P H Y S I C S
A CONTEMPORARY PERSPECTIVE

Volume One
Preliminary Edition

RANDALL D. KNIGHT
California Polytechnic State University—San Luis Obispo

⟁ ADDISON-WESLEY

An imprint of Addison Wesley Longman, Inc.

Reading, Massachusetts • Menlo Park, California • New York • Harlow, England
Don Mills, Ontario • Sydney • Mexico City • Madrid • Amsterdam

Senior Sponsoring Editor   Julia Berrisford
Development Editor   Margy Kuntz
Senior Production Supervisor   Karen Wernholm
Executive Marketing Manager   Kate Derrick
Photo Researcher   Sara Peterson
Text Designer Melinda Grosser
Cover Designer   Diana C. Coe
Prepress Buyer   Caroline Fell
Compositor   Sally Simpson
Art Editor   Susan London Payne
Manufacturing Supervisor   Hugh Crawford

Photo Credits
Page 4, photo courtesy of Digital Instruments, Santa Barbara, CA. Photo captured by a NanoScope® scanning probe microscope. Page 192, ©The Harold E. Edgerton 1992 Trust, courtesy of Palm Press, Inc. Page 196, Richard Megna, *Fundamental Photographs*, New York. Page 270, ©1994, Comstock, Inc. Page 392, The Granger Collection. Page 398, Lick Observatory, University of California, Santa Cruz. Page 418, Astronomical Society of the Pacific, 1983. Page 500, Education Development Center, Newton, MA. Page 510, from *PSSC Physics*, D.C. Heath & Co. Reprinted with permission of Education Development Center, Newton, MA. Page 533, from M. Cagnet et al, *Atlas of Optical Phenomena*, 1962, NY: Springer-Verlag. Reprinted with permission. Page 540, from Jenkins and White, *Fundamental Optics*, 1957, The McGraw-Hill Companies. Reprinted with permission. Page 541, from Carleen Maley Hutchins, "Acoustics of Violins," *Scientific American*, October 1981. Photo by Dr. Karl A. Stetson. Reprinted with permission of C.M. Hutchins, Catgut Acoustical Society. Page 546, Education Development Center, Newton, MA. Page 547, from M. Cagnet et al, *Atlas of Optical Phenomena*, 1962, NY: Springer-Verlag. Reprinted with permission. Page 552, photo courtesy of PASCO Scientific. Page 556, Eugene Hecht. Page 563, from M. Cagnet et al, *Atlas of Optical Phenomena*, 1962, NY: Springer-Verlag. Reprinted with permission. Page 567, General Electric Co. Page 577, from Gerhard Herzberg, *Atomic Spectra and Atomic Structures*, NJ: Prentice-Hall, Inc., 1937. Reprinted with permission. Page 578, Radio Corporation of America. Page 579, from E.R. Huggins, *Physics I*, 1968, Reading, MA: Addison Wesley Longman. Reprinted with permission. Page 582, Culver Pictures. Page 585, Education Development Center, Newton, MA. Page 586, Fig. 18-9, from Paul A. Tipler, *Modern Physics,* NY: Worth Publishers, 1987. Reprinted with permission. Page 586, Fig. 18-10a, from P. Merli and G. Missiroli, *American Journal of Physics*, 1976. Page 586, Fig. 18-10b, from *Reviews of Modern Physics* 60, p. 1067, 1988. Reprinted with permission. Page 587, from the *American Journal of Physics*, 1974, p. 4. Reprinted with permission of the American Institute of Physics.

Text Credit
Page 424, Roszak, *Where the Wasteland Ends*, 1972, Doubleday & Co., New York, 135, 137.

6 7 8 9 10 CRS 06 05 04 03 02

An Invitation to Faculty and Students

Welcome to the Preliminary Edition of *Physics: A Contemporary Perspective*. This is very much a "work in progress," and we invite you to help us shape it into final form. Although these materials have undergone two full years of classroom testing by the author at California Polytechnic State University, where the assessments have been extremely positive, a change in the foundation course of a discipline cannot take place without consulting the community of users. The purpose of this Preliminary Edition is the first of several steps to obtain extensive feedback from both faculty and students prior to the publication of the First Edition.

We expect that concerns will be raised, that omissions and oversights will be discovered, and that many helpful suggestions will be made. Our intentions, and those of the author, are to address the concerns and to heed your suggestions to the fullest extent possible.

Feedback questionnaires for each chapter are provided at the back of the book. We urge instructors to collect these from students and to return them, with their own comments and perspectives, to Addison-Wesley. A more comprehensive survey is also provided as a tear-out card. Please fill out this card at the end of the first term and mail it, postage paid, to Addison-Wesley.

What Is *Physics: A Contemporary Perspective?*

Physics: A Contemporary Perspective is the first comprehensive textbook that

- is based extensively on the results of physics education research, and

- follows the general guidelines of the Introductory University Physics Project.

This is a calculus-based textbook intended for a one-year introductory physics course. Care has been taken from the earliest stages of development to ensure that it meets the needs of a wide variety of classroom settings—large universities, mid-size state universities, liberal arts colleges, and community colleges.

The principal objectives of *Physics: A Contemporary Perspective* are:

- To meet the needs and interests of contemporary science and engineering students in a one-year course.

- To move the results of physics education research into the classroom.
- To replace the encyclopedic scope of existing textbooks with a manageable and coherent group of topics while still covering the essential physics needed by science and engineering students.
- To balance qualitative reasoning and conceptual understanding with quantitative reasoning and problem solving.
- To provide an explicit discussion of and practice with modeling, assumptions, skill development, and multiple representations of knowledge.
- To include a significant component of 20th century physics and modern ideas while keeping in mind the practical needs of much of the audience.
- To develop in students an awareness and appreciation of the fact that physics is a process, not just a list of facts and formulas, and that it has a historical context.

In the course of meeting these objectives, the author has drawn heavily upon the research findings of Arons, McDermott, Hestenes, Van Heuvelen, Hake, Reif, Larkin, Ganiel, Sherwood, Thornton, Laws, Sokoloff, and others. It is very much a research-based textbook that attempts to move the research findings of "what works" into the classroom.

Active learning in the classroom is an essential ingredient for the success of this approach. The text explanations are full and complete, so there is no need to use class time for derivations or for presenting material that the text omitted. The author has designed these materials to support a classroom format based on the effective use of demonstrations, questions and discussions, group activities, and example problem solving.

These materials do not make explicit use of computers or numerical calculations. They will, however, fully support a computer-oriented "workshop" approach to introductory physics. Neither is any specific laboratory component called for. The author, however, suggests that "guided discovery" labs, such as the Thornton/Laws/Sokoloff microcomputer-based labs or Steinberg's batteries-and-bulbs labs, provide the best pedagogical match to the text materials and thus maximize student learning potential.

Major Features of *Physics: A Contemporary Perspective*

A Story Line: Most texts provide little motivation for topics as they are introduced. Therefore, it is not surprising that most students view physics as a collection of loosely-related facts and formulas rather than a few major principles from which many implications are drawn. To combat this fragmentation of knowledge, *Physics: A Contemporary Perspective* adopts the story line of "Physics and the Atomic Structure of Matter." Discovering the link between the microscopic properties of atoms and the macroscopic properties of bulk matter has been a triumph of physics. Further, in this age of nanostructures and quantum well lasers, it is a topic that is relevant to the interests of engineers as well as scientists. The telling of this story is broad enough to visit all the major areas of introductory physics, yet it provides criteria by which specific topics are selected for inclusion.

Works from the Concrete to the Abstract: Arnold Arons, long-time leader in physics education research, stresses the importance of leading the students to answer "How do we know …?" and "Why do we believe …?" questions. The ability to answer such questions represents real knowledge, not just memorized and quickly-forgotten facts. To meet this goal, considerable attention is given throughout to working "from the concrete to the abstract." General principles are arrived at inductively, from evidence, rather than presented *a priori*.

A Workbook: A unique aspect of these materials is a separate Student Workbook. While the text establishes the content and provides the necessary information, there is a limit to how effective a textbook alone can be, no matter how well intentioned, researched, and written. To develop *skills*, such as interpreting graphs, reasoning with ratios, drawing free-body diagrams, drawing electric field maps, or interpreting wave functions, students need opportunities to *practice*. Quantitative, end-of-chapter problems call upon students to assemble and use various skills, but they rarely give students seeing these ideas for the first time an opportunity to develop the individual skills through focused practice.

The Workbook bridges the gap between textbook and problem solving, providing the needed opportunities to learn and practice skills separately—much as a musician practices technique separately from performance pieces. The Workbook exercises are keyed to each section of the text. The exercises are generally graphical or qualitative, letting students practice using the concepts introduced in that section of the text. The Workbook can be used for in-class active learning, in recitation sections, or for assigned homework.

Multiple Representations of Knowledge: *Physics: A Contemporary Perspective* emphasizes multiple representations of knowledge, such as word descriptions, pictorial descriptions, graphical descriptions, and mathematical descriptions. The text provides specific instruction and examples of translating back and forth between different representations, and both the Workbook and the end-of-chapter problems frequently call upon students to do so. This process, called "modeling" by some, is the heart of problem solving in physics. *Physics: A Contemporary Perspective* is unique in the attention it gives to teaching students these skills.

Explicit Problem-Solving Strategies: In conjunction with multiple representation skills, the text develops explicit problem-solving strategies in which students analyze a problem from several qualitative perspectives before approaching it mathematically. Equally important, the worked examples all follow this strategy in detail, with careful explanations of the underlying, and often unstated, reasoning. Students are led to develop a hierarchical knowledge structure, based on general principles, rather than a fragmented knowledge structure of loosely-connected facts and formulas. Optional worksheets are used in some portions of the text to guide students through the difficult process of converting the word representation of a problem statement into intermediate pictorial and graphical representations and, ultimately, to a solvable mathematical representation. Comments from student evaluations find many students saying things such as "I never before realized there is a *method* for solving problems. I wish I had been taught this in my other classes."

Course Content

The textbook has a six-part structure. It can be covered in two semesters or three quarters. The first five parts are generally consistent with a "standard order" to meet the needs of transfer students. The contents of the two volumes are:

Volume One: Part I   Single Particle Dynamics
              Part II   Interacting Particles and Conservation Laws
              Part III   Oscillations and Waves
Volume Two: Part IV: Thermal and Statistical Physics
              Part V: Electric and Magnetic Fields
              Part VI: Quantum Physics and the Structure of Atoms

While all the major concepts of introductory physics are here, the author has endeavored to reduce the encyclopedic scope of other textbooks by omitting some topics and scaling back the level of detail for others. Feedback from users of the Preliminary Edition will be especially critical to judge whether the author's choices meet with widespread agreement.

Part VI: Quantum Physics and the Structure of Atoms places introductory quantum physics on an equal footing with thermal and statistical physics and with electricity and magnetism. The subject matter of quantum physics is rapidly becoming important to engineering students and to scientists in other fields. The goals of Part VI are for students:

- To recognize the experimental basis for quantum physics.

- To understand the primary concepts of energy levels and wave functions.

- To see how one-dimensional quantum mechanics applies to scanning tunneling microscopes, quantum well devices, molecular bonds, radioactivity, and other topics.

- To learn how the atomic shell model, the periodic table of the elements, and the emission and absorption of light are important consequences of quantum mechanics.

These are much less ambitious goals than a full modern physics course, but they are a significant step up from the conventional presentation of modern physics topics. Students who go no farther in physics will have been introduced, in a rigorous rather than a hand-waving fashion, to the basic ideas of quantum physics. And those students who continue to a modern physics course will be better prepared for the abstract formalism of quantum mechanics.

Initial Classroom Testing

These materials have undergone two full rounds of classroom testing by the author at California Polytechnic State University. Student performance, measured using such assessment tools as the Hestenes and Halloun *Force Concept Inventory*, has been outstanding. Details of the initial classroom testing are available upon request.

For more information on classroom test results, or for a sample copy of the text, please contact Addison-Wesley at the address below.

With your help, *Physics: A Contemporary Perspective* will become the new standard for introductory physics.

Addison-Wesley Publishing Company
Physics Editorial
One Jacob Way
Reading, MA 01867
physics@aw.com

Acknowledgments

I have relied upon conversations with and, especially, the written publications of many members of the physics education research community. Those who may recognize their influence include: Arnold Arons, Uri Ganiel, Ibrahim Halloun, David Hestenes, Leonard Jossem, Priscilla Laws, John Mallinckrodt, Lillian McDermott, Edward "Joe" Reddish, Bruce Sherwood, David Sokoloff, Ronald Thornton, Shelia Tobias, and Alan Van Heuvelen. John Rigden, founder and director of the Introductory University Physics Project, provided the impetus that started me down this path.

Valuable review comments have been contributed by Edward Adelson (The Ohio State University), Ronald Bieniek (University of Missouri–Rolla), S. Leslie Blatt (Clark University), David Jenkins (Virginia Polytechnic Institute), John Mallinckrodt (California State Polytechnic University-Pomona), Robert Marchini (Memphis State University), John Risley (North Carolina State University), Cindy Schwarz (Vassar College), and Judy Tavel (Dutchess Community College). My students at Cal Poly also provided vast amounts of feedback as these materials were developed.

I especially want to thank Julia Berrisford, my sponsoring editor; Margy Kuntz, my development editor; Gordon Wong, who developed much of the artwork; and the staff at Addison-Wesley Longman for their enthusiasm toward this project and their efforts to have it published in a timely manner. I am grateful to my wife, Sally, for her encouragement and patience, and to our cats for their complete indifference to this project.

The development of these materials has been supported by the National Science Foundation as the *Physics for the Year 2000* project. Their support is gratefully acknowledged.

Contents

SCENIC VISTA: THE NEWTONIAN SYNTHESIS 418

PART III   OSCILLATIONS AND WAVES

OVERVIEW: OSCILLATIONS AND WAVES 429

Introduction: From Me to You

The most incomprehensible thing about the universe is that it is comprehensible.

Albert Einstein

The day I went into physics class it was death.

Sylvia Plath

Let's you and I talk a little before we start. A rather one-sided talk, admittedly, because you can't respond. But that's OK. I've heard from many of your fellow students over the years, so I have a pretty good idea what's on your mind.

What's your reaction to taking physics? Fear and loathing? Uncertainty? Excitement? All of the above? Probably all of the above, and then some. Let's face it, physics sometimes has a bit of an image problem on campus. You've probably heard that it's difficult, maybe downright impossible unless you're an Einstein. Maybe you've heard other things as well that color your expectations about what this course is going to be like.

Is this a "hard" course? Perhaps. What's easy for one person is hard for another, so there's no way I can know if it's going to be hard for you. But one thing I can say is, easy or hard, it's going to take some effort on your part. There are lots of new ideas, and the pace of this course (like college courses in general) is going to be much faster than science courses you had in high school. I think you could say that it will be an *intense* course. But we can avoid some of the problems and difficulties you might be expecting if we can establish, here at the beginning, what this course is about and what will be expected of you—and of me!

Physics is a mature science. It covers an immense scope of knowledge, with many sub-disciplines and specialties. This, however, is a text for an *introductory* course in physics. Its goal is to convey to you a coherent overview of the major concepts and ideas of physics and to provide you with the tools and skills for using physical concepts in your major and your profession. So this is going to be a mixture of learning *about* physics (the concepts) as well as learning to *do* physics (the skills). Most of you taking this course will need to be able to apply these ideas in your major and your profession, so this is far more than just a "gee whiz" course. We're not going to talk about quarks or black holes or superconductors—they are all fascinating and important topics, but that's not the purpose of this course. However, this course will prepare you to understand those subjects, and many others, in a meaningful fashion.

Just what is physics, anyway? Physics is a way of thinking about nature—specifically the physical characteristics of nature, as opposed to the biological characteristics. Physics is not better than art or poetry or religion—those are also ways to think about nature—it's simply different. One of the things this text will emphasize is that physics is a human endeavor. The information content of this book was not found in a cave or conveyed to us by aliens; it was discovered by real people engaged in a struggle with real issues. I hope to convey to you something of the history and the process by which we have come to accept the principles that form the foundation of today's science and engineering.

Physics, more specifically, is a way of thinking that is based on experiment, that is reproducible, that is quantitative, and that has results that are independent of the person discovering them—that is, objective rather than subjective. This last point is worth noting. It doesn't say that anybody could discover them—that's where scientific creativity and genius come in. But once the theory of relativity was discovered, it makes no difference whether it was Einstein or me or you who discovered it; anyone with the proper training can understand it and verify it. But only Beethoven could write Beethoven's Fifth Symphony and only Leonardo da Vinci could paint the Mona Lisa. Scientific creations are fundamentally different than artistic creations.

Physics is not about "facts." Not that facts are unimportant, but the real heart of physics is about thinking and analyzing—about discovering *relationships* that exist between facts. As a consequence, there's not a lot of memorization when you study physics. There is some—such as definitions to learn—but less than in many other courses. Our emphasis, instead, will be on thinking and reasoning. This is important to factor into your expectations for the course.

Perhaps most important of all, physics is *not* math! Physics is much broader. We're going to be looking for patterns and relationships in nature, we're going to develop the logical chains of inference used to relate different ideas, and we're going to search for reasons *why* things happen as they do. We're going to stress *qualitative* reasoning with words, pictorial and graphical reasoning, and reasoning by analogy. And yes, we will use math, but it's just one tool among many. Many great physicists, in fact, have not been especially good at math.

It's important for you to be aware of this up front. It will save you much frustration if you enter the course with an open mind rather than expecting it to be like your math courses. Many of you, I know, want to find a formula and plug numbers into it—that is, to do a math problem. Maybe that's what you learned in high-school science courses, but it is *not* what this course expects of you. You'll certainly do calculations, but the specific numbers are usually the last and least important step in the analysis.

As you study, you'll often be baffled, puzzled, and confused. That's OK—it's perfectly normal and to be expected. Making mistakes is OK too *if* you're willing to learn from the experience. No one is born knowing how to do physics, any more than they're born knowing how to play the piano or shoot basketballs. Learning comes from practice, repetition, and struggling with the ideas until you "own" them and can apply them yourself in new situations. There's no way to make learning effortless, at least for anything worth learning, so expect to have some difficult moments ahead.

But also expect to have some moments of excitement at the joy of discovery. There will be instants at which the pieces suddenly click into place and you *know* that you understand a difficult idea. There will be times at which you'll surprise yourself by successfully working a difficult problem that you didn't think you could solve. My hope, as author, is that the excitement and sense of adventure will far outweigh the difficulties and frustrations.

Many of you, I suspect, would like to know the "best" way to study for this course. There is no best way. People are too different, and what works for one student is less effective for another. But I do want to stress that *reading the text* is vitally important. Class activities will work on clarifying difficulties and on developing tools for using the knowledge, but your instructor will *not* use class time to repeat information in the text. The basic knowledge for this course is written down right here, and the *number one expectation* is that you will read carefully and thoroughly to find and learn that knowledge.

Despite there being no best way to study, I will suggest *one* way that is successful for many students.

1. Pre-read each chapter before it is discussed in class. I cannot stress too highly how important this is. Your class attendance is largely ineffective if you have not prepared. When you do the first reading of a chapter, you should focus particularly on learning new vocabulary, definitions, and notation. There's a list of important concepts and terms at the end of each chapter—learn them! This is one place where some memorization is required. You won't understand what's being discussed or how the ideas are being used if you don't understand what the terms and symbols mean.

2. Participate actively in class. There is ample scientific documentation that active participation—taking notes, asking and answering questions, working in discussion groups—helps students to learn but that passive listening does not.

3. After class, go back for a *careful* re-reading of the chapter. During your second reading you should pay closer attention to the details and the worked examples. Look for the *logic* behind the example (and I've tried to help make this clear), not just at what formula is being used. Do the workbook exercises for each section as you finish your careful reading of it.

4. Lastly, apply what you have learned to the exercises and problems at the end of each chapter.

Did someone mention a workbook? The companion workbook is an essential part of this course. It contains questions and exercises, not problems, that ask you to reason *qualitatively*, to answer questions and give explanations, and to use graphical information. It is through these exercises that you will learn what the concepts mean (and don't mean!) and will develop the reasoning skills appropriate to that chapter. This is the baseline knowledge that you need *before* trying to apply it in the homework problems. You would never think of performing before you practice in sports or music, so why would you want to in physics? The workbook is where you practice and work on basic skills.

Many of you, I know, prefer to go straight to the homework problems and then thumb through the text looking for a formula that seems like it will work. That strategy will not work in this course, and it's guaranteed to make you frustrated and discouraged. The homework problems are not "plug and chug" problems where you simply put numbers into a formula. You will need a better study strategy—either the one outlined or your own—that helps you understand the concepts and the relationships between the ideas if you're going to work the homework problems successfully. Many of the chapters in this book have features designed to help you develop effective problem-solving skills.

You'll find two kinds of homework problems in most chapters: some called "exercises" and others called "problems." The exercises —like their name implies—are intended to be straightforward, very basic problems for the purpose of practicing the ideas presented in the chapter. Think of them as warm-up exercises for the problems. Learning to work the problems is the real goal of the chapter. The problems are longer, often based on real-world examples, and frequently ask for a blend of algebraic analysis, numerical solutions, and explanation or interpretation *in words* of what the problem is all about. The problems will typically draw upon more than one idea, often requiring you to recall and use ideas from previous chapters. Many of the problems will require you to recognize which information is relevant; others will ask a question but not tell you what calculations are necessary to answer it.

Most of the exercises have answers provided in the back of the book, but only a few of the problems. That's a conscious choice on my part. Part of my rationale is that I want you to be focused on the thinking and reasoning aspects of physics, not just on getting "the answer." Another aspect of my rationale is that the real world doesn't come with an answer manual. Part of what you're being asked to learn, in preparation for a scientific or technical career, is to develop self-confidence in your reasoning abilities. I do *highly* recommend that you join a study group of three or four individuals so that you can talk about the problems and check your solutions against each other.

A traditional rule of thumb in college is that you should study two hours outside of class for every hour spent in class. This text is designed with that expectation. That, of course, is an average. Some chapters are straightforward and will go quickly, but others will likely require much more than two study hours per class. Individual variations in ability and motivation will also play a role in how much study time you need.

Now that you know more about what is expected of you, what can you expect of me? That's a little trickier, because the book is already written! Nonetheless, it was prepared on the basis of what I think my students throughout the years have expected, and wanted, from their physics course. You should know that these materials—the text and the workbook—were developed under a grant from the National Science Foundation. I have incorporated a wide range of recent research about the most effective ways for students to learn physics, and I have also adhered to the guidelines produced by a nationwide consortium of physics educators called the Introductory University Physics Project. In addition, I have tried to make it clear why physics is relevant to you, not just as a technical body of knowledge needed in your profession, but also as an exciting adventure of the human mind.

I hope you'll enjoy the time we're going to spend together.

R.D.K.

PART I

Single Particle Dynamics

PART ■ Overview

Journey into Physics

Said Alice to the Cheshire Cat:
"Cheshire-Puss, ... would you tell me, please, which way I ought to go from here?"
"That depends a good deal on where you want to go," said the Cat.
"I don't much care where –" said Alice.
"Then it doesn't matter which way you go," said the Cat.

Lewis Carroll—*Alice in Wonderland*

Why is the sky blue?

Why is glass an insulator but metal a conductor?

Is it really true that an electron can be in two places at once?

Have you ever wondered about questions such as these? These are the questions of which physics is made. Physics is the science that attempts to answer fundamental questions about the nature of the physical universe in which we live. Physicists try to understand the world by observing the phenomena of nature—such as the sky being blue—and by looking for patterns and principles that explain these phenomena. Some of their discoveries have forever altered the ways in which we live and think.

You are about to embark on a journey into the realm of physics. During the journey, you will gain insight into many physical phenomena and learn to answer questions such as the ones above. You'll also learn how to use physics to analyze and solve many practical problems.

Our ultimate goal is to explore the world of the atom, for an understanding of atoms and their properties is the basis for understanding the nature of the world around us. The word *atom* comes from the Greek root *atmos*, meaning "uncuttable." Until about 1900, which is very recent in the history of science, the essence of atoms was their "uncuttableness." Atoms were thought to be the basic, indivisible building blocks of matter, regardless of whether they were the earth-air-fire-water atoms of the ancient Greeks or the ninety or so periodic table atoms identified by chemists in the nineteenth century. But discoveries

during the very end of the nineteenth century and the beginning of the twentieth century revealed smaller pieces—electrons, protons, neutrons—inside the atoms. Much of today's science and technology, from electronics to lasers to new materials, depends upon our knowledge of this atomic structure.

Physics is an experimental science. We have *evidence* for our theories and beliefs that comes from controlled, repeatable experiments. Although the Greek philosopher Democritus introduced the idea of atoms 2400 years ago, there was no evidence for their reality until the chemical studies of the early 1800s. Even then the evidence was indirect, and many prominent scientists of the nineteenth century refused to accept the reality of atoms. Only with the discovery of the electron and of radioactivity at the end of the nineteenth century, barely one hundred years ago, did the existence of atoms become an accepted fact of science.

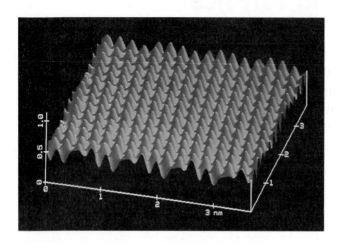

FIGURE OI-1   Image, from a scanning tunneling microscope, of individual atoms on a surface.

As you proceed on your journey you are going to see the way knowledge about the universe is generated—the introduction of new ideas, the experiments that either provide evidence for or disprove those ideas, and the construction of theories based on experimental evidence. The theories and relationships of physics are not arbitrary; they are firmly grounded in experiments and measurements. By the time you finish this text you will be able to *justify* your knowledge by recognizing the evidence upon which it is based.

Which Way Should We Go?

We are rather like Alice in Wonderland, here at the start of the journey, in that we must decide which way to go. Physics is an immense body of knowledge, and without a specific goal it would not much matter which topics we chose for study. But unlike Alice, we *do* have a destination in mind—the world of atomic structure.

Atoms and their properties are described by quantum physics, but we cannot leap directly into that subject and expect that it would make any sense. To reach our destination, we are going to have to study many other topics along the way—rather like having to visit Nebraska if you want to drive from New York to San Francisco. Here, at the beginning, we can survey the route ahead and see how each topic is going to move us a little further along toward our goal.

Parts I and II—*Single Particle Dynamics* and *Interacting Particles and Conservation Laws*—form the basis of what is called *classical mechanics*. Classical mechanics is the study of motion. (It is called *classical* to distinguish it from the *modern* theory of motion at the atomic level, which is called *quantum mechanics*.) Everything that you will study in physics is about change and motion, and the first two parts of this textbook establish the basic language and concepts of motion. These two parts also will establish the twin themes of "force" and "energy," which you will use to study the motion of everything from accelerating sprinters to orbiting satellites.

Part III is titled *Oscillations and Waves*. Waves are ubiquitous in nature, whether they be large-scale oscillations such as ocean waves, the less obvious motions of sound waves, or the subtle undulations of light waves and matter waves that go to the heart of the atomic structure of matter. We are going to emphasize the unity of wave physics, showing that all these wave phenomena can be analyzed with the same concepts and mathematical language. It is here we will begin to accumulate evidence that the theory of classical mechanics is inadequate to explain the observed behavior of atoms. We will end this section with some atomic puzzles that seem to defy understanding.

Part IV, *Statistical Physics and Thermodynamics*, provides much of the traditional evidence for the existence of atoms. It is here that you will explore the connection between the microscopic behavior of large numbers of atoms and the macroscopic properties of bulk matter, such as gases and solids. You will find that some of the properties of gases that you know from chemistry, such as the ideal gas law, turn out to be direct consequences of the underlying atomic structure of the gas. The chapters on thermodynamics will investigate the concept of *energy* and will show how energy is transferred and utilized. These chapters also will provide you with insights into exceedingly basic issues about the nature of time; for example, what distinguishes the past from the future, and will the universe have an end?

Part V, *Electric and Magnetic Fields*, is devoted to the topics of electricity and magnetism. One of the most important forces in nature is the electromagnetic force. It is, in essence, the "glue" that holds atoms together, and it is also the force that makes this the "electronic age." You will learn about electric charges and about the behavior of charges in matter. This will lead to a discussion about electrical conductors and insulators—the basic ideas behind electrical circuits—and to an understanding of magnetism.

Part VI is *Quantum Physics and the Structure of Matter*. Here you will enter the microscopic domain of atoms, where the behavior of light and matter is at complete odds with what our common sense tells us is possible. We will begin by looking at some of the unexpected phenomena of the microscopic realm and seeing the experimental evidence for the quantization of energy. Although the mathematics of quantum theory quickly gets beyond the level of this text, and time will be running out, you will see that the quantum theory of atoms explains many of the things that you learned simply as rules in chemistry.

We will not have visited all of physics on our travels. There just isn't time. However, occasional "Scenic Vistas" along the way will give you a glimpse of other territories waiting to be explored. Even though our journey into physics will have reached its end, you will have the background and the experience to explore new topics in more advanced courses or for yourself.

Model Building: A Vehicle for Understanding

Reality is extremely complicated. If we had to keep track of every little detail of every situation, we would never be able to develop a science. Consider a simple example: throwing a ball. To understand the motion of the ball, is it necessary to keep track of every atom inside? Of every quark inside every proton inside every nucleus inside every atom inside the ball? Do we need to analyze what you ate for breakfast and the biochemistry of how that was translated into muscle power? In principle, the answer to all these questions is "yes." But in practice these are all details that have no influence at all on the measurements you might make on the ball—they exist, but they are just too insignificant to matter.

We can do a perfectly fine analysis if we treat the ball as a round solid and your hand as another solid that exerts a force on the ball. This is now what we call a *model* of the situation. A model is a simplified description of reality that is used to reduce the complexity of a problem to the point where it can be analyzed and understood.

All of physics and engineering is a matter of model building—simplifying the situation, isolating the essential features, and developing a set of equations that provides an adequate, but not perfect, description of reality. Physics, in particular, attempts to strip a phenomena down to its barest essentials in order to illustrate the physical principles involved. Many of the features neglected by the physicist as "unnecessary details" would be considered absolutely essential by an engineer. Both are right: the model each is using has to match the needs each of them is trying to meet.

In building a model, you need to isolate and keep just those features that are essential to the problem. In the case of the ball, for example, you can ignore the ball's atomic structure because it doesn't affect the motion of the ball as a whole. But you cannot ignore gravity. It is an essential feature of the model because it directly influences the motion of the ball. What about air resistance? If you are throwing a very hard, dense ball a very short distance, then ignoring air resistance is probably acceptable. But if you are throwing a ping-pong ball from the Empire State Building, then certainly air resistance is an essential feature of the model. The more details you omit, the simpler the model and the easier it will be to solve the model's equations—but at the expense of accuracy and agreement with reality.

We will pay close attention, especially in the earlier chapters, to model building. One of the major goals of this text is to develop *methods* for solving problems, and model building will turn out to be a major part of the strategy. It is, however, a skill that will take some practice and experience. As we go along, the text will point out where simplifying assumptions are being made, and why. Learning *how* to simplify a situation is the essence of successful modeling. As you begin to apply these ideas in your own homework, you will be gaining experience with model building.

With that said, let us begin.

Chapter **1**

Concepts of Motion

You cannot step twice into the same river, for fresh waters are ever flowing in upon you.

Heraclitus (c. 450 B.C.)

Socrates: The nature of motion appears to be the question with which we begin. What do they mean when they say that all things are in motion? Is there only one kind of motion, or, as I rather incline to think, two? I should like to have your opinion upon this point in addition to my own, that I may err, if I must err, in your company. Tell me, then, when a thing changes from one place to another, or goes round in the same place, is that not what is called motion?

Theodorus: Yes.

Socrates: Here then we have one kind of motion. But when a thing, remaining on the same spot, grows old, or becomes black from being white, or hard from being soft, or undergoes any other change, may not this be properly called motion of another kind?

Theodorus: I think so.

Socrates: Say, rather, that it must be so. Of motion then there are these two kinds, "change" and "motion in place."

Theodorus: You are right.

Plato—*Theaetetus* (c. 375 B.C.)

1.1 Introduction

The universe in which we live is clearly one of change and motion. If you look around, you will see many objects in motion—people walking, leaves rustling, clock hands moving inexorably forward. The pages of this book may look quite still, but a microscopic look would reveal that the pages are made of atoms and their moving electrons. The stars look as stationary as anything, yet astronomers' telescopes reveal the stars to be ceaselessly moving within galaxies that rotate and orbit about yet other galaxies. The Greek philosopher Heraclitus maintained that everything is changing—that permanence is only an illusion. But what is change?

[**Photo suggestion: Time exposure of stars rotating about the north star.**]

The simplest change you might observe is the change of an object's position with time. For example, you could make a mark of some sort on the object and measure the location of that mark every second or every minute or every year. The change in the location of the mark each time you measure it is what we call *motion*. While this sounds straightforward, there are some subtleties. There is clearly change as clouds form, move, and dissipate, but how would we describe it? Or what about the ocean's surface—always in motion, but not describable by making a mark on it? More subtle yet, what about the changes that occur in you as you age—are they motion?

The ancient Greek philosophers puzzled over these different kinds of change. However, they never arrived at a single, clearly defined idea of motion. Much of the reason was their lack of appropriate mathematical concepts. Plato based his arguments on words, which can be rather slippery and ambiguous. Our modern understanding of motion did not begin until Galileo (1564–1642) first formulated the concepts of motion in mathematical form. And it took Newton (1642–1727) and the invention of calculus to finally put the concepts of motion on a firm and rigorous footing.

Throughout this book, the concepts of motion will provide a language for describing our universe of change. A careful and accurate description of motion is harder to achieve than it might at first appear. So rather than jumping immediately into a lot of mathematics and calculations, this chapter is going to provide a *qualitative* introduction to the important concepts. Qualitative reasoning means reasoning with words, rather than numbers or calculations. We will still use mathematical concepts when needed, because they increase the precision of our thoughts; but we will, for now, avoid actual calculations. When using qualitative reasoning, you should articulate your ideas clearly and explain your reasoning in a convincing manner. Doing so leads to true understanding of a subject, which is our goal.

1.2 Basic Types of Motion

As stated so well by Socrates, "The nature of motion appears to be the question with which we begin." For physics purposes we will define **motion** as the change of an object's position with time. For now, we will focus only on the motion of solid objects. Once you have gained an understanding of solid-object motion, it will be straightforward to extend these ideas to more complex motions. The primary objective of this chapter, then, is to determine what concepts are needed to describe the motion of solid objects.

Examples of objects that move are easy to list; you could probably come up with dozens in a minute. For example, bicycles, cars, boats, airplanes, and rockets are all objects that move. These objects move predominantly along a straight line, although it is sometimes necessary to turn a corner! Straight-line motion, whether horizontal, vertical, or at an angle, is called **linear motion**. It is an important class of motion with which we will deal frequently.

Other types of objects move predominantly in circles. A ball rolling around the inside of a roulette wheel, the tip of a fan blade, a rider on a Ferris wheel, and a planet orbiting the sun are all examples of objects that exhibit **circular motion**. (Actually, planetary motion is slightly elliptical, but treating it as circular is a good first approximation.) Some objects follow curved, but not circular, paths. Consider, for example, the path of a baseball or basketball after it is thrown. This type of motion is called **projectile motion.**

A mass bobbing up and down on a spring, a pendulum swinging back and forth in a grandfather clock, and the vibrating reed on a saxophone are all examples of objects that have **oscillatory motion**. And then there are more complex motions yet, such as the motion of all the molecules in a gas or of the electrons in an atom.

All of these moving objects have some things in common—they start and stop, and they speed up and slow down. Our goal is to discover the *concepts* that are needed to give a full and accurate description of *all* of these classes of motion. To avoid falling into some of the word traps that hindered early scientists, our concepts will ultimately need to be defined mathematically. But before we arrive at mathematical definitions, it will first be important to understand qualitatively what the concepts are all about. Our primary tool for achieving a qualitative understanding of motion is called a *motion diagram*. This important topic is the subject of the next section.

1.3 Motion Diagrams

An easy way to study motion is to make a movie of the moving object. A movie camera, as you probably know, takes photographs at a fast rate—typically about 30 every second—and each picture is taken after a set period of time has elapsed. Each separate photo is called a *frame*, and they are all lined up one after the other in a *filmstrip*. When the filmstrip is played back through a projector, your eye and brain perceive the motion of objects to be continuous rather than jerky, even though the individual frames only show the objects at various discrete positions.

Suppose we make a movie of a single object, such as a car, as it moves past a *fixed* movie camera. (A fixed camera does not "pan" to track the object as it moves.) Figure 1-1 shows a few frames from the resulting filmstrip. Not surprisingly, the car is in a somewhat different position in each frame.

Now, suppose we cut the individual frames of the filmstrip apart, stack them on top of each other, and then project the entire stack at once onto a screen for viewing. The resulting picture on the screen, shown in Fig. 1-2, is a composite photo showing the car's position at a series of equally-spaced time intervals. This type of composite "photo" is what we call a **motion diagram**. As simple as motion diagrams seem, they will turn out to be a powerful tool for analyzing motion.

FIGURE 1-1 Frames from the movie of a moving car.

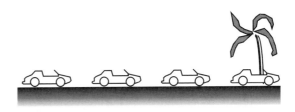

FIGURE 1-2 Motion diagram of a moving car.

FIGURE 1-3 Motion diagram of a rock at rest.

[**Photo suggestion: Several strobe photos illustrating different kinds of motion.**]

Now that you've seen how to make a motion diagram, let's take our camera out into the world and make a few more movies that show different types of motion. We can start by making a motion diagram of a rock as it sits in a garden (Fig. 1-3). This is not a terribly exciting motion diagram because the rock never moves. However, the diagram helps us define one concept of motion: an object that occupies only a *single position* in a motion diagram is **at rest**. Figure 1-4 is a motion diagram of a skateboarder rolling across the driveway. Notice that his positions in adjacent frames of the film are *equally spaced*. This indicates that his speed does not change as he moves. We say that he moves with a **constant speed**.

Next, we are lucky enough to be around when a dog sees a mail carrier and takes out after him (Fig. 1-5). The distance between successive positions in the diagram is *increasing* because

FIGURE 1-4 Motion diagram of a skateboarder at steady speed.

the dog is *speeding up*. Figure 1-6 shows the motion diagram of a woman braking a car just in time to avoid running over a bicycle left in the driveway. Here the distance between successive

FIGURE 1-5 Motion diagram of a dog speeding up.

FIGURE 1-6 Motion diagram of a car slowing down.

positions of the car is *decreasing*, indicating that the car is *slowing down*. Finally, we see some kids playing basketball and film a great jump shot made from center court. The ball's motion diagram as seen from beside the basketball court is two dimensional (Fig. 1-7). The diagram shows aspects of both slowing down (positions getting closer together as the ball rises) and speeding up (positions spreading apart as the ball falls).

Note that we have defined several concepts (at rest, constant speed, speeding up, and slowing down) in terms of how the moving object appears in the motion diagram. These types of definitions are called **operational definitions**,

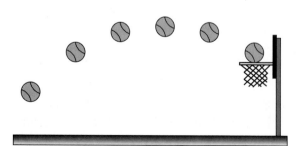

FIGURE 1-7 Motion diagram of a basketball.

meaning that they are defined in terms of a particular procedure or operation. Most of the concepts in physics will be introduced as operational definitions. This reminds us that physics is an experimental science, and it also minimizes the possible ambiguities that might arise if we try to define concepts simply with words.

The Particle Model

In the examples of motion diagrams you have seen so far, we have used representations of the actual objects—a person on a skateboard, a car, a ball—that show both the size and the shape of the objects. However, when the size of an object is much less than the distance it travels, then the object's size and shape become "details" that can be neglected when describing the overall motion. Suppose, for example, you make a motion diagram of a car being driven into a garage. In this case, it's critically important to distinguish between the location of the front bumper and the location of the rear bumper! The distinction is important because the length of the car is comparable to the distance it travels. On the other hand, if you were to make a motion diagram of a car traveling from Los Angeles to Seattle, the distinction between the front and rear bumper would be meaningless. The length of the car is tiny in comparison to the distance between the two cities.

If we state this idea more carefully, we can say that it is all right to treat an object *as if* it were just a single point—without size or shape—as long as the object's size is much less than the distance it travels. We can also treat the object *as if* all of its mass were concentrated into a single point. We call an object treated in this manner a **particle**. In addition to having a mass, a particle can have an electric charge and a magnetic moment (properties that we will discuss in future chapters), but it has no size, no shape, and no distinction between top and bottom or between front and back.

As we discussed in the Overview, a model is a simplification of reality that allows us to focus on the important aspects of a phenomenon while neglecting those aspects which play only a minor role. The **particle model** of motion is a particular simplification in which we treat a moving object as if all of its mass were concentrated at a single point, which can be described by a specific set of coordinates (x, y, z).

Two major conditions must be satisfied for the particle model to be valid:

1. The size of the object must be much less than the distance traveled, and

2. Any internal motions of the object must be of no significance to the problem.

As an example of the second condition, the internal motions of a car's pistons are of no significance if your task is to understand the motion of the car along the road. But the particle model would *not* be appropriate for analyzing an internal combustion engine.

Even with these restrictions, there is an enormous range of situations in which the particle model is an excellent and valid approximation of reality. In fact, the success of the particle model has been so great that the motion of more complex objects, which cannot be treated as a single particle, is often analyzed as if the object were a collection of particles. For this text, however, we will concentrate our efforts on studying the motion of objects which can be treated as a single particle. This model will be implicit throughout the text. Whenever a problem says that "a car moves" or "a rocket moves" or "a person moves," that object will be considered to be a single particle. This is, of course, a simplification of reality, but one that is well justified.

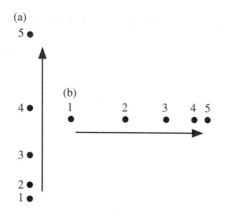

(a)

(b)

FIGURE 1-8 Motion diagrams of a) a rocket taking off and b) a car braking to a halt. The numbers next to the dots indicate the sequence in which the frames were exposed.

The particle model also allows us to make a major simplification in how we draw motion diagrams. If we treat an object as a particle, its size and shape have no significance for the problem. Consequently, we can represent the object in each frame of a motion diagram as a simple dot rather than having to draw a full picture. Figure 1-8 shows how motion diagrams appear when using the particle model. Figure 1-8a shows the motion diagram of a rocket taking off, while Fig. 1-8b shows a car braking to a halt. Note that the dots have been numbered to indicate the sequence of time in the diagram. These diagrams are more "abstract" without the pictures, but they are easier to draw and they still convey our full understanding of the object's motion.

Thought question: How can you distinguish between motion diagrams showing: a) starting from rest, b) stopping, and c) moving at a steady speed?

1.4 Position and Time

To develop our motion diagrams further we need to be able to make measurements. When we look at an object in a motion diagram it would be useful to know *where* the object is (i.e., its *position*) and *when* the object was at that position (i.e., the *time*). These are easy measurements to make. Consider placing a long ruler or tapemeasure horizontally along the ground. (If needed, a second ruler could extend vertically upward from the ground for height measurements.) Also, place a large clock in the foreground of the scene. Then simply start the object moving while the camera is rolling. After you develop the film, every frame will show not only the object but also a clock for measuring the time and a ruler for measuring the position.

Figure 1-9 shows a new motion diagram of the skateboarder that includes the clock and ruler. Now you can see that he was at position 10 feet at the time 4 seconds. These measurements, however, raise some new questions. Ten feet from where? Four seconds after what? To measure the object in one frame as being "10 feet at 4 seconds" does not really

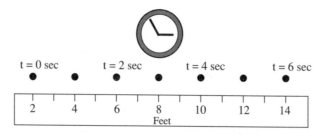

FIGURE 1-9 Motion diagram of a skateboarder, showing position and time.

tell us anything at all unless we also know *relative to where and when.* And that is what a coordinate system is for.

A *coordinate system* is a reference grid that you impose on a situation in order to make measurements. When we measure an object as being "at 4 feet," what we really mean is that it is 4 feet from a specific reference point called the *origin* of the coordinate system. The world does not come with a coordinate system attached. Therefore, where to place the origin is your choice, and different observers of a moving object might all choose to use different origins for their coordinate systems. Likewise, you can choose the orientation of the *x*-axis and *y*-axis. The conventional choice is for the *x*-axis to be horizontal and point to the right while the *y*-axis is vertical and points upward, but there is nothing sacred about this choice. We will soon have many occasions to tilt the axes at an angle.

The point to be made here is that a coordinate system, with a particular origin, is an artificial grid that you place over a problem in order to analyze the motion. The axes and origin are chosen to be convenient and helpful for that particular problem, and they might be different in the next problem. Some choices are better than others, and with experience you will learn how to choose the best coordinate system for a problem. However, no choice is "wrong."

Time is also a coordinate system with an origin, although you may never have thought of time this way. The particular instant that we choose to call "zero seconds" is an arbitrary choice—it is simply the instant you decide to start your clock or stopwatch. Different observers might choose to start their clocks at different moments. A frame showing a clock reading of 4 seconds simply means that it was taken 4 seconds after the clock started. We typically choose $t = 0$ to represent the "start" of our problem, but the object may have been moving before then. Those earlier instants would be measured as negative times, just like objects to the left of the *x*-axis origin have negative values of position. Negative numbers are not to be avoided—they simply locate an event in space or time *relative to an origin.*

For an object moving in a plane, its position in each frame can be described by a pair of numbers (x, y) giving its coordinate values at that instant of time. Alternatively, we can locate the object by drawing an arrow *to* the object *from* the origin and specifying the length and angle of that arrow. Such an arrow is called the **position vector** of the object. More specifically, the position vector $\vec{r}$ is defined as follows:

> The position vector $\vec{r}$ of an object is a vector drawn from the origin to the object.

Figure 1-10 shows the position vector of a basketball at several points in its motion. Note that the position vector does not tell us anything different than the coordinates (x, y). The position vector simply provides the information in a different form, namely a geometric form rather than an algebraic form. But because motion diagrams are geometrical, position vectors provide a "natural" way of locating objects.

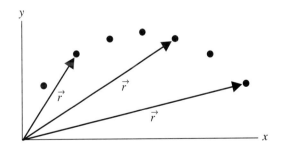

FIGURE 1-10 Position vector $\vec{r}$ of a basketball, shown at several points in its motion.

A Word About Vectors and Notation

Before we continue our discussion, we should take a quick look at what a vector is. (Vectors will be more thoroughly studied in Chapter 2.) Some physical quantities, such as time, mass, and temperature, can be described completely by a single number with a unit. For example, the mass of an object is 6 kg and its temperature is 30°C. Many other quantities, however, have a directional quality and cannot be described by a single number. For example, to describe the motion of a car we must specify not only how fast it is moving, but also the *direction* in which it is moving. When a physical quantity is described by a single number, we call it a **scalar quantity**. A **vector quantity** is a quantity that has both a *size* (the "how far" or "how fast") and a *direction* in space. The size or length of a vector is properly called its **magnitude**.

When we want to represent a vector quantity with a symbol, we need a way to indicate that the symbol is for a vector rather than for a scalar. We do this by drawing an arrow over the letter that represents the quantity. Thus $\vec{r}$ and $\vec{A}$ are symbols for vectors, whereas r and A, without the arrows, are symbols for scalars. In handwritten work you must draw arrows over all symbols that represent vectors. This may seem strange until you get used to it, but it is very important to get in the habit because we will often use both r and $\vec{r}$, or both v and $\vec{v}$, in the same problem and they mean different things! Without the arrow, you will be using the same symbol with two different meanings in your problem, and you will likely end up making a mistake.

In figures and drawings, you first draw the vector arrow itself, and then you place a label beside it (such as $\vec{r}$) that also includes an arrow as part of the symbol. Make sure you do not confuse the arrow over the symbol with the actual vector arrow. Note that the arrow over the symbol always points to the right, regardless of which direction the actual vector points. Thus we write $\vec{r}$ or $\vec{a}$, never $\overleftarrow{r}$ or $\overleftarrow{a}$.

Changes in Position and Time

Consider the following problem: *Sam is standing at a point 50 feet east of a corner. He then walks northeast for 100 feet to a second point. What was Sam's change of position?* Sam's motion is shown in Fig. 1-11. Before Sam starts walking he is at one position, called the **initial position**. When he finishes walking he is at a new position, called the **final position**. In the figure, we have used coordinates (x_i, y_i) to indicate Sam's initial position and (x_f, y_f) to indicate his final position. Sam's *change* from one position to another is called a **displacement**. It is symbolized as $\Delta \vec{r}$. (The Greek letter delta Δ is used to indicate changes in a quantity. This is a standard notation that we will use many times.)

Notice that displacement is a vector quantity—it requires both a length and a direction to describe it. In Fig. 1-11, the quantity $\Delta \vec{r}$ that points *from* Sam's initial position *to* his final position is called the **displacement vector**. This vector describes Sam's change of position. In this case, we can say that Sam's displacement vector is "100 feet, to the northeast."

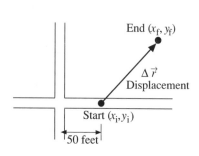

FIGURE 1-11 Sam starts at position (x_i, y_i) and ends at position (x_f, y_f), having undergone a displacement $\Delta \vec{r}$.

The length, or magnitude, of the displacement vector is simply the *distance* between the initial and final positions. It is important to point out that the magnitude of the displacement—that is, the distance between (x_i, y_i) and (x_f, y_f)—is a scalar quantity even though the displacement itself is a vector. The magnitude of Sam's displacement vector is 100 feet, in other words, just the distance without a direction.

Now look at the problem in a slightly different way. Suppose that instead of wanting to find Sam's change in position, the question had been *Where does Sam end up?* As shown in Fig. 1-12, we can use position vectors to answer this question. The position vector $\vec{r_i}$ shows the Sam's starting position relative to the origin (chosen to be at the corner). Position vector $\vec{r_f}$ shows his ending position. The figure seems to suggest that $\vec{r_f}$, the position vector showing where Sam's ends up, is a combination of vector $\vec{r_i}$, where he started, *plus* vector $\Delta \vec{r}$, his displacement. In fact, we *can* find $\vec{r_f}$ by summing the vectors $\vec{r_i}$ and $\Delta \vec{r}$. In mathematical notation this can be written as:

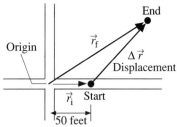

FIGURE 1-12 The vector $\vec{r_f}$, Sam's final position, is equal to the sum of vectors $\vec{r_i}$ and $\Delta \vec{r}$.

$$\vec{r_f} = \vec{r_i} + \Delta \vec{r}. \tag{1-1}$$

Notice, however, that we are adding vector quantities, not scalar quantities. Vector addition is a different process than "regular" addition, and we will study it thoroughly in the next chapter. For now, you can add two vectors $\vec{A}$ and $\vec{B}$ graphically using the following steps:

1. Begin by drawing the first vector $\vec{A}$.

2. Draw vector $\vec{B}$ by placing the "tail" of $\vec{B}$ at the "tip" of $\vec{A}$.

3. Draw an arrow from the tail of $\vec{A}$, where you first started, to the tip of $\vec{B}$, where you finally ended up. This arrow is the vector sum $\vec{A} + \vec{B}$.

Notice in Fig. 1-12 that this is exactly how $\vec{r_i}$ and $\Delta \vec{r}$ are added to give $\vec{r_f}$.

Also notice in Fig. 1-12 that we chose *arbitrarily* to put the origin of the coordinate system at the corner. While this might be convenient, it certainly is not mandatory. Suppose we located it elsewhere, as in Fig. 1-13? Notice something interesting: Vectors $\vec{r_i}$ and $\vec{r_f}$ are different now, because they are position measurements relative to the origin. But the displacement vector $\Delta \vec{r}$ has not changed! The displacement is a quantity that is *independent of the coordinate system*. Graphically this means that the arrow drawn from one position of an object to the next is the same no matter what coordinate system we choose. This gives the displacement $\Delta \vec{r}$ considerably more *physical significance* than the position vectors themselves.

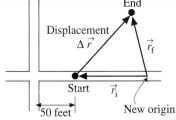

FIGURE 1-13 Sam's motion considered from a new origin. The displacement $\Delta \vec{r}$ is unchanged.

This suggests that the displacement, rather than the actual position, is what we want to focus on as we analyze the motion of an object. Equation 1-1 told us that $\vec{r}_f = \vec{r}_i + \Delta\vec{r}$. This is easily rearranged to give a more precise definition of displacement as:

The displacement $\Delta\vec{r}$ of an object as it moves from an initial position $\vec{r}_i$ to a final position $\vec{r}_f$ is given by $\Delta\vec{r} = \vec{r}_f - \vec{r}_i$ and is independent of the coordinate system.

The displacement is found graphically by drawing an arrow *from* the starting, or initial, position *to* the ending, or final, position.

This definition of $\Delta\vec{r}$ involves *subtracting* vectors. Vector subtraction is a little trickier than vector addition. If you look at Figs. 1-12 and 1-13 you can see that the vector $\Delta\vec{r}$, which is $\vec{r}_f - \vec{r}_i$, is a vector drawn from the *tip* of $\vec{r}_i$ to the *tip* of $\vec{r}_f$. This suggests that the rules for graphically subtracting vector $\vec{B}$ from vector $\vec{A}$ are:

1. Begin by drawing the first vector $\vec{A}$.

2. Draw vector $\vec{B}$ by placing the tail of $\vec{B}$ at the tail of $\vec{A}$.

3. Draw an arrow from the tip of $\vec{B}$ to the tip of $\vec{A}$. This arrow is the vector $\vec{A} - \vec{B}$.

Be sure to notice that vector subtraction requires placing the two vectors "tail-to-tail" whereas vector addition placed them "tip-to-tail."

As a first step in analyzing a motion diagram, you can determine all of the displacement vectors simply by drawing arrows connecting each location of the object to the next. As you do so, make sure you label each arrow with the *vector* symbol $\Delta\vec{r}$! Figure 1-14 shows the earlier Figures 1-5 and 1-6 redrawn with the displacement vectors shown. Note that you do *not* need to show the position vectors because each $\Delta\vec{r}$ is independent of the coordinate system. Now we can conclude, more precisely than before, that displacements which are *increasing* in length as time proceeds represent an object that is *speeding up* (Fig. 1-14a), whereas displacements which are *decreasing* in length represent an object that is *slowing down* (Fig. 1-14b).

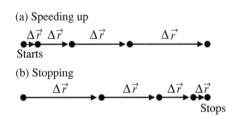

(a) Speeding up

Starts

(b) Stopping

Stops

FIGURE 1-14 Motion diagrams of a) a dog speeding up, and b) a car stopping.

Just as it was more useful to look at change in position rather than the position itself, it is useful to look at *change* in time rather than individual time values. For example, the clock readings of two frames of film might be t_i and t_f. The specific values are arbitrary because they are timed relative to an arbitrary instant that you chose to call $t = 0$. But the **time interval** $\Delta t = t_f - t_i$ is *not* arbitrary. It represents the elapsed time for the object to move from one location to the next, and it will be measured to have the same value by all observers, regardless of when they choose to start their clocks.

The time interval $\Delta t = t_f - t_i$ measures the elapsed time as an object moves from an initial position $\vec{r}_i$ at time t_i to a final position $\vec{r}_f$ at time t_f. Its value is independent of the specific clock used to measure the times.

To summarize the main idea of this section, we have introduced clocks and rulers into our movies in order to make measurements of *when* each frame was exposed and *where* the object was located in each frame. Different observers of the motion, each making their own separate movie, might choose different rulers and different clocks. A particular position $\vec{r}$ or time t is ambiguous because each is measured relative to an arbitrarily chosen origin. However, all independent observers will get the *same* values for the displacements $\Delta\vec{r}$ and the time intervals Δt because these are independent of the specific coordinate system used to measure them. Consequently, it is the displacements $\Delta\vec{r}$ and the time intervals Δt that have physical significance, leading us to conclude that they form a good description of motion.

EXAMPLE 1-1   A mountain climber is climbing straight up a rope when—suddenly—it breaks! He falls, but lands safely in a deep snow bank below. Draw a motion diagram for the climber. Be sure to show and label all displacement vectors.

SOLUTION The motion in which we are interested is from the instant of time when the rope breaks until the instant at which the climber comes to rest in the snow. These two frames of film are labeled in Fig. 1-15. The point at which the climber hits the snow bank is also labeled. Initially the climber is speeding up as he falls, so the displacement vectors are increasing in length from one frame to the next. Once the climber hits the snow he begins slowing, so the displacement vectors then get shorter until he stops. It is reasonable to assume that the climber's stopping distance in the snow is less than the distance he falls to the snow, but we do not want to make his stop *too* quick. Notice that when an object either starts from rest or ends at rest, as is the case here, the initial (or final) dots are *as close together* as you can draw the displacement vector arrow connecting them. In addition, just to be clear, you should write "starts" or "stops" beside the initial or final dot. It is important to distinguish stopping from merely slowing down.

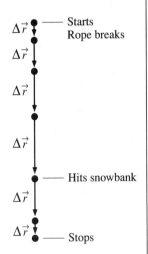

FIGURE 1-15 Motion diagram of the falling mountain climber.

1.5 Velocity

As you know, different objects can move at different speeds, and even one object can move at different speeds at different times. To extend our study of motion, it would be useful to have some way to measure "fastness" or "slowness." The only *direct* measurements that we can make on our motion diagrams are those of position and time. By a "direct measurement" we mean reading a number directly from some measuring instrument, such as a ruler or a clock. Quantities such as the displacement $\Delta\vec{r}$ or the time interval Δt are not found by directly reading instruments but by *calculating* various combinations of direct measurements. The displacement $\Delta\vec{r}$, for example, is the difference between the two directly measured quantities $\vec{r}_f$ and $\vec{r}_i$. These calculated combinations of direct measurements are called *indirect measurements*. Because we do not have a speed-measuring instrument in a motion diagram, a determination of fastness or slowness is going to have to be an indirect measurement made by combining the measured values of $\vec{r}$ and t.

During a given time interval, a speeding bullet has a larger displacement than a speeding snail. The quickness of the bullet, in comparison to the snail, is expressed through its larger value of $\Delta\vec{r}$ during a *fixed* time interval. (Recall that the magnitude of $\Delta\vec{r}$ is just the distance traveled between the initial and final positions.) Suppose, though, that we timed both the bullet and the snail as they each participated in the 100-yard dash. In this case, both have the same value of $\Delta\vec{r}$ but different values of Δt. The faster bullet has a smaller value of Δt as it moves through a *fixed* distance. Therefore, it seems that both the displacement *and* the time interval must enter into a definition of fastness or slowness. What we need is some combination of $\Delta\vec{r}$ and Δt that produces a larger value for faster objects and a smaller value for slower objects.

Consider the ratio $\Delta\vec{r}/\Delta t$. This ratio is a vector, because $\Delta\vec{r}$ is a vector, and it has both a magnitude and a direction. The size, or magnitude, of this ratio will be larger for a fast object than for a slow object. This is because either $\Delta\vec{r}$ will be larger for a faster object than a slower one when Δt is the same, or because Δt will be smaller for the faster object than for the slower one when $\Delta\vec{r}$ is the same. Thus, this ratio meets our goal of having a larger value for faster objects and a smaller value for slower ones. It is convenient to give this ratio a name, so we call it the **average velocity** during the time interval Δt and give it the symbol $\vec{v}$.

The average velocity of an object during the time interval Δt, during which the object undergoes a displacement $\Delta\vec{r}$, is a vector defined as

$$\vec{v} = \frac{\Delta\vec{r}}{\Delta t}.$$

The magnitude of the average velocity vector is called the object's **speed**.

There are several things here worth noting:

1. *Velocity* is a vector quantity, meaning that its direction is an important factor. *Speed*, on the other hand, is simply the *magnitude*, or size, of the velocity vector without regard to its direction. To say that an object is moving at 30 miles per hour is to give its speed, not its velocity. The velocity combines speed with direction, so to specify the velocity we must say that the object is moving at 30 miles per hour to the east, or 10 miles per hour straight up, or whatever is appropriate. In common language we do not make a distinction between speed and velocity, but in physics *the distinction is very important*. You will meet other examples as we go along of how the use of words in physics is more precise than their use in common language.

2. It is possible that the speed of an object will change several times during the time interval Δt. We cannot tell this from a motion diagram because all we have is the rather limited information of the particle's positions $\vec{r}_i$ and $\vec{r}_f$ in two frames of film taken a time interval Δt apart. Thus, all we can learn from the ratio $\Delta\vec{r}/\Delta t$ is the average of

the velocity during this time interval. We will need the mathematics of calculus, which we will introduce in Chapter 3, before we can describe the instantaneous velocity. For now, we will (for convenience) generally omit the word "average" and just refer to "velocity," but you should keep in mind that we are talking about the average of the velocity between two successive frames of the film.

3. We had a specific goal, namely to find a way to measure fastness or slowness. Analysis of the situation led us to consider the ratio $\Delta \vec{r}/\Delta t$. Only after this ratio appeared to be a useful concept did we decide to give it a name. It is the *idea* that is important, not the name! Many people have the impression that science consists of memorizing definitions and formulas. This impression is mistaken. While we do give names to ideas in order to talk about them, science is about the ideas themselves and how they are used to gain understanding and knowledge.

The ratio of a quantity to a time is called a **rate**, and we read or say it as something "per time." Velocity is the *rate of change of position*, usually expressed as distance *per time*. You are probably most familiar with velocity (or, correctly, speed) as being a certain number of "miles per hour." If you live in a country that uses the metric system, your car's speed is measured in "kilometers per hour." In physics, we will most frequently use units of "meters per second." The idea of a rate, though, is more general. There are birth rates (births per year), cancer rates (new cancers per year), and interest rates (dollars per year).

We will look into the mathematics of rates more carefully in Chapter 3; for now, we only need to note that the time unit can be chosen as needed. Our motion diagrams have a very natural unit of time—the *frame*. We will define *one frame* to be the time interval between the exposure of one frame of film and the next. You can think of the movie camera as being a steady and regular clock, with the time between "ticks" being one frame. The velocity, as we have defined it, will then measure the rate of change of position *per frame* of film.

This may seem a bit silly, but it has a very important advantage for motion diagrams. The displacement vectors $\Delta \vec{r}$ that we draw on motion diagrams, such as in Fig. 1-14, are specifically the displacement of the object during an elapsed time interval of one frame. Consequently, the average velocity of the object during that interval is

$$\vec{v} = \frac{\Delta \vec{r}}{\Delta t} = \frac{\Delta \vec{r}}{1 \text{ frame}} = \Delta \vec{r} \text{ per frame.} \tag{1-2}$$

Notice that we haven't specified any units of length. These units could be meters or miles, but they are included *within* the symbol $\Delta \vec{r}$. If $\Delta \vec{r}$ is measured in meters, then the complete units of $\vec{v}$ will be "meters per frame."

Equation 1-2 says that the vector $\vec{v}$ and the vector $\Delta \vec{r}$ both have the same magnitude and the same direction *if* we measure time intervals in frames. Thus $\vec{v}$ and $\Delta \vec{r}$ appear as identical vectors on a motion diagram! Consequently, the vectors previously labeled as $\Delta \vec{r}$ could equally well be labeled as $\vec{v}$. Be careful, though. If you had six apples and six oranges, they would *appear* the same if you wrote their numbers on a piece of paper— namely a "6" for both. But apples are not oranges. Likewise here. Having made a particular choice of units for Δt, the vectors $\vec{v}$ and $\Delta \vec{r}$ *appear* to be the same even though velocity, measured as a length per frame, and displacement, measured simply as a length, are most definitely not the same thing.

Figure 1-16 shows four frames from a motion diagram of a tortoise racing a hare. The vectors connecting each dot to the next, that we previously labeled as displacement vectors $\Delta \vec{r}$, are now identified as velocity vectors $\vec{v}$. That is the consequence of Eq. 1-2. The length of each vector—its magnitude—represents the speed of the object, so we easily see from the diagram that the hare (longer velocity arrows) moves faster than the tortoise (shorter velocity arrows). Notice that the hare's velocity vectors do not change—each has the same length and direction. (Mathematical aside: The location of a vector is not important. A vector is defined as a quantity having size and direction, so two vectors' arrows having the same length and same direction are equal to each other no matter where they are drawn on the page.) The hare's velocity vector is thus unchanging, and we say the hare is moving with **constant velocity**. The tortoise is also moving with its own constant velocity.

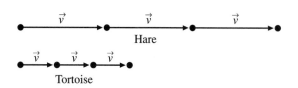

FIGURE 1-16 Motion diagram of a tortoise racing a hare.

To understand a concept, it often helps to look at it from several different perspectives. We have been looking at $\vec{v}$ in terms of the displacement $\Delta \vec{r}$, but let's turn that around. Suppose a rock is at rest, as shown in Fig. 1-17a. The displacement of the rock from one frame to the next is $\Delta \vec{r} = 0$, so the velocity of the rock is $\vec{v} = 0$. (Note that being *at rest* is a special case of constant velocity.) Now what happens if the rock suddenly acquires a velocity? During the time interval of one frame, as you can see in Fig. 1-17b, the rock is displaced from its initial position $\vec{r}_i$ to a new position $\vec{r}_f$. Its displacement $\Delta \vec{r}$ is no longer zero. This implies that *it is the velocity that causes the position to change*. If the velocity is zero, there will not be any change in position, but if the velocity is non-zero, there will be a change in position.

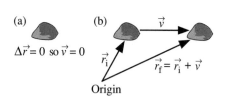

FIGURE 1-17 Motion diagram of a rock. In a) the rock is at rest. In b) the rock acquires a velocity and is displaced from $\vec{r}_i$ to $\vec{r}_f$.

Now continue with the special case, applicable to motion diagrams, where $\Delta t = 1$ frame. Because $\vec{v} = \Delta \vec{r}$ per frame and because $\Delta \vec{r} = \vec{r}_f - \vec{r}_i$, we can combine these to give

$$\vec{r}_f = \vec{r}_i + \vec{v}. \tag{1-3}$$

Equation 1-3 is telling us that an object's velocity is what changes the object's position from $\vec{r}_i$ in one frame to $\vec{r}_f$ in the next. From this perspective we can think of velocity as being the "cause" or "reason" that the position of the object changes.

Caution: Equation 1-3 is valid *only* if $\vec{r}_i$ and $\vec{r}_f$ represent the positions in adjacent frames of the film and if $\vec{v}$ is measured in units of length per frame. This is true for our motion diagrams, so we will use Eqs. 1-2 and 1-3 to help develop intuition about motion. But we will have to find more general expressions, where Δt is measured in seconds, before we can do any real calculations. That will be the subject of Chapter 3.

Figure 1-18 shows two examples where just two frames of film have been isolated. The position of the object in each frame is shown by the dots. As we found previously, the displacement vector from the initial to the final position gives the velocity $\vec{v}$. We have also included the position vectors $\vec{r}_i$ and $\vec{r}_f$, based on an arbitrary choice of origin near the left edge. By referring to the rules for vector addition in the last section, you can see that, indeed, $\vec{r}_f = \vec{r}_i + \vec{v}$. Notice that the objects depicted in Figs. 1-18a and 1-18b start with the same initial position (same $\vec{r}_i$) and are moving with the same speed (same *length* of the $\vec{v}$ vector), but they end up with very different final positions. This emphasizes that velocity is a vector and that the *direction* of $\vec{v}$ is just as important as its size.

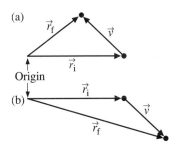

FIGURE 1-18 Parts a) and b) show two objects starting with the same initial position and moving with the same speed. The difference in the *direction* of their velocity vectors leads to different final positions.

EXAMPLE 1-2 A traffic light turns green and a car accelerates, starting from rest, up a 20° hill. What is the motion diagram for the car?

SOLUTION The car's motion takes place along a straight line, but the line is neither horizontal or vertical. Because a motion diagram is made from frames of a movie, it will show the object moving with the correct orientation—in this case, at an angle of 20°. Figure 1-19 shows several frames of the motion diagram, where we see the car speeding up. The displacement vectors have been drawn from each dot to the next, but then they have been identified and labeled as velocity vectors $\vec{v}$. That is the significance of Eq. 1-2. In a motion diagram, $\vec{v}$ is found to have the same length and direction as $\Delta \vec{r}$. Notice that because the car starts from rest, the first arrow is drawn as short as possible and the first dot is labeled "starts."

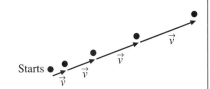

FIGURE 1-19 Motion diagram showing the velocity vectors of a car accelerating up a hill.

1.6 Acceleration

The goal of this chapter is to find a set of concepts with which to describe motion. Position, time, and velocity are clearly important concepts, and at first glance they might appear to be sufficient. But that is not the case. Sometimes an object's velocity is constant, as it was in Fig. 1-16, but more likely the velocity is changing. We need one more motion concept to describe the *change* in velocity.

Because velocity is a vector, it can change in two different ways: either the magnitude can change (equivalent to a change in speed) or the direction can change. Example 1-2 and Fig. 1-19 showed the motion diagram of a car speeding up. This was an example in which the magnitude of the velocity vector changed but not the direction. An example where only the direction changes, but not the magnitude, would be a runner moving around a circular track at constant speed, as shown in Fig. 1-20. Although the runner's speed is not changing (the lengths

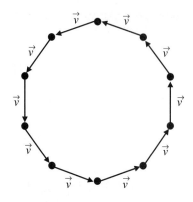

FIGURE 1-20 Motion diagram of a runner on a circular track, showing a changing velocity vector.

of all the velocity vectors are the same), the direction of each velocity vector is different. Thus Fig. 1-20 represents a changing velocity, despite the fact that the speed is constant.

How can we measure the change of velocity in a meaningful way? When we wanted a way to measure changes in position, the ratio $\Delta \vec{r}/\Delta t$, which is the rate of change of $\Delta \vec{r}$, proved to be useful. If either $\Delta \vec{r}$ increases or Δt decreases, the ratio becomes larger and measures "more fastness." By analogy, to measure changes of velocity we might consider the ratio $\Delta \vec{v}/\Delta t$, where $\Delta \vec{v} = \vec{v}_f - \vec{v}_i$ is the *change of velocity*. The ratio $\Delta \vec{v}/\Delta t$ is again a rate, in this case the *rate of change of velocity*. But what does it measure?

Consider two cars—a Volkswagen and a Porsche. Let them start from rest, and then measure their velocities after an elapsed time of 10 seconds. Both have had a change of velocity, having started with $\vec{v}_i = 0$ and having ended with a non-zero velocity. The Porsche, we can assume, will have had a larger $\Delta \vec{v}$, and consequently it will have the larger value of the ratio $\Delta \vec{v}/\Delta t$.

Restart the two cars, but now measure the time interval Δt it takes for each of them to reach a speed of 60 miles per hour. The Porsche will have a smaller value of Δt, which again gives it a larger value of the ratio $\Delta \vec{v}/\Delta t$. This ratio thus appears to measure how quickly the car speeds up, having a larger value for objects that speed up quickly and a smaller value for objects that speed up slowly. This seems to be a useful concept, so let's call it the **average acceleration** during the time interval Δt and give it the symbol $\vec{a}$.

> The average acceleration of an object during the time interval Δt, during which the object undergoes a velocity change $\Delta \vec{v} = \vec{v}_f - \vec{v}_i$, is a vector defined as
> $$\vec{a} = \frac{\Delta \vec{v}}{\Delta t}.$$

Note that acceleration, like position and velocity, is a vector, so both its magnitude and its direction are important pieces of information.

Acceleration is a fairly abstract concept. Position and time are our real hands-on measurements of an object, and they are easy to understand. You can "see" where the object is located and what time the clock reads. Velocity is a bit more abstract, being a relationship between the change of position and the change of time. You cannot directly "see" velocity, but motion diagrams help us visualize velocity as the vector arrows connecting one position of the object to the next. Acceleration, though, is a very abstract idea about changes in the already abstract concept of velocity. We cannot "see" acceleration directly, and we cannot even visualize it directly from the motion diagrams, like we can the velocity. Yet it is essential to develop a good intuition about acceleration because it will be a key concept for understanding why objects move as they do.

Let us again adopt a time unit of *one frame*. As the motion diagram proceeds from one frame to the next, the velocity changes from $\vec{v}_i$ to $\vec{v}_f$. The *change* in the object's velocity between these two adjacent frames is thus $\Delta \vec{v} = \vec{v}_f - \vec{v}_i$. The average acceleration during this interval is

$$\vec{a} = \frac{\Delta \vec{v}}{\Delta t} = \frac{\Delta \vec{v}}{1 \text{ frame}} = \Delta \vec{v} \text{ per frame.} \tag{1-4}$$

Compare this to Eq. 1-2 where we determined the average velocity between two adjacent frames.

Within our special time units, in which $\Delta t = 1$ frame, the acceleration is given by simply

$$\vec{a} = \vec{v}_f - \vec{v}_i \tag{1-5}$$

as the object's velocity changes from $\vec{v}_i$ to $\vec{v}_f$. Note that the Caution following Eq. 1-3 applies here as well. Equation 1-5 is strictly valid *only* if $\vec{a}$ is measured in the special units of $\Delta \vec{v}$ per frame. Our purpose here is to gain a graphical understanding of acceleration. We will introduce a more general expression, where Δt is in seconds, before starting to do numerical calculations.

Vector subtraction is sometimes hard to visualize, so let's turn Eq. 1-5 around and write it as

$$\vec{v}_f = \vec{v}_i + \vec{a}. \tag{1-6}$$

This is exactly analogous to Eq. 1-3, and we can use a similar interpretation: *acceleration is what causes the velocity to change.* That is, if the object has velocity $\vec{v}_i$ in one frame of film, its velocity in the next frame will be changed to $\vec{v}_f = \vec{v}_i + \vec{a}$. If the acceleration is zero, there will not be any change in the velocity, but a non-zero acceleration will result in a change of velocity. From this perspective we can think of acceleration as being the "cause" or "reason" that the velocity of the object changes.

This begins to make a chain of causes. Acceleration causes the velocity to change, and the velocity then causes the position to change. But what causes the acceleration to change? We do not want to get into an infinite chain of changes causing changes causing changes, because that would not explain anything. Fortunately, Newton discovered that the acceleration of an object is determined by the forces exerted on it—as we will see in Chapter 4. This provides a finite chain of causes:

$$\text{forces} \Rightarrow \text{acceleration} \Rightarrow \text{velocity} \Rightarrow \text{position.}$$

Our ultimate goal is to learn how an object moves—that is, how its position varies with time—as a consequence of forces exerted on it. You can now see how acceleration and velocity play essential roles.

Now let's look at how we can determine $\vec{a}$ in our motion diagrams. In Fig. 1-21a there are three frames from the motion diagram of a car that is speeding up. Because the speed is increasing, the car moved further between frames 2 and 3 than it did between frames 1 and 2, so the final velocity vector $\vec{v}_f$ is longer than the initial vector $\vec{v}_i$. Because the velocity is changing, there must be an acceleration vector $\vec{a}$ that, when added to $\vec{v}_i$ in accordance with Eq. 1-6, gives $\vec{v}_f$ as the vector

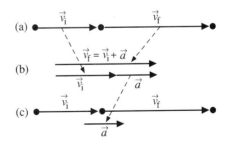

FIGURE 1-21 Construction of the acceleration vector $\vec{a}$ from $\vec{v}_i$ and $\vec{v}_f$.

sum. We can use Eq. 1-5 and the rules of vector subtraction (described in Section 1.4) to determine $\vec{a}$.

First, reposition $\vec{v}_i$ and $\vec{v}_f$ such that their *tails* are located together but they are otherwise unchanged, as shown in Fig. 1-21b. (Note that the two vectors in the figure are slightly separated so that you can see them clearly. They would be placed on top of each other, hiding $\vec{v}_i$, if they were drawn exactly with tails together.) Next, draw a vector from the *tip* of $\vec{v}_i$ to the *tip* of $\vec{v}_f$. According to the rules of vector subtraction, this new vector is $\vec{v}_f - \vec{v}_i$. But, by Eq. 1-5, this is none other than the acceleration vector $\vec{a}$! In the finished motion diagram, as shown in Fig. 1-21c, the acceleration $\vec{a}$ is placed below the midpoint of the two velocity vectors that it links together.

We can summarize the procedure for finding an acceleration vector $\vec{a}$ from two velocity vectors $\vec{v}_i$ and $\vec{v}_f$:

1. Draw the two velocity vectors with their *tails* together.

2. Draw a vector from the *tip* of $\vec{v}_i$ to the *tip* of $\vec{v}_f$. This is $\vec{a}$, because $\vec{v}_f = \vec{v}_i + \vec{a}$.

3. Draw $\vec{a}$ on the motion diagram, locating it at the *midpoint* of $\vec{v}_i$ and $\vec{v}_f$.

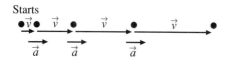

FIGURE 1-22 A complete motion diagram for a car accelerating from rest.

In Fig. 1-21, we focused in on just three frames in order to see how the acceleration was found. A full motion diagram, though, may have more than three frames. Figure 1-22 shows the full motion diagram for a car starting up from rest. A complete motion diagram consists of:

1. The position (shown as a dot) of the object in each frame of the film. You should use five or six dots so that the motion is clear without overcrowding the picture. More complex motions may need more dots.

2. The velocity vectors, showing the average velocity of the object between each frame of film and the next. For clarity, we will now draw the velocity vectors *beside* the position dots, rather than directly between the dots. (This does not change the velocities, because the vectors have the proper length and direction.) There is *one* velocity vector linking each *two* position dots. This emphasizes the idea that the velocity $\vec{v}$ changes position $\vec{r}_i$ to the next position $\vec{r}_f$.

3. The acceleration vectors, showing the average acceleration of the object between each velocity and the next. Because the acceleration $\vec{a}$ changes velocity $\vec{v}_i$ to velocity $\vec{v}_f$, there is *one* acceleration vector linking each *two* velocity vectors. Each acceleration vector is drawn, as in Fig. 1-22, at the midpoint of the two velocity vectors it links.

A complete motion diagram contains all the information that we might want to know about the object's motion. Now let's consider some other examples.

Figure 1-23 shows the motion diagram for the first expedition to land on Mars. The spacecraft descends and slows, using its rockets, until it comes to rest on the surface. Notice that the spacecraft positions get closer together as it slows. The velocity vectors are drawn connecting each position to the next. (Note: If you saw *only* the position dots, you would have no way to tell if the spacecraft is landing or taking off. You must have knowledge as to the *direction* of motion before you can draw the velocity vectors.) To determine the acceleration, two velocity vectors are repositioned to the side, tails together, and a vector is drawn from the tip of $\vec{v}_i$ to the tip of $\vec{v}_f$. This is illustrated in the inset to Fig. 1-23. Note that $\vec{v}_i$ is always the velocity vector that occurs earlier in time. All the other acceleration vectors will be similar, because for each pair of velocity vectors the earlier one ($\vec{v}_i$) is longer than the later one ($\vec{v}_f$). Having thus determined that all of the acceleration vectors point upward, an appropriate acceleration vector is drawn next to the midpoint between each pair of velocity vectors.

Notice something interesting in Figs. 1-22 and 1-23. In Fig. 1-22, where the object is speeding up, the acceleration and velocity vectors point in the *same direction*. But in Fig. 1-23, where the object is slowing down, the acceleration and velocity vectors point in *opposite directions*. This is consistent with thinking of acceleration as the "cause" of the velocity change. For an object to go faster, its velocity vectors have to get longer. The acceleration vector needs to "push on" the velocity vectors in the same direction they are already pointing. To slow an object down, though, the acceleration needs to "push against" the velocity vectors to make them shorter. This result is always true for linear motion:

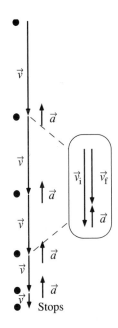

FIGURE 1-23 Motion diagram of a spacecraft landing on Mars. The inset shows how the acceleration $\vec{a}$ is found.

An object is speeding up if and only if $\vec{v}$ and $\vec{a}$ point in the same direction.

An object is slowing down if and only if $\vec{v}$ and $\vec{a}$ point in opposite directions.

An object's velocity is constant if and only if $\vec{a} = 0$.

In everyday language we refer to slowing down as "decelerating." This is not a term that we use in physics. Both speeding up and slowing down are changes in the velocity and consequently, by our definition, *both* are accelerations! In physics, acceleration refers to changing the velocity, no matter what the change is, and not, as it does in everyday language, just to speeding up. We have noted previously that we will be using words more precisely in physics than is the case in ordinary English, and this is another important example.

Notice also, in Figs. 1-22 and 1-23, that when an object is starting from rest or ending at rest the position dots get very close together. This is important for you to keep in mind as you draw your own motion diagrams. Diagrams where the dots get closer together, but not *too* close, would indicate an object that was slowing but not stopping. If the object happens to be at rest at some point in the middle of a motion, make sure that your diagram shows it clearly and labels that point "$\vec{v} = 0$."

EXAMPLE 1-3 A ball rolls along a smooth, horizontal surface at constant speed and then heads down a ramp (Fig. 1-24a). What is the motion diagram for the ball?

SOLUTION: In Fig. 1-24b we have shown the position dots for the ball and connected them with velocity vectors. Notice that we've used five dots for each half of the motion with one dot—exactly at the top of the ramp—common to both the horizontal and downhill motion. This allows us to separate the two halves of the motion cleanly. For the horizontal portion of the motion both the speed *and* direction are constant, so the velocity is constant. The ball then speeds up as it moves down the ramp. This happens gradually, though, so the first velocity vector on the ramp should be only slightly longer than the last horizontal velocity vector. A sudden drastic change in length would not be an accurate portrayal of the motion. (If you're not sure about this, think about *coasting* on your bicycle first across a level surface, then down an incline. Your speed increases only gradually as you start down the incline.)

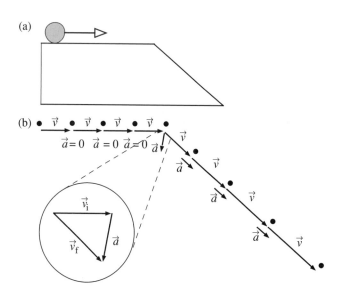

FIGURE 1-24 Motion diagram of a ball on a ramp. The inset shows how the acceleration was determined at the top of the ramp.

The constant velocity during the horizontal motion gives an acceleration $\vec{a} = 0$, because $\vec{v}_i$ and $\vec{v}_f$ are the same. Still, we have written $\vec{a} = 0$ at each location of the acceleration vector; a vector of zero length is still a vector. The downhill motion is a linear acceleration. It is no different than the accelerating car in Fig. 1-22, just tilted at an angle. Because the ball is speeding up, $\vec{v}$ and $\vec{a}$ must point in the same direction. The velocity vectors are increasing in length, but not changing direction, so the acceleration needs to be parallel to the velocity.

We are still left with the position dot at the top of the ramp. The velocity immediately after this dot (the first downhill velocity) differs from the velocity immediately preceding the dot (the last horizontal velocity) in both length and direction. This is a new situation, but we can still use vector subtraction to find $\vec{a}$. As the inset shows, the two velocities are positioned with their tails together and $\vec{a}$ is found as the vector drawn from the tip of $\vec{v}_i$ to the tip of $\vec{v}_f$. We then position this $\vec{a}$ in the motion diagram, as shown.

Notice in Example 1-3 that the acceleration at the top of the ramp is not parallel to either velocity vector! How can we make sense of this? Keep in mind that the acceleration is what changes the velocity. The velocity leading up to this point was horizontal, whereas the velocity afterward is not much different in length but is tilted down. To change the first velocity vector into the second, we need to "push" its tip down and slightly to the left. That is what you see $\vec{a}$ doing—pushing on the nose of vector $\vec{v}_i$ in the direction needed to change it into vector $\vec{v}_f$!

This is *the* key idea for developing an intuition for acceleration when the motion is not linear. It would be a good idea for you to think very carefully about the previous paragraph and to draw the figure inset for yourself. If you can convince yourself that $\vec{a}$ *must* point this direction, rather than just accepting it because the book says so, then you will be well on your way to a successful understanding of this essential concept of motion.

EXAMPLE 1-4   Bill rides the Ferris wheel at an amusement park. What is Bill's motion diagram if the Ferris wheel turns at a constant speed?

SOLUTION   Let's start by using ten frames of film, so we can see one complete revolution of the Ferris wheel, as shown in Fig. 1-25. The seat on the Ferris wheel moves in a circle at a constant *speed*, so the motion diagram will show Bill moving around the edge of a circle with equal distances between one dot and the next. As before, the velocity vectors are found by connecting each dot to the next. Note that the velocity vectors (like all vectors) are *straight lines*, not curves, because the displacement vectors are *not* the same as the actual path followed.

Because the Ferris wheel turns at a constant speed, each of the velocity vectors has the *same length*. However, the *direction* of the vector keeps changing, so the velocity, which includes both size and direction, is *not* constant. Constant velocity can occur *only* along a straight line because neither the size nor the direction of the vector can change. Circular motion can be at constant speed, but it can never occur with constant velocity.

Because the velocity vectors are changing, Bill *must* be accelerating! This is not a "speeding up" or "slowing down" acceleration because the speed is constant, but it is still a change of the velocity with time. The

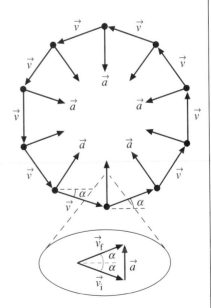

FIGURE 1-25 Motion diagram of a passenger on a Ferris wheel. The inset shows how $\vec{a}$ is found at the bottom of the circle.

inset to Fig. 1-25 shows how to find the acceleration at the dot right at the bottom of the circle. Vector $\vec{v}_i$ is the velocity vector that leads into this dot while $\vec{v}_f$ moves away from it. From the circular geometry of the main figure, the two angles marked α are equal. These angles must be maintained in the inset, where $\vec{v}_i$ and $\vec{v}_f$ are positioned with tails together. Vector $\vec{a}$ is drawn from the tip of $\vec{v}_i$ to the tip of $\vec{v}_f$. The three vectors together make an isosceles triangle because the two velocities have the same length. The sides are tilted at equal angles α above and below the horizontal, so vector $\vec{a}$ must be perpendicular to the horizontal and thus is exactly vertical. When $\vec{a}$ is positioned on the motion diagram, we see that the acceleration vector points directly to the center of the circle!

•

No matter which dot you select on the motion diagram in Example 1-4, the velocities leading to and away from that dot change in such a way as to require an acceleration pointing directly to the center of the circle. You should convince yourself of this by finding $\vec{a}$ at several other points around the circle. Why should this be? Remember that the acceleration is what changes vector $\vec{v}_i$ into $\vec{v}_f$. In this case, the velocity vectors are all tangent to the circle. We need to change their directions *without* changing their lengths. The only way to push the tip of a velocity vector so that its direction changes but not its length is to push at right angles to the circle's perimeter—which is directly toward the center.

An acceleration that always points directly toward the center of a circle is called a *centripetal acceleration*. The word "centripetal" comes from a Greek root meaning "center seeking." We will have a lot to say about centripetal acceleration in later chapters.

EXAMPLE 1-5 The Hulk is an Olympic shot-putter. What is the motion diagram for the shot from the moment it leaves The Hulk's hand until it hits the ground?

SOLUTION: An object moving through the air is a **projectile**. You probably know, from watching baseballs, basketballs, rocks, and so on, that projectiles move in some sort of an arc. Projectiles also slow down as they rise and speed up again as they fall. The top of the arc is where the object has the slowest motion—which is why the ball seems to hesitate there before crashing down again—but it is not at rest. The motion diagram of the shot is shown in Fig. 1-26. Notice that the camera was located off to the side so that we can see the motion as occurring in a vertical plane.

The shot has a curved motion, and we have drawn the velocity vectors on the inside of the curve. Notice that the shot does *not* start from rest—it leaves The Hulk's hand with a non-zero initial velocity. Because the shot slows while rising, the velocity vectors along the ascent are getting shorter. The vectors then lengthen again as the shot falls. The last position is right at ground level, at the instant of impact.

There must be an acceleration of the shot because both the size and the direction of the velocity are changing. The inset shows the determination of $\vec{a}$ at one point along the trajectory. The acceleration must be pointed more-or-less downward, but we cannot determine the exact direction of $\vec{a}$ without further information about the exact shape of the arc. It turns out, for projectile motion, that the acceleration vector at every point is *straight down* and of *constant length*—a conclusion that we will justify later but will simply assert, without proof, for now. This very specific acceleration is called the *acceleration due to gravity*. It will be discussed more thoroughly in Chapter 3. We have used this result, though, to indicate that the acceleration is the same at each point.

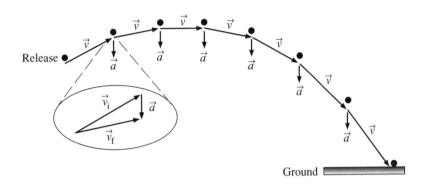

FIGURE 1-26 Motion diagram of the shot thrown by The Hulk.

[**Photo suggestion: Strobe photo of projectile motion for part a of Fig. 1-26.**]

It is a good idea to compare the projectile motion in Example 1-5 to the constant-speed circular motion in Example 1-4. While both objects follow curves, the shapes of the curves are very different as a result of the very different acceleration vectors.

Example 1-5 raises an important aspect of how we interpret problems in physics. The Hulk clearly had to flex his muscles and move various parts of his body to get the shot moving. That part of the motion is complicated. But from the moment the shot leaves his hand until it hits the ground, the shot has a fairly simple motion. The actual impact, and any subsequent motion, again becomes very complicated. In this case, the question explicitly asks for only the motion from the time The Hulk releases the shot until it hits the ground. The complex motions of the throw and the impact are not part of the problem. But many problems will not be this specific. *You* have to supply the interpretation of when the problem "begins" and "ends." The key is to focus on isolating those parts of the problem that are essential while disregarding the non-essential aspects.

EXAMPLE 1-6   What is the motion diagram of a ball tossed straight up in the air?

SOLUTION This problem calls for some interpretation. Should we include the toss itself, or only the motion after the tosser releases the ball? Should we include the ball hitting the ground? It appears that this problem is really concerned with the ball's motion through the air. Consequently, we will focus on the ball's motion from the moment that the tosser releases the ball until a very short time before the ball hits the ground. Neither the toss nor the impact will be considered.

The motion diagram presents a difficulty here because the ball retraces its route as it falls. The upward motion and downward motion will appear on top of each other, leading to confusion. We can avoid this difficulty by making a horizontal separation of the upward

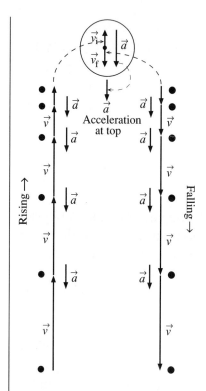

FIGURE 1-27 Motion diagram of a ball tossed straight up in the air. The rising and falling motions are displaced, for clarity, but actually take place along the same vertical line.

motion from the downward motion. This will not affect our conclusions because it will not change any of the vectors. Figure 1-27 shows the motion diagram drawn this way.

As the ball rises, it exhibits linear motion that is slowing down. We already know that such motion has the acceleration vectors pointing opposite the velocity vectors, and they are shown accordingly. Likewise, the falling ball is linear motion that is speeding up, so $\vec{a}$ and $\vec{v}$ point in the same direction. The acceleration vector points downward, regardless of whether the ball is rising or falling. Therefore, both "speeding up" and "slowing down" occur with the *same* acceleration vector. This is an important conclusion, and it is worth pausing to make sure you understand it.

Now let's look at the top of the ball's path. As the ball approaches the top, the velocity vectors are pointing upward but getting shorter and shorter. As the ball starts to fall, the velocity vectors are pointing downward and getting longer. There must be a moment—just an instant as $\vec{v}$ goes from up to down—when the velocity is zero. Indeed, the ball's velocity *is* zero for an instant at the precise top of the motion! But what about the acceleration at the top? The inset shows the acceleration determined from the last upward velocity before the top point and the first downward velocity. From the figure we find that the acceleration is *not* zero at the top but is pointing downward—just as it does elsewhere in the motion.

It may be hard to understand how the ball can be accelerating at the top of its path. Many people expect the acceleration to be zero at that point. The idea to keep in mind is that the velocity at that point *is* changing—from up to down. It is only *instantaneously* zero, it does not stay zero for any finite time. If the velocity is changing, there *must* be an acceleration. To turn the velocity vector from up to down, the acceleration must "push" the tip of the velocity vector downward. This requires a downward-pointing acceleration vector. Another way to think about this is to note that zero acceleration would mean no change of velocity. When the ball reached zero velocity at the top, it would hang there and not fall if the acceleration were also zero! The non-zero acceleration at the top is what causes the ball to move back down.

The motion of the ball in Example 1-6 is really just a special case of projectile motion. The difference between the ball and the shot thrown by The Hulk is that the ball moves through the air with the special starting condition of moving exactly vertically. We saw in Example 1-5 that the acceleration is a constant downward vector for projectile motion. That is exactly what we have found again for the ball!

1.7 From Words to Symbols

Physics is not mathematics. Math problems are precise and clearly stated, such as "What is 2 + 2?" Physics is about the world around us, and to describe that world we must use language. Now language is wonderful—we couldn't communicate without it—but it is also imprecise and sometimes ambiguous. That makes for good poetry but lousy science.

The questions we ask in physics are ones such as, "How much fuel do I need in this rocket to get it into orbit?" or "At what temperature should I run this engine for maximum efficiency?" or "What subatomic particles will be created if two protons collide head-on with an energy of 200 GeV?" Answering such questions requires a lot more than simply typing numbers into your calculator or computer.

Part of the process of answering physics questions is evaluating information. A physics question will carry with it certain information, such as the mass of the rocket and the thrust of its engines. Is the information sufficient to answer the question, or is some item missing? Is all of the information relevant? For example, would knowing that the rocket is red rather than blue affect how you go about answering the question? Textbook problems in most books are rather artificial in that you are given precisely the right amount of information to work the problem—no more and no less. Problems in the "real world" are not this neat and clean, and real science and engineering problems require significant sorting and evaluating of the information to find out what is relevant and what is necessary for a solution. Some of the problems in this book will have unnecessary details or, occasionally, missing data so that you can gain practice at evaluating information.

The real challenge, after the information is evaluated, is to translate the words into symbols that can be manipulated, calculated, graphed, and so on. This translation from words to symbols is the heart of problem solving in physics, the aspect that sets it apart from pure mathematics, and the major stumbling block for many students. This is the point where ambiguous words and phrases must be clarified, where the imprecise must be made precise, and where you arrive at an understanding of exactly what the question is asking. The goal of this section is to begin the development of a strategy for translating words into symbols.

A Problem-Solving Strategy

Expert problem solvers never start a problem by searching for a formula. Instead, they invariably start by reading the problem several times, sketching a picture or two to clarify the situation, defining the symbols they expect to need, and reasoning out the problem *qualitatively*. This reasoning guides them to select one or more general principles that are relevant to the problem. Only then do they consider the mathematics that is needed to apply these principles. After arriving at an answer, an expert problem solver will check to see if the answer has the right units and if it seems to be "reasonable."

We are going to develop a *strategy* for solving physics problems that is similar to the one used by experts. The purpose of a strategy is to guide you in the right direction with minimal wasted effort. A proper strategy will help you to evaluate the information in a problem, to obtain a "feeling" for what is happening in the problem, and ultimately to arrive at a correct mathematical statement of the problem. Learning a strategy will also help you with the task of structuring your knowledge, because you will begin to see the close similarities in many seemingly different kinds of problems.

Our strategy for the first part of this text will have three basic parts:

1. The Pictorial Model: We will use a *pictorial model* to analyze the statement of the problem, evaluate the information, and make initial choices about coordinate systems and variables. This is a *qualitative* analysis of the problem. You can think of the pictorial model as being that portion of problem solving devoted to transforming the word statement of a problem into the symbolic, quantitative statement that will be necessary, in Step 3, for a successful mathematical solution. The difficulty many people experience with "word problems" stems from not knowing how to use a pictorial model. Most of this section is devoted to explaining and demonstrating the pictorial model.

2. The Physical Model: The pictorial model transforms words into symbols, but it is a technique that can be applied equally well to physics, to math, or to any other quantitative subject. We also need to evaluate the *physical* situation, and this is done with a *physical model*. The first piece of the physical model is just the motion diagram, which lets us determine the velocity and acceleration vectors of a moving object. These are essential pieces of information. The second piece of the physical model, which will be introduced in Chapter 4, is called the *free-body diagram*. It is a graphical analysis of the forces acting on an object. You will see, in upcoming chapters, how the information from the physical model can be inserted directly into the mathematical model, eliminating the need for "guessing" which equation to use!

3. The Mathematical Model: Only *after* the first two steps are complete will we introduce and solve specific equations. This last step is the *mathematical model* of the problem. If the first two steps have been successful, they will guide you—almost like magic!—to the proper equations. Once the equations are written, it is then "merely" a mathematics problem to solve them.

Our emphasis throughout this text will be on the first two steps. They represent the "physics" of the problem, as opposed to the mathematics of finding a solution for the resulting equations. Many examples will devote great attention to the details of the pictorial model and the physical model, but will then end with a rather general "Solution of these equations gives ..." This is not to say that the mathematical operations of solving the equations are always easy—in many cases they are not. But we will focus our attention on the *physics* of the problem.

Specific techniques for each step in the strategy will be introduced as needed. The important thing to remember, and one that we will continue to emphasize, is that there *is* a strategy for solving physics problems.

Motion in One Dimension

We have defined the position, velocity, and acceleration of an object with the *vectors* $\vec{r}$, $\vec{v}$, and $\vec{a}$. Chapter 2 will introduce the mathematics of vectors. Our goal in this section is simply to analyze the statement of a problem, without getting into the mathematics. For that purpose we can limit ourselves to straight-line motion in one dimension. In this case, the full machinery of vectors is not needed. We can measure the position, velocity, and acceleration of an object with the *scalar* quantities x, v, and a (or y, v, and a if the motion is vertical).

The symbol v does represent a velocity, though, and not simply the speed. Therefore, we still need to distinguish between motion to the right and motion to the left. We can do this by allowing v to be a positive number for motion toward the right and a negative number for motion toward the left. Similarly, for vertical motion, upward motion and downward motion can be characterized as positive and negative values of the scalar velocity v. We can also do the same with acceleration. An acceleration vector pointing toward the right (or upward) can be represented by a positive value of a, while an acceleration vector pointing to the left (or downward) will have a negative a. Chapters 2 and 3 will consider the mathematics of one-dimensional motion in more detail, but these simple rules are enough to understand the examples and homework problems of this chapter.

The Pictorial Model

In this section we want to focus not on solving physics problems, but simply on *describing* them. This is the first step, where expert problem solvers spend a large fraction of their problem-solving time. A well-described problem is already half solved!

Our initial concern is not with specific numbers, but simply with understanding the *information* we are given. But how do we go about describing the information in a useful way? To begin, if we use the methods of an expert problem solver, we could make a sketch of the situation. This helps us to recognize the different geometrical relationships in the problem. We could then add a coordinate system to our sketch, allowing us to make quantitative measurements. Next, we could list all the quantities mentioned in the problem—such as positions, velocities, and masses—and assign them appropriate symbolic names. Finally, we could identify the unknown quantity or quantities we have been asked to find.

In fact, the combination of all these steps is our starting point for solving a problem. Together they form what we will call the **pictorial model**. The pictorial model is the most literal description of the problem, and it is the starting point for the more abstract models we will use in later stages of the solution.

When you make a pictorial model of a problem, you should include the following:

1. A *sketch* of the situation, showing the object or objects at the beginning of the motion, at the end, and at any points in between where the character of the motion changes.

2. A *coordinate system* that you impose on the problem, including a choice of the origin and the axes. The axes and the origin should be labeled. It is often easiest to have the motion start at the origin, but this is not required and may not even be desirable in all problems.

3. *Symbols*, shown on the sketch, to indicate the quantities of time, position, and velocity for each object. These symbols, with identifying subscripts, should be placed at the beginning of the motion, at the end, and at any points in between where the character of the motion changes. Also show symbols for other relevant quantities, such as the mass of an object. This step *defines* the variables you will be using in the problem. *Every* variable that you use later in the mathematics should be defined at this point of the problem. Some of these variables will have known values, from the statement of the problem, while others are initially unknown. Regardless of whether or not

the values are known, the sketch should show *symbols* for the variables rather than numerical values.

4. *Arrows* indicating the acceleration vector as the object moves from one position to the next in your sketch. Give each vector arrow a symbol $\vec{a}$, with a subscript if there's more than one acceleration in the problem. Write "$\vec{a} = 0$" instead of drawing an arrow if the acceleration is zero during some interval of the motion.

5. A *table* listing the values for all the variables that are known from the statement of the problem or that can be quickly found with simple geometry. Velocities and accelerations should be given appropriate signs. The table should also indicate known relationships, even if the numbers are not known. For example, if you know that the velocity at point x_1 is equal to the velocity at x_2, you could indicate this by writing "$v_1 = v_2$" in the table. All conversions to a consistent set of units should be done at this point.

6. A *list* of the unknown quantity or quantities you have been asked to find.

Let's take a look at how to put these parts together. The following example illustrates how to construct a pictorial model for two problems that are typical of problems you might be asked to solve in the next few chapters.

EXAMPLE 1-7   Create a pictorial model for each of the following problems:

Problem 1: Billy accelerates his bike from rest to 6 m/s in 5 seconds. What distance has he traveled?

Problem 2: Motorcycle Momma Martha is cruising along on her cycle at 120 miles per hour when a small furry mammal runs onto the road. Martha has a 0.5 second response time before she applies the brakes, then she comes to a halt after skidding for 2.5 seconds. What is her total stopping distance?

SOLUTION In Problem 1, Billy starts from rest and accelerates steadily for 5 seconds. This problem has a clear beginning and end. The character of the motion—steady acceleration—is constant throughout. We begin by drawing a sketch, as shown in Fig. 1-28, showing Billy at the beginning and end of his motion. Notice that we have imposed a coordinate system on the problem. This includes choosing the axis we want to use (the *x*-axis in this case), labeling it, and—very importantly—clearly identifying the origin with a "0" next to it. In this case we chose to place the origin at the start of the motion, but nothing forced us to place it there. We could equally well have put the origin at the end of the motion or elsewhere. The important thing is to identify and label your choice clearly.

Figure 1-28 also labels the beginning and ending points in the motion, with x_0, v_0, and t_0 as the initial position, velocity, and time and x_1, v_1, and t_1 as the final values. The acceleration vector is shown and labeled as $\vec{a}$. We know $\vec{a}$ points toward the right because Billy is speeding up. The drawing thus *defines* these symbols. Note that all of these symbols are *lowercase*. We will have other uses later for many of these same letters written in uppercase (i.e., capital letters), so make sure you get in the habit of writing them in the proper case.

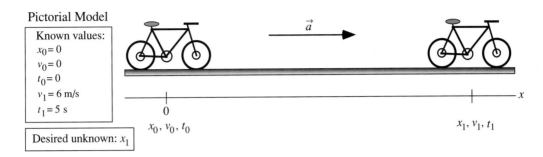

Pictorial Model

Known values:
$x_0 = 0$
$v_0 = 0$
$t_0 = 0$
$v_1 = 6$ m/s
$t_1 = 5$ s

Desired unknown: x_1

FIGURE 1-28 The pictorial model of Billy riding his bike (Example 1-7).

Finally, to complete the pictorial model, the symbols with known values are listed and the quantity x_1 is identified as the unknown.

Problem 2 is somewhat similar, but now there are two separate parts to the motion. Martha cruises at constant velocity for 0.5 seconds before putting on the brakes. We can identify the beginning and end of this segment of the motion with subscripts 0 and 1, respectively, as shown in Fig. 1-29. She then brakes until coming to a stop. This segment of the motion starts at x_1 and ends at x_2. We end up with three labeled points—the beginning, the end, and a point in between (x_1) where the character of the motion changed. Each of these three points is also characterized by a velocity and a time. However, we have only shown *two* acceleration vectors because it takes two velocities to define one acceleration. Thus v_0 and v_1 define a single acceleration $\vec{a}_0$ (which happens to be zero here) that changes v_0 into v_1. Likewise, $\vec{a}_1$ changes v_1 into v_2.

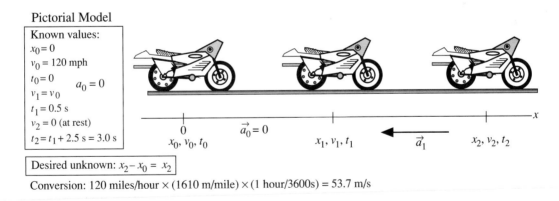

Pictorial Model

Known values:
$x_0 = 0$
$v_0 = 120$ mph
$t_0 = 0$ $a_0 = 0$
$v_1 = v_0$
$t_1 = 0.5$ s
$v_2 = 0$ (at rest)
$t_2 = t_1 + 2.5$ s $= 3.0$ s

Desired unknown: $x_2 - x_0 = x_2$

Conversion: 120 miles/hour $\times$ (1610 m/mile) $\times$ (1 hour/3600s) $= 53.7$ m/s

FIGURE 1-29 The pictorial model of Martha stopping her motorcycle (Example 1-7).

This information is shown with symbols on the figure, thus defining what the symbols represent. The known quantities are then listed separately. We have *interpreted* the question "What is her total stopping distance?" to mean the total distance from when she first sees the furry mammal until she is stopped. This is the quantity $x_2 - x_0$. However, because $x_0 = 0$ due to our choice of origin, we can simply solve for x_2. Notice that a different choice of origin would have required us to compute the full $x_2 - x_0$ to get the total stopping distance. This illustrates that some coordinate system choices are "better" than others in the sense that they minimize the amount of computation needing to be done.

Figures 1-28 and 1-29 illustrate the complete pictorial models for the two problems given above. Verify for yourself that they each have incorporated all six steps for drawing a pictorial model. We have taken a "word problem," interpreted it, and rendered it in a more precise form that is now ready for a quantitative analysis.

•

Using Symbols

It is worth making a few comments about the symbols used in Example 1-7. Symbols are a language that allow us to talk about the relationships in a problem with great precision. As with any language, we all need to agree to use words or symbols in the same way if we want to be able to communicate with each other. Many of the ways we use symbols in science and engineering are somewhat arbitrary, often reflecting ancient historical roots, but practicing scientists and engineers have come to agree on how to use the language of symbols. Learning this language of symbols is part of learning physics.

We frequently use subscripts in physics to designate a particular point in the problem. Scientists generally label the starting point of the problem with the subscript "0," not the subscript "1" that you might expect. Scientists and mathematicians often refer to a zero subscript with the word "naught"—a word that comes from Old English. Thus x_0 is often pronounced "x naught." When using subscripts, make sure that all variables referring to the same point in the problem have the *same numerical subscript*. To have the ending point in a problem characterized by position x_1 but velocity v_2 is guaranteed to lead to confusion!

As you may have already noticed, when writing general equations we use the subscripts i and f, for *initial* and *final*, to represent the end points of a particular segment of the motion. In a sense, these are generic equations. In a real problem, though, it is a good idea *not* to use subscripts i and f. Use specific numbers, starting with zero, and you will be less likely to make mistakes. Whichever numbers are appropriate can then be substituted for i and f when you need to use an equation.

1.8 Working with Motion Diagrams and Pictorial Models

The pictorial model is a tool for translating the words of the problem into symbols and for gaining a full understanding of just what the problem is about. However, it is not our only tool—we also have the motion diagram. Your problem solving should include both the pictorial model and the motion diagram. In fact, you often need to work on both simultaneously as you think about the problem. For example, the instructions for the pictorial model tell you to show the acceleration vector. But how do you know the direction in which $\vec{a}$ points? You find out from the motion diagram! As you gain experience, you will be able to determine the direction of velocity and acceleration vectors directly from the statement of the problem. For now, though, it is usually best to start with the motion diagram, proceed to the pictorial model, then confirm that your two diagrams are consistent with each other. That is, do the velocity and acceleration vectors point the same directions in both diagrams? If not, then you have a problem somewhere and do not yet have a good understanding of the problem statement. Do not skip this comparison step! It only takes a couple of seconds, and these little checks for consistency are part of an expert's problem-solving tactics, helping him or her avoid careless errors. Steps like this will become second nature to you after awhile, but you may have to force yourself to do them in the beginning. You might consider making a checklist of "things to do" when approaching a new problem and keep it beside you while you work.

Let's look at several examples. Keep in mind that our task, at this point, is not to *solve* anything, but rather to focus on "what is happening" in the problem. In other words, our goal is to make the translation from words to symbols in preparation for subsequent mathematical analysis. We will do so, in each case, by drawing the motion diagram and developing the full pictorial model.

EXAMPLE 1-8   A rocket sled accelerates at 50 m/s^2 for 5 seconds, coasts for 3 seconds, then deploys a braking parachute and decelerates at 3 m/s^2 until coming to a halt.

1. What is the top speed of the rocket sled?
2. What is the total distance traveled?

SOLUTION Figure 1-30a shows the motion diagram for the rocket sled. Notice that there are three phases to the motion: first a "speeding up" phase with $\vec{a}$ and $\vec{v}$ pointing to the right, second a "constant velocity" phase (coasting) where $\vec{a} = 0$, and third a "slowing down" phase in which $\vec{a}$ points to the left while $\vec{v}$ is still to the right. We see, from the motion diagram, that there are not only beginning and ending points of the motion, but also two points in between where the character of the motion changes: at the change from acceleration to coasting, and at the change from coasting to braking. The pictorial model thus needs to identify all four points.

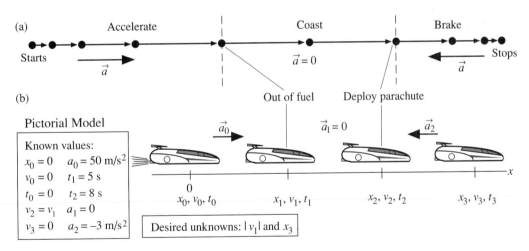

FIGURE 1-30 a) Motion diagram and b) pictorial model for Example 1-8.

Figure 1-30b shows the pictorial model for this problem. You can see that the subscripts 0, 1, 2, and 3 are given to these four points in the motion. We also need three accelerations: a_0 from x_0 to x_1, a_1 from x_1 to x_2, and a_2 from x_2 to x_3. It is essential to keep in mind that these symbols for one-dimensional motion still need to convey information about the direction in which the motion diagram vectors point. Because $\vec{a}_0$ points toward the right it has the positive value $a_0 = +50$ m/s^2. But $\vec{a}_2$ points toward the left, so it has a negative value: $a_2 = -3$ m/s^2. During the coasting phase, we know from the motion diagram that

$a_1 = 0$. Notice that this information about a_1 is not explicitly given in the problem, but instead it comes from our preliminary analysis with the motion diagram.

If the sled starts from rest (not stated—we are making a reasonable interpretation!) and ends at rest, then v_0 and v_3 are both zero. While we do not know v_1 at this time, we do know from the motion diagram that $v_2 = v_1$. As the final piece of our pictorial model we identify the maximum speed, $|v_1|$ and the total distance traveled, x_3 as our desired unknowns. Strictly speaking, the distance traveled is $|x_3 - x_0|$, but you can see that we have conveniently chosen $x_0 = 0$ as the origin so that the distance becomes simply x_3. Note that identifying v_1 rather than $|v_1|$ as the desired unknown in part a) would be incorrect because the problem asked for speed rather than velocity.

●

This completes our qualitative analysis of Example 1-8, and we would now be ready to start the quantitative analysis.

●

EXAMPLE 1-9   A small rocket, such as those used for meteorological measurements of the atmosphere, is launched vertically with an acceleration of 30 m/s². It runs out of fuel after 30 seconds. What is its maximum altitude?

SOLUTION   Here we have to do some interpretation. Common sense tells us that the rocket does not stop the instant it runs out of fuel. Instead, it continues upward, while slowing, until reaching its maximum altitude. This second half of the motion, after the rocket runs out of fuel, is like the first half of Example 1-6. Because the problem does not ask about the rocket's descent, we conclude that the problem ends at the point of maximum altitude.

Figure 1-31 shows the motion diagram and pictorial model for the rocket. The rocket is speeding up during the first half of the motion, so $\vec{a}_0$ points parallel to $\vec{v}$. During the second half, as the rocket slows, $\vec{a}_1$ will point downward, as we discovered in the earlier example. Because this is in the downward direction, a_1 will be a negative number. The initial acceleration $a_0 = 30$ m/s² is given in the problem.

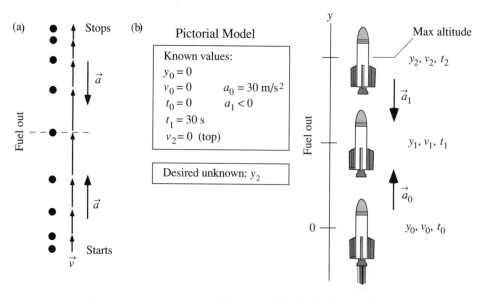

FIGURE 1-31 a) Motion diagram and b) pictorial model for Example 1-9.

This information is then transferred to the pictorial model in Fig. 1-31b. Note that we do know velocity v_2: it must be zero at the very top of the trajectory. Lastly, we have identified y_2 as the desired unknown. This, of course, is not the only unknown in the problem, but it is the one we are specifically asked to find.

Some of you who have had a physics course previously may be tempted to include $a_1 = -9.8$ m/s^2, the acceleration due to gravity. However, that would only be true if there is no air resistance on the rocket. We will need to consider the *forces* acting on the rocket during the second half of its motion before we can determine a value for a_1, and that will be the subject of several upcoming chapters. For now, though, all that we can safely conclude is that a_1 must be negative.

EXAMPLE 1-10   Football hero Fast Fred catches the kick-off while standing directly on the goal line. He immediately runs forward with an acceleration of 6 ft/s^2. At the moment the catch is made, Two-Ton Tommy is crossing the 20-yard line with a steady speed of 15 ft/s, heading directly toward Fred. If neither deviates from his path, where will the tackle occur?

SOLUTION   Here is a problem with two moving objects to consider. Figure 1-31a shows the motion diagram, with Fred accelerating toward the right, making the values for v and a both positive, while Tommy moves toward the left with a *negative* value for v. (Noticing that v_{T0} is negative because the vector $\vec{v}_T$ points left, is *extremely* important.) The ending point of the problem is fairly clear—when Fred and Tommy collide (ouch!).

For the pictorial model in Fig. 1-32b, we have used additional subscripts F and T to distinguish the symbols describing Fred's motion from similar symbols describing Tommy's motion. As expected, v_{T0} is negative while v_{F0} is zero. The time, however, does not have an extra subscript because there is only one clock in the movie from which we made the motion diagram. Thus *both* starting positions are described with the same time symbol t_0.

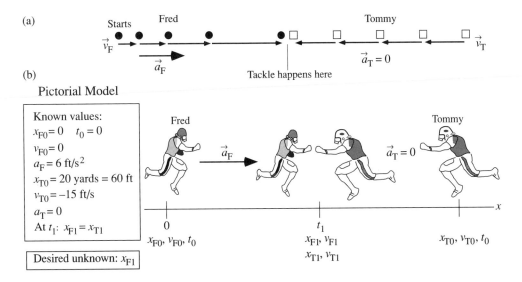

FIGURE 1-32 a) Motion diagram and b) pictorial model for Example 1-10.

Notice that we have chosen to have Fred start at the origin and move in the $+x$-direction. Here is a problem where other choices are equally plausible. For example, we might have let Fred run to the left, or we might have put the origin at Tommy's initial position. We will get the same final answer for any choice of coordinate system.

One important piece of reasoning that we must be careful not to overlook is how to state, in symbols, the condition for the problem to end. We know intuitively how it ends—they collide—but how do we state that more precisely? To collide, they must both reach the same point at the same time (t_1). Therefore, the relation we are looking for is $x_{F1} = x_{T1}$. It is necessary to keep in mind that x_{T1} is Tommy's *position*, measured by the coordinate system, and not the distance that Tommy has traveled.

EXAMPLE 1-11 Sammy Skier steps off the ski lift at an elevation 200 m above the parking lot and heads off down a snow-covered 15° slope. All goes well until he reaches the bottom. Unfortunately, Sammy had skipped the section of his skiing class on stopping, so upon reaching the bottom of the slope he heads out across the horizontal asphalt parking lot. The snow can be treated as frictionless, but the coefficient of friction between expensive fiber-glass skis and asphalt is $\mu = 0.40$. How far does Sammy slide across the parking lot before stopping?

SOLUTION In Figure 1-33a the motion diagram shows Sammy speeding up on the slope and then slowing down on the asphalt. There is clearly a point where the character of the motion changes—as Sammy leaves the snow and moves onto the parking lot. (Note the direction of $\vec{a}$ at that point and its similarity to Example 1-3.) We thus expect the pictorial model to include three points of interest.

This would seem to be a two-dimensional motion problem, and strictly speaking it is. However, the motion diagram shows us that we can consider the motion as being made up of two straight-line segments. This suggests that the imaginative use of coordinate systems can make this a one-dimensional motion problem, and thus much easier than it might at first appear!

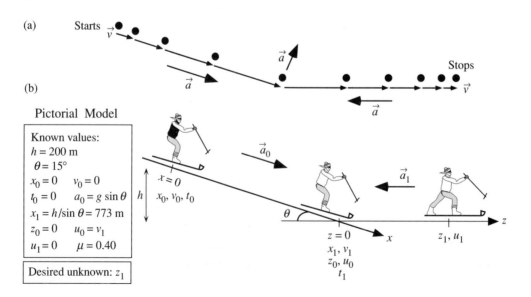

FIGURE 1-33 a) Motion diagram and b) pictorial model for Example 1-11.

The pictorial model, in Fig. 1-33b, shows that we have used two coordinate systems—one for each half of the problem! Furthermore, the coordinate system for the first half has been tilted (with the angle identified and labeled) so as to make the x-axis parallel to the slope. With this choice of coordinate systems, we have turned a two-dimensional motion problem into two one-dimensional problems.

To distinguish the two coordinate systems, we have called the first system the x-axis and the second the z-axis. Thus Sammy starts down the slope at $x_0 = 0$ and reaches the bottom at x_1. He then switches over to the z-axis, with $z_0 = 0$ representing the start of his journey across the parking lot and z_1 his ending point. His velocity along the x-axis is called v, but his velocity along the z-axis is called u, again to distinguish the two segments of motion.

As part of our interpretation of the problem, it is reasonable to assume that Sammy suffers no loss of speed as he moves from the snow to the parking lot. Thus $u_0 = v_1$ is the relation between his final velocity along the x-axis and his initial velocity along the z-axis. This is a problem where the forces on Sammy—gravity and friction—will have to be considered before we can determine the accelerations. Nonetheless, we can see from the motion diagram that a_0 will have to be positive while a_1 will be negative. Notice that we do have another piece of quantitative information in this problem—the coefficient of friction μ—so it has been included in the table of known information. (We will learn in Chapter 5 that μ is the symbol for the coefficient of friction. It is included here just to illustrate its use in a pictorial model, not as anything you need to remember at this point.)   •

You now have some powerful tools for thinking about and analyzing problems. It will take continued practice until using these tools becomes second nature to you; however, you will get plenty of opportunity for that practice in the coming chapters.

Summary

▲ Important Concepts and Terms

motion

linear motion

circular motion

projectile motion

oscillatory motion

motion diagram

at rest

constant speed

operational definition

particle

particle model

position vector

scalar quantity

vector quantity

magnitude

initial position

final position

displacement

displacement vector

time interval

average velocity

speed

rate

constant velocity

average acceleration

projectile

pictorial model

This chapter is exceptionally important and worth extended study. Its purpose has been to establish the basic concepts of motion and to begin building intuition—especially about the more abstract concepts of velocity and acceleration. We will continue to build on and practice these ideas in the next few chapters, but this is the foundation of our future knowledge.

Real understanding is not the memorization of definitions and formulas, but an ability to use ideas and concepts in new situations. To emphasize this, we have introduced the concepts of motion as *operational definitions*. We have given these new concepts names only *after* trying to learn something about how the idea is used.

The list at the beginning of this summary identifies important concepts and terms you should learn in this chapter. The number of concepts has been fairly small: position, time, displacement, time interval, velocity, and acceleration. We have also defined some specific types of motion: linear, circular, and projectile. In addition, we introduced a particular way of characterizing motion called the particle model, and a particular way of analyzing a problem statement called the pictorial model. These could have all been quickly defined in just a few sentences, but your understanding of them would have been slim. The emphasis, instead, has been on how to *use* these concepts to describe motion, and your study should focus on this aspect.

It is worth remembering that the position $\vec{r}$ and the time t are the only quantities that we can directly measure. The other quantities, such as displacement or acceleration, are defined as various combinations of these two measured quantities. Although we have not yet done any calculations, these ideas will form the basis of kinematics in Chapter 3.

Motion diagrams constitute a somewhat simplified and abstract description of a moving object. As such, they are a *model* of the motion and they begin to give a picture of "what is happening" in a problem. They are important as a means of *interpreting* the concepts of motion. In particular, motion diagrams allow us to learn about the acceleration vector. This will turn out to be essential information. Motion diagrams are also the first step toward developing a complete strategy for solving physics problems.

A complete motion diagram for a moving object consists of:

1. The position (shown as a dot) of the object in each frame of the film.

2. The velocity vectors, showing the average velocity of the object between each frame of film and the next. There is *one* velocity vector linking each two position dots. This emphasizes the idea that the velocity $\vec{v}$ changes position $\vec{r}_i$ into position $\vec{r}_f$.

3. The acceleration vectors, showing the average acceleration of the object between each velocity and the next. Because the acceleration $\vec{a}$ changes velocity $\vec{v}_i$ to velocity $\vec{v}_f$, there is *one* acceleration vector linking each two velocity vectors. Each acceleration vector is drawn at the midpoint of the two velocity vectors it links.

The pictorial model is the second step toward a complete problem-solving strategy. The pictorial model is a tool for translating the words of a problem statement into the more precise symbols and quantitative information needed for a mathematical analysis. A pictorial model consists of

1. A sketch, showing the object or objects at the beginning and end of their motion as well as at any point in between where the character of the motion changes.

2. A coordinate system with the origin clearly labeled.

3. Symbols, shown on the sketch, for positions, velocities, times, accelerations, and other quantities. This *defines* the meanings of the symbols that will be used in the mathematical analysis.

4. A table listing the values of known quantities.

5. A list of the unknown quantity or quantities that the problem asks for.

As a final thought for this chapter, it is worth noting that scientific concepts are not "discovered." There was no preexisting definition of $\vec{v} = \Delta\vec{r}/\Delta t$ waiting for someone to find it like a nugget of gold in a river bed. Rather, scientific concepts are abstract, human-created ideas that have proven themselves useful for describing a wide range of phenomena. It is only *because* the ratio $\Delta\vec{r}/\Delta t$ turns out to measure something useful that we give it a name and develop its properties. While there are discoveries in science, they are of phenomena and not of concepts. The discoveries would be pretty uninteresting by themselves without the concepts and theories that allow us to make use of them. Science, like art or literature, is a human-created realm of knowledge and understanding.

Exercises and Problems

1. a. Write a paragraph or two describing the "particle model." What is it and why is it important?
   b. Give two of examples of situations in which the particle model would be used.
   c. Give two of examples of situations in which it would be inappropriate.

2. a. What is an "operational definition?"
   b. Give operational definitions of the concepts of
      i) position
      ii) displacement
      iii) velocity
      iv) acceleration

Your definition should be given mostly in words and sentences, with a minimum of symbols or mathematics.

Problems 3–6 show several complete motion diagrams. For each of these problems, write a one or two sentence "story" about a *real object* that has this motion diagram. Your stories should discuss people or objects by name and say what they are doing.

3. Starts

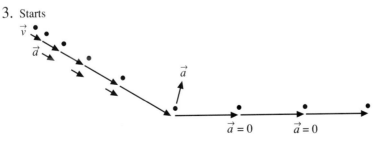

4.

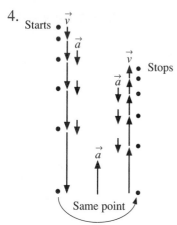

5. Top view of motion in a horizontal plane.

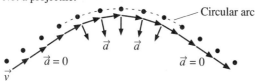

Circular arc

6. Side view of motion in a vertical plane. *Not* a projectile.

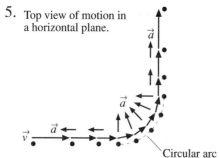

Circular arc

7. Figure 1-25 showed the motion diagram for Bill as he rode a Ferris wheel that was turning at a constant speed. The inset to the figure showed how to find the acceleration vector at the lowest point in his motion. Use a similar analysis to find the acceleration vector at the 12 o'clock, 4 o'clock, and 8 o'clock positions of the motion diagram. Use a ruler so that your analysis is accurate.

8. Consider a pendulum swinging back and forth on a string. At the very top of its arc is the velocity zero or non-zero? Is the acceleration zero or non-zero? Your answer should consist of an *analysis*, using motion diagrams, and a *written explanation*.

For each of the problems 9–15 below, draw a) a motion diagram, and b) a pictorial model. Do not attempt to "solve" these problems or do any mathematics other than units conversion. These questions ask you to analyze the problem statement, but not to solve it.

The ✍ icon in front of these problems indicates that they can be done on a Dynamics Worksheet. If you use a worksheet, place the pictorial model in the "Pictorial Model" section and the motion diagram in the "Physical Model" section. The "Mathematical Model" section is not needed for these problems.

✍ 9. A Porsche accelerates from a stoplight at 8 m/s² for five seconds, then coasts for three more seconds. How far has it traveled?

✍ 10. Billy drops a watermelon from the top of a ten-story building, 30 m above the sidewalk. How fast is it going when it hits?

✍ 11. Sam is recklessly driving 60 mph in a 30 mph speed zone when he suddenly sees the police. He steps on the brakes and slows to 30 mph in three seconds, looking nonchalant as he passes the officer. How far does he travel while braking?

✍ 12. a. You would like to stick a wet spitwad on the ceiling, so you toss it straight up with a speed of 10 m/s. How long does it take to reach the ceiling, 3 m above?
   b. Unfortunately, the spitwad fails to stick and drops off 5 s later. How fast is it going when you catch it?

✍ 13. A ball rolls along a smooth horizontal floor at 10 m/s, then starts up a 20° ramp. How high does it go before rolling back down?

✍ 14. Ice hockey star Bruce Blades is 5 m from the blue line and gliding toward it with a speed of 4 m/s. You are 20 m from the blue line, directly behind Bruce. You want to shoot the puck to Bruce. With what speed should you shoot the puck down the ice so that it reaches Bruce exactly as he crosses the blue line?

✍ 15. You are standing still as Fast Fred runs past you with the football at a speed of 10 m/s. He has only 30 yards left before reaching the goal line to score the winning touchdown.
   a. If you begin running at the exact instant he passes you, what acceleration must you maintain to catch him at the 5-yard line?
   b. What speed will you have as you tackle him?

The following problems each show a partial motion diagram. For each:
   a. Complete the motion diagram by adding acceleration vectors.
   b. Write a problem for which this is the correct motion diagram. Be imaginative! Don't forget to include enough numerical information to make the problem complete and solvable, and to state clearly what is to be found. Do not attempt to actually solve the problem.
   c. Draw a full pictorial model for your problem.

16.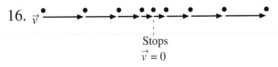

17.

18.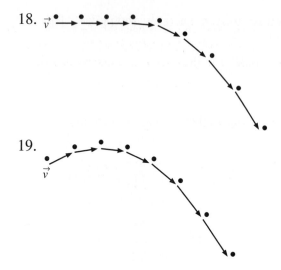

19.

[**Estimated 10 additional problems for the final edition.**]

Chapter **2**

Vectors and Coordinate Systems

2.1 Scalars and Vectors

Many of the quantities that we use to describe the physical world are simply numbers. For example, the mass of an object is 2 kg, its temperature is 21°C, and it occupies a volume of 25 cm^3. A mathematical quantity that is fully described by a single number is called a **scalar quantity**. Mass, temperature, and volume are all scalars. Other scalar quantities include pressure, density, energy, charge, and voltage. Some scalars, such as mass, are inherently positive, but many others, such as charge, can be either positive or negative. Many times, especially when we do not know the actual value, we will represent a scalar by an algebraic symbol. Thus m will represent mass, T temperature, V volume, q charge, and so on. Notice that scalars, in printed text, are shown in italics.

Our universe has three dimensions so it is not surprising that some quantities also need a direction for a full description. If you ask someone the way to the post office, it will not be very helpful if he or she replies only, "Go three blocks." A full description would be, "Go three blocks south." A mathematical quantity having both a size *and* a direction is called a **vector quantity**. You saw several examples of vector quantities in the last chapter: position, displacement, velocity, and acceleration. You will soon make the acquaintance of others, such as force, momentum, and the electric field. Because vector quantities play an essential role in all areas of physics, it is worth spending a little time discussing how they are represented and used.

Suppose, for example, that you are assigned the task of measuring the temperature at various points throughout a large building and then showing the information on a building floor plan. To do this you could put little dots on the floor plan, showing the points at which you made measurements. Then you could write the temperature at that point beside each dot, as shown in Fig. 2-1a. In other words, you can "represent" the temperature at each point simply by writing a number there.

Having done such a good job on your first assignment, you are next assigned the task of measuring the displacements of several employees as they move about in their work. Recall from Chapter 1 that displacement has both a size and a direction. Simply writing the distance each person travels is not sufficient because the number doesn't take into account the direction in which the person moved. After some thought, you conclude that a better way to "represent" the displacement is by drawing arrows that point in the correct directions and whose

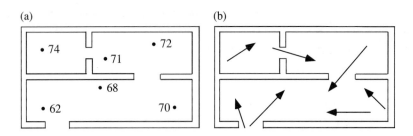

FIGURE 2-1 Measurements of a) temperatures (scalars) and b) displacements (vectors).

lengths are proportional to the distance traveled (Fig. 2-1b). Further, you decide to place the *tail* of an arrow at the point where you first measured an employee's position.

As the preceding example illustrates, we can provide a *geometric representation* of a vector as an arrow, with the tail of the arrow (not its tip!) placed at the point where the measurement is made. The vector then seems to radiate outward from the point. An arrow is a natural representation of a vector because it inherently has both a length and a direction. The proper mathematical term for the length, or size, of a vector is **magnitude**. Therefore, from now on we will say that *a vector is a quantity having a magnitude and a direction*. As an example, Fig. 2-2 shows a representation of vector $\vec{A}$. It is drawn with its tail at the point where the vector quantity was measured.

Arrows are good for pictures, but we will also need an *algebraic representation* of vectors to use in equations. As noted in the last chapter, we do this by drawing an arrow over the letter representing the vector: $\vec{r}$ for position, $\vec{v}$ for velocity, $\vec{a}$ for acceleration, and so on. Note that the arrow symbol *always* points to the right, regardless of which direction the vector itself actually points. The magnitude, or length, of a vector is indicated by placing the vector within absolute value signs: $|\vec{r}|$ indicates the magnitude of the position vector, which is simply a distance. Be sure to note that magnitude, as indicated by the absolute value signs, is *always* a positive scalar number. The vector $\vec{A}$ in Fig. 2-2 has magnitude $|\vec{A}| = 3$ cm.

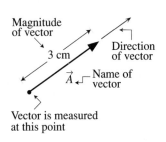

Magnitude of vector

3 cm

Direction of vector

$\vec{A}$ — Name of vector

Vector is measured at this point

FIGURE 2-2 Vector $\vec{A}$ has both a magnitude and a direction. The magnitude of the vector is $|\vec{A}| = 3$ cm.

As was mentioned in Chapter 1, it is extremely important to get in the habit of using the arrow symbol for vectors because we will often use vector symbols and scalar symbols in the same problem. The symbols $\vec{r}$ and r, or $\vec{v}$ and v, do not represent the same thing, so if you omit the vector arrow from vector symbols you will soon have confusion and mistakes.

2.2 Properties of Vectors

Recall from Chapter 1 that a vector drawn from the initial position of an object to the object's position at a later time is called the *displacement vector*. Because displacement is an easy concept to think about, we can use it to discuss some of the properties of vectors. It is important to keep in mind, however, that these properties will apply to *all* vectors, not just to displacement.

Suppose that Sam starts out his front door, walks across the street, and ends up 200 feet to the northeast of where he started. Sam's displacement vector $\vec{S}$ is shown in Fig. 2-3a. It is a straight-line connection from his initial to his final point, not necessarily his actual path. The dotted line indicates a possible route Sam might have taken, but his displacement is the vector $\vec{S}$. To describe a vector we must specify both its magnitude and its direction. You can write Sam's displacement vector as

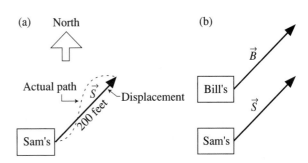

$$\vec{S} = (200 \text{ feet, northeast}).$$

The magnitude of Sam's displacement is $|\vec{S}| = 200$ feet, which is the distance between his initial and final points.

Sam's next-door neighbor Bill also walks 200 feet to the northeast, starting from his own front door. Bill's displacement vector $\vec{B}$ has the same magnitude and direction as Sam's displacement $\vec{S}$. Because vectors are defined only by their magnitude and direction, two vectors are *equal* if they have the same magnitude and direction, regardless of their point of origin. Thus the two displacement vectors in Fig. 2-3b are equal to each other, and you can write $\vec{B} = \vec{S}$. Graphically, this property implies that you can redraw a vector at a different point on a drawing, and the new vector will be equal to the original vector as long as both have the same length and direction.

FIGURE 2-3 a) Sam's displacement vector $\vec{S}$ is the straight-line connection between his initial and final positions. It may not be the same as the actual path he followed. b) Bill's displacement vector $\vec{B}$ is equal to Sam's because both have the same length and direction.

Figure 2-4 shows the displacement of a hiker moving from a starting point at P to a final point S. The hiker first hikes 4 miles to the east, then 3 miles to the north. The first leg of the hike—4 miles east—is described by the displacement vector $\vec{A} = (4 \text{ miles, east})$. The second leg of the hike is vector $\vec{B} = (3 \text{ miles, north})$. By definition, a vector from the initial position P to the final position S is also a displacement vector. This is vector $\vec{C}$ on the figure. $\vec{C}$ is called the *net displacement* because it describes the net result of the hiker undergoing first displacement $\vec{A}$, then displacement $\vec{B}$.

FIGURE 2-4 Net displacement $\vec{C}$ resulting from two displacements $\vec{A}$ and $\vec{B}$.

With scalars, the word *net* implies addition. For example, if you earn $50 on Saturday and $60 on Sunday, your *net* income for the weekend is $110—the sum of $50 and $60. The same is true with vectors. The net displacement $\vec{C}$ is simply an initial displacement $\vec{A}$ *plus* a second displacement $\vec{B}$, or

$$\vec{C} = \vec{A} + \vec{B}. \tag{2-1}$$

The sum of two vectors is also called the **resultant vector**.

Recall from Chapter 1 that the sum of two vectors can be found graphically using the following steps:

1. Draw the first vector $\vec{A}$.
2. Draw the second vector $\vec{B}$ with its tail placed on the tip of $\vec{A}$.
3. Draw the vector from the tail of $\vec{A}$ to the tip of $\vec{B}$. This is the resultant $\vec{C} = \vec{A} + \vec{B}$.

This method for adding vectors is called **graphical addition**. As you can see in Fig. 2-4, this is how vector $\vec{C}$ was found. While this makes sense and is easy to visualize for displacement vectors, keep in mind that *any* two vectors can be added in exactly the same way.

While the graphical method for adding vectors is straightforward, we need to do a little geometry to come up with a complete description of $\vec{C}$. In Fig. 2-4, Vector $\vec{C}$ is defined by its magnitude $|\vec{C}|$ and by its direction, which is measured by the angle θ. (The symbol θ, which we will use frequently for angles, is the Greek letter *theta*.) Because the three vectors $\vec{A}$, $\vec{B}$, and $\vec{C}$ form a right triangle, the magnitude, or length, of $\vec{C}$ is given by the Pythagorean theorem:

$$|\vec{C}| = \sqrt{|\vec{A}|^2 + |\vec{B}|^2} = \sqrt{(4 \text{ miles})^2 + (3 \text{ miles})^2} = 5 \text{ miles.} \qquad (2\text{-}2)$$

Notice we have used absolute value signs because we are using just the magnitudes of the vectors. The angle θ of $\vec{C}$, as defined in Fig. 2-4, is also easily found for a right triangle:

$$\theta = \tan^{-1}\left(\frac{|\vec{B}|}{|\vec{A}|}\right) = \tan^{-1}\left(\frac{3 \text{ miles}}{4 \text{ miles}}\right) = 37°. \qquad (2\text{-}3)$$

Altogether, then, we would describe the hiker's net displacement as

$$\vec{C} = \vec{A} + \vec{B} = (5 \text{ miles, } 37° \text{ north of east}). \qquad (2\text{-}4)$$

You can add vectors in any order that you wish—the final result will be the same. For example, it is often the case that the two vectors to be added are drawn with their tails together, as shown in Fig. 2-5a. How do you evaluate $\vec{D} + \vec{E}$? One way is to "slide" vector $\vec{E}$ over to where its tail is on the tip of $\vec{D}$, then to use the rules of graphical addition (Fig. 2-5b). This gives the vector sum $\vec{F} = \vec{D} + \vec{E}$. But you could also, as in Fig. 2-5c, slide vector $\vec{D}$ up so that its tail is on the tip $\vec{E}$. The sum $\vec{E} + \vec{D}$ is the same as $\vec{D} + \vec{E}$. Thus, vector addition is *commutative*:

$$\vec{D} + \vec{E} = \vec{E} + \vec{D}. \qquad (2\text{-}5)$$

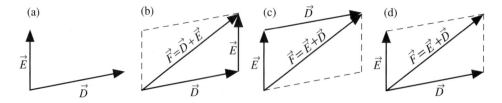

FIGURE 2-5 a) Addition of two vectors with their tails together. The graphical method of b) and c) shows that $\vec{D} + \vec{E} = \vec{E} + \vec{D}$. The resultant can also be found, as in d), by "completing the parallelogram."

Figure 2-5d shows that the vector sum can be considered as the diagonal of the parallelogram formed by vectors $\vec{D}$ and $\vec{E}$. This method for vector addition, which some of you may have learned, is called the *parallelogram rule* of addition.

The concept of vector addition is easily extended to more than two vectors. Figure 2-6 shows a hiker moving from initial position 0 to position 1, then position 2, then position 3, and finally arriving at position 4. These four segments are described by displacement vectors $\vec{D}_1$, $\vec{D}_2$, $\vec{D}_3$, and

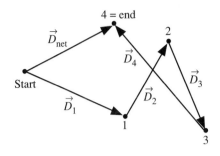

FIGURE 2-6 Net displacement of four individual displacements.

$\vec{D}_4$. The hiker's net displacement, though, is simply from position 0 to position 4, as shown by vector $\vec{D}_{net}$. In this case, we have

$$\vec{D}_{net} = \vec{D}_1 + \vec{D}_2 + \vec{D}_3 + \vec{D}_4 \ . \tag{2-6}$$

EXAMPLE 2-1 A bird flies 100 m due east from a tree, then 200 m northwest (that is, 45° north of west). What is the bird's net displacement?

SOLUTION Let the two displacements be $\vec{A} = (100$ m, east$)$ and $\vec{B} = (200$ m, northwest$)$. These are shown, along with their sum $\vec{C} = \vec{A} + \vec{B}$, in Fig. 2-7. This is the graphical addition of two displacements, with vector $\vec{C}$ pointing from the initial to the final position. But describing $\vec{C}$ is trickier than in the earlier example of the hiker because $\vec{A}$ and $\vec{B}$ are not at right angles. We can find the magnitude of $\vec{C}$ by recalling the law of cosines from trigonometry:

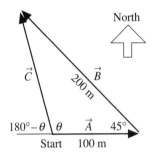

FIGURE 2-7 A bird flies 100 m east, then 200 m northwest. Its net displacement is $\vec{C} = \vec{A} + \vec{B}$.

$$|\vec{C}|^2 = |\vec{A}|^2 + |\vec{B}|^2 - 2|\vec{A}||\vec{B}|\cos(45°)$$

$$= (100 \text{ m})^2 + (200 \text{ m})^2 - 2(100 \text{ m})(200 \text{ m})\cos(45°)$$

$$= 21{,}720 \text{ m}^2$$

$$|\vec{C}| = \sqrt{21{,}720 \text{ m}^2} = 147 \text{ m}.$$

Once $|\vec{C}|$ is known, a second use of the law of cosines can determine angle θ:

$$|\vec{B}|^2 = |\vec{A}|^2 + |\vec{C}|^2 - 2|\vec{A}||\vec{C}|\cos\theta$$

$$\Rightarrow \theta = \cos^{-1}\left[\frac{|\vec{A}|^2 + |\vec{C}|^2 - |\vec{B}|^2}{2|\vec{A}||\vec{C}|}\right] = 106.2°.$$

It's perhaps easier to describe $\vec{C}$ with the angle $180° - \theta = 73.8°$ away from west. The bird's net displacement is $\vec{C} = (147$ m, 73.8° north of west$)$.

Suppose a second bird flies twice as far to the east as the bird in Example 2-1. The first bird's easterly displacement was $\vec{A}_1 = (100 \text{ m, east})$—the subscript "1" has been added to denote the first bird. The second bird's displacement will then certainly be $\vec{A}_2 = (200 \text{ m, east})$. The words "twice as" indicate a multiplication, so we can say

$$\vec{A}_2 = 2\vec{A}_1.$$

The result of multiplying a vector by a scalar is another vector. Consider, for example, the vector $\vec{C} = (147 \text{ m, } 73.8° \text{ north of west})$ that we found for the bird's net displacement. If we multiply $\vec{C}$ by the scalar 5 we get

$$5\vec{C} = (735 \text{ m, } 73.8° \text{ north of west}).$$

Note that the magnitude gets multiplied while the direction remains the same. The arrow of vector $5\vec{C}$ is five times as long as vector $\vec{C}$, but both point in exactly the same direction.

As another example, if we multiply $\vec{C}$ by zero we get

$$0 \cdot \vec{C} = \vec{0} = (0 \text{ m, } 73.8° \text{ north of west}).$$

The result is a vector having zero length or magnitude. This vector is known as the **zero vector**. In this case, the direction of the vector is irrelevant—you cannot describe the direction of an arrow of zero length!

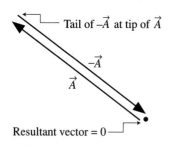

Tail of $-\vec{A}$ at tip of $\vec{A}$

$-\vec{A}$

$\vec{A}$

Resultant vector $= 0$

FIGURE 2-8   Vector $-\vec{A}$ is equal but opposite to $\vec{A}$.

A quantity that we sometimes need to consider is the negative of a vector: $-\vec{A}$. The negative of a vector is equivalent to multiplying the vector by -1. How should this be interpreted? Because $\vec{A} - \vec{A} = \vec{A} + (-\vec{A}) = 0$, the vector $-\vec{A}$ must be such that, when added to $\vec{A}$, the resultant is zero. In other words, the *tip* of $-\vec{A}$ must return to the *tail* of $\vec{A}$. Figure 2-8 shows that a vector that is equal in magnitude to $\vec{A}$, but opposite in direction, will add to $\vec{A}$ to give a resultant of zero. Thus we can conclude that

$$-\vec{A} = \left(|\vec{A}|, \text{ direction opposite } \vec{A} \right). \qquad (2-7)$$

Stated another way, multiplying a vector by -1 yields a vector of the same magnitude but pointing in the opposite direction.

Introduction of the minus sign allows us to consider the meaning of vector subtraction. Figure 2-9a shows two vectors, $\vec{P}$ and $\vec{Q}$. What is $\vec{R} = \vec{P} - \vec{Q}$? We can approach this question in two ways. First, we can write $\vec{P} - \vec{Q} = \vec{P} + (-\vec{Q})$. Because we now know how to form the vector $-\vec{Q}$, we can do so and then add it to $\vec{P}$, using the rules for vector addition to find $\vec{R}$. This is shown in Fig. 2-9b. Alternatively, we could ask what is the vector $\vec{R}$ such that $\vec{Q} + \vec{R} = \vec{P}$? From the rules for vector addition, we know that such a vector must leave the tip of $\vec{Q}$ and arrive at the tip of $\vec{P}$, as shown in Fig. 2-9c. In comparing the results of Figs. 2-9b and 2-9c, keep in mind that a vector is defined by its magnitude and direction, not by its location. Although vector $\vec{R}$ is in different locations in these two figures, it is still the same vector.

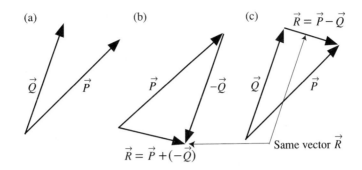

FIGURE 2-9 a) Vectors $\vec{P}$ and $\vec{Q}$. b) $\vec{R} = \vec{P} - \vec{Q} = \vec{P} + (-\vec{Q})$. c) $\vec{R}$ found such that $\vec{Q} + \vec{R} = \vec{P}$ giving $\vec{R} = \vec{P} - \vec{Q}$.

2.3 Component Vectors and Unit Vectors

Our discussion of vectors and their properties has, thus far, not used a coordinate system at all. Vectors do not require a coordinate system. We can add and subtract vectors graphically, as we did in the preceding section. This is a method we will use frequently to clarify our understanding of a situation. But the graphical addition of vectors is not a particularly good way to find quantitative results, as we learned in Example 2-1. In this section, we will look at a method of describing vectors in terms of *coordinates* rather than as a magnitude and direction. The coordinate description will provide us with an easier method for adding and subtracting vectors.

Coordinate Systems

As we noted in the last chapter, the world does not come with a coordinate system attached. A coordinate system is an artificially-imposed grid that you place on a problem in order to make quantitative measurements. It may be helpful to think of drawing a grid on a piece of transparent plastic that you can then overlay on top of the problem. This conveys the idea that you choose: 1) where to place the origin, and 2) how to orient the axes. Different problem solvers may choose to use different coordinate systems—that is perfectly acceptable. Some coordinate systems may make the problem easier, and part of our goal is to show you how to choose the "best" coordinate system. However, no coordinate system is "wrong."

We will generally use **Cartesian coordinates**. This is a coordinate system with the axes perpendicular to each other, forming a rectangular grid. The standard *xy*-coordinate system with which you are familiar is a Cartesian coordinate system. An *xyz*-coordinate system would be a Cartesian coordinate system in three dimensions. There are other possible coordinate systems, such as polar coordinates, but we will not be concerned with those for now.

The placement of the axes is not entirely arbitrary. By convention, the positive *y*-axis is located 90° *counterclockwise* from the positive *x*-axis, as illustrated in Fig. 2-10. Figure 2-10 also identifies the four **quadrants** of the coordinate system, I–IV. Notice that the quadrants are numbered counterclockwise from the positive *x*-axis.

Coordinate axes have a positive end and a negative end, separated by zero at the *origin*. When you draw a coordinate system, it is important to label the axes. This is done by placing *x* and *y* labels at the positive ends of the axes, as was done in Fig. 2-10. The purpose of the labels is to identify which axis is which and to identify the positive ends of the axes. This will be important when you have to determine whether the quantities in a problem should be assigned positive or negative values.

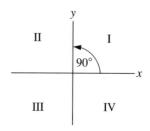

FIGURE 2-10 A conventional Cartesian coordinate system and the quadrants of the *xy*-plane.

Component Vectors

Now let's see how we can use a coordinate system to describe a vector. Figure 2-11 shows a vector $\vec{A}$ with an *xy*-coordinate system imposed. Having established the axes, we can

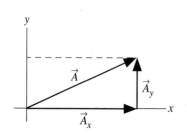

FIGURE 2-11 Component vectors $\vec{A}_x$ and $\vec{A}_y$, with $\vec{A}_x + \vec{A}_y = \vec{A}$.

define two new vectors parallel to those axes. We call these vectors the **component vectors** of $\vec{A}$. Vector $\vec{A}_x$ is called the *x-component vector* and is the projection of $\vec{A}$ along the *x*-axis. Vector $\vec{A}_y$ is the *y-component vector* and is the projection of $\vec{A}$ along the *y*-axis. According to the properties of vector addition, it is apparent that

$$\vec{A} = \vec{A}_x + \vec{A}_y . \qquad (2\text{-}8)$$

Make sure you note that the component vectors are themselves vectors, with the full properties of vectors. Their only requirement is to be *parallel to the coordinate axes*. We have, in essence, broken vector $\vec{A}$ into two pieces parallel to the axes. This process is referred to as the **decomposition** of vector $\vec{A}$ into its component vectors.

Because each component vector lies along a coordinate axis, we need only a single number to describe each one. When the component vector $\vec{A}_x$ points in the +*x*-direction, we define the *scalar A_x* to be the magnitude of $\vec{A}_x$. When the component vector $\vec{A}_x$ points in the –*x*-direction, we define the scalar A_x to be the negative of the magnitude of $\vec{A}_x$. That is, $A_x = |\vec{A}_x|$ if $\vec{A}_x$ points toward the right and $A_x = -|\vec{A}_x|$ if $\vec{A}_x$ points toward the left. We can define the scalar A_y in a similar way. The scalars A_x and A_y are called the **x-component** and the **y-component** of vector $\vec{A}$.

Beware of the somewhat confusing terminology! $\vec{A}_x$ and $\vec{A}_y$ are called component vectors, whereas A_x and A_y are simply called components. The components are scalars—just numbers—so make sure you do not put arrow symbols over the components!

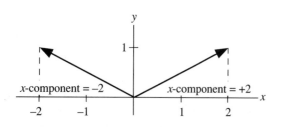

FIGURE 2-12 The importance of signs! The right vector has an *x*-component of +2 while the left vector's *x*-component is –2.

The signs of the components are quite important. Figure 2-12 shows two vectors, one with a positive x-component and one with a negative x-component. The *sign* of the component determines whether the x-component vector points to the right (+x-direction) or to the left (–x-direction). So A_x is *not* the *magnitude* of the x-component vector (which would be just $|\vec{A}_x|$) but includes directional information as well.

It will frequently be the case that we will need to decompose a vector into its components. We will also need, on occasion, to take a vector in component form and express it in terms of a magnitude and a direction. In other words, we need to be able to move back-and-forth between the graphical and the component viewpoints. Doing so is an application of geometry and trigonometry.

Consider first the problem of decomposing a vector into its x- and y-components, as shown in Fig. 2-13a. Here vector $\vec{A}$ has magnitude $|\vec{A}|$ and is at angle θ from the x-axis. It is *essential* to establish, with a picture or diagram, the angle used to describe the direction of a vector. An assertion that "$\theta = 30°$" is meaningless unless angle θ is defined and known, both to you and to anyone else looking at your work. While measuring angles from the x-axis is customary, other choices can be made as long as you are clear about the choice you have made.

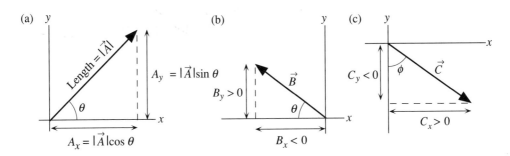

FIGURE 2-13 Moving between the graphical viewpoint (magnitude and angle) and the component viewpoint.

$|\vec{A}|$ is the length of the hypotenuse of a right triangle while the components A_x and A_y are the lengths of the sides of this triangle. Therefore, trigonometry gives

$$A_x = |\vec{A}|\cos\theta$$

$$A_y = |\vec{A}|\sin\theta. \tag{2-9}$$

These equations are the basic means of converting the length and angle description of a vector into the vector's components. But these equations are correct *only* if vector $\vec{A}$ is in the first quadrant, so that the components A_x and A_y are both positive numbers.

Figure 2-13b shows another vector, $\vec{B}$, which is in the second quadrant. In this case, because the vector is pointing toward the left, the x-component B_x will be a *negative* number. The quantity $|\vec{B}|\cos\theta$ gives the length, or absolute value, of the side of the triangle. However, you should realize from the picture that an additional minus sign is needed: $B_x = -|\vec{B}|\cos\theta$. Because the y-component is positive (vector pointing up), we have $B_y = +|\vec{B}|\sin\theta$.

You could also choose, as in Fig. 2-13c, to measure the direction of the vector with an angle ϕ measured from the y-axis. The symbol ϕ is the Greek letter "phi," which is pronounced to rhyme with "pie." In Fig. 2-13c, vector $\vec{C}$ has positive x-component and a negative y-component. The signs and the geometry together lead to $C_x = |\vec{C}|\sin\phi$ and $C_y = -|\vec{C}|\cos\phi$. Notice that the roles of sine and cosine are reversed from that in Eqs. 2-9 because we are using a different angle. The major point here is that there is *not* a formula you can plug into for finding vector components. Each decomposition requires that you pay close attention to the direction in which the vector points and the angles that are defined.

We can also see, from Fig. 2-13, how to go the opposite direction—that is, how to determine the length and angle of a vector from the x- and y-components. Because $|\vec{A}|$, in Fig. 2-13a, is the hypotenuse of a right triangle, its length is given by the Pythagorean theorem as

$$|\vec{A}| = \sqrt{A_x^2 + A_y^2}.$$ (2-10)

Similarly, the tangent of angle θ is the ratio of the far side to the adjacent side, so

$$\theta = \tan^{-1}\left(\frac{A_y}{A_x}\right)$$ (2-11)

where $\tan^{-1}$ is the inverse tangent function, also known as the arc tangent function. Equations 2-10 and 2-11 can be thought of as the "reverse" of Eqs. 2-9.

Equation 2-10 always works for finding the length or magnitude of a vector because the squares eliminate any concerns over the signs of the components. But finding the angle, just like finding the components, requires close attention to how the angle is defined and to the signs of the components. We usually define an angle to lie between 0° and 90°, so the tangent of the angle is a positive number. To find the angle of vector $\vec{B}$ in Fig. 2-13b means that we want just the length $|B_x|$ *without* the minus sign. Thus $\theta = \tan^{-1}|B_y/B_x|$. Without the absolute value signs, your calculator would give you a negative angle! Vector $\vec{C}$ in Fig. 2-13c is even trickier because the angle ϕ is defined differently. Here $\phi = \tan^{-1}|C_x/C_y|$, with the role of x and y reversed from what we used to find angle θ of vectors $\vec{A}$ and $\vec{B}$.

EXAMPLE 2-2   Find the components of vector $\vec{E}$, shown in Fig. 2-14a.

SOLUTION   Vector $\vec{E}$ is described graphically as $\vec{E} = (6\text{ m}, 30°$ below the $-x$-axis). We have redrawn $\vec{E}$ in Fig. 2-14b with component vectors $\vec{E}_x$ and $\vec{E}_y$. Because $\vec{E}_x$ points in

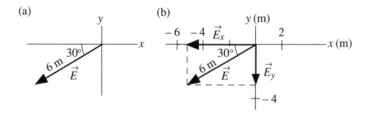

FIGURE 2-14 Example 2-2 describes how to find the component of vector $\vec{E}$.

the –x-direction and $\vec{E}_y$ points in the –y-direction, the components E_x and E_y are both negative. Thus

$$E_x = -|\vec{E}|\cos 30° = -(6 \text{ m})\cos 30° = -5.2 \text{ m}$$

$$E_y = -|\vec{E}|\sin 30° = -(6 \text{ m})\sin 30° = -3.0 \text{ m}.$$

Because $\vec{E}$ has units, so do the components E_x and E_y.

EXAMPLE 2-3   Vector $\vec{F}$ is shown in Fig. 2-15a. Write $\vec{F}$ as a magnitude and a direction.

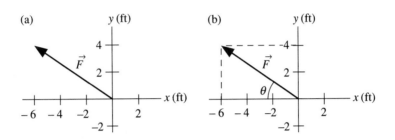

FIGURE 2-15 Example 2-3 describes how to find the magnitude and direction of vector $\vec{F}$.

SOLUTION The vector is redrawn in Fig. 2-15b, and dotted lines have been added to determine the components of $\vec{F}$. We see that $F_x = -6$ feet and $F_y = 4$ feet. This is enough information to find the magnitude of $\vec{F}$:

$$|\vec{F}| = \sqrt{F_x{}^2 + F_y{}^2} = \sqrt{(-6 \text{ feet})^2 + (4 \text{ feet})^2} = 7.2 \text{ feet}.$$

To determine the direction of $\vec{F}$ we must specify an angle by which the direction is measured. Figure 2-15b defines an angle θ. From trigonometry,

$$\theta = \tan^{-1}\left|\frac{F_y}{F_x}\right| = \tan^{-1}\left|\frac{4 \text{ feet}}{-6 \text{ feet}}\right| = 37°.$$

The absolute value signs are necessary because F_x is a negative number whereas θ is defined as an angle whose tangent is positive. The vector $\vec{F}$ can be written $\vec{F} = (7.2 \text{ feet},$ $37°$ above the –x-axis).

Unit Vectors

The x-component vector of a vector $\vec{A}$ can be written as

$$\vec{A}_x = A_x(1, +x\text{-direction}), \tag{2-12}$$

where A_x is the scalar x-component of $\vec{A}$. If A_x is negative, then the component vector $\vec{A}_x$ points in the –x-direction. Similarly,

$$\vec{A}_y = A_y(1, +y\text{-direction}). \tag{2-13}$$

A vector of length 1 in the +x-direction has a special name and symbol:

$$(1, +x\text{-direction}) = \hat{i} = \text{the } \textbf{unit vector} \text{ in the } +x\text{-direction}.$$

The notation $\hat{i}$ (read "i hat"), rather than the usual arrow symbol, indicates specifically a *unit vector* whose magnitude is 1. Unit vectors have no dimensional units. Similarly, the unit vector in the +y-direction is defined to be

$$(1, +y\text{-direction}) = \hat{j} = \text{the } \textbf{unit vector} \text{ in the } +y\text{-direction}.$$

Vectors $\hat{i}$ and $\hat{j}$ are shown in Fig. 2-16a.

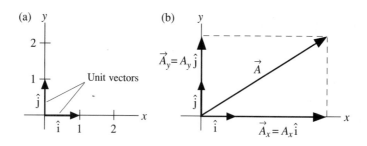

FIGURE 2-16 a) The unit vectors $\hat{i}$ and $\hat{j}$. b) The decomposition of vector $\vec{A}$.

The purpose of unit vectors is to establish the directions of the positive axes of the coordinate system. The vector $\vec{A}$ itself is independent of any coordinate system, but once we decide to place a coordinate system on a problem we need something to tell us "That direction is the +x-direction." This is what the unit vectors do.

With these definitions, Eqs. 2-12 and 2-13 for the component vectors of $\vec{A}$ can be written

$$\vec{A}_x = A_x \hat{i}$$
$$\vec{A}_y = A_y \hat{j}. \tag{2-14}$$

Equations 2-14 separate each component vector into a scalar piece A_x (or A_y) and a directional piece $\hat{i}$ (or $\hat{j}$). The full decomposition of vector $\vec{A}$ is then

$$\vec{A} = \vec{A}_x + \vec{A}_y = A_x \hat{i} + A_y \hat{j}. \tag{2-15}$$

Figure 2-16b illustrates how the unit vectors and the components together form vector $\vec{A}$. In three dimensions, the unit vector along the +z-direction is called $\hat{k}$, and we would include an additional component vector $\vec{A}_z = A_z \hat{k}$.

You may have learned in a math class to think of vectors as pairs or triplets of numbers, such as (4, –2, 5). This is another, and completely equivalent, way to write the components of a vector. Thus, for a vector in three dimensions, we could write,

$$\vec{B} = 4\hat{i} - 2\hat{j} + 5\hat{k} = (4, -2, 5).$$

You will find using unit vectors is more convenient for the types of equations we will use in physics, but rest assured that you already know a lot about vectors if you learned about them as pairs or triplets of numbers.

The importance of vector components and unit vectors becomes apparent when we start doing vector arithmetic. You learned in Section 2-2 how to add two vectors graphically, but it is a tedious problem in geometry and trigonometry to find precise values for the magnitude and length of the resultant. Consider the problem of adding three vectors $\vec{A}+\vec{B}+\vec{C} = \vec{D}$ from a component perspective:

$$\begin{aligned} \vec{D} &= \vec{A}+ \vec{B}+\vec{C} \\ &= (A_x\hat{i} + A_y\hat{j}) + (B_x\hat{i} + B_y\hat{j}) + (C_x\hat{i}+C_y\hat{j}) \\ &= (A_x + B_x + C_x)\hat{i} + (A_y + B_y + C_y)\hat{j}. \end{aligned} \qquad (2\text{-}16)$$

But we can also decompose $\vec{D}$ as

$$\vec{D} = D_x\hat{i} + D_y\hat{j}. \qquad (2\text{-}17)$$

If Eqs. 2-16 and 2-17 are really just two different ways to write $\vec{D}$, they must be equal to each other. This implies that the x-component of Eq. 2-16 must equal the x-component of Eq. 2-17. Likewise for the y-components. Thus we find:

$$\begin{aligned} D_x &= A_x + B_x + C_x \\ D_y &= A_y + B_y + C_y. \end{aligned} \qquad (2\text{-}18)$$

Stated in words, Eq. 2-18 says that we can perform vector addition by adding all of the separate x-components to give the x-component of the resultant and by adding all of the separate y-components to give the y-component of the resultant. This method of vector addition is called **algebraic addition**.

Using components to perform vector subtraction and the multiplication of a vector by a scalar is an easy extension of vector addition. Thus $\vec{R} = \vec{P} - \vec{Q}$ can be written as

$$\begin{aligned} R_x &= P_x - Q_x \\ R_y &= P_y - Q_y. \end{aligned} \qquad (2\text{-}19)$$

Similarly, $\vec{T} = c\,\vec{S}$ is:

$$\begin{aligned} T_x &= cS_x \\ T_y &= cS_y. \end{aligned} \qquad (2\text{-}20)$$

Here both the x-component and the y-component are "stretched" equally by the factor c.

In Section 2.2 we used a graphical approach to see that the vector $-\vec{A}$ can be interpreted as a vector equal in magnitude to $\vec{A}$ but pointing the opposite direction. This interpretation is arrived at easily using components:

$$-\vec{A} = -(A_x\hat{i} + A_y\hat{j}) = (-A_x)\hat{i} + (-A_y)\hat{j}. \qquad (2\text{-}21)$$

Reversing the sign of each component reverses the direction of each component vector, and reversing each component vector causes the direction of $\vec{A}$ to be reversed without changing its magnitude.

We will, throughout this book, make frequent use of vector equations. For example, the equation expressing the force on a falling object in the presence of air resistance is

$$\vec{F} = \vec{W} - b\vec{v}.$$

Using what we have described in this section about vector components, you can see that this equation is really just a shorthand way of writing three simultaneous equations:

$$F_x = W_x - bv_x$$
$$F_y = W_y - bv_y$$
$$F_z = W_z - bv_z.$$

In other words, a vector equation is interpreted as meaning: "Equate the x-components on both sides, then equate the y-components, and then equate the z-components." Vector notation simply allows us to write these three equations in a much more compact form.

EXAMPLE 2-4:   In Example 2-1 we considered a bird that flew 100 to the east, then 200 m to the northwest. Use the algebraic addition of vectors to find the bird's net displacement. Express the resultant in both component form and as a magnitude and a direction.

SOLUTION: Figure 2-17 shows displacement vectors $\vec{A} = (100$ m, east) and $\vec{B} = (200$ m, northwest). We draw vectors tip-to-tail if we are going to add them graphically, but it's usually easiest to draw them all from the origin if we are going to use algebraic addition. The resultant vector $\vec{C}$ is found using the parallelogram rule. To add vectors algebraically we must know their components. From the figure these are seen to be

$$\vec{A} = 100\,\hat{i}\ \text{m}$$

$$\vec{B} = (-200\cos 45°\,\hat{i} + 200\sin 45°\,\hat{j})\ \text{m} = (-141\,\hat{i} + 141\,\hat{j})\ \text{m}.$$

Note that vector quantities still must include units. Also note that $\vec{B}$ has a negative x-component. Adding $\vec{A}$ and $\vec{B}$ gives

$$\vec{C} = \vec{A} + \vec{B} = 100\,\hat{i}\ \text{m} + (-141\,\hat{i} + 141\,\hat{j})\ \text{m} = (-41\,\hat{i} + 141\,\hat{j})\ \text{m}.$$

This would be a perfectly acceptable answer for many purposes. However, we need to translate $\vec{C}$ to the magnitude and direction representation if we want to compare this result

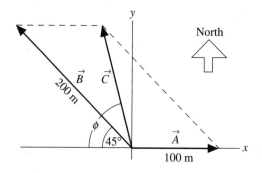

FIGURE 2-17 The net displacement $\vec{C} = \vec{A} + \vec{B}$ of the bird in Example 2-4 can be found by adding the components of vectors $\vec{A}$ and $\vec{B}$.

with our earlier answer. The magnitude of $\vec{C}$ is

$$|\vec{C}| = \sqrt{C_x{}^2 + C_y{}^2} = \sqrt{(-41\ \text{m})^2 + (141\ \text{m})^2} = 147\ \text{m}.$$

The angle ϕ, defined in Fig. 2-17, is

$$\phi = \tan^{-1}\left(\frac{|C_y|}{|C_x|}\right) = \tan^{-1}\left(\frac{141\ \text{m}}{41\ \text{m}}\right) = 73.8°.$$

Thus $\vec{C}$ = (147 m, 73.8° north of west), in perfect agreement with Example 2-1.

Tilted Axes and Arbitrary Directions

As we noted earlier, the coordinate system is entirely your choice. It is a grid that you impose on the problem in whatever manner will make the problem easiest to solve. You will soon meet problems where it will be convenient to tilt the axes of the coor-dinate system, such as shown in Fig. 2-18.

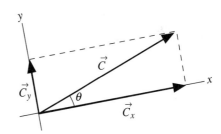

Although you may not have seen such a coordi-nate system before, it is perfectly legitimate. The axes are perpendicular and the y-axis is oriented correctly with respect to the x-axis. While we are used to having the x-axis horizontal, there is no requirement that it has to be that way.

Finding components with tilted axes is no dif-ferent than what we did in the last section: Eqs. 2-9 will apply as long as we correctly measure θ from the x-axis. Note that the unit vectors $\hat{i}$ and $\hat{j}$ corre-spond to the axes, not to "horizontal" and "vertical," so they are also tilted.

FIGURE 2-18 A coordinate system with tilted axes.

Another issue that will arise is to determine component vectors "parallel to" and "per-pendicular to" some line drawn in an arbitrary direction. For example, we will later need to decompose a force vector into component vectors parallel to and perpendicular to a sur-face. Doing so is analogous to using tilted coordinate axes.

Suppose we would like to find the components of vector $\vec{A}$ parallel and perpendicular to the line shown in Fig. 2-19a. Figure 2-19b shows the component vectors, which we label $\vec{A}_\parallel$ and $\vec{A}_\perp$. The subscripts indicate either "parallel to" or "perpendicular to." This

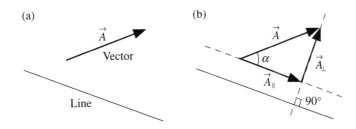

FIGURE 2-19 Vector components parallel and perpendicular to a line.

figure represents a decomposition of $\vec{A}$ because $\vec{A} = \vec{A}_{\parallel} + \vec{A}_{\perp}$. From geometry, the magnitudes of these component vectors are given by

$$|\vec{A}_{\parallel}| = |\vec{A}|\cos\alpha$$

$$|\vec{A}_{\perp}| = |\vec{A}|\sin\alpha$$

(2-22)

where α (Greek *alpha*) is the angle between the desired direction and $\vec{A}$. This is the same result we would get for the *x*- and *y*-components if we tilted the coordinate system to have the *x*-axis parallel to the given line. Note that it is not necessary to have the tail of $\vec{A}$ on the line to find a component of $\vec{A}$ parallel to the line. The line simply indicates a direction, and the component $\vec{A}_{\parallel}$ lies in that direction.

2.4 Significant Figures

Now that we are ready to begin doing calculations, it is necessary to say a few words about a perennial source of difficulty—significant figures. Mathematics is a subject where numbers and relationships can be as precise as desired, but physics deals with a real world of ambiguity and imprecision. It is important in all areas of science and engineering to state clearly what you know about a situation—no less and, especially, no more. Numbers provide one way to specify your knowledge.

If you report that a physical quantity has a value of 6.2, the implication is that the actual value, as best you can determine, falls between 6.15 and 6.25 and thus rounds to 6.2. If that is the case, then reporting a value of simply 6 is saying less than you know—you are withholding information. On the other hand, to report the number as 6.213 is not only wrong, it borders on being fraudulent. Any person reviewing your work—perhaps a client who hired you—would interpret the number 6.213 as meaning that the actual value falls between 6.2125 and 6.2135, thus rounding to 6.213. You are, in this case, claiming to have knowledge and information that you do not really possess. If your client proceeds to make subsequent decisions based upon your report that the value is 6.213, you could be sued! These are the kinds of issues of which real-world law suits are made.

The problem is how to state your numerical knowledge accurately. The solution is the proper use of **significant figures**. A number such as 6.2 is said to have *two* significant figures because the next decimal place—the one-hundredths—is not known and thus cannot be significant. Similarly, 6.213 has four significant figures. A number known to two significant figures is known with an accuracy of roughly 1 part in 10^2, or $\pm 1\%$. Similarly, a number known to four significant figures is known with an accuracy of roughly 1 part in 10^4, or $\pm 0.01\%$. Significant figures thus provide a means of judging the accuracy with which a number is known.

The easiest way to determine how many significant figures a number has is to write it in scientific notation. This helps you see that the location of the decimal point does not in any way affect the number of significant figures. For example, 6200 (6.2×10^3) and 0.00062 (6.2×10^{-4}) both have the same number of significant figures as 6.2—just two. Leading zeros, such as the zeros in 0.00062, simply locate the decimal point but have no influence over the accuracy of a number. Changing units—say from grams to kilograms—

shifts the decimal point but does not alter the number of significant figures. Also, make sure you notice that the number of significant figures is *not* the same as the number of decimal places.

While leading zeros are not significant, trailing zeros are! Thus 0.043 has two significant figures while 0.0430 has three. In the first case we know nothing at all about the fourth decimal place, while in the second case we know specifically that the fourth decimal place is a zero and not any other value. A potentially confusing case is a number like 230; is the zero significant or not? If the number is written as 230, the zero can be interpreted as simply locating the decimal point and not significant. So 230 has two significant figures. But if you know that the ones' place specifically is a zero, so that the zero is significant, the custom is to write the number as "230." with a trailing decimal point. Thus 230. has three significant figures. Likewise, in the example above, 6200 had just two significant figures but 6200. has four.

Arithmetic combinations of numbers follow the "weak link" rule. The saying, which you probably know, is that "a chain is only as strong as its weakest link." Even if nine out of ten links in a chain can support a 1000 pound weight, their strength is meaningless if the tenth link can only support 200 pounds. With numbers, nine out of the ten numbers used in a calculation might be known with an accuracy of 0.01%; but if the tenth number has an accuracy of only 10%, then the result of the calculation cannot possibly be more accurate than 10%. The weak link rules! The point is that the results of a calculation should not have any more significant figures than the *least*-accurately known number used in the calculations. To use more figures would be claiming more knowledge than you really possess about the answer.

We will soon see that the net force F exerted on an object is related to the object's mass m and acceleration a by $F = ma$. If the mass is $m = 63.5$ kg and the acceleration $a = 5.42$ m/s^2, your calculator will give a numerical product $ma = 344.17$ kg m/s^2. To report such an answer would, however, be incorrect. Both numbers used in the calculation have three significant figures and so are accurate to roughly 0.1%. By the weak link rule, it is not possible for the answer to have any more accuracy than this, so it should be reported to just three significant figures: $F = 344$ kg m/s^2. Similarly, $\sqrt{63.5} = 7.97$, not 7.9686887 as given by your calculator.

There are a few exceptions to keep in mind. First, you can unavoidably lose accuracy during subtraction. For example, 65.5 − 63.3 gives an answer of 2.2 (two significant figures), not 2.20 (three significant figures). The original numbers have no information about the one-hundredths column, so there is no way for the answer to generate that knowledge. Second, it is customary to keep one extra significant figure if (and only if) the number starts with a 1. For example, 10.43 could be used in a calculation with other numbers having three significant figures. Third, it is acceptable to keep one or two extra significant figures during intermediate steps of a calculation, as long as the final answer is reported with the proper number. The purpose here is to minimize round-off errors in the calculation. But only one or two extra figures—not the five or six shown in your calculator display— are necessary.

In laboratory work, the proper number of significant figures is determined by the experiment. The least-accurately measured quantity sets the proper number. Textbook problems in science and engineering often use a standard of *three* significant figures for all calculations. Two figures is too imprecise, while four is unnecessary *unless* the problem happens to give very accurate data with which to work. *Three significant figures will be the standard in this text*, unless a specific problem instructs you otherwise.

Caution: Many calculators display only two decimal places as a default setting. This is very dangerous! If you need to calculate 5.23/58.5, your calculator will show a result of 0.09 and it is all too easy to write that number down as an answer. But by doing so, you have reduced a calculation involving two numbers with three significant figures to an answer with only one significant figure. The proper result of this division is 0.0894 or 8.94×10^{-2}. You can avoid these kinds of errors if you keep your calculator set to display numbers in scientific notation with two decimal places.

A frequent concern is with problems that say things such as, "A mass of 1 kg …" Is this not just *one* significant figure? While that appears to be the case, you must learn to interpret this common way of writing problems. In keeping with the accepted standard, such a statement is to be *interpreted* as meaning that the mass is 1.00 kg—known with an accuracy of three significant figures.

One last caution: In math classes, answers like $\pi / \sqrt{3}$ are perfectly acceptable. Not so in science or engineering, where you should *always* evaluate such expressions and give a numerical answer to three significant figures. (An exception is when using angles measured in radians, where answers like "2π radians" are acceptable.)

Learning to use significant figures correctly is part of learning the "culture" of science and engineering. These "cultural issues" are important because you must speak the same language as the "natives" if you wish to communicate effectively. Therefore, we will frequently emphasize them. Most students "know" the rules of significant figures, having learned them in high school, but many fail to apply them. It is important that you understand the reasons for significant figures and that you get in the habit of using them properly.

Summary

Important Concepts and Terms

scalar quantity

vector quantity

magnitude

resultant vector

graphical addition

zero vector

Cartesian coordinates

quadrants

component vectors

decomposition of a vector

x-component

y-component

unit vector

algebraic addition

significant figures

Basic Greek Symbols

α	alpha
β	beta
Δ	delta
π	pi
μ	mu
θ	theta
ϕ	phi

This chapter has been more mathematics than physics, but we have developed some powerful analysis tools that we will use throughout the text. One of these tools is vectors. A vector is a mathematical quantity that is characterized by both a length and a direction. There are two methods for describing vectors: either "graphical," using the vector's magnitude and direction, or "algebraic," using the vector's components in some coordinate system. You should be able to use both of these methods and to switch back and forth between them. The process of finding the components of a vector, known as decomposing the vector into its components, is especially important for our needs. Connected with the algebraic description of vectors is the definition of the unit vectors $\hat{i}$ and $\hat{j}$, which you should review.

Vectors are manipulated and combined using vector arithmetic. Three primary results we have established in this chapter are: how to add two vectors, how to subtract two vectors, and how to multiply a vector by a scalar. You should be able to represent these processes graphically and also be able to perform the operations using components.

The chapter ended with a discussion of the use of significant figures. Significant figures are another aspect of *interpreting* your work and assigning a meaning to your calculations. Your calculator can be a dangerous object; it computes away without regard for whether the computations have any meaning or significance. Interpreting the results of a calculation, which means that you are running the calculator and not vice versa, is an important skill to learn.

Exercises and Problems

Exercises

1. Consider vectors $\vec{A} = 5\,\hat{i} + 2\,\hat{j}$ and $\vec{B} = -3\,\hat{i} - 5\,\hat{j}$. Define vector $\vec{C}$ as $\vec{C} = \vec{A} + \vec{B}$.
   a. Draw a coordinate system and use it to show the vectors $\vec{A}$, $\vec{B}$, and $\vec{C}$.
   b. Write vector $\vec{C}$ in component form.
   c. What are the magnitude and direction of vector $\vec{C}$?

2. Consider the two vectors $\vec{A}$ and $\vec{B}$ from Exercise 1. Define vector $\vec{D}$ as $\vec{D} = \vec{A} - \vec{B}$.
   a. Draw a coordinate system and use it show the vectors $\vec{A}$, $\vec{B}$, and $\vec{D}$.
   b. Write vector $\vec{D}$ in component form.
   c. What are the magnitude and direction of vector $\vec{D}$

3. Consider the two vectors $\vec{A}$ and $\vec{B}$ from Exercise 1. Define vector $\vec{E}$ as $\vec{E} = -2\vec{A} + 3\vec{B}$.
   a. Draw a coordinate system and use it show the vectors $\vec{A}$, $\vec{B}$, and $\vec{E}$.
   b. Write vector $\vec{E}$ in component form.
   c. What are the magnitude and direction of vector $\vec{E}$?

4. Figure 2-20 shows vectors $\vec{A}$ and $\vec{B}$. Let $\vec{C} = \vec{A} + \vec{B}$.
   a. Reproduce the figure as accurately as possible, using a ruler and protractor. Draw vector $\vec{C}$ on your sketch by graphically adding of $\vec{A}$ and $\vec{B}$. Then determine the magnitude and direction of $\vec{C}$ by *measuring* it with a ruler and protractor.

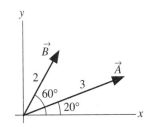

FIGURE 2-20

b. Use geometry and trigonometry to *calculate* exactly the magnitude and direction of $\vec{C}$.

c. Decompose vectors $\vec{A}$ and $\vec{B}$ into components, then use these components to calculate the magnitude and direction of $\vec{C}$.

5. a. In Figure 2-21, what is the angle ϕ between vectors $\vec{E}$ and $\vec{F}$?

b. What is the magnitude and direction of the vector $\vec{G} = \vec{F} - \vec{E}$?

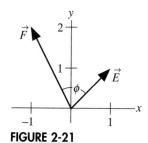

FIGURE 2-21

6. For the three vectors shown in Figure 2-22, $\vec{A} + \vec{B} + \vec{C} = -2\,\hat{\imath}$. What is vector $\vec{B}$? Write your answer in component form, and as a magnitude and a direction.

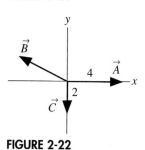

FIGURE 2-22

7. a. Figure 2-23 shows two vectors $\vec{A}$ and $\vec{B}$. What must vector $\vec{C}$ be such that $\vec{A} + \vec{B} + \vec{C} = 0$?

b. Redraw the figure showing vector $\vec{C}$.

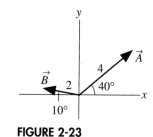

FIGURE 2-23

Problems

8. Bob walks 200 m south, then jogs 400 m southwest, then finally walks 200 m in a direction 30° east of north.
   a. Use a ruler and a protractor to draw an accurate graphical picture, similar to Fig. 2-6, of Bob's motion.
   b. What must Bob do to get back to his starting point by the most direct route?

Determine your answer in two ways: first, by measuring it directly on your sketch from part a), and second, by calculating it and expressing your answer as a magnitude and an angle.

9. Mary needs to row her boat across a 100 m wide river that is flowing to the east at a speed of 1 m/s. Mary can row the boat with a speed of 2 m/s.
   a. If she rows straight north, where will she land?
   b. Draw a picture showing Mary's displacement due to her rowing, her displacement due to the river's motion, and her net displacement.

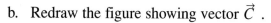

10. Jim's dog Sparky runs 50 m northeast to tree, then 70 m west to a second tree, and finally 20 m south to a third.
    a. Draw a picture of the situation and establish a coordinate system.
    b. Calculate Sparky's net displacement in component form.
    c. Calculate Sparky's net displacement as an magnitude and an angle.

11. The minute hand on a watch is 2 cm in length. What is the displacement vector of the tip of the minute hand:
    a. From 8:00 to 8:20 A.M.?
    b. From 8:00 to 9:00 A.M.?

12. The treasure map shown in Fig. 2-24 gives the following directions to a buried treasure: "Start at the old oak tree, walk due north for 500 paces, then due east for 100 paces. Dig." But when you arrive at the oak tree, you find an angry dragon just north of the tree. To avoid the dragon, you set off along the yellow brick road at an angle 60° east of north. After walking 300 paces you see an opening through the woods.

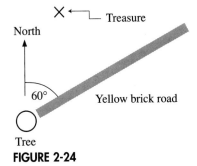
FIGURE 2-24

    a. Use vector subtraction to find the north/south and east/west components of your route to the treasure.
    b. How far are you from the treasure?

13. A jet plane is flying horizontally with a speed of 500 m/s over a hill that slopes upward with a 3% grade (i.e., the "rise" is 3% of the "run"). What is the component of the plane's velocity perpendicular to the ground?

14. Jack and Jill ran up the hill at 3 m/s. If the horizontal component of Jill's velocity vector is 2.5 m/s:
    a. What is the angle of the hill?
    b. What is the vertical component of Jill's velocity?

15. Figure 2-25 shows a car speeding up as it turns a quarter-circle curve from north to east. When exactly halfway around the curve, the car's acceleration vector is $\vec{a}$ = (2 (m/s)/s, 15° south of east).

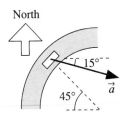
FIGURE 2-25

    a. What is the component of $\vec{a}$ tangent to the circle at that point?
    b. What is the component of $\vec{a}$ perpendicular to the circle at that point?

16. Figure 2-26 shows three ropes tied together in a knot and pulled on in three different directions. How hard and in which direction should the pull on the third rope be to keep the knot from moving?

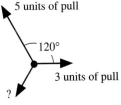
FIGURE 2-26

17. Three forces are exerted on an object placed on a tilted floor, as shown in Figure 2-27. Assuming that forces are vectors:
    a. What is the component of the *net force* $\vec{F}_{net} = \vec{F}_1 + \vec{F}_2 + \vec{F}_3$ parallel to the floor?
    b. What is the component of $\vec{F}_{net}$ perpendicular to the floor?
    c. What is the magnitude and direction of $\vec{F}_{net}$?

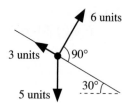

FIGURE 2-27 .

18. Four electrical charges are located on the corners of a rectangle, as shown in Figure 2-28. Like charges, you will recall, repel each other while opposite charges attract. Charge B exerts a repulsive force (directly *away from* B) on charge A of 3 units strength. Charge C exerts an attractive force (directly *toward* C) on charge A of 6 units strength. Finally, charge D exerts an attractive force of 2 units strength on charge A. Assume that the forces exerted on A are vectors. What is the magnitude and the direction of the net force $\vec{F}_{net}$ exerted on charge A?

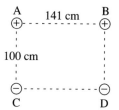

FIGURE 2-28

[**Estimated 10 additional problems for the final edition.**]

Chapter **3**

Kinematics: The Mathematics of Motion

Of all the intellectual hurdles which the human mind has confronted and has overcome in the last fifteen hundred years the one which seems to me to have been the most amazing in character and the most stupendous in the scope of its consequences is the one relating to the problem of motion.

Herbert Butterfield—*The Origins of Modern Science*

Sagredo: From these considerations it appears to me that we may obtain a proper solution of the problem discussed by philosophers, namely, what causes the acceleration in the motion of heavy bodies?

Salviati: The present does not seem to be the proper time to investigate the cause of acceleration of natural motion, concerning which various opinions have been expressed by various philosophers ... At present it is the purpose of our Author merely to investigate and to demonstrate some of the properties of accelerated motion, whatever the cause of this acceleration might be.

Galileo—*Two New Sciences*

LOOKING BACK * | Sections 1.4–1.7; 2.3

3.1 Introduction

We are ready, after several preliminary chapters, to begin a quantitative study of motion. At first glance, such a study would appear to be straightforward. But nothing could be further from the truth. As Butterfield notes in the opening quotation, solving the "problem of motion" was one of the supreme feats of the human imagination. Many of the finest thinkers from Heraclitus to Galileo—a span of roughly 2000 years—pondered the issue of motion, but none succeeded in finding the solution. That honor rests with Newton. Galileo, however, did make two extremely important contributions that Newton was able

* The Looking Back box directs your attention back to specific sections that will be of major significance to the present chapter. This box will be a regular feature of most chapters to come. You need not re-read the sections in detail, but a brief review will improve your study of the present chapter.

to build upon. First, Galileo made a separation between the *cause* of motion and the *description* of motion. Secondly, Galileo was the first major scientist to base his work on *quantifying* nature by making numerical measurements and finding specific mathematical relationships between various measured quantities. Previous scientists and philosophers assumed that "truth" could be found through the power of *reason* alone, but Galileo asserted the then quite radical viewpoint that we can best learn about nature through experiment. In this regard, Galileo was the first modern scientist.

[**Photo suggestion: Portrait of Galileo.**]

Kinematics is the modern name for the mathematical description of motion without regard to the causes of that motion—exactly the separation advocated by Galileo. The term itself comes from the Greek word *kinema*, meaning "movement." You know this word through its English variation: *cinema*—motion pictures! This chapter will be concerned with developing the mathematical tools for describing motion, and we will follow very much in Galileo's footsteps. Only when this has been accomplished will we be ready to take the next step, which Galileo recognized but could not answer, of inquiring as to the cause of motion.

3.2 Measurements and Units

To quantify motion we will need, as did Galileo, to make measurements. As we discussed in Chapter 1, the only quantities that we can measure directly are the position, using a tape measure or ruler, and the time, with a clock or stopwatch. Other quantities, such as velocity, are determined from these basic measurements.

Measurements require *units*. In Chapter 1, for example, we adopted the unit of *one frame* for measuring time. While that was handy at the time, its practical limitations are obvious. One of the important tasks of science is to define and adopt units that can be replicated in any laboratory at any time. The system of units currently used in science is called the *Système Internationale d'Unités*, which was adopted by an international committee in 1960. These are commonly referred to as **SI units**. Older books often referred to *mks units*, which stands for "meter-kilogram-second," or *cgs units*, which is "centimeter-gram-second." For practical purposes, SI units are the same as mks units. In casual speaking we often refer to *metric units*, although this could mean either mks or cgs units. So to be precise, we will follow the internationally accepted custom of calling our system of units *SI units*.

Units of Length and Time

The standard of time, prior to 1960, was based on the *mean solar day*. As time-keeping accuracy and astronomical observations improved, however, it became apparent that the earth's rotation is not perfectly steady. Meanwhile, atomic physicists during the 1950s and 1960s had been developing a device called an *atomic clock*. This instrument is able to measure, with incredibly high precision, the frequency of radio waves absorbed by atoms as they move between two closely-spaced energy levels. These frequencies, as far as we know, have not changed since the beginning of the universe and, furthermore, they can be reproduced with great accuracy at many laboratories around the world. Consequently, the SI unit of time—the second—was redefined in 1967 as: One *second* is the time required for 9,192,631,770 oscillations of the radio wave absorbed by the cesium-133 isotope. The abbreviation for second is the letter s.

[**Photo suggestion: Atomic clock at the National Institute of Standards and Technology.**]

A shortwave radio channel operated (in the United States) by the National Institute of Standards and Technology broadcasts a signal whose frequency is linked directly to the atomic clocks. This signal is the time standard, and any time-measuring equipment you use was calibrated from this time standard.

The SI unit of length—the meter—has also has a long and interesting history. It was originally defined as one ten-millionth of the distance from the North Pole to the equator along a line passing through Paris. There are obvious practical difficulties with implementing this definition, and it was later abandoned in favor of the distance between two scratches on a platinum-iridium bar stored in a special vault in Paris. More recent definitions involved various atomic properties. The present definition was agreed to in 1983 and is as follows: One *meter* is the distance traveled by light in vacuum during 1/299792458 of a second. The abbreviation for meter is the letter m.

This is equivalent to defining the speed of light to be 299,792,458 m/s, exactly. Laser technology is used in various national laboratories to implement this definition and to calibrate secondary standards that are easier to use. These standards ultimately make their way to your ruler or to a meter stick, but it is worth keeping in mind that these measuring devices are only as accurate as the care with which they were calibrated.

We will have occasion to talk about lengths and times that are either very much less or very much greater than the standards of one meter and one second. We will do so by using *prefixes* to denote various powers of ten. Table 3-1 lists the prefixes that will be used frequently throughout this book. Memorize it! There are few things in science which are learned by rote memory, but this is one of them. Although the prefixes make it easier to talk about quantities, the proper SI units are only meters and seconds. Quantities given with prefixed units *must* be converted to meters and seconds before any calculations are done. If all quantities to be used in a formula are in SI units, then you can feel confident that the answer will have the correct SI units. That is why Chapter 1 urged you to make any conversions at the very beginning of a problem, as part of the pictorial model.

TABLE 3-1 Prefixes used in physics.

Prefix	Meaning	Abbreviation	Use
kilo-	10^3	k	km
centi-	10^{-2}	c	cm
milli-	10^{-3}	m	mm or ms
micro-	10^{-6}	μ	μm or μs
nano-	10^{-9}	n	nm or ns

Unit Conversions

Although SI units will be our standard, we cannot entirely forget that the United States still uses English units—feet, pounds, and so on. Even after repeated exposure to metric units in classes, most of us "think" in the English units we grew up with. So it remains important to be able to convert back and forth between SI units and English units. Table 3-2 shows a few frequently-used conversions. These are worth memorizing if you do not

TABLE 3-2 Useful conversions.

1 inch = 2.54 cm	1 m = 39.37 inch
1 mile = 1.61 km	1 km = 0.621 mile

already know them. While the English system was originally based on the length of the king's foot, it is interesting to note that today the conversion 1 inch = 2.54 cm is the definition of the inch. In other words, the English-system unit for length is based on the meter!

There are various techniques for doing unit conversions. One effective method is to write the conversion factor as a ratio equal to 1. For example, using the information in Tables 3-1 and 3-2, we can write the ratio of meters to micrometers as:

$$\frac{10^{-6} \text{ m}}{1 \text{ } \mu\text{m}} = 1.$$

We can also write the ratio of centimeters to inches as:

$$\frac{2.54 \text{ cm}}{1 \text{ in}} = 1.$$

Because multiplying any expression by 1 does not change its value, these ratios are easily used for conversions. For example, to convert 3.5 μm to meters we would compute

$$3.5 \text{ } \mu\text{m} \times \frac{10^{-6} \text{ m}}{1 \text{ } \mu\text{m}} = 3.5 \times 10^{-6} \text{ m}.$$

Similarly, the conversion of 2 feet to meters would be

$$2 \text{ ft} \times \frac{12 \text{ in}}{1 \text{ ft}} \times \frac{2.54 \text{ cm}}{1 \text{ in}} \times \frac{10^{-2} \text{ m}}{1 \text{ cm}} = 0.610 \text{ m}.$$

Notice how units in the numerator and in the denominator cancel until just the desired units remain at the end. You can continue this process of multiplying by 1 as many times as necessary to complete all the conversions.

These examples convert information *into* SI units, as you would need to do in a problem. As we get further into problem solving, we will place emphasis on deciding whether or not the answer to a problem "makes sense." To determine this, at least until you have more experience with SI units, you will probably need to convert from the SI units of your answer back to the English units in which you think. But this conversion does not need to be very accurate. For example, if you are working a problem about automobile speeds and obtain an answer of 35 m/s, all you really want to know for the assessment is whether or not this is a realistic speed for a car. That requires a "quick and dirty" conversion, not a conversion of great accuracy.

Table 3-3 shows a number of approximate conversion factors that can be used to assess the answer to a problem. Using 1 m/s ≈ 2 mph (miles per hour), you find that 35 m/s is roughly 70 mph—a reasonable speed

TABLE 3-3 Approximate conversion factors.

1 cm ≈ 1/2 inch	10 cm ≈ 4 inch
1 m ≈ 1 yard	1 m ≈ 3 feet
1 km ≈ 0.6 mile	1 m/s ≈ 2 mph

for a car. But an answer of 350 m/s—which you might get after making a calculation error—would be an unreasonable 700 mph. Practice with these will allow you to develop intuition for metric units.

3.3 Motion in One Dimension

In this chapter we are going to consider the mathematics of one-dimensional motion, that is, motion along a straight line. You will recall, from Chapter 1, that position $\vec{r}$, velocity $\vec{v}$, and acceleration $\vec{a}$ are all vectors, having a direction as well as a magnitude. Chapter 2 introduced the *components* of a vector. In the special case that a vector is parallel to one of the coordinate axes, then only that component of the vector is non-zero. If an object moves along the x-axis, for example, its velocity vector $\vec{v}$ is parallel to the x-axis. The components v_x and v_y are thus

$$v_x = \begin{cases} +|\vec{v}| & \text{if } \vec{v} \text{ points toward } +x \\ -|\vec{v}| & \text{if } \vec{v} \text{ points toward } -x \end{cases}$$

$$v_y = 0.$$

(3-1)

Thus the x-component v_x contains the full magnitude of the vector, *and* it also includes a *sign* to indicate which direction the vector points along the axis.

We noted, in Section 1.7, that the subscripts are not necessary for one-dimensional motion because only one component of each vector is non-zero. If an object moves parallel to the x-axis, the symbol v will be interpreted as meaning v_x, the only non-zero component of $\vec{v}$. Similarly, the symbol a will mean a_y for an object moving along the y-axis.

Our analysis should be valid regardless of whether the line is the x-axis, the y-axis, or any other straight line. Consequently, we will find it convenient to use a "generic axis" called the s-axis. The position of the object will be represented by the symbol s. This will allow our equations, written in terms of s, to be valid for any one-dimensional motion. In a specific problem, however, you should switch to x or y, whichever is appropriate, rather than s. Vector components s, v, and a all have signs determined as shown in Eq. 3-1. These signs are illustrated in Fig. 3-1. Be sure to note that the sign of s depends on where the object is located relative to the origin, but the signs of v and a are independent of the object's location. Instead, they indicate the direction that the $\vec{v}$ and $\vec{a}$ vectors point.

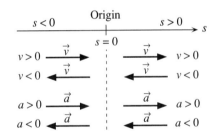

FIGURE 3-1 Signs of the vector components s, v, and a.

Position and Time

When we introduced motion diagrams in Chapter 1, every frame of the film had a specific clock reading and the object was located at a specific place along the ruler. The ruler measurement gives the *position s* at the measured *instant of time t*. We could make a table, based on information measured directly from the motion diagram, showing the position of the object at many different times. Consider, for example, a student walking to school in the morning along a straight street. Her positions at various times are shown in Table 3-4. She leaves home at a time we choose to call $t = 0$ and makes steady progress for awhile. Beginning around $t = 3$ s, there is a period where the distance traveled during each time

TABLE 3-4 Measured position at various times for a student walking to school.

Time t (min)	Position s (m)
0	0
1	60
2	120
3	180
4	200
5	210
6	220
7	320
8	420
9	520

interval becomes less—perhaps the student slowed down to speak with a friend. Then she picks up the pace, and the distances within each interval become longer.

We can see the student's motion more clearly if we make a graph of these measurements. Figure 3-2a shows a graph of s versus t for the student. (A graph of a versus b means that a is the y-variable and b is the x-variable. Saying "graph a versus b" is really a shorthand way of saying "graph a as a function $a(b)$ of b.") Note that the measurements from the motion diagram tell us where the student is at only a few discrete points of time. Hence, a graph of this "data" shows only points—no lines.

However, common sense tells us the following. First, the student—or any other object—was *somewhere specific* at all times. That is, there was never a time when she failed to have a well-defined position, nor could she occupy two positions at one time. (As reasonable as this belief appears to be, it will be severely questioned and found not entirely accurate when we get to quantum physics!) Second, the student moved *continuously* through all intervening points of space. She could not go from $s = 100$ m to $s = 200$ m without passing through every point in between. It is thus quite reasonable to believe that her motion can be shown as a *smooth curve* passing through the measured points, as shown in Fig. 3-2b. A graph such as this is called a *position-versus-time graph*.

It is worth pausing to note that a graph is *not* a "picture" of the motion. The student is walking along a straight line, but the graph itself is not a straight line. Graphs are an *abstract representation* of motion. Consequently, we will place significant emphasis on the process of *interpreting* graphs, and many of the exercises and problems will give you a chance to practice these skills.

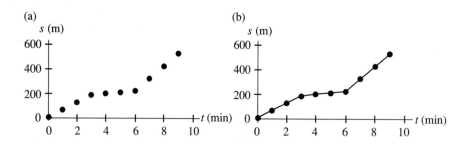

FIGURE 3-2 a) Student's measured position at specific instants of time. b) A position-versus-time graph of the student's motion, with a smooth curve through the data.

EXAMPLE 3-1   Figure 3-3 represents the motion of a car along a road. Describe in words the motion of the car.

SOLUTION Notice that the figure uses the specific symbol x for position rather than the generic s because, in this example, we know the motion is horizontal. At $t = 0$, the car is at $x = 10$ km. Because x is a positive value, this position is 10 km to the *right* of the origin. The value of x steadily *decreases* for 30 minutes, indicating that the car is moving *left*, toward the origin. After 30 minutes has elapsed, the car is located at $x = -20$ km, or 20 km left of the origin. From $t = 30$ min to $t = 40$ min the value of x does not change, so the car must be stopped and not moving. The

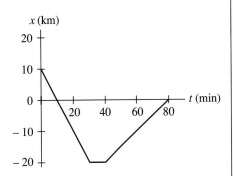

FIGURE 3-3 Position graph of the car of Example 3-1.

value of x begins increasing at $t = 40$ min, so the car is now moving toward the *right*. It reaches the origin at $t = 80$ min, where the graph ends. The car travels the 20 km from $x = -20$ km to $x = 0$ km in 40 minutes, whereas the same 20 km was covered in only 20 min (between $t = 10$ min and $t = 30$ min) when the car was moving toward the left. This leads us to conclude that the motion toward the right is slower than the motion toward the left. We can summarize our *interpretation* of the graph by saying that it represents a car that travels toward the left for 30 minutes, stops for 10 minutes, then travels back toward the right for 40 minutes at a slower speed than it had initially. The car's final position is 10 km to the left of where it started.

   How long does a moving object spend at one particular position? You might think this is a trick question, but it is actually quite important. We have associated each position with a clock reading, so we could restate the question as, How long does a clock reading last? Your answer to both these questions is likely to be, "For an instant." But that answer just defers the problem. How long is an instant? Most people, if pressed for a definition, reply that "an instant is a very, very small interval of time." However, a very, very small interval is still *finite* and could, at least in principle, be measured with a very, very fast and accurate clock. *If* a moving object spends "an instant" at each location, where an instant is a small but finite interval of time, then $\Delta \vec{r} = 0$ during a finite Δt. But then the average velocity during this instant is $\vec{v} = \Delta \vec{r}/\Delta t = 0$. In other words, the object would be at rest ($\vec{v} = 0$) during this very small interval while it is also, by definition, moving!

   The error in our reasoning that led to this contradiction was the assumption that "an instant" has a very small but finite duration. Our only recourse is to accept that "an instant" lasts for zero seconds, or equivalently, that a moving object occupies a particular position for zero seconds. The symbol t stands for *the instant of time* and not for a time interval. By contrast, we will always use Δt to represent an *interval of time*. Thus it is *not* correct to say that s represents the position *for* an instant of time, because this implies that instants of time have duration. Instead, s is the position *at* an instant of time. This is a subtle but crucial distinction. We can call s the **instantaneous position**—the position *at* a particular instant of time. An object spends zero seconds at an instantaneous position before moving on to the next position. We will return to this idea in Section 3.5 when we introduce *instantaneous velocity*.

3.4 Uniform Motion

If you drive your car at a perfectly steady 60 miles per hour, you will cover 60 miles during the first hour, another 60 miles during the second hour, yet another 60 miles during the third hour, and so on. This is an example of what we call *uniform motion*. In this case, 60 miles is not your position, but rather the *change* in your position during each hour, which we defined as being your displacement Δs. Similarly, 1 hour is a *time interval* Δt rather than a specific instant of time. This suggests the following definition:

> Motion in which equal displacements occur during *any* successive equal time-intervals is called **uniform motion**.

The qualifier "any" in this definition is important. If you drive 120 miles per hour for 30 minutes—covering 60 miles—then stop for 30 minutes, and then repeat this each hour, you would cover 60 miles during each successive 1-hour interval. But if you measured your displacement using 30-minute time intervals, then you would *not* find equal displacements during successive intervals. So this motion is not uniform. Your constant 60 mph driving is uniform motion because you will find equal displacements no matter how you choose your successive time intervals.

Consider the two particle motions represented by motion diagrams in Fig. 3-4. In Fig. 3-4a, the displacement between successive frames is always the same. While you cannot be sure what happened between frames, if you assume that the particle was behaving reasonably, then this motion diagram represents uniform motion. (Note the equally spaced points.) Figure 3-4b, though, is clearly not uniform motion because the successive displacements are not equal.

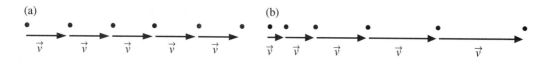

FIGURE 3-4 Motion diagrams for a) uniform motion, and b) nonuniform motion.

Figure 3-5 shows position-versus-time graphs for the two motions of Fig. 3-4. Notice that there is a clear geometric distinction between the graph of uniform motion and that of nonuniform motion—namely, uniform motion is represented by a position-versus-time graph that is a straight line. This follows from the requirement that all Δs corresponding to the same Δt be equal. In fact, an alternative definition of uniform motion is that *motion is uniform if and only if its position-versus-time graph is a straight line.*

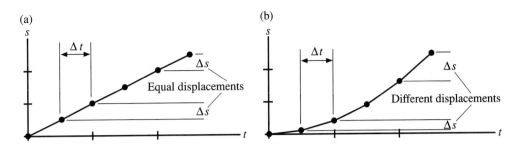

FIGURE 3-5 Position-versus-time graphs for a) uniform motion, and b) nonuniform motion.

A straight line is characterized by its slope, which is defined as the "rise over run," or $\Delta y/\Delta x$. The graphs of Fig. 3-5 have s as the y-variable and t as the x-variable, so their slopes are given by $\Delta s/\Delta t$. Because the slope of a straight line is constant, another way to characterize uniform motion is to say that *motion is uniform if and only if $\Delta s/\Delta t$ is a constant for any time interval Δt*.

In Chapter 1 we defined the **average velocity** to be $\Delta \vec{r}/\Delta t$. This definition is for the vector $\vec{v}$. However, for one-dimensional motion we are only interested in the vector component v along the s-axis. Therefore, it is convenient at this point to give average velocity the special symbol $\bar{v}$, where the bar over the letter is a commonly used method of indicating an average. This will allow us, in the next section, to distinguish between average velocity and instantaneous velocity. In one dimension, the average velocity $\bar{v}$ is:

$$\bar{v} \equiv \frac{\Delta s}{\Delta t} \text{ (average velocity).}$$

Notice the special symbol "$\equiv$" rather than the usual equal sign "$=$". The symbol $\equiv$ stands for "is defined as" or "is equivalent to," which implies more than the two sides simply being equal.

We have *defined* $\bar{v}$ as $\Delta s/\Delta t$. But $\Delta s/\Delta t$ is just the slope of the position-versus-time graph. Thus we can say that *motion is uniform if and only if its average velocity $\bar{v}$ is constant during any time interval Δt*. As a result, uniform motion is also called *motion at constant velocity*. Note that the slope of the position-versus-time graph, which is a *geometrical quantity*, is equivalent to the *physical quantity* that we call the average velocity. This is an extremely important idea that we will use over and over.

EXAMPLE 3-2 Figure 3-6 graphically represents the motion of three people skating on rollerblades. Describe their motion in words.

[**Photo suggestion: Someone on rollerblades.**]

SOLUTION We can make a couple of initial observations. First, all three skaters have the same value of x at $t = 0$; hence, all three are located at the same horizontal position when the clock is started. (Keep in mind, following our earlier discussions, that this is not necessarily the beginning of their motion, but simply where they happened to be when we started the timing clock.) Second, we can conclude that all three are moving uniformly, with a constant velocity, because all three graphs are straight lines. We can get more specific by measuring the slopes in order to determine the velocities. Skater A undergoes a displacement of $\Delta x = 2.0$ m during the first 0.4 second. Thus

$$\bar{v}_A = \frac{\Delta x}{\Delta t} = \frac{2.0 \text{ m}}{0.4 \text{ s}} = 5.0 \text{ m/s}.$$

Note that we are using x rather than the generic s because the motion in this example is horizontal. Similarly skater B, who is displaced 2.0 m during a time interval of 1.0 s, has an average velocity $\bar{v}_B = 2.0$ m/s. We need to be more careful with

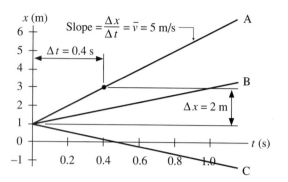

FIGURE 3-6 Position-versus-time graphs for three people on rollerblades.

skater C. Although she covers a distance of 2.0 m during 1.0 second, her *displacement* Δx has a very precise definition: $\Delta x = x_f - x_i$. Here $x_i = 1.0$ m while $x_f = -1.0$ m, so we have $\Delta x = -2.0$ m. As we have emphasized several times, paying attention to the signs is very important! This leads to $\bar{v}_C = -2.0$ m/s. Because $\bar{v}$ is the component of a vector, we interpret the minus sign as indicating that the object is moving toward the negative x-axis—that is, toward the left if we are using a standard axis. Our interpretation of this graph, then, is that three people on rollerblades, each moving with constant velocity, all pass the point $x = 1.0$ m at time $t = 0$. Skater A is moving to the right with a velocity of 5.0 m/s, skater B is also moving to the right with a smaller velocity of 2.0 m/s, while skater C is moving to left with a velocity of -2.0 m/s. One second later, A is at $x_1 = 6$ m, B is at $x_2 = 3$ m, and C, who was moving to the left, is at $x_3 = -1$ m.

•

There are several points worth noting in this example. First, steeper slopes on the position-versus-time graph correspond to faster speeds. Second, negative slopes correspond to negative velocities and, hence, to motion toward the negative end of the coordinate axis. Third, slope is a ratio of *intervals*—$\Delta s/\Delta t$—not a ratio of *coordinates*. That is, the slope is *not* simply s/t. This is a common error—here it would give the incorrect answer $\bar{v}_B = 3.0$ m/s. Fourth, and somewhat more subtle, is that we are distinguishing between the *actual* slope and the *physically-meaningful* slope. If you were to take out your ruler and measure the rise and run of the graphs in the figure, you could compute the *actual* slope—that is, the slope of the actual geometric line as drawn on the page. That is not the slope to which we are referring when we equate $\bar{v}$ with the slope of the line. Instead, we find the *physically-meahingful* slope by measuring the rise and run using the scales along the axes. Thus the line representing skater A has a "rise" $\Delta x = 2.0$ m during the "run" $\Delta t = 0.4$ s. Physically-meaningful rise and run include *units*, and the ratio of these units gives the units of the slope.

You will recall, from math, that a straight-line graph is represented algebraically by the equation $y = b + cx$, where c is the slope of the line and b is its y-intercept (i.e., the value of y when $x = 0$). For a straight-line position-versus-time graph, we have seen that the slope is $\bar{v}$. Let us define s_i to be the *initial value* of the position at $t = 0$. As seen on the graph, this is simply the y-intercept. So the equation for the straight-line graph of uniform motion shown in Fig. 3-5a is:

$$s(t) = s_i + \bar{v}t. \tag{3-2}$$

The notation $s(t)$ indicates explicitly that the position s is a *function* of the time t. Make sure you are convinced that this really is the proper equation to describe the line.

We arrived at Eq. 3-2 through geometry, but we can also get there using algebra. By definition

$$\bar{v} = \frac{\Delta s}{\Delta t} = \frac{s_f - s_i}{t_f - t_i}, \tag{3-3}$$

where i and f represent the initial and final points of some interval of motion. We can choose a particular time interval that starts with $t_i = 0$. Let the time interval end at a clock reading $t_f = t$, and call the position at that time $s_f = s(t)$. Then, in this case, $\Delta s = s(t) - s_0$ and $\Delta t = t - 0 = t$, so Eq. 3-3 becomes

$$\bar{v} = \frac{\Delta s}{\Delta t} = \frac{s_f - s_i}{t_f - t_i},$$

which is easily rearranged to give Eq. 3-2.

Equation 3-2 allows us to find the position of a uniformly-moving particle at any future time t if we know both its initial position and its velocity. There are two conditions, though, that must be satisfied before using this equation. First, the motion *must* be uniform—having constant velocity. Second, s_i *must* be the position at the specific instant $t = 0$. In some problems, however, you may be given an initial value of the position at a time other than zero. Equation 3-2 is not valid in such a case. However, a simple rearrangement of Eq. 3-3 gives a more general result for uniform motion:

$$s_f = s_i + \bar{v}\Delta t \quad \text{(uniform motion)}. \tag{3-4}$$

This applies for any time interval Δt during which the velocity is constant. You can see that Eq. 3-2 is a special case where the time interval happens to start at $t = 0$.

The average velocity of a uniformly-moving particle tells us the amount by which its position changes during each second. A particle with a velocity of 20 meter *per second* changes its position by 20 m during the first second of its motion, by another 20 m during the next second, and so on. If the motion starts at $s = 10$ m, the particle will be at $s = 30$ m after 1 second of motion and at $s = 50$ m after 2 seconds of motion. Thinking of velocity like this will help you develop an intuitive understanding that the velocity of an object is what causes its position to change.

EXAMPLE 3-3 Bob leaves home in Los Angeles at 9:00 A.M. and travels north at a steady 60 mph. Suzy, 400 miles to the north in San Francisco, also leaves at 9:00 A.M. and travels south at a steady 40 mph. Where will they meet for lunch?

SOLUTION Here is a problem where, for the first time, we can use all three aspects of our problem-solving strategy: the pictorial model, the physical model, and, finally, the mathematical model. Figure 3-7 shows the physical model and the pictorial model for the situation. In evaluating the given information, we note that the starting time of 9:00 is not relevant to the problem—we need to find *where* they meet, not *when*. Consequently, the initial time is chosen as $t_0 = 0$ for calculational simplicity. Also notice from the motion diagram that Bob and Suzy are traveling in opposite directions. This implies that one of the velocities must be a negative number. We chose a coordinate system in which Suzy is moving to the left, so she has the negative velocity.

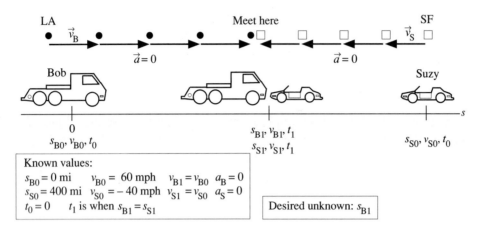

FIGURE 3-7 Motion diagram and pictorial model for Example 3-3.

One purpose of the pictorial model is to establish what we need to find. Bob and Suzy meet when they have the same position at the specific time t_1. That is, $s_{B1} = s_{S1}$ is the mathematical statement of their meeting. It is then the goal of the *mathematical model* to find and use *quantitative* relationships, leading to a mathematical solution of the problem. We can use Eq. 3-4 to find both Bob's position and Suzy's position at time t_1. Using $(s_B)_i = s_{B0}$, $(s_S)_i = s_{S0}$, and $t_i = t_0 = 0$ gives:

$$s_{B1} = s_{B0} + \bar{v}_B t_1$$
$$s_{S1} = s_{S0} + \bar{v}_S t_1$$

where t_1 is the time when they meet. (Note that the symbol $\bar{v}_B$ represents Bob's average velocity, whereas v_{B0} and v_{B1} are his velocities at the specific instants of time t_0 and t_1.) The condition that they meet is given by

$$s_{B1} = s_{S1},$$

so we can equate the two right-hand sides of the previous equations to get

$$s_{B0} + \bar{v}_B t_1 = s_{S0} + \bar{v}_S t_1.$$

Solving this for t_1 gives

$$t_1 = \frac{s_{S0} - s_{B0}}{\bar{v}_B - \bar{v}_S} = \frac{400 \text{ miles} - 0 \text{ miles}}{60 \text{ mph} - (-40) \text{ mph}} = 4.0 \text{ hours}.$$

Finally, inserting this time back into the equation for s_{B1} gives

$$s_{B1} = 0 \text{ miles} + (60 \frac{\text{miles}}{\text{hour}}) \times (4.0 \text{ hours}) = 240 \text{ miles}.$$

While this gives us a number, it is not yet the answer to the question. The phrase "240 miles" taken alone does not say anything meaningful. Because this is the value of Bob's position, and Bob was driving north, the complete answer to the question is, "They meet 240 miles north of Los Angeles."

Before stopping, you should check whether or not this answer seems reasonable. We certainly expected an answer between 0 miles and 400 miles. We also know that Bob is driving faster than Suzy, so we expect that their meeting point will be *more* than halfway from Los Angeles to San Francisco. This assessment tells us that 240 miles seems to be a reasonable answer.   •

It is instructive to look at this example from a graphical perspective. Figure 3-8 shows position-versus-time graphs for Bob and Suzy. Notice the negative slope for Suzy's graph, indicating her negative velocity. The point of interest is the intersection of the two lines—that is where Bob and Suzy have the same position at the same time. The equations for s_{B1} and s_{S1} are the equations for the two lines. Therefore, our method of solution—equating these two equations—is really just solving the mathematical problem of finding the intersection of two lines. Seeing the situation graphically can often help you decide how to proceed with a solution.

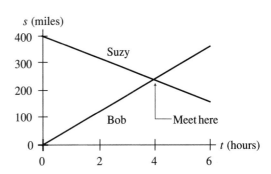

FIGURE 3-8 Position-versus-time graphs for Bob and Suzy.

3.5 Instantaneous Velocity

Not many objects in the universe move at constant velocity—at least not for very long. We need to extend our analysis if we want to describe "real" objects as they move. Consider, for example, what happens when you drive a car. You start from rest, accelerate, perhaps drive at steady speed for awhile (constant v), decelerate, and ultimately stop. This is clearly a more complex problem than we considered in the last section. If we were to graph your car's position at various times, the graph would not be a straight line. So it appears that we need to understand position-versus-time graphs that are curved.

[**Photo suggestion: A rocket being launched.**]

We have previously defined the average velocity $\bar{v} = \Delta s/\Delta t$ as the average during a certain finite time interval Δt. This definition is valid regardless of whether or not the object is moving with a constant velocity. Figure 3-9 shows a position-versus-time graph of an object that is speeding up. Its distance from the origin is increasing more and more rapidly as time goes by. We can still

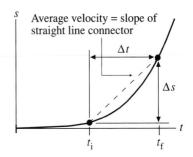

FIGURE 3-9 Position-versus-time graph for an object that is speeding up. The average velocity $\bar{v}$ between the two times t_i and t_f is the slope of the dashed line.

determine $\bar{v}$ between any two times t_i and t_f by selecting those two points on the graph, drawing the straight-line connection between them, measuring Δs and Δt, and finally using these to compute $\bar{v}$. Graphically, $\bar{v}$ is simply the slope of the straight-line connection between the two points. But, because the velocity is changing, the average velocity is not an accurate description of the motion.

There are, in fact, an infinite number of position-versus-time graphs that pass through the two points (t_i, s_i) and (t_f, s_f), and all of them have exactly the same average velocity. The difficulty is that average velocity, by its very definition, refers to a finite interval of time over which the averaging is done, rather than to a single point in time. What we need is a way to determine the velocity of the object at a single *instant* of time.

Suppose that you and Frank and Karen all leave at precisely 8:00 A.M. in your cars. At 9:00 A.M. you have each traveled exactly 60 miles. All that it is possible to discern from this information is that all three of you had the same average velocity $\bar{v} = 60$ miles/hour. Suppose, however, that Frank started out going faster than 60 miles/hour but later slowed down; Karen got a slow start, but later sped up; while you drove at a steady but boring speed the entire hour. What would each of you see if you read your car's speedometer at 8:10? You would see 60 mph, but Frank would see a speed greater than 60 mph, while Karen would see a speed less than 60 mph. Later, at 8:50, Frank's speedometer reads less than 60 mph, Karen's reads more than 60 mph, while yours has not changed. The speedometer reading tells you your velocity *at* an *instant* of time, rather than averaged over the entire hour.

These words should sound familiar—this is much like the discussion in Section 3.3 about how long an instant lasts and about how s is the position *at* an *instant* of time—the instantaneous position. By analogy, we can define the *instantaneous velocity* to be the velocity of the object *at* a single *instant* of time t. When we talk about the instantaneous velocity of an object, we are referring to what the object's speedometer would read, if it had one, at just that instant of time. The length of time an object spends at this particular velocity is zero seconds.

Such a definition, though, raises some troubling issues. Just what does it mean to have a velocity "at an instant?" Suppose a police officer pulls you over and says, "I just clocked you going 80 miles per hour." You might respond, "But that's impossible. I've only been driving for 20 minutes, so I can't possibly have gone 80 miles." Unfortunately for you, the police officer was a physics major. He replies, "I mean that at the instant I measured your velocity, you were moving at a rate such that you *would* cover a distance of 80 miles *if* you were to continue at that velocity without change for 1 hour. That will be a $200 fine."

Here, again, is the idea of velocity as a rate—in this case, the rate at which an object changes its position. Rates tell us how quickly or how slowly things change, and that idea is conveyed by the word "per." A velocity of 80 miles per hour at some particular instant in time means that the rate at which your car's position is changing—at that exact instant—is such that the car would travel a distance of 80 miles in 1 hour *if* it continued at that rate without change. Whether or not it actually does travel at that velocity for another hour, or even for another millisecond, is not relevant.

A position-versus-time graph can help make this clear. Figure 3-9 showed the graph of an object that is accelerating. To find its instantaneous velocity at a particular instant of time t_1, we need to consider how the object would move if the velocity it has at t_1 gets locked in and no longer changes—that is, if it remains *constant* for $t > t_1$. Figure 3-10 shows both the actual motion of the object and the motion that would occur *if* the object were to continue with the velocity at had at time t_1. Notice that the straight-line segment is *tangent* to the curve at t_1. The slope of this straight line is the "as if" velocity we are seeking. In other words:

> The velocity v at a specific instant of time t—which we call the **instantaneous velocity**—is the slope of the straight line that is tangent to the position-versus-time curve at t.

This slope is the reading that a speedometer would have at that instant of time.

It is all well and good to give definitions, but how do we actually find the slope of the tangent line? We could do it graphically, as in Fig. 3-10, but that is not terribly practical. Instead, we would like a way to *compute* the velocity of a moving object from its position function $s(t)$; that is, a way to compute the slope of tangent lines.

To see the situation clearly, let's assume that we have used a movie camera to film an accelerating object. Figure 3-11a shows a motion diagram, made using a normal 30 frames-per-second camera, of the accelerating object. The length of the third velocity vector is the *average* velocity during the time interval Δt it took the object to move the distance Δx from x_2 to x_3. We would like to know the instantaneous velocity v at the specific position X right before x_3. Because the object is accelerating, its velocity at this specific point is going to be a little bit larger than the average velocity between x_2 and x_3. How can we find it?

Suppose that we also have a high-speed camera that takes 300 frames per second, and that we use this camera to film just the segment of motion between x_2 and x_3. This "magnified" motion diagram is shown in Fig. 3-11b. The position X where we wish to find v falls between x_{2c} and x_{2d}. While the velocity changed a lot between x_0 and x_4, causing the length of the arrows in Fig. 3-11a to change greatly, the velocity change between x_2 and x_3 is much less. Consequently, the

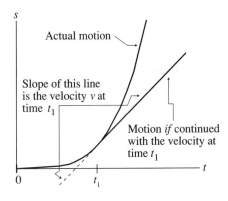

FIGURE 3-10 The straight-line tangent to the curve at t_1 shows the motion that would occur if the velocity did not change after t_1.

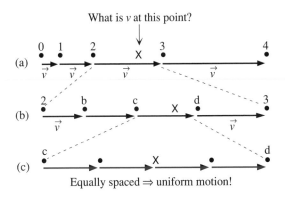

FIGURE 3-11 a) Motion diagram with a 30-frames-per-second camera. b) With a 300-frames-per-second camera. c) At 3000 frames per second.

change in the length of the velocity arrows in Fig. 3-11b is much less. The velocity arrow's length is again the *average* velocity between x_{2c} and x_{2d}. This average is a much closer approximation to the instantaneous velocity because the length of the velocity arrows on either side are nearly the same. Still, though, there is evidence of a changing velocity, so the instantaneous velocity we seek is not the same as the average velocity.

Finally we get out our really high-speed camera that takes 3000 frames per second and film just the interval between x_{2c} and x_{2d}. The result is shown in Fig. 3-11c. Now, at this level of magnification, we no longer see evidence of acceleration—each velocity vector appears to be the same length. On this time scale, the motion near the point in which we are interested appears uniform! But that is exactly what we need. *If* the motion were to continue at a constant velocity after X, then the velocity the object would have is the velocity of Fig. 3-11c. Therefore, the instantaneous velocity v at a particular position X is equivalent to the average velocity $\bar{v}$ when averaged over a *sufficiently small interval* that the average does not change from one frame to the next.

By magnifying the motion diagram, we are using smaller and smaller time intervals Δt. Our procedure, stated in words, says that the instantaneous velocity $v(t)$ at time t can be found as the average velocity $\Delta s/\Delta t$ if averaged over a sufficiently small interval Δt centered on t. Mathematically, we can define the instantaneous velocity $v(t)$ in terms of a limit:

$$v(t) \equiv \lim_{\Delta t \to 0} \frac{\Delta s}{\Delta t} = \frac{ds}{dt} \quad \text{(instantaneous velocity)}. \qquad (3\text{-}5)$$

This is exactly what we did in Fig. 3-11—no more and no less. There we took smaller and smaller intervals Δt (i.e., $\Delta t \to 0$) until the velocity vector $\vec{v} = \Delta \vec{r}/\Delta t$ reached a constant value and was no longer changing. That constant value is what we mean by the *limit*. It is very important to keep in mind that the displacement Δs is also shrinking to zero (that is, $\Delta s \to 0$), even though this is not stated explicitly in Eq. 3-5. What we find is that the *ratio* of Δs to Δt can be a constant, nonzero value while Δs and Δt are each, individually, shrinking to zero.

A Little Calculus

We have reached the point beyond which Galileo could not proceed because he lacked the mathematical tools. Further progress had to await Newton, who invented a new branch of mathematics called *calculus*. (Calculus was invented simultaneously in Germany by Leibniz. Historians of science still debate whether one influenced the other or if the discoveries were entirely independent!) Calculus is designed to deal with instantaneous quantities—in other words, it provides us with the tools for evaluating limits such as that in Eq. 3-5. The notation ds/dt is called "the derivative of s with respect to t," and Eq. 3-5 defines its meaning as the limiting value of a ratio.

EXAMPLE 3-4   What is $v(t)$, the velocity as a function of time, for a particle whose position is given by the function $s(t) = 2t^2$?

SOLUTION To solve this problem we need to "take the derivative" of $s(t)$, which requires evaluating Eq. 3-5. During the time interval Δt, the particle moves from position $s(t)$ to the new position $s(t + \Delta t)$. Its displacement is thus

$$\Delta s = s(t + \Delta t) - s(t)$$
$$= 2(t + \Delta t)^2 - 2t^2$$
$$= 2(t^2 + 2t\Delta t + (\Delta t)^2) - 2t^2$$
$$= 4t\Delta t + 2(\Delta t)^2 .$$

The average velocity during interval Δt is then given by

$$\frac{\Delta s}{\Delta t} = \frac{4t\Delta t + 2(\Delta t)^2}{\Delta t} = 4t + 2\Delta t$$

We can finish by taking the limit $\Delta t \to 0$ to find

$$v(t) = \frac{ds}{dt} = \lim_{\Delta t \to 0} (4t + 2\Delta t) = 4t$$

This answer provides a specific function $v(t) = 4t$ for calculating the velocity at any instant of time. At $t = 3$ s, for example, the particle is located at position $s = 18$ m and it has an instantaneous velocity, at just that instant, of $v = 12$ m/s.   •

Let's look at Example 3-4 graphically. Figure 3-12a shows the position-versus-time graph. Figure 3-12b then shows the velocity-versus-time graph, using the velocity function we just calculated. We see that it is a straight line, as expected for $v(t) = 4t$. It is critically important to understand the relationship between these two graphs. The *value* of the velocity graph at any instant of time, which we can read directly off the vertical axis, is the *slope* of the position graph at that same time. This is illustrated at times $t = 1$ s and $t = 3$ s.

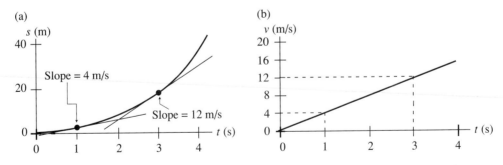

FIGURE 3-12 a) Position-versus-time and b) velocity-versus-time graphs for Example 3-4. Note that the *slope* of the position graph at time t is equal to the *value* of the velocity graph at time t.

Although our example showed how the limits can be evaluated to find to a derivative, the procedure is clearly rather tedious. It would hinder us significantly if we had to do this for every new situation. Fortunately, there are only a few basic derivatives we will need in this text, and we will describe them in the following paragraphs. Learn these, and you do not have to go all the way back each time to the definition in terms of limits.

The only functions we will use in Part I and Part II of this book are polynomials. Consider a generic polynomial function

$$u(t) = ct^n$$

where c and n are constants. It is proven in calculus that the derivative of this function is

$$\frac{du}{dt} = nct^{n-1}. \tag{3-6}$$

Example 3-4 evaluated the derivative of the function $s(t) = 2t^2$. This is a polynomial function with $c = 2$ and $n = 2$. By using Eq. 3-6 you could quickly find the derivative to be

$$\frac{ds}{dt} = 2 \cdot 2t^{2-1} = 4t.$$

Similarly, the derivative of the function $x(t) = 3/t^2 = 3t^{-2}$ is

$$\frac{dx}{dt} = (-2) \cdot 3t^{-2-1} = -6t^{-3} = -\frac{6}{t^3}.$$

The derivative of a function such as $u(t)$, you should notice, is some other *function* of t. It is, in fact, just the function that tells you the *slope* of the $u(t)$ graph at t.

A function with a constant, unchanging value is $u(t) = c = $ constant. This is a special-case polynomial with the exponent $n = 0$. We see from Eq. 3-6 that the derivative is

$$\frac{du}{dt} = 0, \text{ if } u(t) = c = \text{constant}.$$

This makes sense. The graph of the function $u(t) = c$ is simply a horizontal line at height c. The slope of a horizontal line—which is what the derivative du/dt measures—is zero!

The only other information we need about derivatives for now is how to evaluate the derivative of the sum of two or more functions. Let $u(t)$ and $w(t)$ be two separate functions. You will prove in calculus that

$$\frac{d}{dt}(u + w) = \frac{du}{dt} + \frac{dw}{dt}. \tag{3-7}$$

That is, the derivative of a sum is the sum of the derivatives.

EXAMPLE 3-5   A particle moves in one dimension according to the function $s(t) = -t^3 + 3t$. a) What is the particle's position and velocity at $t = 2$ s? b) Draw graphs of $s(t)$ and $v(t)$ during the interval $-3 \text{ s} \leq t \leq 3$ s. c) Draw a motion diagram illustrating this motion.

SOLUTION a) You can compute the position at $t = 2$ s directly from the function $s(t)$. This gives $s(t = 2 \text{ s}) = -8 + 6 = -2$ m. The velocity is given by the derivative $v = ds/dt$. Here the function for s is the sum of two polynomials. Because the derivative of a sum is the sum of the derivatives:

$$v(t) = \frac{ds}{dt} = \frac{d}{dt}(-t^3 + 3t) = \frac{d}{dt}(-t^3) + \frac{d}{dt}(3t).$$

The first derivative is of a polynomial with $c = -1$ and $n = 3$, while the second is of a polynomial with $c = 3$ and $n = 1$. Referring to Eq. 3-6, we can evaluate these to find

$$v(t) = -3t^2 + 3.$$

Evaluating at $t = 2$ s gives $v(t = 2$ s$) = -9$ m/s. The negative sign indicates that the particle, at this instant of time, is moving to the *left* with a *speed* of 9 m/s.

b) Figure 3-13a shows the position-versus-time graph and Fig. 3-13b is the corresponding velocity-versus-time graph. These were created by computing, and then graphing, the values of the $s(t)$ and $v(t)$ functions at several points between -3 s and 3 s. In Fig. 3-13a, the slope of the position-versus-time is -9 m/s at $t = 2$ s. This becomes the *value* that is graphed as the velocity in Fig. 3-13b. Similar measurements are shown at $t = -1$ s.

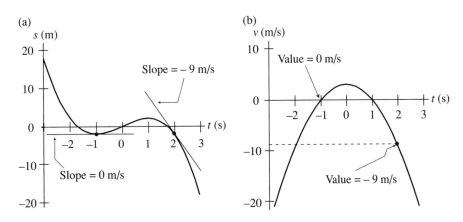

FIGURE 3-13 a) Position-versus-time graph, and b) velocity-versus-time graph for Example 3-5. The *value* of the velocity graph is the *slope* of the position graph at the same instant of time.

c) Finally, we can interpret the graphs of part b) to create a motion diagram. Initially, when $t = -3$ s, the particle is to the right of the origin ($s > 0$) but moving to the left ($v < 0$). Its *speed* is slowing ($|v|$ is decreasing), so the velocity vector arrows will be getting shorter. At $t \approx -1.5$ s, the particle passes the origin but is still moving left. Then, at $t = -1$ s, s reaches a minimum. In other words, the particle is as far left as it is going, so *instantaneously* $v = 0$ and the particle is at rest. This is a point at which the particle is turning around. Between $t = -1$ s and $t = +1$ s, the particle is moving back to the right, with $v > 0$. The particle speeds up until v reaches a maximum at $t = 0$, then it slows back down. The particle turns around again at $t = 1$ s and begins moving left again as the velocity becomes negative. It passes the origin again at $t \approx 1.5$ s, continues to speed up, and disappears off to the left of the origin. This behavior is shown in the motion diagram of Fig. 3-14. The different segments of the motion have been displaced for clarity, but keep in mind that the motion is actually along a single straight line.

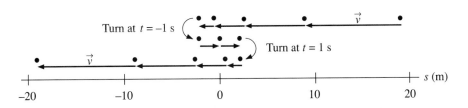

FIGURE 3-14 The motion diagram for Example 3-5. The different segments of motion have been displaced for clarity, but the motion is actually along a single straight line.

EXAMPLE 3-6 Figure 3-15 shows the position-versus-time graph of a moving object, with three points of the motion labeled. a) At which labeled point or points is the object's motion the slowest? b) At which point or points is the object's motion the fastest? c) Sketch an approximate velocity-versus-time graph for the object.

SOLUTION a) The terms "fastest" and "slowest" refer to the object's *speed* |v| rather than its velocity *v*. The *sign* of the velocity is not relevant when we talk about fast or slow. Because the velocity at each point in time is the *slope* of the position-versus-time graph, and because the speed is the absolute value of the velocity, we can look for the points where the magnitude of the slope is a minimum. The slope at both points A and C

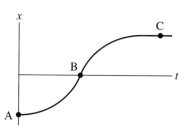

FIGURE 3-15 Position-versus-time graph for Example 3-6.

is zero; hence, the speed is zero. While velocity can be negative, speed cannot. A speed of zero—at rest!—is the slowest the motion can possibly get. Notice, though, that at point A the speed is only *instantaneously* zero, whereas at point C the object seems to have actually stopped, at position x_C, with a speed that *stays* at zero.

b) The fastest point in the motion will occur where the magnitude of the slope is a maximum. The curve reaches its steepest point, with maximum slope, at point B, so this will be the fastest point in the motion. This is the point at which the position *x* is changing most rapidly.

c) We cannot find an exact velocity-versus-time graph without having a function whose derivative we can take. Nonetheless, we can deduce that the slope, and hence *v*, starts at zero, rises to a maximum value at point B, decreases back to zero a little before point C, then remains at zero slope thereafter. Thus Fig. 3-16 represents, at least approximately, the velocity-versus-time graph for this object. Notice that the shape of the velocity graph bears no resemblance to the *shape* of the position graph. You must transfer *slope* information from one graph (position) to *value* information on the other (velocity).

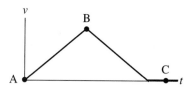

FIGURE 3-16 Velocity-versus-time graph for Example 3-6.

3.6 Are We There Yet? Relating Velocity to Position

Equation 3-5 for $v(t)$ provides a means of finding the velocity *v* at any time *t* if we know the position $s(t)$ at all times. But we frequently need to solve the reverse problem—that is, we want to predict the position *s* at some future time *t* if we know the object's velocity $v(t)$ at all times. Equation 3-4 does exactly this for the case of uniform motion with a constant velocity $\bar{v}$. We need to find a more general expression for the case when *v* is not constant.

Figure 3-17a shows a velocity-versus-time graph for an object that does not have constant velocity. Suppose that we know the position to be s_i at some initial time t_i. How can we find the position s_f at some final time t_f? Because we know how to handle *constant* velocities using Eq. 3-4, let's *approximate* the velocity function of Fig. 3-17a

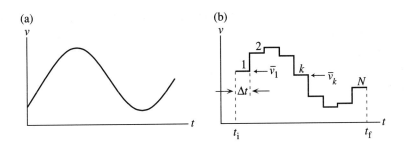

FIGURE 3-17 a) An object with a non-constant velocity. b) Approximating the velocity curve with a series of constant-velocity steps.

as a series of constant-velocity steps, as illustrated in Fig. 3-17b. The first step shows that from time t_i to time $t_i + \Delta t$ the velocity can be approximated with a constant $\bar{v}_1 = v(t_i)$. The second step from time $t_i + \Delta t$ to $t_i + 2\Delta t$ has constant $\bar{v}_2 = v(t_i + \Delta t)$, and so on. Altogether, the $v(t)$ curve has been divided into N steps of equal width Δt, and during each step the velocity is constant and the motion is uniform. Although the approximation shown, with only nine steps, is rather rough, it is easy to imagine that this approximation could be made as precise as desired by having more and more steps of smaller and smaller width: $\Delta t \to 0$.

Now we can use Eq. 3-4 during each step, because the velocity during that step is constant. During the first step the object undergoes a displacement $\Delta s_1 = \bar{v}_1 \Delta t$. The second step has a displacement of $\Delta s_2 = \bar{v}_2 \Delta t$, and so on. Thus during step k, the displacement is $\Delta s_k = \bar{v}_k \Delta t$. The total displacement of the object between t_i and t_f, by definition, is $\Delta s_{\text{total}} = s_f - s_i$. But this total displacement is just the sum of the all the individual displacements Δs_k during each of the N steps. Thus

$$s_f - s_i = \Delta s_{\text{total}} = \sum_{k=1}^{N} \Delta s_k = \sum_{k=1}^{N} \bar{v}_k \Delta t = \sum_{k=1}^{N} v(t_k) \Delta t$$

(3-8)

$$\Rightarrow s_f = s_i + \sum_{k=1}^{N} v(t_k) \Delta t.$$

We have written $\bar{v}_k$, the average velocity during step k, as $v(t_k)$, the function $v(t)$ evaluated at the specific time $t = t_k$.

We can give Eq. 3-8 an important geometric interpretation, as shown in Fig. 3-18. Consider each "step" in the approximation to be a long, thin rectangle of height $\bar{v}_k$ and width Δt. For step k, the product $\bar{v}_k \Delta t$ is the *area* (base $\times$ height) of this small rectangle. The sum in Eq. 3-6 adds up all of these rectangular areas, so Δs_{total} is equal to the total *area* enclosed between the t-axis and the tops of the steps.

If we now let $\Delta t \to 0$, each step's width approaches zero but the total number of steps N

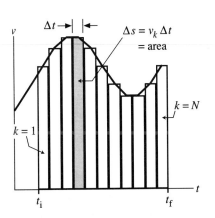

FIGURE 3-18 The displacement Δs is the "area under the curve" of step Δt.

approaches infinity. In this limit, the series of steps becomes a perfect replica of the $v(t)$ curve. Thus

$$s_f = s_i + \lim_{\Delta t \to 0} \sum_{k=1}^{N} v(t_k)\Delta t = s_i + \int_{t_i}^{t_f} v(t)dt. \tag{3-9}$$

The curly-cue symbol is called an *integral*. In this case, the symbol is read as "the integral of $v(t)$ from t_i to t_f." You will see shortly how to find an exact evaluation, but for now it much more useful to have a graphical interpretation of Eq. 3-9:

$$s_f = s_i + \text{ area under the } v(t) \text{ curve between } t_i \text{ and } t_f.$$

Equation 3-9 is the generalization of Eq. 3-4 that we were seeking. Notice that you need not only the $v(t)$ curve but also the initial position s_i in order to find a final position s_f. This differs from the case of finding v from s, where the curve alone was sufficient.

Wait a minute—the displacement Δs is a length. How can a length equal an area? Good question. Recall earlier, when discussing the relation between velocity and the slope of a graph, we made a distinction between the *actual* slope and the *physically-meaningful* slope? The same discussion applies here. The curve described by $v(t)$ does indeed bound a certain area on the page. That is the *actual* area, but it is *not* the area to which we are referring. Here again, we need to measure the quantities we are using—Δt and v—by referring to the scales on the axes. Thus Δt, in this example, is some number of seconds while v is some number of meters per second. When these are multiplied together, the *physically-meaningful* area ends up with units of meters—i.e., a displacement. Note that the units of v *must* be in meters per second if you want s in meters. On the other hand, *if* v is in meters per second, you can be certain that your answer for positions or displacements will be in meters. The examples below should help make all of this clear.

EXAMPLE 3-7   Figure 3-19 shows the velocity-versus-time graph of a moving object. How far does the object move during the first 3 seconds?

SOLUTION The question "how far" indicates that you should look for a displacement Δs rather than a position s. According to our interpretation of Eq. 3-9, the displacement between $t = 0$ s and $t = 3$ s is the area under the curve from $t = 0$ s to $t = 3$ s. The curve in this case is a straight, angled line, so the area is that of a triangle:

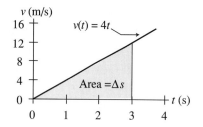

FIGURE 3-19 Velocity-versus-time graph for Example 3-7.

$$\Delta s = \text{ area of triangle}$$
$$= \tfrac{1}{2} \text{ base} \times \text{height}$$
$$= \tfrac{1}{2} \times 3 \text{ s} \times 12 \text{ m}/\text{s}$$
$$= 18 \text{ m}.$$

So this object undergoes a displacement of 18 m during the first 3 seconds.

EXAMPLE 3-8   Find an expression $s(t)$ for the position as a function of time t for the object whose velocity-versus-time graph was shown in Fig. 3-19 and whose initial position is $s(t = 0) = 0$.

SOLUTION In Eq. 3-9, let $s_i = 0$ and $s_f = s(t)$. Also let $t_i = 0$ and $t_f = t$. From Fig. 3-19 we can see that v is a straight-line graph given by the linear function $v(t) = 4t$. Then

$$s(t) = 0 + \int_0^t v(t)dt = \text{ area under triangle between 0 and } t$$

$$= \tfrac{1}{2}(t - 0)(4t - 0)$$

$$= 2t^2.$$

But this is exactly Example 3-4 in reverse! There we found, by taking the derivative, that a particle whose position is described by $s(t) = 2t^2$ has a velocity described by $v(t) = 4t$. Here we have found, by integration, that a particle whose velocity is described by $v(t) = 4t$ has a position described by $s(t) = 2t^2$.

EXAMPLE 3-9   A particle starts at $s_0 = 30$ m and has the velocity shown in Fig. 3-20. a) Draw a motion diagram for the particle. b) Where is the particle instantaneously at rest? c) At what time does the particle reach the origin?

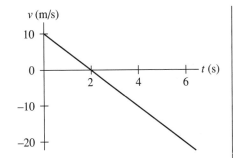

FIGURE 3-20 Velocity-versus-time graph for Example 3-9.

SOLUTION a) Before getting into the mathematics, it is crucial to have a good mental image of the motion. The particle is initially 30 m to the right of the origin and moving *to the right* (because $v > 0$) with a velocity of 10 m/s. So the value of s will be increasing, at least initially, and the particle will get further from the origin. However, v is decreasing so the particle is slowing down. At $t = 2$ s, the velocity—just for an instant—is zero. At that time, the particle is instantaneously at rest. For $t > 2$ s, the velocity is negative—so the particle has reversed direction and is moving back toward the origin, speeding up ($|v|$ increasing) as it goes. At some time, which we want to find, it will pass $s = 0$ and then head into negative positions. Figure 3-21 illustrates this as a motion diagram. The distance scale will be established in parts b) and c) but is shown here for convenience.

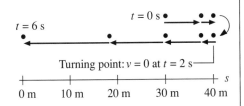

FIGURE 3-21 Motion diagram for the particle whose velocity graph was shown in Fig. 3-20.

b) The particle is instantaneously at rest at time $t = 2$ s, when $v = 0$. To learn *where* it is at that time, we need first to note that $s_i = s_0 = 30$ m and then to find the displacement during the first two seconds. We can do this by finding the area under the curve between $t = 0$ s and $t = 2$ s:

$$s(t = 2 \text{ s}) = s_0 + \int_0^2 v(t)\,dt$$

$$= s_0 + \text{area under the curve between 0 s and 2 s}$$

$$= 30 \text{ m} + \tfrac{1}{2}(2 \text{ s} - 0 \text{ s}) \times (10 \text{ m/s} - 0 \text{ m/s})$$

$$= 40 \text{ m}.$$

The particle instantaneously stops at $s = 40$ m at the time $t = 2$ s, then turns and moves back to the left.

c) The particle needs a displacement $\Delta s = -40$ m to get to the origin from its position $s = 40$ m at $t = 2$ s. Again, this is the area under the curve *from* $t = 2$ s to the desired t. Because the curve is below the axis, with negative values of v, the area to the right of $t = 2$ s is a *negative* area. With a bit of geometry (the details are left as an exercise), you will find that a triangle with a base of 4 s has an "area" of 40 m. Thus the particle will reach the origin at $t = 2$ s $+ 4$ s $= 6$ s. This information is shown on the motion diagram of Fig. 3-21.
•

The particle in Example 3-9a moved out to a maximum distance, then returned. The point in its motion where it reversed direction is called a **turning point**. Because the velocity was positive just before reaching the turning point and negative just after, it had to pass through $v = 0$. Thus a turning point is a point where the instantaneous velocity is zero as the particle reverses direction. We will see many future examples of this.

A Little More Calculus

Evaluating an integral graphically by finding the area is analogous to evaluating a derivative graphically by finding the slope. It is very important method for building intuition about motion, but limited in its practical application. We have found (or rather Newton found!) that derivatives of standards functions can be evaluated and tabulated. The same is true of integrals. The integral seen in Eq. 3-9 is called a *definite integral* because there are two definite boundaries to the area we want to find. These boundaries are called the lower (t_i) and upper (t_f) *limits of integration*.

For a polynomial function $u(t) = ct^n$, the essential result from calculus is that

$$\int_{t_i}^{t_f} u\,dt = \int_{t_i}^{t_f} ct^n\,dt = \left.\frac{ct^{n+1}}{n+1}\right|_{t_i}^{t_f} = \frac{ct_f^{n+1}}{n+1} - \frac{ct_i^{n+1}}{n+1} \qquad (n \neq -1). \qquad (3\text{-}10)$$

In the third step, the vertical bar with subscript t_i and superscript t_f is a shorthand notation from calculus that *means*—as seen in the last step—the integral evaluated at the upper limit t_f *minus* the integral evaluated at the lower limit t_i. We also need to know that for two functions $u(t)$ and $w(t)$:

$$\int_{t_i}^{t_f} (u + w)\,dt = \int_{t_i}^{t_f} u\,dt + \int_{t_i}^{t_f} w\,dt. \qquad (3\text{-}11)$$

That is, the integral of a sum is equal to the sum of the integrals.

EXAMPLE 3-10   Use calculus to solve Example 3-9.

SOLUTION Figure 3-20 is a straight-line graph. We can find its y-intercept as 10 m/s and its slope as –5 m/s. The velocity function is then $v(t) = (10 - 5t)$ m/s. We can find the position function by using Eq. 3-9:

$$s(t) = s_0 + \int_0^t v(t)dt$$

$$= 30 \text{ m} + \int_0^t (10 - 5t)dt$$

$$= 30 \text{ m} + \int_0^t 10 dt - \int_0^t 5t dt.$$

We used Eq. 3-11 for the integral of a sum to get the last line. The first integral is a function of the form $u(t) = ct^n$ with $c = 10$ and $n = 0$, while the second is of the form $u(t) = ct^n$ with $c = 5$ and $n = 1$. Referring to Eq. 3-10:

$$\int_0^t 10 dt = 10t \big|_0^t = 10 \cdot t - 10 \cdot 0 = 10t \text{ m}$$

and

$$\int_0^t 5t dt = \tfrac{5}{2}t^2 \Big|_0^t = \tfrac{5}{2} \cdot t^2 - \tfrac{5}{2} \cdot 0^2 = \tfrac{5}{2}t^2 \text{ m}.$$

Combining the pieces gives:

$$s(t) = (30 + 10t - \tfrac{5}{2}t^2) \text{ m}$$

The particle is instantaneously at rest at $t = 2$ s, and its position at that time is $s(t = 2 \text{ s}) = 40$ m.

For part c) we must set $s = 0$, giving us a quadratic equation to solve for t:

$$30 + 10t - \tfrac{5}{2}t^2 = 0$$

This equation has two solutions: $t = -2$ s or $t = 6$ s.

When we solve a quadratic equation, we cannot just arbitrarily select which root we want. Instead, we must decide on physical grounds which is the *meaningful* root. Here the negative root is a time before the problem began. Because the particle starts at $s = 30$ m and is moving to the right, there very likely was an earlier time when the particle passed the origin $s = 0$. If the velocity function was the same for $t < 0$ as it is for $t > 0$, this must have occurred at $t = -2$ s. But we have no knowledge of what the velocity was doing before $t = 0$ s, so this root is not meaningful. The meaningful one is the positive root, $t = 6$ s, which agrees with the answer we found previously from a graphical solution.

The point to be made with these examples is that there are often many ways to solve a problem. The graphical procedures for finding derivatives and integrals are very simple, but they only work for a limited range of problems where the geometry is simple. Using the techniques of calculus is more demanding, but these techniques allow us to deal with functions whose graphs are quite complex.

As you work on building intuition about motion, you need to be able to move back and forth between four different descriptions of the motion: the motion diagram, the position-versus-time graph, the velocity-versus-time graph, and the description *in words*. Given a description of a certain motion, you should be able to sketch the motion diagram and the position and velocity graphs. Given one graph, you should be able to generate the other. And given position and velocity graphs, you should be able to "interpret" them by describing the motion in words or in a motion diagram.

3.7 Constant Acceleration

This probably seems like enough material already for one chapter, but we still have one more major concept to go—acceleration. As we noted in Chapter 1, acceleration is a very abstract concept. You cannot "see" the value of acceleration, like you can that of position, nor can you even judge it by looking to find if an object is moving fast or slow, like you can do with velocity. Instead, acceleration must be *inferred* from other observations. Nonetheless, acceleration is the linchpin of mechanics because, as you will see very shortly, Newton's laws relate the acceleration of an object to the forces that are exerted on it.

[**Photo suggestion: Two drag race cars at the beginning of a race.**]

To help you visualize acceleration, let's conduct a race between a Volkswagen and a Porsche to see which can achieve a velocity of 60 mph the quickest. Both cars are equipped with computers that will record the speedometer reading ten times each second. This gives a nearly continuous record of the *instantaneous* velocity of each car. Table 3-5 shows some of the data. The velocity-versus-time graph, based on this data, is shown in Fig. 3-22.

TABLE 3-5 Acceleration of a Porsche and a Volkswagen

t (s)	v_{Porsche} (mph)	v_{VW} (mph)
0.0	0.0	0.0
0.1	1.0	0.4
0.2	2.0	0.8
0.3	3.0	1.2
0.4	4.0	1.6
$\vdots$	$\vdots$	$\vdots$

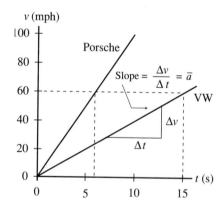

FIGURE 3-22 Velocity-versus-time graphs for the Porsche and the Volkswagen.

How can we describe the difference in performance of the two cars? It is not that one has a different velocity than the other—both achieve every velocity between 0 and 60 miles/hour. The distinction is *how long* it took each to *change* its velocity from 0 to 60 mph. The Porsche changed velocities quickly—in 6 s, while the VW changed slowly—in 15 s. This suggests that the distinction is—once again—a *rate*. In this case, as we compare the two cars, we are looking at the rate at which each object's velocity changes. Because the Porsche changed its velocity $\Delta v = 60$ miles/hour during a time interval $\Delta t = 6$ s, the *rate* at which its velocity changed was

$$\text{rate of velocity change} = \frac{\Delta v}{\Delta t} = \frac{60 \text{ miles / hour}}{6 \text{ s}} = 10 \text{ (miles/hour)/s}.$$

The VW changed its velocity at the slower rate of only 4 (miles/hour)/s.

Notice the units. They are units of "velocity per second." A rate of velocity change of 4 "miles per hour per second" means that the velocity increases by 4 miles/hour during the first second, by another 4 miles/hour during the next second, and so on. In fact, the velocity will increase by 4 miles/hour during any second in which it is changing at the rate of 4 (miles/hour)/s.

As you saw in Chapter 1, *acceleration* is the term we use to describe the rate of velocity change. Acceleration measures how quickly the velocity of an object changes. The Porsche's velocity changed quickly, so it has a large acceleration. The VW's velocity changed more slowly, so its acceleration is much less. In parallel with our treatment of velocity, we need to be precise and call this the **average acceleration** $\overline{a}$ during the time interval Δt:

$$\overline{a} \equiv \frac{\Delta v}{\Delta t} \text{ (average acceleration)}.$$

Since Δv and Δt are the "rise" and the "run" of a velocity-versus-time graph, we see that $\overline{a}$ can be interpreted graphically as the *slope* of a straight-line velocity-versus-time graph. Figure 3-22 shows this for the graph of the Volkswagen's velocity. The VW's average acceleration was

$$\overline{a}_{\text{vw}} = \frac{\Delta v}{\Delta t} = \frac{60 \text{ miles / hour}}{15 \text{ s}} = 4 \text{ (miles/hour)/s}.$$

An object whose velocity-versus-time graph is a straight-line graph has a steady and unchanging acceleration. Such a graph represents motion at *constant acceleration*. Sometimes, again in parallel with velocity, this is called **uniformly accelerated motion**. Galileo described uniformly accelerated motion as:

> We shall call that motion equably or uniformly accelerated which, abandoning rest, adds on to itself equal momenta of swiftness in equal times.

The terminology is a bit antiquated, but his meaning is clear: During equal time intervals Δt, the velocity increases by equal amounts Δv. But this is exactly the condition for the ratio $\Delta v/\Delta t$, and, hence, the slope of the velocity graph, to be a constant!

An important aspect of acceleration is its *sign*—a source of many errors! Like the position s and the velocity v, the acceleration a is the *component* of a vector along the axis of one-dimensional motion. In this case, it is the component of the acceleration vector $\vec{a}$. You should review Section 1.6 carefully to make sure you understand how the direction of $\vec{a}$ is determined.

EXAMPLE 3-11   The velocity graphs for two objects are shown Fig. 3-23. For each object: a) determine its acceleration, b) describe its motion, and c) draw its motion diagram.

SOLUTION a) The average acceleration is the slope of a velocity-versus-time graph. In Fig. 3-23a, the slope of the graph is an unchanging $(10 \text{ m/s})/(2 \text{ s}) = 5$ (m/s)/s. Therefore, this object is undergoing uniformly accelerated motion with the constant acceleration $\bar{a} = +5$ (m/s)/s. That is, the velocity is *changing* at the rate of 5 m/s per second. Figure 3-23b shows a graph with a negative slope, so we expect to find negative acceleration. Measurements from the graph for Fig. 3-23b give the same magnitude as in Fig. 3-23a, but the opposite sign: $\bar{a} = -5$ (m/s)/s.

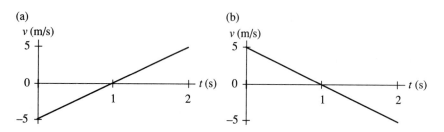

FIGURE 3-23 Two velocity-versus-time graphs for Example 3-11.

b) How are these graphs to be interpreted? In Fig. 3-23a, the initial velocity is negative, hence the object is moving to the left. As time goes by, the *magnitude* of v—that is, its speed—is decreasing. So for the first second the object is slowing down. At $t = 1$ s, the object is *instantaneously* at rest ($v = 0$), but for $t > 1$ s, the velocity immediately begins increasing. Now v is positive, so motion is to the right, and the speed is increasing. What we are seeing, then, is another example of a turning point. The object starts out moving to the left, slows down, reaches the turning point where $v = 0$, and then starts gaining speed as it moves back to the right. The acceleration is positive throughout the entire motion. Figure 3-23b shows a similar situation. The object starts out moving to the right, slows down, reaches a turning point at $t = 1$ s, reverses direction, and then starts moving to the left ($v < 0$) with increasing speed. The acceleration throughout is negative.

c) Figure 3-24 shows the motion diagrams for the two objects represented by the graphs in Fig. 3-23. Here we can see, using the techniques of Chapter 1, that $\vec{a}$ points steadily to the right for the object in part a) and steadily to the left for the object in part b).

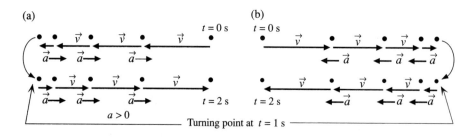

FIGURE 3-24 Motion diagrams for Example 3-11.

Remember that the *sign* of a has *nothing* to do with whether or not the object is speeding up or slowing down. A very common error is to interpret a positive acceleration as speeding up and a negative acceleration as slowing down. The correct interpretation is that a positive acceleration indicates that the acceleration vector $\vec{a}$ points to the right (toward the positive end of the axis), while a negative acceleration indicates that $\vec{a}$ points to the left. Whether or not the object is speeding up or slowing down *cannot* be determined by the acceleration alone. Instead, you have to consider the velocity and the acceleration together! This example is worth very careful study.

EXAMPLE 3-12 a) An object is moving with a velocity of 10 m/s and an constant acceleration of 2 (m/s)/s. What is its velocity one second later? Two seconds later? b) An object is moving with a velocity of –10 m/s and an constant acceleration of 2 (m/s)/s. What is its velocity one second later? Two seconds later?

SOLUTION a) This returns us to our interpretation of just what is meant by "2 (m/s)/s?" It means that "velocity increases by 2 m/s every 1 second." If the initial velocity is 10 m/s, then after one second the velocity will be 12 m/s. After two seconds, which is one additional second later, it will increase by another 2 m/s to 14 m/s. After three seconds it will be 16 m/s. Here a positive $\bar{a}$ is causing the object to speed up.

b) The acceleration changes the velocity, not the speed. If the initial velocity is a *negative* –10 m/s but the acceleration is a positive +2 (m/s)/s, then one second later the velocity will be –8 m/s. After two seconds it will be –6 m/s, and so on. In this case, a positive $\bar{a}$ is causing the object to *slow down* (decreasing |v|).

It is customary to abbreviate the units "(m/s)/s" as simply "m/s^2." The two objects in Example 3-11 had accelerations of +5 m/s^2 and –5 m/s^2, while the object in Example 3-12 had an acceleration of 2 m/s^2. We will use this notation, but it is important to keep in mind the *meaning* of the notation is "(meters per second) per second."

Consider an object whose acceleration a remains constant during the time interval $\Delta t = t_f - t_i$. (We have removed the average symbol from over the a because for a *constant* acceleration the average is just $\bar{a} = a$.) At the beginning of this interval, at time t_i, the object has an initial velocity v_i and an initial position s_i. Note that t_i is often zero, but it does not have to be. Figure 3-25a shows the acceleration-versus-time graph. It is a horizontal line between t_i and t_f, which is what we mean by *constant* acceleration.

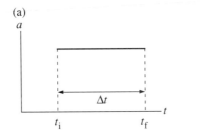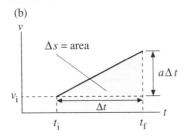

FIGURE 3-25 a) The acceleration-versus-time graph for constant acceleration. b) The velocity graph for motion with constant acceleration.

The object's velocity is changing because the object is accelerating. It is not hard to find the velocity v_f the object has at a later time t_f. By definition,

$$a = \bar{a} = \frac{\Delta v}{\Delta t} = \frac{v_f - v_i}{\Delta t}$$

which is easily rearranged to give

$$v_f = v_i + a\Delta t. \tag{3-12}$$

It is also useful to know the object's position s_f at time t_f. We demonstrated in the last section how to find the position from the velocity. We start by drawing the velocity-versus-time graph of Fig. 3-25b. This is a graph of Eq. 3-12 with Δt allowed to vary continuously from 0 at t_i to $t_f - t_i$ at t_f. Note that v changes by the amount $a\Delta t$ during the interval Δt. From the velocity-versus-time curve we can proceed to determine the position. The final position is:

$$s_f = s_i + \text{ area under the } v(t) \text{ curve between } t_i \text{ and } t_f. \tag{3-13}$$

The area in question, which is shaded in Fig. 3-25b, can be subdivided into a rectangle of area $v_i\Delta t$ and a triangle of area $\frac{1}{2}(a\Delta t)(\Delta t) = \frac{1}{2}(\Delta t)^2$. Combining these areas gives

$$s_f = s_i + v_i\Delta t + \tfrac{1}{2}a(\Delta t)^2$$
$$= s_i + v_i(t_f - t_i) + \tfrac{1}{2}a(t_f - t_i)^2. \tag{3-14}$$

Equations 3-12 and 3-14 tell us the object's exact position and velocity at some future instant of time. They are valid if a) the acceleration a is *constant*, and b) the initial position s_i and the initial velocity v_i are both known. While these equations are very general and very useful, we need one more to complete our set—a direct relation between position and velocity. First, isolate Δt in Eq. 3-12, finding $\Delta t = (v_f - v_i)/a$. Substitute this for Δt in Eq. 3-14, to obtain

$$s_f = s_i + v_i\left(\frac{v_f - v_i}{a}\right) + \tfrac{1}{2}a\left(\frac{v_f - v_i}{a}\right)^2.$$

With a bit of algebra this becomes

$$s_f = s_i + \frac{v_f^2 - v_i^2}{2a}.$$

One last rearrangement puts this into final form:

$$v_f^2 = v_i^2 + 2a\Delta s, \tag{3-15}$$

TABLE 3-6 The kinematic equations for motion with constant acceleration.

$$v_f = v_i + a\Delta t$$

$$s_f = s_i + v_i\Delta t + \tfrac{1}{2}a(\Delta t)^2$$

$$v_f^2 = v_i^2 + 2a\Delta s$$

---

where $\Delta s = s_f - s_i$ is the *displacement* (not the distance). Equations 3-12, 3-14, and 3-15 are the keys to constant acceleration motion, and we will have frequent need of them. Table 3-6 summarizes the results.

We can conclude this section by summarizing the distinction between motion with constant velocity and motion with constant acceleration. This is a critical distinction for many of our applications of kinematics. When the velocity is constant, we have the case of uniform motion. The acceleration is zero, and the position changes linearly with time (Eq. 3-4), giving a position-versus-time graph that is a straight line with slope v. When the acceleration is constant, however, it is the velocity that increases linearly with time—a straight-line graph with slope equal to a—while the position increases quadratically with time (Table 3-6). This causes the position-versus-time graph to have a parabolic shape. These graphs are shown in Fig. 3-26 for the case of a positive acceleration. The case of a negative acceleration is left as an exercise.

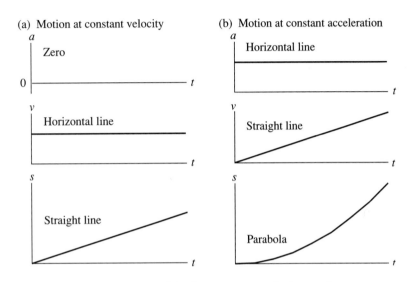

FIGURE 3-26 A summary of the kinematic graphs for motion at a) constant velocity, and b) constant acceleration.

EXAMPLE 3-13   A rocket sled accelerates at 50 m/s^2 for 5 seconds, coasts for 3 seconds, then deploys a braking parachute and decelerates at 3 m/s^2 until coming to a halt. a) What is the top speed of the rocket sled? b) What is the total distance traveled?

SOLUTION a) We previously developed the pictorial model and physical model for this problem as Example 3-1–3-8. The results of that analysis are reproduced here in Fig. 3-27. Now we can proceed with the mathematical model. The top speed is identified in the pictorial model as $|v_1|$, the speed at time t_1 when the acceleration phase ends. The first equation in Table 3-6 gives

$$v_1 = v_0 + a_0(t_1 - t_0) = a_0 t_1 = (50 \text{ m}/\text{s}^2)(5 \text{ s}) = 250 \text{ m}/\text{s}.$$

We started with the complete equation, then simplified by noting that v_0 and t_0 were zero. This gives us the top speed—approximately 500 mph.

b) Finding the total distance the sled travel requires several steps. The sled's position at t_1, when the acceleration ends, is found from the second equation in Table 3-6:

$$x_1 = x_0 + v_0(t_1 - t_0) + \tfrac{1}{2}a_0(t_1 - t_0)^2 = \tfrac{1}{2}a_0 t_1^2$$
$$= \tfrac{1}{2}(50 \text{ m}/\text{s}^2)(5 \text{ s})^2 = 625 \text{ m}.$$

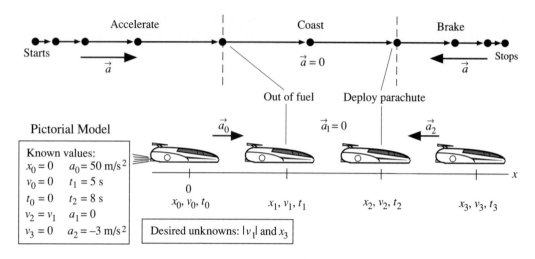

FIGURE 3-27 The pictorial and physical models for Example 3-13.

The same equation can be used during the coasting phase, when $a_1 = 0$, to find:

$$x_2 = x_1 + v_1(t_2 - t_1) + \tfrac{1}{2}a_1(t_2 - t_1)^2 = x_1 + v_1(t_2 - t_1)$$
$$= 625 \text{ m} + (250 \text{ m} / \text{s})(3 \text{ s}) = 1375 \text{ m}.$$

The braking phase is a little different because we don't know how long it lasts. But we do know that the sled ends with $v_3 = 0$, so we can use the third equation in Table 3-6 to find $v_3{}^2$:

$$v_3{}^2 = v_2{}^2 + 2a_2(x_3 - x_2).$$

Rewriting this in terms of x_3 gives:

$$x_3 = x_2 + \frac{v_3{}^2 - v_2{}^2}{2a_2}$$

$$= 1375 \text{ m} + \frac{0 - (250 \text{ m} / \text{s})^2}{2(-3 \text{ m} / \text{s}^2)} = 9040 \text{ m}.$$

The total distance traveled is ≈ 9 km ≈ 5.5 miles. This is reasonable because it takes a very long distance to stop from a top speed of 500 mph! You should note that we have used explicit numerical subscripts throughout the mathematical model, each referring to a symbol that was defined in the pictorial model. The subscripts i and f in the Table 3-6 equations are just generic "place holders" and don't have unique values. During the acceleration phase we had i = 0 and f = 1, but then during the coasting phase these became i = 1 and f = 2. The numerical subscripts have a clear meaning and are less likely to lead to confusion.

EXAMPLE 3-14 Football hero Fast Fred catches the kick-off while standing directly on the goal line. He immediately runs forward with an acceleration of 6 ft/s^2. At the moment the catch is made, Two-Ton Tommy is crossing the 20-yard line with a steady

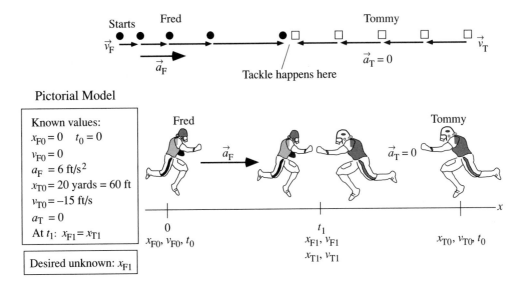

FIGURE 3-28 The pictorial and physical models for Example 3-13.

speed of 15 ft/s, heading directly toward Fred. If neither deviates from his path, where will the tackle occur?

SOLUTION This problem was analyzed in Example 3-1–3-10, and the pictorial and physical models are shown again in Fig. 3-28. We want to find *where* Fred and Tommy have the same position. The pictorial model designates time t_1 as *when* they meet. The axes have been chosen so that Fred starts at $x_{F0} = 0$ and moves to the right while Tommy starts at $x_{T0} = 60$ feet and runs to the *left* with a *negative* velocity. The first equation of Table 3-6 allows us to find their positions at time t_1. These are:

$$x_{F1} = x_{F0} + v_{F0}(t_1 - t_0) + \tfrac{1}{2} a_F(t_1 - t_0)^2 = \tfrac{1}{2} a_F t_1^2$$

$$x_{T1} = x_{T0} + v_{T0}(t_1 - t_0) + \tfrac{1}{2} a_T(t_1 - t_0)^2 = x_{T0} + v_{T0} t_1.$$

Notice that Tommy's position equation contains the term $+ v_{T0} t_1$. The fact that he is moving to the left has already been considered in assigning a *negative value* to v_{T0}, so we don't want to add any additional negative signs in the equation. If we now set x_{F1} and x_{T1} equal to each other, indicating the point of the tackle, we can solve for t_1:

$$\tfrac{1}{2} a_F t_1^2 = x_{T0} + v_{T0} t_1$$

$$\Rightarrow \tfrac{1}{2} a_F t_1^2 - v_{T0} t_1 - x_{T0} = 0$$

$$\Rightarrow 3t_1^2 + 15t_1 - 60 = 0.$$

Notice that we added the numerical information only in the final step, *after* all the algebraic manipulations were completed. This is a quadratic equation for t_1 whose solutions are found to be $t_1 = (-7.62$ s, $+2.62$ s$)$. The negative time is not physically meaningful in this problem, so the time of the tackle is $t_1 = 2.62$ s. Using this to compute x_{F1} gives

$$x_{F1} = \tfrac{1}{2} a_F t_1^2 = 20.6 \text{ feet} = 6.9 \text{ yards}.$$

Tommy makes the tackle at just about the 7-yard line!

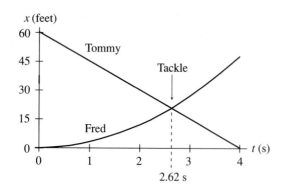

FIGURE 3-29 Position-versus-time graphs for Fred and Tommy.

It is worth exploring this example graphically. Figure 3-29 shows position-versus-time graphs for Fred and Tommy. Tommy's graph is a straight line with negative slope because he is running to the left with a constant speed. But Fred is accelerating, so his graph is a parabola with an increasing slope. The curves intersect at $t = 2.62$ s, and that is where the tackle occurs. You should compare this problem to Example 3-3 and Fig. 3-8 for Bob and Suzy to notice the similarities and the differences.

3.8 Instantaneous Acceleration

After defining $\bar{v}$ as the slope of a straight-line position-versus-time graph, we noted that not all position-versus-time graphs are straight lines. This led us to define the instantaneous velocity $v(t)$ as the slope of the tangent to the $s(t)$ curve at t. It is equally true that not all velocity-versus-time graphs are straight lines. Figure 3-30 shows a velocity that first increases with time, reaches a maximum, then decreases with time. We can still define an average acceleration, but the concept of average acceleration just does not contain enough information to give a complete description of a nonlinear velocity-versus-time curve.

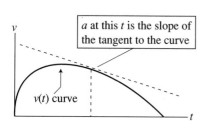

FIGURE 3-30 A nonuniform acceleration. The acceleration is the slope of the velocity-versus-time curve at each point.

We need the acceleration *at* a single *instant* of time, by which we mean (as before) the amount by which the velocity would change in one second if it continued at that rate. By analogy with our analysis of Section 3.5, this is just the slope of the tangent to the $v(t)$ curve, as shown in Fig. 3-30. So we can define

> The acceleration *a* at a specific instant of time *t*—which we call the **instantaneous acceleration**—is the slope of the straight line which is tangent to the velocity-versus-time curve at time *t*.

Mathematically, this becomes

$$a(t) \equiv \lim_{\Delta t \to 0} \frac{\Delta v}{\Delta t} = \frac{dv}{dt} \quad \text{(instantaneous acceleration).} \tag{3-16}$$

The reverse problem—to find the velocity v if the acceleration $a(t)$ is known—is also important. The analogy with velocity continues to hold. There we took a $v(t)$ curve, divided it into N steps, and found that Δs_k during each step was the area $\bar{v}_k \Delta t$ inside the step. Then we added all the steps (i.e., integrated) to find s_f. We can do the same with acceleration. An $a(t)$ curve can be divided into N very narrow steps so that during each step the acceleration is an essentially constant $\bar{a}_k$. During that interval, Eq. 3-12 gives a velocity change as $\Delta v_k = \bar{a}_k \Delta t = a(t_k)\Delta t$, which is the area inside the step. The total velocity change, between t_i and t_f is then found by summing all the small Δv_k. In the limit $\Delta t \to 0$, we have

$$v_f = v_i + \lim_{\Delta t \to 0} \sum_{k=1}^{N} a(t_k)\Delta t = v_i + \int_{t_i}^{t_f} a(t)dt. \tag{3-17}$$

We want to give this a graphical interpretation, similar to Eq. 3-9. In this case:

$$v_f = v_i + \text{ area under the } a(t) \text{ curve between } t_i \text{ and } t_f.$$

The constant-acceleration equation, $v_f = v_i + a\Delta t$, is a special example of this. As can be seen in Fig. 3-25a, the quantity $a\Delta t$ is the area of the rectangular area under the horizontal $a(t)$ curve.

Let us immediately apply these ideas to some examples.

EXAMPLE 3-15   Figure 3-31a shows a velocity-versus-time graph for an object whose velocity is given by the function $v(t) = [10 - (t - 5 \text{ s})^2]$ m/s. a) Find a and sketch the acceleration-versus-time graph, and b) describe the motion.

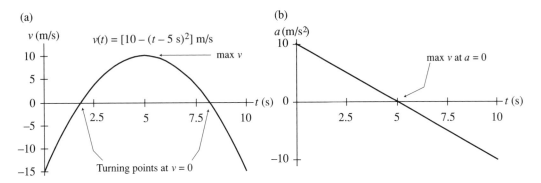

FIGURE 3-31 a) Velocity and b) acceleration-versus-time graphs for Example 3-15.

SOLUTION a) The velocity function is given as $v(t) = [10 - (t - 5 \text{ s})^2]$ m/s, which describes a parabola centered at $t = 5$ s with an apex at $v_{max} = 10$ m/s. Before getting quantitative, it is worth noting what we expect the acceleration to look like. For $t < 5$ s, the slope of v is positive but decreasing in magnitude. The slope is zero at $t = 5$ s, and it is negative and increasing in magnitude for $t > 5$ s. So the $a(t)$ graph should start positive and decrease steadily, pass through zero at $t = 5$ s, and end up negative. To confirm this, we can take the derivative of $v(t)$. First, expand the square in v to give:

$$v(t) = [-t^2 + 10t - 15] \text{ m/s}.$$

Then use the information in Eqs. 3-6 and 3-7 to evaluate the derivatives:

$$a(t) = \frac{dv}{dt} = [-2t + 10] \text{ m/s}^2.$$

This is a linear equation, which is graphed in Fig. 3-31b. The graph meets our expectations.

b) This is a complex motion, so how would we describe it? The initial velocity is negative, so the object is initially moving to the left with a speed of 15 m/s. There is a positive acceleration causing the *velocity* to become *more positive* and the *speed* to *decrease*. The velocity becomes zero just before $t = 2$ s—a turning point! Between the turning point and $t = 5$ s, the object is moving to the right ($v > 0$) and picking up speed (positive acceleration making v more positive). The velocity reaches a maximum at $t = 5$ s (when $a = 0$) and then starts to decrease (a goes negative). The object, though, is still moving toward the right (v is still positive)! That is, $a = 0$ is *not* a turning point—it is simply a point where the object switches from speeding up to slowing down. The slowing continues until just past $t = 8$ s, when another turning point occurs. Beyond this time, the object is again moving to the left and picking up speed as a negative a makes the velocity ever more negative.

EXAMPLE 3-16 Figure 3-32 shows the acceleration-versus-time graph for an object which has an initial velocity of 10 m/s. What is the velocity at $t = 8$ s?

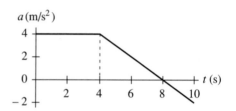

FIGURE 3-32 Acceleration graph for Example 3-16.

SOLUTION Our interpretation of Eq. 3-17 is that

$$v_f = v_i + \text{area under the } a(t) \text{ curve between } t_i \text{ and } t_f.$$

The area under the curve can be subdivided into a rectangle and a triangle, each of width 4 s. Those areas can be easily computed. Thus

$$v(t = 8 \text{ s}) = 10 \text{ m/s} + (4 \text{ m/s}^2)(4 \text{ s}) + (\tfrac{1}{2})(4 \text{ m/s}^2)(4 \text{ s}) = 34 \text{ m/s}.$$

3.9 Free Fall and Inclined Planes

A scientific problem of great antiquity is that of so-called natural motion—that is, the motion of objects moving freely through the air. This includes objects that are dropped or that are thrown upwards and fall back to the ground. It also includes a problem of great practical interest, namely the trajectory of a projectile that is shot upwards at an angle and returns to the ground some distance away.

[**Photo suggestion: Strobe photo of rock and feather falling together in a vacuum.**]

Aristotle asserted that heavier objects "fall faster" than lighter objects. This seems fairly clear if you drop a rock and a feather side-by-side. Galileo was the first to challenge

Aristotle and to put Aristotle's assertion to a rigorous experimental test. Galileo wrote about this in his *Two New Sciences*:

> *Simplicio:* As I recall it, Aristotle makes two assumptions; one concerning moveables differing in heaviness but moving in the same medium, and the other concerning a moveable moved in different mediums. As to the first, he assumes that moveables differing in heaviness are moved in the same medium with unequal speeds, which maintain to one another the same ratio as their weights. Thus, for example, a moveable ten times as heavy as another is moved ten times as fast.

> *Salviati:* I think it is possible to go against his assumptions and deny them. As to the first one, I seriously doubt that Aristotle ever tested whether it is true that two stones, one ten times as heavy as the other, both released at the same instant to fall from a height, say, of one hundred braccia (about 200 feet), differed so much in their speeds that upon the arrival of the larger stone upon the ground, the other would be found to have descended no more than ten braccia.

> *Simplicio:* But it is seen from his words that he appears to have tested this, for he says "We see the heavier ..." Now this "We see" suggests that he had made the experiment.

> *Sagredo:* But I, Simplicio, who have made the test, assure you that a cannonball that weighs one hundred pounds does not anticipate by even one span the arrival on the ground of a musket ball of no more than half an ounce, both coming from a height of two hundred braccia.

> *Simplicio:* Truly, your reasoning goes along very smoothly; yet I find it hard to believe that a birdshot must move as swiftly as a cannonball.

> *Salviati:* You should say "a grain of sand as fast as a millstone." But I don't want you, Simplicio, to do what many others do, and divert the argument from its principal purpose, attacking something I said that departs by a hair from the truth, and then trying to hide under this hair another's fault that is a big as a ship's hawser (the large rope for tying a ship to the dock). Aristotle says, "A hundred pound iron ball falling from the height of a hundred braccia hits the ground before one of just one pound has descended a single braccio." I say that they arrive at the same time. You find, on making the experiment, that the larger anticipates the smaller by two inches; that is, when the larger one strikes the ground, the other is two inches behind it. And now you want to hide, behind those two inches, the ninety-nine braccia of Aristotle, and speaking only of my tiny error, remain silent about his enormous one.

> (There follows a lengthy discussion about motion in mediums of different density. Finally ...)

> *Salviati:* What now? Surely a gold ball at the end of a fall through one hundred braccia will not have outrun one of copper by four inches. This seen, I say, I came to the opinion that *if one were to remove entirely the resistance of the medium, all materials would descend with equal speed.*

This is truly a remarkable passage! Galileo, showing no mercy, uses his spokesman Salviati to demolish 2000 years of accepted wisdom. He utilizes careful measurement and planned experiments—the cornerstone of modern science. He is also not deterred by "tiny errors," realizing that there is a general principle lurking nearby although certain aspects of the experiment may not be ideal. He correctly identifies the resistance of the air as the source of his tiny errors, and so forms his general principle for an *idealized situation* of motion in a vacuum. (This was prior to the invention of the vacuum pump, and followers of Aristotle even denied that a vacuum could exist.) In this regard, Galileo developed a *model* of motion—namely motion in the absence of air resistance—that could only be approximated by real motion. He focused on what is *common* to the motion of all objects, and then explained small differences in terms of the local conditions, rather than focusing on the *differences* in the motion of objects, which was the prevailing attitude.

This passage is the basis for the well-known story of Galileo dropping different weights from the leaning bell tower at the cathedral in Pisa. Historians cannot confirm whether or not Galileo ever conducted experiments in Pisa, but bell towers of "one hundred braccia" (about 200 feet) were common in the Italy of Galileo's day, so he had ample opportunity to make the measurements and observations that he describes. Notice, as Galileo emphasized in the quotation that opened this chapter, that he was not inquiring or speculating as to the *cause* of the motion but was merely describing its character—that is, the kinematics of freely-falling objects.

We call the motion of objects moving under the influence of gravity only, and no other forces, **free fall**. Strictly speaking, this motion would occur only in a vacuum so as to avoid the forces of air resistance. Nonetheless, as Galileo noted, the errors involved are "tiny" if we treat "heavy" objects that move through distances that are "not too big" *as if* they were in free fall. This assumption is valid for most of the objects with which we have common experience and which we will study in this section. For very light objects, such as a feather or a piece of paper, or for objects that fall through very large distances and gain very high speeds, the effect of air resistance is *not* negligible. We will need the more complete analysis of Newton before we can attack those problems.

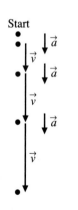

Figure 3-33 shows the motion diagram of a freely-falling object. As Galileo discovered, and as we can confirm today with great precision, the acceleration $\vec{a}$ is the *same for all falling objects*, independent of their mass or size, in the ideal case of no air resistance. At this point, Galileo's discovery is simply an observation; the *reasons* for this behavior will be investigated in future chapters. Careful measurements show that the value of this acceleration varies slightly at different places on the earth, due to the slightly non-spherical shape of the earth and to the effects of local mountains or high-density mineral deposits. A global average, at sea level, is

FIGURE 3-33 Motion diagram of an object in free fall.

$$\vec{a}_{\text{free fall}} = (9.80 \text{ m/s}^2, \text{ vertically downward}).$$

Because $\vec{a}_{\text{free fall}}$ is a vector, we have given both its magnitude and its direction. It is customary to speak of just the magnitude of the free-fall acceleration as the **acceleration due**

to gravity, and to give it a special symbol g:

$$g = |\vec{a}_{\text{free fall}}| = 9.80 \text{ m/s}^2 \quad \text{(acceleration due to gravity)}.$$

There are two things worthy of note here. First, g is, by definition, *always* positive. There will *never* be a problem that will use a negative value for g. But, you say, objects fall when you release them rather than rise, so how can g be positive? This is a common point of confusion. The key is that g is *not* the acceleration $\vec{a}$, but simply its magnitude. To work a problem, you will need $\vec{a}$. For the usual case of a y-axis pointing vertically upward, a downward pointing acceleration vector $\vec{a}$ will require a one-dimensional component $a = -g$. So the minus sign, which is used to indicate direction, goes with a and not with g. Second, it is important to learn and to use the proper name. The symbol g is not called "gravity." "Gravity" is not an acceleration. So learn to refer to g by its correct name of "the acceleration due to gravity."

Free fall is a case of motion with constant acceleration. We can use the kinematic equations of Table 3-6 with the acceleration due to gravity, $a = -g$. The kinematic equations specific to free fall are:

$$v_f = v_i - g\Delta t$$

$$y_f = y_i + v_i\Delta t - \tfrac{1}{2}g(\Delta t)^2 \quad \text{(free fall)}. \tag{3-18}$$

$$v_f^2 = v_i^2 - 2g\Delta y$$

Let us apply these equations to some examples. We will use the full problem-solving strategy of pictorial model, physical model, and mathematical model.

EXAMPLE 3-17   A rock is released from rest at the top of a 100 m tall building. a) How long does the rock take to fall to the ground? b) What is its impact velocity?

SOLUTION Figure 3-34 shows the pictorial model and the motion diagram of the physical model. We have chosen the origin at the ground, which makes $y_0 = 100$ m, but we could equally well have chosen the origin to be where the rock is released and had $y_1 = -100$ m. a) This question involves a relation between time and distance, so only the second of Eqs. 3-18 is relevant. Using $v_0 = 0$ and $t_0 = 0$, we get:

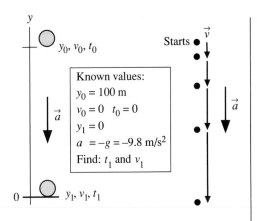

FIGURE 3-34 Pictorial model and motion diagram for Example 3-17.

$$y_1 = y_0 + v_0\Delta t - \tfrac{1}{2}g\Delta t^2 = y_0 - \tfrac{1}{2}gt_1^2.$$

We can now solve for t_1, finding:

$$t_1 = \sqrt{\frac{2(y_0 - y_1)}{g}} = \sqrt{\frac{2(100 \text{ m} - 0 \text{ m})}{9.80 \text{ m}/\text{s}^2}} = \pm 4.52 \text{ s}.$$

The ± indicates that there are two mathematical solutions; therefore, we have to use physical reasoning to choose between them. In this case, a negative t_1 would be a time before we dropped the rock—not *physically* meaningful. So we select the positive root: $t_1 = 4.52$ s.

b) Knowing the fall time from part a), we can use the first of Eqs. 3-18 to solve for v_1:

$$v_1 = v_0 - g\Delta t = -gt_1 = -(9.80 \text{ m}/\text{s}^2)(4.52 \text{ s}) = -44.3 \text{ m}/\text{s}.$$

Alternatively, we could work directly from the last of Eqs. 3-18:

$$v_1 = \sqrt{v_0{}^2 - 2g\Delta y} = \sqrt{-2g(y_1 - y_0)}$$

$$= \sqrt{-2(9.80 \text{ m}/\text{s}^2)(0 \text{ m} - 100 \text{ m})} = \pm 44.3 \text{ m}/\text{s}.$$

This method is useful if you don't know Δt. However, we must again choose the correct sign of the square root. Because we know that the velocity vector points downward, the sign of v, the y-component of $\vec{v}$, has to be negative! Thus $v_1 = -44.3$ m/s. The importance here of careful attention to the signs cannot be overemphasized.

A common error would be to say "The rock fell 100 m, so $\Delta y = 100$ m." This would have you trying to take the square root of a negative number. The problem here is that Δy is not a distance—it is a *displacement*, with a carefully defined meaning of $y_f - y_i$. In this case, $\Delta y = y_1 - y_0 = -100$ m. The negative sign here cancels that of the -2 under the square root sign, so everything comes out all right.

Lastly, let us assess our result. Do we believe the answers? Are they reasonable? Well, 100 m is about 300 feet, which is about the height of a 30-floor building. How long does it take something to fall 30 floors? Four or five seconds seems pretty reasonable. How fast would it be going at the bottom? Referring back to Table 3-3, where we find that 1 m/s ≈ 2 mph, we see that 44.3 m/s ≈ 90 mph. That also seems pretty reasonable after falling 30 floors. Had we misplaced a decimal point, though, and found 443 m/s, we would be suspicious when we converted this to ≈900 mph! So the answers all seem reasonable.

EXAMPLE 3-18   A ball is shot straight upward with an initial speed of 100 m/s. How high does it go?

SOLUTION Figure 3-35 shows the pictorial and physical models for the ball's motion. This *is* a free fall problem, even though the ball was shot upward, because the ball (after being launched) is moving under the influence of gravity *only*. Because we are given the initial *speed*, rather than the initial velocity, we have to choose whether $v_0 = +100$ m/s or $v_0 = -100$ m/s. The upward direction of $\vec{v}$ is the clue. Another key piece of setting up a problem like this is knowing where it ends—how do we put "how high" into symbols? Here the clue is that the very

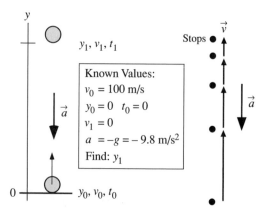

FIGURE 3-35 Pictorial model and motion diagram for Example 3-18.

top of the trajectory is a *turning point*. The object is moving upward before it reaches the top, and it will move downward after passing the top. Recall that at a turning point the *instantaneous* velocity $v = 0$. So we can characterize the "top" of the trajectory by $v_1 = 0$. This was not explicitly stated in the problem, instead it is our "knowledge input" to the problem. With this information, we see that we are looking at a relationship between distance and velocity, without knowing the time interval. This relationship is described mathematically by the third of Eqs. 3-18:

$$v_1{}^2 = v_0{}^2 - 2g\Delta y.$$

Solving this for Δy gives

$$\Delta y = y_1 = \frac{v_0{}^2 - v_1{}^2}{2g} = \frac{(100 \text{ m}/\text{s})^2}{2(9.80 \text{ m}/\text{s}^2)} = 510 \text{ m}.$$

Is this answer reasonable? A speed of 100 m/s ≈ 200 mph—pretty fast! The calculated height is 510 m ≈ 1500 ft. In Example 3-17 we found that an object dropped from 100 m is going 44 m/s when it hits the ground, so it is reasonable that an object shot upward at 100 m/s will go significantly higher than 100 m. While we cannot say that 510 m is necessarily better than 400 m or 600 m, we can say that it is not unreasonable. The point of the assessment is not to prove that the answer *has* to be right, but to find answers that are obviously wrong.

EXAMPLE 3-19   A rock is tossed straight up with a speed of 20 m/s. When it returns, it falls into a hole 10 m deep. a) What is its velocity as it hits the bottom of the hole? b) How long is it in the air?

SOLUTION   Figure 3-36 shows the pictorial model and the motion diagram for the rock. You might be tempted to include the top of the trajectory as y_1 and label the bottom of the hole as y_2. Before you do this, you should ask the question, "Does the character of the motion change at the top?" While the top is indeed a turning point, the character of the motion does *not* change—it is just free fall throughout the entire trip! So we only need the starting and ending points of the motion.

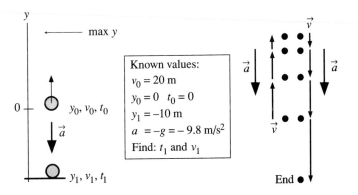

FIGURE 3-36 Pictorial model and physical model for Example 3-19.

a) We once again have a relationship between displacement and velocity, so

$$v_1 = \sqrt{v_0{}^2 - 2g\Delta y} = \sqrt{(+20 \text{ m}/\text{s})^2 - 2(9.80 \text{ m}/\text{s}^2)(-10 \text{ m})} = -24.4 \text{ m}/\text{s}.$$

As was the case earlier, Δy is negative and we chose the negative root for v_1.

b) Either of the two time equations from Eqs. 3-18 can be used for part b). The velocity equation is the easier of the two to use:

$$v_1 = v_0 - g\Delta t = v_0 - gt_1$$

$$\Rightarrow t_1 = \frac{v_0 - v_1}{g} = \frac{20 \text{ m}/\text{s} - (-24.4) \text{ m}/\text{s}}{9.80 \text{ m}/\text{s}^2} = 4.53 \text{ s}.$$

If we had selected the distance equation instead, we would have:

$$y_1 = y_0 + v_0\Delta t - \tfrac{1}{2}g\Delta t^2 = v_0 t_1 - \tfrac{1}{2}gt_1{}^2$$

$$\Rightarrow \tfrac{1}{2}gt_1{}^2 - v_0 t_1 + y_1 = 4.9t_1{}^2 - 20t_1 - 10 = 0.$$

This is a quadratic equation for t_1. Application of the quadratic formula yields

$$t_1 = (-0.45 \text{ s}, +4.53 \text{ s}).$$

As before, a negative value of t_1 is not physically meaningful, so we choose $t_1 = 4.53$ s.
•

Inclined Planes

A problem closely related to free fall is that of an object moving down a straight, but frictionless, inclined plane. The ideal case of no friction, like that of no air resistance, eliminates any resistive forces of the "medium." We can come very close to this ideal by rolling a ball on a very smooth surface. In fact, most of Galileo's precise research into the nature of kinematics utilized motion on an inclined plane. Freely-falling objects are nice, but they move too quickly for precise measurement using the crude timing devices available to Galileo. It was the much slower motion of a ball on an inclined plane that allowed Galileo to discover the properties of uniformly accelerated motion.

[**Photo suggestion: Ball rolling on an inclined plane.**]

Figure 3-37 shows an ball accelerating down a frictionless inclined plane. The motion of the ball is confined to a line parallel to the surface. In Section 2-3, we considered the problem of finding the components of a vector parallel and perpendicular to a given line. Here, the perpendicular component $\vec{a}_\perp$ of the acceleration $\vec{a}$ is "blocked" by the surface, and is therefore ineffective. But the parallel component $\vec{a}_\parallel$ is unhindered. It is this component of $\vec{a}$ that accelerates the ball along the surface.

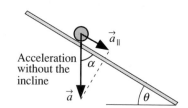

FIGURE 3-37 Acceleration on an inclined plane.

According to Eq. 2-22, the magnitude of the acceleration is

$$|\vec{a}_\parallel| = |\vec{a}|\cos\alpha$$

where α is the angle between $\vec{a}$ and the surface. It is generally more common to refer to inclined planes and slopes by their angle θ above the horizontal. From Fig. 3-37,

$\alpha = 90° - \theta$, so $\cos\alpha = \sin\theta$. What is $|\vec{a}|$? If $\theta = 90°$, a completely vertical surface, then clearly the acceleration is just that of free fall—namely g. So it must be that $|\vec{a}| = g$. Thus the magnitude of the acceleration along a frictionless plane tilted at an angle θ above the horizon is

$$|\vec{a}_{\parallel}| = g\sin\theta \quad \text{(frictionless inclined plane).} \qquad (3\text{-}19)$$

This is the *magnitude* of $\vec{a}_{\parallel}$ only. The *component* of $\vec{a}_{\parallel}$ along the axis of motion will thus be $a = \pm g\sin\theta$, with the sign depending upon the direction of the axis and whether the ramp is tilted up or down. The two major points to be made here are: 1) the motion is in one dimension, but we have to make a proper choice of coordinate system to exploit this fact, and 2) this is again a problem of *constant* acceleration.

EXAMPLE 3-20   Figure 3-38a shows a ball moving horizontally at 2 m/s. What angle θ should a ramp 10 m long have for the ball's speed at the bottom to be 10 m/s?

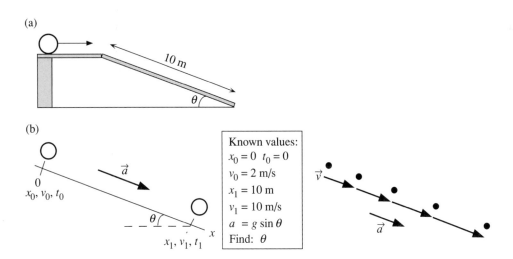

FIGURE 3-38 a) A ball rolling down an inclined ramp. b) The pictorial model and motion diagram for Example 3-20.

SOLUTION The problem really starts at the "top" of the ramp and looks at motion along an inclined plane. We assume that the ball loses no speed as it moves onto the downward-sloping ramp. Motion along an inclined surface is a problem where the use of a *tilted* axis allows the easiest solution. With an x-axis tilted parallel to the surface, as shown in the pictorial model of Fig. 3-38b, the motion is one-dimensional with constant acceleration. Had we chosen conventional horizontal and vertical axes, the motion would have had components in both dimensions and would be much harder to describe. With the axis shown, $v_0 = 2$ m/s, $v_1 = 10$ m/s, and $\Delta x = 10$ m. The acceleration is in the $+x$-direction and thus is $a = +g\sin\theta$. The third equation from Table 3-6 then gives

$$v_1^2 = v_0^2 + 2a\Delta x = v_0^2 + 2g\sin\theta\Delta x.$$

Solving for $\sin \theta$ gives:

$$\sin \theta = \frac{v_1^2 - v_0^2}{2g\Delta x} = \frac{(10 \text{ m}/\text{s})^2 - (2 \text{ m}/\text{s})^2}{2(9.80 \text{ m}/\text{s}^2)(10 \text{ m})} = 0.490.$$

Then

$$\theta = \sin^{-1}(0.490) = 29.3°.$$

EXAMPLE 3-21 In an amusement park ride, a car is shot up a 30° inclined track, comes momentarily to rest near the top, then rolls back down—as shown in Fig. 3-39a. If the track has a height of 20 m, what is the maximum allowable speed with which the car can start?

SOLUTION The problem "starts" as the car begins to move up the incline. We will assume that the track is frictionless, so this is just an inclined-plane problem. The *maximum* starting speed is that for which the car goes right to the very edge! If we use an inclined x-axis, as shown in the Fig. 3-39b, then the displacement Δx at maximum initial speed is related to the height h by

$$\Delta x = x_1 - x_0 = \frac{h}{\sin 30°} = \frac{20 \text{ m}}{\sin 30°} = 40 \text{ m}.$$

The condition for the top turning point is $v_1 = 0$. The last key step is to recognize that the acceleration vector $\vec{a}_\parallel$ points *downhill*, in the $-x$-direction, and so $a = -g\sin \theta$. Then

$$v_1^2 = 0 = v_0^2 - 2g\sin 30° \Delta x$$

$$\Rightarrow v_0 = \sqrt{2g\sin 30° \Delta x} = \sqrt{2(9.8 \text{ m}/\text{s}^2)(0.500)(40 \text{ m})} = 19.8 \text{ m}/\text{s}.$$

This is the maximum speed, because a car going any faster will run off the top. Make sure you understand why the sign of a was negative here but positive in Example 3-20.

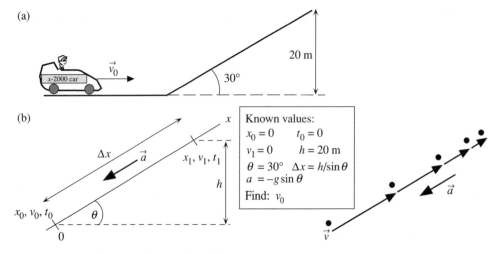

FIGURE 3-39 a) Car going up inclined track. b) Pictorial model and motion diagram for the car on the way *up* the track.

3.10 Pulling It All Together

Kinematics is the language of motion. You will spend the rest of this course studying moving objects—from baseballs to electrons—and the concepts we have developed in this chapter will be used extensively. One of the most important ideas has been the *relationships* that exist between the position, velocity, and acceleration of an object. Those relationships are most easily expressed graphically, so that you can visualize distances, slopes, and areas under curves.

A good way to solidify your new-found knowledge of kinematics is to consider the motion of a hard, smooth ball rolling on a smooth track. The track is made up of several straight segments that are connected together. Each segment may be either horizontal or inclined. Your analysis of the ball's motion will be purely qualitative—graphs of position, velocity, and acceleration that describe the motion—requiring you to reason about, rather than calculate, the relationships between s, v, and a.

There are two variations to this type of problem. In the first, you are given a picture of a track and the initial condition for the ball. The problem is then to draw graphs of s, v, and a. In the second, you are given the graphs, and the problem is to deduce the shape of the track on which the ball is rolling. We will do a couple of examples of each in this section, then there will be several more for you to try as problems.

There are a small number of "rules" to follow for all of these problem:

1. Assume that the ball passes smoothly from one segment of the track to the next, with no loss of speed.

2. The position, velocity, and acceleration graphs should be drawn so that they are stacked vertically. They should each have the same horizontal scale so that a vertical line drawn through all three will connect points describing the same instant of time.

3. Although the graphs have no numbers, they should show the correct *relationships*. For example, the slope of one portion of the position-versus-time graph should be steeper than the slope of a second portion if the velocity is greater during the first portion than the second. Similarly, longer-lasting motions should span a greater horizontal range than shorter-lasting motions.

4. The position s is the position measured *along* the track (not simply the horizontal position x). Similarly, v and a are the components of velocity and acceleration along the direction of the track.

EXAMPLE 3-22   Figure 3-40 shows a ball on a frictionless track. Draw position, velocity, and acceleration graphs for the motion up to the point that the ball leaves the track.

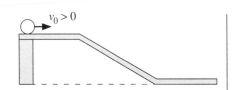

FIGURE 3-40 A ball rolling along a frictionless track.

SOLUTION It is often easiest to start a problem such as this by considering the velocity. Here we see that the ball has an initial velocity v_0 that is not zero—it is already rolling. For frictionless motion along a horizontal surface, the acceleration is $a = 0$ (see Eq. 3-19 for $\theta = 0$). Consequently the velocity remains constant. Eventually, the ball reaches the slope and starts down. Now we have a case of motion on an inclined plane, which, we have learned, has constant acceleration. During constant acceleration motion, the velocity increases linearly with time. After reaching the bottom segment, the ball returns to constant-velocity motion. This is shown in the middle graph of Fig. 3-41.

We already have enough information to deduce the acceleration graph. We noted that the acceleration is zero while the ball is on the horizontal segments and is a constant positive value on the slope. This is consistent with the slope of the velocity-versus-time graph during the three segments of the motion: zero slope, then positive slope, then a return to zero slope. This acceleration graph is shown at the bottom of Fig. 3-41. The acceleration cannot *really* change instantly from zero to a nonzero value, but the change can be so quick that we do not see it on the time scale of the graph. That is what the vertical dotted lines imply.

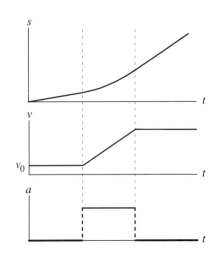

FIGURE 3-41 Motion graphs for the ball in Example 3-22.

Lastly we need to find the position-versus-time graph. (You might want to refer back to Fig. 3-26 to see how the position graph looks for constant-velocity and constant-acceleration motion.) During the first segment at constant velocity, the position will increase *linearly* with time. This will also be true during the third segment of motion, but with a steeper slope indicating the faster velocity. In between, while the acceleration is non-zero but constant, the position graph will have a parabolic shape. The top section of Fig. 3-41 shows the appropriate position-versus-time graph.

•

There are a couple of points worth noting about this example. First, we have drawn dotted vertical lines at the instants of time when the ball moves from one segment of the track to the next. As noted in Rule 1, the speed does not change *abruptly* at these points—it is just a gradual change. Second, the parabolic section of the position-versus-time graph blends *smoothly* into the straight lines on either side. This is really a consequence of Rule 1: If the velocity immediately after a segment change is the same as immediately before, then the *slope* of the position-versus-time graph will be the same immediately before and after the segment change. An abrupt change of slope would indicate an abrupt, discontinuous change in velocity, and would thus violate Rule 1.

• **EXAMPLE 3-23** Figure 3-42 shows a somewhat more complex track. This track has a "switch," so that a ball moving left-to-right passes through and heads up the incline, while a ball rolling down the incline will proceed straight through and continue downhill. The arrows next to "switches" indicate how balls pass through from the different sides. Draw position, velocity, and acceleration graphs of the motion.

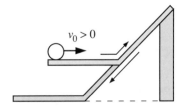

FIGURE 3-42 The ball and track for Example 3-23.

SOLUTION Let us again start by considering the velocity. It will remain constant at v_0 while the ball is on the level segment. As the ball moves onto the uphill incline, the velocity will *decrease linearly* with time (constant *a* implies linear change of *v*). At some point, *v* reaches zero (a turning point), then the ball starts rolling back down. Rolling downhill will be a *negative* velocity, but the velocity will still change linearly with time. Upon reaching the bottom, the ball will roll across the horizontal segment with constant negative velocity (moving toward the left). It is important to note that the *speed* will be greater on the lower horizontal segment than it was on the upper horizontal segment, so |v| at the end will be larger than v_0. Figure 3-43 shows the velocity graph in the middle.

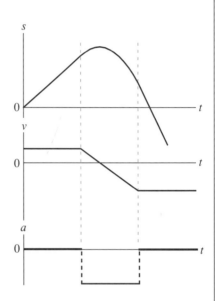

FIGURE 3-43 Figure 3-43 Motion graphs for the ball of Fig. 3-42.

Turning to the acceleration, it is zero on the two horizontal segments. Along the inclined segment, the situation is the same as in Example 3-21; that is, *a* is *negative* because the vector $\vec{a}_\parallel$ points downhill. The component *a* is constant the whole time that the ball is on the incline, *regardless* of whether it is moving uphill or downhill! "Moving uphill" or "moving downhill" is determined by the velocity, not by the acceleration. The acceleration determines how the velocity vector will change, but not which direction the velocity vector points. Two things can reinforce this idea for you if you still are not quite sure: 1) Draw a motion diagram and determine the acceleration vector directly, as we did in Chapter 1. You should find that the vector $\vec{a}$ is the same when the ball rolls downhill as it was when the ball rolled uphill. 2) Consider the slope of the velocity-versus-time graph during this segment; the slope is a constant negative value. The change of sign of *v* shows that motion changes direction—with the turning point at *v* = 0—but *nothing happens to the acceleration at that point*! Figure 3-43 shows the acceleration graph at the bottom.

The position is analyzed in a manner similar to the previous example. It changes linearly while the velocity is constant, and parabolically while the acceleration is constant. In this case, the position reaches a *maximum value* at the turning point, which is the top of the parabola. The constant negative velocity on the last segment implies a straight-line position graph with negative slope. The position-versus-time graph is shown at the top of Fig. 3-43.   •

EXAMPLE 3-24   Figure 3-38 shows a set of motion graphs for a ball moving on a track. Draw a picture of what the track looks like and describe the ball's initial condition. Each segment of the track is *straight*, but the segments may be tilted.

SOLUTION As with the last two examples, let's start by examining the velocity graph. In this case, the ball starts with initial velocity $v_0 = 0$. Then the velocity immediately increases linearly with time. This indicates that the ball was released from rest while on a slope, and it started rolling downhill. The acceleration graph confirms this—the initial acceleration is positive. Because v is positive, the motion is toward the right. This is seen also from the position graph, which has s becoming increasingly positive. After a while, the acceleration drops to zero and the velocity holds constant—a horizontal segment! Then the acceleration becomes negative and the velocity starts decreasing. However, the velocity is still *positive*, so the ball is still moving toward the right but it is slowing down. Thefore, the ball must be going uphill.

Notice that the acceleration has a smaller magnitude in this third segment than it had during the first segment, and that the magnitude of the velocity's slope is less. This indicates that the uphill tilt is less than the initial downhill tilt. Lastly, the acceleration is again zero and the velocity is constant, so the final segment is horizontal. Because the position has increased throughout and v was never negative, there are no turning points in the motion; it is purely left-to-right. Figure 3-45 shows the track and the initial conditions that are responsible for the graphs of Fig. 3-44.

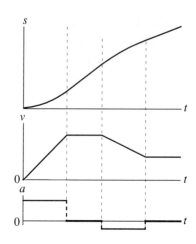

FIGURE 3-44 Motion graphs of a ball rolling on a track of unknown shape.

FIGURE 3-45 Track responsible for the motion graphs of Fig. 3-44.

EXAMPLE 3-25 Figure 3-46 shows a set of motion graphs. Determine the track and the initial conditions that created these graphs.

SOLUTION In this example, the initial velocity v_0 is negative, indicating an initial motion toward the left. Further, the *magnitude* of the velocity (i.e., the ball's speed) is decreasing, so the ball is slowing down. Eventually, the velocity passes through zero—but only instantaneously—and then becomes positive. We have seen several examples of this before; it indicates a turning point where the motion reverses direction. The position graph, indeed, shows that s steadily decreases (moving toward the left) until the instant of time when $v = 0$, then begins increasing (moving back toward the right). This would be explained by a ball that has been given an initial *uphill* push toward the left. After going part of the way up the incline, the ball turns and starts rolling back down. The acceleration vector would then point downhill, in the direction of $+s$, and we indeed see a constant positive acceleration throughout this period.

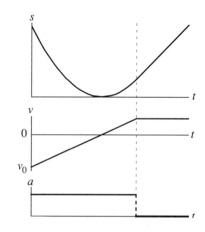

FIGURE 3-46 Motion graphs for Example 3-25.

Lastly, the ball reaches a steady velocity with no acceleration—a horizontal segment. Because the ball's speed |v| here is less than v_0, and the time the ball spent rolling back downhill was less than the time spent going uphill, it must not have rolled all the way back to its initial position. This is confirmed by the position graph at the time when the acceleration drops to zero. We can thus deduce that there must be a "switch" that intercepts the ball part way down the incline and diverts it to a horizontal segment of track. Our final picture of the track is shown in Figure 3-47.

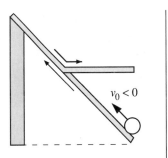

FIGURE 3-47 Track responsible for the motion graphs of Fig. 3-46.

Summary

Important Concepts and Terms

kinematics

SI units

instantaneous position

uniform motion

average velocity

instantaneous velocity

turning point

average acceleration

uniformly accelerated motion

instantaneous acceleration

free fall

acceleration due to gravity

In this chapter we have developed the mathematical basis of kinematics. In doing so, we have combined the concepts of position, velocity, and acceleration—which we defined operationally in Chapter 1—with the properties of vectors from Chapter 2. We have also added some new mathematical concepts from calculus. The result is a very powerful set of mathematical relationships between the different concepts of motion.

We can look at the flow of ideas in two directions: from position to acceleration, or from acceleration to position. Given the position $s(t)$ as a function of time, we can find first the velocity

$$v(t) = \frac{ds}{dt} = \text{ slope of the } s(t) \text{ graph at } t$$

and then the acceleration

$$a(t) = \frac{dv}{dt} = \text{ slope of the } v(t) \text{ graph at } t.$$

Alternatively, given the acceleration $a(t)$ as a function of time, we can first find the velocity

$$v(t) = v_i + \int_{t_i}^{t} a(t)dt$$

$$= v_i + \text{ area under the } a(t) \text{ curve between } t_i \text{ and } t$$

and then the position

$$s(t) = s_i + \int_{t_i}^{t} v(t)dt$$

$$= s_i + \text{area under the } v(t) \text{ curve between } t_i \text{ and } t.$$

These equations are true for any motion. There are, however, two special kinds of motion that occur sufficiently often that it becomes worthwhile to do the derivatives and integrals and thus arrive at explicit results. The first is motion with constant acceleration (also called uniformly accelerated motion). For this case:

$$v_f = v_i + a\Delta t$$

$$s_f = s_i + v_i\Delta t + \frac{1}{2}a(\Delta t)^2$$

$$v_f^2 = v_i^2 + 2a\Delta s.$$

The second is motion with constant velocity (also called uniform motion). In this case we have simply

$$a = 0$$

$$v(t) = v_i = \bar{v} = \text{ constant}$$

$$s(t) = s_i + \bar{v}\Delta t.$$

While these two special cases are important and do occur frequently, *do not* make the common error of trying to apply them indiscriminately to every problem. You will meet many situations where neither the velocity nor the acceleration are constant, and you will then need to rely on the more general relationships between s, v, and a.

It would be a serious mistake to focus only on the mathematical aspect of kinematics. As you have seen repeatedly throughout this chapter, a full understanding of kinematics requires not only mathematics but also an ability to use and interpret graphs as well as an ability to transform information from words and symbols to graphs and equations.

Exercises and Problems

▲ Exercises

The ✍ icon indicates that these problems can be done on a Dynamics Worksheet.

✍ 1. Bob leaves Los Angeles at 8:00 A.M. to drive to San Francisco, 400 miles away. He travels at a steady 50 mph. Bill leaves at 9:00 A.M. and drives a steady 60 mph.
a) Who gets to San Francisco first? b) How long does the first to arrive have to wait for the second?

✍ 2. It is 100 miles between San Luis Obispo and Santa Barbara. On her way to Santa Barbara, Jane drives half the *time* at 40 mph and half the time at 60 mph. On her return trip, she drives half the *distance* at 40 mph and half the distance at 60 mph.
a. How long does it take for Jane to drive to Santa Barbara from San Luis Obispo?
b. What is her average speed during that trip?
c. How long does it take Jane to return from Santa Barbara to San Luis Obispo?
d. What is her average speed on the return trip?

✍ 3. a. What acceleration, in SI units, must a car have in order to go from zero to 60 mph in 10 s?
   b. What fraction of "*g*" is this?
   c. How far has the car traveled when it reaches 60 mph?

✍ 4. Ball bearings are made by letting spherical drops of molten metal fall inside a tall tower—called a "shot tower"—and solidify as they fall.
   a. If a bearing takes 4 s to solidify enough for impact, how high must the tower be?
   b. What is the bearing's impact velocity for a tower this high?

✍ 5. A jet plane is cruising at 300 m/s when, suddenly, the pilot turns the engines up to full throttle. After traveling 4 km the jet is moving with a speed of 500 m/s.
   a. What is the jet's acceleration, assuming it to be a constant acceleration?
   b. Is your answer reasonable? Explain.

✍ 6. A car traveling at 30 m/s runs out of gas while traveling up a 20° slope. How far up the hill will the car coast before starting to roll back down?

✍ 7. a. How many days will it take a spaceship to accelerate to the speed of light (3×10^8 m/s) with the acceleration *g*?
   b. How far will it travel during this interval?
   c. What fraction of a light year is your answer to part b)? A light year is the *distance* light travels in one year.

(**Note:** We now know, from Einstein's theory of relativity, that no material object can travel at the speed of light. So this problem, while interesting and instructive, is not realistic.)

8. A particle moving along the *x*-axis has its position as a function of time described by the function $x(t) = 2t^3 - t + 1$ meters.
   a. What is the particle's position at *t* = 2 s?
   b. What is the particle's velocity at *t* = 2 s?
   c. What is the particle's acceleration at *t* = 2 s?

9. Figure 3-48 shows the velocity-versus-time graph of a particle moving along the *x*-axis. Its initial position is $x(t = 0) = 2$ m.
   a. What is the particle's position at *t* = 2 s?
   b. What is the particle's velocity at *t* = 2 s?
   c. What is the particle's acceleration at *t* = 2 s?

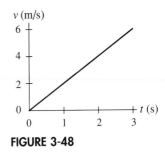

FIGURE 3-48

▲ Problems

10. An object moves along the *x*-axis according to the function $x(t) = -3t^3 + 9t^2 + 72t$.
   a. Make a position-versus-time graph for the interval -5 s $\leq t \leq 5$ s, either by calculating and graphing 21 values of the position at different times (every 0.5 s) or by using appropriate computer software.
   b. Determine the object's velocity at *t* = 0 two ways: by drawing the tangent line on your graph and measuring its slope, and by evaluating the derivative at *t* = 0.

c. At what position or positions does the object have a velocity $v = -63$ m/s?

d. Are there any turning points in the object's motion? If so, at what position or positions? Determine your answer two ways: by locating the turning point(s) on your graph and reading the position, and by solving for their exact location.

e. At what time or times does the object have zero acceleration?

f. Locate and label the zero-acceleration points on your graph. What can you say about the shape of your graph at those points?

11. The three graphs in Fig. 3-49 describe the motion of a particle by using a) a position-versus-time graph, b) a velocity-versus-time graph, or c) an acceleration-versus-time graph. For each, determine the particle's velocity at $t = 7$ s if the particle has an initial velocity $v_0 = v(t = 0) = 10$ m/s. In doing these, you are not to use any kinematics formulas but are to work with the geometry of the graphs.

(**Note:** Parts a), b), and c) are independent of each other. That is, the velocity graph in part b) does *not* correspond to the same particle whose motion is graphed in parts a) or c).)

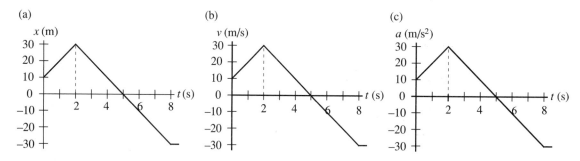

FIGURE 3-49

12. Figure 3-50 is a velocity-versus-time graph for a particle having initial position $x_0 = x(t = 0) = 0$. At what time or times is the particle located at $x = 35$ m? Work directly from the graph, using the graphical relationship between velocity and position, and not from any kinematics formulas.

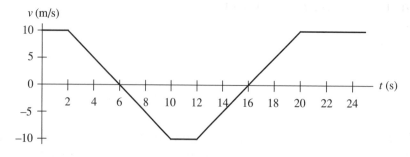

FIGURE 3-50

13. The lower graph in Fig. 3-51 shows the acceleration-versus-time graph for a particle moving along a straight line. There are periods of uniform acceleration with very abrupt jumps between them. Reproduce this graph on your page along with the set

of "empty" graph axes shown at the top of the figure. (You can, if you wish, just photocopy this figure and tape it onto your page.)

a. Use the empty graph axes to plot the velocity-versus-time graph for this particle, assuming that it starts from rest at $t = 0$.

b. Describe, in words, how the velocity-versus-time graph would look if the particle had an initial velocity of 2 m/s.

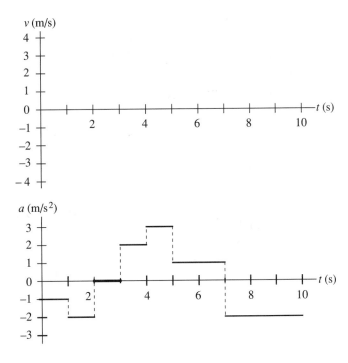

FIGURE 3-51

14. Figure 3-52 shows the acceleration-versus-time graph for an object that starts from rest at $t = 0$. Determine the object's velocity at times $t = 0$ s, 2 s, 4 s, 6 s, and 8 s both by:

a. Calculating the values graphically, from the figure, and

b. Using calculus.

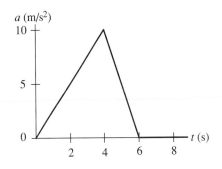

FIGURE 3-52

15. Home experiment: Drop a rubber ball from a height of about 1 m and watch carefully as it bounces. Think about how both the position and the velocity change with time. Then sketch a) a position-versus-time graph, b) a velocity-versus-time graph, and c) an acceleration-versus-time graph for the time interval from the instant you drop the ball until it reaches its maximum height after the *second* bounce. Stack your three graphs vertically so that the time axes are aligned with each other.

16. An object starts from rest at position $x = 0$ at time $t = 0$. Five seconds later, at $t = 5$ s, the object is observed to be at $x = +40.0$ m and to have instantaneous velocity $v = +11$ m/s.
    a. Was the object's acceleration uniform or nonuniform? Explain your reasoning.
    b. Are the kinematic equations of Table 3-6 a valid description of this motion?
    c. Sketch the shape of the velocity-versus-time graph implied by this data. Is the graph a straight line or curved? If curved, is it concave upward or downward?

17. Figure 3-53 shows a motion diagram of a ball rolling along a track, made with a movie camera that exposes two frames of film per second. A meter stick is shown below the track.
    a. Measure, directly from the motion diagram, the x-value of the center of the ball at each position. Place your data in a table, similar to Table 3-4, showing each position and the instant of time at which it occurred.
    b. Make a position-versus-time graph for the ball. Because you have data at only certain instants of time, your graph should consist of dots that are not connected together.
    c. What is the *change* in the ball's position from $t = 0$ s to $t = 1.0$ s?
    d. What is the *rate of change* of the ball's position from $t = 0$ s to $t = 1.0$ s?
    e. What is the *change* in the ball's position from $t = 2.0$ s to $t = 4.0$ s?
    f. What is the *rate of change* of the ball's position from $t = 2.0$ s to $t = 4.0$ s?
    g. What is the ball's velocity on the lower horizontal segment?
    h. What is the ball's velocity on the upper horizontal segment?
    i. Determine the ball's acceleration on the inclined section of the track. Explain how you made this determination.

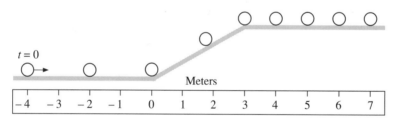

FIGURE 3-53

18. Figure 3-54 shows a ball that rolls along a frictionless track. Each segment of the track is straight, and you can assume that the ball passes smoothly from one segment to the next. For both a) and b) draw three vertically-stacked graphs showing position, velocity, and acceleration versus time. Each graph should have the same time axis and the proportions of the graphs should be qualitatively correct.
    a. The ball has enough speed to reach the top.
    b. The ball does *not* have enough speed to reach the top.

$v_0 > 0$

FIGURE 3-54

19. Draw position, velocity, and acceleration graphs for the ball shown in Fig. 3-55, as described in Problem 18.

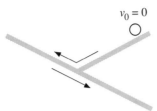

FIGURE 3-55

20. Draw position, velocity, and acceleration graphs for the ball shown in Fig. 3-56, as described in Problem 18.

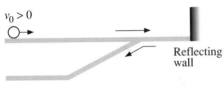

FIGURE 3-56

21. Figure 3-57 shows a set of kinematic graphs for a ball rolling on a track. All segments of the track are straight lines, but some may be tilted. Draw a picture of the track and also indicate the ball's initial condition.

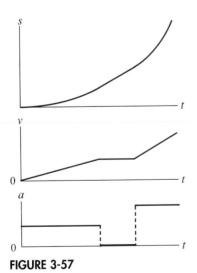

FIGURE 3-57

22. Figure 3-58 shows a set of kinematic graphs for a ball rolling on a track. All segments of the track are straight lines, but some may be tilted. Draw a picture of the track and also indicate the ball's initial condition.

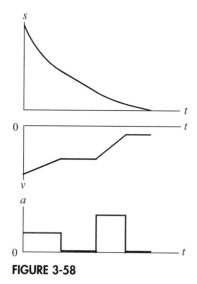

FIGURE 3-58

For Problems 23 and 24 you are given the kinematical equation or equations that are used to "solve" a problem. For each of these, you are to

   a.   Write a *realistic* problem for which this is the correct kinematic equation(s). Make sure that the answer your problem requests is consistent with the equation(s) given.
   b.   Draw the pictorial model and the motion diagram for your problem.
   c.   Finish the solution of the problem.

23. $(10 \text{ m/s})^2 = v_0^2 - 2(9.8 \text{ m/s}^2)(10 \text{ m} - 0 \text{ m})$

24. $v_1 = 0 \text{ m/s} + (20 \text{ m/s}^2)(5 \text{ s} - 0 \text{ s})$

   $x_1 = 0 \text{ m} + (0 \text{ m/s})(5 \text{ s} - 0 \text{ s}) + \frac{1}{2}(20 \text{ m/s}^2)(5 \text{ s} - 0 \text{ s})^2$

   $x_2 = x_1 + v_1(10 \text{ s} - 5 \text{ s}) + \frac{1}{2}(0 \text{ m/s}^2)(10 \text{ s} - 5 \text{ s})^2$

25. A car traveling at a speed of 30 m/s can stop in a minimum distance of 60 m, including the distance traveled during the driver's reaction time of 0.5 s.
   a.   What is the minimum stopping distance for the same car traveling at a speed of 40 m/s? Assume that the maximum deceleration during braking is constant.
   b.   Draw a position-versus-time graph for the motion of the car in part a). Assume the car is at $x = 0$ when the driver first sees the emergency situation ahead that calls for a rapid halt.

26. A 200 kg weather rocket is loaded with 100 kg of fuel and is fired straight up. It accelerates upward at 30 m/s² for 30 s, then runs out of fuel. Ignore any air resistance effects.
   a.   What is the rocket's maximum altitude?
   b.   How long is the rocket in the air?
   c.   Draw a velocity-versus-time graph for the rocket from liftoff until it hits the ground.

27. Figure 3-59 shows the motion diagram for a subway train that has an acceleration and deceleration of 1 m/s^2 and a top speed of 30 m/s. The train stops at each station for 30 s. If the train leaves station A at exactly 8:00:00 A.M., at what time will it arrive at station D?

●   1.0 km   ● 0.6 km ●   2.0 km   ●
A   B   C   D

FIGURE 3-59

28. Ann and Carol are driving their cars along the same straight road. Carol is located at $x = 2.4$ miles at $t = 0$ and drives at a steady 36 mph. Ann, who is traveling in the same direction, is located at $x = 0.0$ miles at $t = 0.50$ hours and drives at a steady 50 mph.
   a. At what time does Ann overtake Carol?
   b. What is their position at this instant?
   c. Draw a position-versus-time graph showing the motion of both Ann and Carol.

29. Bob and Bill are standing on a bridge 50 m above a river. Bob throws a rock straight down with a speed of 20 m/s while Bill, at exactly the same instant of time, throws a rock straight up with the same speed. Ignore air resistance.
   a. How much time elapses between the first splash and the second splash?
   b. Which rock has the faster speed as it hits the water?

30. A ball rolls along the smooth frictionless track shown in Figure 3-60 with an initial speed of 5 m/s. Assume that the ball turns all the corners smoothly with no loss of speed.
   a. What is the ball's speed as it goes over the top?
   b. What is its speed when it reaches the level track on the right side?
   c. By what percentage does its final speed differ from its initial speed? Is this surprising?

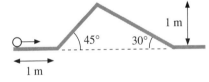

FIGURE 3-60

31. Challenge Problem: The Starship Enterprise returns from warp drive to ordinary space with a forward speed of 50 km/s. To the crew's great surprise, a Klingon ship is 100 km directly in front of them, traveling in the same direction at a mere 20 km/s. Without evasive action, the Enterprise will overtake and collide with the Klingons in just slightly over 3 s. The Enterprise's computers react instantly to brake the ship. What is the magnitude of the acceleration with which the Enterprise must slow to just barely avoid a collision with the Klingon ship?

(**Hint:** As part of your initial qualitative analysis you should draw a single position-versus-time graph showing the motions of both the Enterprise and the Klingon ship. Let $x = 0$ be the location of the Enterprise as it returns from warp drive. How do you show graphically the situation in which the collision is "barely avoided?" Once you decide what it looks like graphically, how do you then express that situation mathematically?)

[**Estimated 10 additional problems for the final edition.**]

Chapter 4

Force and Motion

L O O K I N G
B A C K
| Sections 2.2–2.3 |

4.1 Dynamics

The science of motion is called **mechanics**. In the first three chapters you have seen how to *describe* an object's motion with words, pictures, graphs, and equations. We can now turn our attention to the more fundamental task of inquiring as to the *cause* of motion. This aspect of mechanics is called **dynamics**. The principles of dynamics were first formulated by Isaac Newton. Today we call these principles **Newton's laws of motion**, and the entire subject of how an object moves in response to forces is called **Newtonian mechanics**. We will begin our study of dynamics qualitatively in this chapter, then quantitatively in the next. These two chapters will establish the foundations of the subject. Much of the rest of this text will then be devoted to working out the multitude of implications that flow from Newton's laws of motion.

This chapter will introduce a significant new concept: experimental evidence. All of kinematics is a matter of *definition*. We defined what we meant by position, velocity, and acceleration, and then dressed those definitions in mathematical clothing. Kinematics is thus a language; it describes the "how" of motion. Kinematics is neither "true" nor "false"—it is not a theory. It is just as capable of describing "unrealistic" motion as it is "realistic" motion.

Our overall goal in physics is to understand how and why the universe behaves as it does. We need to be able to distinguish realistic motion from unrealistic motion. Or stated another way, we want to go beyond simply describing *how* something moves; we also want to explain *why* it moves as it does. Descriptions are embodied in language, even if that language is mathematical, as is the case with kinematics. But explanations are embodied in *theories*. Further, theories themselves are based on experimental evidence of how the universe actually behaves. Newtonian mechanics is a theory about how motion occurs as a consequence of forces. No amount of pure thinking would lead us to Newton's laws—we must instead appeal to the evidence of experiments.

A difficulty in learning physics is that a textbook is not an experiment. The book can assert that a certain experiment will have a certain outcome, but you may not be convinced unless you see or do the experiment yourself. Newton's laws are frequently contrary to "common sense," and so unfamiliarity with the actual evidence for them is a major source of difficulty for many students. You will have an opportunity through lecture demonstrations and in the laboratory to see for yourself much of the evidence supporting Newtonian mechanics. This is an essential activity. Physics is not an arbitrary collection of definitions and formulas, but a consistent theory as to how the universe really works. It is only with experience and evidence that we learn to separate physical fact from fantasy.

4.2 Force

[**Photo suggestion: Person pushing or pulling on a heavy object.**]

You know, from long years of experience, that if you shove a box it will slide across the floor. If you pull a string, whatever is attached to the string will start to move. There is a sense that some sort of *force* is required to move these objects. The dictionary definition of force refers to "an exertion of strength or power," and pushing boxes or pulling strings would certainly qualify as force. This definition, as it stands, is much too vague to form the basis of a scientific theory, yet the observation that force and motion are connected suggests that this would be a good direction to explore. The two major issues at the heart of dynamics, issues that this chapter will begin to examine, are:

1. What is force?
2. What is the connection between force and motion?

Our common sense idea of a **force** is that it is a *push* or a *pull*. While we will refine this idea as we go along, it is an adequate starting point. We can say that an object that experiences a push or a pull has a force exerted on it. Notice that we have chosen the wording here carefully, referring to "a force," rather than simply "force." The undifferentiated word "force" is too vague for scientific purposes. It might mean "military force," or it might simply be a synonym for "powerful" (as in "she gave a forceful speech"). We want to think of a force as a very specific *singular action*, so we can talk about a single force or perhaps about two or three individual forces that we can clearly distinguish. Hence, the specific idea of "a force" acting on an object.

Implicit here, as part of our concept of force, is that a force *acts on an object*. From the object's perspective, it experiences a force *exerted* on it. In other words, pushes and pulls are applied *to* something—an object. They do not exist in isolation from the object that experiences them.

If you push an object, you can either push gently or very hard. Similarly, you can either push left or right, or up or down. To quantify a push, we need to specify both a magnitude *and* a direction. It should thus come as no surprise that a force is a vector quantity. Our symbol for a force will be the vector symbol $\vec{F}$.

We can identify two basic classes of forces: contact forces and long-range forces. **Contact forces** are forces that act on an object only by touching it at a point of contact. A string must be tied to an object before it can pull that object. A compressed spring must be in contact with an object before it can push that object. The majority of forces that we will examine are contact forces. **Long-range forces** are forces that act on an object without physical contact. You have undoubtedly held a magnet over a paper clip and seen the paper clip leap up to the magnet. If you bring two magnets together, you find that they either want to repel each other or else they are pulled together. The pushes and pulls of magnetism operate through space, without physical contact between the magnet and the object on which the magnetic force is exerted. Magnetism is an example of a long-range force. The most common long-range force is the force of gravity. A coffee cup released from your hand is somehow pulled to the earth despite the fact that it is not in contact with the earth.

An important aspect of a force, regardless of whether it is a contact force or a long-range force, is that a force has an **agent**. An agent is "something that acts or exerts power," so what we are saying here is that forces have specific, identifiable *causes*. As you throw a ball, it is your hand in contact with the ball that is the agent, or the cause, of the force exerted on the ball. *If* an object has a force being exerted on it, then you must be able to identify a specific cause (i.e., the agent) of that force. Conversely, a force is not exerted on an object *unless* you can identify a specific cause or agent. While this idea may seem like it is stating the obvious, you will find it to be a powerful tool for avoiding some common misconceptions about what is and is not a force.

A somewhat more subtle concept is that the agent of a force is the *immediate* cause, as seen from the perspective of the object experiencing the force. Consider an example to make this clear. Suppose Mike attaches a rope to a crate and then uses the rope to drag the crate across the floor. What is the cause of the force acting on the crate? You might be tempted to say that it is Mike. But Mike is not in contact with the crate—only the rope is! From the crate's perspective, the force is the *contact force* of the rope on the crate. The response of the crate will be the same regardless of whether the rope is pulled by Mike or by Michelle or by a mechanical winch. So we are distinguishing the *immediate cause* (the rope attached to the crate) from the *final cause* (Mike). This distinction can help clarify a common misconception about force, namely that only animate objects (people, animals, etc.) can exert forces. For physics purposes, we define force as being that push or pull exerted by the *immediate* agent, which could be either animate (your hand) or inanimate (the rope).

You can use a pictorial method to show how forces are exerted on objects. Because we are using the particle model, in which objects are treated as very small particles, the process of drawing force diagrams is fairly straightforward. Here is how it goes:

1. Represent the object as a particle.

2. Draw the force vector as an arrow pointing in the proper direction and with a length proportional to the size of the force.

3. Place the tail of the force vector on the particle.

Rule 3 may seem contrary to what a "push" should do. Keep in mind, though, that moving a vector around on the page does not change the vector as long as the length and angle do not change. The vector $\vec{F}$ is the same regardless of whether the tail or the tip is placed on the particle. The reason for our choice of using the tail will become clear when we consider how to combine several forces. For now let's look at a few examples of force diagrams.

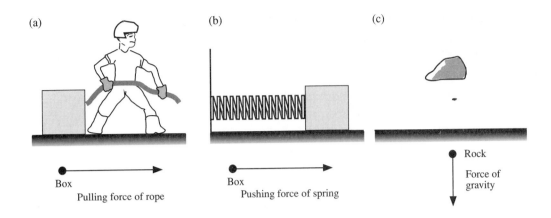

FIGURE 4-1 Three force diagrams showing a) the force of a rope pulling on a box, b) the force of a spring pushing on a box, and c) the force of gravity pulling down on a rock.

Figure 4-1 shows three objects experiencing a force. In Fig. 4-1a, the rope (not the person!) is the immediate cause in contact with the box, so the force diagram shows the box as a particle and the force of the rope. Figure 4-1b is similar to Fig. 4-1a except that the force is a pushing force from a compressed spring. Even though this force is a pushing force, the force diagram has the *tail* of the force vector on the object. Finally, Fig. 4-1c shows the long-range force of gravity pulling down on a rock.

A word of caution: In the particle model, objects do *not* exert forces on themselves. A force on an object will always occur because of an agent or cause external to the object. Now there are certainly objects that have internal forces (think of all the forces inside the engine of your car), but the particle model is not valid if you need to consider those internal forces. If you are going to treat your entire car as a particle and look only at the overall motion of the car *as a whole*, that motion will be a consequence of external forces acting on the car, not of internal forces within the car. Stated another way, forces act *on* an object, they are not properties *of* an object. Gravity and air resistance are forces that act on your car, but they are not properties of your car. Objects do possess certain properties, such as mass, velocity, temperature, and charge, that characterize the object itself. Force, however, does not fall into this category. Forces are external influences acting on the object, not characteristics that the object possesses.

4.3 Force, Motion, and Newton's Second Law

The fundamental question we want to answer is: "How does an object move when a force is exerted on it?" The only way to find the connection between force and motion is to do experiments. To do such experiments, however, we need a way to reproduce the same amount of force again and again. We also need to compare two forces to each other. In other words, we need a *scale* of forces.

Let's start with a scale you can visualize. Imagine putting a rubber band over your finger and stretching the rubber band a certain distance that you can measure—say 2 inches. Your finger "feels" the pulling force exerted on it by the rubber band. This is a reproducible force, because you can always use a ruler to stretch the rubber band the same distance. We can call the amount of force exerted by one rubber band stretched this standard distance "one rubber band" of force. For the time being, this is our unit of force. Let's abbreviate it as "1 r.b."

Now attach the rubber band to some other object and stretch it the same distance. This object must be experiencing the same amount of force—1 r.b.—as did your finger. So you can apply a known and reproducible force to any object you wish.

Suppose you put two identical rubber bands around your finger and pull the standard distance. What happens to the force you feel? How much force would you feel if you use three rubber bands, or four? If you are not sure, find a few rubber bands and try this—the experience will give you a much better understanding of force. You will discover that the more rubber bands you use, the stronger the pulling force you feel. This result is not surprising—it makes sense that if two rubber bands are each pulling equally hard on your finger, the "net pull" is twice that of one rubber band. The same will be true for rubber bands pulling on any other object. You can apply a force of N r.b. to an object by attaching N identical rubber bands to it, side-by-side, and stretching each the standard distance that you can measure with your ruler.

Now that we have a way of applying and measuring a known force, let's apply a single *constant* force to an object and then measure the object's motion in response to that force. You have to be a bit careful, because you know that the force of gravity may be acting on our object to pull it in the vertical direction. If you pull the object vertically up or down with rubber bands, the object will respond to both the rubber-band force *and* to the force of gravity, so we might not be able to distinguish the effects of the two forces. It makes more sense to pull horizontally.

To keep the motion horizontal, we will need to support the object on a table or other horizontal surface. There might be some friction between the object and the surface. Is friction a force? That is a question we will return to later, but to be on the safe side (because we want to study the motion in response to only our rubber-band force) let's eliminate friction. You could, for example, pull a smooth block of ice over a smooth ice surface. Or you could support an object on a cushion of air. These situations are not absolutely frictionless, but they would make the effects of friction sufficiently small for us to conduct a valid experiment.

Once we have arranged our experiment to measure just the horizontal motion of our test object under the influence of just the rubber-band force, we must make sure that the applied force is held *constant*. That means that you need to keep a constant amount of stretch in the rubber band. If you stretch the rubber band and then release the object, it moves toward your hand. But as it does so, the stretch in the rubber band diminishes and the pulling force decreases. To keep the force constant, you must *move your hand forward* at just the right speed to keep the stretch in the rubber band from changing!

[**Photo suggestion: Strobe photo of an object accelerating under a constant force.**]

Let's start the experiment and see what happens. Attach one rubber band to the object and stretch it the standard distance so that the force is 1 r.b. Next, release the object, moving your hand forward so as to keep the force constant at 1 r.b., and measure the motion of the object using a movie camera. This is illustrated in Fig. 4-2. You will find that the object moves with *constant acceleration*. This could not have been anticipated in advance, but nonetheless that is what happens. Once you know that the motion is one of constant acceleration, all the constant-acceleration kinematics you learned in Chapter 3 comes into play: the velocity will increase linearly with time, the position will increase quadratically with time, and so on.

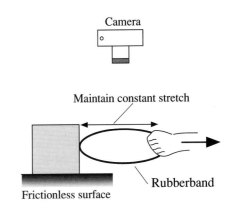

Next consider what will happen if you increase the force by using several rubber bands. Measure the acceleration produced by a 1 r.b. force and call it a_1. Then add a second rubber band to increase the applied force to 2 r.b. What happens? Because the force is constant, you will find that the object's acceleration is again constant. Call the acceleration produced by a 2 r.b. force a_2. You can similarly measure a_3 produced by a 3 r.b. force, a_4 produced by a 4 r.b. force, and so on. When you do so, you find another important result: $a_2 = 2a_1$, $a_3 = 3a_1$, $a_4 = 4a_1$, and so on. In other words, doubling the force produces twice the acceleration, tripling the force produces three times the acceleration, and so on.

FIGURE 4-2 Measuring the motion of an object experiencing a force of 1 r.b.

Figure 4-3 shows a graph of force-versus-acceleration. It demonstrates that force and acceleration are *proportional* to each other. We write this mathematically as $F \propto a$, where the symbol $\propto$ means "is proportional to." This can also be written as $F = ca$, where c is called the *proportionality constant*. Proportionality is a stronger relationship than merely being "linear" because proportionality indicates the specific linear relationship that has zero *y*-intercept (i.e., the graph passes through the origin). The proportionality constant c is the slope of a linear graph that passes through the origin.

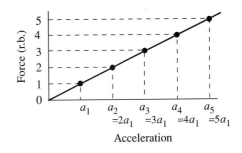

FIGURE 4-3 Graph of force-versus-acceleration for forces of 1 r.b. to 5 r.b.

Recall that force is a vector. When we write the scalar symbol F, we mean the vector *component* of the force vector along the direction of motion. This is consistent with a being the vector component of acceleration along the axis of motion. Both a and F can be negative—which they would be if the force vector pointed toward the left, resulting in an acceleration toward the left. The *magnitude* of the force vector $\vec{F}$ will be designated, as usual, by $|\vec{F}|$. It is always positive. Make sure you understand the distinction between component F and magnitude $|\vec{F}|$.

We have established, *via experiment*, that the force is proportional to the acceleration for forces applied to one single object. What happens if a 1 r.b. force is applied to a different object, having a different size? To find out, let us continue our experiment by doubling the size of the object. You can do this by taking two copies of the original object and gluing them together. Then measure the acceleration of this object for the same forces of 1 r.b., 2 r.b., and so on. You can also glue three objects together to create an object three times the size of the original, and you can cut an object in half to create one that is half the size. After doing several such experiments, you would find the following results:

$$a_F(\tfrac{1}{2} \text{ object}) = 2a_F(1 \text{ object})$$

$$a_F(2 \text{ objects}) = \tfrac{1}{2} a_F(1 \text{ object})$$

$$a_F(3 \text{ objects}) = \tfrac{1}{3} a_F(1 \text{ object})$$

$$a_F(n \text{ objects}) = \tfrac{1}{n} a_F(1 \text{ object})$$

where a_F means "the acceleration as a result of force F." A "double object" has only half the acceleration of the original object when both experience the same force. A "half object," by contrast, has twice the acceleration of the original. If we add this new data to Fig. 4-3, we get the graph shown in Fig. 4-4.

It is an experimental fact that the force applied to an object is proportional to the object's acceleration. It is also an experimental fact, as seen in Fig. 4-4, that the proportionality constant c between force and acceleration—the slope of the line—is a property *of the object*. This property, which we have loosely called the "size" of the object, is properly the called the *inertial mass* of the object, where

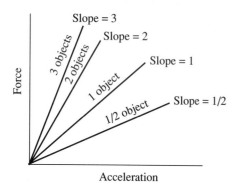

FIGURE 4-4 Force-versus-acceleration graphs for different sized objects. Notice that the slope is equal to the size of the object.

inertial mass = m = slope of the force-versus-acceleration graph.

More generally, we usually just refer to the inertial mass as "the mass." Mass is an *intrinsic* property of an object. It is the property that determines the object's acceleration in response to a force applied to the object. Mass is *not* the same thing as weight.

Because the slope of the force-versus-acceleration graph for a double object is twice the slope for a single object, it must be the case that $m(2 \text{ objects}) = 2m(1 \text{ object})$. Similarly, $m(3 \text{ objects}) = 3m(1 \text{ object}) = m(1 \text{ object}) + m(1 \text{ object}) + m(1 \text{ object})$. We find—experimentally!—that the total mass of several combined objects is just the arithmetic sum of their individual masses. Although this is common sense, it is worth noting that there is a *reason* why mass has this additive property.

Having now defined the inertial mass as $m = F/a$, where F/a is the slope of the force-versus-acceleration graph, we can summarize in a single equation the results of our experiments to find how an object moves in response to a single applied force F:

$$F = ma.$$

This result is known as **Newton's second law.** Newton himself did no experiments to find this law, but he relied on the experimental work of Galileo and others. The reasons it is called his second law are historical; it does not indicate second in importance. We can see directly from this equation that acceleration and force are proportional to each other, and that the proportionality constant is none other than the object's mass. If we rewrite the second law in the form

$$a = \frac{F}{m},$$

then we see that it correctly describes the experiments in which the force was kept constant but the mass was varied: For constant force, acceleration is inversely proportional to mass.

Force and acceleration are both *defined* quantities. There was no reason to suspect that there should be any simple relationship between force and acceleration. Yet there it is—$F = ma$—a simple but exceedingly powerful equation relating the two. This result came not from definitions, not from mathematical manipulations, but from experiment. It says that "this is the way the world actually works." Newton's work, preceded to some extent by Galileo's, marks the beginning of a highly successful period in the history of science during which it was learned that the behavior of physical objects can be described and predicted by mathematical relationships. While some relationships are found to apply only in special circumstances, others seem to have universal applicability. These universally-valid equations, which appear to apply at all times and under all conditions, have come to be called "laws of nature." Newton's second law is our first law of nature; you will meet others as you go through this book.

Combining Forces

It is not obvious what happens if we apply more than one force to an object. Figure 4-5, for example, shows a box being pulled by two ropes, each exerting a force on the box. How will it move? Experimentally, we find that when several individual forces $\vec{F}_1$, $\vec{F}_2$, $\vec{F}_3$, ... are exerted on an object, they combine to form a **net force** given by the *vector* sum of the individual forces:

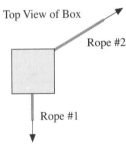

FIGURE 4-5 Two forces applied to an object.

$$\vec{F}_{net} = \sum_{i=1}^{N} \vec{F}_i = \vec{F}_1 + \vec{F}_2 + ... + \vec{F}_N. \qquad (4\text{-}1)$$

Mathematically, this is called the **superposition of forces**. The net force is also frequently called the **resultant force**.

When several forces are exerted on an object of inertial mass m, the object experiences an acceleration

$$\boxed{\vec{a} = \frac{\vec{F}_{net}}{m}. \quad \text{(Newton's second law)}} \qquad (4\text{-}2)$$

Equation 4-2 is the full statement of Newton's second law. We will postpone discussing the mathematical aspects of the second law until the next chapter, but we will note that there are two implications of the fact that this is a *vector* equation. First, the *magnitude* of the acceleration $|\vec{a}|$ is given by the *magnitude* of the net force $|\vec{F}_{net}|$ divided by the mass: $|\vec{a}| = |\vec{F}_{net}|/m$. Second, and equally as important, the acceleration vector $\vec{a}$ points in the *same direction* as the net force vector $\vec{F}_{net}$.

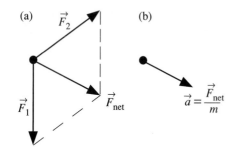

FIGURE 4-6 a) Force diagram showing the net force as the superposition of forces $\vec{F}_1$ and $\vec{F}_2$. b) The direction of the acceleration $\vec{a}$ is the same as the direction of $\vec{F}_{net}$.

Figure 4-6 completes the analysis of the box in Figure 4-5. First we represented the box as a particle, then we graphically add the two force vectors to find the net force. Finally, we can deduce that the box will accelerate in the direction of $\vec{F}_{net}$ with an acceleration of magnitude

$$|\vec{a}| = \frac{|\vec{F}_{net}|}{m} = \frac{|\vec{F}_1 + \vec{F}_2|}{m}.$$

You can begin to see now why we put so much emphasis in Chapters 2 and 3 on understanding vectors and acceleration. The entire strategy of dynamics can be summarized as follows:

Identify all forces on an object
$\Downarrow$
Determine the object's acceleration from

$$\vec{a} = \frac{\vec{F}_{net}}{m}$$

$\Downarrow$
Deduce the subsequent motion from the equations of kinematics

We will be following this strategy for many chapters to come.

Force and Mass Scales

So far we have been using "rubber bands" as our unit of force. Because the inertial mass of an object is the slope of its force-versus-acceleration graph, the units of mass must be force units divided by acceleration units. If force is measured in units of "rubber bands," then mass will have units of "(r.b.)/(m/s^2)." These are obviously not practical units for mass. If we define the unit of force, as we did, then we have no control over the units of mass. It makes more sense to define the unit of mass and to adopt a consistent unit of force. The SI unit of mass is defined as: One *kilogram* is the mass of a platinum-iridium alloy cylinder stored in a special vault in Paris. The abbreviation for kilogram is kg.

This definition is rather old-fashioned compared to the definitions of the meter and the second. Physicists have hopes that we will someday base the unit of mass on the masses of atoms, but we do not yet have the technology to do this with sufficient accuracy. Note that micrograms (μg), milligrams (mg), and grams (g) are all commonly used units, but you must convert all masses to kilograms if you want the results of your calculations to have SI units.

With acceleration and mass having defined units, we can now define the basic unit of force to be "the force that causes a 1 kg mass to accelerate at 1 m/s^2." From the second law, this force is

$$1 \text{ basic unit of force} \equiv (1 \text{ kg}) \times (1 \text{m/s}^2) = 1 \frac{\text{kg m}}{\text{s}^2}.$$

This basic unit of force is called a *newton*, and it is defined as:

> One **newton** is the force that must be exerted on a 1 kg mass to cause it to accelerate at 1 m/s^2. The abbreviation for newton is N. Mathematically, 1 N = 1 kg·m/s^2.

The newton is a *secondary unit*, defined in terms of the *primary units* of kilograms, meters, and seconds. We will introduce other secondary units as needed.

It is hard to acquire an "intuition" about how big a 1 N force is. Nonetheless, it is important to have a feeling for what the appropriate size of forces should be. Table 4-1 shows some typical forces. As you can see, "typical" forces on "typical" objects are likely to be in the range 1 N–10,000 N. Forces less than 1 N are too small to consider unless you are dealing with very small objects. Forces greater than 10,000 N would only make sense if applied to very massive objects.

TABLE 4-1 Approximate magnitude of some typical forces.

Force	Approx. magnitude (newtons)
weight of a U.S. quarter	0.05 N
weight of a 1 pound object	5
weight of a 110 pound person	500
propulsion force of a car	5000
thrust force of a rocket motor	5,000,000

4.4 Inertia and Newton's First Law

As we have remarked in earlier chapters, Aristotle and his contemporaries were very interested in motion. One question they asked was, "What is the 'natural state' of an object if left to itself?" It does not take a multimillion-dollar research program to see that every moving object on earth, if left to itself, eventually comes to rest. Aristotle concluded that the "natural state" of an object is one of being at rest. An object at rest requires no explanation—it is doing precisely what comes naturally to it. A moving object, though, is not in its natural state and thus requires an explanation: "Why is this object moving? What is keeping it going and preventing it from being in its natural state?"

Galileo, as part of his research on the motion of spheres rolling along inclined planes, reopened the question of the "natural state" of objects. He suggested focusing on the *limiting case* in which resistance to the motion (i.e., friction or air resistance) is zero. This is an idealization that may not be realizable in practice, but Galileo had asserted previously, with great success, that the idealized case can establish a *general principle*. After many careful experiments in which he minimized the influence of friction, Galileo came to a conclusion about the "natural state" that was in sharp contrast to Aristotle.

For the case of spheres rolling along a track, Galileo had the following to say:

> It may also be noted that whatever degree of speed is found in the moveable, this is by its nature indelibly impressed on it when external causes of acceleration or retardation are removed … From this it likewise follows that motion in the horizontal is eternal, since it is indeed equable if it is not weakened or remitted.

What is he saying here? The acceleration of an object has an "external cause"—we would say "a force." When the force is removed, the object continues to have a velocity (it is "indelibly impressed" on the object). Furthermore, if the motion is *horizontal* then the object will continue to move with that velocity forever ("eternal"). In other words, the "natural state" of an object, if left alone, is *uniform motion* with constant ("equable") velocity! This does not happen in practice because friction or air resistance are "retardations" that prevent the object from being left alone. Galileo greatly extended the concept of the "natural state," and in the process the concept of "at rest" lost its special status.

Galileo restricted his discussion to motion along horizontal surfaces. It was left to Newton to generalize this result. Today we call this generalization Newton's first law of motion:

> **Newton's first law**: An object that is at rest will remain at rest, or an object that is moving will continue to move in a straight line with constant speed, if and only if the net force acting on that object is zero.

Newton's first law is also called the law of inertia. **Inertia** is the tendency of an object to resist change. If the object is at rest, it has a tendency to remain at rest. If it is moving, it has a tendency to continue moving with the *same velocity*.

Notice the "if and only if" nature of Newton's first law. If the object moves with constant velocity, then we can conclude that there is no net force acting on it. Conversely, if there is no net force acting, we can conclude that the object will have constant velocity (not just constant speed—the direction remains constant, too!). The first law completes our concept and definition of force. It answers the question, "What is a force?" If an "influence" on an object causes that object's velocity to change, then the influence is a force. It is very important to note that the first law refers to *net* force. There may very well be forces exerted on the object, but all that matters is the net force when the individual forces are summed.

Newton's first law changes the question the ancient Greeks were trying to answer. They wanted to know, "What causes an object to move?" The first law says *no "cause" is needed for an object to move!* Instead, uniform motion is the object's natural state. Nothing at all is required for it to remain in that state. The proper question, according to Newton, is "What causes an object to *change* its motion?" Newton also gave us the answer: a *force* is what causes an object to change its motion.

The preceding paragraph contains the essence of Newtonian mechanics. This new perspective on motion, however, is totally contrary to our common experience. We all know perfectly well that you must keep pushing an object—exerting a force on it—to keep it moving. Newton is asking us to change our point of view and to consider motion *from the object's perspective* rather than from our personal perspective. As far as the object is concerned, our push is just one of several forces acting on it. Others might include friction, air resistance, or gravity. Only by knowing the *net* force can we determine the object's motion.

An object on which the net force $\vec{F}_{net}$ is equal to zero is said to be in **equilibrium**. This can occur either if the object is at rest (static equilibrium) or if it is moving with constant velocity (dynamic equilibrium). We will deal frequently with equilibrium situations in which the equation $\vec{F}_{net} = 0$ establishes a relationship that must be obeyed among the various forces. Newton's first law can be rephrased as "An object that is in equilibrium will either remain at rest (static equilibrium) or continue to move with constant velocity (dynamic equilibrium)."

It may seem as if Newton's first law is just a special case of Newton's second law. After all, the equation $\vec{F}_{net} = m\vec{a}$ tells us that an object stays at rest or continues to move with constant velocity, both cases of $\vec{a} = 0$, if and only if $\vec{F}_{net} = 0$. The difficulty is that the second law *assumes* that we already know what forces are. The purpose of the first law is to *identify* a force as something that disturbs a state of equilibrium. The second law then goes on to describe how the object responds to this force. Thus from a *logical* perspective, the first law really is a separate statement that must precede the second law. But this is a rather formal distinction. From a pedagogical perspective it is better—as we have done—to use a common-sense understanding of force and start with Newton's second law.

4.5 A Short Catalog of Forces

There are many forces we will deal with over and over. The purpose of this section is to introduce you to some of them and to describe situations where they are—or are not—important. Many of these forces have special symbols. As you learn the major forces, make sure you also learn the appropriate symbol for each. Whether the symbol is uppercase (a capital letter) or lowercase is also important: t stands for time but T stands for tension.

Gravity

Probably the most common force of all, present in nearly every problem, is the force of gravity. For objects near the surface of the earth (or another planet), the force of gravity is directed vertically downward, toward the earth's center. We call the gravitational force on an object its **weight**. Note, and this is extremely important, that *weight is not mass*. Mass (a scalar) is a measure of how an object accelerates in response to a force. Weight (a vector) is a force acting on the object due to the gravitational pull of the planet. There is a connection between weight and mass, which we will explore in Chapter 5, but they are most definitely not the same thing. Confusing the two is a common error, so make a special note of this. Figure 4-7a shows a situation where weight would be a force acting on an object, and Fig. 4-7b shows the appropriate force diagram. The symbol for weight is $\vec{W}$, and the vector *always* points straight down. Weight is a long-range force.

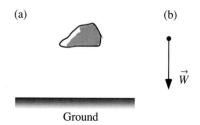

(a)

(b)

$\vec{W}$

Ground

FIGURE 4-7 a) The force of gravity acts on an object. b) A force diagram shows the weight force $\vec{W}$.

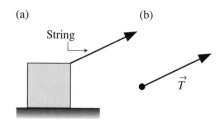

(a)

(b)

String

$\vec{T}$

FIGURE 4-8 a) A string exerts a tension force on a object. b) A force diagram for the tension force $\vec{T}$.

Tension

[**Photo suggestion: An object suspended by a cable under tension.**]

When a string or rope or wire pulls on an object, it exerts a contact force that we call **tension**. Tension is *always* a pulling force, never a pushing force. (Imagine trying to push something with a string!) Further, the direction of the tension force is always *parallel* to the string or rope. Figure 4-8 shows a situation where tension is present. The symbol for tension is a capital $\vec{T}$.

Springs

[**Photo suggestion: A compressed spring pushing an object.**]

Springs provide one of the most commonly used contact forces in physics and engineering. The spring force differs from tension in that springs can both pull (when stretched) *and* push (when compressed). While we often think of a spring as a metal coil that can be stretched or compressed, this is only one type of spring. Hold a ruler or any other thin piece of wood or metal by the ends and bend it slightly—it flexes. When you let go, it "springs" back to its original shape. This is just as much a spring as is a metal coil. We will have a great deal to say about springs later in the course, because they are the basis of *oscillations*. Figure 4-9a shows a compressed spring that is exerting a pushing force on an object, while Fig. 4-9b shows the pulling force of a stretched spring. In both cases, the *tail* of the force vector is placed on the particle in the force diagram. The force, it is important to note, is exerted *on* the object *by* the spring; the spring does not exert a force on itself. There is not a special symbol for a spring force, so we simply use a subscript label: $\vec{F}_{sp}$.

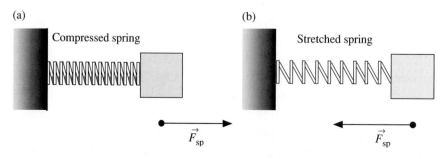

(a)

Compressed spring

$\vec{F}_{sp}$

(b)

Stretched spring

$\vec{F}_{sp}$

FIGURE 4-9 The spring force and force diagrams for a) a compressed spring and b) a stretched spring.

EXAMPLE 4-1 Consider a brick resting on top of a vertically-oriented spring, as shown in Fig. 4-10a. Is the magnitude of the spring force on the brick greater than, less than, or equal to the magnitude of the brick's weight force?

SOLUTION To answer this question, we need to utilize two ideas from this chapter: combining forces and relating the net force to the acceleration. The brick has two forces exerted on it—the pushing force of the spring upward and the weight force downward. These are shown, using the particle model, in Fig. 4-10b. The net force is the *vector* sum of these:

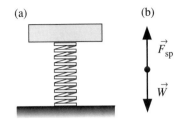

FIGURE 4-10 a) The brick and spring for Example 1. b) A force diagram for the brick shows the upward push of the spring and the downward pull of gravity.

$$\vec{F}_{net} = \vec{F}_{sp} + \vec{W}.$$

Now we can use Newton's laws. The brick is at rest—in static equilibrium—so the first law tells us that the net force acting on the brick is zero: $\vec{F}_{net} = 0$. (This also follows from the second law, because an object at rest has $\vec{a} = 0$, which implies $\vec{F}_{net} = 0$.) Because the two forces point in opposite directions, they will add to give a net force of zero *if* they are equal in magnitude. So we conclude that $\vec{F}_{sp} = -\vec{W}$ and that $|\vec{F}_{sp}| = |\vec{W}|$.

Normal Force

Example 4-1 illustrated a fairly obvious idea—if you place an object on a vertically-oriented spring, the spring *compresses* and, as a consequence of the compression, exerts an upward force on the object. If we were to use a stiffer spring, it would show less compression but still exert a spring force of magnitude equal to the object's weight. If the spring were extremely stiff, you might not even notice the compression unless you used sensitive measuring instruments. Nonetheless, the spring would compress ever-so-slightly and cause an upward spring force to be exerted on the object.

Figure 4-11 shows a similar situation: a flexible bar (a meter stick is a good example) is supported at both ends and then a heavy object placed on its center. The bar flexes a little and, as a result, exerts an upward *spring force* on the object. If you are not sure about this, take a meter stick, or even a thin ruler, and support each end on a book. Then press down on the center with your fingertip. As you flex the meter stick, you can *feel* the force it is exerting upward on your finger. If you push down harder to increase the flex, you feel a larger force pushing back up on your fingertip. This is the spring force that you feel, exactly as if you had pressed down on the top of a regular coil spring and felt it pushing your finger upward. This same upward force is exerted on any object that flexes the meter stick.

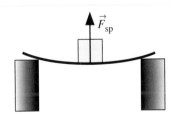

FIGURE 4-11 Upward spring force of a flexible bar.

Now take the same object an place it on top of a sturdy table. You may not *see* the table flex, but does it? If you were to use a very powerful microscope to look at the table, you would "see" that it is made of *atoms* that are joined together by *molecular bonds*. These bonds hold the atoms together as a solid; without them, the table would be a gas. Molecular bonds are not rigid, solid connections between the atoms. A better model of a solid, which we will be able to justify later in this book, is to think of molecular bonds as *springs* holding the atoms together, as in Fig. 4-12a. These are *very* stiff springs, to be sure, but they can indeed compress or stretch. If you stretch a bond too far, it will break—that is what happens when you cut something with a knife or bend it until it breaks.

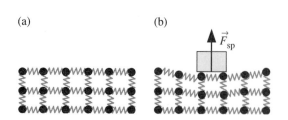

FIGURE 4-12 a) Atomic-level view of a solid, with molecular bonds as springs. b) Compression of the molecular springs, causing an upward force on the object.

When we place a large object on a table the object deforms the molecular springs in the table, as shown in Fig. 4-12b. This causes the table to flex. As a consequence, the molecular springs *push upward* on the object. The size of the flex may be very small, because molecular springs are so stiff, but it is not zero. Optical techniques, using laser beams reflected from the table surface, demonstrate that the table really does bend slightly. We say that "the table" exerts the upward force, but it is important to understand that the springy molecular bonds are what *really* do the pushing. Similarly, an object resting on the ground deforms the molecular springs holding the ground together and, as a consequence, the ground *pushes up* on the object.

[**Photo suggestion: Someone leaning against a wall.**]

We can extend this idea. Suppose you place your hand on a wall and lean against it. Does the wall exert a force on your hand? As you lean, you deform the molecular springs in the wall and, as a consequence, they *push outward* against your hand. So the answer is "yes," the wall does exert a force on you that is really no different than the force exerted by a table on an object resting on the table.

In one case (the object on table) the force exerted by the surface is vertical. In the other case (the hand on wall), the force exerted by the surface is horizontal. But in both cases, the force exerted on an object that is pressing against a surface is in a direction *perpendicular* to the surface. Mathematicians refer to a line that is perpendicular to a surface as being *normal* to the surface. In keeping with this terminology, we define the **normal force** as that force exerted by a surface against an object that is pressing against the surface. The symbol for the normal force is $\vec{N}$. It is *always* perpendicular (i.e., normal) to the surface.

Make a special note not to confuse the symbol $\vec{N}$ used to represent the normal force with the very similar N used to abbreviate newtons. A recurring difficulty in science and engineering is that we use the same letter or symbol to represent different things. These two different uses of the letter N are our first encounter with this, but there will be others in the near future. When you see a symbol, you must observe the *context* in which it is used to understand its meaning.

We have spent a lot of time describing the idea behind the normal force because it is a force that many students have a difficult time understanding. The normal force is a very real force arising from the very real stretching and compressing of molecular bonds. It is, in essence, just a spring force, but one exerted by a vast number of microscopic springs acting at once. For example, it is the normal force that prevents you from passing right through the chair you are sitting in. As you press against the chair, the molecular bonds deform and exert the upward spring force that supports your weight—just like the spring supported the brick in Example 4-1. The normal force is also responsible for the "solidness" of solids. If you bang your head into a door, it is the normal force that causes the pain and the lump! Your head can then tell you that the force exerted on it by the door was very real.

Figure 4-13a shows an object on an inclined surface. The deformation of the molecular bonds can only press *outward* from the surface, not parallel to the surface, so Fig. 4-13b shows a normal force $\vec{N}$ that is perpendicular to the surface.

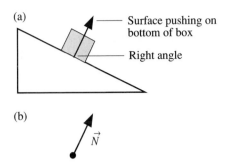

FIGURE 4-13 a) A surface exerts a normal force on an object. b) A force diagram for the normal force $\vec{N}$.

Friction

Galileo arrived at the concept of inertia by considering the ideal case of motion with no friction. Without friction, an object moving across a horizontal surface would continue forever at constant velocity. In the real world, though, objects given an initial velocity slow down and eventually stop. Why? An object that is slowing has an acceleration which is non-zero. In fact, the acceleration vector points opposite the velocity vector. From a Newtonian perspective, the presence of an acceleration indicates the presence of a net force that points in a direction opposite the direction of travel. If the motion is that of an object rolling or sliding over a surface, we call this force **friction**. The symbol for friction is a lowercase $\vec{f}$.

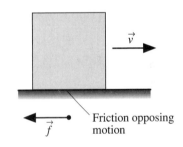

FIGURE 4-14 The friction force.

Figure 4-14 shows the friction force on a moving object. The main thing to notice is that $\vec{f}$ is *opposite to* the direction of motion across the surface.

On a microscopic level, friction arises from electric forces between atoms of the object and atoms of the surface as they pass each other. The rougher the surface is, the more these atoms are forced into close proximity and, as a result, the larger the forces. A detailed treatment of friction is quite complex. We will develop a simple model of friction in the next chapter that will be sufficient for our needs.

Friction at a surface is one example of a resistive force. In general, a force that opposes or resists motion is called a *resistive force*. Another important resistive force is the force due to air resistance. This force is often called a **drag force**, and is symbolized as $\vec{F}_{drag}$. Like friction, $\vec{F}_{drag}$ points in the direction opposite the direction of motion. Air resistance is a contact force, although the contact is across the object's entire surface rather than at a single point. In the particle model, however, the object's dimensions are neglected and the air resistance is just another force acting on the particle.

[**Photo suggestion: A falling leaf.**]

The air resistance can be a large force for objects moving at high speeds. Hold your arm out the window as you ride in a car and feel how the resistance of the air against your arm increases rapidly at high speeds. However, for objects that are heavy and compact (small, hard spheres) and whose speed is not too great, the air-resistance force is fairly small. In order to keep things as simple as possible, you can neglect air resistance unless a problem explicitly asks you to include it. The error introduced into calculations by this approximation will generally be pretty small. More advanced courses will study the implications of air resistance and other drag forces; their inclusion in the problem often results in equations that can only be solved numerically on a computer.

Propulsion Forces

A jet airplane obviously has a force that propels it forward during takeoff. Similarly, when you accelerate your car forward from a stop sign there must be, according to Newton's second law, a forward-directed force on your car. Even something as simple as walking or running requires a force to accelerate you from rest. These forces are called **propulsion forces**. Understanding how propulsion comes about is actually rather tricky, and we will have to postpone that discussion until we have introduced Newton's third law. For now, we will treat propulsion forces as forces that point in the direction of acceleration. For example, in a problem about a rocket being launched, there must be a propulsion force in the upward direction that we could call $\vec{F}_{\text{thrust}}$.

External Forces

Lastly, a common physics problem is to consider the motion of an object under the influence of some unspecified "external" force. Such forces are quite real, even if we are not told their cause, so they need to be included on force diagrams. We generally label such forces $\vec{F}_{\text{ext}}$.

4.6 Identifying Forces

Force and motion problems generally have two basic steps: 1) Identify all of the forces acting on an object, then 2) use Newton's laws and kinematics to determine the motion. The second step in this procedure is mathematical, and it will be the subject of Chapter 5. The first step is the topic for the rest of this chapter.

A typical problem will describe an object that is being pushed and pulled in various directions. Some forces are given explicitly, but many are only implied. In order to proceed, it is necessary to determine what forces act on the object and what the direction and magnitude of those force are (i.e., to identify the force *vectors*). It is also necessary to avoid including forces that do not really exist. Now that you have learned the properties that characterize forces, and seen a catalog of typical forces, we can proceed to develop a step-by-step method for identifying each force in a problem.

Identifying Forces

1. First, divide the problem into "the system" and "the environment." The system is just the object whose motion you wish to study, and the environment is everything else.
2. Draw a picture of the problem, showing the object and everything in the environment. Ropes, springs, surfaces, and so on should all be considered parts of the environment.
3. Draw a closed curve around the system, with the object inside the curve and everything else outside the curve.
4. Locate *every* point on the boundary of this curve where the environment touches or contacts the system. These are the points where the environment exerts *contact forces* on the object. Do not leave any out!
5. Identify by *name* the contact force or forces (there may be more than one) at each point of contact, and label each force with the appropriate symbol. When necessary, use numerical subscripts to distinguish forces of the same type.
6. Identify any long-range forces acting on the object. Name the force and write its symbol beside your picture. For now, the only long-range force we have is weight; other long-range forces, such as electrical forces, will be introduced later.

This procedure is part of the physical model of the problem, and it will be used in Section 4.7 to develop a *free-body diagram*. Let us look at a couple of examples.

EXAMPLE 4-2 A car with a dead battery is being towed up a 30° hill. What forces are being exerted on the car?

SOLUTION First, we can identify the car as the system of interest in this problem. Next, as shown in Fig. 4-15, we draw a picture of the problem, including the rope that is towing the car and the road surface. We then draw a closed curve around the car. After doing so, we see that the car contacts the environment at two points. There is a tension force $\vec{T}$ at the point where the rope contacts the car. At the contact between the surface and the car there are *two* forces: a normal force $\vec{N}$ and a friction force $\vec{f}$. (In accordance with information given in the last section, we neglect air resistance.) Lastly, we include $\vec{W}$, the car's weight due to the long-range force of gravity. We have now identified all of the forces acting on the car.

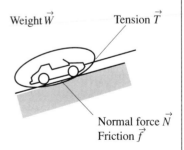

Weight $\vec{W}$ Tension $\vec{T}$

Normal force $\vec{N}$
Friction $\vec{f}$

FIGURE 4-15 Identifying the forces on a car being towed up a hill.

EXAMPLE 4-3   A jet airplane accelerates down the runway for takeoff. Air resistance is not negligible. What are the forces acting on the plane?

SOLUTION: We proceed as in the last example. First, we identify the plane as the system, draw a picture, and draw a closed curve around the plane. There are three points of contact. The normal force $\vec{N}$ and friction $\vec{f}$ at the ground surface are easy to identify. The air contacts the plane at the front, providing force $\vec{F}_{drag}$. (As we noted in the Section 4.5, the contact between the air and the plane is not a "point" of contact, but we can treat it as such in the particle model.) The thrust of the engines is the tricky force. A jet engine burns fuel and produces an exhaust. The molecules of the exhaust are left behind—that is, the exhaust is part of the environment, not part of the system. So the point where the closed curve passes through the engine exhaust, is a point of contact between the system and the environment. The force exerted on the system at that point is the propulsion force $\vec{F}_{thrust}$. Lastly, we again include the weight force $\vec{W}$. These forces are illustrated in Fig. 4-16.

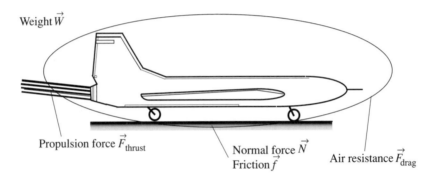

FIGURE 4-16   Identifying the forces on a jet airplane during takeoff.

Some Misconceptions About Force

It is important to identify correctly all the forces acting on an object. It is equally important to not include forces that do not really exist. We have established a number of criteria for identifying forces, of which the two critical ones are: 1) A force has an agent, something tangible and identifiable that causes the force; and 2) forces exist at the point of contact between the cause of the force and the object experiencing the force (except for the few special cases of long-range forces).

We all have had many experiences suggesting that a force is necessary to keep something moving. If we see a moving object, it seems "obvious" that there must be a force responsible for the motion. Consider a bowling ball rolling along on a smooth floor. It is very tempting to assign a horizontal "force of motion" to it in the forward direction. But if we draw a picture and put a closed curve around the ball, *nothing is contacting the ball except the floor*. We cannot identify an agent giving the ball a forward push. According to our definition, then, there is *no* forward "force of motion" acting on the ball. So what keeps it going? Recall our discussion of the first law: *no* cause is needed to keep an object moving at constant velocity. It continues to move forward simply because of its inertia.

Part of the reason for wanting to include a "force of motion" is that we try to view the problem from our perspective as one of the agents of force. If we push a crate across the floor at constant velocity, we certainly have to keep pushing. Newton's laws, though, require that we adopt the object's perspective. The crate experiences *both* our pushing force in the forward direction *and* a friction force in the backward direction. For the crate to move at constant velocity, the *net* force must be zero. Therefore, our pushing force exactly cancels the friction force.

If we stop pushing, the crate moves forward for some distance and stops. It begins decelerating the instant we stop pushing—it is *not* steady speed for a time, followed by an abrupt stop. A steady deceleration until $v = 0$ is exactly what we have learned to expect for a velocity vector in the forward direction and an acceleration vector, due to the friction force, in the backward direction. No "force of motion" is required to maintain the forward motion as it decelerates.

A related problem occurs when we consider a ball that has been thrown. A forward pushing force was indeed required to accelerate the ball *as it was thrown*. But the force is gone the instant the ball loses contact with the thrower's hand—it does not stick with the ball as the ball travels through the air. The ball has no "memory" of previous forces. Once the ball has acquired a horizontal velocity, *nothing* is needed to keep it moving forward with that velocity.

Another common misconception is to consider inertia to be a force. As we noted in Section 4.4, inertia is the tendency of an object to resist change. This is a consequence of Newton's first law, which states that a force is necessary to change what the object is doing. Inertia itself is *not* a force.

Yet another difficulty for many students is concern over forces due to air pressure, which are sometimes confused with gravitational forces. The difficulty here is knowing "too much," but not knowing it quite correctly. You have likely learned in high school that the pressure of the air, like any fluid, exerts forces on objects. Perhaps you learned this idea as "the air presses down with a weight of 15 pounds on every square inch." There is only one error here, but it is a serious one—the word "down." Air pressure, at sea level, does indeed exert a force of 15 pounds per square inch, but in *all* directions. It presses down on the top of an object, but also inward on the sides and upward on the bottom of the object. The *net* force, for all practical purposes, is zero! The top of your forearm has a surface area of about 40 square inches, so the air presses down on the top of your arm with a force of about 600 pounds! It would break your arm except for the fact that the air also pushes up from the bottom with 600 pounds of force. The *net* force on your arm is zero.

But, some of you will ask, what happens if we place an object on a flat surface so that there is no air under it? Try this experiment: Place your arm on a flat table top. Do you now feel the 600 pounds of force pressing on it? Of course not. It is not possible to remove the air and the upward air pressure force from under an object. The only way to experience the air-pressure force is to form a seal around one side of the object and to remove the air from that side—creating a *vacuum*. When you press a suction cup against the wall, you press the air out and the rubber forms a seal that prevent the air from returning. Now the air-pressure force pushing the cup against the wall *is* more than the air-pressure force pushing away from the wall, and thus the air pressure holds the suction cup in place! Vacuum-sealed jars used for bottling things such as fruit jams and jellies are similar. But these are unusual situations. You will not need to be concerned with air pressure in any of the problems in this part of the text.

Lastly, a common error is to attempt to include a force called "$m\vec{a}$." After all, the second law says $\vec{F}_{net} = m\vec{a}$, so isn't $m\vec{a}$ a force? This is a matter of understanding what the second law is saying. It is not *defining* $m\vec{a}$ to be a force. $\vec{F}_{net}$ is a vector that can be found from the definition of what a force is and from the rules of vector addition. Likewise, $m\vec{a}$ is a vector that is found by analyzing an object's motion to find $\vec{a}$ and then multiplying by the scalar m. What Newton's second law says—and this is really the essence of the second law—is that the vector $\vec{F}_{net}$ and the vector $m\vec{a}$ are experimentally found to be equal to each other. We will use this operationally, starting in Chapter 5, to *find* the acceleration from the forces. But neither the acceleration itself nor $m\vec{a}$ is a force.

4.7 Free-Body Diagrams

Having discussed, at length, what is and is not a force, we are ready to assemble our knowledge about force and motion into a single diagram—the **free-body diagram**. You will learn in the next chapter how to write the equations of motion directly from the free-body diagram. Solution of the equations is a mathematical exercise—possibly a difficult one, but nonetheless an exercise that could be done by a computer. The *physics* of the problem, as distinct from the purely calculational aspects, is the steps that lead to the free-body diagram.

A free-body diagram is a combination of the particle model with a coordinate system and a force diagram showing *all* of the forces acting on the object. The free-body diagram and the motion diagram together form the full *physical model* of the problem. To draw a free-body diagram you should use the following steps:

Drawing Free-Body Diagrams

1. Identify all forces acting on the object, as described in Section 4.6.
2. Reproduce the same coordinate axes as used in your pictorial model. If those axes were tilted, then the axes of the free-body diagram should be similarly tilted.
3. Represent the particle as a "dot" at the origin of the coordinate axes.
4. Draw arrows representing each of the identified forces. Each arrow should
   a. have its *tail* placed on the dot representing the particle,
   b. point in the proper direction for the force,
   c. have a length proportional to the magnitude of the force (if known), and
   d. have an angle shown if the arrow is not parallel to a coordinate axis.
5. Draw a single *net-force* vector that is the vector sum of all forces. You might want to draw the net-force vector in a different color than the individual forces to avoid confusion. If the net force is zero, write "$\vec{F}_{net} = 0$" beside the diagram.

Once the free-body diagram is complete, you know the direction of $\vec{F}_{net}$. You should already know the direction of the acceleration $\vec{a}$ from the motion diagram. $\vec{F}_{net}$ and $\vec{a}$ should, according to the second law, have the same direction! Do they? Verifying that they have the same direction is the critical checkpoint in the solution of a problem. If they do not, then one of them (or perhaps both) is wrong and you have an error somewhere in your reasoning. There is no point in proceeding until you locate and correct the error, because you are assured that any mathematical answer will be wrong. While agreement is no guarantee of a correct final answer, it at least suggests that you are on the right track.

EXAMPLE 4-4   An elevator, suspended by a cable, accelerates upward. What is the physical model of this motion?

SOLUTION   First, let us analyze the kinematics with a motion diagram, as shown in Fig. 4-17a. The main conclusion is that the acceleration vector $\vec{a}$ points upward. Now we can proceed to analyze the forces. In Fig. 4-17b we have identified the forces by drawing a picture and isolating the system from the environment. From this picture we can see that there is only one contact force, the cable tension $\vec{T}$. There is also the long-range weight force $\vec{W}$. Figure 4-17c then shows the free-body diagram for this situation. Notice that the free-body diagram includes a coordinate system. The coordinate axes, with a vertical y-axis, are the ones we would have used in a pictorial model. The elevator has been shown as a dot at the origin, and the two force vectors have been drawn. Because the motion is accelerating upward, we suspect that the magnitude of $\vec{T}$ is larger than the magnitude of $\vec{W}$, so the diagram has been drawn with the $\vec{T}$ vector longer than the $\vec{W}$ vector. Lastly, we have drawn $\vec{F}_{net}$, which is upward. Having completed the model, we then check to see if $\vec{F}_{net}$ and $\vec{a}$ point the same direction. They do, which gives us confidence that we have correctly understood the physics.

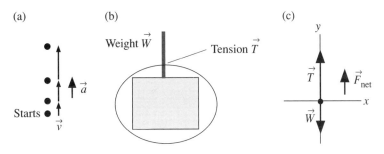

FIGURE 4-17 a) Motion diagram, b) identification of forces, and c) free-body diagram for an elevator accelerating upward.

EXAMPLE 4-5   Bobby straps a small rocket to a block of ice and shoots it across the smooth surface of a frozen lake. What is the physical model of its motion?

SOLUTION   Figure 4-18a shows a motion diagram for an accelerating ice block, with $\vec{a}$ pointing toward the right. Figure 4-18b then identifies the forces on the block. There is a normal force at the contact with the surface of the frozen lake, and a propulsion force where the boundary around the system crosses the rocket exhaust. Because of the wording of the problem, and our knowledge about how slippery ice is, we interpret the problem to mean that friction can be neglected. The long-range force of the block's weight is, of course, also present.

Figure 4-18c shows the free-body diagram. In drawing this diagram, we have made use of our knowledge that the motion is purely horizontal. This knowledge tells us that the y-component of acceleration is zero (it is not accelerating upward or downward, so $a_y = 0$).

The second law then implies that the y-component of the net force is also zero: $(F_{net})_y = 0$. This can only be true, according to the rules of vector addition, if the upward-pointing $\vec{N}$ and the downward-pointing $\vec{W}$ are equal in magnitude and thus cancel each other ($W_y = -N_y$). The vectors have been drawn accordingly. This leaves the net-force vector pointing toward the right—in agreement with $\vec{a}$ from the motion diagram.

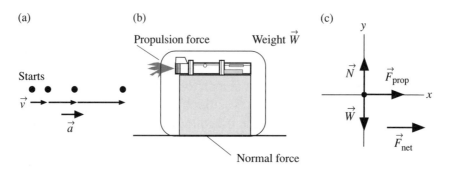

FIGURE 4-18 a) Motion diagram, b) identification of forces, and c) free-body diagram for a block of ice being shot across a frictionless frozen lake.

Note that in the previous example we could equally well justify our conclusion that $W_y = -N_y$ by using Newton's first law. The y-component of the block's motion is at rest. According to the first law, this can only be true if there is no net force in the y-direction. Notice that the first law does not say *no* forces but simply no *net* force. Thus the two vertical forces must add to zero. This type of analysis is one we will use frequently.

EXAMPLE 4-6   Your dead car is being towed up a 30° slope by an accelerating tow truck. What is the physical model of the motion?

SOLUTION This is Example 4-2 again with the additional information that the tow truck is accelerating. Figure 4-19a shows the motion diagram for a car accelerating up an incline. We have already identified the forces in Example 4-2 as tension $\vec{T}$ in the rope, a normal force $\vec{N}$ that is *perpendicular* to the surface, a friction force $\vec{f}$ that is *opposite* the direction

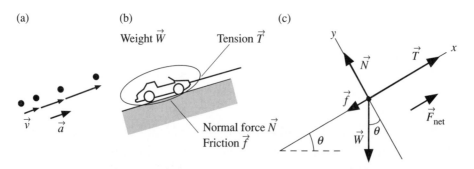

FIGURE 4-19 a) Motion diagram, b) identifying forces, and c) free-body diagram for a car being towed up a hill.

of motion and thus parallel to the surface, and a weight $\vec{W}$. If we were to draw a pictorial model it would use a tilted coordinate system, so we use these same tilted coordinate axes for the free-body diagram, as shown in Fig. 4-19c. In adding the force vectors, we have shown $\vec{T}$ pulling parallel to the slope and $\vec{f}$, which opposes the direction of motion, pointing down the slope. $\vec{N}$ is perpendicular to the surface and thus along the +y-axis. Finally, and this is an important step, the weight $\vec{W}$ is *vertically* downward, which in this case is *not* along the negative y-axis. In fact, you should convince yourself from the geometry that the angle θ between the $\vec{W}$ vector and the −y-axis is the same as the angle θ of the incline above the horizontal. (We will do many problems of motion along a tilted surface, so make a special note of this geometry of the angles.)

The fact that there is no acceleration in the y-direction allows us to conclude that the y-component of $\vec{F}_{net}$ must be zero. Thus we have drawn the vectors such that the y-component of $\vec{W}$ is equal in magnitude to $\vec{N}$. The net force vector will be in the x-direction, parallel to the slope, but *three* of the vectors have x-components. We suspect that the net force will point up the slope, toward the +x-direction, so $\vec{T}$ must be sufficiently large to compensate for the negative x-components of both $\vec{f}$ and $\vec{W}$. The direction of $\vec{F}_{net}$ then agrees, as we expected, with that of $\vec{a}$. •

Free-body diagrams will be our major analysis tool for the next several chapters, and they will continue to be used even later in the course when we learn about the forces of oscillating systems and electromagnetic forces. Careful practice with the exercises and homework in this chapter will pay immediate benefits in the next chapter. Indeed, it is not too much to assert that a problem is half solved, or even more, when you complete the free-body diagram.

Summary

Important Concepts and Terms

mechanics	resultant force
dynamics	newton
Newton's laws of motion	Newton's first law
Newtonian mechanics	inertia
force	equilibrium
contact force	weight
long-range force	tension
agent	normal force
inertial mass	friction
Newton's second law	drag force
net force	propulsion force
superposition of forces	free-body diagram

This chapter has begun the analysis of forces: how to identify them, how to combine them, and how they are related to motion. The boxed diagram in Section 4.3 summarizes the relationship of force and motion, and it will be the basis of our problem-solving strategy for many chapters to come.

Newton's first law states that an object either at rest or moving with constant velocity will remain that way if and only if the net force on the object is zero: $\vec{F}_{net} = 0$. This tendency for an object to resist change is called inertia. When $\vec{F}_{net} = 0$, the object is in equilibrium. We distinguish between two types of equilibrium: static equilibrium (at rest) and dynamic equilibrium (motion at constant velocity).

Newton's first law implies that no "cause" is needed for an object to move. Uniform motion is an object's natural state, and no cause is required to remain in that state. The proper question of dynamics is "What causes an object to *change* its motion—that is, to accelerate?" Newton's second law answers this question by relating acceleration—change of motion—to force:

$$\vec{a} = \frac{\vec{F}_{net}}{m}$$

where $\vec{F}_{net}$ is the vector sum of all the individual forces acting on the object and m is the object's inertial mass. This relationship between force and motion is an experimental discovery. Many of the ideas associated with this point of view seem contrary to our common sense experiences, and it will take some practice to develop a Newtonian intuition about motion.

There are two major procedures to follow in a force analysis. The first is the step-by-step method of Section 4.6 for the identification of forces. The second is the drawing of a free-body diagram, as described in Section 4.7. The free-body diagram will be an essential tool throughout this course and is also widely used in engineering.

Exercises and Problems

Exercises

1. Write a short essay of about two paragraphs' length on the topic "What Is a Force?" Your essay should include explanations of how to recognize a force as well as a description of the properties of forces. The essay should *not* talk about motion. Motion is a consequence of force and is not one of the defining characteristics of force.

2. Write a short essay of about two paragraphs' length on the topic "Force and Motion." Your essay should include a description of the connection between force and motion. Include any evidence you can cite supporting your statements.

Problems

Problems 3–6 show a free-body diagram. For each of these:

a. Identify, if possible, the direction of the acceleration vector $\vec{a}$. If not possible, state why not.

b. Identify, if possible, the direction of the velocity vector $\vec{v}$. If not possible, state why not.

c. Is the object of the force diagram speeding up, slowing down, or moving at a constant speed? Explain how you can tell.

d. Write a description of a real object, similar to the descriptions of Examples 4-4, 4-5, and 4-6, for which this is the correct free-body diagram.

3.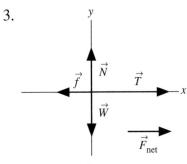

4.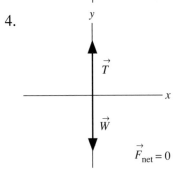

5.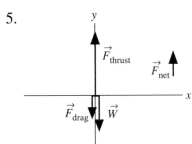

6.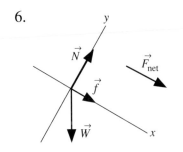

Problems 7–10 each describe a situation. For each:
  a. Draw a picture and identify all the forces acting on the object, then
  b. Draw a complete free-body diagram for the object.

7. An ice hockey puck slides across frictionless ice.

8. An elevator, suspended by a single cable, has just left the tenth floor and is accelerating downward.

9. A rocket is being launched upward. Air resistance is *not* negligible.

10. You're braking your car and slowing down while going down a 30° hill.

11. If a rubber ball is dropped and hits the ground, it bounces. Explain *how* the ball bounces. Your explanation should consist of the following:
  a. A motion diagram of the ball *just* during the brief time interval that it is in contact with the ground. (Use a high-speed movie "camera" for your motion diagram so that you'll have 8 or 10 frames showing the ball as it compresses, then re-expands). What is the direction of $\vec{a}$ during the time that the ball is touching the ground?
  b. A picture of the ball during its contact with the ground, on which you identify all forces acting on the ball.
  c. A free-body diagram of the ball during its contact with the ground. Is there a net force? If so, in which direction?
  d. A one-paragraph essay in which you describe what you have learned from a)–c) and in which you answer the question "How does a ball bounce?"

12. When a car stops suddenly, you feel "thrown forward." Similarly, you feel "thrown backward" if a bus in which you are standing suddenly accelerates forward from rest. Are there really forces pushing you forward or backward? Use the following steps to answer this question:
  Consider a person sitting on a *very* slippery bench (no friction and no seat back) inside a car. He is not holding on to anything.
  a. Draw a picture of the person on the bench and use it to identify all of the forces acting on the person during the three cases when the car moves at a steady speed, the car suddenly stops, and the car suddenly accelerates forward from rest.
  b. Draw a free-body diagram of the person for each of the three cases.
  c. In what direction, if any, is the net force acting on the person in each case?
  d. Describe what happens to the person in each case.
  e. Now explain, basing your explanation on Newton's laws, why you seem to be "thrown forward" in one case and "thrown backward" in another. Are there forces pushing on you in those directions?
  f. Suppose now that the bench is not slippery, so the person accelerates forward with the car when it starts from rest. What force is responsible for his acceleration? In which direction does this force point? Include a free-body diagram as part of your answer.

[**Estimated 5 additional problems for the final edition.**]

Chapter **5**

Dynamics I: Newton's Second Law

Nature and Nature's laws lay hid in night;
God said, let Newton be! and all was light.

Alexander Pope

LOOKING BACK | Sections 2.3; 3.7–3.9; 4.3–4.7

5.1 A Strategy for Force and Motion Problems

The last chapter developed the idea of what a force is and is not. We also introduced the fundamental relationship between force and motion: Newton's second law. As we continue to emphasize, this relationship is not just a definition, nor is it something arrived at purely by thinking. It is, instead, an empirical discovery of a connection between the force exerted on an object and the acceleration of that object. That relationship is

$$\vec{F}_{\text{net}} = \sum_i \vec{F}_i = m\vec{a}, \tag{5-1}$$

where the summation indicates a vector sum of all the individual forces $\vec{F}_i$ acting on the object. Because this relationship was thought by Newton to hold for *all* motions, he called it a "law" of nature.

Discoveries of the twentieth century have shown that Newton's second law fails to hold for motion at the atomic level or for motion with velocities near that of light (3×10^8 m/s). These motions, though, were far beyond the realm that Newton was able to study in the late seventeenth century—indeed, special instruments and techniques are needed even today to study motions that fail to adhere to Newton's laws. All of our everyday experience and the vast majority of engineering applications are extremely well described by the physical laws that Newton postulated 300 years ago.

We want, in this chapter, to develop a *strategy* for dealing with force and motion problems. Our strategy will be a set of procedures to be learned, not a set of equations to be memorized. In the last four chapters we have discussed several different methods for analyzing motion and forces: creating pictorial models, making motion diagrams, identifying forces, and drawing free-body diagrams. We've also discussed how to describe motion mathematically. Now we want to put all these ideas together, along with Newton's second law, into a complete problem-solving strategy.

Strategy for Force and Motion Problems

1. Analyze the problem statement by preparing a *pictorial model*. The pictorial model translates words to symbols, clarifies information that is known, and identifies what the problem is trying to find.

2. Analyze the motion by preparing a *motion diagram*. The motion diagram will determine the acceleration vector $\vec{a}$. It will also provide useful information to be used later in the kinematical analysis.

3. Identify all forces acting on the object and show them on a *free-body diagram*.

4. Invoke Newton's second law in operational form: $\vec{a} = \dfrac{1}{m} \sum_i \vec{F}_i$.

5. Determine the vector sum of the forces *directly* from the free-body diagram.

6. Solve for the subsequent motion, using appropriate results from kinematics.

7. Assess the results.

Let's take a closer look at Step 4, because this one you have not seen before. We noted in the last chapter that Newton's second law is a *vector* equation. Recall, from Chapter 2, that the equation in Step 4 is a short-hand way of writing three simultaneous equations:

$$
\begin{aligned}
a_x &= \frac{1}{m} \sum_i (F_i)_x = \frac{1}{m} (F_{\text{net}})_x \\
a_y &= \frac{1}{m} \sum_i (F_i)_y = \frac{1}{m} (F_{\text{net}})_y \\
a_z &= \frac{1}{m} \sum_i (F_i)_z = \frac{1}{m} (F_{\text{net}})_z .
\end{aligned}
\tag{5-2}
$$

Each *component* of the acceleration vector is obtained by summing the *components* of the force vectors along the appropriate axis, then dividing by the mass. Fortunately, we will restrict ourselves in this text to motion in a plane and, thus, will only need the first two of Eqs. 5-2. So the phrase "invoke Newton's second law" in Step 4 means to write down—literally—the first two of Eqs. 5-2.

5.2 Using Newton's Second Law

Let's look at some examples of how to use this strategy to solve problems.

EXAMPLE 5-1 A 500 g model rocket with a weight of 4.9 N is launched straight up. The small rocket motor burns for 5 s and has a thrust of 20.0 N. What is the maximum height reached by this rocket? Assume that the mass loss of the burned fuel is negligible.

SOLUTION The *procedure* we wish to follow is the seven-step strategy of Section 5.1. We begin by preparing a pictorial model, which we have shown in Fig. 5-1a. Note that we have established a coordinate system, identified several points of interest in the problem with symbols, and listed known and unknown information. This is an essential preliminary step in problem solving. Second, we have drawn a motion diagram in Fig. 5-1b. From it, we see that the acceleration vector $\vec{a}$ points upward while the fuel burns and points downward as the rocket "coasts" from the burn-out point to the maximum altitude. This will be important information later.

Next, we identify the forces acting on the object and draw a free-body diagram. As you learned in the previous chapter, we identify the forces by isolating the system and looking at the points of contact with the environment. Notice in Fig. 5-1c that the rocket's exhaust contacts the environment, giving the thrust force $\vec{F}_{thrust}$, but there are no other contact forces. In addition, there is the long-range weight force $\vec{W}$. (Air-resistance forces have been neglected because they were not explicitly called for.) These forces are shown on the free-body diagram of Fig. 5-1d. Once the fuel is exhausted, the thrust force vanishes and the weight remains as the only force on the rocket.

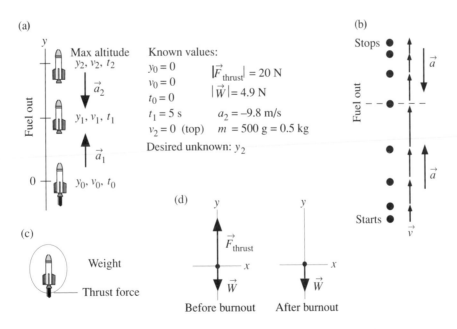

Known values:
$y_0 = 0$ $|\vec{F}_{thrust}| = 20$ N
$v_0 = 0$ $|\vec{W}| = 4.9$ N
$t_0 = 0$
$t_1 = 5$ s $a_2 = -9.8$ m/s
$v_2 = 0$ (top) $m = 500$ g $= 0.5$ kg

Desired unknown: y_2

FIGURE 5-1 a) Pictorial model, b) motion diagram, c) force identification, and d) free-body diagrams for the rocket of Example 5-1.

At this point we know what the problem is asking, have established relevant symbols and coordinates, and know what the forces and accelerations are. The mathematical analysis is now straightforward. We begin the mathematical model, by writing down Newton's second law. Because this is a *vector* equation, we want to use the component form of Eqs. 5-2. Thus

$$a_x = \frac{(F_{net})_x}{m} = \frac{1}{m}\left[(F_{thrust})_x + W_x\right]$$

$$a_y = \frac{(F_{net})_y}{m} = \frac{1}{m}\left[(F_{thrust})_y + W_y\right].$$

(5-3)

Note that the net force, by definition, is the *sum* of the forces $\vec{F}_{thrust}$ and $\vec{W}$. Thus the plus sign on the right-hand side of Eq. 5-3. The fact that vector $\vec{W}$ points downward, which might have tempted you to use subtraction in the y-equation, will be taken into account when we *evaluate* the components.

Referring to the motion diagram and free-body diagram, we see that *none* of the vectors in this problem have an x-component. This is a purely one-dimensional problem, so only the y-component of the second law will be needed. Now we can use the free-body diagram to determine that, while the rocket motor is burning fuel, $(\vec{F}_{thrust})_y = +|\vec{F}_{thrust}|$ while $W_y = -|\vec{W}|$. This is the point at which the directional information about the force vectors enters. Writing $a_y = a_1$, the acceleration while thrusting, the y-component of Eq. 5-3 becomes

$$a_1 = a_y = \frac{1}{m}\left[|\vec{F}_{thrust}| - |\vec{W}|\right]$$

$$= \frac{20.0\ N - 4.9\ N}{0.500\ kg} = 30.2\ m/s^2.$$

Notice that we converted the mass to SI units of kilograms before doing any calculations and that the division of newtons by kilograms, because of the definition of newtons, automatically gives the correct SI units of acceleration.

We now know that the acceleration of the rocket, until it runs out of fuel, is 30.2 m/s². We can next solve for the subsequent motion (Step 6 of the strategy). Because this is a constant acceleration, we can use constant-acceleration kinematics to find the height and velocity at burnout ($t_1 = 5$ s):

$$y_1 = y_0 + v_0(t_1 - t_0) + \tfrac{1}{2}a_1(t_1 - t_0)^2 = 377\ m$$

$$v_1 = v_0 + a_1(t_1 - t_0) = 151\ m/s.$$

Following burnout, the only force acting on the rocket is that of gravity, so we have a free-fall situation with $a_2 = -g$. While we do not know the time interval to the top, we do know that the final velocity is $v_2 = 0$. We can use this to find the final altitude, or maximum height, to be

$$v_2^2 = 0 = v_1^2 - 2g(y_2 - y_1)$$

$$\Rightarrow y_2 = y_1 + \frac{v_1^2}{2g} = 1540\ m = 1.54\ km.$$

Finally, we want to *assess* whether this is a reasonable, believable result. The maximum height reached by this rocket is 1.54 km, or just slightly under one mile. While that is fairly high, the upward acceleration of 30.2 m/s² is ≈3g—a quite rapid acceleration. Sustaining such an acceleration for 5 s will, indeed, create a very high velocity by burnout and a significant final height. Under these circumstances, a 1.54 km height is a believable result. •

We have described this example in vastly more detail than upcoming examples will receive. Our purpose has been to show how the seven-step strategy is put into practice. While the details will change from problem to problem, the basic *procedure* will remain the same.

EXAMPLE 5-2   A 1000 kg car is pulled by a tow truck. The tension in the tow rope is 2500 N, and a 500 N friction force acts in the backward direction. If the car starts from rest, what is its velocity after 5 seconds?

SOLUTION We need to apply Steps 1–7 of the problem-solving strategy. Figure 5-2a shows a pictorial model while Fig. 5-2b shows a motion diagram. The acceleration $\vec{a}$ points to the right. Note that the information we are given is the *magnitude* of the forces: $|\vec{T}| = 2500$ N and $|\vec{f}| = 500$ N.

There are four forces acting on the car: the tension force of the rope, the friction force of the ground, the normal force exerted by the ground, and the car's weight. Although the problem statement only mentioned $\vec{T}$ and $\vec{f}$, a force-identification analysis reveals the presence of the other two. Figure 5-2c shows a free-body diagram, including the all-important coordinate system. The net-force vector, as shown on the free-body diagram, agrees with the acceleration vector from the motion diagram. Note that force vectors $\vec{T}$ and $\vec{f}$ have only x-components while $\vec{N}$ and $\vec{W}$ have only y-components.

Having drawn the free-body diagram, we begin the mathematical model by invoking the second law:

$$a_x = \frac{(F_{net})_x}{m} = \frac{1}{m}(T_x + f_x + N_x + W_x)$$

$$a_y = \frac{(F_{net})_y}{m} = \frac{1}{m}(T_y + f_y + N_y + W_y).$$

(5-4)

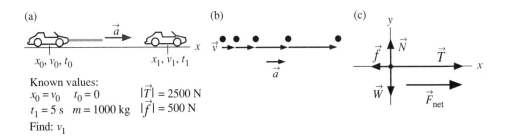

(a)

x_0, v_0, t_0 x_1, v_1, t_1

Known values:
$x_0 = v_0$ $t_0 = 0$ $|\vec{T}| = 2500$ N
$t_1 = 5$ s $m = 1000$ kg $|\vec{f}| = 500$ N
Find: v_1

(b)

(c)

FIGURE 5-2 a) Pictorial model, b) motion diagram, and c) free-body diagram for Example 5-2.

Reference to the free-body diagram allows us to determine the vector components:

$$T_x = +|\vec{T}| \quad N_x = 0$$
$$T_y = 0 \quad\quad N_y = +|\vec{N}|$$
$$f_x = -|\vec{f}| \quad W_x = 0 \quad\quad\quad\quad (5\text{-}5)$$
$$f_y = 0 \quad\quad W_y = -|\vec{W}|.$$

Notice that Eq. 5-4 is perfectly general, with plus signs everywhere, depending only on the fact that four vectors are *added* to give $\vec{F}_{net}$. Equations 5-5 then refer *explicitly* to the free-body diagram and have the correct signs. Substituting these components into Eq. 5-4 yields

$$a_x = \frac{1}{m}(|\vec{T}| - |\vec{f}|) = \frac{1}{1000 \text{ kg}}(2500 \text{ N} - 500 \text{ N})$$

$$= 2.00 \frac{\text{m}}{\text{s}^2} \quad\quad\quad\quad (5\text{-}6)$$

$$a_y = \frac{1}{m}(|\vec{N}| - |\vec{W}|).$$

Newton's second law has allowed us to determine a_x exactly, but it has given only an algebraic expression for a_y. However, we know *from the motion diagram* that $a_y = 0$! That is, the motion is purely along the x-axis, so there can be *no* acceleration along the y-axis. The condition $a_y = 0$ allows us to conclude, from Eqs. 5-6, that $|\vec{N}| = |\vec{W}|$. While we do not need $|\vec{N}|$ for this problem, it will be important in many future problems.

Having found the car's acceleration to be a constant $a = a_x = 2.00$ m/s², we can use kinematics to find the velocity:

$$v_1 = v_0 + a\Delta t = 0 + (2.00 \text{ m}/\text{s}^2) \cdot (5 \text{ s}) = 10.0 \text{ m}/\text{s},$$

where $a = a_x$. To assess the result, we note that 10 m/s ≈ 20 mph, a quite reasonable speed after 5 s of acceleration. We thus confirm that the answer is plausible.

●

EXAMPLE 5-3   A hockey puck can slide over the ice with negligible friction. Figure 5-3a shows a top view of three *horizontal* forces applied to a puck. What must the magnitude and angle of force $\vec{F}_3$ be to prevent the puck from moving?

SOLUTION This example, unlike the first two, is an *equilibrium problem* with the puck at rest, and thus $\vec{a} = 0$. A static equilibrium problem doesn't need a motion diagram because nothing is moving, but we still need a pictorial model and a free-body diagram. Notice that angle θ of $\vec{F}_3$ has been

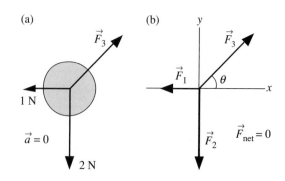

FIGURE 5-3 a) Top view of three forces applied to a hockey puck. b) Free-body diagram for hockey puck.

defined in the free-body diagram of Fig. 5-3b. Newton's second law is:

$$a_x = \frac{(F_{\text{net}})_x}{m} = \frac{1}{m}(F_{1x} + F_{2x} + F_{3x})$$

$$a_y = \frac{(F_{\text{net}})_y}{m} = \frac{1}{m}(F_{1y} + F_{2y} + F_{3y}).$$

(5-7)

The components of the force vectors can be identified from the free-body diagram as

$$F_{1x} = -|\vec{F_1}| \qquad F_{1y} = 0$$

$$F_{2x} = 0 \qquad F_{2y} = -|\vec{F_2}|$$

$$F_{3x} = +|\vec{F_3}|\cos\theta \qquad F_{3y} = +|\vec{F_3}|\sin\theta.$$

We know, from Chapter 2, that a vector is zero if and only if *all* its components are zero. Because $\vec{a} = 0$ in equilibrium, we must have $a_x = 0$ and $a_y = 0$. Thus Eqs. 5-7 become

$$\frac{1}{m}\left(-|\vec{F_1}| + |\vec{F_3}|\cos\theta\right) = 0$$

$$\frac{1}{m}\left(-|\vec{F_2}| + |\vec{F_3}|\sin\theta\right) = 0.$$

(5-8)

The $1/m$ is not relevant because both equations equal zero. We can thus rewrite Eqs. 5-8 as

$$|\vec{F_3}|\cos\theta = |\vec{F_1}|$$

$$|\vec{F_3}|\sin\theta = |\vec{F_2}|.$$

(5-9)

Equations 5-9 are two simultaneous equations for the two unknowns $|\vec{F_3}|$ and θ. We will encounter equations of this form on many occasions, so make a special note of the method of solution. First, divide the second equation by the first to isolate $\tan\theta$:

$$\left[\frac{|\vec{F_3}|\sin\theta}{|\vec{F_3}|\cos\theta} = \frac{\sin\theta}{\cos\theta} = \tan\theta\right] = \frac{|\vec{F_2}|}{|\vec{F_1}|}$$

$$\Rightarrow \theta = \tan^{-1}\left(\frac{|\vec{F_2}|}{|\vec{F_1}|}\right) = \tan^{-1}\left(\frac{2\text{ N}}{1\text{ N}}\right) = 63.4°.$$

(5-10)

Now use θ in the first of Eqs. 5-9 to find

$$|\vec{F_3}| = \frac{|\vec{F_1}|}{\cos\theta} = \frac{1\text{ N}}{\cos 63.4°} = 2.24\text{ N}.$$

The force that maintains equilibrium and prevents the puck from moving is thus

$$\vec{F_3} = (2.24\text{ N}, 63.4°).$$

You should note, in these three examples, how we are using the seven-step strategy of Section 1 rather than searching for a formula. This will become our standard procedure.

5.3 Mass and Weight

Gravity is a ubiquitous force. It acts on *every* object in the universe. We introduced weight in Section 4.5 as the force of gravity. More specifically, it is the gravitational force exerted by planet-size masses on objects at or near their surface. We will later extend the concept of gravitational force to arbitrarily large distances, but for now we will restrict our study to objects moving on or very near the surface of a planet.

We do not make a large distinction in our ordinary use of language between the terms *weight* and *mass*, but in physics their distinction is of critical importance. Mass is a scalar quantity that first appeared as the proportionality constant between force and acceleration. It measures the inertia of an object, which is why we called it the *inertial mass*. Loosely speaking, it also measures the amount of matter in an object. (We will make this statement more precise when we later introduce the idea of *gravitational mass*.) Mass is measured in kilograms, and it is a constant property of an object, regardless of the object's location.

Weight, on the other hand, is a *force*. Specifically, it is the gravitational force exerted *on* an object *by* a planet. As a force, weight is a vector rather than a scalar. The magnitude of the weight is measured in newtons, and the vector's direction is always straight down. In fact, the direction of the weight force can be thought of as an operational definition of the term "vertical." An implicit assumption here is that the planet's surface is flat, not curved, so that the weight vectors of separated objects are exactly parallel. We know, of course, that planets are really spherical (approximately), but this is not something of which we are aware unless we rise a distance above the surface that is a reasonable fraction of the planet's radius. We will have to treat planets as spheres when we come to the motion of satellites, but for the time being, as we stay close to the surface, a "flat earth approximation" is extremely good.

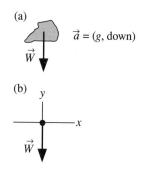

FIGURE 5-4 a) The weight force on a rock in free fall. b) The corresponding free-body diagram.

If weight is the gravitational force on a mass, it seems likely that there must be a connection between the amount of mass and the size of the weight force. Indeed there is, and we can use Galileo's discovery about free fall to make the connection. Figure 5-4a shows an object of mass m in free fall. The *only* force acting on the object is the weight force—the downward pull of gravity—so $\vec{F}_{net} = \vec{W}$. This is shown as a free-body diagram in Fig. 5-4b. Newton's second law says that the net force on this mass is related to its acceleration by

$$\vec{F}_{net} = \vec{W} = m\vec{a}. \tag{5-11}$$

You will also recall Galileo's discovery that *any* object in free fall, regardless of its mass, has the same acceleration:

$$\vec{a}_{\text{free fall}} = (9.80 \text{ m/s}^2, \text{downward}) = (g, \text{downward}),$$

where $g = 9.80 \text{ m/s}^2$ is the acceleration due to gravity at the earth's surface. Keep in mind that g is *always* positive. Equation 5-11 thus implies that the weight-force vector is:

$$\vec{W} = (mg, \text{downward}) \qquad \text{(weight force).} \tag{5-12}$$

We see that the magnitude of the weight is directly proportional to the mass, with g as the constant of proportionality:

$$|\vec{W}| = mg.$$

The value of g, however, varies from planet to planet. (We will see later that g depends on both the mass and the radius of the planet.) Because weight depends upon g, weight is not a fixed, constant property of an object. The value of g on the surface of the moon is about one-sixth its earthly value, so an object on the moon would have a weight only one-sixth its weight on Earth. The object's weight on Jupiter would be larger than its weight on Earth. Its mass, however, would be the *same* in all locations; the amount of matter has not changed, only the gravitational force exerted on that matter.

If we use a conventional coordinate system in which the y-axis points vertically upward, as was used in Fig. 5-4, we can write the vector $\vec{W} = -mg\,\hat{j}$. That is, the *components* of $\vec{W}$ are $W_x = 0$ and $W_y = -mg$. The minus sign in W_y indicates the direction of the vector $\vec{W}$.

Measuring Mass and Weight

Two common devices are used to measure mass and weight. Figure 5-5 shows a *pan balance*, such as those used in chemistry to "weigh out" a predetermined mass of a chemical. Each pan is pulled downward by the force of gravity. This causes the beam to rotate about the pivot if the weight $|\vec{W}_u|$ of the unknown differs from $|\vec{W}_s|$ of the standard mass. Balance of the pans is achieved only when the magnitudes $|\vec{W}_u|$ and $|\vec{W}_s|$ are equal:

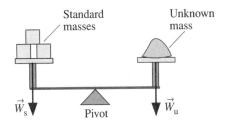

FIGURE 5-5 A pan balance, used to measure masses by comparing an unknown mass to a standard mass.

$$|\vec{W}_u| = |\vec{W}_s|.$$

Because the acceleration due to gravity g cancels out, the masses must be equal:
$$m_u = m_s.$$

[**Photo suggestion: A pan balance being used.**]

The important thing to notice is that the pan balance only *compares* two forces, and it balances when those forces are equal. The *magnitude* of the force is not measured or relevant. Consequently, the pan balance would work equally well on another planet. It is a devise for measuring *masses*, not for measuring weights.

Figure 5-6 shows two *spring scales*. These scales are used to weigh items in the grocery store or the hardware store (Fig. 5-6a). The bathroom scale on which you weigh yourself (Fig. 5-6b) is also a spring scale. Both have a spring—stretched in one case, compressed in the other—and both can be understood on the basis of Newton's first law: The object being weighed is at rest, in equilibrium, so the net force on it must be zero. The scale exerts a force on the object—an upward tension $\vec{T}$ *pulling* up (as shown in Fig. 5-6a) or an upward normal force $\vec{N}$ *pushing* up (shown in Fig. 5-6b). In order to have $\vec{F}_{net} = 0$, the upward tension force of Fig. 5-6a must *exactly* cancel the downward weight force, and thus $|\vec{T}| = |\vec{W}|$. Similarly, in Fig. 5-6b, the upward normal force exerted on the object by the scale's surface exactly cancels the downward weight, so $|\vec{N}| = |\vec{W}|$.

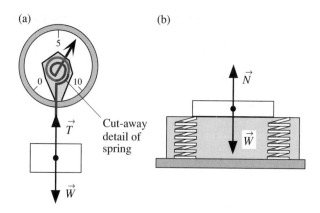

FIGURE 5-6 Two spring scales, used to measure weight:
a) Grocery store scales and b) bathroom scales.

[**Photo suggestion: A grocer's scale being used.**]

The reading of a spring scale is the magnitude of the force—either $|\vec{T}|$ or $|\vec{N}|$—that the scale is exerting on the object. (This is a very important point that we will return to when we discuss apparent weight in Section 5.5.) The scale does not "know" the weight of the object. All it can do is to measure the extent to which its spring is deformed and to produce a reading that indicates how much force the deformed spring exerts on the object. If the object is in equilibrium, so that $\vec{F}_{\text{net}} = 0$, then the magnitude of the spring force $|\vec{T}|$ or $|\vec{N}|$ is exactly equal to the object's weight $|\vec{W}|$. A spring scale thus measures weight, not mass. On a different planet, with a different value for g, the expansion or compression of the spring would be different and thus the scale's reading would be different.

A word of caution: Mass, in SI units, is measured in kilograms and weight is measured in newtons. A common English unit is *pounds*. Do pounds measure mass or weight? Pounds actually are the English unit of force, not mass. The well-known "conversion" that 1 kg = 2.2 pounds is thus, strictly speaking, not true. The correct statement would be, "A mass of 1 kg has a weight *on Earth* of 2.2 pounds." On another planet, the weight of a 1 kg mass would be other than 2.2 pounds.

5.4 Equilibrium

In Section 4.4 we defined *equilibrium* as the condition of no net force on an object. According to Newton's laws, this can occur only if the object's acceleration $\vec{a} = 0$, which means the object is either at rest or moving with constant velocity. An object at rest is said to be in **static equilibrium**, while an object moving with constant velocity is in **dynamic equilibrium**. Despite the difference in names, both are identical from a Newtonian perspective in which "at rest" does not have any special distinction. The primary condition that applies to both is $\vec{F}_{\text{net}} = 0$, and both, according to the first law, will continue without change.

[**Photo suggestion: A suspension bridge.**]

A slight modification of the seven-step strategy is appropriate for equilibrium problems. We already know $\vec{a} = 0$ in equilibrium, so a motion diagram isn't necessary. In addition Newton's first law can be used in place of Newton's second law. The simplified strategy for equilibrium problems is:

Strategy for Equilibrium Problems

1. Analyze the problem statement by preparing a *pictorial model*. The pictorial model translates words to symbols, clarifies information that is known, and identifies what the problem is trying to find.
2. Identify all forces acting on the object and show them on a *free-body diagram*.
3. Invoke Newton's first law: $\vec{F}_{net} = \sum_i \vec{F}_i = 0$.
4. Determine the vector sum of the forces *directly* from the free-body diagram.
5. Solve for needed information.

We have already looked at two examples of equilibrium—Example 5-3, with three forces exerted on a hockey puck, and also our analysis of the spring scale in the previous section. Because equilibrium problems occur fairly frequently, especially in engineering applications, it is worth looking at a couple more examples.

EXAMPLE 5-4   An object is suspended *between* two spring scales, as shown in Fig. 5-7a. The upper scale reads 200 N while the lower reads 290 N. What is the object's mass?

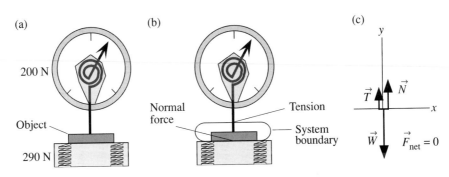

FIGURE 5-7 a) Pictorial model of an object suspended between two scales. b) Force identification. c) Free-body diagram.

SOLUTION This is an equilibrium problem because the object is at rest. The pictorial model, which is shown in Fig. 5-7a, only needs to establish a coordinate system and any relevant geometry—no position or time variables are needed. In Fig. 5-7b we identify the forces on the object by isolating the object as the system. These forces are the tension force exerted upward by the upper scale, the normal force exerted upward by the lower scale, and the downward weight force. These three forces are shown on the free-body diagram in Fig. 5-7c. (Notice how two parallel forces are shown on a free-body diagram.) Newton's first law then gives:

$$(F_{net})_x = \sum F_x = T_x + N_x + W_x = 0$$
$$(F_{net})_y = \sum F_y = T_y + N_y + W_y = 0 .$$

(5-14)

This problem is particularly straightforward because all of the x-components are zero and the first equation is satisfied automatically. The y-components can be determined directly from the free-body diagram to be $T_y = |\vec{T}|$, $N_y = |\vec{N}|$, and $W_y = -|\vec{W}| = -mg$. (Note how we're using the explicit relationship $|\vec{W}| = mg$ to determine the weight.) Thus

$$|\vec{T}| + |\vec{N}| - mg = 0$$

$$\Rightarrow m = \frac{|\vec{T}| + |\vec{N}|}{g} = \frac{200 \text{ N} + 290 \text{ N}}{9.8 \text{ m}/\text{s}^2} = 50 \text{ kg} .$$

(5-15)

Here we have used the fact, as noted in the previous section, that the reading of a spring scale is the magnitude of the force it is exerting. Therefore, in this example $|\vec{T}| = 200$ N and $|\vec{N}| = 290$ N. This example also emphasizes that the reading of a spring scale is *not necessarily* the weight of the object, which here is $|\vec{W}| = mg = 490$ N. Spring scales read a correct weight only if a) the spring scale's force is the only contact force exerted on the object, *and* b) the system is in equilibrium. Condition a) is not satisfied here for either spring scale due to the contact force of the other spring scale.

•

EXAMPLE 5-5   A 1500 kg car is being towed up a 30° frictionless slope at constant velocity. The towrope is "rated" at 8000 N maximum force. Will it break?

SOLUTION   This problem asks for a yes or no answer, not a number. To answer the question, however, will require a quantitative analysis. Because we are not told what to compute, part of our initial analysis of the problem statement will have to be a determination of which quantity or quantities will allow us to answer the question. In this case the answer is clear—we need to calculate the tension in the rope.

Here we have an example of dynamic equilibrium, because the car is moving at constant velocity. This example is an extension of Examples 4-2 and 4-6 from Chapter 4, which you may wish to review. There we found the forces on the car to be its weight $\vec{W}$, the normal force $\vec{N}$, the tow rope tension $\vec{T}$, and friction $\vec{f}$. In the present problem, the friction force is not present. Figure 5-7 shows the free-body diagram; note the tilted coordinate axes because the motion is one dimensional. We noted in the previous examples that the angle between $\vec{W}$ and the negative y-axis is the same as the angle of the slope, and that fact has

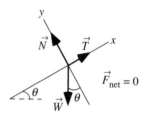

FIGURE 5-8 Free-body diagram for the car of Example 5-5.

been used. We can, for an equilibrium problem, use the first law:

$$(F_{\text{net}})_x = \sum F_x = T_x + N_x + W_x = 0$$

$$(F_{\text{net}})_y = \sum F_y = T_y + N_y + W_y = 0 . \tag{5-16}$$

We can see directly from the free-body diagram that ($T_x = |\vec{T}|$, $T_y = 0$) and ($N_x = 0$, $N_y = |\vec{N}|$). (This simplification is a major reason for the tilted axes.) The weight has both x- and y-components, both of which are negative due to the direction of the vector $\vec{W}$:

$$\vec{W} = (-mg\sin\theta)\,\hat{i} + (-mg\cos\theta)\,\hat{j} . \tag{5-17}$$

Note that in this case W_x is *not* $-mg\cos\theta$ because the angle θ in this problem is measured from the y-axis rather than from the x-axis. Using Eq. 5-17 in Eq. 5-16 gives (with an algebraic rearrangement):

$$|\vec{T}| = mg\sin\theta$$

$$|\vec{N}| = mg\cos\theta . \tag{5-18}$$

We can then evaluate the tension in the rope to be

$$|\vec{T}| = (1500 \text{ kg}) \cdot (9.8 \text{ m}/\text{s}^2) \cdot \sin 30° = 7350 \text{ N} .$$

Because $|\vec{T}| < 8000$ N, we conclude that the rope will *not* break. We have not used the second of Eqs. 5-18 in this problem. However, once we introduce friction we will need this second equation even for motion in one dimension.   •

Let's look again at Eqs. 5-18. Are these equations reasonable? A good way to find out is to examine *limiting cases*. Consider a flat surface with $\theta = 0°$, for which Eqs. 5-18 give $|\vec{T}| = 0$ and $|\vec{N}| = mg$. This is what we expect on a horizontal surface—the rope is slack (no tension) and the normal force is equal to the weight. What about the other limiting case of $\theta = 90°$? Then we find $|\vec{T}| = mg$ and $|\vec{N}| = 0$. The surface is irrelevant in this case, and the full weight of the car is hanging from the rope so the tension is equal to the weight. The limiting cases of a horizontal surface and a vertical surface convince us that Eqs. 5-18 make sense.

5.5 Apparent Weight

Recall from the discussion in Section 5.3 that the reading of a spring scale is the magnitude of the force it is exerting *on* the object. As we saw in Example 5-4, if the spring scale's force is the only contact force exerted on the object, *and* the system is in equilibrium, then the scale's reading is the correct weight of the object. Example 5-4 failed to meet the first condition, which is why the scale readings differed from the weight. Here we want to consider what happens if the second condition is not met.

Figure 5-9a shows a man weighing himself by standing on a spring scale in an elevator that has upward acceleration $\vec{a}$. The system (i.e., the man) is *not* in equilibrium because it is accelerating. We want to ask two questions: "What does the scale read?" and "How does the scale reading correspond to how heavy the man *feels*?"

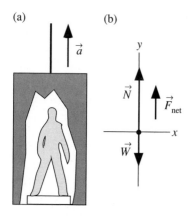

FIGURE 5-9 a) A man weighing himself in an elevator that is accelerating. b) Free-body diagram.

The man's only point of contact with the environment is where his feet touch the scale. Consequently, there is a contact force there—the upward normal force $\vec{N}$ of the scale pushing against his feet. He has, in addition, his weight force $\vec{W}$. These are the only forces acting on him. Figure 5-9b shows his free-body diagram, with a net force $\vec{F}_{net}$ pointing upward. There *has* to be a net force, because the system is accelerating. This leads us to an important conclusion: Because there is a net force $\vec{F}_{net}$ pointing upward, the magnitude of the normal force must be *greater* than the magnitude of the man's weight. That is, $|\vec{N}| > |\vec{W}|$. This result has major implications.

As we saw in Example 5-1, when the motion *and* all force vectors are along a single axis, only that component of the second law is relevant. The y-component of the second law in this case is

$$\left[(F_{net})_y = N_y + W_y\right] = ma_y \qquad (5\text{-}19)$$

where m is the man's mass. We can drop the subscript from a_y, as we did in one-dimensional kinematics, because there is no chance of confusion. From the free-body diagram, $N_y = +|\vec{N}|$ and $W_y = -|\vec{W}| = -mg$. Eq. 5-19 then becomes

$$|\vec{F}_{net}| = |\vec{N}| - mg = ma. \qquad (5\text{-}20)$$

Now the number showing on the scale—the scale reading—is the value $|\vec{N}|$. This is the magnitude of the force that the scale is exerting on the man. Solving Eq. 5-20 for $|\vec{N}|$ gives

$$|\vec{N}| = \text{scale reading} = mg + ma = mg\left(1+\frac{a}{g}\right) = |\vec{W}|\left(1+\frac{a}{g}\right). \qquad (5\text{-}21)$$

If the elevator is either at rest or moving with constant velocity, then $a = 0$ and the man is in equilibrium. In that case, we see from Eq. 5-21 that $|\vec{N}| = |\vec{W}| = mg$, so the scale correctly reads his weight. If, however, $a \neq 0$, the scale's reading will not be the man's true weight.

We have answered our first question of what the scale reads—it reads the value $|\vec{N}|$ given by Eq. 5-21. But what is the significance of this result? What does it mean, and how does it correspond to what the man feels as his *sensation* of weight?

While the *weight* of an object, the force of gravity on that object, has magnitude mg, gravity is not a force that you can sense directly. Your *sensation* of weight—how heavy you feel—is due to contact forces exerted *on* you by surfaces pressing against you. As you read this, your sensation of weight results from the upward force—the normal force—exerted on you by the chair in which you are sitting. When you stand, you *feel* the contact force of the floor against your feet. If you hang from a rope, your sensation of weight results from the upward force—the tension force in this case—exerted on you by the rope.

This idea, that your *sensation* of weight results from contact forces against you, is an important one that you may want to think about for a minute. If you are standing at rest, with $a = 0$, then we see from Eq. 5-21 that the magnitude of the force you *feel* is exactly equal in magnitude to your weight. This would be a trivial conclusion if objects were always at rest, but what happens if $a \neq 0$?

Recall the sensations of how you feel while being accelerated. You feel "heavy" when an elevator suddenly accelerates upward or when an airplane accelerates for takeoff. This heavy sensation vanishes as soon as the elevator or airplane reaches a steady cruising speed. This apparent increase in your weight must be associated with *acceleration* rather than velocity. As the upward-moving elevator brakes to a halt, or a roller coaster goes over the top, your stomach seems to rise a little and you feel lighter than normal. Your true, or actual, weight $|\vec{W}| = mg$ has not changed during these events, but your *apparent weight* has.

Going back to the man in the elevator, we call the scale reading $|\vec{N}|$ his **apparent weight**. This is the magnitude of the force pressing against the man to create his *sensation* of weight. Furthermore, $|\vec{N}|$ is what the scale actually reads. That is, he *appears* to weigh whatever the scale reads although, as the saying goes, appearances can be deceiving. If we call the apparent weight W_{app}, then Eq. 5-21 can be rewritten

$$W_{app} = |\vec{N}| = |\vec{W}|\left(1 + \frac{a}{g}\right) = mg\left(1 + \frac{a}{g}\right). \tag{5-22}$$

As the elevator accelerates upward, $a > 0$ and thus $W_{app} > mg$. The man *feels* heavier than normal and the scale actually reads more than his "true" weight. As the elevator brakes, the acceleration vector $\vec{a}$ points downward so the component $a < 0$. In this case, $W_{app} < mg$ and the man feels lighter than normal. In both cases, however, his true weight remains $|\vec{W}| = mg$.

Note that the spring scale need not be present, and usually is not, for an object to have an apparent weight different from its true weight. An object's apparent weight is just the magnitude of the contact force supporting it. A scale would read this value if present, but the force is exerted on the object whether the scale is there or not. Note also that the concept of apparent weight applies equally to animate and inanimate objects. We "personalized" our analysis so that you can relate the concept to your experience, but we could equally well discuss the apparent weight of a crate in the elevator.

The idea of apparent weight has important applications. Astronauts are nearly crushed by their apparent weight, which can be several times their true weight, during the high acceleration ($a \gg g$) launch of a rocket. Much of the thrill of amusement park rides, such as roller coasters, comes from rapid changes in your apparent weight. The concept of apparent weight will even help us to understand, in Chapter 6, how you can swing a bucket of water over your head without the water falling out! Apparent weight is an idea we will use rather extensively because it will focus your attention on many of the difficult issues surrounding forces and motion.

One last issue before leaving this topic: Suppose the elevator cable breaks and the elevator, along with the man and his scale, plunges straight down in free fall. What will the scale read as they fall? In free fall, the acceleration is $a = -g$. The negative sign is critical—make sure you understand it. When this acceleration is used in Eq. 5-22, we find that $W_{app} = 0$! In other words, the man has *no sensation* of weight.

Think about this carefully. Suppose, as the elevator falls, the man inside releases a ball from his hand. In the absence of air resistance, both he and the ball would continue to accelerate with exactly the same free-fall acceleration −g. From the man's perspective, the ball would appear to "float" beside him—getting neither closer nor further away. The scale floats beneath his feet, not exerting any normal force, and reads "0." He is what we call **weightless**.

The condition we call *weightless* does *not* mean "no weight." Instead, it means "no *apparent* weight." The distinction is significant. The man in the elevator still has weight *mg*—that has not changed—but he has no *sensation* of weight as he free falls. The term "weightless" is a very poor one, almost guaranteed to cause confusion because it seems to imply that objects have no weight. But, as we see, that is not the case.

But isn't this exactly what happens to astronauts orbiting the earth? When you see films from space, an astronaut releases an object and it floats beside her. If she tries to stand on a scale, it does not exert any normal force against her feet and thus it reads "0." She is said to be weightless. But if the criterion to be weightless is to be in free fall, and if astronauts orbiting the earth are weightless, does this mean that they and their spacecraft are in free fall? This is a very interesting question that we shall examine in the next chapter.

5.6 Friction

Friction is everywhere. It is great to think about idealized experiments where friction is absent, but it is equally necessary to connect Newton's laws to our everyday experience where friction is present. Friction, in the "real world," is not just an occasional nuisance—it is absolutely essential for many things we do. Without friction you could not walk, drive, or even sit down (you would slide right off the chair!). Your pencil would not write, and when you set it down it would probably slide off the table. Although friction is a complicated force, many aspects of friction can be described with a simple model.

Think about your experience pushing things when friction is significant. For example, consider pushing a heavy box across the floor. You have likely noticed an important fact: it is harder to get the box started than it is to keep it moving after it has started. To account for this behavior, let us divide friction forces into two categories: **static friction**, which is the force on a static object that keeps it from moving, and **kinetic friction**, which is the drag force that a surface exerts on a moving object. We will give these the symbols $\vec{f}_s$ and $\vec{f}_k$.

Figure 5-10 shows a person pushing against a box that is at rest, "stuck" to the floor. He is applying the horizontal contact force $\vec{F}_{push}$ to the box. But the box is in static equilibrium, so the *net* force on it must be zero. This can be true only if the box also experiences a force to the left that balances $\vec{F}_{push}$. The leftward force is the static friction force. It is a contact force exerted *by* the floor *on* the box. (There are, of course, vertical forces $\vec{W}$ and $\vec{N}$ on the box, but they will not enter into the *x*-equation of Newton's laws so we have

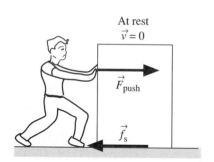

At rest
$\vec{v} = 0$

$\vec{F}_{push}$

$\vec{f}_s$

FIGURE 5-10 A static friction force resists your pushing force.

not shown them.) Because the box is at rest, Newton's first law is:

$$(F_{net})_x = (F_{push})_x + (f_s)_x = |\vec{F}_{push}| - |\vec{f}_s| = 0$$

$$\Rightarrow \quad |\vec{f}_s| = |\vec{F}_{push}|.$$

(5-23)

The size of the static friction force must *exactly* equal the size of the pushing force if the forces are to balance each other and give no net force.

This is a curious result. The harder the person pushes, the harder the floor pushes back. If the pushing force is large (but not so large as to make the box move), the static friction force will be large. Reduce the pushing force, and the static friction force will automatically be reduced to match. The static friction force acts only in *response* to other applied forces. Another aspect of static friction is its direction. The direction of $\vec{F}_{push}$ is determined by the person doing the pushing. Friction *responds* by exerting a force in the appropriate direction to prevent the box from moving. That direction is obvious with only one applied force, such as $\vec{F}_{push}$, but determining the direction of $\vec{f}_s$ can be tricky if there are several applied forces. The procedure is to ask what direction the object move in *if* there were no friction. The static friction force $\vec{f}_s$ has to point in the *opposite* direction to prevent the motion.

If you push sufficiently hard, eventually the box slips and starts to accelerate. This implies that the static friction force has a *maximum* possible size which we call $f_{s\,max}$. The magnitude of the static friction force must satisfy $|\vec{f}_s| \le f_{s\,max}$. If $|\vec{F}_{push}|$ exceeds $f_{s\,max}$, then Eq. 5-23 can no longer be satisfied. A net force appears and the box begins to accelerate.

Once the box starts to move, the static friction force is replaced by the kinetic friction force $\vec{f}_k$, as shown in Fig. 5-11. Experiments reveal that the kinetic friction force, unlike static friction, has a nearly *constant* value. Because it is easier to keep the box moving than it is to start it moving, the kinetic friction force is smaller than the maximum static friction force, or $|\vec{f}_k| < f_{s\,max}$. The direction of $\vec{f}_k$ is always *opposite* to the direction in which the object moves across the surface.

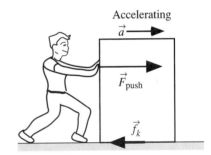

FIGURE 5-11 The kinetic friction force is opposite the direction of motion and is smaller than the static friction force.

Further experiments with friction (these were first done by Leonardo di Vinci!) show that both $f_{s\,max}$ and $|\vec{f}_k|$ are proportional to the magnitude of the normal force $|\vec{N}|$. That is,

$$f_{s\,max} = \mu_s |\vec{N}|$$

$$|\vec{f}_k| = \mu_k |\vec{N}|,,$$

(5-24)

where μ_s is called the **coefficient of static friction** and μ_k is called the **coefficient of kinetic friction**. The coefficients are constants that depend on the materials of which both object and surface are made. Because $|\vec{f}_k| < f_{s\,max}$, we find that $\mu_k < \mu_s$. Table 5-1 shows some typical values of coefficients of friction. It is to be emphasized that these are very approximate. The exact value of the coefficients depends on the roughness, cleanliness, and dryness of the surfaces.

TABLE 5-1 Typical coefficients of friction.

Materials	μ_s	μ_k
rubber on concrete	1.00	0.80
steel on steel (dry)	0.80	0.60
steel on steel (lubricated)	0.15	0.06
wood on wood	0.50	0.20
wood on snow	0.12	0.06
ice on ice	0.10	0.03

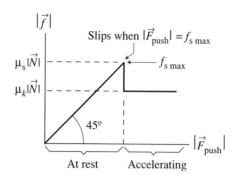

FIGURE 5-12 Measured value of the size of the friction force in response to an increasing pushing force.

Consider an experiment in which an object on a rough surface is pushed by a force $\vec{F}_{push}$ that steadily increases in magnitude. At first the object doesn't move, so there must be a friction force $\vec{f}_s$ whose magnitude also increases to match $\vec{F}_{push}$. When $|\vec{F}_{push}|$ reaches $f_{s\,max}$, the object slips and begins to accelerate. At that point the friction force becomes kinetic friction with magnitude $|\vec{f}_k|$. Figure 5-12 shows the magnitude of the friction force as $|\vec{F}_{push}|$ increases. The graph starts as a 45° line (slope of 1) because $|\vec{f}| = |\vec{f}_s| = |\vec{F}_{push}|$ as long as the object remains stuck in place. At the point of slippage, when $|\vec{F}_{push}| = f_{s\,max} = \mu_s|\vec{N}|$, the friction force drops in size to $|\vec{f}| = |\vec{f}_k| = \mu_k|\vec{N}|$. It remains *constant* at this value even if the pushing force grows stronger.

We can summarize our model of friction as follows:

$$\vec{f}_s = (0 \text{ to } \mu_s|\vec{N}|, \text{ direction as necessary to prevent motion})$$
$$\vec{f}_k = (\mu_k|\vec{N}|, \text{ direction opposite the motion}). \qquad (5\text{-}25)$$

The maximum value of static friction, $|\vec{f}_s| = \mu_s|\vec{N}|$, occurs at the point where the object slips and begins to move.

Note that we call Eq. 5-25 a "model" of friction and not a "law" of friction. These equations provide a reasonably accurate, but not perfect, description of how friction forces act. You might notice that according to this model, the friction force is independent of the surface area of the object. Two objects with the same normal force but different surface areas might be expected to experience different amounts of friction. Surprisingly, its turns out that the surface area has little effect, so our model makes the approximation of ignoring surface area. Likewise, our model assumes that the friction forces are independent of the speed of the object across the surface. This is again an approximation—fairly good, but not perfect. By making such approximations, we are able to find a model of friction that is simple to use and that provides a reasonable description of the effects of friction. More realistic models of friction are much more complex, generally requiring computers to solve the resulting Newtonian equations. We will leave such calculations to engineering courses and be satisfied with this simple model. But keep in mind that it has its limitations and is not a "law" of nature.

EXAMPLE 5-6   Bill pushes a 10 kg wooden box across a wooden floor at a steady speed of 2 m/s. a) How much force is he exerting on the box? b) If he stops pushing, how far will the box move before coming to rest?

SOLUTION   Figure 5-13a shows the pictorial model for this problem, focusing on the motion *after* Bill stops pushing and the box comes to rest at x_1. The coefficient of kinetic friction for wood on wood was taken from Table 5-1. Figure 5-13b shows motion diagrams both for Bill pushing, when $\vec{a} = 0$, and after he stops, when $\vec{a}$ points to the left. We have previously identified the forces on the box, so we can proceed to draw the free-body diagrams—a different one for each part of the problem—as shown in Fig. 5-13c. They differ only in the presence or absence of $\vec{F}_{push}$. You might think that static friction enters the problem because the box ends up at rest. Static friction would come into play *after* it is at rest, but here the box is moving up until the very *instant* that the problem ends, so only kinetic friction is relevant.

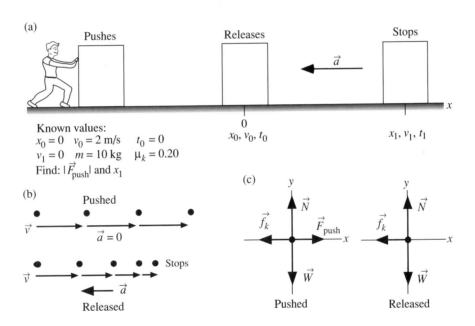

FIGURE 5-13 a) Pictorial model of Bill pushing crate. b) Motion diagrams for when crate is pushed and after it is released. c) Free-body diagrams.

a) For the first part of this problem we have dynamic equilibrium ($\vec{a} = 0$) and can use Newton's first law:

$$\sum F_x = |\vec{F}_{push}| - |\vec{f}_k| = 0 \tag{5-26}$$

$$\sum F_y = |\vec{N}| - |\vec{W}| = |\vec{N}| - mg = 0.$$

In addition, we also have our model of kinetic friction:

$$|\vec{f}_k| = \mu_k |\vec{N}|. \tag{5-27}$$

Equations 5-26 and 5-27 together are three simultaneous equations in the three unknowns $|\vec{F}_{push}|$, $|\vec{f}_k|$, and $|\vec{N}|$. Fortunately, these equations are easy to solve. The second of Eqs. 5-26 tells us that $|\vec{N}| = mg$. Substituting this into Eq. 5-27 gives $|\vec{f}_k| = \mu_k mg$. Now we can use this in the first of Eqs. 5-26:

$$|\vec{F}_{push}| = |\vec{f}_k| = \mu_k mg = (0.20)(10 \text{ kg})(9.80 \text{ m / s}^2) = 19.6 \text{ N} .$$

b) The crate is not in equilibrium after Bill stops pushing it. The basic strategy, as always, is to use the second law to find the acceleration, then do the necessary kinematics. We see, from the motion diagram, that the motion is one-dimensional with $a_x = a$ and $a_y = 0$. The second law, applied to the second free-body diagram of Fig. 5-13c, is

$$a_x = a = \frac{-|\vec{f}_k|}{m}$$

$$a_y = 0 = \frac{|\vec{N}| - mg}{m} , \tag{5-28}$$

and we also continue to have our model of friction

$$|\vec{f}_k| = \mu_k |\vec{N}| . \tag{5-29}$$

The negative sign occurs in the first of Eqs. 5-28 because $\vec{f}_k = -|\vec{f}_k| \hat{i}$ (the vector points to the left). Thus the *component* is negative: $(f_k)_x = -|\vec{f}_k|$. As in part a), we see that $|\vec{N}| = mg$, so $|\vec{f}_k| = \mu_k mg$. Using this in the first of Eqs. 5-28 gives

$$a = \frac{-\mu_k mg}{m} = -\mu_k g = -(0.20)(9.80 \text{ m / s}^2) = -1.96 \text{ m / s}^2 . \tag{5-30}$$

Why the minus sign? Because the acceleration vector $\vec{a}$ points to the left, as we see from the motion diagram, so the vector *component a* is negative. Now we are faced with a problem of kinematics at *constant* acceleration. Because we are interested in a distance, rather than a time interval, the easiest way to proceed is

$$v_1^2 = 0 = v_0^2 + 2a\Delta x$$

$$\Rightarrow \Delta x = \frac{-v_0^2}{2a} = \frac{-(2 \text{ m / s})^2}{2(-1.96 \text{ m / s}^2)} = 1.02 \text{ m} .$$

Note that we get a positive answer because the two negative signs cancel. Is our answer reasonable? Bill was pushing at 2 m/s ≈ 4 mph—fairly quickly. The box slides 1.02 m, which is slightly over 3 feet. That sounds pretty reasonable.

•

It is worth noting that both the horizontal and the vertical components of Newton's second law were needed in Example 5-6, even though the motion was entirely horizontal. This is typical when friction is involved because we must find the normal force to evaluate the friction force.

EXAMPLE 5-7   A steel block is placed on a steel table that can be tilted to various angles. a) If the tilt is gradually increased, at what angle will the block begin to slide? b) Is there an angle at which the block will slide down the incline at constant speed, after getting a push to start it? If so, what is that angle?

SOLUTION a) The first part is a static equilibrium problem with $\vec{a} = 0$ and we can apply Newton's first law: $\vec{F}_{net} = 0$. Figure 5-14a shows a picture of the block and labels the angle as θ. The forces on the block are its weight $\vec{W}$ and the contact forces $\vec{N}$ and $\vec{f}_s$. The weight has to point straight down, while $\vec{N}$ has to be perpendicular to the surface and $\vec{f}_s$ has to be parallel to the surface. To keep the block from slipping, the static friction force must point up the slope and must adjust its value to balance the x-component of the weight force that is pointing downhill. As the angle increases, $|\vec{f}_s|$ will have to increase in value. We want to find the maximum angle of the table before the block slips. This is the angle at which the static friction force reaches its maximum value. For any angle above this, $|\vec{f}_s|$ will not be able to assume a large enough value to maintain the equilibrium. Figure 5-14b shows the free-body diagram.

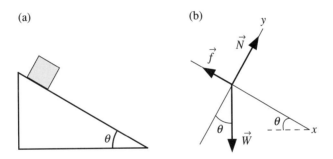

(a)

(b)

FIGURE 5-14 a) Block sliding down an inclined plane. b) Free-body diagram. The friction force is static in part a) and kinetic in part b).

In applying Newton's laws, we see that $\vec{f}_s$ has only a *negative x*-component and that $\vec{N}$ has only a y-component. The weight can be written $\vec{W} = +mg\sin\theta\,\hat{i} - mg\cos\theta\,\hat{j}$, where θ is both the angle of the slope and the angle between $\vec{W}$ and the y-axis. The first law then gives

$$\sum F_x = mg\sin\theta - |\vec{f}_s| = 0$$

$$\sum F_y = |\vec{N}| - mg\cos\theta = 0.$$

(5-31)

At the angle where the block slips, the static friction force will reach its maximum value

$$|\vec{f}_s| = f_{s\,max} = \mu_s|\vec{N}|.$$

(5-32)

From the second of Eqs. 5-31 we see that $|\vec{N}| = mg\cos\theta$. From this, $|\vec{f}_s| = \mu_s mg\cos\theta$, which can be substituted into the first of Eqs. 5-32 to give

$$mg\sin\theta - \mu_s mg\cos\theta = 0$$

$$\Rightarrow \frac{\sin\theta}{\cos\theta} = \tan\theta = \mu_s \qquad (5\text{-}33)$$

$$\Rightarrow \theta = \tan^{-1}\mu_s = \tan^{-1}(0.80) = 38.7°.$$

It is worth noting in this example that $f_{s\,max} = \mu_s|\vec{N}| = \mu_s mg\cos\theta$. A common error is to use simply $f_{s\,max} = \mu_s mg$. This is often true on a horizontal surface, but that is a special case. Be sure to evaluate the normal force within the context of each specific problem.

b) Is there an angle at which the block will slide down at *constant* velocity, without accelerating? If so, then we have a dynamic equilibrium situation with $\vec{F}_{net} = 0$. The only change when the block slides is that $\vec{f}_s$ is replaced by $\vec{f}_k$, so the free-body diagram remains exactly the same as in Fig. 5-14b. All we need to do is determine the angle θ for which the forces sum to zero. The analysis proceeds exactly as in part a), because it is also an equilibrium problem, except that $\vec{f}_s$ and μ_s are replaced by $\vec{f}_k$ and μ_k. So instead of Eqs. 5-33, we end with

$$\theta = \tan^{-1}\mu_k = \tan^{-1}(0.60) = 31.0°.$$

The block will not start by itself at this angle, because $31.0° < 38.7°$, but if started by a push it will then slide at constant velocity. If the angle is larger then $31.0°$, the block will *accelerate* rather than slide at constant velocity. Conversely, if $\theta < 31.0°$ you can start the block, by pushing it, but it will then *decelerate* and stop.

 ●

The angle at which slippage begins is called the *angle of repose*, and it is just the inverse tangent of the coefficient of static friction. The angle of repose is very important in geology because it is the angle at which loose materials (gravel, sand, snow, etc.) begin to slide on a mountainside, leading to landslides and avalanches. Water can act as a lubricant, lowering the coefficient of friction and thus the angle of repose. That is why mudslides are a common rainy-season problem in the steep coastal mountains of California.

Causes of Friction

It is worth a brief pause to look at the *cause* of friction. All surfaces, even those quite smooth to the touch, are very rough on a microscopic scale, as suggested in Fig. 5-15. When two objects are placed in contact, they do not make a smooth fit. Instead, the high points on one surface become jammed against the high points on the other surface, while the low points are not in contact at all. Only a very small fraction (typically 10^{-4}) of the surface area is in actual contact. The amount of actual contact depends on how hard the surfaces are pushed together, which is why the friction forces are proportional to $|\vec{N}|$.

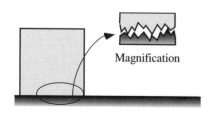

FIGURE 5-15 Magnified view of the area of contact between two surfaces.

[**Photo suggestion: Microscope photo of a surface.**]

At the points of actual contact, the atoms in the two materials are pressed closely together and molecular bonds are established between them. These bonds are ultimately understood as *electrostatic forces* between the atoms, and they are the "cause" of the static friction force. For an object to slip, you must push it hard enough to break these molecular bonds between the surfaces. Once they are broken, and the two surfaces are sliding against each other, there are still attractive forces between the atoms on each side as the high points of the materials push past each other. These attractive electrostatic forces cause kinetic friction. However, the atoms move past each other so quickly that they do not have time to establish the tight bonds of static friction. That is why the kinetic friction force is smaller.

Occasionally, in the course of sliding, two high points will be forced together so closely that they do form a tight bond. As the motion continues, it is not this surface bond that breaks but weaker bonds at the *base* of one of the high points. When this happens, a small piece of the object is left behind, "embedded" in the surface over which the object is sliding. This is what we call *abrasion*. Abrasion is what causes materials to wear out as a result of friction, be they the piston rings in your car or the seat of your pants. In machines, abrasion is minimized with lubrication—a very thin film of liquid between the surfaces that allows the surfaces to "float" past each other with many fewer points in actual contact.

Friction, as seen from the atomic level, is a very complex phenomenon. A detailed understanding of friction is at the forefront of engineering research today, where it is especially important for designing highly miniaturized machines.

5.7 More Examples of the Second Law

We will finish this chapter with several additional examples of using the second law.

EXAMPLE 5-8 A 1500 kg car traveling to the left at a speed of 30 m/s ($\approx$60 mph) on a slope of angle θ slams on its brakes and skids to a halt. Determine the stopping distance for $\theta = 20°$ (traveling downhill), $\theta = 0$ (level), and $\theta = -20°$ (traveling uphill).

SOLUTION This problem, shown as a pictorial model in Fig. 5-16a, is similar to Example 5-7; however, the slope is tilted the other direction. Notice the negative sign of v_0, because

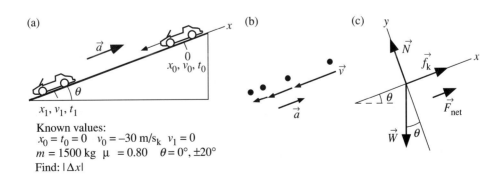

FIGURE 5-16 a) Pictorial model, b) motion diagram, and c) free-body diagram for car skidding to a halt on a slope.

the car is moving in the negative x-direction. The acceleration vector $\vec{a}$ is found, from the motion diagram (Fig. 5-16b), to be opposite to $\vec{v}$ because the car is slowing. The free-body diagram of Fig. 5-16c is like that of Example 5-7, with $\vec{f}_k$ pointing to the right, opposite to $\vec{v}$. In this coordinate system, $\vec{W} = -mg\sin\theta\,\hat{i} - mg\cos\theta\,\hat{j}$. The friction force is here sufficiently large to make the net force point uphill, in agreement with $\vec{a}$. Invoking the second law and the model of friction gives

$$a_x = a = \frac{1}{m}(|\vec{f}_k| - mg\sin\theta)$$

$$a_y = 0 = \frac{1}{m}(|\vec{N}| - mg\cos\theta) \tag{5-34}$$

$$|\vec{f}_k| = \mu_k|\vec{N}|.$$

The second equation gives $|\vec{N}| = mg\cos\theta$. When this is incorporated into the friction model, we find $|\vec{f}_k| = \mu_k mg\cos\theta$. Inserting this result back into the a_x equation, we find

$$a = \frac{1}{m}(\mu_k mg\cos\theta - mg\sin\theta) = g(\mu_k\cos\theta - \sin\theta).$$

From kinematics at constant acceleration:

$$v_1^2 = 0 = v_0^2 + 2a\Delta x$$

$$\Rightarrow |\Delta x| = \frac{v_0^2}{2a} = \frac{v_0^2}{2g(\mu_k\cos\theta - \sin\theta)}.$$

In the last step we have taken notice of the fact that the *displacement* Δx is going to be negative because the car is moving in the negative x direction. Because the questions only ask about the distance, not the displacement, we have taken an absolute value. Evaluating our result at the three different angles, using μ_k for rubber against concrete from Table 5-1, gives the stopping distances.

$$|\Delta x| = \begin{cases} 112 \text{ m} & \theta = 20° & \text{downhill} \\ 57 \text{ m} & \theta = 0° & \text{level} \\ 42 \text{ m} & \theta = -20° & \text{uphill.} \end{cases}$$

The implications are clear about the potential danger of driving too fast down hills!

•

This is a good example for pointing out the advantages of working problems *algebraically* until the very end. If you had starting plugging in numbers early, you would not have found that the mass eventually cancels out and you would have done several needless calculations. In addition, it is very easy to calculate just the final answer for different angles. Had you been computing numbers, rather than algebraic expressions, you would have had to go all the way back to the beginning for each angle.

EXAMPLE 5-9   David drags a heavy 100 kg wooden crate at steady speed across a wooden floor, using a rope at an angle of 30° (Fig. 5-17a). What is the tension in the rope?

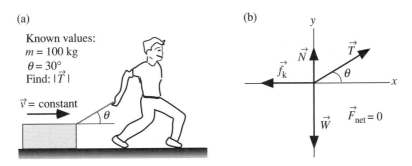

(a)

Known values:
$m = 100$ kg
$\theta = 30°$
Find: $|\vec{T}|$

$\vec{v} =$ constant

(b)

$\vec{F}_{net} = 0$

FIGURE 5-17 a) Pictorial model and b) free-body diagram for David pulling a crate.

SOLUTION "Steady speed" means $\vec{a} = 0$, so this is a dynamic equilibrium problem and we can use the first law: $\vec{F}_{net} = 0$. If we isolate the crate, we see that the forces on it are its weight and three contact forces: the normal force, the tension force, and the kinetic friction force. Because the motion is horizontal and to the right, $\vec{f}_k$ is horizontal and to the left. Remember that David is not in contact with the crate, so he exerts no force on it. This information is shown in the free-body diagram of Fig. 5-17b. The first law and the model of friction give:

$$\sum F_x = |\vec{T}|\cos\theta - |\vec{f}_k| = 0$$

$$\sum F_y = |\vec{N}| + |\vec{T}|\sin\theta - mg = 0$$

$$\Rightarrow |\vec{N}| = mg - |\vec{T}|\sin\theta \qquad (5\text{-}35)$$

$$|\vec{f}_k| = \mu_k|\vec{N}| = \mu_k(mg - |\vec{T}|\sin\theta).$$

Make sure you understand where all the terms in the two force equations come from, including their signs. This is an example where $|\vec{N}|$ is *not* equal to mg, and that has a big effect on the friction force. Because the rope is supporting part of the weight, the floor does not press against the bottom of the crate as hard as it would otherwise, thus the friction force is reduced. Inserting the friction result back into the first equation gives

$$|\vec{T}|\cos\theta - \mu_k(mg - |\vec{T}|\sin\theta) = |\vec{T}|(\cos\theta - \mu_k\sin\theta) - \mu_k mg = 0$$

$$\Rightarrow |\vec{T}| = \frac{\mu_k mg}{\cos\theta - \mu_k\sin\theta} = \frac{(0.20)(100 \text{ kg})(9.80 \text{ m/s}^2)}{\cos 30° - (0.20)(\sin 30°)} = 256 \text{ N}.$$

EXAMPLE 5-10 A 100 kg box of dimensions 50 cm × 50 cm × 50 cm sits in the back of a truck. The coefficients of friction between the box and the bed of the truck are $\mu_s = 0.8$ and $\mu_k = 0.4$. What is the maximum acceleration the truck can have without the box slipping?

SOLUTION This is a rather different problem than any we have looked at thus far, and it will take some careful thought. Let the box be the system. It contacts its environment only at the bottom surface where it touches the truck bed, so only the truck exerts contact forces on the box. If the box is *not* slipping, then there is no motion of the box *relative to the truck*. Therefore, only static friction enters the problem. The fact that the box, along with the truck, is moving relative to the ground is not relevant because there is no contact between the ground and the box. Because the surface area does not play a role in our model of friction, it would appear that neither the box dimensions nor the kinetic coefficient of friction are needed in this problem.

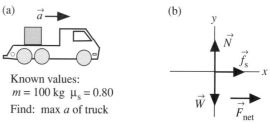

Known values:
$m = 100$ kg $\mu_s = 0.80$
Find: max a of truck

FIGURE 5-18 a) Pictorial model and b) free-body diagram for crate in truck.

Figure 5-18a shows a pictorial model and a motion diagram. As long as the box does not slip, its acceleration has to be exactly the same as that of the truck: $a_{box} = a_{truck}$! This is an important piece of information, because our analysis is of the box whereas the question asked about the truck. But their accelerations are the same. The forces on the box, shown in the free-body diagram of Fig. 5-18b, are the contact forces $\vec{N}$ and $\vec{f}_s$ exerted by the truck and the box's weight $\vec{W}$. The direction of $\vec{f}_s$ is likely to cause some difficulties. Why have we shown $\vec{f}_s$ pointing to the right rather than the left?

Suppose the truck bed were a frictionless surface. What happens as the truck accelerates? We would say that the box "slides out of the back" of the truck, but that is from the perspective of a rider in the truck. From the perspective of an observer on the ground, the truck simply drives out from underneath the box! Without any friction, the inertia of the box would keep it fixed as the truck leaves, until it finally falls vertically downward after the truck bed's normal force is removed. How does static friction keep this from happening? Because static friction acts in the direction necessary to prevent slipping, it must point in the *forward* direction, toward the right, so that the box accelerates *with* the truck. For the box to accelerate with the truck there must be, according to Newton's second law, a net force in the forward direction. Of the three forces on the box, only $\vec{f}_s$ has a horizontal component. Thus static friction is the force with which the truck bed surface *pushes* the box in the forward direction! This is a somewhat subtle but extremely important point, and one that we will meet again in Chapter 7.

Newton's second law and the model of friction provide the mathematical relationships:

$$a_x = a = \frac{1}{m}|\vec{f}_s|$$

$$a_y = 0 = \frac{1}{m}(|\vec{N}| - |\vec{W}|) = \frac{1}{m}(|\vec{N}| - mg) \tag{5-36}$$

$$|\vec{f}_s| \le \mu_s|\vec{N}|.$$

The second equation yields, $|\vec{N}| = mg$, which allows us to conclude from the third equation that $|\vec{f}_s| \leq \mu_s mg$. We can then use this in the first equation to deduce

$$a \leq \mu_s g$$

$$\Rightarrow a_{max} = \mu_s g = 7.84 \text{ m}/\text{s}^2.$$

This is the maximum possible acceleration of the crate, and the truck can accelerate no faster if slipping is to be avoided. Thus the truck's maximum acceleration is 7.84 m/s². •

It is worth noting that the mathematical analysis of this problem was quite straight-forward. The challenge was in the analysis that preceded the mathematics—that is, in the physics of the problem rather than the mathematics. It is here that our analysis tools—motion diagrams, force identification, and free-body diagrams—prove their value.

Summary

Important Concepts and Terms

static equilibrium	static friction
dynamic equilibrium	kinetic friction
apparent weight	coefficient of static friction
weightless	coefficient of kinetic friction

We have looked at numerous applications of Newton's laws in this chapter. Now you can begin to appreciate and understand all of the tools and techniques we developed during the first four chapters! The most important thing to notice from the examples is that every single one of them:

1. Began with *qualitative* analysis tools: the pictorial model, the motion diagram, force identification, and a free-body diagram,

2. Started the mathematical analysis from Newton's laws, and

3. Inserted the proper forces into Newton's laws using the free-body diagram.

This is our strategy for solving dynamics problems! No long list of formulas to memorize, but instead a *procedure* that works in every single problem.

We have now developed Newton's first and second laws into major problem-solving tools. These apply both to equilibrium problems, where $\vec{F}_{net} = \Sigma \vec{F}_i = 0$, and to dynamics problems, where $\vec{a} = \vec{F}_{net}/m$. It is important to remember that these are *vector* equations and that, in using them, you must write out each component of the equations separately. The mathematical portion of a problem solution, beginning with these equations, is usually straightforward *if* your analysis of the problem statement and your free-body diagram are correct.

This chapter has also developed specific information about two important forces:

Weight: $\vec{W} = (mg, \text{downward}) = -mg\,\hat{\jmath}$ (conventional coordinate system)

Friction: $\vec{f}_s = (0 \text{ to } \mu_s |\vec{N}|, \text{ direction as necessary to prevent motion})$

$\vec{f}_k = (\mu_k |\vec{N}|, \text{ direction opposite the motion}).$

Our use of the friction force is based on a model that works reasonably well but is not perfect. Do not confuse a *model*, such as that for friction, with a *law*, such as Newton's laws.

We have also discussed the relationship between the true weight and the apparent weight of an object. Apparent weight provides your sensation of weight and is what a scale would read, but it equals your true weight only in equilibrium, with $\vec{a} = 0$. Apparent weight and true weight can differ significantly for an accelerating object. This will have important implications for understanding the idea of being "weightless."

Exercises and Problems

The ✍ icon indicates that these problems can be done on a Dynamics Worksheet.

▬ Exercises

✍ 1. Three ropes are tied to a small, very light ring, as shown in Fig. 5-19. Two of these ropes are anchored to walls, forming right angles, with the tensions indicated in the figure. What is the magnitude and direction of the tension $\vec{T}$ in the third rope?

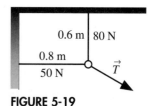

FIGURE 5-19

✍ 2. A 1000 kg car traveling at a speed of 30 m/s skids to a halt on dry level concrete. How long are the skid marks?

✍ 3. A 2 kg steel block at rest on a steel table is pulled on by a string, as shown in Fig. 5-20.
   a. What is the minimum string tension needed to move the block?
   b. If the string tension is 20 N, what is the block's speed when it reaches the edge of the table 1 m away?
   c. If the string tension is 20 N and the table is coated with oil, what is the block's speed when it reaches the edge of the table?

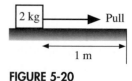

FIGURE 5-20

✍ 4. A 20,000 kg rocket has a rocket motor that generates 3×10^5 N of thrust.
   a. What is the rocket's initial upward acceleration?
   b. At an altitude of 5000 m the rocket's acceleration has increased to 6.00 m/s². What mass of fuel has it burned?

5. Air pressure is used to fire a 50 g ball vertically upward from a 1 m long tube (Fig. 5-21). The air pressure exerts an upward force of 2 N on the ball as long as it is in the tube. How high does the ball go above the top of the tube?

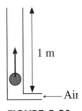

FIGURE 5-21

6. Figure 5-22 shows four free-body diagrams. In each diagram, several forces act on a 2 kg object. All forces are given in newtons. For each diagram find the values of a_x and a_y, the x- and y-components of the acceleration.

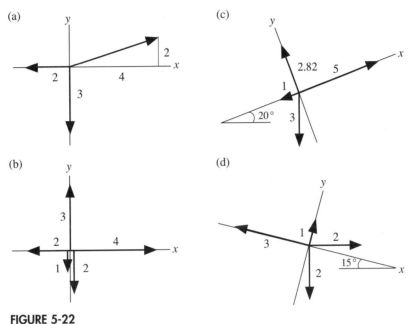

FIGURE 5-22

◣ Problems

7. A rifle with a barrel length of 60 cm fires a 10 g bullet with a horizontal speed of 400 m/s. The bullet strikes a block of wood and penetrates to a depth of 12 cm.
   a. What frictional force (assumed to be constant.) does the wood exert on the bullet?
   b. How long does it take the bullet to come to rest?
   c. Draw position-versus-time and velocity-versus-time graphs for the bullet in the wood.

8. Figure 5-23 shows a 500 kg piano being lowered into position by a crane while two men steady it with ropes pulling to the side. Bob's rope pulls to the left, 15° below horizontal, with 500 N of tension while Sam's rope pulls toward the right, 25° below horizontal.
   a. What tension must Sam maintain in his rope to keep the piano descending at a steady speed?
   b. What is the tension in the main cable supporting the piano?

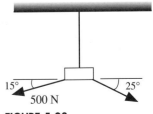

FIGURE 5-23

9. Little Johnny jumps off a swing, lands sitting down on a grassy 20° slope, and slides 3.5 m before stopping (Fig. 5-24). The coefficient of kinetic friction between grass and the seat of Johnny's pants is 0.5. What was his initial speed on the grass?

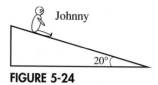

FIGURE 5-24

10. A 2 kg wooden block is launched up a long 30° wooden incline with a speed of 10 m/s.
    a. What vertical height does it reach above its starting point?
    b. What speed does it have when it slides back down to its starting point?

11. In an electricity experiment, a 1 g plastic ball is suspended from a support on a 60 cm string and then charged. An electric field is then generated that exerts a horizontal electrical force $\vec{F}_{elec}$ on the ball, causing it to swing out to a 20° angle and remain there. This situation is shown in Fig. 5-25.
    a. What is the magnitude of the electrical force?
    b. What is the tension in the string?

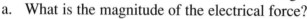
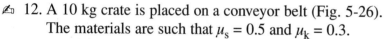

FIGURE 5-25

12. A 10 kg crate is placed on a conveyor belt (Fig. 5-26). The materials are such that $\mu_s = 0.5$ and $\mu_k = 0.3$.
    a. If the conveyor belt runs at constant speed, what force or forces act on the crate. Draw a free-body diagram.
    b. If the belt accelerates gently, what force or forces act on the crate. Draw a free-body diagram.
    c. What is the maximum acceleration the belt can have without the crate slipping?

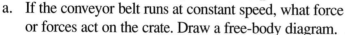
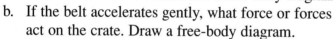

FIGURE 5-26

13. A 2 kg wooden box slides down a vertical wooden wall while a person pushes on it at a 45° angle, as shown in Fig. 5-27. What magnitude force does the person need to apply for the box to slide down at constant speed?

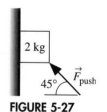

FIGURE 5-27

14. A 1 kg wooden block is pressed against a vertical wooden wall by the 12 N force shown in Fig. 5-28. If the block is initially at rest, will it move upward, move downward, or stay at rest?

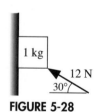

FIGURE 5-28

15. A 5 kg object initially at rest at the origin is subjected to the time-varying force shown in the graph in Fig. 5-29. Use graphical means to determine the object's velocity at time $t = 6$ s.

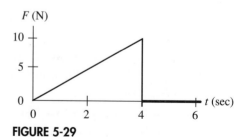

FIGURE 5-29

16. Henry gets into an elevator on the 50th floor of a building. The elevator begins moving at $t = 0$. The graph in Fig. 5-30 shows Henry's apparent weight over the next 12 s.

    a. What is the elevator's initial direction, up or down? Explain how you can tell.
    b. What is Henry's mass?
    c. How far has Henry traveled at $t = 10.5$ s?
    d. Describe the elevator's motion from $t = 0$ s to $t = 12$ s.

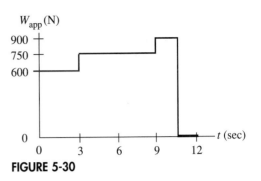

FIGURE 5-30

17. A mechanical testing laboratory wants to determine if a prototype of a new widget can withstand large accelerations and decelerations. To find out, they glue a widget, which has a mass of 5 kg, to a test stand that will drive it vertically upward (Fig. 5-31a). The test stand starts from rest, at $y = 0$, and its acceleration ($a = a_y$) during the first second is given by the acceleration-versus-time graph in Fig. 5-31b. This problem will analyze the kinematics of the widget's motion, then Problem 18 will analyze the forces.

    a. Write an equation for the widget's acceleration as a function of time, $a(t)$. Use this equation to find an equation for the velocity as a function of time, $v(t)$. (**Hint:** This is a calculus problem; the acceleration is *not* constant.)
    b. Compute v at several instants of time and use the results to draw a velocity-versus-time graph for the first second of motion. Does the slope of this graph agree with the acceleration graph? Explain.
    c. Use your answers from part a) to find an equation $y(t)$ for the position as a function of time.
    d. Compute y every 0.1 s, and then draw a position-versus-time graph for the widget. Does the slope of this graph agree with your velocity graph? Explain.
    e. Write a description of the widget's motion during the first second. Your description should include both a motion diagram *and* a written description. Within the written description, identify the time or times at which
       i) the acceleration is maximum,
       ii) the velocity is maximum,
       iii) the velocity is zero, and
       iv) the displacement is maximum.

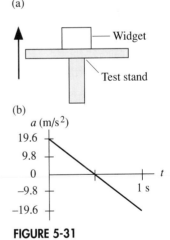

FIGURE 5-31

18. Continue your analysis of the situation described in Problem 17.

    a. Identify the forces acting on the widget and draw a free-body diagram.
    b. Determine the y-component N_y of the normal force acting on the widget during the first second of motion. Give your answer as a graph of N_y-versus-t. (**Hint:** The normal force does not remain constant. Use the second law and the a-versus-t graph of Problem 17.)
    c. Your answer to b) should show an interval of time during which N_y is negative. How can this be? What does it mean physically for the normal force to be negative? Explain.

    d. At what time is the apparent weight of the widget a maximum? What is the acceleration at that time?

    e. Is the apparent weight of the widget ever zero? If so, at what instant of time does this happen? What is the acceleration at that instant of time?

    f. Suppose the technician forgets to glue the widget to the test stand, so it is not held in place. Will the widget remain on the test stand throughout the first second, or will it fly off the stand at some instant of time? If so, at what time will this occur?

19. Consider a 1 kg ball on a 1 m long string hanging from the ceiling of a truck, as shown in Fig. 5-32. The back of the truck, where you are riding with the ball, has had all the windows completely sealed so that you cannot see outside.

    a. Initially the truck is either at rest or is moving forward with a steady speed of 10 m/s on a very smooth road. Can you determine which it is? If so, how? If not, why not?

    b. Later the truck is either at rest or accelerating forward with a steady acceleration of 1 m/s² on a very smooth road. How can you use the hanging ball to determine which it is?

    c. If the truck accelerates forward, is there a force on the ball pushing it in the backward direction?

    d. If the truck has been accelerating forward at 1 m/s² long enough for the ball to achieve a steady position, does the ball have an acceleration? If so, what is the magnitude and direction of the ball's acceleration?

    e. What force or forces are acting on the ball?

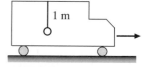

FIGURE 5-32

    f. Draw a free-body diagram for the ball while the truck is accelerating forward. Is the net force on the ball, as determined by your free-body diagram, in the same direction as the ball's acceleration?

    g. Suppose the ball is observed to make a 10° angle with the vertical. What is the truck's acceleration?

✍ 20. Figure 5-33 shows three free-body diagrams with the force magnitudes labeled. For each of these write a dynamics problem for which the diagram is the correct free-body diagram. Your problem should refer to a specific, realistic object; should ask a question that will be answered with a value of position or velocity (such as "How far?" or "How fast?"); and should give sufficient information to allow a solution. Then solve your problem.

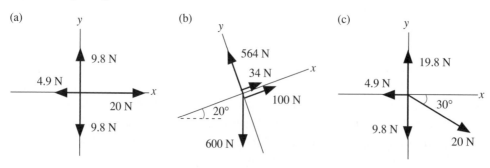

FIGURE 5-33

[**Estimated 10 additional problems for the final edition.**]

Chapter **6**

Dynamics II: Motion in a Plane

LOOKING BACK | Sections 1.4–1.6; 3.7, 3.9; 5.1; 5.5

6.1 Kinematics in Two Dimensions

[**Photo suggestion: Airplane taking off.**]

We have limited ourselves, thus far, to motion along a straight line. Although one-dimensional motion contains a lot of interesting physics and applications, motion in the "real world" is often in two dimensions. A car turning a corner, a fly ball to center field, or a planet orbiting the sun are all examples of two-dimensional motion. Restricting ourselves to one-dimensional motion has allowed us to concentrate on basic physics principles, but the time has come to broaden our horizons and consider a wider variety of motion.

Newton's laws are "laws of nature," meaning that they describe all motion, not just motion in a straight line. Consequently, this chapter will not introduce any new physics but will, instead, extend the application of Newton's laws to new situations. In the process, we will continue to build upon our understanding of the basic nature of force and its relationship to motion.

Consider a particle moving along a *curved* path in the *xy*-plane, as shown in Fig. 6-1. Such a curved path is what we call a **trajectory**. The particle's position at one instant of time is given by the position vector $\vec{r}_i$. We can write $\vec{r}_i$ in unit vector notation as

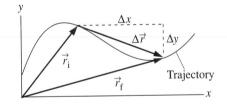

FIGURE 6-1 A particle moving along a curved trajectory in the *xy*-plane.

$\vec{r}_i = x_i\hat{i} + y_i\hat{j}$. At a later instant the position vector is $\vec{r}_f = x_f\hat{i} + y_f\hat{j}$. The vector connecting these two points on the trajectory is the *displacement vector*

$$\Delta\vec{r} = \Delta x\,\hat{i} + \Delta y\,\hat{j}$$

where $\Delta x = x_f - x_i$ and $\Delta y = y_f - y_i$.

Recall from Chapter 1 that the average velocity of a particle moving through a displacement $\Delta \vec{r}$ in a time interval Δt is:

$$\vec{v}_{avg} = \frac{\Delta \vec{r}}{\Delta t} = \frac{\Delta x}{\Delta t}\hat{i} + \frac{\Delta y}{\Delta t}\hat{j}. \tag{6-1}$$

The *instantaneous velocity* is the limit of $\vec{v}_{avg}$ as $\Delta t \rightarrow 0$. Taking the limit of Eq. 6-1, the instantaneous velocity in two dimensions is:

$$\begin{aligned} \vec{v} &= \lim_{\Delta t \to 0} \frac{\Delta \vec{r}}{\Delta t} = \frac{d\vec{r}}{dt} \\ &= \frac{dx}{dt}\hat{i} + \frac{dy}{dt}\hat{j} \\ &= v_x\hat{i} + v_y\hat{j}. \end{aligned} \tag{6-2}$$

We find that the velocity is a *vector* $\vec{v}$ *with x- and y-components* given by:

$$\begin{aligned} v_x &= \frac{dx}{dt} \\ v_y &= \frac{dy}{dt}. \end{aligned} \tag{6-3}$$

That is, the velocity vector has an *x*-component that is the rate dx/dt at which the particle's *x*-coordinate is changing, and similarly for the *y*-component.

As you can see in Fig. 6-1, the displacement vector $\Delta \vec{r}$ becomes tangent to the curve (i.e., tangent to the trajectory) as $\Delta \vec{r} \rightarrow 0$. Because $\vec{v}_{avg}$ has the same direction as $\Delta \vec{r}$, we can conclude that the *instantaneous* velocity vector $\vec{v}$ is, at each point, *tangent* to the trajectory. Figure 6-2 illustrates this for several points along the trajectory. Figure 6-2 also illustrates another important feature of the velocity vector: If the vector's angle θ is measured from the +*x*-direction, the velocity vector components are

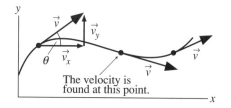

The velocity is found at this point.

FIGURE 6-2 The particle's velocity vector at each point is tangent to the trajectory.

$$\begin{aligned} v_x &= \frac{dx}{dt} = |\vec{v}|\cos\theta \\ v_y &= \frac{dy}{dt} = |\vec{v}|\sin\theta \end{aligned} \tag{6-4}$$

where

$$|\vec{v}| = \sqrt{v_x{}^2 + v_y{}^2} \tag{6-5}$$

is the particle's *speed* at that point. Speed is always a positive number, whereas the components are *signed* quantities that convey information about the direction of the velocity vector.

For curved motion along a trajectory, we have to consider the rate at which both the horizontal position changes *and* the vertical position changes. These two rates, dx/dt and dy/dt, are two separate quantities, and they form the components v_x and v_y of the velocity vector $\vec{v}$. If we know the speed $|\vec{v}|$ and the direction θ of motion, we can use Eq. 6-4 to determine these two components. Conversely, we can use the two components to determine the direction of motion using the equation

$$\tan\theta = \frac{v_y}{v_x}. \tag{6-6}$$

Geometrically, you can see from Figs. 6-1 and 6-2 that the velocity *vector* is an arrow in the *xy*-plane that, at any instant of time, is tangent to the trajectory at that point.

It is important to make a distinction between two different kinds of graphs that you have seen. In Chapter 3 you learned that the *value* of the velocity at an instant of time *t* is given by the *slope* of the position-versus-time graph at time *t*. Now we see that the *direction* of the velocity vector is given by the *tangent* to the *y*-versus-*x* graph of the trajectory. While these two ideas may seem similar, they are in fact two different interpretations of the tangent lines in two different graphs. The *y*-versus-*x* graph of the trajectory, such as the one shown in Fig. 6-2, is an actual picture of the motion. Notice that the trajectory does *not* tell us anything about how fast the particle is moving, but only its direction. A position-versus-time graph, on the other hand, is an abstract representation of the motion—*not* a picture of the motion. Furthermore, a position-versus-time graph can only represent one component of the position vector—not a problem for one-dimensional motion, but less useful for curved motion along a trajectory. A common error is to confuse one type of graph with the other, so think about this carefully and make sure you understand the distinction.

Now let's look at a case where a particle's velocity changes from $\vec{v}_i$ to $\vec{v}_f$ in a time interval Δt. The particle's average acceleration is

$$\vec{a}_{\text{avg}} = \frac{\Delta\vec{v}}{\Delta t} = \frac{\vec{v}_f - \vec{v}_i}{\Delta t}.$$

If we now take the limit $\Delta t \to 0$, the instantaneous acceleration is

$$\vec{a} = \lim_{\Delta t \to 0} \frac{\Delta\vec{v}}{\Delta t} = \frac{d\vec{v}}{dt}$$

$$= \frac{dv_x}{dt}\hat{i} + \frac{dv_y}{dt}\hat{j} \tag{6-7}$$

$$= a_x\hat{i} + a_y\hat{j}.$$

Figure 6-3 shows the velocity vectors $\vec{v}_i$ and $\vec{v}_f$ at two points along a trajectory. Note how each is tangent to the trajectory. As we did in Chapter 1, we can find the change in the velocity $\Delta\vec{v}$ such that $\vec{v}_i + \Delta\vec{v} = \vec{v}_f$. The direction of $\vec{a}$ at the midpoint between the two velocity vectors is the same as the direction of $\Delta\vec{v}$. In the $\Delta t \to 0$ limit, the two points on the trajectory converge toward a single point and $\vec{a}$ is the instantaneous acceleration at that point.

Recall from Chapter 3 that a particle's velocity v and acceleration a are defined as derivatives ds/dt and dv/dt of the position and the velocity. These definitions have been extended, in Eqs. 6-2 and 6-7, to the *x*- and *y*-components of the velocity $\vec{v}$ and

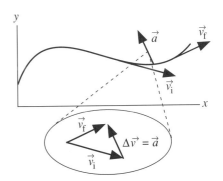

FIGURE 6-3 The acceleration vector at a point on a curved trajectory.

acceleration $\vec{a}$ vectors. You also learned in Chapter 3 that the velocity and the position of a particle are integrals of the acceleration and velocity, respectively. For motion in a plane, these integrals become

$$v_x(t) = (v_x)_i + \int_{t_i}^{t} a_x(t)dt \quad x(t) = x_i + \int_{t_i}^{t} v_x(t)dt$$

$$v_y(t) = (v_y)_i + \int_{t_i}^{t} a_y(t)dt \quad y(t) = y_i + \int_{t_i}^{t} y_x(t)dt.$$

(6-8)

Therefore, if the motion has a *constant* acceleration along one axis (i.e., one component of $\vec{a}$ is a constant), then the constant-acceleration kinematic equations of Table 3-6 apply along *just* that axis. If the motion happens to have *zero* acceleration along one axis (i.e., one component of $\vec{a}$ is zero), then the motion along just that axis proceeds with constant velocity. These ideas will be illustrated in the following examples.

EXAMPLE 6-1   The up thrusters on a shuttlecraft of the Starship Enterprise give it an upward acceleration of 5 m/s^2. Its forward thrusters provide a forward acceleration of 10 m/s^2. As the shuttle takes off, the shuttle's crew turns on the up thrusters at $t = 0$ s, adds the forward thrusters at $t = 2$ s, then turns off the up thrusters at $t = 4$ s. Plot a trajectory for the shuttlecraft for its first six seconds of flight.

SOLUTION Figure 6-4 shows a pictorial model for this situation. The coordinate system has been chosen with the y-axis upward, relative to the shuttlecraft, and with the journey beginning at the origin. The craft will move vertically upward for two seconds, then begin to acquire a forward motion after the forward thrusters are fired. There are four points in the motion—the beginning, the end, and two points at which the acceleration changes. The acceleration during the first two seconds is $\vec{a}_0 = 5\,\hat{j}$ m/s^2.

During the next two seconds it is $\vec{a}_1 = (10\hat{i} + 5\hat{j})$ m/s^2, and finally, during the last two seconds it is $\vec{a}_2 = 10\hat{i}$ m/s^2 . The acceleration is constant *during* each of these intervals, so we can use constant-acceleration kinematics.

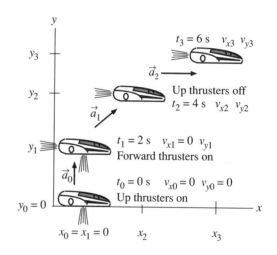

FIGURE 6-4 Pictorial model for Example 6-1.

During the first two seconds, when $a_{x0} = 0$ and $a_{y0} = 5$ m/s^2, the motion is

$$x(t) = x_0 + v_{x0}(t - t_0) + \tfrac{1}{2}a_{x0}(t - t_0)^2 = 0$$

$$y(t) = y_0 + v_{y0}(t - t_0) + \tfrac{1}{2}a_{y0}(t - t_0)^2 = 2.5t^2.$$

At the end of two seconds, $x_1 = 0$, $v_{x1} = 0$, $y_1 = 10$ m, and $v_{y1} = 10$ m/s. During the next two seconds, when $a_{x1} = 10$ m/s^2 and $a_{y1} = 5$ m/s^2, the motion is

$$x(t) = x_1 + v_{x1}(t - t_1) + \tfrac{1}{2}a_{x1}(t - t_1)^2 = 5(t - 2 \text{ s})^2$$

$$y(t) = y_1 + v_{y1}(t - t_1) + \tfrac{1}{2}a_{y1}(t - t_1)^2 = 10 + 10(t - 2 \text{ s}) + 2.5(t - 2 \text{ s})^2.$$

The velocity during this interval is $v_x = 10(t - 2 \text{ s})$ and $v_y = 10 + 5(t - 2 \text{ s})$. At $t = 4$ s, the shuttlecraft has reached $x_2 = 20$ m, $y_2 = 40$ m and has velocity $v_{x2} = 20$ m/s, $v_{y2} = 20$ m/s. The motion during the last interval, when $a_{x2} = 10$ m/s^2 and $a_{y2} = 0$, is

$$x(t) = x_2 + v_{x2}(t - t_2) + \tfrac{1}{2}a_{x2}(t - t_2)^2 = 20 + 20(t - 4 \text{ s}) + 5(t - 4 \text{ s})^2$$

$$y(t) = y_2 + v_{y2}(t - t_2) + \tfrac{1}{2}a_{y2}(t - t_2)^2 = 40 + 20(t - 4 \text{ s}).$$

The trajectory of Fig. 6-5 was plotted from these three sets of equations for x and y, each valid for a two-second interval. The position is indicated every half second. You can see how the shuttlecraft lifts off from the Enterprise during the first two seconds, then begins to accelerate forward. Note that the upward motion *continues* with a constant $v_y = 20$ m/s after the up thrusters are turned off. No upward force is necessary to sustain this motion.

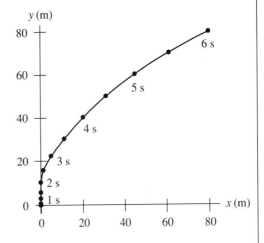

FIGURE 6-5 Trajectory followed by the shuttlecraft of Example 6-1.

6.2 Dynamics in Two Dimensions

Newton's second law, $\vec{a} = \vec{F}_{\text{net}}/m$, makes no distinction between linear motion and nonlinear motion. It is the law that determines an object's acceleration. The x- and y-components of the acceleration vector are given by:

$$a_x = \frac{(F_{\text{net}})_x}{m}$$

$$a_y = \frac{(F_{\text{net}})_y}{m},$$

(6-9)

Our previous strategy for dynamics problems is still valid: Identify the forces, draw a free-body diagram, determine the net force, calculate *both* components of the acceleration, and finally calculate position and velocity information (i.e., the trajectory) from kinematics. We have identified the plane of motion as the xy-plane, but bear in mind that this is only a notation issue—the plane could be horizontal, vertical, or even tilted.

EXAMPLE 6-2 Alice tapes a small rocket motor (from a model rocket) to an ice-hockey puck. She orients the puck so the rocket's exhaust points in the $-y$-axis. Alice then pushes the puck in the $+x$-direction across frictionless ice with speed v_0. The rocket motor is ignited by remote control as it crosses a point on the ice designated as the origin. Find an equation for the puck's trajectory after the motor is ignited.

SOLUTION This is a more difficult problem than you have seen before. There are no numbers in the problem, but you are asked to "find an equation" for the puck's trajectory. What equation? Because the puck follows a curved path, you need to identify this curve. In other words, you need to find the *function* $y(x)$ that describes the trajectory. The coordinate system was already defined in the problem statement. The problem starts at the origin, so $(x_0, y_0) = (0, 0)$. The initial velocity components are $(v_{x0}, v_{y0}) = (v_0, 0)$. Figure 6-6a shows the pictorial model of this situation.

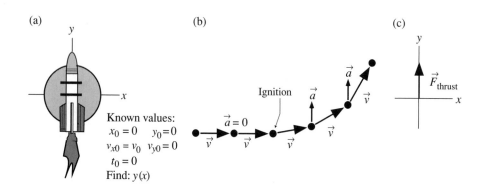

FIGURE 6-6 a) Pictorial model of hockey puck with rocket; b) motion diagram; and c) free-body diagram.

Now let's examine the forces acting on the puck. We know that it has a weight pulling vertically downward and that the normal force of the ice pushes vertically upward. However, these are not in the plane of motion! In fact, the puck is in static equilibrium (at rest) along the z-axis. Therefore, we can ignore these forces. Now let's draw a closed curve around the puck in the xy-plane to find the forces acting in this plane. Doing so shows that there is only one contact force—$\vec{F}_{thrust}$, the thrust of the rocket motor, which points in the $+y$-direction. This is the *net* force in the xy-plane, so $\vec{F}_{net} = |\vec{F}_{thrust}|\hat{j}$.

Let us think about this qualitatively before proceeding with a mathematical analysis. The x-component of $\vec{F}_{net}$ is zero because the force points along the y-axis. Therefore from Eq. 6-9 we know that $a_x = 0$. As we noted above, if one component of $\vec{a}$ is zero, then the motion along that axis continues with constant velocity. We already know the initial value of v_x to be v_0, and we can conclude that it is not going to change. Thus $v_x(t) = v_0 =$ constant. The rocket thrust along the y-axis causes no change in the x-component of the puck's velocity.

$\vec{F}_{net}$ does have a y-component but it is of constant magnitude, so the puck will have a constant acceleration ($a_y =$ constant) in the $+y$-direction. Thus v_y will steadily increase from its original value of zero. The distance traveled in the y-direction, which is just the value y itself, will increase quadratically with time: $y = \frac{1}{2}at^2$. Shortly after ignition, the y-component of velocity will still be small and the puck will still be moving nearly parallel to the x-axis. As time goes by, however, v_y will keep increasing until eventually $v_y > v_x$. We can conclude that the trajectory will start parallel to the *x*-axis and then curve upward in the *xy*-plane. This is shown in the motion diagram of Fig. 6-6b. Notice that the horizontal separation Δx between any two adjacent points remains constant (v_x = constant) while the vertical separation Δy steadily increases (v_y increasing).

We can use the free-body diagram of Fig. 6-6c to find the acceleration:

$$a_x = \frac{(F_{\text{net}})_x}{m} = 0$$

$$a_y = \frac{(F_{\text{net}})_y}{m} = \frac{|\vec{F}_{\text{thrust}}|}{m},$$

where *m* is the combined mass of the puck and the rocket. The *x*-motion is one of constant velocity, because $a_x = 0$, starting from $x_0 = 0$. The *y*-motion, however, is one of constant acceleration, starting from $y_0 = 0$ with $v_{y0} = 0$. Thus the position at time *t* is given by

$$x(t) = x_0 + v_{x0}(t - t_0) + \tfrac{1}{2}a_x(t - t_0)^2 = v_0 t$$

$$y(t) = y_0 + v_{y0}(t - t_0) + \tfrac{1}{2}a_y(t - t_0)^2 = \frac{1}{2}a_y t^2 = \left(\frac{|\vec{F}_{\text{thrust}}|}{2m}\right)t^2.$$

This is a solution, giving both *x* and *y* as explicit functions of time, but it is not the *y*(*x*) solution we were looking for. The important thing to notice is that whatever value *t* might have, it is the same in both equations. That is, our solution gives the simultaneous values of *x* and *y* at one instant of time. To express *y* in terms of *x*, we need to eliminate the time variable by finding $t = x/v_0$ and to substitute this into the *y*-equation. We then arrive at the trajectory equation

$$y(x) = \left(\frac{|\vec{F}_{\text{thrust}}|}{2mv_0^2}\right)x^2.$$

The trajectory is an equation of the form $y = cx^2$, which is the equation of a parabola. Figure 6-7 shows a possible trajectory which, as we had already surmised, curves upward away from the *x*-axis. The amount of curvature depends on the value of the constant multiplying x^2—the larger the value of *c*, the larger the curvature. Notice that the curvature could be increased either by increasing the rocket thrust or by decreasing the mass—both have the effect of increasing the *y*-acceleration. We could also increase the curvature by decreasing the initial velocity v_0, which would "compress" the *x*-motion while having no effect on the *y*-motion.

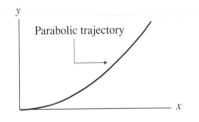

FIGURE 6-7 Parabolic trajectory of the hockey puck.

Example 6-2 illustrates the general principles of analyzing motion in a plane. Although the problem was stated in terms of a rocketing hockey puck, our analysis and conclusion would apply equally well to *any* problem in which one component of acceleration is zero while the other is non-zero. The primary finding is that such an object follows a *parabolic trajectory*. This is an important class of problems that includes not only projectile motion, which we will study more thoroughly in the next section, but also such applications as the "deflection plates" that steer an electron beam through a cathode ray tube.

6.3 Projectile Motion

Baseballs and tennis balls flying through the air, divers leaping from diving boards, dare-devils shot from cannons, and bullets shot from guns are all examples of *projectile motion*.

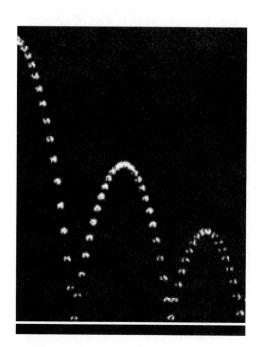

FIGURE 6-8 Parabolic trajectory of a bouncing ball.

Projectiles are objects that move in two dimensions under the influence of only the gravitational force, or weight $\vec{W}$. Projectile motion is an extension of the free-fall motion we studied in Chapter 3. As we examine projectile motion we will continue to neglect the influence of air resistance. The results will be a good approximation of reality for objects moving relatively slowly over relatively short distances.

[**Photo suggestion: A diver or gymnast in the air.**]

If the only force acting on an object is its weight, pointing in the $-y$-direction, then we have a situation where $a_x = 0$ but $a_y \neq 0$. As we noted in Example 6-2, these are the conditions for a *parabolic trajectory* and, indeed, projectiles do move along parabolic paths. Figure 6-8 shows a multiple exposure strobe photograph of a bouncing ball; the parabolic trajectory between bounces is quite obvious. Notice, from the distance between images, how the ball looses speed as it rises and gains speed as it falls.

The start of a projectile's motion, be it thrown by hand or shot from a gun, is called the *launch*. The angle θ of the initial velocity $\vec{v}_0$ above the ground (and thus above the x-axis) is called the **launch angle**. Figure 6-9 illustrates the relationship between the initial velocity vector $\vec{v}_0$ and the initial values of the components v_{x0} and v_{y0}:

$$v_{x0} = |\vec{v}_0|\cos\theta$$

$$v_{y0} = |\vec{v}_0|\sin\theta. \tag{6-10}$$

The vector components v_{x0} and v_{y0} are not always positive. In particular, a projectile launched at an angle *below* the horizontal (such as a ball thrown downward from the roof of a building) has *negative* values for both θ and v_{y0}. The *speed* $|\vec{v}_0|$, however, is always a positive quantity.

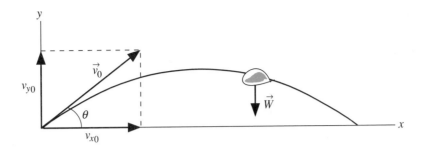

FIGURE 6-9 The trajectory of a projectile that is launched with initial velocity $\vec{v}_0$ at angle θ.

Because a projectile has $\vec{F}_{net} = \vec{W} = -mg\,\hat{j}$, Newton's second law is:

$$a_x = \frac{(F_{net})_x}{m} = 0$$

$$a_y = \frac{(F_{net})_y}{m} = \frac{-mg}{m} = -g.$$

(6-11)

In other words, the vertical acceleration is just the familiar $-g$ of free fall, while the horizontal acceleration is zero. Because $a_x = 0$, the horizontal component of the velocity *will not change*,

$$v_{x1} = v_{x0} = |\vec{v}_0|\cos\theta = \text{ constant}$$

$$\Rightarrow x_1 = x_0 + v_{x0}\Delta t.$$

(6-12)

That is, the horizontal component of projectile motion is just uniform motion at constant velocity v_{x0}. Vertically there is an acceleration $a_y = -g = $ constant. Thus the vertical component of the velocity will *decrease* by g m/s every second:

$$v_{y1} = v_{y0} - g\Delta t$$

$$\Rightarrow y_1 = y_0 + v_{y0}\Delta t - \frac{1}{2}g(\Delta t)^2.$$

(6-13)

Equation 6-13 is exactly what we found before for one-dimensional free-fall along the y-axis (Eqs. 3-18), but now we must add to it an x-component of *uniform* motion. Equations 6-12 and 6-13 are *parametric equations* for the trajectory of a projectile. The resulting motion, as seen in Fig. 6-9, is a parabola.

Figure 6-10 shows a projectile launched from $(x_0, y_0) = (0, 0)$ with an initial velocity $\vec{v}_0 = (9.8\,\hat{i} + 19.6\,\hat{j})$ m/s. The velocity vector is then shown every 1.0 s. Notice that the value of v_x never changes because $a_x = 0$. But v_y decreases by 9.8 m/s² every second. This is what it means to have an acceleration of -9.8 m/s² $=$ $(-9.8$ m/s) per second. No force *pushes* the

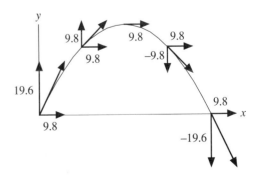

FIGURE 6-10 The velocity vector shown every 1 s for a projectile.

projectile along the curve; instead, the downward weight force causes a downward acceleration that *changes* the velocity vector as shown.

You should note several significant features of the parabolic trajectory, as illustrated in Fig. 6-10. Most important is to recognize that the x- and y-components of the motion are independent of each other! The vertical motion is free-fall, while horizontally the motion is uniform. Both proceed simultaneously, but neither influences the other. Second, the magnitude $|\vec{v}|$ decreases as the particle rises and increases again as it falls, but the magnitude is *not* zero at the top. The particle still maintains its horizontal motion, so $|\vec{v}_{top}| = v_{x0}$. This differs from the one-dimensional problems we considered previously, where the top of the trajectory was characterized by $v = 0$. Third, when the projectile returns to its original level $v_y = -19.6$ m/s $= -v_{y0}$. The relationship $(v_y)_f = -(v_y)_i$ is often useful for problems in which the projectile lands *at the same level* from which it was launched.

EXAMPLE 6-3   A stunt man drives a car at speed $|\vec{v}_0|$ off a horizontal cliff of height h. How far does the car land from the base of the cliff?

SOLUTION This is a problem in which you are given only symbols, not numbers. Thus the question "How far?" cannot be answered numerically. A problem such as this is asking for an *algebraic* expression for the distance from the cliff. You can interpret the speed $|\vec{v}_0|$ and height h as known information, so your answer should be given in terms $|\vec{v}_0|$ and h as well as, possibly, constants such as g.

Figure 6-11 shows the pictorial model for this problem. The number of symbols grows rapidly in trajectory problems, so it is especially important to use a pictorial model to define them and their values. The horizontal nature of the cliff dictates that $\theta = 0°$, which simplifies the problem a bit by making $v_{x0} = |\vec{v}_0|$ and $v_{y0} = 0$. We have chosen to put the origin at the base of the cliff, making $y_0 = h$ and $y_1 = 0$. It is customary in projectile problems to call the total horizontal distance the **range R**, so we have defined $x_1 = R$.

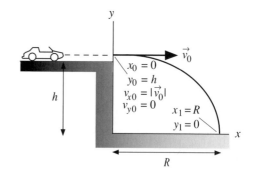

FIGURE 6-11 Pictorial model for Example 6-3.

In this problem, the horizontal motion is uniform and we know the horizontal velocity. To find the distance R, we will need the time interval $\Delta t = t_1 - t_0 = t_1$ that the motion lasts. This time interval, though, is simply the time it takes the car to *fall* a distance h from rest (because $v_{y0} = 0$). Thus we can use the vertical motion to find the flight time Δt and then to use that result to determine the horizontal motion. This is an additional piece of strategy common to projectile problems.

As always, we want to start with the second law. In this case, $\vec{F}_{net} = -mg\,\hat{j}$, so

$$a_x = \frac{(F_{net})_x}{m} = 0$$

$$a_y = \frac{(F_{net})_y}{m} = \frac{-mg}{m} = -g.$$

This is the acceleration of a projectile, so the trajectory is described by Equations 6-12 and 6-13. Applied to this situation, these are

$$\left(x_1 = R\right) = x_0 + v_{x0}(t_1 - t_0) = |\vec{v}_0| t_1$$

$$\left(y_1 = 0\right) = y_0 + v_{y0}(t_1 - t_0) - \tfrac{1}{2}g(t_1 - t_0)^2 = h - \tfrac{1}{2}g t_1^2.$$

As already noted, we first want to use the vertical equation to determine the time of flight:

$$\Delta t = t_1 = \sqrt{\frac{2h}{g}}\ .$$

We then insert this value into the horizontal equation to find the range:

$$R = |\vec{v}_0| \sqrt{\frac{2h}{g}}\ .$$

This result says that the range depends, not surprisingly, on both the car's speed and the height of the cliff. However, it does indicate that speed is the more important factor because v_0 enters as the first power while h enters as a square root. In other words, to double R, you would only have to double $|\vec{v}_0|$ if h is fixed, but you would have to *quadruple* h if $|\vec{v}_0|$ is fixed. To get a feeling for actual values, if $h = 30$ m (≈100 feet) and $|\vec{v}_0| = 20$ m/s (≈40 mph), calculation gives $R = 49.5$ m (≈160 feet).

Figure 6-12 illustrates another way to think about the parabolic trajectory of a projectile. The upper trajectory shows how a particle would move in the *absence* of gravity—simply a straight-line motion, in agreement with the first law. Because of gravity, however, the particle has "fallen" a distance $\tfrac{1}{2}gt^2$ below this line at time t, as shown at two different times. This separation between the straight line and the actual trajectory, a separation that grows according to $\tfrac{1}{2}gt^2$, is the cause of the parabolic shape. This diagram can help you to understand a couple of "classic" problems in physics.

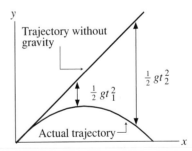

FIGURE 6-12 A projectile trajectory seen as "falling" a distance $\tfrac{1}{2}gt^2$ below a straight-line trajectory.

First, consider a rifle that fires a bullet exactly horizontally, a height h above the ground. At the exact instant that the first bullet is fired, a second bullet is dropped from directly beside the rifle. Which bullet hits the ground first? (Assume that the ground in front of the rifle is perfectly level.) In the absence of gravity, the trajectory of the first bullet would be perfectly horizontal and the bullet would not fall. Because gravity does exist, the bullet falls $\tfrac{1}{2}gt^2$ beneath the horizontal line. But the dropped bullet also falls distance $\tfrac{1}{2}gt^2$ in time t. Both bullets fall the same distance during equal intervals of time, so they will strike the ground *simultaneously*. The key to understanding this problem is to remember that the horizontal and vertical components of the first bullet's motion are *independent* of each other. Both bullets have an initial value $v_{y0} = 0$, so the two bullets will have *identical y-motions* and

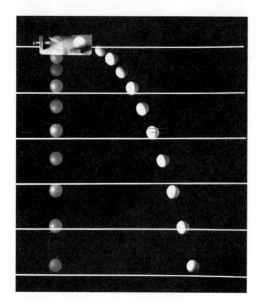

FIGURE 6-13 A projectile launched horizontally falls in the same time as a projectile that is released from rest.

will hit the ground at the *same time*! The horizontal motion of the first bullet has *no* effect on the time it takes to fall vertically through a distance h. Figure 6-13 shows a multiple-exposure photograph made of two balls—one shot horizontally and the other released from rest at the same instant. The *vertical* motions of the two balls are identical, and they hit the floor simultaneously.

Here is another "classic" motion problem. A hunter in the jungle wants to shoot a monkey who is hanging from the branch of a tree. He aims the gun directly at the monkey, but the monkey lets go of the branch at the *exact* instant the hunter pulls the trigger. Does the hunter succeed in shooting the monkey? To answer this, we have to keep in mind that the bullet goes in a parabolic trajectory, not a straight line, because of the downward weight force on it. In the absence of gravity, the bullet would follow the straight line shown in Fig. 6-14. Because of gravity, however, the bullet falls a distance $\frac{1}{2}gt^2$ during the time $t = D/v_{x0}$ it takes the bullet to move a horizontal distance D. But $\frac{1}{2}gt^2$ is also the distance the monkey falls during time t while the bullet is in transit. Thus the bullet and the monkey will fall the same distance and will meet at the same point!

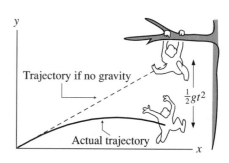

FIGURE 6-14 The Hunter and the Monkey problem.

• **EXAMPLE 6-4**   A baseball is hit with velocity $\vec{v}_0$ at angle θ. a) If the ball is caught at the same height at which it was hit, what is the horizontal distance traveled? b) If the ball is hit at a 30° angle, with what speed must it leave the bat to travel 100 m? c) What angle will yield the maximum distance the ball can travel?

SOLUTION   a) Figure 6-15 shows the pictorial model for this problem, where the horizontal distance is the range R. The height above the ground at which the ball was hit and caught is not relevant because they are the same, so we have located the origin to give both $y_0 = 0$ and $y_1 = 0$. The solution is identical to Example 6-3 up to the point where we develop the horizontal and vertical kinematic equations for

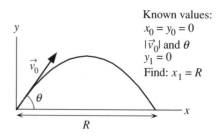

FIGURE 6-15 Trajectory of a baseball.

x and y. In this case, the equations are the same, but the initial values differ:

$$\left(x_1 = R\right) = x_0 + v_{x0}(t_1 - t_0) = |\vec{v}_0|\cos\theta\, t_1$$

$$\left(y_1 = 0\right) = y_0 + v_{y0}(t_1 - t_0) - \frac{1}{2}g(t_1 - t_0)^2 = |\vec{v}_0|\sin\theta\, t_1 - \frac{1}{2}g t_1^2 \,.$$

As in Example 6-3, we want to use the vertical equation to find the time of flight:

$$0 = |\vec{v}_0|\sin\theta\, t_1 - \frac{1}{2}g t_1^2 = (|\vec{v}_0|\sin\theta - \frac{1}{2}g t_1)\cdot t_1$$

$$\Rightarrow t_1 = 0 \text{ or } \frac{2|\vec{v}_0|\sin\theta}{g}.$$

Both values are legitimate solutions—the first corresponding to the time when $y = 0$ at the beginning of the trajectory and the second to when $y = 0$ at the end. Clearly, though, we want the second solution and can use it to find the range equation for R:

$$R = \frac{2|\vec{v}_0|^2\sin\theta\cos\theta}{g} = \frac{|\vec{v}_0|^2\sin(2\theta)}{g},$$

where we have used $2\sin\theta\cos\theta = \sin(2\theta)$.

b) A typical fly ball travels about 100 m. If it is hit at a 30° angle, we can solve the range equation to find that the ball must leave the bat with speed $|\vec{v}_0| = 33.6$ m/s (≈ 70 mph).

c) To find the maximum range, we need to examine the form of the range equation: What value of θ will maximize R? As you know, the sine function has a maximum value of 1 at an angle of 90°. Because the range equation contains the expression $\sin(2\theta)$, it will reach a maximum for $\theta_{max} = 45°$. If the batter hitting the ball with a speed of 33.6 m/s had hit it at a 45° angle, the ball would have traveled 115 m and just cleared the left field wall (at 110 m)—winning the World Series and bringing fame and fortune to our hero. Instead, though, the player hit at a mere 30° angle, the ball was caught, and history has forgotten his name. •

Calculating the range of projectiles obviously has many applications other than baseballs. In fact, the air resistance on a fairly light object such as a baseball is *not* entirely negligible, and real baseballs fall somewhat short of the calculated range. But a heavy projectile—say a cannon ball—follows the parabolic trajectory quite accurately. Our analysis that maximum range is achieved at a specific angle of 45° represents a significant understanding of the nature of projectile motion. Galileo phrased it well in *Two New Sciences*:

> *Sagredo:* The force of rigid demonstrations such as occur only in mathematics fills me with wonder and delight. From accounts given by gunners, I was already aware of the fact that in the use of cannons and mortars the maximum range is obtained when the angle of elevation is 45°; but to understand why this happens far outweighs the mere information obtained by the testimony of others.

The range equation has yet more information that we can uncover. Because $\sin(180° - x) = \sin x$, it follows that $\sin(2[90° - \theta]) = \sin(2\theta)$. This tells us we will achieve the same range R if we launch the projectile either at angle θ *or* at angle $90° - \theta$. All that

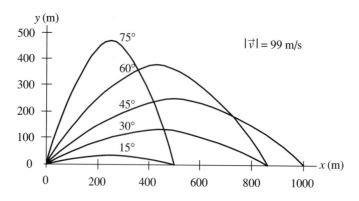

FIGURE 6-16 Trajectories of a projectile launched at different angles with a speed of 99 m/s.

will differ is the maximum *height* of the trajectories (which it is left as a homework problem to find). Figure 6-16 shows projectiles launched, all with the same initial speed, in 15° increments. (This figure is computed directly from the *x*- and *y*-equations given previously.) You can see clearly that maximum range is achieved at 45°, which is the unique angle where 90° − θ = θ, and that any shorter range can be achieved by two different launch angles.

As you compare Examples 6-3 and 6-4, you should notice that the way we solved the problems was nearly identical. In fact, we could go so far as to assert that there is really only *one* projectile problem, and that we have solved it by finding $a_x = 0$ and $a_y = -g$. The rest of any specific problem is then just kinematics, so the details will vary depending upon the information you are given and the result you wish to find. You will find projectile problems much easier if you recognize that they are all just variations on a basic problem, which we have now solved, rather than thinking of each one as a brand new problem.

6.4 Uniform Circular Motion

Projectile motion is just one example of motion in a plane. Equally important, and with a vast number of applications, is circular motion. Nearly every mechanical device has parts, such as wheels or gears, that rotate. As you drive your car around curves on the highway you are moving along the arc of a circle. And the orbital motions of satellites around the earth, planets around the sun, or electrons around atoms are, at least to a first approximation, circular. In this text, we are going to restrict our analysis to particles that move in a circle at *constant speed*. This motion is called **uniform circular motion**.

[**Photo suggestion: A roller coaster doing a loop-the-loop.**]

As you will recall, we looked at an example of uniform circular motion in Chapter 1 (Example 1-4). The motion diagram from that example is shown again in Fig. 6-17.

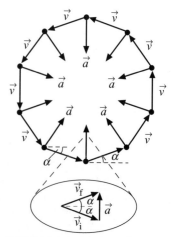

FIGURE 6-17 Motion diagram for uniform circular motion.

Even though the object moves with constant speed, it still has an acceleration. This is because the object's velocity $\vec{v}$—which has a *direction* as well as a magnitude—is always changing even though its speed $|\vec{v}|$ remains constant. Acceleration is the rate of change of velocity, not the rate of change of speed, so a changing velocity direction requires an acceleration. The motion diagram analysis shows that an acceleration pointing *toward the center* is needed to turn the velocity vector around the circle without changing its length.

We call this particular type of acceleration a **centripetal acceleration**, a term from a Greek root meaning *center seeking*. We will use the symbol $\vec{a}_c$ to represent a centripetal acceleration. Note that all we are doing is *naming* an acceleration that happens to act in a particular way; we are *not* introducing a new concept or a new type of acceleration.

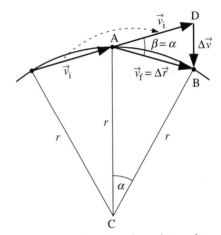

While we know the direction of $\vec{a}_c$, we still need a quantitative relationship between the centripetal acceleration and the velocity. We can find this relationship from a motion diagram analysis. Figure 6-18 shows three points from a motion diagram as a particle moves around a circle at constant speed $|\vec{v}|$. Velocity vector $\vec{v}_i$ joins the first two points while $\vec{v}_f$ joins the second two. If we slide $\vec{v}_i$ forward so that the tails of the two velocity vector are together, as shown, the change in the velocity $\Delta\vec{v} = \vec{v}_f - \vec{v}_i$ is then the vector that points straight down from the tip of $\vec{v}_i$ to the tip of $\vec{v}_f$.

FIGURE 6-18 Finding the relationship between velocity and acceleration for circular motion.

Because the motion is at constant speed, $|\vec{v}_i| = |\vec{v}_f|$. Therefore triangle ABD is an isosceles triangle of angle β. Triangle ABC is also isosceles, of angle α, because sides AC and BC are both of length r. It is not hard to show, although we will skip the geometrical details, that the angles are equal: $\alpha = \beta$. Consequently, ABC and ABD are *similar triangles*. Recall from geometry that the ratios of the sides of similar triangles are equal, so:

$$\frac{|\Delta\vec{v}|}{|\vec{v}|} = \frac{|\Delta\vec{r}|}{r}, \qquad (6\text{-}14)$$

where the speed is $|\vec{v}| = |\vec{v}_i| = |\vec{v}_f|$. The acceleration is, by definition, $\vec{a} = \Delta\vec{v}/\Delta t$. Using $|\Delta\vec{v}| = |\vec{v}||\Delta\vec{r}|/r$ from Eq. 6-14, the magnitude of $\vec{a}$ is

$$|\vec{a}| = \frac{|\Delta\vec{v}|}{\Delta t} = \frac{|\vec{v}||\Delta\vec{r}|}{r\Delta t} = \frac{|\vec{v}|^2}{r}. \qquad (6\text{-}15)$$

In the last step we made use of $|\Delta\vec{r}|/\Delta t = |\vec{v}|$. Strictly speaking, this analysis has been in terms of the *average* velocities and the *average* acceleration. If we consider the limit $\Delta t \to 0$, the three points on the motion diagram of Fig. 6-18 move together but the geometrical relationships do not change. Thus Eq. 6-15 remains valid when we take the limit, and it applies equally well to instantaneous velocities and accelerations.

The particular acceleration we have found is a centripetal acceleration, one pointing to the center of a circular trajectory. Our conclusion from this analysis is that the centripetal acceleration for a particle with uniform circular motion of radius r at speed $|\vec{v}|$ is

$$\vec{a}_c = \left(\frac{|\vec{v}|^2}{r}, \text{ toward center of circle} \right) \quad \text{(centripetal acceleration).} \quad (6\text{-}16)$$

For a particle moving in a circle, the time interval needed to complete one revolution is called the **period**, which we give the symbol T and define as

$$T = \text{the period} = \Delta t \text{ (one complete circle).} \quad (6\text{-}17)$$

The period measures the number of seconds required to go once around the circle.

A related concept is **frequency**, which is the number of revolutions that are completed in one second. The symbol for frequency is f. Clearly, period and frequency must be related to each other. If a particle moves around a circle f times in 1 second, the amount of time required for each revolution is $\Delta t_{\text{per rev}} = 1/f$ second. That is, the one second is divided into f equal parts. But this time interval for one revolution is just the period T, so we have

$$T = \frac{1}{f}. \quad (6\text{-}18)$$

Frequency is measured as "revolutions per second" or "cycles per second." Cycles and revolutions, however, are not actual units (like seconds or kilograms) because you are simply *counting* rather than *measuring*. If a particle revolves 5 times in one second, we may *say* that its frequency is "5 revolutions per second." But the proper SI units are simply 5 s^{-1}. A fairly common unit for frequency is "revolutions per minute," or "rpm's." This is not an SI unit, so you must convert any frequency given in rpm's to s^{-1} before doing calculations.

It is important to distinguish carefully between the scalar period T and the vector tension $\vec{T}$. Likewise, to distinguish between frequency f and the friction force $\vec{f}$. This is another situation of potentially confusing notation.

Consider a particle moving around a circle of radius r at a constant speed $|\vec{v}|$, as shown in Fig. 6-19. As the particle moves one time around the circle at speed $|\vec{v}|$, it travels a distance of one circumference during a time interval of one period. These must be related by

$$|\vec{v}| = \frac{1 \text{ circumference}}{1 \text{ period}} = \frac{2\pi r}{T} = 2\pi r f. \quad (6\text{-}19)$$

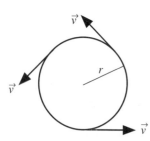

Speed = $|\vec{v}|$ = constant

FIGURE 6-19 A particle moving in a circular trajectory.

This is an important kinematic relationship for many problems. But rather than memorize it, it is helpful to remember how we derived it: speed = (distance/time) for one revolution. Then you can always find the relationship quickly when you need it.

EXAMPLE 6-5 You will later learn about the *Bohr atom*, which is a model of the hydrogen atom in which an electron orbits a proton with a frequency of 6.59×10^{15} revolutions per second at a radius of 5.29×10^{-11} m. What is the magnitude of the electron's centripetal acceleration?

SOLUTION This is a straightforward computation:

$$|\vec{v}| = 2\pi r f = 2\pi (5.29 \times 10^{-11} \text{ m})(6.59 \times 10^{15} \text{ s}^{-1}) = 2.19 \times 10^{6} \text{ m/s}$$

$$\Rightarrow |\vec{a}_c| = \frac{|\vec{v}|^2}{r} = \frac{(2.19 \times 10^6 \text{ m/s})^2}{5.29 \times 10^{-11} \text{ m}} = 9.07 \times 10^{22} \text{ m/s}^2.$$

This was not intended as a profound problem, but merely illustrates how a centripetal acceleration is computed. In addition, it demonstrates the unbelievably enormous accelerations and frequencies that take place at the atomic level. It should then come as no surprise that atomic particles may behave in ways that our intuition, trained by accelerations of only a few m/s^2, cannot easily grasp.

EXAMPLE 6-6 A car tire is 60 cm in diameter. If the car travels at a steady 20 m/s ($\approx$40 mph), what is the magnitude of the acceleration of a rock stuck in the treads?

SOLUTION Although the car itself has no acceleration, the edge of the tire has a centripetal acceleration. During one period of the tire's revolution, lasting $\Delta t = T$, the car rolls forward a distance Δx = one circumference of the tire = $2\pi r$. The car's velocity is thus

$$v_{\text{car}} = \frac{\Delta x}{\Delta t} = \frac{2\pi r}{T}.$$

But according to Eq. 6-19, $2\pi r/T$ is also the rotational speed of the rock as it moves with period T around a circle of radius r. That is, $|\vec{v}_{\text{rock}}| = v_{\text{car}} = 20$ m/s, where $|\vec{v}_{\text{rock}}|$ is the rotational speed of the rock. The magnitude of its centripetal acceleration is thus

$$|\vec{a}_c| = \frac{|\vec{v}_{\text{rock}}|^2}{r} = \frac{(20 \text{ m/s})^2}{0.3 \text{ m}} = 1330 \text{ m/s}^2.$$

Because a particle moving in a circle is undergoing an acceleration, Newton's second law tells us that there *must* be a net force responsible for that acceleration. In fact, the second law tells us exactly how much force is needed to cause a particle to accelerate in circular motion:

$$\vec{F} = m\vec{a}_c = \left(\frac{m|\vec{v}|^2}{r}, \text{ toward center of circle} \right) \qquad \text{(circular motion).} \qquad (6\text{-}20)$$

In other words, a particle needs a force of magnitude $(m|\vec{v}|^2)/r$ pointing toward the center of the circle to keep it moving in a circle. According to the first law, without such a force the particle would move off in a straight line that is tangent to the circle.

It is very important to emphasize that the force described by Eq. 6-20 is not a *new* force! Our rules for identifying forces have not changed. What we are saying is that a particle moves with uniform circular motion *if and only if* a net force always points toward the center of the circle. That force (or superposition of forces) will cause the particle to have a centripetal acceleration and to move with circular motion. Consider, for example, a ball on a string swinging in a circle. The string's tension acting on the ball *has* to point along the direction of the string. Because the tension force is pointing to the center of a circle, tension is the force causing the ball to have a centripetal acceleration and to move in a circle. Equation 6-20 simply tells us *how much* tension is needed if the ball has speed $|\vec{v}|$ and radius r. If you cut the string, so that the tension force is no longer present, then the ball moves off in a straight line. Under other circumstances the tension force could cause the ball to move in a straight line. It is *how* the force acts, not the type of force, that determines the object's path of motion. The same forces, under different circumstances, will cause very different kinds of motion.

It will be useful to have a special coordinate system for drawing free-body diagrams in circular motion problems. Figure 6-20a shows an *edge* view of a circular trajectory—the particle is moving in and out of the plane of the page—with a coordinate axis drawn at the particle's position as it passes through the page. Figure 6-20b provides a "picture" to show this, but we will use the edge view of Fig. 6-20a as we work problems. First, we define the z-axis as being perpendicular to the plane of the orbit. Then we define the r-axis along a *radius* of the circle from the particle (at the origin) through the center of the circle. The location of the center is marked with a C (for center) and the symbol ^. The crucial property of the r-axis is that the positive end of the axis is the end *toward* the center of the circle, and the negative end is the end *away from* the center of the circle. The r-axis in Fig. 6-20a has the positive end on the left, toward the center.

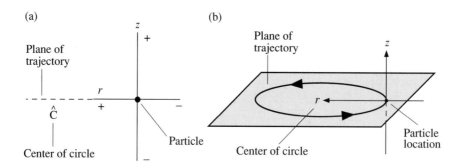

FIGURE 6-20 a) Coordinate system appropriate for circular motion, with r- and z-axes. b) Picture of circular motion, showing the r- and the z-axes.

Because there is no motion along the z-axis, $a_z = 0$. Therefore, the z-axis is an equilibrium axis. Along the r-axis, however, a_r is *not* zero. Because the particle is moving in a circle, a_r *has* to be

$$a_r = +|\vec{a}_c| = +|\vec{v}|^2/r.$$

We have explicitly included the + sign because vector $|\vec{a}_c|$ always points toward the center of the circle and, hence, by the way we defined the r-axis, the r-component of $|\vec{a}_c|$ will always be positive. Newton's second law for circular motion is thus

$$(F_{net})_z = \sum_i (F_i)_z = ma_z = 0$$

$$(F_{net})_r = \sum_i (F_i)_r = ma_r = \frac{m|\vec{v}|^2}{r}$$

(6-21)

where we have used the known values of a_r and a_z on the right-hand side. It is time for some examples to clarify these ideas.

EXAMPLE 6-7   An energetic father places his 20 kg child on a 5 kg cart to which a 2 m long rope is attached. He then holds the end of the rope and spins the cart and child around in a circle, keeping the rope parallel the ground. If the tension in the rope is 100 N, how many revolutions per minute (rpm) does the cart make? Assume friction between the cart's wheels and the ground is negligible.

SOLUTION Figure 6-21a shows how to draw the pictorial model for a circular-motion problem. Remember that the main idea of the pictorial model is to illustrate the relevant geometry and to define the symbols that will be used. A circular-motion problem does not have starting and ending points like a projectile problem, so numerical subscripts (such as x_1 and y_2) are usually not needed. Here we need to define the cart's speed $|\vec{v}|$ and the radius r of the circle. A motion diagram is shown in Fig. 6-21b, with the centripetal acceleration pointing toward the center of the circle. We will consider (cart+child) as a single particle of mass 25 kg.

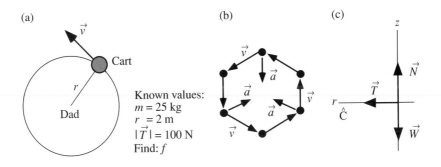

FIGURE 6-21 a) Pictorial model of man spinning a cart in a circle;
b) motion diagram; and c) free-body diagram.

The free-body diagram in Fig. 6-21c shows the edge view described previously, with axes r and z. The contact forces acting on the particle are the normal force of the ground and the tension force of the rope. The direction of $\vec{T}$ is determined by the statement that the rope is parallel to the ground. Because of the way we defined the r-axis, $\vec{T}$ points in the $+r$ direction and thus has an r-component $T_r = +|\vec{T}|$. In addition, there is the long-range weight force $\vec{W}$.

Newton's second law, in the form of Eq. 6-21, gives us the equations of motion:

$$\sum F_z = |\vec{N}| - |\vec{W}| = 0$$

$$\Rightarrow |\vec{N}| = |\vec{W}|$$

$$\sum F_r = |\vec{T}| = \frac{m|\vec{v}|^2}{r}$$

$$\Rightarrow |\vec{v}| = \sqrt{\frac{r|\vec{T}|}{m}} .$$

The z-equation is not needed in this case. Using this result for the speed, Eq. 6-19 gives the frequency:

$$f = \frac{|\vec{v}|}{2\pi r} = \frac{1}{2\pi}\sqrt{\frac{|\vec{T}|}{rm}} = \frac{1}{2\pi}\sqrt{\frac{100\ \text{N}}{(2\ \text{m})(25\ \text{kg})}} = 0.225\ \text{s}^{-1}.$$

Keep in mind that the results of calculations are in SI units, in this case s^{-1}. We must convert to the desired units of rpm's:

$$f = (0.225\ \frac{1}{\text{s}}) \times \frac{60\ \text{s}}{1\ \text{min}} = 13.5\ \frac{1}{\text{min}} = 13.5\ \text{min}^{-1} = 13.5\ \text{rpm}.$$

This has been a very typical circular-motion problem. You might want to think how the solution would change if the rope is *not* parallel to the ground or if the ground itself is not horizontal—that is, if the father were standing on top of a hill.

EXAMPLE 6-8   What is the maximum speed with which a 1500 kg car can turn a curve of radius 50 m on a level (unbanked) road without sliding?

SOLUTION This question, of obvious practical importance, raises all kinds of interesting issues. First, as shown in the pictorial model and motion diagram of Figs. 6-22a and

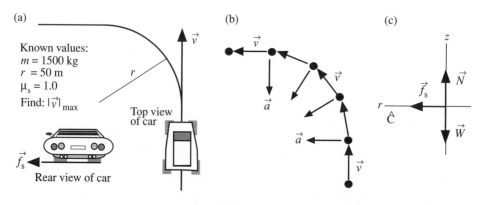

FIGURE 6-22 a) Pictorial model of car turning a curve; b) motion diagram; and c) free-body diagram. Inset in part a) shows the static friction force on the tires.

6-22b, we need to note that a particle can engage in circular motion without having to complete a full circle. In this case, the car moves along a circular arc at constant speed for the quarter-circle necessary to complete the turn. The motion before and after the turn is not relevant to the problem.

Having established that a circular-motion analysis is valid, we need to ask *how* it is that cars turn corners? What force or forces are responsible for the change of their velocity vector? Consider what would happen on a frictionless road—say a very icy road. The car would have no success attempting to turn the corner, it would just slide straight ahead in accordance with both the first law and the experience of anyone who has ever driven on icy roads! It must, then, be a *friction* force acting on the car that causes it to turn the corner instead of sliding straight ahead. *Not*, however, the friction force that slows and eventually stops the car if it runs out of gas—that force acts opposite the direction of motion and, hence, points toward the rear of the car. The friction force responsible for allowing the car to turn the corner pushes *sideways* on the tires, toward the center of the circle, as shown in the free-body diagram of Fig. 6-22c. This force appears only when the wheels are turned, which brings the *outside* edge of the tire around toward the direction of motion. Turning the wheels causes a friction force to push against the edge of the tire to keep the tire from sliding sideways on the road. By pushing against the edge of the tire, this force is *perpendicular* to the wheel and points toward the center of the circle. Furthermore, this is a *static* friction force, because it prevents the tire from sliding across the road surface. Thus static friction provides the force that causes the car to turn!

The size of the required friction force is proportional to $|\vec{v}|^2$, according to Eq. 6-20. Because the static friction force has a maximum size, there will be a maximum speed with which the car can turn without sliding. The maximum speed is reached when the static friction force reaches its maximum of $\mu_s|\vec{N}|$. If the car enters the curve at a speed higher than the maximum, the static friction force will not be large enough to provide a sufficient centripetal acceleration and the car will slide.

The free-body diagram, drawn from behind the car, shows the static friction force pointing toward the center of the circle. Vector $\vec{f}_s$ is in the $+r$-direction. Newton's second law gives:

$$\sum F_z = |\vec{N}| - |\vec{W}| = 0$$

$$\Rightarrow |\vec{N}| = |\vec{W}|$$

$$\sum F_r = |\vec{f}_s| = \frac{m|\vec{v}|^2}{r}$$

$$\Rightarrow |\vec{v}| = \sqrt{\frac{r|\vec{f}_s|}{m}}.$$

The speed will be a maximum when $|\vec{f}_s|$ reaches its maximum value of

$$|\vec{f}_s| = f_{s\,max} = \mu_s|\vec{N}| = \mu_s|\vec{W}| = \mu_s mg.$$

At that point,

$$|\vec{v}|_{max} = \sqrt{\frac{rf_{s\,max}}{m}} = \sqrt{\mu_s rg} = \sqrt{(1.00)(50\text{ m})(9.8\text{ m}/\text{s}^2)} = 22.1\text{ m}/\text{s},$$

where we have used μ_s from Table 5-1. Notice that the car's mass cancels out and that the final equation for $|\vec{v}|_{max}$ is quite simple. This is another example of why it pays to work algebraically until the very end. In assessing our result, we note that 22.1 m/s ≈ 45 mph—a very reasonable answer.

It is worth commenting on the significance of the result. Because μ_s depends on road conditions, the maximum speed of turns can vary dramatically. Wet roads, in particular, lower the value of μ_s and thus lower the speed of turns. A car that handles normally during straight-line motion can suddenly slide out of control when turning a corner on a wet road. Icy conditions are even worse—the corner you turn every day at 45 mph would require a speed of no more than 15 mph if the coefficient of friction on ice drops to 0.1.

[**Photo suggestion: Race cars on a steeply-banked curve.**]

EXAMPLE 6-9 a) A highway curve of radius 70 m is banked at a 15° angle. At what speed $|\vec{v}_0|$ can a car turn this curve without assistance from friction? b) Explore *qualitatively* the role of friction for speeds above and below $|\vec{v}_0|$.

SOLUTION Having just discussed the role of friction in turning corners, it is perhaps surprising to suggest that the same turn can also be accomplished without friction. Example 6-8, however, considered only a level roadway. Real curves are *banked*; that is, tilted up at the outside edge of the curve as shown in Fig. 6-23a. The angle is modest on ordinary highways, but it can be quite large on high-speed racetracks. The purpose of banking becomes clear if you look at the free-body diagram of Fig. 6-23b, which shows that the normal force $\vec{N}$ now has a component along the r-axis. This is rather like an inclined-plane problem, *but* make a special note that we are *not* using a tilted coordinate system. The center of the circle is in the same horizontal plane as the car, and for circular-motion problems we need the r-axis to pass through the center. Tilted axes are for *linear* motion along inclines.

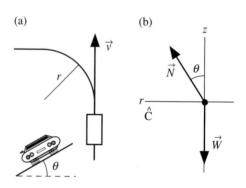

FIGURE 6-23 a) Pictorial model of a car on a banked curve, and b) the corresponding free-body diagram.

a) In the absence of friction, N_r is the only radial component of force, so it must be providing the inward force on the car that causes it to turn the corner. By tilting the road, the normal force acquires a component that is directed toward the center of the circle. Newton's laws applied to the free-body diagram of Fig. 6-23b give:

$$\sum F_z = |\vec{N}|\cos\theta - |\vec{W}| = 0$$

$$\Rightarrow |\vec{N}| = \frac{|\vec{W}|}{\cos\theta} = \frac{mg}{\cos\theta}$$

$$\sum F_r = |\vec{N}| \sin \theta = \frac{m|\vec{v}_0|^2}{r}$$

$$\Rightarrow |\vec{v}_0|^2 = \frac{r|\vec{N}| \sin \theta}{m} = \frac{rmg \sin \theta}{m \cos \theta} = rg \tan \theta$$

$$\Rightarrow |\vec{v}_0| = \sqrt{rg \tan \theta} = 13.6 \text{ m / s}.$$

This is ≈ 27 mph, a reasonable speed, so our assessment of the result is positive. Only at this very specific speed can the turn be negotiated without reliance on friction forces.

b) To explore the role of friction at other speeds, let us modify the free-body diagram to include a static friction force, as shown in Fig. 6-24a. It is very important to realize that $\vec{f}_s$, like any friction force, must be parallel to the surface. Therefore, $\vec{f}_s$ is tilted downward at angle θ. Because $\vec{f}_s$ has a component in the +r-direction, the *net* radial force is larger than that provided by $\vec{N}$ alone, thus providing a larger force directed toward the center. This is exactly what is needed to take the curve at $|\vec{v}| > |\vec{v}_0|$. Note that both the normal force *and* the friction force are contributing to the net force. We could, in a quantitative analysis, proceed to determine the maximum speed on a banked curve by analyzing Fig. 6-24a for the case that $|\vec{f}_s| = f_{s \, max}$, which is similar to Example 6-8.

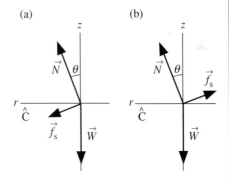

FIGURE 6-24 Free-body diagrams showing the static friction force if a) $|\vec{v}| > |\vec{v}_0|$ and b) $|\vec{v}| < |\vec{v}_0|$.

Perhaps more interesting, however, is to examine the case of taking the curve slowly: $|\vec{v}| < |\vec{v}_0|$. In this situation, the r-component of the normal force is too big—not that much center-directed force is needed. How can the size of the net force be reduced? By having $\vec{f}_s$ point *up* the slope, as in Fig. 6-24b, rather than down! This may seem very strange at first, but consider the limiting case of $|\vec{v}| = 0$, with the car parked on the banked curve. Were it not for a static friction force pointing *up* the slope, the car would slide down the incline. In fact, for any speed less than $|\vec{v}_0|$ the car will slip to the inside of the curve unless it is prevented from doing so by a static friction force pointing up the slope. Our analysis thus shows three divisions of speed. At $|\vec{v}_0|$, the car turns the corner with no assistance from friction. At greater speeds, the car will slide out of the curve unless an inward-directed friction force increases the size of the net force. And lastly, at lesser speeds the car will slip down the incline unless an outward-directed friction force prevents it from doing so.

EXAMPLE 6-10 A stone-age hunter places a 1 kg rock in a sling and swings it in a horizontal circle around his head on a rope 1 m long. If his rope breaks at a tension of 200 N, what is the maximum frequency with which he can swing the rock?

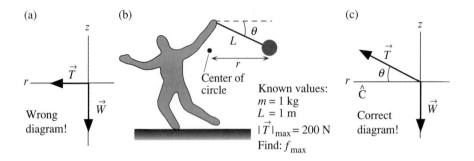

FIGURE 6-25 a) Incorrect view of forces; $\vec{T}$ is not horizontal. b) Pictorial model, showing that string is angled. c) Correct free-body diagram.

SOLUTION This problem appears, at first, to be the same as Example 6-7. However, a closer look reveals that the lack of a normal force from a supporting surface makes a *big* difference. In this case, the *only* contact force on the rock is the tension $\vec{T}$. Because the rock moves in a horizontal circle, you may be tempted to draw a free-body diagram like that in Fig. 6-25a where $\vec{T}$ is directed along the r-axis. You will quickly run into trouble, however, because the rock now has a net force in the z-direction. This make it impossible to satisfy $\Sigma F_z = 0$. The weight $\vec{W}$ certainly points vertically downward, so the difficulty must be with $\vec{T}$. Try tying a weight to a string, swing it over your head, and check the *angle* of the string. You will quickly discover that the string is *not* horizontal but, instead, is angled down as shown in Fig. 6-25b. The correct free-body diagram is thus the one in Fig. 6-25c. Notice that the downward weight force is now balanced by an upward component of the tension, leaving the radial component of the tension to cause the centripetal acceleration. Newton's laws give

$$\sum F_z = |\vec{T}|\sin\theta - |\vec{W}| = |\vec{T}|\sin\theta - mg = 0$$

$$\Rightarrow \quad \sin\theta = \frac{mg}{|\vec{T}|}$$

$$\sum F_r = |\vec{T}|\cos\theta = \frac{m|\vec{v}|^2}{r}$$

$$\Rightarrow \quad |\vec{v}| = \sqrt{\frac{r|\vec{T}|\cos\theta}{m}}.$$

The frequency is given by

$$f = \frac{|\vec{v}|}{2\pi r} = \frac{1}{2\pi}\sqrt{\frac{|\vec{T}|\cos\theta}{rm}}.$$

It is important to note that the radius r of the circle is *not* the length L of the rope. As shown in Fig. 6-25b, the center is not at the hunter's hand but is a horizontal distance

$r = L\cos\theta$ from the rock. Thus

$$f = \frac{1}{2\pi}\sqrt{\frac{|\vec{T}|}{Lm}}$$

$$\Rightarrow f_{max} = \frac{1}{2\pi}\sqrt{\frac{|\vec{T}|_{max}}{Lm}} = 2.25 \text{ revolutions per second.}$$

It turns out that we did not need the z-equation in Example 6-10, although you might not have been able to tell that when we started. Just for fun, though, let's use the z-equation to calculate the angle of the rope when the rock is swung at its maximum frequency: $\theta = \sin^{-1}(mg/|\vec{T}|_{max}) = 2.81°$. The rope's angle of inclination is small, but not zero.

A Nonexistent Force

Suppose you are riding in a car which makes a sudden stop. You may feel as if a force "throws" you forward toward the windshield. But there is no such force. What actually happens is that you try to continue forward with constant velocity—obeying Newton's first law—while the car that surrounds you is decelerating. Relative to the car, you do *seem* to be hurled forward, but an observer watching from beside the road would simply see you continuing forward while the car body is stopping. In an extreme case, say in an accident, the car stops so suddenly that you collide with the windshield if you are not wearing a seat belt. While we *say* that you were "thrown through the windshield," what really happened is that the car accelerated in the backward direction until the windshield ran into you! You cannot identify any forces "throwing" you forward—they are nonexistent.

A somewhat similar situation occurs when a car turns a corner quickly. Suppose you are the passenger and that the seats are very slippery. If the driver makes a quick left-hand turn, you find yourself "thrown" against the door of the car. If the door were not closed, you could even be "thrown" from the car. In amusement park rides that whirl in circles, you feel "thrown" to the outside of the car and pressed firmly against the wall. What force is "throwing" you toward the outside of the circle? Or is there such a force?

Figure 6-26 shows the top view of a passenger sitting on a very slippery bench in the back of a flatbed truck with nothing to hold on to. Suddenly the truck makes a left turn, and the passenger finds himself "thrown" over the side and into the road. But from a bird's eye perspective, we see that the passenger's motion really was quite simple—he just continued straight ahead at constant velocity while the truck *turned out from beneath him*! After all, there are no horizontal forces acting on him if the seat is very slippery (i.e., frictionless), so the first law says that he will continue to move in a straight line at constant speed. *Relative to the truck*, the passenger seems to have been thrown out, but the truck is *accelerating* as it turns, and accelerating coordinate systems are not proper places from which to view the situation. The bird's eye view, from a fixed coordinate system, gives the proper perspective of what happens.

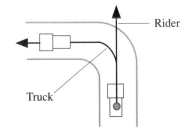

FIGURE 6-26 Bird's eye view of a passenger in the back of a truck as it turns a corner.

The situation is no different if you are in the passenger's seat: you try to continue moving in a straight line when the door suddenly turns right in front of you and runs into you! You do, indeed, then feel the force of the door pressing against you because it is now the normal force of the door, pointing inward toward the center of the curve, that is causing you to turn the corner. But you were not "thrown" into the door; it ran into you! If the seat has sufficient friction, the static friction force the seat exerts on you becomes the force that turns you through the curve along with the car. In that case you do not collide with the door.

The bottom line is that there is not any force that throws you or pushes you to the outside of a circle, any more than there is a force that "throws" you forward when the car stops. It is nonexistent. Such forces only *seem* to exist when you use an inappropriate coordinate system—one that is accelerating—to look at the situation.

As you no doubt know, the "force" that supposedly pushes an object to the outside of a circle is called the "centrifugal force." But what we see is that *there is no such thing as a "centrifugal force."* An object moving in a circle must have a net force pointing inward toward the center of the circle, and in the absence of such a force it will move in a straight line. This sometimes produces a conflict with the object's surroundings if those surroundings continue to move in a circle, and it is this conflict that is the origin of the "centrifugal force" idea. But Newtonian physics has no need for such a force—it is nonexistent. Other than in this one paragraph, you will never again see or hear the term "centrifugal force" in this text. Likewise, you should erase it from your memory banks!

6.5 Circular Motion and Apparent Weight

The idea of apparent weight that we introduced in Chapter 5 has many interesting applications to circular motion. Recall that apparent weight is the magnitude $|\vec{N}|$ of the normal force exerted on an object by the surface that supports it. If the surface is a spring scale, then $|\vec{N}|$ will be the reading of the scale—what the object *appears* to weigh—but the reading will differ from the true weight if the object is accelerating. Because an object moving in a circle at constant speed is always accelerating, there are many opportunities for the apparent weight to differ from the true weight. This is the source of much of the thrill of amusement park rides.

Consider a passenger on a Ferris wheel of radius r turning with constant frequency f, as shown in Fig. 6-27a. At the bottom of the circle she feels heavy and pressed into the seat—a case of her apparent weight exceeding her true weight. At the top she feels light and nearly "thrown out" of the seat because her apparent weight is less than her true weight. The only forces acting on the passenger are the normal force $\vec{N}$ of the seat pushing upward and her weight $\vec{W}$ pulling downward. Together, these must provide the centripetal acceleration $\vec{a}_c$ that keeps her moving in a circle. Because her speed is constant, the magnitude of the centripetal acceleration is also constant: $|\vec{a}_c| = |\vec{v}|^2/r = 4\pi^2 r f^2$, where Eq. 6-19 has been used to relate $|\vec{v}|$ to f.

Figure 6-27b shows a free-body diagram at the *bottom* of the circle. Because the woman is moving in a circle, we know that her acceleration *has* to point upward toward the center of the circle. The y-component of the net force on her is

$$(F_{\text{net}})_y = N_y + W_y = |\vec{N}| - |\vec{W}|$$

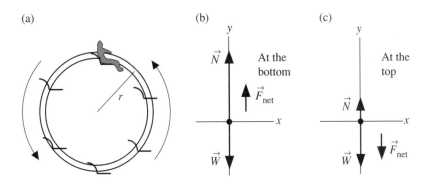

FIGURE 6-27 a) Ferris wheel; b) free-body diagram of a passenger at the bottom of the circle; and c) free-body diagram of a passenger at the top of the circle.

while the y-component of her acceleration is

$$a_y = +|\vec{a}_c| = \frac{|\vec{v}|^2}{r} = 4\pi^2 rf^2 .$$

Because $(F_{net})_y = ma_y$, we have

$$|\vec{N}| - |\vec{W}| = 4\pi^2 mrf^2$$

$$\Rightarrow W_{app} = |\vec{N}| = |\vec{W}| + 4\pi^2 mrf^2 \qquad (6\text{-}22)$$

$$= mg + 4\pi^2 mrf^2 .$$

The woman's apparent weight at the bottom is *larger* than her true weight mg by an amount $4\pi^2 mrf^2$.

Now consider what happens as the woman passes over the top of the circle. This situation is shown in Fig. 6-27c. The net force is still $(F_{net})_y = |\vec{N}| - |\vec{W}|$. Now, however, her acceleration vector—which always points toward the center—points *downward* and has a *negative y*-component:

$$a_y = -|\vec{a}_c| = -4\pi^2 rf^2 .$$

This time, $(F_{net})_y = ma_y$ gives

$$|\vec{N}| - |\vec{W}| = -4\pi^2 mrf^2$$

$$\Rightarrow W_{app} = |\vec{N}| = |\vec{W}| - 4\pi^2 mrf^2 \qquad (6\text{-}23)$$

$$= mg - 4\pi^2 mrf^2 .$$

We see that the woman's apparent weight, as expected, is *reduced* at the top of the circle.

Equation 6-23, however, presents a bit of a problem. What happens if the frequency f is increased to the point where $W_{app} < 0$? A negative apparent weight does not make physical sense. We need to keep in mind the source of the apparent weight: it is the force the woman experiences of the seat pushing up against her. As the frequency increases, it reaches a critical value f_0 where, according to Eq. 6-23, $|\vec{N}| = W_{app} = 0$. At that point the seat is *not* pushing against her any more—she and the seat are moving independently of each other as they cross the top of the circle and she is, momentarily, "weightless."

It might help to think about this situation in the following way. For small frequencies of revolution, the upward normal force at the top of the circle is only slightly less than the weight force, giving a net force downward that is just the right size to create a small centripetal acceleration. As the frequency increases, the necessary centripetal acceleration also increases. Because $|\vec{W}|$ cannot change, $|\vec{N}|$ has to decrease. When f reaches f_0, the weight *by itself* provides sufficient downward force for the woman to turn the top of the circle—so $|\vec{N}| = 0$. What happens if the frequency exceeds f_0? Now the woman needs *more* force directed downward, toward the center, than her weight can provide. If she is going to continue moving in the circle, the normal force needs to *pull* her down rather than to push her up. Thus, a *negative* y-component of $\vec{N}$ would be required. But surfaces can only push, they cannot pull! So what happens? The woman flies out of the seat! She simply *cannot*, with the forces available, move in a circle of radius r at a frequency $f > f_0$.

When a solution becomes physically impossible, it usually indicates that the analysis of the problem is incorrect. In this case, we *assumed* that the motion was circular when we set up the problem. But if we find that $|\vec{N}| < 0$, it tells us that our assumption is no longer valid—the woman is no longer moving in a circle but is off flying through space! The critical point where she loses contact with the seat is when $|\vec{N}| = W_{app} = 0$. In practice, the safety bar would prevent this by exerting a *downward* force on her—just the downward force needed to supplement her weight force so that she keeps moving in a circle.

EXAMPLE 6-11   A roller-coaster car shoots through a vertical loop-the-loop of diameter 10 m, as shown in Fig. 6-28a. a) What is the *minimum* speed the car must have at the top of the loop so that the car does not fall off the track? b) What is the apparent weight of a 100 kg passenger if the car reaches the bottom of the loop at twice this minimum speed?

SOLUTION a) Although the circular motion in this problem is not at constant speed, the only points we need to consider are the very top and very bottom of the circle. The roller-coaster car differs from the Ferris wheel because the contact force on it comes from the track beneath its wheels. Thus the force pushes up when the car is at the bottom, but it *presses down* when the car is at the top and the track is above the car. Figure 6-28b shows

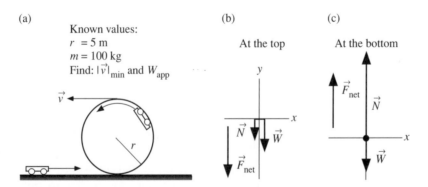

FIGURE 6-28   a) Roller coaster doing a loop-the-loop; b) free-body diagram at the top of the loop; and c) free-body diagram at the bottom of the loop.

a free-body diagram at the top of the loop. The net force, which points in the $-y$-direction, is

$$(F_{\text{net}})_y = N_y + W_y = -|\vec{N}| - |\vec{W}|.$$

The centripetal acceleration points downward, toward the center of the circle, so

$$a_y = -|\vec{a}_c| = -\frac{|\vec{v}|^2}{r}.$$

Make sure you understand why all the signs are negative. Thus $(F_{\text{net}})_y = ma_y$ gives

$$-|\vec{N}| - |\vec{W}| = -\frac{m|\vec{v}|^2}{r}$$

$$\Rightarrow |\vec{N}| = \frac{m|\vec{v}|^2}{r} - |\vec{W}| = m\left(\frac{|\vec{v}|^2}{r} - g\right).$$

For the car to be in contact with the track, it *must* have $|\vec{N}| > 0$. We see that high speeds are not a problem, but if $|\vec{v}|$ is decreased there will come a point where $|\vec{N}|$ goes negative and the solution becomes physically impossible. At this speed the car cannot turn the full loop but instead goes part way up, then comes off the track and begins a parabolic projectile motion! The critical point of minimum speed is defined by $|\vec{N}| = 0$, which happens when

$$|\vec{v}|_{\text{min}} = \sqrt{rg} = 7.00 \text{ m / s}.$$

Note that the problem statement gave the *diameter* as being 10 m, so we had to use $r = 5$ m in the calculation.

b) At the bottom of the loop, both the normal force of the track and the acceleration point upward. Figure 6-28c shows the free-body diagram at the bottom. Analysis of this diagram gives

$$\left[(F_{\text{net}})_y = +|\vec{N}| - |\vec{W}|\right] = \left[+m|\vec{a}_c| = \frac{m|\vec{v}|^2}{r}\right]$$

$$\Rightarrow W_{\text{app}} = |\vec{N}| = \frac{m|\vec{v}|^2}{r} + |\vec{W}| = m\left(\frac{|\vec{v}|^2}{r} + g\right).$$

Evaluation for $|\vec{v}| = 2|\vec{v}|_{\text{min}} = 14.0$ m/s, using $m = 100$ kg, gives $W_{\text{app}} = 4900$ N. This is five times the passenger's true weight of $|\vec{W}| = mg = 980$ N!

EXAMPLE 6-12 It is often proposed that future space stations create an artificial gravity by rotating. Suppose a space station was constructed as a large cylinder 200 m in diameter, with the inside surface of the cylinder serving as the "deck," as shown in Figure 6-29. What rotation frequency would be necessary to provide "normal" gravity?

SOLUTION The only force acting on each space station passenger is the normal force $\vec{N}$ of the deck, which acts as the center-directed net force moving the passengers in a circle. The space-station occupants will feel "normal" if this force gives an apparent weight equal to their true weight: $|\vec{N}| = mg$. The normal force when rotating with speed $|\vec{v}|$ has to be

$$|\vec{N}| = m|\vec{a}_c| = \frac{m|\vec{v}|^2}{r} = 4\pi^2 mrf^2$$

where we've used $|\vec{v}| = 2\pi rf$. Equating $|\vec{N}|$ to the desired value of mg gives

$$4\pi^2 mrf^2 = mg$$

$$\Rightarrow f = \frac{1}{2\pi}\sqrt{\frac{g}{r}} = 0.050 \text{ revolutions per second.}$$

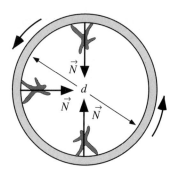

FIGURE 6-29 Space station with "gravity" created by rotation.

This is equivalent to $f = 3$ rpm, or to a period $T = 20$ s. A rotation this fast is likely to be objectionable for other reasons, so full-strength artificial gravity is unlikely for space stations of a size that might be constructed within the next few decades (although a slower spin to give a weak gravity is feasible). If future civilizations were to construct a large space colony of diameter 10 miles ≈ 16,000 m, the necessary rotation period would be $T = 180$ s = 3 min, which could easily be achieved.

6.6 Orbits

Let us return, for a moment, to projectile motion. We made an assumption in our analysis of projectiles that the earth is flat and that the acceleration due to gravity is always straight down. This is an acceptable approximation for projectiles of limited range, such as baseballs or cannon balls, but there clearly comes a point where we can no longer ignore the curvature of the earth.

Consider a perfectly smooth, spherical, airless earth with one tower of height h, as shown in Fig. 6-30. A projectile is launched from this tower. The motion is parallel to the ground ($\theta = 0°$) with speed $|\vec{v}_0|$. We want to know where it lands. If $|\vec{v}_0|$ is very small, as in trajectory 1, the flat-earth approximation is valid, and the problem is identical to Example 6-3 of the car driving off a cliff. The projectile simply falls to the ground along a parabolic trajectory of range R.

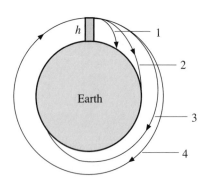

FIGURE 6-30 Projectiles being launched at increasing speeds from height h.

As the initial speed $|\vec{v}_0|$ is increased, the range reaches a point where we notice that the ground is curving out from beneath the projectile. It is still falling the entire time, getting ever closer to the ground, but the distance that it travels before finally reaching the ground—that is, its range—increases rapidly. This is because the projectile needs to "catch up" with the ground that is curving away from it. Trajectories 2 and 3 are of this type. The actual calculation of these trajectories is beyond the scope of this course, but you should be able understand the factors that influence the trajectory.

Because both the trajectory and the surface of the earth curves, there might be a situation where they have the same curve. In that case, the projectile would "fall," but it would never get any closer to the ground! This is the situation for trajectory 4. A circular trajectory around a planet or star, such as trajectory 4, is called an **orbit**.

To analyze this situation, consider a projectile traveling a horizontal distance d in a time interval $\Delta t = d/|\vec{v}_0|$. During this interval, as shown in Fig. 6-31, it falls a distance

$$\Delta y_{\text{proj}} = \frac{1}{2}g\Delta t^2 = \frac{gd^2}{2|\vec{v}_0|^2}. \qquad (6\text{-}24)$$

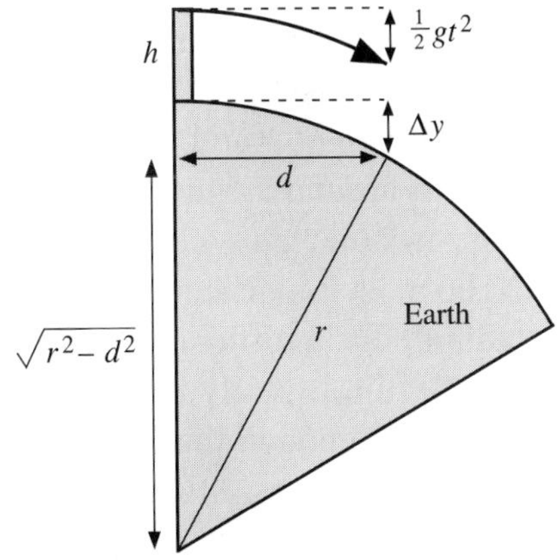

FIGURE 6-31 Finding the condition for which the particle "falls" the same distance that the earth curves.

Over this same horizontal distance, the surface of the earth curves downward by a distance that can be determined from geometry:

$$\Delta y_{\text{earth}} = r - \sqrt{r^2 - d^2} = r\left[1 - \sqrt{1 - \frac{d^2}{r^2}}\right]. \qquad (6\text{-}25)$$

Equation 6-24 is based on the flat-earth approximation. It is valid only if the distance d is very small. (For clarity, the figure greatly exaggerates the distance and the curvature.) Now r is the radius of the earth, about 4000 miles, so $r \gg d$. This means the ratio d^2/r^2 is a *very* small number. For example, if $d = 100$ miles then $d/r = 0.025$ and $(d/r)^2 = 0.000625$. Thus Eq. 6-25 takes the square root of $(1 - x)$ where x is a very small number. There is an approximation that you will learn about in calculus, but that we will just state for now, that

$$\sqrt{1 - x} \approx 1 - \frac{1}{2}x, \text{ if } x \ll 1. \qquad (6\text{-}26)$$

As an example, $\sqrt{0.99} = 0.99499$. But we can use Eq. 6-26 to approximate $\sqrt{0.99}$ as

$$\sqrt{0.99} = \sqrt{1 - 0.01} \approx 1 - \frac{1}{2}(0.01) = 0.99500.$$

The approximation differs from the exact value by only 0.00001—pretty good! If we use the approximation of Eq. 6-26 in Eq. 6-25, with $x = d^2/r^2$, we get

$$\Delta y_{\text{earth}} \approx r\left[1 - \left(1 - \frac{1}{2}\frac{d^2}{r^2}\right)\right] = \frac{d^2}{2r}. \qquad (6\text{-}27)$$

To find the conditions under which the projectile's trajectory and the earth's surface have the same curve, we need to equate Δy_{proj} from Eq. 6-24 to Δy_{earth} from Eq. 6-27:

$$\frac{gd^2}{2|\vec{v}_0|^2} = \frac{d^2}{2r} \qquad (6\text{-}28)$$

$$\Rightarrow |\vec{v}_0| = \sqrt{rg}.$$

This result says that for a very specific velocity, namely $|\vec{v}_0| = \sqrt{rg}$, the projectile falls but it never gets any closer to the earth! In the absence of air resistance, the projectile will still be at height h and traveling at speed $|\vec{v}_0|$ after completing one full revolution. It will then start around again. In other words, the projectile will *orbit* the earth in a circular trajectory if launched with this speed!

The most important point in this analysis is that an orbiting projectile is in free fall. This is, admittedly, a difficult idea to understand, but one worth some careful thought. An orbiting projectile is really no different than a baseball or a car driving off a cliff; it is simply moving fast enough that the curvature of its trajectory matches the curvature of the earth. When this happens, the projectile "falls" under the influence of gravity but never gets any closer to the surface, which curves away beneath it.

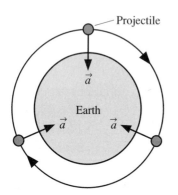

FIGURE 6-32 The acceleration $\vec{a}$ due to gravity as a centripetal acceleration.

This has been an analysis of orbital motion based on kinematics. Centripetal acceleration gives us another way to look at orbits. In the flat-earth approximation, the acceleration due to gravity is

$$\vec{a}_{\text{free fall}} = (g, \text{ vertically downward}) \quad \text{(flat earth)},$$

where $g = 9.80 \text{ m/s}^2$ on earth but has a different value on other planets. If we consider a spherical earth, however, we know that the force of gravity at any point is actually directed toward the *center* of the earth, as shown in Fig. 6-32. In this case the acceleration is

$$\vec{a}_{\text{free fall}} = (g, \text{ toward center}) \quad \text{(round earth)}. \quad (6\text{-}29)$$

As we have learned, an acceleration that always points toward the center of a circle provides a centripetal acceleration for circular motion at constant speed. A projectile moving in a circle of radius r at speed $|\vec{v}_0|$ will have the centripetal acceleration $\vec{a}_c = (g, \text{ toward center})$ if

$$|\vec{a}_c| = \frac{|\vec{v}_0|^2}{r} = g.$$

Solving this for $|\vec{v}_0|$ we find:

$$|\vec{v}_0| = \sqrt{rg}. \quad (6\text{-}30)$$

That is, *if* a projectile moves with the speed $|\vec{v}_0| = \sqrt{rg}$, then the acceleration due to gravity provides exactly the centripetal acceleration needed for a circular orbit. This result, of course, is exactly the same as Eq. 6-28, but you should think carefully about the two different methods we have used to derive it.

The earth has a radius $r = R_e = 6.37 \times 10^6$ m, from which we can calculate the orbital speed of a projectile just skimming the surface of our airless, bald earth: $|\vec{v}_0| = 7900$ m/s, or approximately 16,000 mph. Even if there were not problems such as trees and mountains, a real projectile moving at this speed would burn up from the friction of air resistance. Suppose, however, that we launched the projectile from a tower of height $h = 140$ miles $\approx 2.3 \times 10^5$ m, above the earth's atmosphere. This is approximately the height of low-earth-orbit satellites. Note that $h \ll R_e$, so the radius of the projectile's orbit is $r = R_e + h = 6.60 \times 10^6$ m—only 3.6% greater than the earth's radius. Most people have a mental image that satellites orbit far "above" the earth, but in fact they really come pretty close to skimming the surface. Our calculation of $|\vec{v}_0|$ for a projectile skimming the surface thus turns out to be quite a good estimate of the speed of satellites in low-earth orbit. (We will see later that the value of g differs slightly at a satellite's altitude, but not by much. Our calculation remains a good estimate.)

We can also use Eq. 6-19 to calculate the period of an orbit:

$$T = \frac{2\pi r}{|\vec{v}_0|} = 2\pi\sqrt{\frac{r}{g}}. \qquad (6\text{-}31)$$

For an earth-skimming orbit with $r = R_e$, we can calculate $T = 5070$ s $= 84$ minutes. This is again a very good estimate for satellites in low-earth orbit. (The Space Shuttle orbiting at an altitude of 200 miles has a period of 90 minutes.) So we have achieved an initial understanding of how satellite orbits work; it is not nearly as "exotic" as most people believe!

When we discussed "weightlessness" in Chapter 5, we discovered that it occurs during free fall. We asked the question, at the very end of Section 5.5, of whether astronauts and their spacecraft were in free fall. We can now give an affirmative answer to that question—they are, indeed, in free fall. They are falling continuously around the earth, under the influence of only the gravitational force, but never getting any closer to the ground because the earth's surface curves beneath them. The apparent "weightlessness" in space is thus no different than the "weightlessness" in a free-falling elevator. It does *not* occur from an absence of weight or an absence of gravity, but because the astronaut, the spacecraft, and everything in it (including any scales on which an astronaut attempts to weigh himself) are all falling together.

Gravity is the one long-range force with which we have dealt so far. Somehow, the earth can "reach out" through the vacuum of space and pull on an object. Because that pull is always directed toward the center of the earth, it causes the object to move in an orbit about the earth. We have seen that the orbit is circular if the object has just the right speed. A more extended analysis, which is beyond the scope of this text, shows that the orbit for other speeds is *elliptical*. Because gravity is the most ubiquitous force in the universe, it should come as no surprise that orbital motion is extremely common. The moon orbits the earth, the earth and other planets orbit the sun, binary star systems orbit their common center of mass, the entire solar system orbits the center of the galaxy, and giant clusters of galaxies orbit each other. Computers can be used to predict these orbits with incredible accuracy. They can predict, for example, the exact time of eclipses hundreds of years from now. The equations that these computers solve are just Newton's laws—they are calculating trajectories, exactly as you are learning to do, with just a few extra details (like the finite size of planets) thrown in.

We find the same ideas if we turn inward from the cosmos to the microscopic, with electrons pursuing their unimaginably fast orbits around the nucleus of each atom. In this case, it is the electromagnetic force, rather than the gravitational force, providing the inward-directed force that keeps the electrons orbiting. The electromagnetic force, like gravity, is a long-range force, spanning the inner void of the atom to pull the oppositely-charged electrons and protons together. Once we learn how to specify the electromagnetic force, however, Newton's laws take over and the physics presented in this chapter will determine the electron's speed and period—at least to the extent that the electron can be considered a classical particle.

We can leave this chapter with a peek ahead, where we will look at the gravitational force more closely. If a satellite is simply "falling" around the earth, using the gravitational force as a centripetal force, then what about the moon? Is it obeying the same laws of physics, or do celestial objects obey laws that we cannot discover by experiments here on earth? The radius of the moon's orbit around the earth is $r = R_m = 3.84 \times 10^8$ m. If we calculate the moon's period—the time it takes to circle the earth once, from Eq. 6-31, we get $T = 39,300$ s $= 655$ min $= 10.9$ hours. This is clearly wrong, because you know that the full moon occurs roughly once a month. More exactly, we know from astronomical measurements that the period of the moon's orbit is $T = 27.3$ days $= 2.36 \times 10^6$ s—a factor of 60 longer than we calculated it to be.

Should we conclude that our theory of orbits does not apply to something like the moon? Newton thought that it should. But, he reasoned, why should we assume that the magnitude of the acceleration due to gravity g be the same at the distance of the moon as it is on or near the earth's surface. If gravity is the force of the earth pulling on an object, it seems plausible that the size of that force, and thus the size of g, would diminish at larger distances away from the earth. Then g would be a function of the distance r away from the center of the earth. What value of g would be needed to explain the moon's observed period? Using $r = R_m$, we can calculate from Eq. 6-31 that $g_{\text{at moon}} = 4\pi^2 R_m/T^2 = 0.00272$ m/s², where we used data supplied in the previous paragraph. This is much less than the earth-bound value of 9.8 m/s². Now notice something interesting that we can find by playing with ratios:

$$\frac{g_{\text{on earth}}}{g_{\text{at moon}}} = \frac{9.80 \text{ m/s}^2}{0.00272 \text{ m/s}^2} = 3600$$

$$\left(\frac{R_m}{R_e}\right)^2 = \left(\frac{3.84 \times 10^8 \text{ m}}{6.37 \times 10^6 \text{ m}}\right)^2 = 3630$$

where R_e is the radius of the earth's surface, where $g = 9.8$ m/s², and where R_m is the radius of the moon's orbit, where $g = 0.00272$ m/s². These two ratios differ by less than 1%. This is unlikely to be purely a coincidence, particularly because we know that various factors, such as an earth that is not a perfect sphere and an orbit that is not a perfect circle, will make our calculations less than perfect. Let us tentatively conclude that these two ratios would be exactly equal for perfect spheres moving in perfect circles. What does this discovery imply?

$$\frac{g_{\text{on earth}}}{g_{\text{at moon}}} = \frac{R_m^2}{R_e^2} = \frac{1/R_e^2}{1/R_m^2}$$

$$\Rightarrow g \propto \frac{1}{r^2}.$$

(6-32)

In other words, if g has the value 9.80 m/s² at the earth's surface, and if g decreases in size depending upon the "inverse square" of the distance from earth, then g will have exactly the value it needs at distance $r = R_m$ to give the moon its observed orbital period. This reasoning forms the background for Newton's universal law of gravitation, which states that the gravitational force between two masses depends on the "inverse square" of the distance between them. This is a topic to which we will return in Chapter 12.

Summary

▲ Important Concepts and Terms

trajectory

projectile

launch angle

range

uniform circular motion

centripetal acceleration

period

frequency

orbit

This chapter has expanded our application of Newton's laws to a wider range of motions. After finishing this chapter, you should be able to work with particles in equilibrium as well as particles moving in straight lines, curved trajectories, and circles. As we keep emphasizing, the strategy to be followed for *all* dynamics problems is that described at the beginning of Chapter 5: draw pictures to establish the geometry and the symbols, identify the forces, draw a free-body diagram, then invoke Newton's laws. With this strategy, along with a few supplemental ideas like the model for friction and the speed-period-frequency relationships, you are prepared to solve a wide array of problems. Our emphasis in this text has been, and will continue to be, on learning and using thinking tools—not on memorizing formulas.

Newton's second law, applied to motion in a plane, is

$$a_x = \frac{(F_{\text{net}})_x}{m}$$

$$a_y = \frac{(F_{\text{net}})_y}{m}$$

where, in general, neither a_x nor a_y are zero. From the forces we can determine the components of the acceleration vector $\vec{a}$, and from the acceleration we can then deduce the kinematics of the motion. The only difference from one-dimensional motion is that we end up with two simultaneous equations $x(t)$ and $y(t)$ for the trajectory. We can, if we wish, eliminate t to express the result as an equation $y(x)$ of the trajectory.

Most of the attention in this chapter has been given to two particular types of motion in a plane. For *projectile motion* we have a simple force, $\vec{F}_{\text{net}} = -mg\,\hat{\jmath}$, from which we find the acceleration to be $\vec{a} = -g\,\hat{\jmath}$. Because $a_x = 0$, we have constant velocity motion along the x-axis simultaneous with constant acceleration motion along the y-axis. The kinematics are:

$$x_1 = x_0 + v_{x0}\Delta t$$

$$y_1 = y_0 + v_{y0}\Delta t - \tfrac{1}{2}g(\Delta t)^2.$$

The elapsed time Δt is the same in both equations, and it is common in problem solving to use the motion along one axis to determine the value of Δt, which is then used in the other component equation. Care must be given to determining the correct initial and final conditions.

Uniform circular motion is a case of a non-constant force. Although the force has a constant magnitude, it has a continuously changing direction. When the net force always points inward to a central point, the acceleration it causes is called a centripetal acceleration $\vec{a}_c$ and the motion is circular about the central point. Kinematics determine that $a_r = |\vec{a}_c| = |\vec{v}|^2/r$, where the r-axis points toward the center of the circle of radius r. Because there is no motion perpendicular to the plane (the z-axis), Newton's second law for circular motion becomes

$$(F_{net})_z = ma_z = 0$$

$$(F_{net})_r = ma_r = \frac{m|\vec{v}|^2}{r}.$$

It is important to realize that the forces causing circular motion are not new forces. They are, instead, familiar forces (such as tension) applied in new ways. The use of a free-body diagram to analyze the forces remains a critical part of the overall problem solving strategy.

Circular-motion kinematics relate the speed to the object's frequency f and period T:

$$|\vec{v}| = 2\pi rf = \frac{2\pi r}{T},$$

where the period is the inverse of the frequency. It is important to remember that constant acceleration kinematics, from Chapter 3, are *not* relevant to circular motion.

Exercises and Problems

Exercises

1. A physics student on Planet Exidor throws a ball, and it follows the parabolic trajectory shown in Fig. 6-33. The ball's position is shown at one-second intervals until $t = 3$ s. At $t = 1$ s, the ball's velocity is $\vec{v} = (2\hat{i} + 2\hat{j})$ m/s.
   a. Determine the ball's velocity at $t = 0$ s, 2 s, and 3 s.
   b. What is the value of g on Planet Exidor?
   c. What was the ball's launch angle?

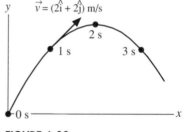

FIGURE 6-33

2. A rifle is aimed horizontally at a target 50 m away. The bullet hits the target 2 cm below the aim point.
   a. What was the bullet's flight time?
   b. What was the bullet's speed as it left the barrel?

3. Five balls are released simultaneously from the same height h above the ground, as shown in Fig. 6-34. Balls 1–4 all have the same initial speed but are launched at, respectively, angles of 45°, 0°, –45°, and –90°. Ball 5 is released from rest. What is the order, from fastest to slowest, of the speeds with which they hit the ground?

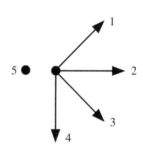

FIGURE 6-34

4. a. Find a general expression for the maximum height of a projectile launched with speed $|\vec{v}_0|$ and angle θ.

   b. A baseball is hit with a speed of 33.6 m/s. Calculate its maximum height if hit at angles of 30°, 45°, and 60°.

5. A *conical pendulum* is formed by attaching a 500 g mass to a 1 m long string, then allowing the mass the move in a horizontal circle of radius 20 cm. As shown in Fig. 6-35, the string traces out the surface of a cone, hence the name.

   a. What is the tension in the string?

   b. What is the mass's frequency in rpm?

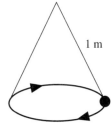

FIGURE 6-35

6. A highway curve of radius 500 m is designed for traffic moving at a speed of 90 km/hr. What is the correct banking angle of the road?

▲ Problems

7. A ball is thrown at a speed of 30 m/s at an angle of 60° toward a cliff of height h (Fig. 6-36). It lands on the cliff 4 s later.

   a. How high is the cliff?

   b. What was the maximum height of the ball?

   c. What is the ball's impact velocity?

   d. What is the ball's impact speed?

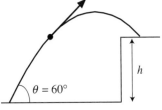

FIGURE 6-36

8. A stunt man drives a car at a speed of 20 m/s off a cliff 30 m high . The road leading to the cliff is inclined upward at an angle of 30°. a) How far from the base of the cliff does the car land? b) What is the car's impact speed?

9. A 100 g ball on a string 60 cm long is swung in a *vertical* circle about a point 200 cm above the floor. The tension in the string when the ball is at the very bottom of the circle is 5.0 N. During one revolution, a very sharp knife is inserted, as shown in Fig. 6-37 cutting the string directly below the point of support. Where does the ball hit the floor?

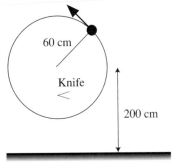

FIGURE 6-37

✍ 10. A rocket-powered hockey puck on frictionless ice is sliding along the y-axis in the positive direction with speed v_0 (Fig. 6-38). The front of the rocket is tilted an angle θ away from the x-axis. The rocket motor is ignited by remote control as the puck crosses the origin.

a. Find an algebraic expression $y(x)$ for the puck's trajectory.

b. Let the rocket have a thrust of 2 N, the total mass of the puck and rocket be 1 kg, and the initial speed be 2 m/s. Make a graph of your function $y(x)$ from $x = 0$ to $x = 20$ m for the three cases $\theta = 45°, \theta = 0°$, and $\theta = -45°$. (Calculate the positions for $x = 0, 2, 4, 6, \ldots, 20$ m and plot them on graph paper. Alternatively, use computer software to generate the graphs.) Show all three trajectories as different curves on a *single* set of graph axes.

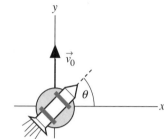

FIGURE 6-38

✍ 11. The rocketing hockey puck from Problem 10 b) is released from rest on a horizontal, frictionless surface, as shown in Fig. 6-39. It is 16 m from the edge of a 3 m drop. The front of the rocket is pointed directly toward the edge. How far will the puck land from the base of the cliff?

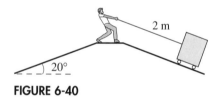

FIGURE 6-39

✍ 12. In Fig. 6-40, the father of Example 6-7 is standing on the top of a 20° hill as he spins his 20 kg child around on a 5 kg cart with a rope 2 m long. He again keeps the rope parallel to the ground, and friction is negligible. What rope tension will allow the cart to spin with the same 13.5 rpm it had in the example?

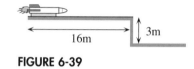

FIGURE 6-40

✍ 13. A concrete highway curve of radius 70 m is banked at a 15° angle. What is the maximum speed with which a 1500 kg rubber-tired car can turn this curve without sliding?

✍ 14. A 5 g coin is placed 15 cm from the center of a turntable. The coin has static and kinetic coefficients of friction with the turntable surface of $\mu_s = 0.80$ and $\mu_k = 0.50$. The turntable is very slowly sped up to a frequency of 60 rpm. Does the coin slide off?

✍ 15. In an old-fashioned amusement park ride, passengers stand inside a 5 m diameter hollow steel cylinder with their backs against the wall (Fig. 6-41). The cylinder begins to rotate and gradually speeds up to frequency f. Then the floor on which the passengers are standing suddenly drops away! If all goes well, the passengers will "stick" to the wall and not slide. Clothing has a static coefficient of friction against steel in the range 0.6 to 1.0 and a kinetic coefficient in the range 0.4 to 0.7. A sign next to the entrance says "No children under 50 kg allowed." What is the minimum rotation frequency f, in rpm, for which the ride is safe?

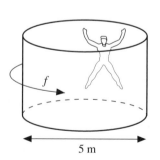

FIGURE 6-41

16. As shown in Fig. 6-42, a 10 g steel marble is spun so that it rolls around the *inside* of a vertically-oriented steel tube at 150 rpm. The tube is 12 cm in diameter. While "rolling resistance" will cause the marble to gradually slow down, assume that the rolling resistance is small enough for the marble to maintain ≈150 rpm for several seconds. Will the marble spin in a horizontal circle, at constant height, or will it spiral down the inside of the tube with a decreasing height?

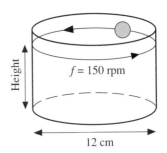

FIGURE 6-42

17. Suppose you swing a ball on a string 1 m long in a vertically-oriented circle at constant speed. As you probably know from experience, there is a *minimum* frequency of rotation f_{min} you must maintain if you want the ball to complete the full circle. If you try a frequency $f < f_{min}$, then the string "collapses" and goes slack as the ball crosses the top of the circle. What is f_{min}? Express your answer in rpm.

18. A ball of mass m is shot through a frictionless plastic tube having a circular cross section slightly larger than the diameter of the ball. The tube has been bent into a circular arc of radius r and it is oriented vertically (Fig. 6-43).
    a. If the ball crosses the top of the arc slowly, it will roll along the lower inside surface of the tube. Identify the forces on the ball at the top of the arc. Show the forces on a free-body diagram.
    b. If the ball is shot through the tube very rapidly, it will be pressed against the *upper* inside surface of the tube as it crosses the top of the arc. In this case, what forces act on the ball at the top of the arc? Show the forces on a free-body diagram.

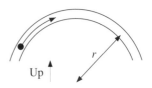

FIGURE 6-43

    c. At what speed or speeds will the ball have each of the following apparent weights as it crosses the top of the arc?
       i)   0
       ii) $\frac{1}{2}mg$
       iii) mg
       iv) $2mg$.

19. How does water stay in a bucket if you swing it rapidly over your head? To answer this, consider a somewhat simpler situation. Imagine placing a small ball of mass m in a bucket, then swinging the bucket in a *vertical* circle from your knees to over your head and back down. If you swing the bucket quickly enough, the ball stays in the bucket.
    a. If you swing the bucket fast enough for the ball to stay in, what force or forces act on the ball at the very top of the circle? Show these force on a free-body diagram.
    b. Give an explanation, in words, of *why* the ball stays in the bucket. Your explanation should be precise and it should be given in terms of physical laws and specific forces.
    c. If you swing the bucket too slowly, the ball falls out at the top of the swing. What force or forces are acting on it at the top of the circle in this case? Draw a free-body diagram.

   d. *Why* does the ball fall out when the speed is too slow? Your explanation should be precise and it should be given in terms of physical laws and specific forces.

   e. Measure the length of your arm. Then calculate the minimum frequency, in rpm, with which you must swing a bucket of water to avoid getting wet.

20. Communications satellites are placed in a circular orbit where they stay directly over a fixed place on the equator as the earth rotates. These are called *geosynchronous orbits*. The radius of a geosynchronous orbit is 4.22×10^7 m ($\approx 26,000$ miles).

   a. What is the period of a satellite in a geosynchronous orbit?

   b. Use the radius and period of the orbit to determine the value of g at the given radius.

   c. Is this value of g consistent with Newton's idea that $g \propto 1/r^2$? Why or why not?

   d. What is the apparent weight of a 2000 kg satellite in a geosynchronous orbit?

21. Planet X is in an alternate universe, with different laws of physics. The planet has a radius $R_X = 6.000 \times 10^6$ m and an acceleration due to gravity $g = 10.00$ m/s^2. Planet X has one moon, which orbits with radius $R_M = 4.000 \times 10^8$ m and has a period of 6009 hours. The great scientist Dr. Fig Newton suspects that the moon's orbit can be understood if g varies with distance from the center of the planet as some inverse power of r: $g \propto 1/r^n$. Analyze the data provided and determine the "law of gravitation" in this universe. (**Hint:** Your answer will be analogous to Eq. 6-32, but not the same.)

22. A satellite of mass m travels in a circular orbit of radius r about a center of force (Fig. 6-44). The magnitude of the attractive force exerted on the satellite by the center is a some power p of the radius of the orbit: $|\vec{F}| \propto r^p$. Experiments with different satellites show that the orbital period T is independent of r—that is, satellites orbiting in circles of *different* radii r have the *same* period T. (Note: This is *not* true for satellites orbiting the earth due to the gravitational force, where $p = -2$.) What is the value of p?

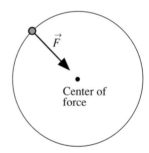

FIGURE 6-44

[**Estimated 10 additional problems for the final edition.**]

PART I Scenic Vista

The Forces of Nature

This is an opportune moment for a short rest break to look back over where we have come and to take a glimpse ahead. The major goal of Part I was to establish the connection between force and motion. Table SVI-1 summarizes what you have learned about the dynamics of single-particle motion.

Table SVI-1 represents the *knowledge structure* for single-particle dynamics. A knowledge structure summarizes the general physical principles, categories, and equations of a theory. The first section of the table indicates that we have introduced only one *general principle*, Newton's second law, and that principle has been the starting point for every dynamics problem we have solved. The second section lists the three *categories* of motion that we have studied: linear motion, trajectories, and circular motion. The specific form of the second law depends upon the category. Finally, the third section lists the kinematic *equations* that relate the acceleration to the position and velocity.

You use this knowledge structure by working your way through it, from top to bottom. Once you recognize a problem as a single-particle dynamics problem, you immediately know to start with Newton's second law. You can then determine the appropriate category of motion and apply the second law in that form. Lastly, you must determine which set of kinematic equations are appropriate and apply them to reach the solution you seek.

The structure represented by Table SVI-1 provides the *procedural knowledge* for solving single-particle dynamics problems, but it does not represent the total knowledge required. That is, it provides you with a set of procedures to follow. But you must add to it knowledge about what position and velocity are, about how forces are identified, about drawing and using free-body diagrams, and so on. These are *tools* for problem solving. The strategy we developed in Chapter 5 combines the procedures and the tools into a powerful method for thinking and for solving problems.

Looking Ahead

While we have shown how particles move in response to forces, we have not yet really explored the nature of the forces themselves. How do forces originate, and why do they have one magnitude and direction rather than another? In two cases, the normal force and

TABLE SVI-1 The knowledge structure for single-particle dynamics.

General Principle: Newton's second law $\vec{F}_{\text{net}} = m\,\vec{a}$

Force Law: Relate forces to acceleration

Linear Motion (x-direction)	Trajectory Motion	Circular Motion
$\sum F_x = ma_x$	$\sum F_x = ma_x$	$\sum F_r = m\lvert\vec{v}\rvert^2/r$
$\sum F_y = 0$	$\sum F_y = ma_y$	$\sum F_z = 0$

Kinematics: Use to find position and velocity

Linear and Trajectory Motions (applies to both x and y)   Circular Motion

Uniform motion:

$$a = 0$$
$$v = \text{constant}$$
$$x_{\text{f}} = x_{\text{i}} + v\Delta t$$

$$a_{\text{c}} = \lvert\vec{v}\rvert^2/r$$
$$\lvert\vec{v}\rvert = 2\pi r/T = 2\pi r f$$
$$s = \lvert\vec{v}\rvert\Delta t$$

Uniform acceleration:

$$a = \text{constant}$$
$$v_{\text{f}} = v_{\text{i}} + a\Delta t$$
$$x_{\text{f}} = x_{\text{i}} + v_{\text{i}}\Delta t + \frac{1}{2}a(\Delta t)^2$$
$$v_{\text{f}}^2 = v_{\text{i}}^2 + 2a\Delta x$$

General:

$$a \neq \text{constant}$$
$$v_{\text{f}} = v_{\text{i}} + \int_{t_{\text{i}}}^{t_{\text{f}}} a(t)\,dt$$
$$x_{\text{f}} = x_{\text{i}} + \int_{t_{\text{i}}}^{t_{\text{f}}} v(t)\,dt$$

---

the friction force, we have "explained" the behavior of these forces in terms of atomic-level forces. Can we explain *all* of our forces in terms of atomic-level forces, or are some of them fundamental forces that have no further level of explanation? And what about those atomic-level forces—are they fundamental forces, or can we understand atomic-level forces in terms of yet smaller subatomic-level forces?

These are very significant questions, and they have been a central topic of physics throughout the 20th century as the atomic structure of matter has been probed and unfolded. At the very heart of the issue is the question, "What are the fundamental forces of nature?" That is, what set of distinct, irreducible forces can explain everything we know about nature? Friction is clearly not a fundamental force of nature because it can be reduced to electrical forces between atoms. What about other forces?

Physicists since the time of Newton have recognized three basic forces: the gravitational force, the electric force, and the magnetic force. These are all long-range forces. The gravitational force is an inherent attraction between any two masses. It is, however, an extremely weak force, only becoming large enough to be of significance when one or both of the masses is planet-size or larger. As we have already seen, the gravitational force is responsible for objects having weight, for planets orbiting the sun, for the shapes and motions of galaxies, and ultimately for the large-scale structure of the entire universe. The cosmological significance of the gravitational force, despite its weakness, is that it extends to infinite distances.

Electric forces between charges are, in many ways that we will see later, analogous to gravitational forces between masses. There is, however, a difference between electric forces and gravitational forces that is of critical importance: charges come in two varieties, positive and negative, that allow for both attractive and *repulsive* electric forces. The electric force is vastly stronger than the gravitational force on the atomic scale and, sometimes, even on the human scale (think of clothes sticking together because of "electrostatic cling" when you remove them from the drier, "defying" the force of gravity by not falling).

The magnetic force acts between magnets and certain magnetic materials, such as iron. It is responsible for compass needles pointing north and for holding your shopping list on the refrigerator door. If you have played with magnets, you know that the magnetic force can be either attractive or repulsive. It is also a stronger force than gravity—you can use even a small magnet to lift paper clips from the table.

A series of discoveries in the early 1800s suggested that there is a connection between the electric force and the magnetic force; that they are not entirely independent of each other. By 1870, the British physicist James Clerk Maxwell had developed a theory which *unified* the electric and magnetic forces into a single *electromagnetic force*. Where there had appeared to be two separate forces, Maxwell's theory showed that, in fact, there is a single force that, under appropriate conditions, can exhibit "electric force behavior" or "magnetic force behavior." His theory provides the precise conditions under which the electric or magnetic aspects of the force act alone as well as conditions under which the unified electromagnetic force acts as a single entity. Maxwell's theory is the basis for modern electrical engineering. The most dramatic prediction of Maxwell's theory was the existence of *electromagnetic waves*. His prediction was confirmed a few years later with the first artificial generation and transmission of radio waves. Our entire communications industry today (radios, televisions, cellular phones, satellite communications, lasers, etc.) is testimony to Maxwell's genius.

At the beginning of the 20th century, Maxwell's electromagnetic force was found to be the "glue" holding atoms together, forming molecular bonds, and condensing molecules into liquids and solids. With the exception of the gravitational force, every other force we have considered can be understood at the atomic level as a manifestation of the electromagnetic force. We have seen this already for the normal force and the friction force. Similarly, tension in strings and ropes is viewed at the atomic level as an electric force forming the molecular bonds between the atoms in the rope. As those bonds are stretched, the electric forces pull harder to keep the bonds from breaking. It is this electrical pull from billions of molecular bonds that is seen, at the human scale, as "tension."

With the success of atomic theories, there seemed to be just two fundamental forces of nature—the gravitational force and the electromagnetic force. Neither, however, was capable of explaining the subatomic behavior of the atomic nucleus that was first found when nuclear physics experiments were begun in the 1930s. The atomic nucleus, as you may have learned in chemistry, is an unimaginably dense and compact ball of protons and neutrons. But what holds it together against the repulsive electric forces between the protons? There must be an attractive force between *nucleons* (protons and neutrons) that is stronger than the repulsive electric force. This force is called the *strong force*, and it is the force that binds atomic nuclei together. Unlike the gravitational force and the electromagnetic force, the strong force is a *short-range* force, extending only about 10^{-14} m, which is roughly the size of the nucleus. At the distance of the orbiting electrons, a mere 10^{-10} m away, the strong force is completely negligible. Exploring the characteristics of the strong force is still a very active area of physics research. The subatomic particles called *quarks*, of which you have likely heard, are part of our current understanding of how the strong force works.

Sometimes, in some nuclei, a neutron spontaneously turns into a proton and an electron (and also a very unusual particle called a *neutrino*). The electron is ejected from the nucleus at very high speed. This behavior is a form of radioactivity called *beta decay*. Neither the electromagnetic force nor the strong force could explain this nuclear phenomenon. Many careful experiments established that a previously undiscovered force was acting in the nucleus. The strength of this force is much less than either the strong force or the electromagnetic force. The new force, in fact, was named the *weak force*. Like the strong force, its actions are confined to the atomic nucleus.

By 1950, the recognized fundamental forces of nature had expanded to four: the gravitational force, the electromagnetic force, the strong force, and the weak force. Physicists, very aware of Maxwell's success at unifying the electric and the magnetic forces, were understandably curious whether all four of these forces were truly fundamental or if some or all of them could be further unified. Albert Einstein spent much of his later years trying to find a theory that would unify the gravitational force and the electromagnetic force. This type of theory is often called a *unified field theory*. Einstein did not succeed. However, very innovative work in the 1960s and 1970s by three separate physicists—Steven Weinberg, Sheldon Glashow, and Abdus Salam—produced a theory that unified the electromagnetic force and the weak force.

Although many physicists were initially skeptical, experiments during the 1980s at some of the world's largest particle accelerators have amply confirmed predictions of the Weinberg-Glashow-Salam theory, and we can now speak of the *electroweak force*. Under appropriate conditions, the electroweak force can exhibit "electromagnetic force" behavior or "weak force" behavior. But under other conditions, new phenomena appear that are consequences of the full electroweak force. These conditions appear on earth only in the largest and most energetic particle accelerators, which is why we were not previously aware of the unified nature of the electromagnetic and weak forces. However, the earliest moments of the Big Bang also provided the right conditions for the electroweak force to play a significant role. Thus a theory that was developed to help us understand the workings of nature on the smallest subatomic scale has unexpectedly given us powerful new insights into the origin of the universe!

The success of the electroweak theory has prompted intense research to take the next step and to unify the electroweak force with the strong force—an effort known as the search for the *Grand Unified Theory*, or GUT. These efforts have thus far failed to produce a successful GUT, although many physicists are confident that they are on the right track. Only time will tell if the strong force and the electroweak force are really just two different aspects of a single force, or if they are truly distinct. Some physicists even envision a day when all the forces of nature will be unified into a single theory—the so-called *Theory of Everything*! For today, however, our understanding of the forces of nature is in terms of three fundamental forces: the gravitational force, the electroweak force, and the strong force. Figure SVI-1 shows a historical progression of the recognized forces of nature.

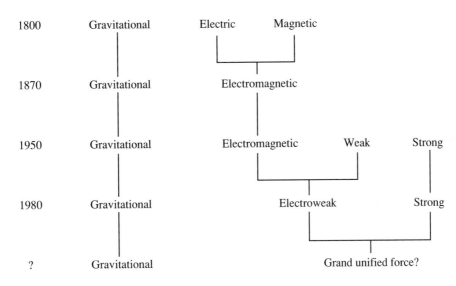

FIGURE SVI-1   An historical progression of our concept of the fundamental forces of nature.

As our concept of "force" has evolved since the time of Newton, one aspect has become increasingly clear: Force is not a "thing." It does not have an independent existence of its own. A more modern idea is that force is an *interaction* between two objects. In Part II we will concentrate on seeing how the concept of interactions can help us understand more complex motions in which two or more objects move in response to forces that they exert mutually on each other.

Well, this has been a short pause to think about where we have been and to see a little of what lies ahead. But now it's time to continue our journey.

Interacting Particles and Conservation Laws

PART **II** Overview

Interacting Particles and Conservation Laws

Consider the stars in a galaxy, the molecules in a gas, or the electrons in an atom. These *systems of particles* exhibit very complex motions. They consist of many objects—stars, molecules, or electrons—that are *interacting* with each other. For example, each of the electrons in an atom is attracted toward the positive nucleus at the center but is repelled by all the other negative electrons. The electrons are all in motion, and we cannot analyze the motion of any one electron independently of the others. The simple model of single-particle motion that we developed in Part I is inadequate to describe this system of interacting particles. Our goal in Part II is to study the interactions of particles, thereby expanding our understanding of motion.

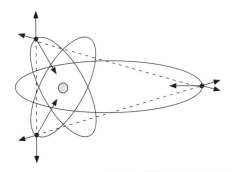

FIGURE OII-1 The many complex interactions between the electrons and the nucleus of an atom.

Newton recognized that force is not a "thing." It does not have an independent existence of its own. There is an old saying, "It takes two to tango," that aptly describes situations in which forces act. Suppose you and a friend engage in a tug-of-war with a rope. The tension in the rope pulls your friend toward you, but at the same time it pulls you toward your friend. You have to pull in one direction *and* your friend has to pull in the opposite direction if there is to be tension in the rope.

When we introduced the concept of force we noted that force requires an agent. This made a distinction between the agent that causes the force and the object that responds. But that distinction is artificial. Your friend may be pulling on you, but at the same time you are pulling on him or her. Each of you, in a sense, is an agent acting on the other. There's no obvious distinction as to who is the "agent" and who is the "object."

Rather than make an artificial distinction between agent and object, Newton adopted the more comprehensive idea of force being an *interaction* between objects. From this viewpoint, you and your friend are interacting with each other. The electrons in an atom are all interacting with each other as well as with the nucleus. Viewing forces as interactions removes the temptation to think of forces as "things" with their own existence.

Newton was able to give an accurate description of the interactions between objects when he formulated his third law of motion. You may have heard the popular version of this law: "For every action there is an equal and opposite reaction." The ease with which we repeat this phrase tends to obscure the profound discovery about force and motion that lies behind the words. The fact that actions are paired with reactions is hardly surprising, but the fact that they turn out to be *equal* is by no means obvious.

The third law was perhaps Newton's boldest and most innovative discovery. Newtonian mechanics would not have progressed far on his first two laws alone. Most situations of real interest have to be viewed as interactions, and it is the third law that provides their correct description. Newton's *three* laws, taken together, are a scientific theory of immense power and influence.

Newton's third law will be the starting point for our study of interactions. The third law will lead us naturally to investigate a quantity called *momentum*. There we will investigate our first conservation law, the *law of conservation of momentum*. Momentum conservation is particularly useful for analyzing *collisions*, in which two objects come together, and *explosions*, where two objects fly apart. Collisions and explosions are clearly *interactions* of particles.

[**Photo suggestion: Two cars colliding.**]

A conservation law tells us that only certain outcomes of an interaction are possible—namely, those for which some quantity remains unchanged. We say that the unchanged quantity is *conserved*. Consider pouring 10 g of chemical A and 10 g of chemical B into a box, then sealing it tightly. The atoms in the two chemicals can interact with each other, producing many new compounds as reaction products, but no atoms can enter or leave the box. As a result, we feel certain that at any later time the total mass of all the compounds in the box will be 20 g. As obvious as this seems, it is only through experiments that we know that this is true. This behavior of masses is called the *law of conservation of mass*. It tells us that certain *conceivable* reactions—such as one that would produce 21 g of products—simply don't happen. Only those interactions are possible that conserve mass.

The law of conservation of mass isn't really surprising. It seems so obvious, due to our everyday experience with mass, that we hardly recognize it as a law. But it is, and it tells us something significant about the nature of interactions. (Interestingly, Einstein's theory of relativity predicted, and it was later confirmed, that there are exceptions to the law of conservation of mass. Mass can be lost in some circumstances if it is transformed into an equivalent amount of energy in accordance with Einstein's famous formula $E = mc^2$.)

We don't have everyday experience with momentum, so we have no expectation that it should be conserved. But under the proper conditions—analogous to having the box tightly sealed—only those interactions can occur that do not change the momentum of a system. Conservation laws provide us with a tool for analyzing what can and, equally important, what cannot happen as a system of particles interacts.

Now there's another curious property of systems of interacting particles: Each system is characterized by a certain number, and no matter how complex the interactions or how long they last, the value of this number never changes. This number is called the *energy* of the system, and the fact that it never changes is called the *law of conservation of energy*. It is, perhaps, the single most important physical law ever discovered.

The law of conservation of energy is more fundamental than Newton's laws of motion. Newton's laws are a very good description of motion under many circumstances, but we now know that they fail for objects moving at extremely high speeds or for objects that are the size of atoms. But the law of conservation of energy is, as far as we know, valid under all circumstances. No violations of this law have ever been observed. Energy will be *the* fundamental concept upon which atomic physics is built. The introduction of energy in Part II is an important steppingstone in our quest to understand atoms and the atomic structure of matter.

But what is energy? How do you determine the energy number for a system? These are not easy questions. Energy is an abstract idea, not as tangible or easy to picture as a force. Our modern concept of energy wasn't fully formulated until the middle of the nineteenth century, nearly two hundred years after Newton. The long delay was a consequence of the difficulty in answering the question "What is energy?" Conservation of energy was not recognized as a law until the relationship between *energy* and *heat* was understood. This is a topic we will take up in Part IV, where the full concept of energy will be found to be the basis of thermodynamics. In Part II, however, we will simply introduce the concept of energy and show how energy gives us an important new perspective on interacting particles.

We will first consider the energy of single objects, then pairs of interacting objects, and finally of systems of objects. You will learn how energy is transferred into and out of system, and also how the energy within a system can be transformed from one form into another. These are subtle issues, so we will approach them one at a time. When we are finished, we will find that what started as a simple idea has blossomed into the entire science of thermodynamics. Thermodynamics, in its broadest sense, is about how to use sources of energy to do useful things. But all that in due time. First we have to see how just two particles interact with each other.

Chapter **7**

Dynamics III: Newton's Third Law

7.1 Interacting Systems

In Part I, you learned that particles respond to forces with an acceleration given by Newton's second law: $\vec{a} = \vec{F}_{net}/m$. So far, however, we have not investigated what *causes* forces. The goal of this chapter is to look more closely at the nature of forces. Recall that a force must have an *agent*—something doing the pushing or pulling. Often the agent itself is another object that experiences forces and is in motion. For example, consider a golf club hitting a golf ball. From the ball's perspective, the club exerts a force on the ball that accelerates it rapidly during the short time in which they are in contact, leaving the ball with a high forward velocity. But the golf club itself is also in motion, responding to the various forces exerted on it—its weight, the golfer's hands, and the force of the golf ball striking it. If the ball's motion is in response to the force of the club, while the club's motion is in response (partly) to the force of the ball, how could we ever hope to understand what happens? The second law, powerful though it may be, is not sufficient to explain what happens in situations where two objects *interact* with each other.

[**Photo suggestion: Close-up of golf club hitting a golf ball.**]

In this example, the heart of the problem is to understand the interaction between the golf club and the golf ball. The club exerts a force on the ball, while at the same time the ball exerts a force on the club. If you are not sure that the ball exerts a force on the club, imagine using your hand instead of the club. If you hit a golf ball with your hand hard enough to knock it 200 yards or more, as a golf club can, you would have a *very* sore hand! Why? Because of the force of the ball on your hand.

If you stop to think about it, *any* time that object A pushes or pulls on object B, object B pushes or pulls back on object A. As you sit in your chair, it pushes upward on you (the normal force) while at the same time you push down on the chair. If a tow rope pulls your car forward, your car is pulling backward on the rope. These are examples of what we will call **interaction forces**. We will label these forces as $\vec{F}_{A\,on\,B}$ and $\vec{F}_{B\,on\,A}$. These labels will help you avoid getting the different forces confused.

The examples of the golf club and golf ball, or of the rope and car, showed interactions that are contact forces. The same idea holds true for long-range forces. Most of you have played with kitchen magnets or bar magnets. If you hold two magnets such that they are attracted toward each other, you can feel with your fingertips that *both* have forces pulling on them. If magnet A pulls magnet B toward the right, you can feel that magnet B simultaneously pulls magnet A toward the left. Long-range forces also act as interaction forces.

If you release a ball, it falls because the earth's gravity exerts a downward force on it—its weight. But does the ball also pull upward on the earth? Newton was the first to recognize that, indeed, the ball *does* pull upward on the earth and that, likewise, the moon pulls on the earth in response to the earth's gravity pulling on the moon. Newton's evidence was the tides. Scientists and astronomers have studied and timed the ocean's tides since antiquity, and it was long known that the tides depend on the position of the moon. But no one understood the *mechanism* by which the tides function. Newton recognized that the tides were nothing more than the ocean's response to the gravitational pull of the moon on the earth—the flexible water bulges toward the moon while the inflexible crust of the earth remains stationary. Even so, our everyday observation is that balls fall downward when released rather than the earth "falling" upward. Understanding why this is so is a major goal of this chapter.

Recall that in earlier chapters we considered forces acting on a single object that we called the *system*. Figure 7-1a shows a representation of this. We now want to extend our model to include situations where two or more objects interact with each other. Figure 7-1b, for example, shows three objects with interaction forces between them. The interactions can be written as force pairs, such as $\vec{F}_{1\,\text{on}\,2}$ and $\vec{F}_{2\,\text{on}\,1}$.

Often we are interested in the motion of Systems 1 and 2 but not that of System 3. For example, Systems 1 and 2 might be the golf club and golf ball while System 3 is the earth. The earth does interact with both the ball and the club, but in a practical sense the earth remains "at rest." It will be convenient to separate "systems of interest" from "systems in the environment." Then, as shown in Fig. 7-1c, we can analyze interaction forces between the systems of interest while identifying forces from the environment as *external forces*.

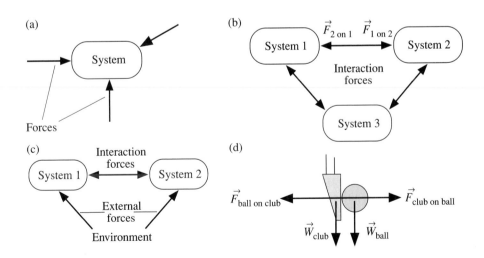

FIGURE 7-1 a) Model of single-particle dynamics. b) Model of interacting systems. c) Forces originating from systems in the environment are considered to be external forces. d) The interaction forces and external forces of hitting a golf ball.

If object B pushes or pulls back on object A whenever A pushes or pulls on B, then ultimately *every* force is one member of an interaction pair. There is no such thing as a truly "external force." What we will call an external force is an interaction between a system of interest and a system in the environment whose motion is not of interest. Figure 7-1d illustrates this for the example of hitting a golf ball. $\vec{F}_{\text{club on ball}}$ and $\vec{F}_{\text{ball on club}}$ are interaction forces while the weights $\vec{W}_{\text{ball}}$ and $\vec{W}_{\text{club}}$ are external forces of the earth on the ball and the club. The ball and club *do* exert forces back on the earth, but in this example we only want to know how the earth affects the ball and club and not how the ball and club affect the earth.

Newton's second law continues to apply to *each* of Systems 1 and 2 in Fig. 7-1c:

$$\text{System 1:} \quad \vec{a}_1 = \frac{1}{m_1} \sum \vec{F}_{\text{on 1}} = \frac{1}{m_1}\left(\vec{F}_{2 \text{ on } 1} + \sum \vec{F}_{\text{ext on 1}} \right)$$

$$\text{System 2:} \quad \vec{a}_2 = \frac{1}{m_2} \sum \vec{F}_{\text{on 2}} = \frac{1}{m_2}\left(\vec{F}_{1 \text{ on } 2} + \sum \vec{F}_{\text{ext on 2}} \right). \tag{7-1}$$

The sum is over *all* forces acting on System 1 (and similarly for System 2). It will be useful, though, to divide this sum into the interaction force plus a sum over external forces. Notice that the equation for System 1 contains *only* forces that act *on* System 1. Forces that are exerted *by* System 1, such as $\vec{F}_{1 \text{ on } 2}$, do *not* appear in the equation of motion for System 1. ($\vec{F}_{1 \text{ on } 2}$ does appear in the equation for System 2 because that is the system *experiencing* this force.) We will emphasize repeatedly throughout this chapter that objects change their motion in response to forces exerted *on* them, not by forces exerted *by* them. One of the most common errors in problems involving interacting systems is to forget this vital aspect of forces.

7.2 Identifying Interaction Forces

Pairs of interacting forces are often called **action/reaction pairs**, although the name can be misleading. For example, we cannot say which force is the "action" and which is the "reaction." Neither is there any implication about cause and effect; the action does *not* cause the reaction. An action/reaction pair of forces exists *as a pair*, or not at all. In identifying action/reaction pairs, the labels are the key. The reaction to force $\vec{F}_{\text{A on B}}$ is the force $\vec{F}_{\text{B on A}}$. Forces do not come with labels, of course, so part of what you have to do is identify "A" and "B" for each force.

The following examples will help you to recognize and to distinguish between interaction forces and external forces. In these examples we will:

1. Draw the interacting objects separately.

2. Find and label *every* action/reaction pair.

3. Connect the two members of each pair with a dotted line.

4. Identify which objects are systems and which are in the environment.

5. Identify which forces must be treated as an action/reaction pair and which can be considered as external forces.

EXAMPLE 7-1 Analyze the forces involved in pushing a large crate across a rough surface, as shown in Fig. 7-2a.

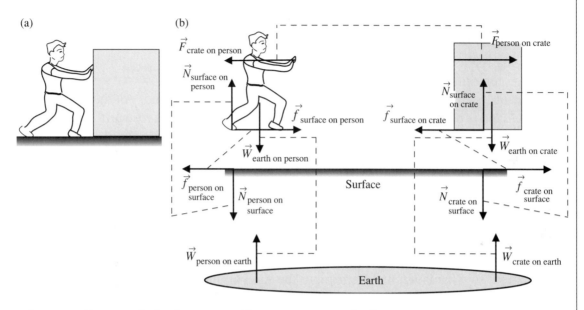

FIGURE 7-2 Force analysis of person pushing a crate across the floor.

SOLUTION The first step in a successful analysis of action/reaction pairs is to redraw the figure with *every* object physically separated from all other objects. In this example, two obvious objects are the person and the crate. The surface is also an object that exerts/experiences forces. Figure 7-2b shows these objects drawn separately, but still with the correct spatial relationships.

We can now start identifying and labeling forces, as you learned to do in Chapter 4. Starting with the crate, we know that it is being pushed with force $\vec{F}_{\text{person on crate}}$, that it has an upward normal force $\vec{N}_{\text{surface on crate}}$, and that it has a friction force $\vec{f}_{\text{surface on crate}}$ in the direction opposite the motion. It also has a downward weight force $\vec{W}_{\text{earth on crate}}$, which is exerted on it by the earth as a whole. It will be important to distinguish between forces exerted by the surface and forces exerted by the earth as a whole.

The person experiences similar forces: there is the force $\vec{F}_{\text{crate on person}}$ of the crate pushing back against the person's hands, the upward normal force $\vec{N}_{\text{surface on person}}$, a friction force $\vec{f}_{\text{surface on person}}$, and the downward weight force $\vec{W}_{\text{earth on person}}$. The surface also experiences several forces. Both the crate and the person press down on the surface with forces $\vec{f}_{\text{crate on surface}}$ and $\vec{N}_{\text{person on surface}}$. Friction forces $\vec{f}_{\text{crate on surface}}$ and $\vec{f}_{\text{person on surface}}$ are exerted on the surface by the sliding crate and the person's feet. Notice that in a problem with multiple systems we will have multiple versions of forces such as weight forces, normal forces, and friction forces. Subscript labels are *essential* to distinguish these forces from each other!

Now let's identify the action/reaction pairs. We know that the person pushes on the crate and that the crate pushes back on the person. So $\vec{F}_{\text{person on crate}}$ and $\vec{F}_{\text{crate on person}}$ are an action/reaction pair. The figure shows these two forces connected with a dotted line. The forces $\vec{N}_{\text{surface on crate}}$ and $\vec{N}_{\text{crate on surface}}$ are also an action/reaction pair; the crate pushes down on the surface while the surface pushes up on the crate. Similarly, $\vec{N}_{\text{surface on person}}$ and $\vec{N}_{\text{person on surface}}$ are an action/reaction pair. Notice how the labels work in these examples: The reaction to the force $\vec{F}_{\text{A on B}}$ is the force $\vec{F}_{\text{B on A}}$.

The friction forces have to be analyzed carefully. The force $\vec{f}_{\text{surface on crate}}$ is a kinetic friction force retarding the crate's motion, so it points to the left. How do we identify the direction of $\vec{f}_{\text{crate on surface}}$? Suppose the floor is covered with sand. As the crate slides along the sand is pushed toward the right. This is a consequence of a *forward*-directed friction force of the crate on the surface. So $\vec{f}_{\text{crate on surface}}$ points toward the right and forms an action/reaction pair with $\vec{f}_{\text{surface on crate}}$.

What about friction between the person and the surface? It is very tempting to place a left-pointing friction force $\vec{f}_{\text{surface on person}}$ on the person because, after all, friction forces are supposed to be in the direction opposite the motion. But if we did so, the person would have two forces directed to the left ($\vec{F}_{\text{crate on person}}$ and $\vec{f}_{\text{surface on person}}$) and none to the right, giving a net force to the *left* and causing a *backward* acceleration! That is clearly not the situation, so what is wrong? Remember the sand on the floor? In what direction does the sand under the person's feet move? In this case, the sand under the person's feet moves to the *left*, and would be kicked backwards if the person's feet slipped. This can happen only if the friction force of the person against the surface is to the *left*. In reaction, the force of the surface against the person is a friction force to the *right*, providing exactly the force necessary for the person to accelerate in the forward direction! So $\vec{f}_{\text{surface on person}}$ points to the right while $\vec{f}_{\text{person on surface}}$ points left. These forces form another action/reaction pair.

The friction force $\vec{f}_{\text{surface on person}}$ is an example of a *propulsion force*. It is worth a careful look because propulsion forces are the source of much confusion. Imagine trying to walk across a frictionless floor—very smooth ice, for example. As you take a step, your foot *slips* and slides in the *reverse* direction across the floor. In order for you to walk, the floor needs to have some friction so that your foot *sticks* to the floor while you straighten your leg, moving your body forward. In order for your foot to be stuck to the floor during this time, there must be a *static* friction force acting on it that prevents it from slipping. Static friction, you will recall, acts in a direction to prevent slipping. Because your foot it as rest relative to the floor but would like to slip backward, the *static* friction force $\vec{f}_{\text{surface on person}}$ *has* to point in the forward direction to prevent the slip. It is this forward-directed static friction force that propels you forward! The force of your foot on the floor, which would kick sand backward, is in the reverse direction. This is the other half of the action/reaction pair.

The distinction between you and the crate is that you have an internal source of energy that allows you to straighten your leg by pushing backward against the ground. The ground responds by pushing you forward. These are static friction forces. All the crate can do, however, is slide. Because of this, *kinetic* friction opposes the motion of the crate.

The two weight forces are still not paired. The force $\vec{W}_{\text{earth on crate}}$ is the gravitational force *by* the earth *on* the crate. By "earth" we mean the *entire* earth, not just the surface, because the earth as a whole is the object that exerts the weight force on the crate. To identify the reaction force, simply reverse the words in the label: The reaction to $\vec{W}_{\text{earth on crate}}$ is $\vec{W}_{\text{crate on earth}}$. Similarly, the reaction to $\vec{W}_{\text{earth on person}}$ is $\vec{W}_{\text{person on earth}}$. Now you see why we included the earth as an object in Fig. 7-2b! It has the two upward forces $\vec{W}_{\text{crate on earth}}$ and $\vec{W}_{\text{person on earth}}$ exerted on it. We can complete our identification of action/reaction pairs with two more dotted lines. Altogether we have identified fourteen forces and seven action/reaction pairs.

As you can see, *every* force is one member of an action/reaction pair. As you work similar problems, make sure you are not left with any unpaired forces. But surely fourteen forces is excessive for such a simple problem! While they all exist, it is not necessary to list every single force to understand how the crate moves. This is the point where we can distinguish between interaction forces and external forces. In this problem, we need to analyze the motion of the *crate* and the *person*, so we will call those objects System 1 and System 2. The surface and the earth do not move, so we can locate them in the environment. The six forces acting on the surface and on the earth have *no* effect on the motion of the crate or the person, so we can omit them from further consideration. Only the four forces on the crate and the four on the person are necessary in this problem. Of these, we can identify $\vec{F}_{\text{crate on person}}$ and $\vec{F}_{\text{person on crate}}$ as interaction forces. The remaining forces—normal, weight, and friction forces—all originate in the environment and are classified as external forces.

EXAMPLE 7-2   Analyze the forces involved in a tow truck pulling a car along a horizontal road, as shown in Fig. 7-3a.

SOLUTION This example has many similarities to Example 7-1. We start by drawing the objects separately but with the correct spatial relationships, as shown in Fig. 7-3b. We now know to include both the surface *and* the earth as objects. Notice that we have included the rope as a separate object. The normal forces and weight forces, with their reactions, are identical to those described in Example 7-1. Make sure you avoid the common error of considering $\vec{N}$ and $\vec{W}$ to be an action/reaction pair—they are both forces on the *same*

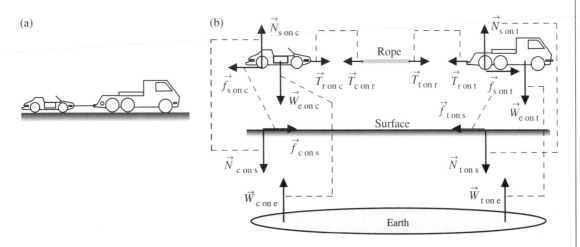

FIGURE 7-3 Force analysis of a tow truck pulling a car.

object, whereas the two forces of an action/reaction pair are always on two *different* objects that are interacting with each other.

What about the friction forces? The car is an inert object rolling along. It would slow and stop if the rope were cut, so there must be a friction force $\vec{f}_{\text{surface on car}}$ to the left and a reaction $\vec{f}_{\text{car on surface}}$ to the right. The truck, however, has an internal source of energy that turns the drive wheels. The drive wheels would like to slip and throw loose sand and gravel backwards, but they are prevented from doing so by a *static* friction force on the wheels in the *forward* direction. The force $\vec{f}_{\text{surface on truck}}$ is a propulsion force that accelerates the truck in the forward direction while the reaction force $\vec{f}_{\text{truck on surface}}$ points to the left.

Finally, we need to identify the forces between the car, the truck, and the rope. What pulls on what in the horizontal direction? The rope pulls on the car with a tension force $\vec{T}_{\text{rope on car}}$. You might be tempted to put the reaction force on the truck, because we say that "the truck pulls the car," but the truck is not in contact with the car. The reaction is a force on the *rope*: $\vec{T}_{\text{car on rope}}$. At the other end, the truck pulls on the rope and the rope pulls back on the truck. This is a tension force $\vec{T}_{\text{rope on truck}}$ on the truck, with reaction force $\vec{T}_{\text{truck on rope}}$. Whether or not $\vec{T}_{\text{truck on rope}}$ is equal to $\vec{T}_{\text{rope on car}}$ is an interesting question that we will examine in Section 7-4. For now, we just want to *identify* forces.

In a practical problem we would again place the surface and the earth in the environment, with normal forces, weight forces, and friction forces considered as external forces. Notice we have *three* systems (the truck, the rope, and the car) and two different pairs of interaction forces (the tension pairs at each end of the rope).

7.3 Newton's Third Law

It seems likely that the two members of an action/reaction pair of forces are related to each other. Newton was the first to recognize the exact nature of this relationship, which we know today as **Newton's third law**. This law can be defined as follows:

Newton's Third Law:

1. *Every* force occurs as one member of a pair of interaction forces;

2. The two members of a pair of interaction forces act on two *different* objects; and

3. The two members of an interaction pair are equal in magnitude but opposite in direction: $\vec{F}_{\text{A on B}} = -\vec{F}_{\text{B on A}}$.

We have deduced most of the third law from our analyses in Section 7-2. There we found that the two members of an action/reaction pair were always opposite in direction (review Figs. 7-2 and 7-3). According to the third law, this will always be true. But the most significant statement of the third law, which is by no means obvious, is that the two members of an action/reaction pair have *equal* magnitudes: $|\vec{F}_{\text{A on B}}| = |\vec{F}_{\text{B on A}}|$. This is the quantitative relationship between forces that will allow you to solve problems of interacting systems.

Newton's third law extends and completes our conception of "force." It is especially significant because it allows us to recognize force as an *interaction* between objects rather than as some "thing" with an independent existence of its own. The third law is frequently stated in the form: "For every action there is an equal but opposite reaction." While this is indeed a catchy phrase, it lacks the preciseness of our preferred version. In particular, it fails to capture an essential feature of action/reaction pairs: that they each act on a *different* object.

EXAMPLE 7-3   If you release a ball, it falls down rather than the earth rising up. But if the ball and the earth exert equal and opposite forces on each other, why don't you see the earth move too? Use Newton's third law to explain this.

SOLUTION The key, surprisingly, is that the forces the ball and the earth exert on each other *are* equal in magnitude. The force exerted on the ball is $\vec{W}_{\text{earth on ball}} = -m_{\text{ball}}g\,\hat{\jmath}$, as shown in Fig. 7-4. According to Newton's second law, this force gives the ball an acceleration

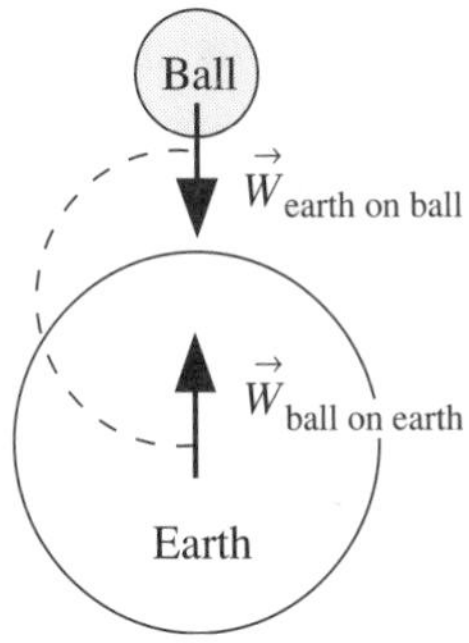

FIGURE 7-4 Third law analysis of a ball falling to earth.

$$\vec{a}_{\text{ball}} = \frac{\vec{W}_{\text{earth on ball}}}{m_{\text{ball}}} = -g\,\hat{\jmath}.$$

This is just the free-fall acceleration due to gravity. In the meantime, the earth of mass m_{earth} undergoes an acceleration also given by the second law:

$$\vec{a}_{\text{earth}} = \frac{\vec{W}_{\text{ball on earth}}}{m_{\text{earth}}}.$$

Here is where the third law enters the picture. Because $\vec{W}_{\text{earth on ball}}$ and $\vec{W}_{\text{ball on earth}}$ are an action/reaction pair:

$$\vec{W}_{\text{ball on earth}} = -\vec{W}_{\text{earth on ball}} = +m_{\text{ball}}g\,\hat{\jmath}.$$

That is, the force exerted on the earth by the ball is the same magnitude (mg) but opposite direction (hence the +) as the force exerted on the ball by the earth. If we assume a 1 kg ball in order to estimate the size of $\vec{a}_{\text{earth}}$, the earth as a whole undergoes an acceleration

$$\vec{a}_{\text{earth}} = \frac{\vec{W}_{\text{ball on earth}}}{m_{\text{earth}}} = \frac{m_{\text{ball}}}{m_{\text{earth}}}g\,\hat{\jmath}$$

$$\Rightarrow |\vec{a}_{\text{earth}}| \approx \frac{1\text{ kg}}{6\times10^{24}\text{ kg}}g \approx 2\times10^{-24}\text{ m}/\text{s}^2.$$

With this acceleration, it would take the earth 8×10^{15} years, approximately 500,000 times the age of the universe, to reach a speed of 1 mph! So we certainly would not expect to "feel" the earth move by dropping a ball.

This example illustrates an important point: Newton's third law equates the size of two forces, *not* two accelerations. The acceleration continues to depend on the mass, as stated in the second law. In an interaction between two very different masses, the lighter mass will do essentially all of the accelerating even though the forces exerted on the two objects are equal.

Acceleration Constraints

Newton's third law gives us one quantitative relationship to use in solving interacting system problems, but that often is not enough. We do, though, frequently have other information about the motion in a problem. Return, for a moment, to Fig. 7-2 of Example 7-1. As long as the person is pushing the crate, the crate *has* to have exactly the same acceleration as the person. If they were to accelerate differently, either the crate would take off on its own or it would suddenly slow down and the person would run over it! It is an implicit assumption of our problem that neither of these is happening. Thus the two accelerations are *constrained* to be equal: $\vec{a}_{\text{crate}} = \vec{a}_{\text{person}}$. This is the final relationship we need to solve the motion problem.

Example 7-2 involves a similar situation. As long as the truck is towing the car and the rope is under tension, the accelerations are constrained to be equal: $\vec{a}_{\text{car}} = \vec{a}_{\text{truck}}$. Be careful, however, not to make the assumption that a problem involving interacting systems A and B will always have the constraint $\vec{a}_A = \vec{a}_B$. In many problems, because of the coordinate systems chosen, the sign of one acceleration will be opposite the other and the constraint will be $\vec{a}_A = -\vec{a}_B$. In problems involving pulleys, it is possible to have constraints such as $\vec{a}_A = 2\vec{a}_B$. The point is that you need to examine the geometry and the coordinate systems of each problem to determine the mathematical form of any acceleration constraints.

A further bit of explanation may be worthwhile about the relationship $a_A = -a_B$, which causes confusion for some students. This relationship does *not* say that a_A is a negative number. It is simply a relational statement, saying that a_A is (-1) times whatever a_B happens to be. If a_B is positive, then a_A will be negative. However, it could equally well be true that a_B is negative and a_A positive. No assumption at all is made about what the actual signs of a_A and a_B are—that may not even be known until the problem is solved, but the *relationship* is known from the beginning.

7.4 Interactions with Strings and Ropes

In Example 7-2 we analyzed a situation in which two systems—the car and the truck—were connected by a rope. This is a common situation in physics. We frequently deal with systems that are connected together by strings, ropes, cables, bungee cords, and so on. For convenience, we usually will refer to all of these simply as "strings." In single particle dynamics, we defined tension as the force exerted on an object by a string. Now, however, we need to think more carefully about the string itself. Just what do we mean when we talk about the tension "in" a string?

[**Photo suggestion: A cable-and-pulley arrangement in a robot arm.**]

Figure 7-5a shows an object hanging vertically on a string, placing the string under tension. If we cut the string, the object and the lower portion of the string will fall—no big surprise. So *before* the string was cut, there must have been a force *within* the string by which the upper portion of the string pulled upward on the lower portion of the string, preventing it from falling. This is the force $\vec{T}_{\text{upper on lower}}$ shown in Fig. 7-5b. In reaction to this force, the lower portion of the string must pull downward on the upper portion of the string with force $\vec{T}_{\text{lower on upper}}$. According to Newton's third law, the sizes of these forces are equal: $|\vec{T}_{\text{lower on upper}}| = |\vec{T}_{\text{upper on lower}}|$. These forces arise from the stretching of millions of molecular bonds in the string, and they are called the *tension* in the string. At any point in the string the tension forces are pulling *equally* in *both* directions.

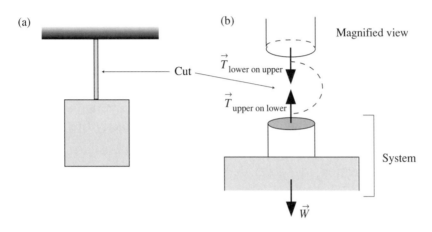

FIGURE 7-5 a) Object hanging by a string. b) Tension forces within the string if a cut is made through the string.

Now suppose we consider the object and the lower portion of the string to be the system, while the upper portion of the string is part of the environment. We can do this by making an imaginary slice through the string, as in Fig. 7-5b. If we apply Newton's first law to the object and the lower portion of the string, which are in equilibrium, we find:

$$(F_{\text{net}})_y = |\vec{T}_{\text{upper on lower}}| - |\vec{W}| = 0$$

$$\Rightarrow |\vec{T}_{\text{upper on lower}}| = |\vec{W}|$$

(7-2)

where $|\vec{W}|$ is the combined weight of the object and the lower portion of the string. This is exactly the result we found in single particle dynamics—that the upward tension force and the downward weight force balance each other when in equilibrium. When we bring the third law into the picture, we also see that $|\vec{T}_{\text{lower on upper}}| = |\vec{T}_{\text{upper on lower}}| = |\vec{W}|$. In other words the tension at any point "in" the string, which is pulling the string straight, is just the weight hanging beneath it at that point. This analysis gives us insight into how tension works by viewing tension as a force *at all points* within the string that holds the string together.

EXAMPLE 7-4 Figure 7-6a shows a man pulling horizontally with a 100 N force on a rope that is attached to a wall, while Fig. 7-6b shows two men in a tug-of-war pulling on opposite ends of a rope with 100 N each. How does the tension in the second rope compare to that in the first rope? Is it larger, smaller, or the same?

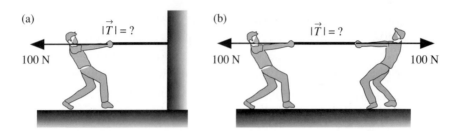

FIGURE 7-6 a) Pulling on one end of a rope with a force of 100 N.
b) Pulling on both ends with forces of 100 N.

SOLUTION The analysis of the situation shown in Fig. 7-6a is really no different than the situation of Fig. 7-5b—it is the same problem rotated 90°, with the pulling force replacing the weight force. We can make an imaginary slice through the rope, as shown in Fig. 7-7a. Because the left portion of the rope is in equilibrium, the force with which the right half of the rope pulls on the left, $\vec{T}_{\text{right on left}}$, has to balance exactly the 100 N pulling force to the left. That is, $|\vec{T}_{\text{right on left}}| = |\vec{F}_{\text{pull on left}}|$. But $\vec{T}_{\text{right on left}}$ is also equal in magnitude, though opposite in direction, to $\vec{T}_{\text{left on right}}$ because these are an action/reaction pair. Thus

$$|\vec{T}_{\text{left on right}}| = |\vec{T}_{\text{right on left}}| = |\vec{F}_{\text{pull on left}}| = 100 \text{ N}.$$

The tension in the rope is a constant 100 N.

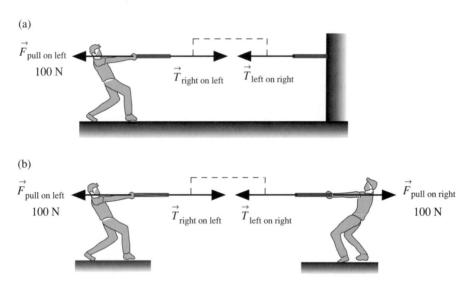

FIGURE 7-7 Analysis of tension forces for a) a man pulling on a rope tied to a wall, and b) a rope pulled by men at each end.

If pulling on the rope from one end causes 100 N of tension, then surely pulling on it from both ends will produce 200 N of tension. Right? Before jumping to conclusions, let's analyze the situation carefully. Again we can make an imaginary cut through the center of the rope, as shown in Fig. 7-7b. The left half of the rope experiences the forces $\vec{T}_{\text{right on left}}$ and $\vec{F}_{\text{pull on left}}$. The left half of the rope is again in equilibrium because the rope is at rest.

The first law thus insists that we still have $|\vec{T}_{\text{right on left}}| = |\vec{F}_{\text{pull on left}}| = 100$ N. Similarly, the right half of the rope experiences forces $\vec{T}_{\text{left on right}}$ and $\vec{F}_{\text{pull on right}}$. But these forces are also in equilibrium and so $|\vec{T}_{\text{left on right}}| = |\vec{F}_{\text{pull on right}}| = 100$ N. Thus the tension forces within the rope are still

$$|\vec{T}_{\text{left on right}}| = \vec{T}_{\text{right on left}}| = 100 \text{ N},$$

exactly as in Fig. 7-7a. The tension has not changed!

You may have assumed that the second man pulling from the right in Fig. 7-6b is doing something to the rope that the wall in Fig. 7-6a does not do. Is this really true? Figure 7-8 shows a detailed view of the point at which the rope of Fig. 7-6a is tied to the wall. Because the rope pulls on the wall with force $\vec{F}_{\text{rope on wall}}$, the wall must pull back on the rope (action/reaction pair) with force $\vec{F}_{\text{wall on rope}}$. Because the rope is in equilibrium, the wall's pull to the right must exactly balance the man's pull to the left: $|\vec{F}_{\text{wall on rope}}| = |\vec{F}_{\text{man on rope}}| = 100$ N. Therefore, the wall in Fig. 7-6a pulls the right end of the rope with a force of 100 N. In Fig. 7-6b a man pulls the right end with 100 N. The forces are the same in both situations! The rope does not care whether it is pulled by a wall or

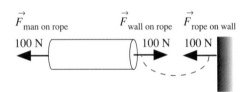

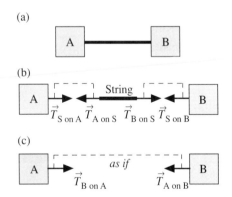

FIGURE 7-8 A closer look at the forces on the rope. The wall pulls on the rope with a force of 100 N.

by a hand—it experiences the same forces in both cases, so the tension is the same 100 N in both.

Figure 7-9 summarizes this information about tension. It shows two objects, or systems, connected by a string under tension. Figure 7-9b shows the separated systems—treating the string itself as a system—and the forces acting on each system. The dotted lines identify the action/reaction pairs. For example, the string pulls on A with $\vec{T}_{\text{S on A}}$, while A pulls back on the string with $\vec{T}_{\text{A on S}}$. These two forces must, according to Newton's third law, be equal in magnitude.

If the rope is in equilibrium, having $\vec{a} = 0$, then Newton's second law requires that $|\vec{T}_{\text{A on S}}| = |\vec{T}_{\text{B on S}}|$. The tension is the same at any point in the string. These tension forces are then applied to objects A and B as forces $\vec{T}_{\text{S on A}}$ and $\vec{T}_{\text{S on B}}$ of equal magnitude but opposite directions. This is a consequence of the *second law*. But the *third law* tells us that $|\vec{T}_{\text{A on S}}| = |\vec{T}_{\text{S on A}}|$ and $|\vec{T}_{\text{B on S}}| = |\vec{T}_{\text{S on B}}|$, so it follows that the two forces $\vec{T}_{\text{S on A}}$ and $\vec{T}_{\text{S on B}}$ are equal in magnitude but opposite in direction. Notice how we used the second law to compare the two

FIGURE 7-9 a) Two systems connected by a rope. b) Force analysis. c) When the systems are in equilibrium, it seems *as if* A and B act directly on each other.

forces on the string and the third law to compare the forces on the string to the forces acting on A and B.

So when the systems are in equilibrium, we find that $\vec{T}_{\text{S on A}} = -\vec{T}_{\text{S on B}}$. Thus the pair of forces exerted directly on the objects A and B act *as if* they are an action/reaction pair. It is common to draw a simplified diagram, such as Fig. 7-9c, with the "action/reaction pair" $\vec{T}_{\text{B on A}}$ and $\vec{T}_{\text{A on B}}$. We can say that A and B are systems that interact with each other *through the string*, and thus that the force of A on B is paired with the force of B on A.

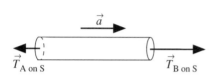

FIGURE 7-10 Forces on a string that is accelerating.

This is not literally true because A and B are not in contact. Nonetheless, it is a common practice to omit the string when all it does is transmit a force from one object to the other without changing the magnitude of that force.

Suppose, however, that the string is *not* in equilibrium, that it is accelerating as shown in Fig. 7-10? Here $\vec{T}_{\text{A on S}}$ and $\vec{T}_{\text{B on S}}$ are the tensions at the left and right end of the string, respectively. Are they still equal in magnitude? That is, is there *constant* tension at all points in the string, as was the case for the equilibrium situation of Fig. 7-9? Let's apply the second law to the string:

$$(F_{\text{net}})_x = |\vec{T}_{\text{B on S}}| - |\vec{T}_{\text{A on S}}| = m_s a, \tag{7-3}$$

where m_s is the mass of the string. *If* the string has a mass, then the tensions at the two ends *cannot* be the same. Thus the accelerating string does not have a constant tension. In fact, we see that

$$|\vec{T}_{\text{B on S}}| = |\vec{T}_{\text{A on S}}| + m_s a, \tag{7-4}$$

so the tension at the "front" of the string is higher than the tension at the "back." This difference in the tensions is necessary to accelerate the string!

Massless Strings

It is frequently the case in physics and engineering problems that the mass of the string or rope is much less than the masses of the objects that it connects. In that case, we can adopt the **massless string** approximation. In the limit $m_s \to 0$, Eq. 7-4 tells us that

$$|\vec{T}_{\text{B on S}}| = |\vec{T}_{\text{A on S}}| \quad \text{(massless string)} \tag{7-5}$$

and we can treat the string tension as a constant. For problems in this book, you can assume that any strings are massless unless the problem specifically states otherwise. Under these conditions, the "simplified view" of Fig. 7-9c is acceptable even if A and B are accelerating. All the string does is transmit an undiminished force from A to B. *But* if the string has a mass, it must be treated as a separate system, as in Fig. 7-9b. These points will become clearer when you study the following examples.

EXAMPLE 7-5 Blocks A and B are connected by a massless string 2 and pulled by massless string 1 with constant tension $|\vec{T}_1|$ across a frictionless table (Fig. 7-11a). Mass B is larger than mass A ($m_B > m_A$). Is the tension $|\vec{T}_2|$ in string 2 larger, smaller, or equal to the tension $|\vec{T}_1|$ in string 1? Explain.

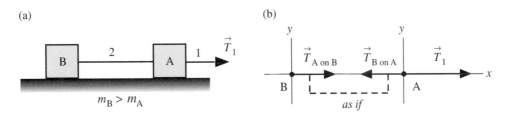

FIGURE 7-11 a) Blocks A and B being pulled across a table. b) Free-body diagram.

SOLUTION Both blocks have the same acceleration ($a_A = a_B$) and B has the larger mass. Thus it may be tempting to conclude, on the basis of Newton's second law, that $|\vec{T}_2|$ must be greater than $|\vec{T}_1|$. The error here is that the second law only tells us about the *net* force. It is true that $|\vec{F}_{\text{net on B}}| > |\vec{F}_{\text{net on A}}|$, due to the larger mass of B, but the net force on A is *not* just the tension $\vec{T}_1$ in the forward direction. The tension in string 2 also pulls *backward* on A!

Figure 7-11b shows the free-body diagram for this frictionless situation. Because A and B are connected by a massless string, the forces $\vec{T}_{\text{A on B}}$ and $\vec{T}_{\text{B on A}}$ act as if they are an action/reaction pair. These two forces are just the tension of string 2 pulling *in both directions*! From the third law we have $|\vec{T}_{\text{A on B}}| = |\vec{T}_{\text{B on A}}| = |\vec{T}_2|$. Therefore, the net force on A (*x*-component) is

$$(\vec{F}_{\text{net on A}})_x = |\vec{T}_1| - |\vec{T}_{\text{B on A}}| = |\vec{T}_1| - |\vec{T}_2| > 0$$

where $F_{\text{net on A}} > 0$ because A is accelerating to the right. The net force on A is the *difference* between the string 1 tension pulling forward and the string 2 tension pulling backward. For this force to be positive, we must have $|\vec{T}_1| > |\vec{T}_2|$—that is, the tension in string 2 is *less* than the tension in string 1. While we have based this conclusion on an analysis of each block separately, we could also have noted that $\vec{T}_1$ pulls *both* blocks, of combined mass ($m_A + m_B$), whereas $\vec{T}_2$ pulls only block B. Thus string 1 must have the larger tension. This is not an intuitively obvious result, and careful study of this example, noting the interplay between the second and third laws, is worthwhile. •

Pulleys

Strings and ropes often are used over pulleys. For example, ropes and pulleys are used to lift heavy weights, and internal cable-and-pulley arrangements move and precisely manipulate robotic arms. The physical principles are the same in both. Let's take a look at the forces involved as a string passes over a pulley.

Figure 7-12a shows a string passing over a pulley to connect two blocks. For simplicity, we will assume the table to be level and frictionless. In Fig. 7-12b we have separated the objects, treating the pulley as part of the table, and identified all the action/reaction pairs. The string is considered as a separate system. Notice particularly the action/reaction pair $\vec{F}_{\text{pulley on S}}$ and $\vec{F}_{\text{S on pulley}}$. It is force $\vec{F}_{\text{pulley on S}}$ that "balances" $\vec{T}_{\text{A on S}}$ and $\vec{T}_{\text{B on S}}$, allowing the string's tension to, in essence, "turn the corner."

Look more closely at the string passing over the pulley, as shown in Fig. 7-12c. The force that we called $\vec{F}_{\text{pulley on S}}$ is really just the normal force of the pulley surface on the

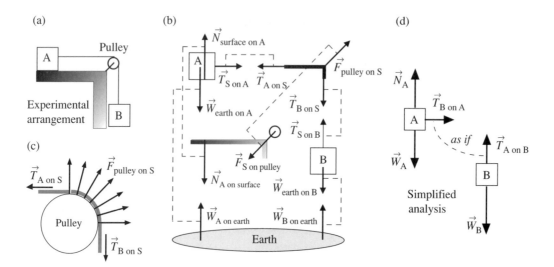

FIGURE 7-12   a) A block on a frictionless table connected by a string over a pulley to a hanging block. b) Force analysis. c) Closer look at the string passing over the pulley. d) Simplified diagram for the case of massless strings and pulleys.

string. But the normal force is *perpendicular* to the string *at every point*, so it has no influence over motion *along* the direction of the string. The string's acceleration is caused—as was the case for a linear string—simply by the *difference* $|\vec{T}_{A \, on \, S}| - |\vec{T}_{B \, on \, S}|$ in the tensions at the two ends.

As the string accelerates, it also—via friction forces—turns the pulley. This makes the situation more complicated than the case of a linear string connecting two objects. But if the string *and* the pulley are both massless, and if the pulley turns on frictionless bearings, then *no* net force is needed to accelerate the string and turn the pulley. Thus for a massless string and a massless, frictionless pulley

$$|\vec{T}_{A \, on \, S}| = |\vec{T}_{B \, on \, S}| \qquad (7\text{-}6)$$

In this text we will, for simplicity, always assume that massless strings pass over massless, frictionless pulleys. In this case, the tension in the string remains constant as it passes over the pulley.

From Newton's third law we know that $|\vec{T}_{S \, on \, A}| = |\vec{T}_{A \, on \, S}|$ and that $|\vec{T}_{S \, on \, B}| = |\vec{T}_{B \, on \, S}|$. Substituting these into Eq. 7-6 allows us to conclude that

$$|\vec{T}_{S \, on \, A}| = |\vec{T}_{S \, on \, B}| \qquad (7\text{-}7)$$

Because of this, it is common to draw a simplified diagram where the string is omitted, as shown in Fig. 7-12d. Forces $\vec{T}_{A \, on \, B}$ and $\vec{T}_{B \, on \, A}$ act *as if* they were an action/reaction pair, even though they are not opposite in direction. We can again say that A and B are systems that interact with each other *through the string*, and thus that the force of A on B is paired with the force of B on A. As a consequence of the pulley, the tension force gets "turned," which is why the two forces are not opposite each other. It is important to keep in mind that this is a simplification valid for the idealized case of a massless string and a massless, frictionless pulley. True action/reaction pairs *have* to be opposite each other.

Figure 7-13 shows free-body diagrams and the acceleration constraint for the example of Fig. 7-12. We have drawn the two diagrams in the correct spatial orientation and have connected $\vec{T}_{\text{A on B}}$ and $\vec{T}_{\text{B on A}}$ as if they were an action/reaction pair. So in addition to the second law equations that we can write for blocks A and B, we also have the relationship $|\vec{T}_{\text{A on B}}| = |\vec{T}_{\text{B on A}}|$. Lastly, because A and B are connected by a string, it must be true that $|\vec{a}_{\text{A}}| = |\vec{a}_{\text{B}}|$. But we must be careful here! Block A is accelerating in the $+x$-direction whereas B is accelerating in the $-y$-direction. While the magnitudes of the accelerations may be equal, they

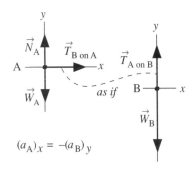

FIGURE 7-13 Free-body diagram for the blocks of Fig. 7-12.

have different components of opposite signs. Thus the acceleration constraint reads $(a_{\text{A}})_x = -(a_{\text{B}})_y$.

7.5 A Revised Strategy for Interacting-Particle Problems

Interacting-particle problems require a few modifications to the basic problem-solving strategy we developed in Chapter 5. The following is the revised strategy.

Revised Strategy for Force and Motion Problems

1. Identify which objects are "systems" and which are part of the "environment."
2. Analyze the problem statement by preparing a pictorial model. Each system should be given a *separate* coordinate system as well as clearly defined labels and symbols. Include acceleration constraints as part of the pictorial model.
3. Identify all forces acting on each system and all action/reaction pairs, then show them on *separate* free-body diagrams for each system. Arrange the diagrams to show the proper spatial orientation between systems. Connect the force vectors of action/reaction pairs with dotted lines. Use subscript labels to distinguish forces, such as $\vec{N}$ and $\vec{W}$, that act independently on more than one system.
4. Invoke Newton's second law for each system, using the information shown on the free-body diagrams.
5. Invoke Newton's third law to equate the magnitudes of action/reaction pairs.
6. Use the acceleration constraints, the friction model, and other quantitative information relevant to the problem.
7. Solve for the subsequent motion.
8. Assess the results.

Notice that we have dropped the motion diagram from our problem-solving strategy. It served a useful function throughout Part I of helping you recognize the velocity and acceleration vectors. But now that you have practiced motion analysis extensively, you should be able to determine the directions of the velocity and acceleration in most problems without the need for an explicit diagram. But if you are uncertain—use one!

You might be a little puzzled that Step 5 calls for the use of the third law to equate just the *magnitudes* of action/reaction forces. What about the "opposite in direction" part of the third law? You have already used it! Your free-body diagrams of Step 3 should show the two members of an action/reaction pair to be opposite in direction, and that information will have been used in writing the second law equations of Step 4. (Note that there is an exception: Problems that omit a massless string may have an action/reaction pair that is not opposite in direction if, as in Fig. 7-12, the string direction is turned by a pulley.) Because the directional information has already been used, all that is left for Step 5 is the magnitude information.

Two steps are especially important when drawing free-body diagrams for interacting-particle problems. First, draw a *separate* diagram for each system. The diagrams do not have to have the the same coordinate system; use whatever coordinates are most appropriate for each system. Second, show only the forces acting *on* that system. The force $\vec{F}_{\text{A on B}}$ goes on the free-body diagram of System B, but $\vec{F}_{\text{B on A}}$ goes on the diagram of System A.

7.6 Examples of Third Law Problems

We will conclude this chapter with several extended examples. Although the mathematics will become more involved than in any of our work up to this point, our emphasis will continue to be on the *reasoning* used in approaching problems such as these. We will use the revised strategy outlined in the last section. In fact, these problems are now reaching a level of complexity that it becomes, for all practical purposes, impossible to work them unless you are following a well-planned strategy. Our earlier emphasis upon identifying forces and using free-body diagrams will now really begin to pay off!

EXAMPLE 7-6 A 90 kg mountain climber is suspended from ropes as shown in Fig. 7-14a. If the maximum tension that rope 3 can withstand before breaking is 1500 N, how small can angle θ become before the rope breaks and the climber falls into the gorge?

SOLUTION This is a static equilibrium problem, so we can use Fig. 7-14a as a pictorial model. Because there is no acceleration, there are no acceleration constraints. The main issue in a

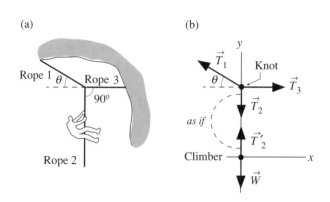

FIGURE 7-14 a) Pictorial model of a mountain climber. b) Free-body diagrams of the knot and of the climber.

problem such as this is to decide what the systems are. The climber has forces being exerted on him, so he is one system. The other significant point where forces are exerted is the knot, where the three ropes are tied together. Because the knot is stationary, the three tension forces are exactly balanced at that point. If we consider the knot to be a second system, we can draw two free-body diagrams as shown in Fig. 7-14b. Rope 2 exerts a downward tension force $\vec{T}_{2\,\text{on knot}}$ on the knot and an upward tension force $\vec{T}_{2\,\text{on climber}}$ on the climber. The full subscript labels are useful for force identification but unwieldy in problem solving, so we will call these forces $\vec{T}_2$ and $\vec{T}_2'$. These are *not* the same force, even though both are exerted by rope 2, hence the prime to distinguish $\vec{T}_2 = \vec{T}_{2\,\text{on knot}}$ from $\vec{T}_2' = \vec{T}_{2\,\text{on climber}}$.

Forces $\vec{T}_2$ and $\vec{T}_2'$ are not, strictly speaking, an action/reaction pair because the climber is not in contact with the knot. But if the ropes are massless, which we will treat them as being, then $\vec{T}_2$ and $\vec{T}_2'$ act as if they are an action/reaction pair. We have thus shown them connected by a dotted line. We can proceed to write Newton's second law for each system, noting that $\vec{a}_{\text{climber}} = 0$ and $\vec{a}_{\text{knot}} = 0$:

Climber:
$$\sum F_y = |\vec{T}_2'| - |\vec{W}| = 0 \tag{7-8}$$

$$\sum F_x = |\vec{T}_3| - |\vec{T}_1|\cos\theta = 0$$

Knot:
$$\tag{7-9}$$

$$\sum F_y = |\vec{T}_1|\sin\theta - |\vec{T}_2| = 0.$$

We can also invoke Newton's third law for the action/reaction pair:

$$|\vec{T}_2'| = |\vec{T}_2|. \tag{7-10}$$

From Eqs. 7-8 and 7-10 we can conclude that $|\vec{T}_2| = |\vec{W}| = mg$. Using this in Eqs. 7-9, we get:

$$|\vec{T}_1|\sin\theta = mg$$

$$|\vec{T}_1|\cos\theta = |\vec{T}_3|.$$

Dividing the first of these by the second gives

$$\tan\theta = \frac{mg}{|\vec{T}_3|}$$

which in turn gives:

$$\theta_{\min} = \tan^{-1}\left(\frac{mg}{|\vec{T}_3|_{\max}}\right)$$

$$= \tan^{-1}\left(\frac{(90\text{ kg})(9.8\text{ m}/\text{s}^2)}{1500\text{ N}}\right) = 30.5°.$$

EXAMPLE 7-7 Figure 7-15a shows two blocks connected by a massless string that passes over a frictionless pulley. Such a device is called an **Atwood machine**. Find the tension in the string and the acceleration of the 2 kg block.

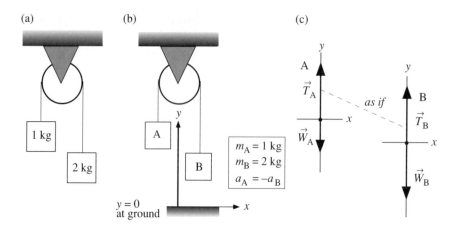

FIGURE 7-15   a) An Atwood machine. b) Pictorial model, with coordinate systems and acceleration constraints. c) Free-body diagrams for blocks A and B.

SOLUTION There are three moving objects in this problem—both blocks and the string. Because the string is massless, we can use the simplified view that the string simply transmits the interaction force between A and B. If we choose blocks A and B as our two systems, we can draw the pictorial model shown in Fig. 7-15b. The acceleration constraint in this problem is $\vec{a}_A = -\vec{a}_B$. The negative sign appears because A is accelerating in the +y-direction while B is accelerating in the –y-direction. This is one-dimensional motion for each block, so we will only need the y-components of forces and accelerations. In that case, the acceleration constraint couples together the two y-components of acceleration and is $a_A = -a_B$.

Noting that each block experiences two forces—a weight force and a tension force—we can proceed to draw the two free-body diagrams shown in Fig. 7-15c. Here $\vec{T}_A$ and $\vec{T}_B$ are not literally an action/reaction pair, but they act as if they are because the string is massless. In this case the pulley has "turned" the tension force so that $\vec{T}_A$ and $\vec{T}_B$ are *parallel* to each other rather than opposite, as members of a true action/reaction pair would have to be. Using the second law we find:

$$a_A = \frac{1}{m_A}\left(|\vec{T}_A| - |\vec{W}_A|\right) = \frac{1}{m_A}\left(|\vec{T}_A| - m_A g\right) = \frac{|\vec{T}_A|}{m_A} - g$$

$$a_B = \frac{1}{m_B}\left(|\vec{T}_B| - |\vec{W}_A|\right) = \frac{1}{m_B}\left(|\vec{T}_B| - m_B g\right) = \frac{|\vec{T}_B|}{m_B} - g \, .$$

(7-11)

Our other pieces of information are the acceleration constraint and the third law relationship $|\vec{T}_A| = |\vec{T}_B|$. We can define $a = a_A = -a_B$ and $|\vec{T}| = |\vec{T}_A| = |\vec{T}_B|$. With these, Eqs. 7-11 become

$$a = \frac{|\vec{T}|}{m_A} - g$$

$$-a = \frac{|\vec{T}|}{m_B} - g \, .$$

(7-12)

Notice how the acceleration constraint was used. We now have two simultaneous equations in the two unknowns a and $|\vec{T}|$. Adding the equations to eliminate a gives

$$0 = |\vec{T}|\left(\frac{1}{m_A} + \frac{1}{m_B}\right) - 2g$$

$$\Rightarrow \quad |\vec{T}| = \frac{2m_A m_B g}{m_A + m_B}.$$

(7-13)

Inserting the result of Eq. 7-13 back into the first of Eqs. 7-12, we find a to be:

$$a = \left(\frac{2m_B}{m_A + m_B} - 1\right)g = \left(\frac{m_B - m_A}{m_A + m_B}\right)g.$$

Because $a = -a_B$:

$$a_B = -a = \left(\frac{m_A - m_B}{m_A + m_B}\right)g.$$

(7-14)

For the given values of m_A and m_B, we find $|\vec{T}| = 13.1$ N and $a_B = -3.27$ m/s^2. Because we were asked for the acceleration of block B, it is important to remember that it is *negative*.   •

Let's take another look at whether Eqs. 7-13 and 7-14 make sense. To do so, we need to consider some special cases. Consider the case where the blocks have equal masses: $m_A = m_B = m$. We find, in this case, $a = 0$ and $|\vec{T}| = mg$. This does make sense. If the masses were equal, they would balance each other and the blocks would hang motionless in equilibrium—thus $a = 0$. Further, we just saw in the Example 7-6 that *stationary* weights hanging from ropes cause a tension $|\vec{T}| = mg$. So our equations give the expected results for equal masses. The other special case of interest is to let $m_A = 0$. This would correspond physically to having the rope over the pulley without being tied to anything. When we let go of block B, it would simply be in free fall ($a_B = -g$) with a loose rope ($|\vec{T}| = 0$) trailing behind. And indeed, Eqs. 7-13 and 7-14 give $a_B = -g$ and $|\vec{T}| = 0$ if we set $m_A = 0$. Having verified that the algebraic equations give the expected results in special cases, we can feel confident in our answers. Note that in the general case the tension will be somewhere between 0 and mg while the acceleration will be somewhere between 0 and g. *Both* masses are involved in determining the final results because of their interaction through the rope.

EXAMPLE 7-8   A 1 kg block on a table is connected by a massless string and a frictionless pulley to a 2 kg block hanging 1 m above the ground, as shown in Fig. 7-16a. If the blocks are released from rest, find the speed of the 2 kg block when it hits the ground in the case that a) The table surface is frictionless, and b) The coefficient of kinetic friction between the table surface and the block is 0.5.

SOLUTION   This is a continuation of the situation that we analyzed in Figs. 7-12 and 7-13. We will consider the table and pulley as part of the environment, because they do not

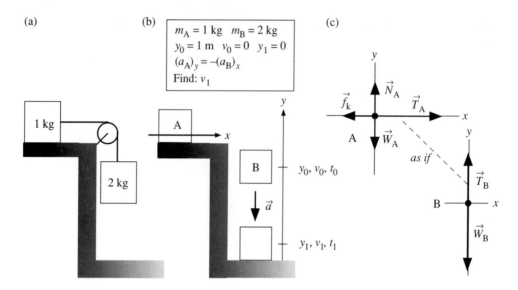

FIGURE 7-16 a) Two connected blocks. b) Pictorial model. c) Free-body diagrams.

move. The fact that the string is massless allows us to use the simplified view that the two blocks interact directly with each other. The pictorial diagram shown in Fig. 7-16b establishes separate coordinate systems for each block and the necessary symbols to carry out the kinematical analysis for block B. Pay particular attention to the acceleration constraint in this problem. While $|\vec{a}_A| = |\vec{a}_B|$, when written in terms of the components of the accelerations this becomes $(a_B)_y = -(a_A)_x$.

We could work the problem once without friction then a second time with friction included, but it is easier to include friction from the beginning and then set $\mu_k = 0$ for part a). The free-body diagram of Fig. 7-16c is modeled after Fig. 7-13, but with a kinetic friction force now included for block A. Forces $\vec{T}_A$ and $\vec{T}_B$ act as if they are an action/reaction pair because the string is massless, so they have been connected with a dotted line. We can write the second law directly from the free-body diagrams:

Block A:
$$(a_A)_x = \frac{1}{m_A}\sum(F_{\text{on A}})_x = \frac{|\vec{T}_A| - |\vec{f}_k|}{m_A}$$

$$(a_A)_y = \frac{1}{m_A}\sum(F_{\text{on A}})_y = \frac{|\vec{N}| - |\vec{W}_A|}{m_A} = 0 \qquad (7\text{-}15)$$

$$\Rightarrow \quad |\vec{N}| = |\vec{W}_A| = m_A g.$$

Block B:
$$(a_B)_y = \frac{1}{m_B}\sum(F_{\text{on B}})_y = \frac{|\vec{T}_B| - |\vec{W}_B|}{m_B} = \frac{|\vec{T}_B| - m_B g}{m_B}. \qquad (7\text{-}16)$$

There are a lot of unknowns here, but we have three additional pieces of information:

Friction:
$$|\vec{f}_k| = \mu_k |\vec{N}| = \mu_k m_A g \qquad (7\text{-}17)$$

Third law:
$$|\vec{T}_A| = |\vec{T}_B| = |\vec{T}| \qquad (7\text{-}18)$$

Acceleration constraint:
$$(a_A)_x = a = -(a_B)_y. \qquad (7\text{-}19)$$

Equation 7-18 defines $|\vec{T}|$ as the tension in the string, which is constant because $|\vec{T}_A| = |\vec{T}_B|$. In Eq. 7-19 we have defined a so that it will be positive because we know that $(a_A)_x$ is positive. Thus $(a_B)_y$ will be negative. Substituting the results from Eqs. 7-17 through 7-19 into the first part of Eq. 7-15 and Eq. 7-16 gives:

$$(a_A)_x = a = \frac{|\vec{T}| - \mu_k m_A g}{m_A}$$

$$(a_B)_y = -a = \frac{|\vec{T}| - m_B g}{m_B}.$$

(7-20)

We have succeeded in reducing our knowledge to two simultaneous equations in the two unknowns a and $|\vec{T}|$. If we add these two equations, to eliminate a, we find (after a bit of algebra) that the tension is:

$$|\vec{T}| = \frac{m_A m_B}{m_A + m_B}(1 + \mu_k)g.$$

(7-21)

Inserting this into either of Eqs. 7-20, we can, at last, find the acceleration:

$$a = \left(\frac{m_B - \mu_k m_A}{m_A + m_B}\right)g.$$

(7-22)

Now we need to perform the kinematic analysis for block B. Because the time of the fall is not known or needed, we can use

$$v_1^2 = v_0^2 + 2(a_B)_y \Delta y = 0 + 2(-a)(0 - y_0) = 2ay_0$$

Solving this for $|v_1|$ gives:

$$|v_1| = \sqrt{2ay_0} = \sqrt{2gy_0 \frac{m_B - \mu_k m_A}{m_A + m_B}}$$

$$= \begin{cases} 3.61 \text{ m/s} & \mu_k = 0 \\ 3.13 \text{ m/s} & \mu_k = 0.5. \end{cases}$$

Notice how the negative sign in $(a_B)_y = -a$ cancels the negative sign in Δy. The value of v_1 itself should be negative, because the motion is in the $-y$-direction, but we only needed to find the speed so we took the absolute value. •

Once again we should consider if the equations in the example make sense. Looking back at Eq. 7-22, consider the table to be frictionless ($\mu_k = 0$). Then

$$a = \left(\frac{m_B}{m_A + m_B}\right)g < g \quad (\mu_k = 0).$$

(7-23)

You might be surprised to find that $a \neq g$ in the frictionless case because, after all, this situation looks a lot like free fall. But free fall, you will recall, is the motion of an object under the influence of *only* the gravitational force. Block B is falling; however, it is acted upon not only by the gravitational force (its weight) *but also* by the tension force in the string. So this is *not* free fall! In fact, you could think about it as follows: Gravity does indeed pull down on block B with force of magnitude $m_B g$. But because the two blocks are constrained to move together, the total mass that has to accelerate is $m_A + m_B$. The second law then tells us that the combined system of both blocks will have acceleration $a = F/m_{tot} = m_B g/(m_A + m_B)$—exactly Eq. 7-23. If $m_B \gg m_A$, then $a \approx g$, as we would expect. On the other hand, if $m_B \ll m_A$, then $a \approx 0$.

When friction is included, as in Eq. 7-22, we see that a sufficiently large value of μ_k will cause a to be negative. This is not a physically realistic solution—a large friction force will *not* cause block A to accelerate to the left, pulling block B upward! If μ_k is increased to the point where $a = 0$, the blocks never move—block A is "stuck" to the table. Our analysis *assumed* that the blocks were moving and that a *kinetic* friction force was exerted on block A. If we find unphysical solutions, as we find here for a sufficiently large μ_k, it indicates that the assumptions in our analysis have failed. Thus our analysis is only valid if $\mu_k \leq m_B/m_A$. We could, though, turn this around and use it to find the maximum mass that block A can have and still move: $(m_A)_{max} = m_B/\mu_k$. A mass larger than this will stick.

[**Photo suggestion: Students engaged in a tug-of-war with a rope.**]

EXAMPLE 7-9   A major attraction at the flea circus is the flea tug-of-war, shown in Fig. 7-17a. Two fleas, Alpha and Beta, each with a mass of 1 mg, are tied together by a 2 mg thread and pull in opposite directions. The static and kinetic coefficients of friction between flea feet and the table are 1.0 and 0.5, respectively. Alpha is winning the current contest. By applying her maximum pulling force, she is dragging Beta across the table. a) What is Alpha's acceleration? b) What is the tension at each end of the thread?

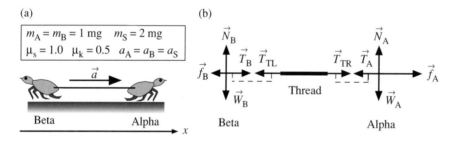

FIGURE 7-17   a) Two fleas having a tug-of-war. b) Free-body diagrams for the fleas and the thread that connects them.

SOLUTION   Here is a problem where the string is *not* massless, so we will need to treat it as a third system. The coordinate axes are straightforward because all the systems are moving along the same line, which we will call the x-axis. All three systems are tied together and moving in the same direction, so the acceleration constraint has to be $a_A = a_S = a_B$, where a_S is the acceleration of the string. We have written the acceleration constraint in terms of the x-components, rather than as vectors, because the motion is one dimensional.

Figure 7-17b shows the free-body diagrams for both fleas and the thread. The *static* friction force $\vec{f}_A$ on Alpha is a propulsion force exerted on her in the forward direction by the table, just like the person pushing the crate had back in Example 7-1. Beta, who is simply being dragged, has the usual kinetic friction force $\vec{f}_B$ opposing his motion. $\vec{T}_{TR}$ and $\vec{T}_{TL}$ are the tension in the thread at the right and left ends, respectively. Because the thread is accelerating to the right, the net force on it must be greater than zero. For this to be true, the tension $|\vec{T}_{TR}|$ at the right end of the thread must be greater than the tension $|\vec{T}_{TL}|$ at the left end. In this problem the pair $\vec{T}_{TR}$ and $\vec{T}_A$ and the pair $\vec{T}_{TL}$ and $\vec{T}_B$ are true action/reaction pairs. We can write Newton's second law for each system as:

Alpha:
$$\sum (F_{on\ A})_x = |\vec{f}_A| - |\vec{T}_A| = m_A(a_A)_x = m_A a$$

$$\sum (F_{on\ A})_y = |\vec{N}_A| - |\vec{W}_A| = |\vec{N}_A| - m_A g = m_A(a_A)_y = 0 \qquad (7\text{-}24)$$

$$\Rightarrow \quad |\vec{N}_A| = m_A g$$

Thread:
$$\sum (F_{on\ thread})_x = |\vec{T}_{TR}| - |\vec{T}_{TL}| = m_S(a_S)_x = m_S a \qquad (7\text{-}25)$$

Beta:
$$\sum (F_{on\ B})_x = |\vec{T}_B| - |\vec{f}_B| = m_B(a_B)_x = m_B a$$

$$\sum (F_{on\ B})_y = |\vec{N}_B| - |\vec{W}_B| = |\vec{N}_B| - m_B g = m_B(a_B)_y = 0 \qquad (7\text{-}26)$$

$$\Rightarrow \quad |\vec{N}_B| = m_B g.$$

We utilized the acceleration constraint in these equations when we replaced all of the a_x with the common value a. Other information that we have is:

Third law:
$$|\vec{T}_{TR}| = |\vec{T}_A| \text{ and } |\vec{T}_{TL}| = |\vec{T}_B| \qquad (7\text{-}27)$$

Kinetic friction:
$$|\vec{f}_B| = \mu_k |\vec{N}_B| = \mu_k m_B g. \qquad (7\text{-}28)$$

The last thing we need to consider is how to interpret the statement that Alpha is pulling with her maximum force. The force she has control over is $\vec{f}_A$, so it must be that she is pushing against the table with the maximum possible force and the table, in reaction, is pushing back equally hard to provide a maximum possible $\vec{f}_A$. The force with which Alpha can push against the table surface is limited by the point at which her feet slip. We noted in our earlier discussion of propulsion forces that these forces are *static* friction forces. We also know that the static friction force has a maximum possible value—the object slips if the static friction force tries to exceed $f_{s\ max}$. So the maximum force $\vec{f}_A$ is given by the maximum value of the static friction force on Alpha's feet. Stated quantitatively, this is

Static friction:
$$|\vec{f}_A|_{max} = f_{s\ max} = \mu_s |\vec{N}_A| = \mu_s m_A g. \qquad (7\text{-}29)$$

We can now insert the information from Eqs. 7-27 through 7-29 into Eqs. 7-24 through 7-26. Because the fleas have equal masses, we will also simplify notion with $m_A = m_B = m$:

$$\mu_s m g - |\vec{T}_A| = ma$$

$$|\vec{T}_A| - |\vec{T}_B| = m_S a \qquad (7\text{-}30)$$

$$|\vec{T}_B| - \mu_k m g = ma.$$

This is a pretty drastic simplification from the original equations, but we still have three equations in three unknowns: a, $|\vec{T}_A|$, and $|\vec{T}_B|$. Notice that we can eliminate *both* tensions by adding the three equations. Upon doing so, we are left with

$$(\mu_s - \mu_k)mg = (2m + m_S)a$$

$$\Rightarrow a = \frac{(\mu_s - \mu_k)mg}{(2m + m_S)}. \tag{7-31}$$

This makes sense when you stop to think about it. The numerator is the *net* force on the combined system of both fleas *and* the string—the *difference* between the forward static friction force on Alpha and the backward drag force on Beta. Were it not for the difference between μ_s and μ_k, Alpha would not be able to generate a force sufficient to overcome Beta's resistive friction force. The denominator is the total mass of the combined system. Thus Eq. 7-31 is just the second law F_{net}/m_{total} applied to the combined system. We can now use Eq. 7-31 back in the first and third of Eqs. 7-30 to find the tensions at the ends of the thread:

$$|\vec{T}_{TR}| = |\vec{T}_A| = \mu_s mg - m\left[\frac{(\mu_s - \mu_k)mg}{2m + m_S}\right] = \left[\frac{(\mu_s + \mu_k)m + \mu_s m_S}{2m + m_S}\right]mg$$

$$|\vec{T}_{TL}| = |\vec{T}_B| = \mu_k mg + m\left[\frac{(\mu_s - \mu_k)mg}{2m + m_S}\right] = \left[\frac{(\mu_s + \mu_k)m + \mu_k m_S}{2m + m_S}\right]mg. \tag{7-32}$$

Evaluation of Eqs. 7-31 and 7-32 gives

$$a = 1.22 \text{ m/s}^2$$

$$|\vec{T}_{TR}| = 8.57 \times 10^{-6} \text{ N}$$

$$|\vec{T}_{TL}| = 6.13 \times 10^{-6} \text{ N}.$$

The acceleration is quite respectable even though the forces are extremely small. Note that, as expected, the tension is larger at the right end of the thread than at the left end. The difference of 2.44×10^{-6} N is, in fact, exactly the net force needed to accelerate the 2 mg thread at 1.22 m/s². Be careful not to make a very common mistake when you perform the numerical evaluation in a problem such as this. We were given the masses in *milli*grams, 1×10^{-3} g. Neither milligrams nor grams are SI units, so we had to convert all masses to *kilograms* before doing calculations.

●

●   **EXAMPLE 7-10**   In Fig. 7-18a, a rope attached to a 20 kg wooden sled is pulling the sled up a 20° snow-covered hill with tension force $\vec{T}$. A wooden box with a mass of 10 kg is placed on top of the sled. What is the maximum tension in the rope such that the box does not slip?

SOLUTION The answer to this question will depend critically on identifying the action/reaction pairs. Let us start by defining coordinate systems as shown in the pictorial model of Fig. 7-18b. As previously, tilted axes will make the problem more tractable. As long as the box does not slip, the acceleration constraint must be $(a_B)_x = (a_S)_x = a$.

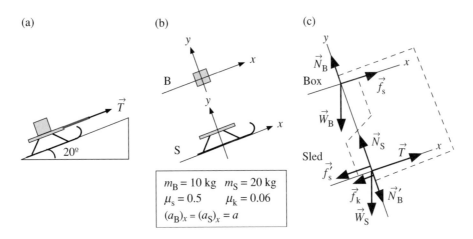

FIGURE 7-18 a) Box on sled being pulled up a hill. b) Pictorial model. c) Free-body diagrams for box and sled, with action/reaction pairs identified.

What forces act on the box? It is in contact with the top surface of the sled, so it experiences a normal force $\vec{N}_B$. This is the force of the *sled* on the box, not the force of the ground. The ground doesn't contact the box. Neither does the rope contact the box, so there is no tension force. The sled, however, does more than simply press upward on the box—there is also friction at the sled-box surface. Because there is no motion between the two, the friction force must be a static friction. A common error is to place a static friction force $\vec{f}_s$ on the box that points *down* the slope, opposite the motion. Why is that wrong? Keep in mind that a static friction force acts in such a direction as needed to *prevent slipping*. Were it not for friction, the box would slide *downhill* until it fell off of the *back* of the sled. To prevent this from happening, $\vec{f}_s$ must point *up* the slope! This is the force—a force of the sled on the box—that will accelerate the box up the incline. Because $|\vec{f}_s|$ has a maximum value, there will be a maximum acceleration a the sled can have without the box slipping. Somehow, in a way yet to be determined, this is going to be related to the maximum tension in the rope. The upper part of Fig. 7-18c shows the free-body diagram for the box.

What about the sled? It has a typical normal force $\vec{N}_s$ and kinetic friction force $\vec{F}_k$ exerted on its bottom surface by the ground. This friction force does point downhill, opposite the motion. The tension force $\vec{T}$ points uphill and is, for this problem, an external force. But what goes on at the *top* surface of the sled? We already noted that the normal force $\vec{N}_B$ on the box is a force of type $\vec{f}_{\text{sled on box}}$. This force must have a reaction force $\vec{F}_{\text{box on sled}}$—that is, the box presses downward on the sled! We can call this force $\vec{N}'_B$. The static friction force $\vec{f}_s$ on the box was also a force of the sled on the box. Consequently there must be a reaction force $\vec{f}'_s$ on the sled. This reaction force has to be opposite $\vec{f}_s$, so it points downhill.

Force $\vec{f}'_s$ on the sled is hard for many students to understand, even though Newton's third law tells us it must be there. One way to visualize this force is to notice that the presence of the box does make it harder to pull the sled up the hill—the heavier the box is, the more tension will be needed. How does the box make its presence known? The only way it can do so is by exerting a *retarding* force on the sled, which is exactly what $\vec{f}'_s$ does. With this information, we can draw the lower free-body diagram shown in Fig. 7-18c and link the two pairs of action/reaction forces.

This is an excellent illustration of how crucial it is to focus on the *qualitative* analysis of identifying forces and drawing free-body diagrams. The rest of the problem is not trivial, but we can work our way through it with confidence after having found the free-body diagrams. It would be hopeless—even for an experienced expert—to try to go directly to Newton's laws without this analysis. But now we can proceed with the second law for each object:

Box:

$$\sum (F_{\text{on box}})_x = |\vec{f_s}| - |\vec{W}_B| \sin\theta$$

$$= |\vec{f_s}| - m_B g \sin\theta$$

$$= m_B (a_B)_x = m_B a$$

$$\sum (F_{\text{on box}})_y = |\vec{N}_B| - |\vec{W}_B| \cos\theta \qquad (7\text{-}33)$$

$$= |\vec{N}_B| - m_B g \cos\theta = 0$$

$$\Rightarrow \quad |\vec{N}_B| = m_B g \cos\theta.$$

Sled:

$$\sum (F_{\text{on sled}})_x = |\vec{T}| - |\vec{W}_S| \sin\theta - |\vec{f_s'}| - |\vec{f_k}|$$

$$= |\vec{T}| - m_S g \sin\theta - |\vec{f_s'}| - |\vec{f_k}|$$

$$= m_S (a_S)_x = m_S a$$

$$\sum (F_{\text{on sled}})_y = |\vec{N}_S| - |\vec{W}_S| \cos\theta - |\vec{N}_B'| \qquad (7\text{-}34)$$

$$= |\vec{N}_S| - m_S g \cos\theta - |\vec{N}_B'| = 0$$

$$\Rightarrow \quad |\vec{N}_S| = m_S g \cos\theta + |\vec{N}_B'|.$$

In addition to this mess of equations, we also have the following information:

Third law:

$$|\vec{f_s'}| = |\vec{f_s}| \text{ and } |\vec{N}_B'| = |\vec{N}_B| = m_B g \cos\theta$$

$$\Rightarrow \quad |\vec{N}_S| = (m_S + m_B) g \cos\theta \qquad (7\text{-}35)$$

Kinetic friction:

$$|\vec{f_k}| = \mu_k |\vec{N}_S| = \mu_k (m_S + m_B) g \cos\theta \qquad (7\text{-}36)$$

Static friction:

$$|\vec{f_s}| = f_{s\,\text{max}} = \mu_s |\vec{N}_B| = \mu_s m_B g \cos\theta. \qquad (7\text{-}37)$$

We have used the *maximum* value of $|\vec{f_s}|$ because we are interested in the point at which the box slips. We have already used the two y-equations from the second law to find the friction forces. If we insert the information from Eqs. 7-35 through 7-37 into the second law x-equations, we finally end up with

$$\mu_s m_B g \cos\theta - m_B g \sin\theta = m_B a$$

$$|\vec{T}| - m_S g \sin\theta - \mu_s m_B g \cos\theta - \mu_k (m_S + m_B) g \cos\theta = m_S a. \qquad (7\text{-}38)$$

We can solve the first of these directly for the acceleration at which the box starts to slip:

$$a = (\mu_s \cos\theta - \sin\theta)g = 1.25 \text{ m}/\text{s}^2,$$

where we found the coefficient of friction from Table 5-1. Inserting this into the second of Eqs. 7-38 gives, after some algebraic rearrangement,

$$|\vec{T}| = (\mu_s + \mu_k)(m_S + m_B)g\cos\theta = 155 \text{ N}.$$

It is worth noting that the kinetic friction force in Eq. 7-36 depends on the *total* mass $m_S + m_B$ because, after all, the total mass is pressing down on the hill. However, we did not put this information *into* the equations, it came *out* of the equations. The friction force depended only on the normal force of the snow on the sled because that is the point of contact where the friction occurs. You should trace through the steps to make sure you understand how the box's mass ended up in Eq. 7-36 as a consequence of both the second and the third laws.

These are not easy problems. But you *can* solve problems such as these if you follow the strategy and the step-by-step approach to analyzing forces and using Newton's laws that we have developed in these first seven chapters.

We have reached the pinnacle of Newton's view of forces and motion. It is possible to solve extremely complex dynamics problems using the physics principles and tools that you now have learned. Rather than pursue ever more complex problems, however, our road is now going to turn in a new direction as we begin to look at the new concepts of momentum and energy. These ideas, developed largely by Newton's successors, will give us a different perspective on the problem of motion. Forces will remain with us, always forming the most basic level of our understanding of dynamics, but they will diminish in importance as they begin to share the stage with newer ideas.

Summary

Important Concepts and Terms

interaction forces

action/reaction pairs

Newton's third law

massless string

Atwood machine

We have introduced a new concept, that of action/reaction pairs of forces, and a new physical law, Newton's third law, to describe them. These, in turn, led us to revise our basic problem-solving strategy for interacting-particle problems. We have not introduced new equations or formulas but, instead, have increased the sophistication with which we carry out a Newtonian analysis of forces. The second law remains the link between force and motion, but when objects interact we need to understand and include the third law as part of our overall analysis.

Strings and the tension in strings received an extended analysis in this chapter because strings and cables are one of the most common ways by which two objects interact. The significance of the massless string approximation is that it allows us to view the forces on two objects *as if* they are an action/reaction pair. The logic and reasoning in behind this, as explained in Section 7-4, is worth careful study.

Exercises and Problems

The ✍ icon indicates that these problems can be done on a Dynamics Worksheet.

◣ Exercises

✍ 1. A 1 kg block tied to the wall with a rope is sitting on top of a 2 kg block (Fig. 7-19). The lower block is being pulled to the right with a second rope under a tension of 20 N. The coefficient of kinetic friction at both the lower and upper surfaces of the 2 kg block are $\mu_k = 0.40$.

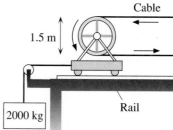
FIGURE 7-19

   a. What is the tension in the rope holding the 1 kg block to the wall?
   b. What is the acceleration of the 2 kg block?

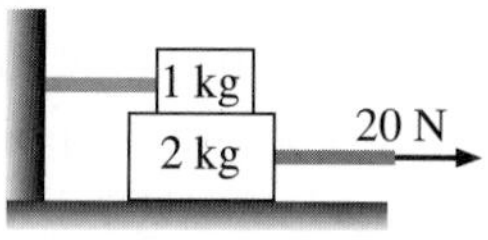

✍ 2. The cable cars in San Francisco are pulled along their tracks by an underground steel cable that moves along at 9.5 mph. The cable is driven by large motors at a central power station and extends, via an intricate pulley arrangement, for several miles beneath the city streets. The length of the cable stretches up to 100 feet during the cable's lifetime. To keep the tension constant, the cable passes over a 1.5 m diameter pulley that can roll back and forth on rails as shown in Fig. 7-20. A 2000 kg block is attached to the pulley's cart, via a rope and another pulley, and is suspended in a deep hole. What is the tension in the cable car's cable?

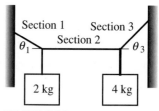
FIGURE 7-20

✍ 3. Two objects having masses of 2 kg and 4 kg are suspended from ropes, as shown in Fig. 7-21. The points where the rope contacts the walls are adjusted until the center section 2 is perfectly horizontal. It is then found that $\theta_1 = 20°$.

FIGURE 7-21

   a. What is angle θ_3?
   b. What is the tension in section 3 of the rope?

◣ Problems

✍ 4. A block of mass m rests on a 20° slope and has coefficients of friction $\mu_s = 0.80$ and $\mu_k = 0.50$ with the surface. The block is connected via a massless string over a frictionless pulley to a hanging block of mass 2 kg as shown in Fig. 7-22.

   a. What is the minimum mass m that the block can have so that it will stick and not slip?
   b. If a block with this minimum mass is nudged ever so slightly, it will start being pulled up the incline. What acceleration will it have?

FIGURE 7-22

5. Figure 7-23 shows two 1 kg blocks connected by a rope having a mass of 250 g. A second 250 g rope hangs beneath the lower block. The entire assembly is accelerated upward at 3 m/s^2 by force $\vec{F}_{ext}$.

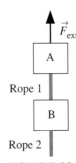

FIGURE 7-23

   a. What is $|\vec{F}_{ext}|$? (**Hint:** Consider the combined systems as one object.)
   b. What is the tension in rope 1 where it contacts block A?
   c. What is the tension in rope 1 where it contacts block B?
   d. What is the tension in rope 2 where it contacts block B?
   e. What is the tension in rope 2 at its lower end?
   f. What is the net force on block A?
   g. What is the net force on block B?
   h. Why should your answers to f) and g) have such a simple relationship? Discuss.

6. The lower block in Fig. 7-24, which has a mass of 2 kg, is pulled on by a rope with a tension force of 20 N. The coefficient of kinetic friction between both the lower block and the surface and between the lower block and the 1 kg upper block is 0.30. What is the acceleration of the 2 kg block?

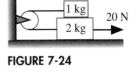

FIGURE 7-24

7. A Federation starship (2×10^6 kg) uses its tractor beam to pull a shuttlecraft (2×10^4 kg) aboard from a distance of 10 km away (Fig. 7-25). The tractor beam exerts a constant force of 4×10^4 N on the shuttlecraft.

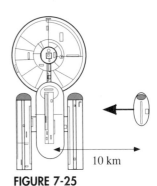

FIGURE 7-25

   a. What is the acceleration of the shuttlecraft?
   b. How far does the starship move as it pulls the shuttlecraft aboard?

8. A 2000 kg cable car descends a hill 200 m high at a 30° angle, as shown in Fig. 7-26. In addition to using its brakes, the cable car controls its speed by pulling an 1800 kg counterweight up the other side of the hill, which has an angle of 20°. The rolling friction of both the cable car and the counterweight are negligible.

   a. How much braking force does the cable car need for it to descend at constant speed?
   b. One day the brakes fail just as the cable car leaves the top on its downward journey. What is the runaway car's speed at the bottom of the hill?

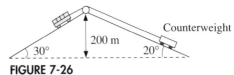

FIGURE 7-26

9. A 4 kg box is on a frictionless 35° slope and is connected by a massless string over a frictionless pulley to a hanging 2 kg weight. The picture for this situation is the same as Fig. 7-22.

   a. If the 4 kg box is held in place, so that it cannot move, what is the tension in the string?
   b. If the box is then released, which way will it move on the slope?
   c. What is the tension in the string once the box begins to move?

✍ 10. Mass m_1 on a frictionless table is connected by a string through a hole in the table to a hanging mass m_2 (Fig. 7-27). With what speed must m_1 rotate in a circle of radius r if m_2 is to remain hanging at rest?

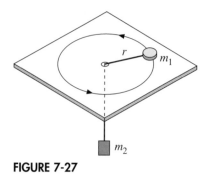

FIGURE 7-27

✍ 11. Figure 7-28 shows a block of mass m placed on a lower block of mass M and slope θ. The lower block, in turn, rests on a scale.

a. Initially, static friction is sufficient to keep m from moving. In this case, the two blocks are effectively a single block of mass $M + m$ and the scale should read a weight $(M + m)g$. Show, by treating the two blocks as *separate* systems and analyzing the forces, that this is the case.

b. An extra-fine lubricating oil, having $\mu_s = \mu_k = 0$, is sprayed on the upper surface of M, allowing m to begin sliding down. However, the friction between M and the scale is large enough that the lower block does *not* slip on the scale. Now what does the scale read?

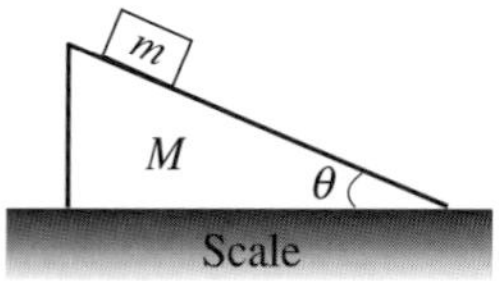

FIGURE 7-28

✍ 12. Bob, who has a mass of 75 kg, can throw a 500 g rock with a speed of 30 m/s. The distance through which his hand moves as he accelerates the rock forward, from rest until he releases it, is 1 m.

a. What constant force must Bob exert on the rock to throw it with this speed?

b. If Bob is standing on frictionless ice, what is his recoil speed after releasing the rock?

c. Give an explanation of why Bob has a recoil velocity.

13. a. Describe how a car accelerates from rest. Your explanation should be given in terms of forces and physical laws. You should include and use a free-body diagram.

b. Why can a car accelerate but a house cannot? What is the role played by the engine and the fuel? Explain. Again, your explanation should be in terms of forces and their properties.

c. Two-thirds of the weight of a 1500 kg car rests on the drive wheels. What is the maximum acceleration of this car on a concrete surface?

14. A 100 kg basketball player can leap straight up in the air to a height of 80 cm as shown in Fig. 7-29. How? Answer this question by analyzing the situation as follows:

a. The player bends his legs until the upper part of his body has dropped by 60 cm, then he begins his jump. Explain in terms of forces what the player is doing from the time he begins his jump until his feet leave the ground.

b. Draw separate free-body diagrams for the player and for the floor *as* he is jumping but before his feet leave the ground.

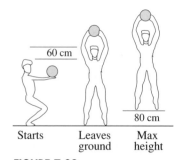

FIGURE 7-29

   c. Is there a net force on the player as he jumps (before his feet leave the ground)? How can that be? Explain.

   d. What is the speed with which the player leaves the ground?

   e. What was the player's acceleration, assumed to be constant, as he jumped?

   f. Suppose the player jumps while standing on metric bathroom scales that read in N. What do they read before he jumps? As he is jumping? After his feet leave the ground?

[**Estimated 6 additional problems for the final edition.**]

Chapter **8**

Momentum and Its Conservation

L O O K I N G B A C K | Sections 3.7; 7.3

8.1 Conservation Laws: Connecting "Before" and "After"

In Chapter 7 you learned that Newton's third law describes the interactions between objects. An interesting and important class of interactions are those that take place "quickly." Consider a locomotive that is rolling along railroad tracks when it collides and couples with a boxcar. There are certainly some complicated interaction forces occurring during the collision, but it would be very difficult to know exactly what they are. If we know the locomotive's speed before the collision, is it possible to find the speed of the combined locomotive and boxcar after the collision *without* knowing the interaction forces? Or consider two balls in a pipe with an explosive charge between them, as in Fig. 8-1. After the explosive is detonated, the two balls fly out of the pipe with velocities v_1 and v_2. The explosion itself is extremely complex, but do we really have to understand all of its details to find v_1 and v_2?

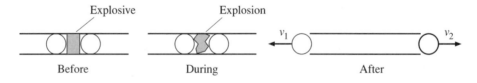

FIGURE 8-1 Explosion shooting two balls out of a pipe.

These examples, and we could easily supply many more, have two basic characteristics in common. First, as previously noted, the interaction happens quickly. Second, the event can be divided into three phases: *before*, *during*, and *after* the interaction, as shown in Fig. 8-1. What we would like to find is some relationship between the "before" and the "after" that does not require us to know all of the messy details of what goes on during the interaction.

It would be very useful to find some quantity Q that is not changed by the interaction. If such a quantity exists, it will have the same value *after* the interaction that it did *before* the interaction. We could write this as:

$$Q_f = Q_i \text{ (Conservation of } Q\text{)},$$

indicating that Q does not change in value during the interaction. A quantity that does not change its value during a physical process is said to be **conserved**. If it turns out that Q is conserved in just one or two processes, we might find that to be interesting but hardly profound. If, on the other hand, we find an entire *class* of processes where Q is conserved, then we have discovered an important insight into how nature behaves. The significance of such a discovery would be comparable to the significance of Newton's laws. We would be justified in calling such a discovery the "law of conservation of Q."

Conservation laws—and we will find some—tell us that nature is *not* free to act in arbitrary ways. There are an infinite number of ways that we could imagine the two balls of Fig. 8-1 emerging from the pipe, but only one of these ways actually occurs—namely, the one that satisfies the proper conservation laws. This is a very powerful statement about nature. It says that we *can* connect "before" and "after," and thus can make predictions about the outcome of events without having to know the interaction details, *if* we can find a conservation law that governs the process.

You already know one conservation law that we have tacitly assumed to be valid. For a system consisting of several particles of masses m_1, m_2, ..., we have assumed that the total mass $M = m_1 + m_2 + ...$ does not change with time. The statement $M_f = M_i$ is a conservation law. This seems perfectly obvious, but in fact it is only *because* you have repeatedly observed it to be true—empirical evidence!—that it has achieved the status of "common sense." We can state our knowledge about mass in a more formal way as the **law of conservation of mass**. This law states the following:

Law of Conservation of Mass: The total mass in a closed system is constant.

The law of conservation of mass is a critical underpinning of all of chemistry, where the mass of the reaction products is set equal to the mass of the reactants. Despite common sense, it was only after many long and precise experiments that this law was accepted as being true. (Surprisingly, and at complete odds with common sense, Einstein's theory of relativity has shown that mass actually is *not* conserved but can, under some circumstances, be converted to energy in accordance with his famous formula $E = mc^2$. Nonetheless, conservation of mass is an exceedingly good approximation in nearly all applications of science and engineering.)

An important aspect of conservation laws is that they give us a new and different *perspective* on how motion changes. This is not insignificant. You probably have seen optical illusions where a figure appears one way from one perspective but quite different from another perspective, even though the basic information in the figure has not changed. Likewise with motion. We will soon see that there are some situations most easily analyzed from a Newton's laws perspective, but others that make much more sense when analyzed from a conservation law perspective. One of your tasks is to learn which perspective is best for a given problem. The examples we looked at earlier provide a strong hint: If you need to relate "before" to "after" in a situation but do not care about the details of the interaction, then a conservation law perspective is likely the best. If, on the other hand, you need to understand the details of the interaction itself, then you will probably need to use Newton's laws.

8.2 Impulse and Momentum

The epitome of a *sudden* change of motion is a *collision*. During a collision, two objects approach each other, interact, then quickly move apart. The interaction usually lasts for a small but *finite* amount of time. For example, a collision of a ball with a tennis racket may seem instantaneous to your eye, but that is a limitation of your perception. Figure 8-2 shows a high-speed photograph of a tennis ball colliding with a tennis racket. What you should notice, in particular, is that the right side of the ball is no longer round—it is flattened and pressed up against the strings of the racket. The compression is the result of the interaction between the racket and the ball.

FIGURE 8-2 High-speed photo of a tennis ball being hit. Notice the flattened right side.

The actual contact time for collisions such as a tennis ball and racket or a baseball and bat is typically 1–10 ms (0.001–0.01 s). This is too short a time to perceive, but easily measured with simple equipment. The harder the surfaces, the shorter the contact time. For example, two steel balls bouncing off each other are probably in contact for less than 1 ms. During this time the molecular bonds compress a little and then re-expand. No matter how hard the materials, they are in contact for a small but finite amount of time.

When a tennis ball hits a racket, the ball and the racket exert forces on each other during the short time interval while they interact. First there is a period during which the ball is being compressed and the strings are stretching. The force is increasing during this interval. Eventually, the ball reaches maximum compression—the interaction forces $\vec{F}_{\text{ball on racket}}$ and $\vec{F}_{\text{racket on ball}}$ are maximum at this instant of time. Then the ball expands and "springs" away from the racket during a period of decreasing force. This is shown graphically in Fig. 8-3. Forces such as this that act quickly and then vanish are called **impulsive forces**. Impulsive forces describe not only collisions but also events such as explosions.

Consider an object of mass m that is traveling in a straight line with one-dimensional velocity v_i. It suddenly and quickly interacts with another object, experiencing an impulsive force $F(t)$ from time t_i to time t_f, after which it emerges with velocity v_f. How is v_f related to v_i? Notice that we have explicitly allowed the force to be a function of time $F(t)$, consistent with the picture of a rapidly-changing collision force that we see in Fig. 8-3. Keep in mind that both v and F are vector components and thus have *signs*, depending upon which way the vector points.

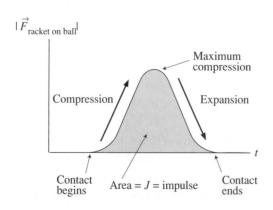

FIGURE 8-3 The rapidly-changing force during a collision of a tennis racket with a tennis ball.

We can use Newton's second law to answer the question of how v_f is related to v_i. Recall that acceleration is defined (in one dimension) as $a = dv/dt$. We can write the second law as

$$ma = m\frac{dv}{dt} = F(t) \tag{8-1}$$
$$\Rightarrow m\,dv = F(t)\,dt.$$

Equation 8-1 tells us how to find the infinitesimal change in velocity dv that occurs during the infinitesimal time interval dt at time t. The *total* change in velocity during a finite time interval can be found by adding up all of the infinitesimal changes dv—that is, by integrating. Because we are examining a special class of problems where the force $F(t)$ is non-zero only during a small interval of time, we only need to integrate Eq. 8-1 over the time interval from t_i to t_f, during which the velocity changes from v_i to v_f. Doing so gives

$$m\int_{v_i}^{v_f} dv = mv_f - mv_i = \int_{t_i}^{t_f} F(t)dt. \tag{8-2}$$

Notice that Eq. 8-2 contains the product of the particle's mass and velocity. This product is called the *momentum p* of the particle. Momentum is defined as

$$\textbf{momentum} = p = mv. \tag{8-3}$$

The plural of *momentum* is *momenta*, from its Latin origin.

Momentum is another example of a term that we use in everyday speech without a precise definition. In physics and engineering, however, it is a technical term whose meaning is defined as the product of mass and velocity. An object can acquire momentum either by having a small mass but a large velocity (a bullet fired from a rifle) or a small velocity but a large mass (a large truck rolling at a slow 1 mph).

Momentum, strictly speaking, is a vector quantity: $\vec{p} = m\vec{v}$. Because we are only looking at one-dimensional motion in this chapter, Eq. 8-3 defines the *component* of the momentum vector along the axis of motion. As a vector component, p has a *sign*—exactly the same sign as v. Newton actually formulated his second law in terms of momentum rather than acceleration. Equation 8-1, you will notice, can be written in terms of momentum as

$$F = \frac{dp}{dt}. \tag{8-4}$$

This statement of the second law, saying that force is the rate of change of momentum, is more general than our earlier version $F = ma$. It allows for the possibility that the mass of the object might change—such as a rocket that is losing mass as it burns fuel. We will not consider variable-mass problems in this course, because the mathematics becomes complex, but many of you will see such problems in more advanced courses.

The question we are trying to answer is how a particle's final velocity v_f is related to its initial velocity v_i when it experiences an impulsive force. Equation 8-2 tells us that the particle's change in velocity can be expressed in terms of the time-integral of the force. Let's define a quantity called the **impulse** J to be

$$\textbf{impulse} = J = \int_{t_i}^{t_f} F(t)dt = \text{area under the } F(t) \text{ curve between } t_i \text{ and } t_f. \tag{8-5}$$

The units of both momentum and impulse are kg m/s. (Impulse, strictly speaking, would have units of N s, but you should be able to show that N s are equivalent to kg m/s.)

Equation 8-2 can be rewritten in terms of impulse and momentum as

$$p_f - p_i = \Delta p = J \quad \text{(Impulse-Momentum Theorem).} \tag{8-6}$$

Equation 8-5 is called the **impulse-momentum theorem**, and it is what we have been looking for to connect "before" and "after." This theorem tells us that an impulse delivered to an object causes a *change* of the object's momentum. The momentum p_f "after" a brief interaction, such as a collision or an explosion, is equal to the momentum p_i "before" the interaction *plus* the impulse that arises from the interaction: $p_f = p_i + J$. Furthermore, we have a *graphical* interpretation of J as the area under the $F(t)$ curve, as shown on Fig. 8-3. This indicates that we do *not* need to know the exact force function $F(t)$, which gives the full details of the interaction, but only its integral.

EXAMPLE 8-1 A baseball of mass 150 g is thrown with a speed of 20 m/s and is then hit straight back toward the pitcher with a speed of 40 m/s. If the interaction force between the ball and the bat has the shape shown in Figure 8-4a, what is the maximum force F_{max} exerted on the ball by the bat?

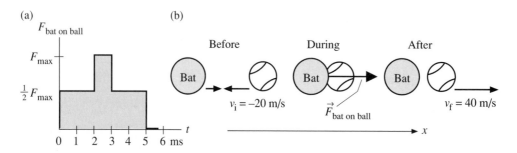

FIGURE 8-4 a) Interaction force of baseball and bat, showing a contact time of 5 ms. b) Pictorial model showing the situation before, during, and after the collision.

SOLUTION We need a new approach to solve a problem such as this. Our earlier strategy of pictorial models and free-body diagrams was oriented toward the use of Newton's laws and a subsequent kinematical analysis. Now we are interested in making a connection between "before" and "after," using impulse rather than forces. Figure 8-4b shows a new type of pictorial model more suited to this problem: a sketch showing the situation before the collision, during the collision, and after the collision. Note that a coordinate axis is still needed.

Rather than starting the mathematical model with the second law, as we have done consistently until now, we want to start with the impulse-momentum theorem of Eq. 8-6:

$$\Delta p = p_f - p_i = mv_f - mv_i = J.$$

We know the mass of the ball and both its initial and final velocity. Notice that we converted the problem statement about *speeds* into information about *velocities* in Fig. 8-4b, with one velocity (and thus one momentum) being negative. We chose to show the ball initially moving toward the left so that the force component F, and thus the impulse, will be positive. You could equally well have started with the ball moving toward the right, but doing so would cause many of the quantities (including J) to be negative and would increase the chances of making sign errors. This is a nice illustration of how you can maximize your opportunity for success with just a little thought as you set up a problem.

The impulse J is the area under the force curve of Fig. 8-4a. The area can be subdivided into two parts: a lower rectangle of height $(1/2)F_{max}$ and width 5 ms, and a smaller upper rectangle of height $(1/2)F_{max}$ and width 1 ms. We can then find the total area to be

$$J = \text{area} = (\tfrac{1}{2}F_{max} \times 0.005 \text{ s}) + (\tfrac{1}{2}F_{max} \times 0.001 \text{ s})$$

$$= \tfrac{1}{2}F_{max} \times 0.006 \text{ s}$$

$$= (F_{max})(0.003 \text{ s}).$$

Using this result for J in the impulse-momentum theorem gives

$$mv_f - mv_i = m(v_f - v_i) = J = (F_{max})(0.003 \text{ s})$$

$$\Rightarrow F_{max} = \frac{m(v_f - v_i)}{(0.003 \text{ s})} = \frac{(0.15 \text{ kg})(40 \text{ m/s} - (-20) \text{ m/s})}{(0.003 \text{ s})} = 3000 \text{ N}.$$

This is a very large force, but actually quite typical of the size of impulsive forces during collisions. The main thing to focus on is our new perspective of how an impulse changes the momentum of an object.   ●

Often other external forces are acting on an object during a collision or other brief interaction. In Example 8-1, for instance, the baseball also has a weight force acting on it. However, it is usually the case that these other external forces are *much* smaller than the interaction forces. For example, the weight of the ball is 1.5 N, which is vastly less than the 3000 N force of the bat on the ball. It is a reasonable approximation to neglect such external forces *during* the brief time of the interaction—they simply are not large enough to cause a noticeable change of motion during an interval of a few milliseconds. This is called the **impulse approximation**. When we use the impulse approximation, it is important to note that p_i and p_f are then the momenta *immediately* before and after the collision. The velocities in Example 8-1 are not necessarily the velocities with which the pitcher throws the ball or the second baseman catches it. Those velocities are influenced by weight and air resistance forces acting on the ball during the much longer times that the ball moves through the air. Instead, v_i and v_f are the velocities of the ball *immediately* before and after it collides with the bat. We could then do a follow-up problem to find the ball's trajectory if we know it leaves the bat with a velocity of 40 m/s.

8.3 Momentum in Two-Particle Collisions

[**Photo suggestion: Automobile collision.**]

In the last section, our analysis of a particle's response to an impulsive force was from the perspective of single particle dynamics—much like the application of Newton's second law to a single particle in Chapters 4–6. It is true that we used the concepts of impulse and momentum, rather than force and acceleration, but we still treated the impulse as an external influence. After having introduced the third law, however, we have developed a more sophisticated understanding of forces as *interactions* between systems. This has some very interesting implications for collisions between two particles.

Figure 8-5 shows a collision between two particles that move together along a straight line with velocities v_{1i} and v_{2i}, collide, then bounce apart with final velocities v_{1f} and v_{2f}. The forces responsible for the collision are the interaction forces $\vec{F}_{1 \text{ on } 2}$ and $\vec{F}_{2 \text{ on } 1}$. The forces have vector components $F_{1 \text{ on } 2}$ and $F_{2 \text{ on } 1}$ along the axis of motion, which cause the two particles to experience impulses J_1 and J_2. Using the impulse approximation, we can write Newton's second law (in the form of Eq. 8-4) for each particle *during* the collision as

$$\frac{dp_1}{dt} = F_{2 \text{ on } 1}$$

$$\frac{dp_2}{dt} = F_{1 \text{ on } 2} = -F_{2 \text{ on } 1}.$$

(8-7)

We have made explicit use of Newton's third law in the second of Eqs. 8-7 to relate the two members of the action/reaction pair of forces. It is important to keep in mind that the p's and F's are vector components, so they do have signs.

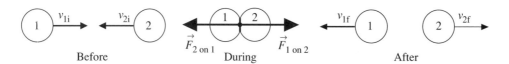

FIGURE 8-5 A collision between two particles. The force that each exerts on the other during the time of contact causes each particle to experience an impulse that changes its momentum.

Although Eqs. 8-7 are for two different particles, suppose—just to see what happens—we were to *add* these two equations. If we do, we find that:

$$\frac{dp_1}{dt} + \frac{dp_2}{dt} = \frac{d}{dt}(p_1 + p_2) = F_{2 \text{ on } 1} + (-F_{2 \text{ on } 1}) = 0.$$

(8-8)

But if the derivative of the sum $p_1 + p_2$ is zero, it must be the case that

$$p_1 + p_2 = \text{constant}.$$

(8-9)

In other words, the *sum* of the momenta of the two particles is a constant that does not change its value during the collision! We can use the unchanging value of $p_1 + p_2$ to infer that the sum of the momenta *before* the collision will equal the sum of the momenta *after* the collision, or:

$$p_{1i} + p_{2i} = p_{1f} + p_{2f}$$

$$\Rightarrow m_1 v_{1i} + m_2 v_{2i} = m_1 v_{1f} + m_2 v_{2f}.$$

(8-10)

We have discovered a *conservation law*—a quantity Q that is not changed by the interaction. In this case, the quantity Q is the sum of the momenta: $p_1 + p_2$. Equation 8-10 says that the value of this sum is not changed by the collision. Its value immediately before the collision is equal to its value immediately after the collision. (We have to insert the word *immediately* because we are using the impulse approximation. The value of the sum long before or long after the collision may differ as a consequence of other forces that we are neglecting during the brief collision.) We have succeeded in connecting "before" and "after" without having to know *anything* about the interaction forces $F_{1 \text{ on } 2}$ and $F_{2 \text{ on } 1}$. This result is called the **law of conservation of momentum**.

Although we have made a significant discovery about two-particle collisions, we still cannot use Eq. 8-10 to predict the final velocities v_1 and v_2 for an arbitrary collision—it is just one equation but it has two unknowns. We will soon find another conservation law— the law of conservation of energy—that we can combine with Eq. 8-10 to find a complete solution. There is, however, one class of collisions that we can fully understand on the basis of Eq. 8-10.

Collisions can have different possible outcomes. A rubber ball dropped on the floor bounces, but a ball of clay sticks to the floor without bouncing. A golf club hitting a golf ball causes the ball to rebound away from the club, but a bullet striking a block of wood embeds itself in the block. A collision in which the two objects stick together and move with a *common* final velocity is called a **perfectly inelastic collision**. The clay hitting the floor and the bullet embedding itself in the wood are examples of perfectly inelastic collisions. Other examples include railroad cars coupling together upon impact and darts hitting a dart board.

Mathematically, the condition for a perfectly inelastic collision is that the two objects have a common final velocity: $v_{1f} = v_{2f} = v_f$. If we apply Eq. 8-10 to a perfectly inelastic collision, we find

$$m_1 v_{1i} + m_2 v_{2i} = (m_1 + m_2)v_f \quad \text{(perfectly inelastic collision).} \quad (8\text{-}11)$$

Even though we have written Eq. 8-11 in terms of masses and velocities, keep in mind that this equation is really just stating the momentum conservation law: $p_1 + p_2 = $ constant.

EXAMPLE 8-2   Bob (75 kg) is jogging along with a speed of 4 m/s when he jumps onto a stationary cart (25 kg) and rolls away. What is his speed just after landing on the cart?

SOLUTION You might not have thought of Bob jumping onto a cart as a "collision," but it is from a physics perspective. The interaction forces between Bob and the cart—friction forces in this case—act only over a very short time interval, just the fraction of a second it takes him to become stuck to the cart. It is reasonable to neglect any other forces during this brief interval (impulse approximation). Because Bob becomes "stuck" to the cart, and they roll away with a common velocity, this is a perfectly inelastic collision. Other forces, such as friction with the ground, may cause the cart to slow down and eventually stop, but the question asks about Bob's speed on the cart *immediately* after the "collision." That is what our newly found conservation law is designed to answer.

Figure 8-6 shows a pictorial model consistent with our new strategy of looking at "before" and "after." We note that $v_{2i} = 0$ because initially the cart is at rest, which will make the mathematics somewhat easier. The physical principle that forms the basis for our

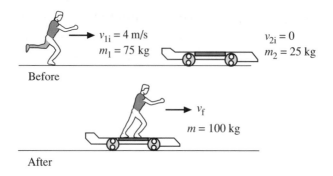

FIGURE 8-6 Pictorial model showing the "before" and "after" of Bob jumping onto the cart.

mathematical model is the momentum conservation law of Eq. 8-11. Applying it to this problem gives

$$m_1 v_{1i} + m_2 v_{2i} = m_1 v_{1i} = (m_1 + m_2) v_f$$

$$\Rightarrow v_f = \frac{m_1}{m_1 + m_2} v_{1i} = \frac{75 \text{ kg}}{100 \text{ kg}} (4 \text{ m/s}) = 3 \text{ m/s}.$$

Notice how easy this was! No forces, no acceleration constraints, no simultaneous equations—why didn't we think of this before? While conservation laws are indeed powerful, they can only answer certain questions. Had we wanted to know how far Bob slid across the cart before sticking to it, how long the collision took, or what the cart's acceleration was during the collision, we would not have been able to answer such questions on the basis of the conservation law. There is a price to pay for finding a simple connection between "before" and "after," and that price is the loss of all information about the details of the interaction. If we are satisfied with only knowing about "before" and "after," then the conservation law is a simple and straightforward way to proceed. But there are frequently problems where we *do* need to understand the interaction, and in that case there is no avoiding Newton's laws and all that they entail.

EXAMPLE 8-3   A 2000 kg Cadillac had just started forward from a stop sign when it was struck from behind by a 1000 kg Volkswagen. The cars' bumpers become entangled, and the two cars skidded forward together until they come to rest. Fortunately, both cars were equipped with airbags and the drivers were using seat belts, so no one was injured. Officer Tom, responding to the accident, measured the skid marks from the point of impact to the cars' final location to be 3 m ($\approx$10 feet) in length. He also took testimony from the Cadillac driver that the Cadillac's speed at the moment of impact was 5 m/s. Officer Tom charged the Volkswagen driver with reckless driving. Should the Volkswagen driver also be charged with exceeding the 50 km/hr ($\approx$30 mph) speed limit? The judge calls you as a "expert witness" to analyze the evidence. What is your conclusion?

SOLUTION Where should we begin a problem like this? The important observation to make is that this is really *two* problems. First, there is a collision problem—in this case, because the cars stick together, a perfectly inelastic collision. Equation 8-11 will apply to the collision, but not to the subsequent skidding motion. All that we can learn from the conservation law is the velocity *immediately* after the impact. While that is the end of the collision problem, it is just the beginning of a dynamics problem of how far an object— the two cars combined—will slide before coming to rest. The *final* velocity of the collision problem is the *initial* velocity for the dynamics problem. Our two-part strategy will be first to analyze the collision, then to use the results of the collision analysis as the starting point for a dynamics analysis.

Figure 8-7a shows an extended pictorial model, covering both the before and after of the collision as well as the more familiar picture for the dynamics of the skidding. We do not need to consider forces explicitly for the collision because we will use the law of conservation of momentum. However, we do need a free-body diagram of the forces during the subsequent skid. This is shown in Fig. 8-7b. Note that we have converted 50 km/hr to 13.9 m/s. We have also set the *final* velocity of the collision, v_f, equal to the *initial* velocity v_0 of the skidding problem. Our ultimate goal is to find v_{1i}.

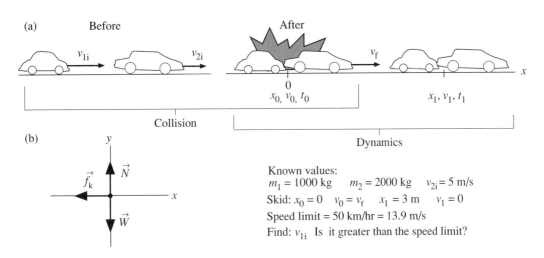

FIGURE 8-7 a) Pictorial model showing both "before" and "after" of the collision as well as the kinematics of the cars skidding. b) Free-body diagram of the skid.

Let's begin with the collision analysis. The law of conservation of momentum for a perfectly inelastic collision is Eq. 8-11:

$$m_1 v_{1i} + m_2 v_{2i} = (m_1 + m_2)v_f$$

$$\Rightarrow v_{1i} = \frac{(m_1 + m_2)v_f - m_2 v_{2i}}{m_1}.$$

Because we know v_{2i}, from the driver's testimony, we can evaluate v_{1i} *if* we can find v_f, the velocity *immediately* after the collision as the cars begin to slide. This, we hope, will come out of the dynamics analysis of the skid.

Dynamics analysis: The forces on the combined cars as they skid are shown on the free-body diagram. We can invoke Newton's second law to find

$$a_x = a = \frac{1}{(m_1 + m_2)} \sum F_x = \frac{-|\vec{f}_k|}{(m_1 + m_2)}$$

$$a_y = 0 = \frac{1}{(m_1 + m_2)} \sum F_x = \frac{|\vec{N}| - |\vec{W}|}{(m_1 + m_2)}$$

$$\Rightarrow |\vec{N}| = |\vec{W}| = (m_1 + m_2)g,$$

where we have noted that $\vec{f}_k$ points to the left (negative component) and that the total mass is $m_1 + m_2$. Our model of friction gives

$$|\vec{f}_k| = \mu_k |\vec{N}| = \mu_k (m_1 + m_2)g$$

$$\Rightarrow a = \frac{-|\vec{f}_k|}{(m_1 + m_2)} = -\mu_k g.$$

With the acceleration determined, we can move on to the kinematics. Constant acceleration relationships are valid in this case. Our information is about velocity and distance, so

$$v_1^2 = 0 = v_0^2 + 2a(x_1 - x_0) = v_0^2 + 2ax_1 = v_0^2 - 2\mu_k g x_1$$

Solving this for v_0 gives:

$$v_0 = \sqrt{2\mu_k g x_1} = \sqrt{2(0.8)(9.8 \text{ m}/\text{s}^2)(3 \text{ m})} = 6.86 \text{ m}/\text{s},$$

where we have used the coefficient of kinetic friction for rubber on concrete from Table 5-1. This is the initial velocity of the skid but, as we have already noted, it is the final velocity of the collision: $v_f = 6.86$ m/s. Inserting this back into the momentum conservation equation, we can calculate that $v_{1i} = 10.6$ m/s. On the basis of your testimony, the Volkswagen driver is *not* charged with speeding!

•

It is perhaps worth noting that both cars were traveling in the *same* direction when they collided. This differs from the picture of Fig. 8-5, but nothing in the derivation of Eq. 8-10 or Eq. 8-11 required the initial directions to be opposite. As long as the only forces are $F_{1 \text{ on } 2}$ and $F_{2 \text{ on } 1}$, which is still true when the initial directions are the same, then Eqs. 8-10 and 8-11 remain valid.

8.4 Conservation of Momentum

The discovery of a conservation law involving momentum seems quite valuable, but we discovered it as part of an analysis of two-particle collisions. Is the conservation law restricted to such situations, or does it have broader applicability? This section will develop a more general law of conservation of momentum.

Consider a quite general system consisting of N particles. The particles might be large entities (cars, baseballs, etc.), or they might be the many microscopic atoms in a container of gas or a solid crystal. We can identify each particle by an identification number $i = 1 - N$. Every particle in the system is subjected to possible external forces $F_{\text{ext on } i}$ that originate *outside* the system. In addition, every particle has the possibility that it *interacts* with every other particle via pairs of interaction forces $F_{i \text{ on } j}$ and $F_{j \text{ on } i}$.

Each particle in the system has a momentum defined by $p_i = mv_i$. Let us define a new concept, the **total momentum,** as:

$$P = total\ momentum = p_1 + p_2 + p_3 + \dots + p_N = \sum_i p_i,$$

where the sum extends over all particles. The total momentum, in other words, is the sum of all the individual momenta. We can find how the total momentum of a system changes with time if we take the time derivative of P:

$$\frac{dP}{dt} = \sum_i \frac{dp_i}{dt} = \sum_i F_i = \sum_i F_{\text{ext on } i} + \sum_i \sum_{j \neq i} F_{j \text{ on } i}. \tag{8-12}$$

To arrive at this result we have used Newton's second law for each particle in momentum form: $F_i = dp_i/dt$. In addition, we have divided the forces acting on particle i into external forces and interaction forces due to all the other particles. Because i does not exert a force on itself, the sum over all the other particles is expressed as $j \neq i$.

The double sum on $F_{j \text{ on } i}$ adds *every* interaction force within the system. If we group the forces in terms of action/reaction pairs and use Newton's third law ($F_{i \text{ on } j} = -F_{j \text{ on } i}$), we find

$$\sum_i \sum_{j \neq i} F_{j \text{ on } i} = \sum_{\text{all pairs } ij} (F_{j \text{ on } i} + F_{i \text{ on } j}) = \sum_{\text{all pairs } ij} (F_{j \text{ on } i} - F_{j \text{ on } i}) = 0. \tag{8-13}$$

In other words, the sum of *all* the internal forces is zero because all the action/reaction pairs cancel. (If you are not sure of Eq. 8-13, you might want to verify it by hand for the case of $N = 4$.) Equation 8-12 then becomes

$$\frac{dP}{dt} = \sum_i F_{\text{ext on } i} = F_{\text{tot ext}}. \tag{8-14}$$

But this is just Newton's second law written for the system as a whole! That is, the rate of change of the total momentum of the whole system is equal to the net force applied to the whole system. This has two very important implications. First, it tells us that we do not need to know about or consider interaction forces between the particles *if* we are only interested in the motion of the entire system. We have, in fact, been using this idea all along as an *assumption* of the particle model. When we have treated cars and rocks and baseballs as particles, we have assumed that the internal forces between the atoms, the forces that hold the object together, do not affect the motion of the object as a whole. Now we have *justified* that assumption and shown that it is valid.

The second implication of Eq. 8-14, and the more important one from the perspective of this chapter, is with regard to what we will call an **isolated system**. An isolated system is defined to be a system for which the *net* external force is zero: $F_{\text{tot ext}} = 0$. For an isolated system, Eq. 8-14 tells us

$$\frac{dP}{dt} = 0 \implies P = \text{constant} \quad \text{(isolated system).} \tag{8-15}$$

The *total* momentum of an isolated system does not change. It remains constant *regardless* of whatever interactions are going on *inside* the system. The importance of this result, which extends even beyond Newtonian mechanics, is such as to elevate it to a *law of nature*, alongside Newton's laws. Equation 8-15 is the full statement of the law of conservation of momentum:

> **Law of Conservation of Momentum**: The total momentum of an isolated system is a constant.

It is worth emphasizing the critical role of Newton's third law in the derivation of Eq. 8-15. The title of this entire part of the text is *Interacting Particles and Conservation Laws*. These two topics initially seemed quite unrelated, but we now see that the law of conservation of momentum is, in fact, a direct consequence of the specific characteristics of the interactions between the particles within an isolated system.

8.5 Recoil, Radioactivity, and Rockets

Our analysis of two-particle collisions in Section 8.3 was simply an application of the law of conservation of momentum—the total momentum after the collision equaled the total momentum before the collision. Another interesting application of momentum conservation is the opposite of collisions—namely *explosions*, or systems where pieces fly apart from each other. The explosive forces, which could be from an expanding spring or from expanding hot gases, are *internal* forces. If the entire system is isolated, its total momentum during the explosion will be conserved.

EXAMPLE 8-4   A 10 g bullet is fired from a 3 kg rifle with a speed of 500 m/s. What is the recoil velocity of the rifle?

SOLUTION In a simple analysis we might say that the rifle exerts a forward force on the bullet and that the bullet, by the third law, thus exerts a backward force on the rifle, causing it to recoil. This is, however, a little *too* simple. The rifle, after all, has no means by which to exert a force on the bullet. Instead, the rifle causes a small mass of gunpowder to explode, releasing energy stored in the gunpowder. The expanding gases exert forces on *both* the bullet and the rifle. If we assume the gun to be fired horizontally, the total momentum along the horizontal axis is

$$P = p_{\text{bullet}} + p_{\text{rifle}} + p_{\text{gases}}.$$

The momentum of the expanding gases is actually the sum of the momenta of all the molecules in the gas. For every molecule moving in the forward direction with speed v and momentum mv there is, on average, another molecule of equal mass moving in the backward direction with speed v and thus momentum $-mv$. When summed over the enormous number of molecules in the gas, we will be left with $p_{\text{gases}} \approx 0$. Thus we are justified in considering the total momentum to be $P = p_{\text{bullet}} + p_{\text{rifle}}$.

Figure 8-8 shows a pictorial model of the before and after situations. Because both the bullet and the rifle are at rest prior to pulling the trigger, we have $P_i = 0$. Is momentum conserved in the firing of a rifle? There are no horizontal forces during the explosion other than the internal forces of the expanding gases. In addition, any friction forces between the bullet and the rifle as the bullet travels down the barrel are also internal forces. So this is, indeed, an isolated system and the law of conservation of momentum will apply. Using momentum conservation gives

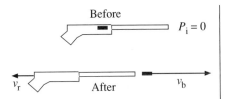

FIGURE 8-8 Before and after a bullet is fired from a rifle.

$$[P_f = m_b v_b + m_r v_r] = [P_i = 0]$$

$$\Rightarrow v_r = -\frac{m_b}{m_r} v_b = -\frac{0.010 \text{ kg}}{3 \text{ kg}}(500 \text{ m/s}) = -1.67 \text{ m/s}.$$

The minus sign is necessary, of course, because the rifle's recoil is to the left.

Example 8-4 is a simple problem when approached from the before/after perspective of a conservation law. We would not even know where to begin to solve a problem such as this using Newton's laws because we do not know the force $F(t)$ exerted on the rifle by the expanding gases. Those details of the interaction are not needed to connect "before" with "after." Note, however, that the selection of bullet+gases+rifle as the system was a critical step. For momentum conservation to be a useful principle, we had to select a system such that the complicated forces of expanding gases and friction were all internal forces. The rifle by itself is *not* an isolated system—the expanding gases and friction forces would be external forces in that case—so its momentum is *not* conserved.

EXAMPLE 8-5 A ^{238}U uranium nucleus is radioactive and spontaneously disintegrates into a small fragment that is ejected with a measured speed of 1.50×10^7 m/s and a *daughter nucleus* that recoils with a measured speed of 2.56×10^5 m/s. What are the atomic masses of the ejected fragment and the daughter nucleus?

SOLUTION The notation ^{238}U indicates the isotope of uranium with an atomic mass of 238, meaning 92 protons (uranium is atomic number 92) and 146 neutrons. The disintegration of a nucleus is, in essence, an explosion that depends only on *internal* forces. Thus the total momentum is conserved. Figure 8-9 shows before and after pictures of the process, calling the mass of the daughter nucleus m_1 and the mass of the ejected fragment m_2. Notice that the initial momentum is $P_i = 0$. For the total momentum to remain zero after the disintegration, the two pieces will have to move away back-to-

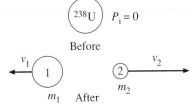

FIGURE 8-9 Before and after the radioactive decay of a ^{238}U nucleus.

back with momenta equal in magnitude but opposite in sign. Because the ejected fragment has a *much* higher speed than the daughter nucleus, it will have to have a much smaller mass in order to conserve momentum. Momentum conservation gives

$$\left[P_f = m_1 v_1 + m_2 v_2\right] = \left[P_i = 0\right].$$

Although we know both velocities, this is not enough information to find two unknown masses. However, we also have another conservation law—conservation of mass. This law implies

$$m_1 + m_2 = 238 \text{ amu}$$

where we have used *atomic mass units*. Combining these two conservation laws gives

$$m_1 v_1 + (238 - m_1) v_2 = 0.$$

Solving this equation for m_1 and then using the result to find m_2 gives:

$$m_1 = \frac{v_1}{v_1 - v_2}(238) = \frac{1.50 \times 10^7 \text{ m/s}}{[1.50 \times 10^7 - (-2.56 \times 10^5)] \text{ m/s}}(238 \text{ amu}) = 234 \text{ amu}$$

$$m_2 = 238 - m_1 = 4 \text{ amu}.$$

Make sure you notice that we converted the speed information to velocity information, with v_1 and v_2 having opposite signs. The masses are all we can learn from a momentum analysis. Further chemical analysis of the daughter nucleus shows that it is the element thorium, with atomic number 90—two protons less than the uranium nucleus had. This implies that the ejected fragment carried away two protons as part of its mass of 4 amu, so it must be a particle consisting of two protons and two neutrons. This is nothing other than the nucleus of a helium atom, ^{4}He, which in nuclear physics is called an *alpha particle* α. Thus the radioactive decay of ^{238}U can be written as ^{238}U $\rightarrow$ ^{234}Th $+ \alpha$.

•

[**Photo suggestion: Rocket being launched.**]

Much the same reasoning allows us to understand how a rocket or jet aircraft accelerates forward. Figure 8-10 shows a rocket with a parcel of fuel on board. The fuel is converted, by burning, to hot gases that are expelled from the rocket motor. The burning and expulsion are all internal forces. The total momentum of the rocket+gases system must therefore be conserved, and the rocket gains forward velocity and momentum as the exhaust gases are shot out the back. The *total* momentum remains zero. Many people find it hard to understand how a rocket can accelerate in the vacuum of space because, after all, there is nothing to "push against." What we see is that the rocket does not push against anything *external*, but only against the gases that it pushes out the back. In return, in accordance with Newton's third law, the gases push forward on the rocket. The details of rocket propulsion are more complex than we want to handle, because the mass of the rocket is changing, but you should at least be able to understand and explain the principle by which propulsion occurs.

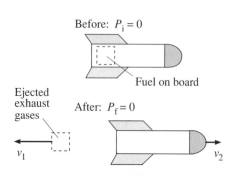

FIGURE 8-10 Rocket propulsion is an example of conservation of momentum.

8.6 Collisions in Two Dimensions

[**Photo suggestion: Time-exposure photo of colliding billard balls.**]

It is worth a brief detour to look at collisions in two dimensions, although we will not do any quantitative analysis or calculations.

Figure 8-11 shows a hard sphere, such as a billiard ball, being shot toward a second sphere at rest. The incoming sphere is going to bounce off at an angle while the stationary sphere is going to recoil at a different angle. Because the motion of m_1 is in the x-direction and m_2 is initially at rest, the initial total momentum *vector* is:

$$\vec{P}_i = m_1 \vec{v}_{1i} + m_2 \vec{v}_{2i} = m_1 v_{1i} \hat{1}.$$

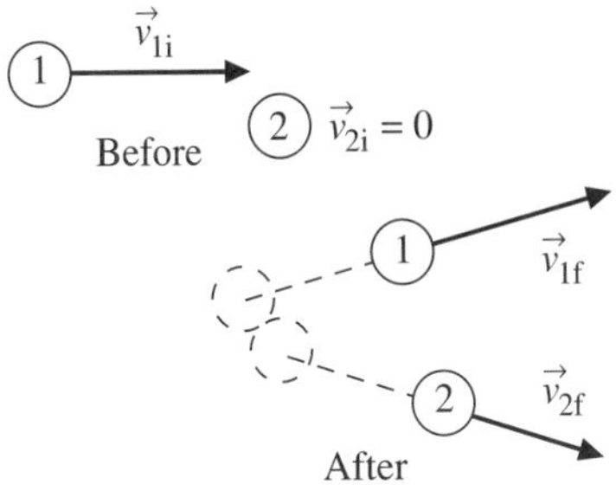

FIGURE 8-11 A collision of two hard spheres in two dimensions.

Our one-dimensional analysis of a two-particle collision can be extended to show that the total momentum *vector* is conserved: $\vec{P}_f = \vec{P}_i$. We can analyze the two-dimensional collision of Fig. 8-11 in terms of its x- and y-components:

$$[(P_f)_x = (p_{1f} + p_{2f})_x] = [(P_i)_x = m_1 v_{1i}]$$
$$[(P_f)_y = (p_{1f} + p_{2f})_y] = [(P_i)_y = 0].$$

Note that the two spheres move in such as way that the y-component of their total momentum remains zero. Questions you might want to ask are: What are the velocities of the two spheres after the collision? What is the angle between the two trajectories after the collision? What influence does the radius of the spheres have?

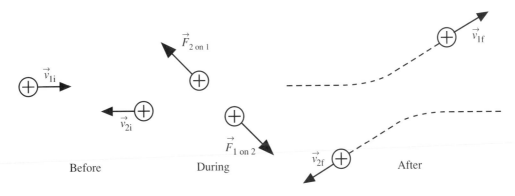

FIGURE 8-12 A "collision" of two positively-charged particles that repel each other via long-range interaction forces.

Another particularly interesting kind of collision is a collision due to long-range forces. All of the collisions we have examined so far have been "hard collisions" due to contact forces, but consider a collision such as that shown in Fig. 8-12. Here two positively charged particles are moving toward each other. As the particles get closer together, they each experience a repulsive force along the line joining them. (Recall that "like" charges repel.) Because the system of two charges is subject to no external forces, this "soft collision" is again a process characterized by the conservation of total momentum. The net result of such a collision, as shown in the after view of Fig. 8-12, is that each particle is deflected along a curved trajectory and then continues, after the collision, along a straight line in a new direction.

We will not proceed any further in looking at two-dimensional collisions—we merely wanted to point out that you *do* know enough physics that you *could* now calculate the details of such a collision.

8.7   Rutherford and the Structure of Atoms

Finally, before leaving this chapter, let's take a look at how the results of atomic collision experiments led to a better understanding of the structure of atoms. The existence of atoms was known from chemistry long before it was discovered that atoms have an internal structure. You may have learned in high school, or even earlier, that an atom consists of a heavy positive nucleus with a swarm of light negative electrons orbiting about it—rather like a miniature solar system. This is the basis for the electron "shells" that you learned about in chemistry. But how do we *know* that this is the structure of atoms? We cannot see atoms, so our knowledge of their structure must be inferred from experiments that are able to probe inside the atom.

An extremely important event in the history of science was the J. J. Thompson's discovery of the electron in 1897. Thompson was investigating what were called "cathode rays." He found that these rays were material particles having both a negative charge and a very small mass. We now call these particles *electrons*, but their earlier name lives on when we refer to the picture tube in a television or computer display as a *cathode-ray tube*. Thompson correctly deduced that these particles were part of the atom, and that they were separated from a positive part that was left behind. Furthermore, the very small mass of the electron implied that nearly all (>99.9%) of an atom's mass resided in the positive part.

With this limited knowledge about the constituents of the atom, Thompson proposed a model of atomic structure. He suggested that atoms consist of small negatively-charged electrons embedded in a heavier positively-charged fluid-like substance, as suggested by Fig. 8-13. This came to be known as the "plum pudding" model of atomic structure, with the electrons seeming like small "plums" stuck within a positive "pudding."

FIGURE 8-13 Thompson's "plum pudding" model of the atom.

The end of the nineteenth century also saw the discovery of *radioactivity*, wherein certain materials, such as radium or uranium, were found to eject charged particles with very high velocities. One particular type of radiation was called *alpha radiation*, which we met in Example 8-5. This type of radiation was found to consist of positively-charged particles that were significantly more massive than an electron. These particles were later found to be the nuclei of helium atoms and are now called *alpha particles*. The English scientist Ernest Rutherford and his students developed a technique for forming a *beam* of alpha particles. As Fig. 8-14 shows, they did this by aligning a source of alpha particles with a series of

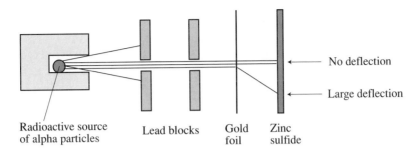

Radioactive source
of alpha particles   Lead blocks   Gold
foil   Zinc
sulfide

No deflection

Large deflection

FIGURE 8-14 Rutherford's experiment to measure the size of atomic nuclei by collisions with a beam of alpha particles.

small holes through lead blocks. Having produced a controlled source of alpha particles, they began directing it toward targets of various kinds to see what would happen.

In 1909, Rutherford's student Hans Geiger (who would go on to later fame as the inventor of the Geiger counter) was shooting alpha particles at a very thin gold foil. Nearly all of the alpha particles passed straight through the foil and impinged on a zinc sulfide screen, where each caused a small flash of light that could be seen and recorded. On occasion, however, Geiger would see a flash of light off to the side that indicated an alpha particle had been deflected through a large angle. Further investigations even found rare "events" in which an alpha particle was deflected almost straight back toward the source!

Rutherford and Geiger realized that the alpha particles were undergoing collisions with atoms in the foil and being deflected, much like the charged particles in Fig. 8-12. The deflection is due entirely to the massive positively-charged part of the atom, because an alpha particle will no more be deflected by a lightweight electron than a bowling ball would be by a pea—it just pushes the electrons aside. However, Rutherford and his students could not reconcile their measurements with Thompson's model of the atom. If the positive charge is evenly distributed throughout the atom, and if the atoms in the foil are spaced close together, then essentially every alpha particle will come close enough to an atom's positive core—even in a very thin foil—to "collide" and be deflected. But that is not what was found. The majority of the alpha particles were not deflected at all, indicating that they "missed" the atoms and had no collision.

Furthermore, a collision with a Thompson-like atom would be very "soft," because the atom's positive charge is all spread out, and the deflection angles of the very fast alpha particles should all be fairly small. Most of Geiger's alpha particles did not seem to have a collision at all, but those that did were deflected through very large angles. It was like throwing a ball through a fence made of small, widely spaced, but very hard bars. Most balls would pass through without deflection, but a few would hit a bar and bounce at large angles—occasionally even straight back! Apparently the positive part of the atom was *much* smaller than the atom itself.

In 1911, after two years of further experiments and analysis, Rutherford proposed a new model of atomic structure that consisted of a very small (but massive), positively-charged *nucleus* orbited by the light, negatively-charged electrons. This solar-system model of the atom (Fig. 8-15) came to be known as the **Rutherford atom**, and it marked a great step forward in our understanding of atomic structure. Further collision experiments showed that the radius of the nucleus is roughly $r_{nuc} \approx 10^{-14}$ m, vastly less than the

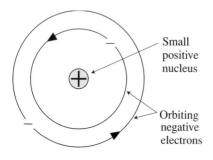

FIGURE 8-15 The Rutherford model of the atom.

radius of the atom itself, which is $r_{atom} \approx 5 \times 10^{-11}$ m. The nucleus takes up only about 1 part in 10^{11} of the volume of an atom: $V_{nuc}/V_{atom} \approx 10^{-11}$. Thus the *overwhelming* majority of an atom is empty space!

To put this in context, consider the period at the end of this sentence. It is about 0.5 mm in diameter. If this were an atomic nucleus, the diameter of the atom would be about 2.5 m $\approx$ 8 feet. This is the ceiling height in a typical house. So to make a model atom, suspend the period in the center of a room and imagine little dust specks of electrons (remember, they are much lighter than the nucleus) swirling around the walls, floor, and ceiling. There you have it; an atom is mostly nothingness!

Summary

Important Concepts and Terms

conserved	impulse approximation
law of conservation of mass	law of conservation of momentum
impulsive force	perfectly inelastic collision
momentum	total momentum
impulse	isolated system
impulse-momentum theorem	Rutherford atom

This chapter has introduced a new perspective on motion. The first seven chapters were devoted to developing a view of dynamics based on the conceptual tools of *force* and *acceleration*. Now we have introduced the two new concepts of *impulse* and *momentum*, and we have found that they give us a very different, but equally valid, understanding. The momentum of a particle is $p = mv$, while the impulse exerted on the particle by a time-varying force is $J = \int F(t)dt$. The impulse in interpreted geometrically as the area under the $F(t)$ curve.

The motion of a single particle, even if that particle has internal structure, was found to obey both Newton's second law, written in momentum form

$$F_{ext} = \frac{dp}{dt},$$

and the impulse-momentum theorem

$$J = \Delta p.$$

The significance of the subscript "ext" on F_{ext} is that the motion of a complex object as a whole is determined only by any *external* forces acting on the object. Internal forces do not affect the object's motion. This finding helps to validate the particle model.

We earlier introduced the idea that force is the physical quantity responsible for a *change* of velocity. Now, in analogous fashion, we see that impulse is the physical quantity responsible for a *change* of momentum. In the absence of an impulse, the momentum of an object is unchanged, or *conserved*.

The primary importance of this new perspective, however, is not so much for single particles as it is for *systems* of particles. We found that the *total momentum* of an *isolated system* is a *conserved* quantity:

$$P = \sum_i p_i = \text{constant for an isolated } (F_{ext} = 0) \text{ system.}$$

It is this mathematical statement that connects "before" with "after," because $P = $ constant implies that $P_f = P_i$. The critical role of Newton's third law in momentum conservation cannot be overemphasized. While the momentum of a single isolated *particle* is indeed conserved, the true significance of momentum conservation is to *systems* of *interacting* particles.

Exercises and Problems

The ✍ icon indicates that these problems can be done on a Conservation Worksheet.

◖ Exercises

✍ 1. A 50 kg archer, standing on frictionless ice, shoots a 100 g arrow at a speed of 100 m/s.
   a. What is the "recoil velocity" of the archer?
   b. The arrow impales a 10 kg target resting on the ice. What happens to the target?

✍ 2. The parking brake on a 5000 pound Cadillac has failed, and it is rolling slowly, at 1 mph, toward a group of small, innocent children. As you see the situation, you realize there is just time for you to drive your 2000 pound Volkswagen head-on into the Cadillac and thus to save the children. With what speed should you impact the Cadillac to bring it to a halt?

✍ 3. A 300 g bird flying along at 6 m/s sees a 10 g insect heading straight toward it with a speed of 30 m/s. The bird opens his mouth wide and enjoys a nice lunch. What is the bird's speed immediately after swallowing?

◖ Problems

✍ 4. A 200 g ball is dropped from a height of 2 m and bounces on a hard floor. The impulse received from the floor is shown in Fig. 8-16. How high does the ball rebound?

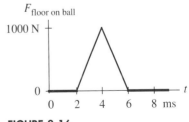

FIGURE 8-16

✍ 5. It is thought by many geologists that the dinosaurs became extinct 65 million years ago when a large comet or asteroid struck the earth and threw up so much dust that the sun was blocked out for a period of many months. The asteroid would have had a diameter of about two miles and a mass of $\approx 4\times 10^{10}$ kg, and it would have hit the earth with an impact speed of $\approx 4\times 10^{4}$ m/s.

    a. What is the earth's recoil velocity from such a collision?

    b. What percentage is this of the earth's speed around the sun? (Use the data inside the front cover.)

6. In Problem 12 of Chapter 7 you found the recoil speed of Bob as he throws a rock while standing on frictionless ice. Bob has a mass of 75 kg and can throw a 500 g rock with a speed of 30 m/s. Re-analyze the problem using momentum, and again find Bob's recoil speed.

7. A spaceship of mass 2×10^{6} kg is cruising along at a speed of 5×10^{6} m/s when the antimatter reactor fails, blowing the ship into three pieces. One section, having a mass of 5×10^{5} kg, is blown straight backwards with a speed of 2×10^{6} m/s. A second piece, of mass 8×10^{5} kg, continues forward at 1×10^{6} m/s.

    a. Determine the direction of the third piece. Explain your reasoning.

    b. What is the velocity of the third piece?

8. An alpha particle (atomic mass of 4) is shot toward a gold target (atomic mass of 197) with a speed of 1.5×10^{7} m/s. It is deflected by the target straight back toward the source with 80% of its initial speed. What is the recoil velocity of the gold nucleus with which it collided?

9. Fred Fingers, 60 kg, is running upfield with the football at a speed of 6 m/s when he is met head-on by Brutus, 120 kg, who is moving a speed of 4 m/s. Brutus grabs Fred in a tight grip, and they fall to the ground. Which way do they slide, and how far? The coefficient of kinetic friction between football uniforms and Astroturf is 0.3.

10. a. A neutron is an electrically neutral particle having a mass just slightly greater than that of a proton. In the nuclei of atoms, where neutrons are usually found, the neutron is stable. A free neutron, however, is radioactive and decays after a few minutes into other subatomic particles. Consider a neutron (1.68×10^{-27} kg) that decays while at rest. It is experimentally observed that an electron (9.11×10^{-31} kg) and a proton (1.67×10^{-27} kg) are shot out back-to-back along the same line. The proton speed is measured to be 1.0×10^{5} m/s, while that of the electron is 3.0×10^{7} m/s. No other decay products are detected. Is momentum conserved in the decay of this neutron?

    b. Experiments such as the one described in part a were first performed in the 1930s and seemed to indicate a failure of the law of conservation of momentum. Physicists, however, were reluctant to give up a conservation law that had worked so well in every other circumstance. In 1933, Wolfgang Pauli postulated that there might be a *third* decay product of the neutron, in addition to the proton and electron, that is so light and interacts so weakly with other matter that it is virtually impossible to detect. Even so, this new particle can carry away just enough momentum to keep the total momentum conserved. This proposed particle was named the *neutrino* by Enrico Fermi, meaning "little neutral one." Neutrinos were, indeed, discovered nearly 20 years later. They are now well understood as an important aspect of the weak force. If a neutrino were emitted in the neutron decay described in part a, in which direction would it have to travel to conserve momentum? Explain your reasoning. How much momentum would this neutrino "carry away" with it?

11. Ann (50 kg) is standing at one end of a 500 kg cart that is 15 m long. The cart has frictionless wheels and rolls on a frictionless track. Initially both Ann and the cart are at rest. Suddenly, Ann starts running along the cart at a speed of 5 m/s relative to the cart.

    a. How will the cart respond? Give a qualitative response.

    b. How far will Ann have run *relative to the ground* when she reaches the far end of the cart?

12. You are a world-famous physicist-lawyer defending a client who has been charged with murder. It is alleged that your client, Mr. Smith, shot the victim, Mr. Wesson. The detective who investigated the scene of the crime found a second bullet, from a shot that missed Mr. Wesson, embedded in a chair. You arise to cross examine the detective:

You: In what type of chair did you find the bullet?

Det: A wooden chair.

You: How massive was this chair?

Det: It had a mass of 20 kg.

You: How did the chair respond to being struck with a bullet?

Det: It slid across the floor.

You: How far?

Det: Three centimeters. The slide marks on the dusty floor are quite distinct.

You: What kind of floor was it?

Det: A wood floor, very nice oak planks.

You: What was the mass of the bullet you retrieved from the chair?

Det: Its mass was 10 g.

You: And how far had it penetrated into the chair?

Det: A distance of 4 cm.

You: Have you tested the gun you found in Mr. Smith's possession?

Det: I have.

You: What is the muzzle velocity of bullets fired from that gun?

Det: The muzzle velocity is 450 m/s.

You: And the barrel length?

Det: The gun has a barrel length of 62 cm.

With that, you turn confidently to the jury and proclaim, "My client's gun did not fire these shots!" How are you going to convince the jury and the judge?

[**Estimated 10 additional problems for the final edition.**]

Chapter 9

Concepts of Energy I:
Work and Energy

L O O K I N G
B A C K

Sections 2.3; 5.2–5.4; 5.6; 8.2

9.1 Introduction

[**Photo suggestion: A photo collage showing the sun, a windmill, a power plant, and an electric transmission line.**]

The concept of *energy* is one of the most important concepts in all of science. Automobiles, rockets, computers, biological organisms, and ecosystems all use and transform energy in a myriad of ways. Energy is also vital to our daily lives. We use chemical energy to heat our homes and bodies, electrical energy to power our lights and computers, and solar energy to grow our crops and forests. Energy—its characteristics, transformations, and conservation—will be a major new theme in this text. Energy will give us a new perspective on motion. In addition, energy will give us a new tool for problem solving

It is difficult, however, to define in a general way just what energy is. The concept of energy has grown and changed with time, and you will come to understand energy better through seeing many examples of how energy is used than by formal definitions. We will start by discussing what seems to be a completely unrelated topic—money. As you will discover, monetary systems have much in common with energy.

The Parable of the Lost Penny

Jose was a hard worker. His only source of income was the paycheck he received each month. Even though most of each paycheck had to be spent on basic necessities, Jose managed to keep a respectable balance in his checking account. He even saved enough to occasionally buy a few stocks and bonds. Jose never cared much for pennies. To him they seemed more trouble than they were worth. So Jose kept a jar by the door and dropped all of his pennies into it at the end of each day. Eventually, he reasoned, his saved pennies would be worth taking to the bank and converting into crisp new dollar bills.

Jose found it fascinating to keep track of these various sources of money. He noticed, somewhat to his dismay, that the amount of money in his checking account did not spontaneously increase overnight. Neither did the number of pennies in his jar. Furthermore, there seemed to be a definite correlation between the size of his paycheck and the amount of money he had in the bank. So Jose decided to embark on a systematic study of money.

He began, as would any good scientist, by using his initial observations to formulate a hypothesis. Because this was a fairly involved hypothesis involving the relationships between various sorts of money, Jose called it a *model* of his monetary system. He found that he could represent the monetary model as a flow chart, as shown in Fig. 9-1.

As the chart shows, Jose found that it was convenient to divide his money into two basic types—liquid assets and saved assets. The *liquid assets L*, which included his checking account and the cash in his pockets, were moneys available for immediate use. His *saved assets S*, which included his stocks and bonds as well as his jar of pennies, had the *potential* to be converted into liquid assets, but they were not available for immediate use. Jose decided to call the sum total of assets his *wealth*: $W = L + S$.

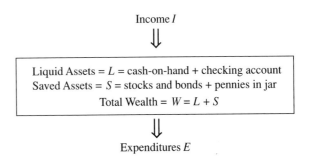

FIGURE 9-1 Jose's model of the monetary system.

Jose's assets were, more or less, simply definitions. The more interesting question, he thought, was how his wealth depended on the *monetary transfers* of *income I* and *expenditures E*. These transfers represented both money transferred *to* him by his employer and money transferred *by* him to stores and bill collectors. After painstaking collection and analysis of data, Jose finally deduced the *quantitative* relationship between the monetary transfers and his assets to be

$$I - E = \Delta L + \Delta S = \Delta W.$$

Jose interpreted this equation to mean that the *net* monetary transfer to him, given by $I - E$, was numerically equal to the *change* in his wealth, ΔW. (His data clearly refuted the competing hypothesis of his next door neighbor Bubba, who had asserted that $I - E = W$. That is, that the net monetary transfer *equals* net wealth. After all, Jose noted, his wealth did not drop to zero on days when he had neither income nor expenses.)

During a one-week period when Jose stayed home sick, *isolated* from the rest of the world, he had neither income nor expenses: $I - E = 0$. Amazingly, but in grand confirmation of his hypothesis, he found that his wealth W_f at the end of the week was identical to his wealth W_i at the week's beginning: $W_f = W_i$, or $\Delta W = 0$. This occurred despite the fact that he had moved pennies from his pocket to the jar and also, by telephone, had sold some stocks and transferred the money to his checking account. In other words, Jose found that he could make all of the *internal* conversions of assets from one form to another that he wanted, but his total wealth remained constant (W = constant) as long as he was isolated from the world. This seemed such a remarkable rule that Jose named it the law of conservation of wealth.

One day, however, Jose added up both his income and expenditures for the day and the changes in his various assets—and he was 1¢ off! $I - E = \Delta L + \Delta S - 1$ ¢. Jose quickly verified that it wasn't just a math error, but that some money really, and inexplicably, seemed to have vanished. Jose was devastated. All those years of careful research, and now it seemed that his monetary hypothesis was not true, that under some circumstances—yet to be determined—$I - E \neq \Delta W$. Off by a measly penny. A wasted scientific life....

But wait! Jose realized, in a flash of inspiration, that perhaps there were other types of assets that he had not discovered and that, if *all* assets were included, his monetary hypothesis would still be valid. Weeks went by as Jose, in frantic activity, searched fruitlessly for previously *hidden* assets. While his life disintegrated, his girlfriend gave him an ultimatum: "Knock off this manic behavior and clean house, or I'm moving out." As Jose, his spirits low, lifted the cushion off the sofa to vacuum out the potato chip crumbs—lo and behold!—there it was! The missing penny!

Jose raced to complete his theory, now including the sofa (as well as the washing machine) as previously unknown forms of saved assets that needed to be included in S. Other researchers soon discovered other types of assets—particularly the remarkable find of the "cash in the mattress"—that were included in Jose's hypothesis. To this day, when all known assets are included, monetary scientists have never found a violation of Jose's simple hypothesis that $I - E = \Delta W$. Jose was last seen sailing for Stockholm to collect the Nobel Prize for his theory of wealth.

9.2 Energy

Jose, despite his diligent efforts, did not discover a "law of nature." The monetary system is a human construction that, by design, obeys Jose's "laws." Monetary system laws, such as those that say you cannot print money in your basement, are enforced by society, not by nature. Suppose, however, that physical objects possessed a "natural money" that was governed by a theory, or model, similar to Jose's. An object might have several different forms of this "natural money" that could be converted back-and-forth, but the total amount of an object's "natural money" would only *change* as a result of this natural money being *transferred* in or out of the object. Two of the key words here, as in Jose's model, are *transfer* and *change*.

One of the greatest and most significant discoveries of science is that there is such a "natural money," which we call **energy**. A difficulty for many students, however, is that the concept of energy is rather abstract. Force is very tangible, something you can feel, and acceleration is a concept you can visualize. So thinking of motion in terms of forces and acceleration seems, at least after some practice, fairly straightforward. Energy is a more abstract and more subtle concept than is force.

But, if you think about it, money is also a very abstract idea. Money is *not* the pieces of paper or objects of metal you carry in your pocket or purse. The real and tangible "value" of that piece of paper with $100 written on it is essentially zero—the same as any other piece of paper the same size. However, the fact that someone would give you a pair of shoes or large stack of CDs in exchange for that piece of paper implies that the paper has some *intangible* value. This *monetary* value of the paper is an idea that is useful only because we, as a society, have agreed to certain "laws" about how money behaves and about how one form of monetary value (e.g., a piece of paper with numbers on it) is exchanged for another (e.g., real goods). So despite its abstract nature, money is quite "real." Energy is much the same—abstract, intangible, but nonetheless "real." The distinction is that energy obeys natural laws, which we discover, rather than human laws, which we invent.

There is not a single definition of energy. Energy is a concept that has developed over a long span of time, and there are many forms, or types, of energy. You have heard of some of these, such as solar energy or nuclear energy, but others may be new to you. These forms of energy can differ as much as a checking account differs from loose change in the sofa. Much of our study is going to be focused on the *transformation* of energy. A large fraction of modern technology is concerned with transforming energy from one form (e.g., the chemical energy of oil molecules) to another (e.g., electrical energy or the kinetic energy of your car). We will continue to expand the concepts of energy and its transformation properties for the rest of this text.

As we use energy concepts we will be "accounting" for energy that is transferred in or out of a system, or that is converted from one form to another without loss. It is this characteristic of energy that makes the analogy with money so useful. The fact that nature "balances the books" for energy is one of the most profound discoveries of science. The possible behaviors of physical systems are sharply constrained by having to maintain this balance. Behaviors that might otherwise seem plausible are simply found not to occur in nature if the energy cannot be accounted for. The implications of this true "law of nature" extend well beyond physics. Chemistry, biology, engineering, and ecology are all significantly influenced by the laws of energy use. In the long run, the laws of energy are far more wide-ranging and important in other disciplines than are Newton's laws of mechanics.

Even though certain forms of energy were recognized quite early, the *law of conservation of energy* was not recognized until the mid-nineteenth century, long after Newton. The reason, similar to the situation with Jose's "lost penny," was that it took scientists a long time to realize how many types of energy there are and the various ways that energy can be converted from one form to another. The ideas involved go well beyond Newtonian mechanics to include new concepts about heat, about chemical energy, and about the energy of the individual atoms and molecules that comprise a system. The complete statement about energy and its transformation properties is known as the *first law of thermodynamics*. It is a new statement about nature, having more content and meaning than can be deduced from Newton's laws alone. We will defer the full first law until our study of thermodynamics, but these next few chapters will begin the important task of introducing the concepts of energy.

It is worth emphasizing that the law of conservation of energy is a scientific *hypothesis* about nature. Simply writing down a quantitative relationship between various concepts does not make the relationship true. Like Jose, we must first postulate a relationship and *then* seek evidence for its validity. As of today, with 150 years of experimental evidence, we know of no violations of the law of conservation of energy. It has become one of the firmest principles of science. Many scientific discoveries, in fact, have been made as a consequence of experiments where there seemed to be some "missing energy." Rather than believe they had discovered a violation of energy conservation (although that could happen!), the scientists believed so firmly in energy conservation that they searched until they discovered the source of the missing energy.

9.3 The Basic Energy Model

Figure 9-2 provides a basic model of energy that is analogous to Jose's model of money. As a *basic* model it is certainly not complete, and we will add significant new features to our model as we need them. Nonetheless, it is a good starting point.

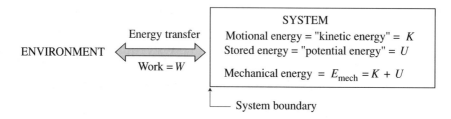

FIGURE 9-2 The basic energy model of a system interacting with its environment.

Once again we are distinguishing between the *system* that we wish to study and its surrounding *environment*. The system can be characterized by two quantities that we call the **kinetic energy** and the **potential energy**. All we need to know about these for now is that kinetic energy (symbol K) is an "energy of motion," while potential energy (symbol U) is a "stored" energy that has the "potential" to be converted to kinetic energy. (Note that such a conversion would be an **energy transformation**.) Kinetic energy is analogous to Jose's liquid assets, while potential energy is analogous to his saved assets. The sum of kinetic and potential energy (analogous to wealth) is called the **mechanical energy** $E_{mech} = K + U$. The term *mechanical energy* designates this form of energy as being due to motion and mechanical effects (like stretching springs) rather than chemical effects or heat effects, which are other forms of energy that we will introduce later. For now, we will omit the subscript and use the symbol E to mean mechanical energy. Later, when it becomes important, we'll use E_{mech} to distinguish mechanical energy from other forms of energy.

Kinetic energy, as you will learn in the next section, depends upon the *speed* $|\vec{v}|$ of the system. Because the speed can be zero but never negative, we require $K \geq 0$. The fact that kinetic energy can never be negative is one of its important characteristics. Potential energy is a bit harder to understand, but a good prototype of potential energy is a stretched rubber band. A stretched rubber band, as you know, is just waiting to be released in order to launch a paper wad. In other words, it has the "potential" for producing kinetic energy. The rubber band's stretch represents "saved" or stored energy, and that is what potential energy is all about. The energy stored in a rubber band depends on how far it is stretched—that is, on the *position* of the ends of the rubber band. So while kinetic energy depends on *speed*, potential energy depends on *position*. As you will see, we can use energy ideas to relate an object's speed to its position.

A system, unless it is completely isolated, has the possibility of exchanging energy with its surrounding environment. There are two primary processes by which this can occur. The first, which is the only one we are going to be concerned with for now, is as a result of forces—pushes and pulls—exerted on the system by the environment. This *mechanical* energy transfer goes by the name **work**. It is also possible for energy to be transferred between the system and its environment, if they are at different temperatures, by a *nonmechanical* energy transfer process called *heat*. Heat is a significant aspect of the energy model that we will add when we get to the study of thermodynamics, but for the time being we will concentrate on the mechanical transfer of energy via work.

The symbol for work is W. (The possibility for confusion once again rears its ugly head, because now you have to make sure that you do not confuse work W with weight $\vec{W}$ or its magnitude $|\vec{W}|$). Notice that the arrow labeled *work* in Fig. 9-2 is bi-directional, rather different than the single-direction arrows of "income" and "expenditures" in Fig. 9-1. This is because work is a quantity that can be either positive or negative, with the interpretation that:

$W > 0 \Rightarrow$ the environment does work on the system and the system's energy increases,

$W < 0 \Rightarrow$ the system does work on the environment and the system's energy decreases.

This is equivalent to considering expenditures—money out—to be a negative income. In fact, that is just how accountants really do handle incomes and expenditures.

Having established our basic quantities, what is the relationship between them? Our hypothesis, which is confirmed by experiment, is:

$$W = \Delta E = \Delta K + \Delta U. \tag{9-1}$$

In words, Eq. 9-1 says that the energy transferred *to* a system via work changes the total mechanical energy *of* the system. Further, the system's change of energy might be a change of kinetic energy, a change of potential energy, or both. Equation 9-1 gives no information about how the total energy change is divided up between kinetic energy and potential energy.

Now consider what happens if you push on an object. As you push, you exert a force on the object and do work on it. That is, you mechanically transfer energy to the object. What happens to the object as a result of this push? One possibility is that the object will accelerate and have a higher speed at the end of the push than it had at the beginning—an increase of kinetic energy. The energy you transferred *to* the system via work ends up, in this case, as an increase *of* the system's kinetic energy. It is also possible that the push causes a rubber band inside the system to be stretched—an increase of potential energy. Here the energy transferred *to* the system via work ends up as an increase *of* the system's potential energy.

The situation could go the other way as well. Suppose a rubber band inside the system is initially stretched and that it is used to pull an object in the environment closer. Now the system is doing work on the environment, by pulling on it. In this case the work is a negative quantity. But the system is also *losing* potential energy as the rubber band retracts, so ΔU is also negative. So in this example the energy *of* the system decreases and that amount of energy is transferred *to* the environment by doing work on it.

It is not just poor typing that has led us to emphasize the terms *of* and *to* in the last three paragraphs. We are making a very significant point about work and energy. Kinetic energy and potential energy are properties that *characterize* the system. Like mass or charge, we could say that they are properties *of* the system. We will often talk about the **state** of the system, by which we mean the specific characteristics of the system at a particular time. In addition to mass, charge, kinetic energy, and potential energy, quantities such as pressure and temperature also characterize the state of the system. They are *of* the system.

Work, on the other hand, is *not* a property of the system. It is a *process*, or an interaction, between the system and the environment. It is something done *to* the system in order to *change the state* of the system. That is why Eq. 9-1 has Δ's on the right side but *not* on the left. The system had a certain amount of kinetic energy K_i and potential energy U_i before work was done on it, and it ends up with some other amount K_f and U_f after the work was done. We can measure the change of state of the system in terms of its change of kinetic energy $\Delta K = K_f - K_i$ and its change of potential energy $\Delta U = U_f - U_i$. But work does not have a before and after—it is simply a measurement of something that was done *to* the system.

So as we interpret Eq. 9-1, we want to say that a *process*, namely doing work on a system, causes a *change* in the state of the system. The work done does not tell us anything about how much total energy E the system has (recall Bubba's mistaken conjecture in the parable), but only by how much the total energy *changes*.

9.4 Kinetic Energy

In the previous section we began to develop the idea that the state of a system can be changed by an external influence or process acting on the system. This is rather vague, so let us see if we can make the idea more precise by associating "state of the system" and "external influence" with specific quantities that we can measure or calculate. We will concentrate, for now, on the motion of a single particle—the simplest possible system. Chapters 10 and 11 will expand these ideas to more complex systems of multiple particles.

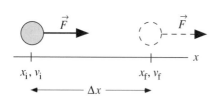

FIGURE 9-3 A particle moving from x_i to x_f under the influence of a *constant* force $\vec{F}$.

Figure 9-3 shows a particle of mass m that moves along the x-axis from an initial position x_i to a final position x_f under the influence of a *constant* force F. The force acts steadily on the particle as it moves— that is, the force is not an impulse force that acts briefly on the particle at x_i, but is a force that is applied throughout the particle's motion. Such a force will cause a constant acceleration $a = F/m$. Recall, from one-dimensional kinematics, that a particle accelerating from initial velocity v_i to final velocity v_f with constant acceleration a obeys

$$v_f{}^2 = v_i{}^2 + 2a(x_f - x_i)$$
$$\Rightarrow v_f{}^2 - v_i{}^2 = 2a\Delta x. \tag{9-2}$$

Substituting $a = F/m$ and doing a bit of rearranging gives

$$v_f{}^2 - v_i{}^2 = \frac{2F\Delta x}{m}$$
$$\Rightarrow \tfrac{1}{2}mv_f{}^2 - \tfrac{1}{2}mv_i{}^2 = F\Delta x. \tag{9-3}$$

We can rewrite Eq. 9-3 in the form

$$\Delta(\tfrac{1}{2}mv^2) = F\Delta x. \tag{9-4}$$

What does Eq. 9-4 tell us? The basic quantity on the left, $\tfrac{1}{2}mv^2$, is a characteristic *of* the particle. It depends, at any particular point of the motion, only on the particle's mass and velocity. Thus the quantity $\frac{1}{2}mv^2$ measures the *state* of the system. The right-hand side, however, measures something being done *to* the system. A force, from an agent somewhere in the environment, reaches in and pushes the particle through a displacement Δx. The quantity $F\Delta x$ thus represents a *process* that happens. With these ideas, we can interpret Eq. 9-4 as saying that the *process* of a force F pushing the particle through a displacement Δx causes a *change* in the *state* of the system, as measured by the quantity $\frac{1}{2}mv^2$. This is exactly the idea behind the basic energy model.

It is worth noting that the quantity $F\Delta x$ tells us nothing at all about the *value* of $\frac{1}{2}mv^2$, but only about how $\frac{1}{2}mv^2$ *changes*. That change could be either positive—if the particle speeds up—or negative—if it slows down. How could it be negative? The quantities F and Δx, as in our earlier analyses of one-dimensional motion, are vector components and, accordingly, have signs. They are the "force component along the axis of motion" and the "displacement," both of which are signed quantities. They are *not* the magnitude of the force $|\vec{F}|$ or the distance $|\Delta x|$, which are always positive. If, for example, the force in Fig. 9-3 points to the *left* as the particle moves toward the right, it would act as a braking force that slows the particle. The component F would be negative, because the vector $\vec{F}$ points in the $-x$-direction, while Δx would continue to be positive. This would make the product $F\Delta x$ negative, exactly what is needed to match the negative sign of $\Delta(\frac{1}{2}mv^2)$ for a particle slowing down. (The *change* in a quantity, recall, is *always* the final value minus the initial value.)

The quantity $\frac{1}{2}mv^2$ is the kinetic energy K of the particle. It is defined as

$$K = \tfrac{1}{2}m|\vec{v}|^2 , \tag{9-5}$$

where m is the mass of the particle. In terms of vector components,

$$K = \begin{cases} \frac{1}{2}m({v_x}^2 + {v_y}^2 + {v_z}^2) & \text{general formula} \\ \frac{1}{2}mv^2 & \text{one-dimensional motion.} \end{cases} \tag{9-6}$$

By its definition, kinetic energy can never be a negative number: $K \geq 0$. The *change* in kinetic energy can, of course, be negative if K decreases in value. If you find, in the course of solving a problem, that K is negative—stop! You have made an error somewhere. Don't just "lose" the minus sign and hope that everything turns out OK.

One of the most important characteristics of kinetic energy is that it is a scalar, rather than a vector. It depends on the speed $|\vec{v}|$, but not on the velocity's direction. The kinetic energy of a particle will be the same regardless of whether it is moving up or down, or left or right. Consequently, the mathematics of an energy solution to a problem is often much easier than the vector mathematics required by a force and acceleration solution.

The unit of energy is that of mass times velocity squared. In the SI system of units, this is $\text{kg}\,\text{m}^2/\text{s}^2$. The unit of energy is so important that is has been given its own name: the **joule**. We define:

$$1 \text{ joule} = 1 \text{ J} \equiv 1 \text{ kg}\,\text{m}^2/\text{s}^2.$$

To give you an idea about the size of a Joule, consider a 0.5 kg mass (weight on earth of ≈ 1 pound) moving with a speed of 2 m/s (≈ 4 mph). The mass's kinetic energy is

$$K = \tfrac{1}{2}mv^2 = \tfrac{1}{2}(0.5 \text{ kg})(2 \text{ m}/\text{s})^2 = 1 \text{ J}.$$

This suggests that ordinary-sized objects moving at ordinary speeds will have kinetic energies of a fraction of a Joule up to, perhaps, a few thousand joules (running person $K \approx 1000$ J = 1 kJ). A high-speed truck might get up to around 10^6 joules. For ordinary-sized objects, problem answers that are much smaller or much larger than these values are indicative of an error in your work. Keep in mind that you must have masses in kg and velocities in m/s before doing calculations.

9.5 Work

Having defined the kinetic energy as a means of measuring the state of a system, we next want to look at what causes the state of the system to change. That is, what process results in a *change* of kinetic energy ΔK? Equation 9-4 gives us a big hint—the process must have something to do with a force acting on the particle as it moves through a displacement. We will start with the simplest case and then gradually extend this idea to more complex and general cases.

Constant Force in the Direction of Motion

Consider a particle that moves along a linear axis, which we can call the x-axis, under the influence of a constant force $\vec{F}$ which points either toward $+x$ or toward $-x$. This is exactly the situation we considered in Fig. 9-3. The force has only an x-component F, which can be either positive or negative. We have analyzed examples of such situations from the force-acceleration perspective: pushing a object, pulling (or lifting) an object with a rope under constant tension, an object sliding to a halt under the influence of kinetic friction, and so on.

As the particle moves through a displacement Δx, its kinetic energy changes by

$$\Delta K = F \Delta x, \tag{9-7}$$

where we have used Eq. 9-4, the result of our earlier analysis, along with Eq. 9-6 for the definition of kinetic energy for one-dimensional motion.

The product of force and displacement is called the work W done *by* the force *on* the object. It is defined as:

$$work = W \equiv F \Delta x \quad \text{(First definition of work)}. \tag{9-8}$$

We call this the *first definition* because it *only* applies to the particular case of a constant force in one dimension. Make a note of this! With this definition, Eq. 9-7 becomes simply

$$\Delta K = W, \tag{9-9}$$

which states that the work done on a particle changes its kinetic energy.

Work is a common word in the English language, with many meanings. In one dictionary work is defined as:

1. Physical or mental effort; labor.

2. The activity by which one makes a living.

3. A task or duty.

4. Something produced as a result of effort, such as a *work of art*.

5. Plural *works*: A factory or plant where industry is carried on, such as *steel works*.

6. Plural *works*: The essential or operating parts of a mechanism.

7. The transfer of energy to a body by application of a force.

When you first think of work, you probably think of the first definition, namely "physical effort." After all, we talk about "working out," or we say, "Boy, that was a lot of work" after completing a physically demanding task. But that is *not* the type of work we are referring to with our definition of Eq. 9-8. If, for example, you were to hold a 200-pound weight over your head, you might break out in a sweat, your arms would tire, and by the time you dropped the weight you would "feel" that you had done a lot of work. However, you would have done *zero* work in the physics sense because the weight was not displaced while you were holding it ($\Delta x = 0 \Rightarrow W = 0$) and you transferred no energy to it. So set aside your prior ideas about what *work* is and concentrate on learning a new definition. We are making a very specific, technical definition: *work* is "the transfer of energy to a body by application of a force."

Work, as a *process*, is something done *to* a particle to change its state. Positive work is a *transfer of energy to* the particle, increasing its kinetic energy. Negative work is a transfer of energy *from* the particle, decreasing its kinetic energy. We will continue to emphasize the idea that work is a *transfer* of energy to or from the particle, causing the particle's kinetic energy to either increase or decrease.

The units of work, by its definition, are N m. Recalling the definition of the newton, this is equivalent to $(kg\,m/s^2)\,m = kg\,m^2/s^2$. But these are just the units of energy—which they of course have to be if work is a transfer of energy. So rather than use N m, we will measure work, just like we do kinetic energy, in joules.

EXAMPLE 9-1 A 500 g ice-hockey puck is sliding across the ice at a speed of 2 m/s. A compressed air gun exerts a forward force on the puck of 1 N. This force is continuously applied as the puck moves 50 cm. What is the puck's final speed?

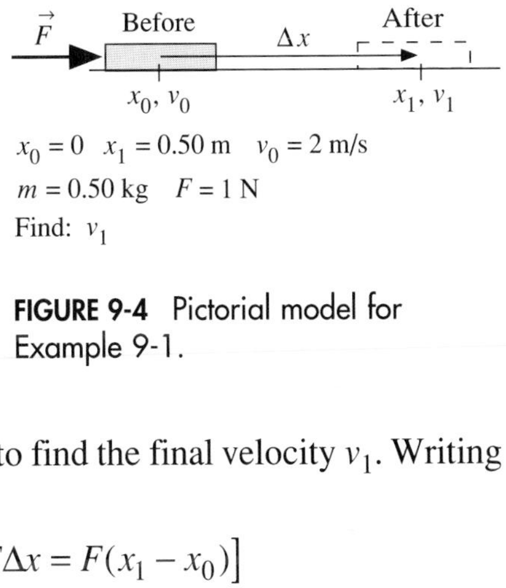

SOLUTION Figure 9-4 shows a simple pictorial model in which we have shown the before and after state of the puck. This is much like we did for momentum problems. We can use the relationship between work and kinetic energy, as given by Eq. 9-9, to find the final velocity v_1. Writing out the definitions of ΔK and W, we get

$x_0 = 0$ $x_1 = 0.50\,m$ $v_0 = 2\,m/s$
$m = 0.50\,kg$ $F = 1\,N$
Find: v_1

FIGURE 9-4 Pictorial model for Example 9-1.

$$\left[\Delta K = \tfrac{1}{2}mv_1^2 - \tfrac{1}{2}mv_0^2\right] = \left[W = F\Delta x = F(x_1 - x_0)\right]$$

$$\Rightarrow v_1 = \sqrt{v_0^2 + \frac{2}{m}F(x_1 - x_0)} = 2.45\,m/s.$$

While we could have used Newton's laws and kinematics to solve Example 9-1, notice how straightforward the work-energy method is.

EXAMPLE 9-2   The puck of Example 9-1 is again moving at 2 m/s. This time, however, the air gun exerts a *backward* force of 1 N while the puck travels 50 cm. What is v_1?

SOLUTION The before/after picture for this situation is the same as that shown in Fig. 9-4. The only difference is that we must use $F = -1$ N. This is because F is a vector component, and the force *vector* now points to the left (toward $-x$) instead of to the right. With this change, we can calculate v_1 in exactly the same way as in Example 9-1. This gives the result

$$v_1 = \sqrt{v_0^2 + \frac{2}{m}F(x_1 - x_0)} = 1.41 \text{ m/s}.$$

Force Perpendicular to the Direction of Motion

Consider, as shown in Fig. 9-5a, an ice-hockey puck sliding across frictionless ice with no horizontal forces applied to it. Because $F_x = 0$, the puck slides with an unchanging velocity. Thus, $\Delta K = 0$. There *are* forces acting on the puck, namely the normal force $\vec{N}$ and the weight force $\vec{W}$, but these forces are perpendicular to the direction of motion.

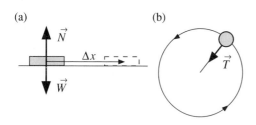

(a)   (b)

FIGURE 9-5 a) The normal force and the weight force are perpendicular to the direction of motion of a hockey puck. b) The tension force in a string is perpendicular to motion in a circle.

The relationship $\Delta K = W$ is still valid for this motion if the work done on the puck is zero. This will be true either if the *net* work done by the two forces is zero (one positive and the other negative) or if the work done by each individual force is zero. Which is it? Neither force can cause a horizontal acceleration, so neither force can transfer kinetic energy to the puck. Therefore, it seems to make sense that the work done by a force perpendicular to the direction of motion is zero.

That this is the correct definition of work is confirmed if you tie a string to the puck and whirl it in a horizontal circle on the ice, as shown in Fig. 9-5b. The puck's *speed* doesn't change, even though its velocity vector does, so its kinetic energy remains unchanged: $\Delta K = 0$. In this case, however, there *is* a net force on the puck—the tension $\vec{T}$ in the string. But the tension force is, at every point on the circle, perpendicular to the direction of motion. The tension force is accelerating the puck but *not* changing its kinetic energy. This is possible if we define work in such a way that no work is done, $W = 0$, by a force that is perpendicular to the direction of motion

Constant Force in an Arbitrary Direction

Next, consider a particle that undergoes a displacement Δx along a straight line under the influence of a constant force $\vec{F}$ that makes an angle θ with the x-axis. Figure 9-6 shows the situation. The force $\vec{F}$ clearly cannot be the only force acting on the particle because the particle is not accelerating in the direction of $\vec{F}$. Other unseen forces, such as a normal force or a weight force, must be balancing the y-component of $\vec{F}$ so that the particle can move horizontally. But these additional forces, if perpendicular to the direction of motion do no work on the particle. Force $\vec{F}$, we will assume, is the only force with an x-component: $F_x = |\vec{F}|\cos\theta$.

As before, it is important to keep in mind that $\vec{F}$ is acting with a constant magnitude and at a constant angle throughout the entire displacement Δx. The x-component of the particle's acceleration is

$$a_x = \frac{F_x}{m}.$$

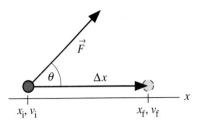

FIGURE 9-6 A particle moving under the influence of a constant force at an arbitrary angle.

We can calculate the particle's kinetic energy from constant-acceleration kinematics:

$$v_f^2 = v_i^2 + 2a_x\Delta x = v_i^2 + \frac{2F_x\Delta x}{m}$$

$$\Rightarrow \quad \tfrac{1}{2}mv_f^2 - \tfrac{1}{2}mv_i^2 = \Delta K = F_x\Delta x.$$

Thus the relationship $\Delta K = W$ remains valid if we broaden our definition of work to be

$$W \equiv F_x\Delta x \quad \text{(Second definition of work).} \tag{9-10}$$

Equation 9-10 is consistent with the first definition of Eq. 9-8, but it is more general in that it allows the force to be at an angle with respect to the direction of motion. It is also consistent with the work being zero for a force perpendicular to the motion because, in that case, $F_x = 0$.

Work is not always a positive quantity. Figure 9-7 shows two examples where the work done by force $\vec{F}$ is negative. In Fig. 9-7a the work is negative because Δx is negative. In Fig. 9-7b the work is negative because the force *component* F_x is negative.

It is worth reiterating that only the component of force *along the axis of motion* does work on the particle. The force component perpendicular to the motion, F_y in this case, does no work and consequently cannot change the particle's kinetic energy. This makes sense from a force-acceleration perspective because only F_x causes an x-component of acceleration a_x that, in turn, changes the x-component of velocity v_x. Nonetheless, it is sometimes easy to lose sight of this when using the work-energy perspective.

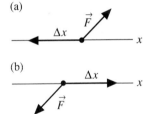

FIGURE 9-7 Examples of negative work. a) Δx is negative. b) F_x is negative.

EXAMPLE 9-3 The 500 mg ice-hockey puck of Examples 9-1 and 9-2 is again sliding along at 2 m/s. The compressed air gun exerts a 1 N force against the front of the puck. The compressed air flow is aimed 30° below the horizontal, and this force is applied continuously as the puck moves 50 cm. What is the puck's final speed?

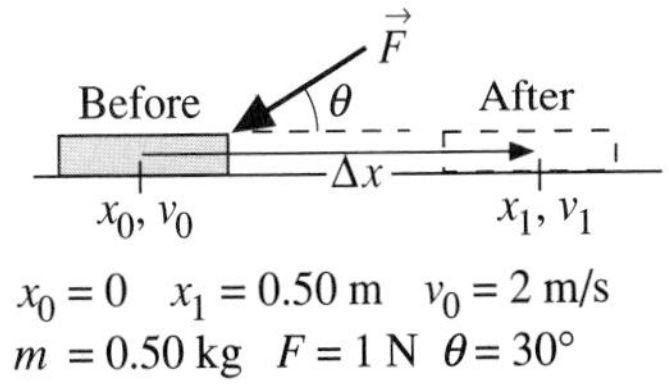

Before $\quad$ After

$x_0, v_0 \qquad x_1, v_1$

$x_0 = 0 \quad x_1 = 0.50\text{ m} \quad v_0 = 2\text{ m/s}$
$m = 0.50\text{ kg} \quad F = 1\text{ N} \quad \theta = 30°$

FIGURE 9-8 Pictorial model for Example 9-3.

SOLUTION Three forces act on the puck: Its weight $\vec{W}$, the normal force $\vec{N}$, and the force $\vec{F}$ of the compressed air. The weight and the normal force are perpendicular to the direction of motion, so they do no work on the puck. We can find the puck's final speed by using the relationship $\Delta K = W$, where the only work W is done by the compressed air force $\vec{F}$. The work, according to Eq. 9-10, is given by

$$W = F_x \Delta x = \left(-|\vec{F}| \cos 30° \right) \Delta x = -(1 \text{ N}) \cos 30° (0.5 \text{ m}) = -0.433 \text{ J}.$$

It was important to notice that the x-component of the force is negative. We've also made use of the fact that $1 \text{ N m} = 1 \text{ J}$. Now we can compute the change in speed to be

$$\Delta K = \tfrac{1}{2} m v_1^2 - \tfrac{1}{2} m v_0^2 = W$$

$$\Rightarrow \quad v_1 = \sqrt{v_0^2 + \frac{2W}{m}} = \sqrt{(2 \text{ m/s})^2 + \frac{2(-0.433 \text{ J})}{0.5 \text{ kg}}} = 1.51 \text{ m/s}.$$

Not surprisingly, the puck slows a bit less than in Example 9-2, where the compressed air force was directed exactly opposite the direction of motion.

•

9.6 Calculating Work: The Vector Dot Product

Our first two definitions of work considered only situations where the particle moved along the x-axis. Suppose, however, that we were faced with a more general situation, such as the one shown in Fig. 9-9. Here a *constant* force $\vec{F}$ is *continuously* applied to a particle as it moves through a quite general displacement $\Delta \vec{r}$. Let the angle between the force vector $\vec{F}$ and the displacement vector $\Delta \vec{r}$ be called α.

How much work does this force do on the particle, and by how much is the particle's kinetic energy changed as a result? To answer this question we need to introduce some new vector mathematics.

Newton's second law tells us the three components of the acceleration vector:

$$\vec{a} = a_x \hat{i} + a_y \hat{j} + a_z \hat{k} = \frac{F_x}{m} \hat{i} + \frac{F_y}{m} \hat{j} + \frac{F_z}{m} \hat{k}. \qquad (9\text{-}11)$$

Because we are assuming $\vec{F}$ to be a *constant* vector, all three components of the acceleration are also constant. We can thus use constant acceleration kinematics to deduce

$$(v_x)_f^2 = (v_x)_i^2 + 2a_x \Delta x = (v_x)_i^2 + \frac{2F_x}{m} \Delta x$$

$$(v_y)_f^2 = (v_y)_i^2 + 2a_y \Delta y = (v_y)_i^2 + \frac{2F_y}{m} \Delta y \qquad (9\text{-}12)$$

$$(v_z)_f^2 = (v_z)_i^2 + 2a_z \Delta z = (v_z)_i^2 + \frac{2F_z}{m} \Delta z.$$

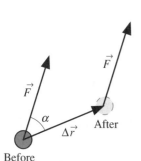

FIGURE 9-9 A more general case of a force acting on a particle during a displacement. How much work is done by the force?

The three-dimensional definition of kinetic energy, from Eq. 9-6, gives for ΔK:

$$\Delta K = K_f - K_i = \tfrac{1}{2}m\left[(v_x)_f^2 + (v_y)_f^2 + (v_z)_f^2\right] - \tfrac{1}{2}m\left[(v_x)_i^2 + (v_y)_i^2 + (v_z)_i^2\right]$$

$$= \left[\tfrac{1}{2}m(v_x)_f^2 - \tfrac{1}{2}m(v_x)_i^2\right] + \left[\tfrac{1}{2}m(v_y)_f^2 - \tfrac{1}{2}m(v_y)_i^2\right] + \left[\tfrac{1}{2}m(v_z)_f^2 - \tfrac{1}{2}m(v_z)_i^2\right] \quad (9\text{-}13)$$

$$= F_x\Delta x + F_y\Delta y + F_z\Delta z.$$

The result of this somewhat messy calculation is that ΔK is given by a *multiplicative* combination of the components of the vectors $\vec{F} = F_x\hat{i} + F_y\hat{j} + F_z\hat{k}$ and $\Delta\vec{r} = \Delta x\,\hat{i} + \Delta y\,\hat{j} + \Delta z\,\hat{k}$. In other words, we seem to be multiplying the two vectors together.

Now for the mathematics. Given two vectors $\vec{A} = A_x\hat{i} + A_y\hat{j} + A_z\hat{k}$ and $\vec{B} = B_x\hat{i} + B_y\hat{j} + B_z\hat{k}$, we define the **scalar product** or the **dot product** (both names are used) to be

$$\vec{A}\cdot\vec{B} = A_xB_x + A_yB_y + A_zB_z. \quad (9\text{-}14)$$

The dot product is the sum of the products of like-components of the two vectors. Note that the dot symbol between the vectors is *required* to indicate a dot product. The notation $\vec{A}\vec{B}$, without the dot, is *not* the same thing as $\vec{A}\cdot\vec{B}$.

An important characteristic of the dot product, which we will assert without proof, is that the value is *independent* of the coordinate system. We can make use of this fact to learn some interesting things about the dot product of two vectors. Figure 9-10a shows a situation in which the vectors $\vec{A}$ and $\vec{B}$ are parallel to each other. Because the dot product $\vec{A}\cdot\vec{B}$ can be evaluated in any coordinate system, consider coordinate axes such that the x-axis is parallel to $\vec{A}$ and $\vec{B}$. In this case, $A_x = |\vec{A}|$ and $B_x = |\vec{B}|$ while the other components are zero. Using Eq. 9-14, we can evaluate the dot product to be:

$$\text{Parallel vectors: } \vec{A}\cdot\vec{B} = A_xB_x = |\vec{A}||\vec{B}|. \quad (9\text{-}15)$$

That is, for parallel vectors (only!) the dot product is simply the product of the magnitudes.

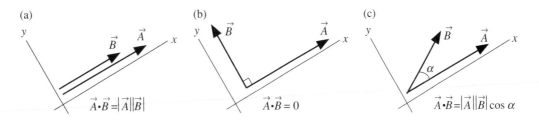

FIGURE 9-10 Using a tilted coordinate system allows us to evaluate $\vec{A}\cdot\vec{B}$ when a) $\vec{A}$ and $\vec{B}$ are parallel, b) $\vec{A}$ and $\vec{B}$ are perpendicular, and c) $\vec{A}$ and $\vec{B}$ have angle α between them.

Figure 9-10b shows another "special" situation, this time with the vectors $\vec{A}$ and $\vec{B}$ perpendicular to each other. Again choose the x-axis to be parallel to $\vec{A}$. In this case, the vectors are $\vec{A} = (|\vec{A}|, 0, 0)$ and $\vec{B} = (0, |\vec{B}|, 0)$. Because $A_y = 0$ and $B_x = 0$, Eq. 9-14 gives

$$\text{Perpendicular vectors: } \vec{A}\cdot\vec{B} = (A_x)(0) + (0)(B_y) = 0. \quad (9\text{-}16)$$

You can see that the dot product of two vectors depends on the orientation between the vectors.

Finally, consider the "general" case, shown in Fig. 9-10c. Here the two vectors $\vec{A}$ and $\vec{B}$ are oriented with angle α between them. Once again, let the x-axis be parallel to $\vec{A}$ so that $\vec{A} = (|\vec{A}|, 0, 0)$. In this case we have $\vec{B} = (|\vec{B}|\cos\alpha, |\vec{B}|\sin\alpha, 0)$. From Eq. 9-14, the dot product is

$$\text{General case: } \vec{A} \cdot \vec{B} = A_x B_x + (0)(B_y) = |\vec{A}||\vec{B}|\cos\alpha. \tag{9-17}$$

Equations 9-14 and 9-17 are alternative, but equivalent, ways of evaluating the dot product. It is important to note that α has to be the angle *between* the two vectors if you are going to use Eq. 9-17.

EXAMPLE 9-4   Compute the dot product of the two vectors shown in Fig. 9-11.

SOLUTION We can evaluate the dot product using either Eq. 9-14 or Eq. 9-17. Using vector components, we can write

$$\vec{A} = 5\cos 50° \,\hat{i} + 5\sin 50° \,\hat{j} = 3.21\,\hat{i} + 3.83\,\hat{j}$$

$$\vec{B} = 4\cos 20° \,\hat{i} + 4\sin 20° \,\hat{j} = 3.76\,\hat{i} + 1.37\,\hat{j}.$$

The dot product in terms of component is given by Eq. 9-14 as

$$\vec{A} \cdot \vec{B} = A_x B_x + A_y B_y = (3.21)(3.76) + (3.83)(1.37) = 17.32.$$

Neither vector has a z-component, so the $A_z B_z$ term was omitted. Alternatively, we can use Eq. 9-17. The angle *between* the vectors is $\alpha = 30°$, so

$$\vec{A} \cdot \vec{B} = |\vec{A}||\vec{B}|\cos\alpha = (5)(4)\cos 30° = 17.32.$$

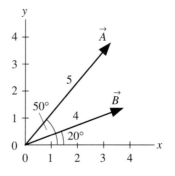

FIGURE 9-11 Vectors $\vec{A}$ and $\vec{B}$ of Example 9-4.

Now let's return to Eq. 9-13, where we calculated ΔK. The right side of the equation, as you can see by comparison with Eq. 9-14, is nothing other than the dot product of the force vector $\vec{F}$ with the displacement vector $\Delta\vec{r}$. That is:

$$\Delta K = F_x \Delta x + F_y \Delta y + F_z \Delta z = \vec{F} \cdot \Delta\vec{r}. \tag{9-18}$$

Equation 9-18 for the change in kinetic energy is again in the form $\Delta K = W$ if we broaden our definition of work to be

$$W \equiv \vec{F} \cdot \Delta\vec{r} \quad \text{(Third definition of work)}. \tag{9-19}$$

If we assume that the motion takes place in the xy-plane, with $\Delta z = 0$, then Eq. 9-19 can be written as

$$W = \begin{cases} F_x \Delta x + F_y \Delta y \\ |\vec{F}||\Delta\vec{r}|\cos\alpha \end{cases} \tag{9-20}$$

where α is the angle between the $\vec{F}$ and the $\Delta\vec{r}$ vectors. This is now the most general definition of work for a *constant* force, and our earlier two definitions are seen to be special cases where either the force or the displacement (or both) happened to be along the x-axis.

9.7 The Work-Kinetic Energy Theorem

There are very few situations, as you know, where an object is acted on by only a single force. More typical is a superposition of forces:

$$\vec{F}_{\text{net}} = \sum_i \vec{F}_i \quad \Rightarrow \quad \vec{a} = \frac{1}{m} \sum_i \vec{F}_i. \tag{9-21}$$

In other words, the acceleration is determined by the vector sum of *all* the forces acting on the object—a fact that you have used many times in Newton's second law problems. If you use this more general expression for $\vec{a}$ in Eq. 9-12, you can see that it leads to

$$\Delta K = \sum_i (F_i)_x \Delta x + (F_i)_y \Delta y + (F_i)_z \Delta z = \sum_i \vec{F}_i \cdot \Delta \vec{r} = \sum_i W_i \tag{9-22}$$

where W_i is the work done by force $\vec{F}_i$. Let's define the **net work** done *on* the object to be the sum of the work done by each force:

$$W_{\text{net}} = \sum_i W_i. \tag{9-23}$$

It is the *net* work, due to all forces acting on an object, that causes an object's kinetic energy to change. The change in kinetic energy is given by

$$\Delta K = W_{\text{net}}. \tag{9-24}$$

This basic idea—that a force or forces acting on a particle as it moves through a displacement $\Delta \vec{r}$ causes the particle's kinetic energy to change—has continued to hold up as we have broadened the picture to include multiple forces pushing and pulling in arbitrary directions. The relationship is beginning to look like a general principle, rather than just a special case of one-dimensional forces and motion. This principle, called the work-kinetic energy theorem, is defined as follows:

> **The Work-Kinetic Energy Theorem**: When one or more forces $\vec{F}_i$ act on a particle as it is displaced from an initial position to a final position, the net work done on the particle by these forces causes a *change* in the particle's kinetic energy given by $\Delta K = W_{\text{net}}$.

You might note that this has a strong similarity to the impulse-momentum theorem of Chapter 8, a point we will explore later. First, however, let us look at some examples.

EXAMPLE 9-5 An 70 kg ice skater on frictionless ice grabs hold of a tow rope that pulls him forward with a constant tension of 30 N. What is the skater's speed after he has been pulled for 10 m?

SOLUTION Figure 9-12a shows the pictorial model for this problem. The diagram, similar to those used for momentum problems, shows the situation before and after the interaction. Note that no time variable is defined in a work/energy problem; time does not enter into work and energy descriptions of motion. The skater's displacement, as the figure shows, is chosen to be along the *x*-axis. Figure 9-12b shows the free-body diagram. Although you don't need the free-body diagram for setting up Newton's second law, you still *do* need it to calculate the work done by the various forces acting on the skater.

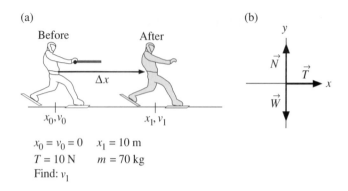

$x_0 = v_0 = 0$ $x_1 = 10$ m
$T = 10$ N $m = 70$ kg
Find: v_1

FIGURE 9-12   The pictorial model for Example 9-5 with a) the before and after pictures of the skater, and b) the free-body diagram for the skater.

The work-kinetic energy theorem applied to the skater tells us that

$$\Delta K = W_{net}.$$

However, to use this equation we must evaluate each side of the equation in terms of how it is defined. Because there are three forces, W_{net} contains three terms:

$$W_{net} = \sum_i W_i = W_{tension} + W_{gravity} + W_{normal}.$$

Forces $\vec{W}$ and $\vec{N}$ are both perpendicular to the direction of motion, so these forces do zero work on the skater. The tension force, on the other hand, is a constant force along the x-axis, so it does work on the skater given by $W_{tension} = |\vec{T}|\Delta x$. Using Eq. 9-6 for K in one-dimensional motion, the work-kinetic energy theorem becomes

$$\left[\Delta K = \tfrac{1}{2}mv_1{}^2 - \tfrac{1}{2}mv_0{}^2\right] = \left[W_{net} = |\vec{T}|(x_1 - x_0)\right]$$

We can now solve this for v_1:

$$\tfrac{1}{2}mv_1{}^2 = |\vec{T}|x_1$$

$$\Rightarrow v_1 = \sqrt{\frac{2|\vec{T}|x_1}{m}} = \sqrt{\frac{2(30 \text{ N})(10 \text{ m})}{70 \text{ kg}}} = 2.93 \text{ m / s}.$$

EXAMPLE 9-6   After the skater from Example 9-5 has been towed 10 m he is at the top of a 10° icy hill with a 50 m long down slope. If he lets go of the rope at the top of the hill, what will his speed be at the bottom?

SOLUTION   Figure 9-13 shows the pictorial model and the free-body diagram for the skater on the hill. We have used a tilted coordinate system so that the skater's displacement is still along the x-axis. We see, by inspection, that the normal force $\vec{N}$ does not have an x-component and will do no work on the skater. But the weight force is no longer perpendicular to the direction of motion, so the force of gravity *will* do work on the skater as he glides downhill. The work done on the skater by the force of gravity is

$$W = \vec{W} \cdot \Delta\vec{r} = W_x \Delta x = |\vec{W}|\sin\theta\Delta x = mg\sin\theta\Delta x.$$

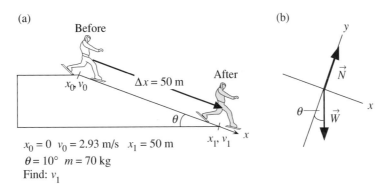

FIGURE 9-13 a) Pictorial model of skater. b) Free-body diagram.

(Make sure you agree that $W_x = |\vec{W}|\sin\theta$.) The work-kinetic energy theorem then gives

$$\Delta K = W_{net}$$

$$\Rightarrow \quad \tfrac{1}{2}mv_1^2 - \tfrac{1}{2}mv_0^2 = mg\sin\theta\,(x_1 - x_0)$$

$$\Rightarrow \quad v_1 = \sqrt{v_0^2 + 2gx_1\sin\theta}\,.$$

The skater's initial speed at the top of the hill v_0 is his *final* speed after being towed 10 meters, which we found in Example 9-5 to be 2.93 m/s. Thus

$$v_1 = \sqrt{(2.93 \text{ m/s})^2 + 2(9.8 \text{ m/s}^2)(50 \text{ m})\sin 10°} = 13.4 \text{ m/s}.$$

Notice in this example that it is the work done by the force of gravity that causes the skater to speed up.   •

EXAMPLE 9-7   A 10 kg wooden crate is being pulled by a rope at 5 m/s up a 10° wooden slope when suddenly the rope breaks. How far will the crate slide up the hill before stopping?

SOLUTION Figure 9-14 shows the pictorial model and free-body diagram *after* the rope breaks. The rope is responsible for the crate's initial speed, but it will play no role in the crate sliding to a halt. The motion is again linear, with displacement Δx, and we can again see by inspection that $\vec{N}$ will do no work on the crate because it has no x-component. The work-kinetic energy theorem then gives us

$$\Delta K = W_{net} = W_{gravity} + W_{friction}.$$

To determine $W_{friction}$ we will need $\vec{f}_k$, and to determine the friction force we need to do just a little Newton's laws analysis. We know that $|\vec{f}_k| = \mu_k|\vec{N}|$, and if we apply the second law along just the y-axis we have

$$\sum F_y = |\vec{N}| - |\vec{W}|\cos\theta = |\vec{N}| - mg\cos\theta = ma_y = 0$$

$$\Rightarrow |\vec{N}| = mg\cos\theta$$

$$\Rightarrow |\vec{f}_k| = \mu_k mg\cos\theta.$$

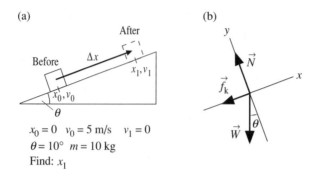

FIGURE 9-14 Pictorial model and free-body diagram of a wooden crate sliding uphill.

To calculate the work done by the forces we need their x-components. We see, from the free-body diagram, that the x-components of both $\vec{W}$ and $\vec{f}_k$ will be negative:

$$W_x = -mg\sin\theta$$

$$(f_k)_x = -|\vec{f}_k| = -\mu_k mg\cos\theta.$$

The net work done by these forces is thus:

$$W_{net} = \vec{W}\cdot\Delta\vec{r} + \vec{f}_k\cdot\Delta\vec{r} = W_x\Delta x + (f_k)_x\Delta x$$

$$= -mg\sin\theta\Delta x - \mu_k mg\cos\theta\Delta x$$

$$= -mg\Delta x(\sin\theta + \mu_k\cos\theta).$$

The work done in this example is *negative* because the x-components of both forces are *opposite* the crate's displacement. Inserting this result for W_{net} into the work-kinetic energy theorem gives

$$\tfrac{1}{2}mv_1^2 - \tfrac{1}{2}mv_0^2 = -\tfrac{1}{2}mv_0^2 = -mg\Delta x(\sin\theta + \mu_k\cos\theta)$$

$$\Rightarrow \Delta x = \frac{v_0^2}{2g(\sin\theta + \mu_k\cos\theta)} = \frac{(5\text{ m}/\text{s})^2}{2(9.8\text{ m}/\text{s}^2)(\sin 10° + 0.2\times\cos 10°)} = 3.44\text{ m}.$$

The coefficient of friction was taken from Table 5-1. The important point of this example has been the calculation of *negative* work, which *decreases* the particle's kinetic energy.

•

All of these examples, it is true, could have been solved by a straightforward application of Newton's second law. As we continue to develop the idea of work and energy, you will soon find problems where the work-energy perspective will allow you to solve problems that are *not* easily solved by Newton's laws.

One thing you will have noticed from these examples is that the work-kinetic energy theorem $\Delta K = W$ is deceptively simple when viewed as simply a mathematical statement. You have to keep in mind that both sides of this equation contain a significant amount of physics. To *use* the work-kinetic energy theorem, you have to recall how the terms K and W are defined and do the necessary analysis to evaluate them. But do not lose sight, in the midst of the mathematics, of what the work-kinetic energy theorem is all about. It is a statement about *energy transfer*, saying that work is energy transferred to or from the system due to forces exerted on the system by the environment. As a consequence of this energy transfer, the system's kinetic energy either increases or decreases.

A word of caution: The work-kinetic energy theorem, as we have developed it thus far, applies *only* to a one-particle system. You know, from experience, that objects sliding under the influence of friction forces get *hot*—a thermal effect. This suggests that our analysis of Example 9-7 has not included everything that is going on. The difficulty is that a crate is *not* a single particle. The idea of temperature is meaningless for a single particle that occupies only a single point in space and has no internal structure. Real objects, which are systems of very large numbers of atomic particles, have other forms of energy in addition to the bulk kinetic energy of the entire system. We will see later that these other forms of energy are associated with the object's temperature. So our analysis of Example 9-7, in which we treated the crate *as if* it were a particle, is not entirely valid. We will return to this example in Chapter 11, where we examine the energy properties of systems of particles. It will turn out that our answer is, in fact, correct, but it is not the full story.

9.8 Work Done by a Continuously-Varying Force

The previous sections have shown how to calculate the work done on a particle by a force that has a constant magnitude and direction throughout the particle's displacement. But what about a force that changes in either magnitude or direction as the particle moves—a force whose value is different at every point? Our first three definitions of work, which were for *constant* forces, clearly do not apply to a variable force. Does the work-kinetic energy theorem apply to such variable forces? And if so, how can we compute the work done by such a force?

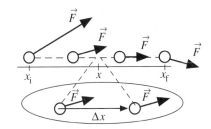

FIGURE 9-15 A particle moving under the influence of a variable force. The inset show a small segment Δx over which the force is constant.

Figure 9-15 shows a particle moving in one dimension from initial position x_i to final position x_f while being acted on by a continuously-varying force. In this case, the force varies in both magnitude *and* direction. Because the force is *not* constant, our earlier definition $W = F_x \Delta x$ will *not* give the correct value for the work done by this force.

Suppose, instead, we were to divide the particle's trajectory into many very small segments Δx, such that the force $\vec{F}(x)$ acting on the particle over this small length Δx, centered at the position x, is essentially constant. Then Eq. 9-10 is valid for each small interval, and we can find the kinetic energy change during one such interval to be

$$\Delta K_{\text{during interval } \Delta x} = K(x + \Delta x) - K(x)$$
$$= F_x(x)\Delta x \tag{9-25}$$
$$= \delta W$$

where $\delta W = F_x(x)\Delta x$ is the small amount of work done by the force during this interval. The symbol δ is a lowercase Greek delta. We are going to use it, rather than Δ, so that δW means a *small quantity* of work while ΔK means the corresponding *change* in the kinetic energy. The notation $F_x(x)$ indicates the x-component of the force $\vec{F}$ acting on the particle when it is at position x.

Having computed the small amount of work δW done on the particle as it moves through the small distance Δx, and its corresponding change of kinetic energy, we now want somehow to "add up" these quantities over all of the intervals Δx in the large interval from x_i to x_f. To do this, let's divide the full interval into N small intervals of width Δx:

First small interval:	x_i	to	$x_i + \Delta x$
Second small interval:	$x_i + \Delta x$	to	$x_i + 2\Delta x$
Third small interval:	$x_i + 2\Delta x$	to	$x_i + 3\Delta x$
$\vdots$			$\vdots$
Nth (last) interval:	$x_i + (N-1)\Delta x$	to	$x_i + N\Delta x = x_f$

where we have noted that the last interval ends at x_f. Now add the ΔK for all these intervals:

$$\sum_{\text{all } \Delta x} \Delta K_{\text{during interval } \Delta x} = [K(x_i + \Delta x) - K(x_i)] + [K(x_i + 2\Delta x) - K(x_i + \Delta x)]$$
$$+ [K(x_i + 3\Delta x) - K(x_i + 2\Delta x)] + \cdots \quad (9\text{-}26)$$
$$+ [K(x_f) - K(x_i + (N-1)\Delta x)].$$

If you examine this sum carefully you will notice that all of the terms cancel except the initial $K(x_f)$ and the final $K(x_i)$, leaving only

$$\sum_{\text{all } \Delta x} \Delta K_{\text{during interval } \Delta x} = K(x_f) - K(x_i) = \Delta K_{x_i \to x_f}. \quad (9\text{-}27)$$

In other words, the sum of all the small ΔK's is just the net change in kinetic energy ΔK as the particle moves from x_i to x_f.

To find the total work done by force $\vec{F}$ as the particle moves from initial position x_i to final position x_f, we next want to add up the small amounts of work δW that are done during each of the N small intervals. During interval j, of length Δx at position x_j, the small amount of work δW_j is given by the right-hand side of Eq. 9-25. Adding these, we find the total work done is

$$W = \sum_j \delta W_j = \sum_j F_x(x_j)\Delta x. \quad (9\text{-}28)$$

Not surprisingly, we want to consider the limit $\Delta x \to 0$. In that case, the sum of Eq. 9-28 becomes an integral and the work is

$$W \equiv \lim_{\Delta x \to 0} \sum_j F_x(x_j)\Delta x = \int_{x_i}^{x_f} F_x(x)dx \quad \text{(Fourth definition of work).} \quad (9\text{-}29)$$

This is our fourth, and last, definition of work. You should convince yourself that the first two definitions of work, Eqs. 9-8 and 9-10, can be derived from Eq. 9-29 for the case that the force is a constant and can be taken outside the integral. By summing both sides of Eq. 9-25, we have now shown that the work-kinetic energy theorem is valid for the case of a continuously-varying force *if* the work is properly computed. We end up with

$$\Delta K_{x_i \to x_f} = W, \text{ where } W = \int_{x_i}^{x_f} F_x(x)dx. \qquad (9\text{-}30)$$

Keep in mind that we chose the x-axis simply for convenience, and that you could equally well use the y-axis if that is more appropriate. Let us see how this can be useful.

EXAMPLE 9-8   A 1500 kg car accelerates from rest. The net force on the car (propulsion force minus any drag forces) is shown in Fig. 9-16 as a function of the car's position. What is the car's speed after having traveled a) 100 m, and b) 200 m?

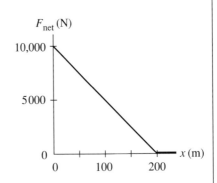

FIGURE 9-16 Force-versus-position graph for the car of Example 9-8.

SOLUTION A car's acceleration is high as it starts but decreases as it picks up speed and drag forces increase. Figure 9-16 is a more realistic portrayal of the net force on a car than was our earlier model of a constant acceleration. But we cannot use constant-acceleration kinematics with a variable force. Instead, we can use the work-kinetic energy theorem. Because $v_i = 0$, we have

$$\left[\Delta K = K_f - K_i = \tfrac{1}{2}mv_f^2\right] = W = \int_{x_i}^{x_f} F_x(x)dx.$$

The x-component of the force is just F_{net}, because the net force points in the +x-direction. Using $x_i = 0$, the work is

$$W = \int_0^{x_f} F_{net}\,dx = \text{area under the } F_{net}\text{-versus-}x \text{ curve from 0 to } x_f.$$

Once again we've given a graphical interpretation to an integral.

a) The area under the curve of Fig. 9-16 out to $x_f = 100$ m can be divided into a rectangle of width 100 m and height 5000 N and a triangle of base 100 m and height 5000 N. Thus

$$W(\text{to 100 m}) = \text{area} = (5000 \text{ N})(100 \text{ m}) + \tfrac{1}{2}(5000 \text{ N})(100 \text{ m}) = 750,000 \text{ J}.$$

Using this result for the work in the work-kinetic energy theorem gives

$$\tfrac{1}{2}mv_f^2 = W$$

$$\Rightarrow \ v_f = \sqrt{\frac{2W}{m}} = \sqrt{\frac{2(750,000 \text{ J})}{1500 \text{ kg}}} = 31.6 \text{ m}/\text{s}.$$

Notice that the speed has SI units of m/s because the work and the mass both had SI units.

b) When the car reaches $x_f = 200$ m the curve is a full triangle of width 100 m. Thus

$$W(\text{to } 200 \text{ m}) = \text{area} = \tfrac{1}{2}(10\,000 \text{ N})(200 \text{ m}) = 1{,}000{,}000 \text{ J},$$

which gives:

$$v_f = \sqrt{\frac{2W}{m}} = 36.5 \text{ m / s}.$$

Very little additional speed was gained in the second 100 meters because little additional work was done.

•

An Analogy with Impulse-Momentum

You have likely noticed that there is a similarity of the work-kinetic energy theorem to the impulse-momentum theorem of Chapter 8:

Work-kinetic energy theorem: $\Delta K = W = \displaystyle\int_{x_i}^{x_f} F_x \, dx$

Impulse-momentum theorem: $\Delta p = J = \displaystyle\int_{t_i}^{t_f} F_x \, dt.$

(9-31)

In both cases, a force acting on a particle causes the state of the system to change. If the force acts over a time interval $t_i \rightarrow t_f$, it creates an *impulse* that changes the particle's momentum. If the force acts over a spatial interval $x_i \rightarrow x_f$, it does *work* that changes the particle's kinetic energy. This is *not* to suggest that a force *either* creates an impulse *or* does work, but not both. Quite the contrary—a force acting on a particle *both* creates an impulse *and* does work, changing both the momentum and the kinetic energy of the particle. The issue of whether to use the work-kinetic energy theorem or the impulse-momentum theorem depends on the question you are trying to answer. Note that the geometric interpretation of impulse as the area under the F-versus-t graph can be applied equally well to an interpretation of work as the area under the F-versus-x graph.

We can, in fact, express the kinetic energy in terms of the momentum as

$$K = \tfrac{1}{2}mv^2 = \frac{(mv)^2}{2m} = \frac{p^2}{2m}.$$

(9-32)

This makes it quite clear that you cannot change a particle's kinetic energy without also changing its momentum.

9.9 Restoring Forces and Hooke's Law

[**Photo suggestion: A stretched slingshot getting ready to shoot a projectile.**]

If you stretch a rubber band, a force appears that tries to pull the band back to its equilibrium, or unstretched, length. This force did not exist before the rubber band was stretched. A force that restores a system to an equilibrium position is called a **restoring force**. Systems that exhibit restoring forces are called **elastic**. The most basic examples of elasticity are things such as springs and rubber bands. If you stretch a spring, there is a tension-like force trying to pull it back. Similarly if you compress a spring, it tries to re-expand back to its equilibrium length. Other examples of elasticity and restoring forces abound. If you press your finger on a drum head (or against any stretched material), it deflects a little and exerts a force back against your finger. As soon as you remove your finger, this restoring force returns the material to its equilibrium position. If you drive your car over a bridge, the steel beams sag downward slightly—but they are restored to equilibrium after your car passes by. Nearly everything that stretches, compresses, flexes, bends, or twists exhibits a restoring force and can be called elastic.

One observation you have probably made about elastic objects is that they *oscillate* a little if you pull them away from equilibrium and then release them. Pull down a spring and release it—it oscillates. Flex a metal ruler in your hands and then let go of one end—the ruler "quivers" (oscillates) for a few seconds. Bang a drum head—it oscillates. An entire section of this text—Part 3—will be devoted to the physics of oscillations and waves. There we will encounter restoring forces over and over. In this section we want to introduce some of the basic characteristics of restoring forces. We will find that work and energy are useful concepts for understanding elasticity.

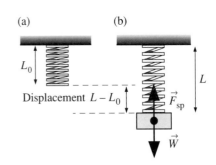

FIGURE 9-17 a) A spring hangs at its equilibrium length L_0. b) A mass hung on the spring stretches it to length L.

If an elastic system is displaced from its equilibrium position, the restoring force tries to push or pull it back. We would like to know how the magnitude of the restoring force depends on the displacement from equilibrium. This is an experimental question, so suppose we attach a spring to a firm support where we can measure it. Its equilibrium length, as shown in Fig. 9-17a, is L_0. The spring in this position is neither pushing nor pulling. Now attach a mass m to the spring and let the spring stretch out to length L, as shown in Fig. 9-17b. The mass is in static equilibrium, so the weight force pulling down on it must be balanced by an upward pull of the spring with force $\vec{F}_{sp}$. These forces add to zero, according to Newton's first law, so we must have

$$|\vec{F}_{sp}| = |\vec{W}| = mg.$$

In other words, the weight mg measures the restoring force of the spring when it is stretched to length L. By using different masses to produce different lengths, we can determine how the size of the restoring force $|\vec{F}_{sp}|$ depends on the length L.

Figure 9-18 shows some actual data of the restoring force of a real spring and also of a rubber band. Notice that the quantity graphed along the horizontal axis is $L - L_0$. This is the distance that the end of the spring has moved, which we call the **displacement from equilibrium**. The graphs of Fig. 9-18 describe a *variable* force. The force is zero if $L = L_0$, and it steadily increases as the stretching distance increases. The graph for the high-quality spring indicates that the restoring force is proportional to the displacement—the data falls along the straight line $|\vec{F}_{sp}| = k(L - L_0)$. The rubber band exhibits a linear variation of force with displacement only for very small displacements ($L - L_0 < 0.04$ m). For larger displacements, this linear relationship fails.

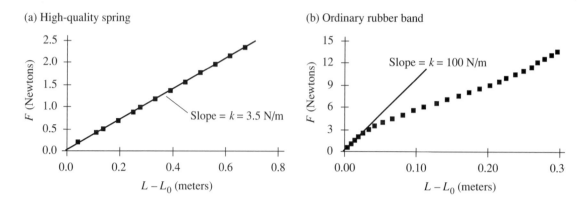

FIGURE 9-18 Measured data of restoring force versus displacement for a) a high-quality metal spring, and b) an ordinary rubber band.

Figure 9-18 shows that for "small" displacements from equilibrium, the magnitude of the restoring force is *directly proportional* to the displacement from equilibrium $|L - L_0|$. We call this relationship **Hooke's law**, and write it as:

$$|\vec{F}_{sp}| = k|L - L_0| \quad \text{(Hooke's law)}, \tag{9-33}$$

where k, which is the *slope* of the force-versus-displacement graph, is called the **spring constant**. This "law" about the restoring force of springs was first suggested by Robert Hooke, a contemporary (and sometimes bitter rival) of Newton. Hooke's law is not a "law of nature" in the way that Newton's laws are, but is actually just a *model* of a restoring force. It works extremely well for some springs, as in Fig. 9-18a, but less well for others, such as the rubber band of Fig. 9-18b. In this regard, Hooke's law is more closely related to our "law" of friction, which is a model of the friction force that is based on experiment and, as we noted earlier, has definite limitations.

The spring constant k is a property that characterizes the spring, just like mass m characterizes a particle. If k is large, a large weight is needed to cause a significant stretch. In this case we refer to the spring as a "stiff" spring. If k is small, the spring can be stretched with very little force. A small-k spring is called a "soft" spring. Every spring has its own, unique, value of k that remains constant for that spring. The units of the spring constant, as suggested by Eq. 9-33, are N/m. The spring constants for the two springs in Fig. 9-18 can be determined from the slopes of the straight lines: $k = 3.5$ N/m for the metal spring (pretty soft) and a stiffer $k = 100$ N/m for the rubber band (at least for small displacements).

Springs can be compressed, as well as stretched. Experiments with compressed springs show that the restoring force is proportional to the distance compressed, as long as the displacement is small. In this case, however, the quantity $L - L_0$ is negative. Consequently, we have written Hooke's law, in Eq. 9-33, with the absolute value of $L - L_0$. Hooke's law will then apply equally well to both elongation or compression of the spring.

Although we "discovered" Hooke's law for a spring, it actually applies to small displacements of nearly any elastic object. As long as the displacements from equilibrium are small, the object is said to have a *linear restoring force*, indicating that the magnitude of the restoring force is linearly proportional to the displacement from equilibrium. But, as we saw in Fig. 9-18, Hooke's law fails if the displacement gets too large.

In this text, we are going to adopt the idealization of a *massless spring*. This is similar to our use of massless strings. While not a perfect description, it will be a sufficiently good approximation if any masses that are hung on springs are much larger than the mass of the spring itself. Although the spring's mass *can* be included in the equations, doing so greatly increases the mathematical complexity while obscuring the physical ideas that we want to stress. Engineers will learn, in more advanced courses, how to deal with a spring's mass because clearly it is going to be important in some applications.

Hooke's law, as expressed in Eq. 9-33, gives only the magnitude of the restoring force. Because the restoring force is a vector, we will frequently have need of its direction as well as its magnitude. We need to rewrite Eq. 9-33 in a slightly different form that will give us an appropriate sign for the vector component F_{sp}.

Consider the situation depicted in Fig. 9-19. In this case, the spring lies along a generic s-axis and the equilibrium position of the end of the spring is denoted s_0. Note, and this is important, that s_0 is *not* the equilibrium length L_0 of the spring. It is, instead, the *position*, or coordinate, of the free end of the spring. It *might* be equal to the length *if* the fixed end happens to be at $s = 0$, but this is usually not the case.

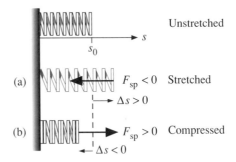

In Fig. 9-19a, where the spring is stretched, we see that the displacement from equilibrium $\Delta s = s - s_0$ is positive whereas the s-component of the restoring force, which points to the left, is negative. In the compression of Fig. 9–19b, on the other hand, the displacement from equilibrium Δs is negative while the s-component of $\vec{F}_{sp}$, which now points to the right, is positive. No matter which way the spring is displaced from equilibrium, the *sign* of the displacement Δs is opposite to the *sign* of the force component F_{sp}. We can state this mathematically as

FIGURE 9-19 Determining the sign of the spring force for a) a stretched spring and b) a compressed spring.

$$F_{sp} = -k(s - s_0) = -k\Delta s \quad \text{(mathematically-correct Hooke's law)}. \qquad (9\text{-}34)$$

Equation 9-34, with its minus sign, is the mathematically correct way to express Hooke's law for the vector component of the spring force.

A word of caution: Some of you, in an earlier physics course, may have learned to write Hooke's law as $F = -kx$, rather than as $-k\Delta x$ (assuming the spring is along the x-axis). This can be misleading, and it is a common source of errors. The restoring force will be $-kx$ *only* if the coordinate system in the problem is chosen such that the origin is at the equilibrium position of the free end of the spring. This makes $x_0 = 0$ and $\Delta x = x$. This is often done in simple problems, but in more complex problems it will frequently be most convenient to locate the origin of the coordinate system somewhere else. If you have memorized $F = -kx$ and try to use it after the origin has moved elsewhere—big trouble! So erase the earlier version from your memory banks and replace it with the more correct statement of Hooke's law as $F = -k\Delta s$.

EXAMPLE 9-9 A spring-loaded gun is used to launch a 10 g plastic ball. The spring, which has a spring constant of 10 N/m, is compressed by 10 cm when the ball is pushed into the barrel. When the trigger is pulled, the spring is released and shoots the ball back out. With what speed does the ball exit the barrel? Assume that friction is negligible.

SOLUTION Figure 9-20 shows a pictorial model of the ball being pushed out by the spring from initial position x_1 to final position x_2. The spring will continue to push on the ball until the spring has returned to its equilibrium, unstretched length. We have chosen to put the origin of the coordinate system at the equilibrium position x_0 of the free end of the spring, making $x_2 = 0$ and $x_1 = -10$ cm. The force exerted on the ball by the spring is given by Hooke's law (including the minus sign!) as

$$F_x(x) = -k(x - x_0) = -kx,$$

where we've used $x_0 = 0$ for this case. Notice that the ball's position x will be negative while it is in contact with the spring (left of the origin). When a negative x is combined with the minus sign in Hooke's law, F_x will be positive—properly indicating a force vector pointing to the right.

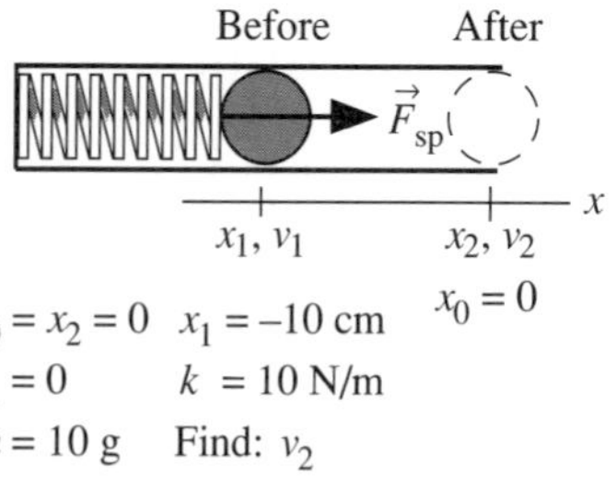

The force of the spring is going to do work on the ball, changing its kinetic energy. No work will be done by gravity or by the normal force of the barrel on the ball because these forces have no x-component. Because the spring force is a variable force, we will need Eq. 9-29—the definition of work as an integral—to compute the work. The work-kinetic energy theorem, using $v_1 = 0$, gives us

$x_0 = x_2 = 0 \quad x_1 = -10$ cm $\quad x_0 = 0$
$v_1 = 0 \quad\quad k = 10$ N/m
$m = 10$ g $\quad$ Find: v_2

FIGURE 9-20 A spring-loaded gun shooting a plastic ball.

$$\left[\Delta K = \tfrac{1}{2}mv_2{}^2 - \tfrac{1}{2}mv_1{}^2 = \tfrac{1}{2}mv_2{}^2 \right] = \left[W = \int_{x_1}^{x_2} F_x(x)dx \right].$$

In this example we know the functional form of $F_x(x)$, so we can do the integral explicitly:

$$\begin{aligned}
W &= \int_{x_1}^{x_2} F_x(x)dx \\
&= \int_{x_1}^{x_2} (-kx)dx = -k \int_{x_1}^{x_2} x\,dx \\
&= -\tfrac{1}{2}kx^2 \Big|_{x_1}^{x_2} = -\tfrac{1}{2}k(x_2{}^2 - x_1{}^2) \\
&= \tfrac{1}{2}kx_1{}^2.
\end{aligned}$$

Thus

$$\tfrac{1}{2}mv_2{}^2 = \tfrac{1}{2}kx_1{}^2,$$

which we can solve for v_2:

$$v_2 = \sqrt{\frac{kx_1{}^2}{m}} = \sqrt{\frac{(10 \text{ N}/\text{m})(-0.10 \text{ m})^2}{(0.010 \text{ kg})}} = 3.16 \text{ m}/\text{s}.$$

Notice that Example 9 is a problem that we could *not* have solved with Newton's laws. We could, it is true, write the acceleration $a(x) = F(x)/m$, but this is a non-constant acceleration and we do not know how to handle the kinematics in this situation. (It can be done, but the mathematics gets significantly more advanced.) Although we did have to do an integral to compute the work, we were able to solve the problem fairly easily using work and energy.

EXAMPLE 9-10   A wooden crate of mass 10 kg sliding along a wooden floor collides with a spring of spring constant 100 N/m. If the crate's speed at the time of impact is 3 m/s, what is the maximum compression of the spring?

SOLUTION   The pictorial model of Fig. 9-21a shows that we have again chosen to place the origin at the equilibrium position of the free end of the spring, so the crate contacts the spring at $x_1 = 0$ and compresses the spring until it reaches maximum compression at x_2. These are the before and after of the interaction. The point at which the spring is fully compressed has $v_2 = 0$—it is a *turning point* for the crate, which is instantaneously at rest. In this problem we can easily calculate ΔK, so we want to find a final position x_2 such that $W(\text{for } x_1 \rightarrow x_2) = \Delta K$.

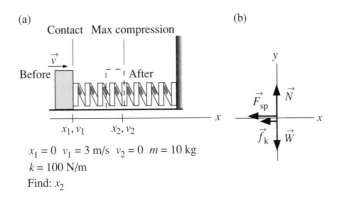

FIGURE 9-21   a) A wooden crate colliding with a fixed spring. b) The free-body diagram.

We can again employ the work-kinetic energy theorem, but we need to realize, as shown in Fig. 9-21b, that there are now *two* forces acting on the crate and thus two amounts of work being done. The friction force we can easily find, from much past experience, to be

$$|\vec{f}_k| = \mu_k |\vec{N}| = \mu_k mg$$
$$\Rightarrow (f_k)_x = -\mu_k mg.$$

The friction force *vector* points to the left, so its x-component is $-|\vec{f}_k|$. This is a constant force, so we can use Eq. 9-10 to evaluate the work it does on the crate. The work done by the spring force, however, will again require an integration. Thus

$$\Delta K = \tfrac{1}{2}mv_2{}^2 - \tfrac{1}{2}mv_1{}^2 = -\tfrac{1}{2}mv_1{}^2 = W_{\text{friction}} + W_{\text{spring}}$$

$$= (-\mu_k mg)(x_2 - x_1) + \int_{x_1}^{x_2}(-kx)dx$$

$$= -\mu_k mgx_2 - k\int_{0}^{x_2} x\,dx$$

$$= -\mu_k mgx_2 - \tfrac{1}{2}kx_2{}^2,$$

which is rearranged to give

$$\tfrac{1}{2}kx_2{}^2 + \mu_k mgx_2 - \tfrac{1}{2}mv_1{}^2 = 0.$$

This is a quadratic equation for x_2, with solutions $x_2 = (-1.16 \text{ m}, +0.773 \text{ m})$. The negative solution is clearly unphysical, showing the crate stopping before it hits the spring, so our answer is $x_2 = 0.773$ m.

●

9.10 Power

[**Photo suggestion: A sprinter coming out of the blocks.**]

As you have seen, work is a transfer of energy from the environment to a particle. It is a particular type of energy transfer, in which the environment pushes (or pulls) the particle through some displacement, thereby changing the particle's speed and kinetic energy. There are many situations, especially in engineering applications of physics, where we would like to know *how fast* the energy is transferred. Does the force act quickly and transfer the energy very rapidly, or is it a slow and lazy transfer of energy? If you need to lift 2000 pounds up 50 feet, which requires work from a motor, it makes a *big* difference whether the motor has to do this in 3 hours or 3 seconds!

The question "how fast?" implies that we are talking about a *rate of change*. For example, the velocity of an object—how fast it is going—is the rate of change of position. So when we raise the issue of how fast the energy is changing as a result of energy transfer, we must be talking about the *rate of change of energy*. This rate of change is called the **power** P, and it is defined to be:

$$P \equiv \frac{dE}{dt}. \tag{9-35}$$

The unit of power is the **watt**, which is defined as 1 watt = 1 W ≡ 1 J/s.

A particle gaining energy at the rate of 3 J/s is said to "consume" a power of 3 W. The source of that power is the environment that is doing work on the particle by "generating" 3 W of power. Common prefixes used with power are mW (milliwatts), kW (kilowatts), and MW (megawatts). Note that we have yet *another* use of the symbol "W"—be careful!

The English unit of power is the *horsepower*—a pretty antiquated unit today, when few of us have ever seen a horse power anything! Nonetheless, the conversion factor to watts is 1 horsepower = 1 hp = 746 W. Common appliances, such as motors, are rated in hp.

The idea of power as a rate of energy transfer applies regardless of the form of energy being considered. For example, a 100 W light bulb is using electrical energy at a rate of

100 Joules per second, transforming it to light and heat. A well-trained athlete can "put out" a sustained power of ≈ 0.5 hp ≈ 370 W by transforming the chemical energy of glucose and fat ("burning" fat) into mechanical energy at a rate of 370 Joules per second. The gas furnace in my home has a sticker on it that rates it at "43,000 BTU/hr," a set of units even more arcane than horsepower but that converts to ≈ 12 kW. This says that the gas flame delivers heat energy—another major form of energy transfer that we will study later—at the rate of 12,000 Joules per second. In all of these cases the important thing is not the amount of energy that is transferred, but the *rate*. You will not win the race if your body takes 2 s to convert 370 W of chemical energy into mechanical energy while your competitor's body can convert 370 W in 1 s!

Although power as the rate of energy transfer is a quite general concept, we want to focus on *work* as the source of energy transfer. Within this more limited scope, *power* is simply the *rate of doing work* by a force: $P = dW/dt$. If a particle moves through a small displacement dx while acted on by force $\vec{F}$, the force does a small amount of work dW given by Eq. 9-25:

$$dW = F_x dx. \tag{9-36}$$

Dividing both sides by dt, to give a *rate* of change, yields

$$\frac{dW}{dt} = F_x \frac{dx}{dt} = F_x v \tag{9-37}$$

$$\Rightarrow \quad P = F_x v_x.$$

This can be stated more generally as

$$P = \vec{F} \cdot \vec{v}. \tag{9-38}$$

In other words, the power delivered to a particle by a force acting on it is the dot product of the force with the particle's velocity. Because v is the instantaneous velocity, P is the instantaneous power. The power delivered will change if either the force or the velocity changes. Our use of power in this text will be limited to one-dimensional examples, in which case we can use a somewhat simpler $P = Fv$. These ideas will become clearer with some examples.

EXAMPLE 9-11 What horsepower motor is needed to lift a 1000 kg elevator at 3 m/s?

SOLUTION Figure 9-22 shows a free-body diagram of the elevator. The net force is zero, because the elevator moves at constant speed. The tension in the cable *does* do work on the elevator while gravity does an equal but opposite amount of work such that $W_{net} = 0$. Because the cable is pulled by the motor, we say that the motor does the work of lifting the elevator. The lifting force provided by the motor is $\vec{F}_{lift} = \vec{T}$, so the power output of the motor is $P = (F_{lift})_y v = |\vec{T}|v$.

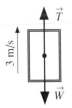

FIGURE 9-22 Free-body diagram for Example 9-11.

From Newton's first law, for constant velocity motion, $|\vec{T}| = |\vec{W}| = mg$. Thus

$$P = |\vec{T}|v = mgv = 29,400 \text{ W} = 39.4 \text{ hp}.$$

EXAMPLE 9-12 An automobile braking system brings a car to a halt at constant deceleration. a) What is the maximum power required to stop a 1500 kg car traveling at 25 m/s in 3 s? b) Draw a graph of the braking power as a function of time.

SOLUTION a) The car's deceleration is found from linear kinematics:

$$v_1 = 0 = v_0 + a\Delta t$$

$$\Rightarrow a = -\frac{v_0}{\Delta t} = -\frac{25 \text{ m / s}}{3 \text{ s}} = -8.33 \text{ m / s}^2.$$

The braking force on the car, from the second law, has to be

$$F_{\text{brake}} = ma = (1500 \text{ kg})(-8.33 \text{ m / s}^2) = -12{,}500 \text{ N}.$$

The minus sign simply indicates a force *vector* directed opposite the motion. The question about "braking power," however, is not concerned with the sign—it generally makes most sense to think of power as a positive quantity. So we will consider just the *magnitude* of the braking force: $|F_{\text{brake}}| = 12{,}500$ N. The power generated by this force at time t is

$$P_{\text{brake}} = |F_{\text{brake}}v_{\text{car}}| = |F_{\text{brake}}|(v_0 + at)$$

$$= (12{,}500 \text{ N})(25 \text{ m / s} - 8.33t)$$

$$= (312{,}500)\left(1 - \frac{t}{3}\right) \text{ W}.$$

Notice that the braking power is a function of time. $P = Fv$ is a maximum at $t = 0$ when the velocity is a maximum. Its value at that time is 312,500 W = 419 hp. This tells us that, as the brakes are first applied, kinetic energy is being removed from the car *at a rate* of 312,500 joules per second. Most of this energy is going to end up in the car's brake drums and brake pads, raising their temperature.

b) A graph of the braking power as a function of time is shown in Fig. 9-23. It decreases linearly with time from P_{max} at $t = 0$ to zero at $t = 3$ s, when the car comes to rest. The point here is that the power decreases, even while the braking *force* stays constant, because the car's velocity decreases.

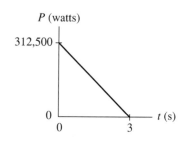

FIGURE 9-23 Braking power as a function of time.

EXAMPLE 9-13 A cable attached to a motor is used to pull a 100 kg steel box across a steel surface at constant speed. What is the tension in the cable and the power output of the motor if the box speed is a) 1 m/s and b) 3 m/s?

SOLUTION Figure 9-24 shows the pictorial model and free-body diagram for this problem. The force applied by the motor, through the cable, is the tension force $\vec{T}$. The motor does work $W = |\vec{T}|\Delta x$ that requires a power output $P = |\vec{T}|v$. The box is in dynamical equilibrium, because the motion is at constant velocity. You can see from the free-body diagram that $|\vec{N}| = |\vec{W}| = mg$ and $|\vec{T}| = |\vec{f}_k|$. Because $|\vec{f}_k| = \mu_k|\vec{N}|$, these pieces combine to give $|\vec{T}| = \mu_k mg$. The required tension is independent of the box's speed. This may be somewhat surprising, but it is a consequence of the friction model being independent of speed. The motor's power output is then

$$P = Tv = \mu_k mgv.$$

The coefficient of friction is found from Table 5-1 to be $\mu_k = 0.6$. Evaluating:

a) For $v = 1$ m/s: $T = 588$ N $P = 588$ W $= 0.788$ hp

b) For $v = 3$ m/s: $T = 588$ N $P = 1764$ W $= 2.36$ hp

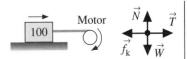

FIGURE 9-24 a) Box pulled by a motor and cable. b) The free-body diagram.

EXAMPLE 9-14   A 50 kg sprinter accelerates 50 m from rest in a time of 7 s at constant acceleration. What is her power output as a function of time?

SOLUTION Equation 9-37 gives the power at any instant of time as $P = Fv$. We must, however, be careful. The force exerted on the runner, in the particle model, is a friction force—the static friction force of the track on the soles of her shoes. Strictly speaking, it is the track that is doing work on her as she accelerates. But we know from Newton's third law that $|\vec{F}_{\text{track on runner}}| = |\vec{F}_{\text{runner on track}}|$. The track exerts a force on her only because she is able to exert a force against the track by transforming stored chemical energy into the mechanical energy of her legs. That is why we can ask about *her* power output rather than the track's power output.

From kinematics, $v = at$ and $\Delta x = \frac{1}{2}at^2$. From the latter, using $\Delta x = 50$ m and $t = 7$ s, we find $a = 2.04$ m/s². We can then find the sprinter's instantaneous velocity at any t. We also have Newton's second law, giving us $F = ma$. Combining these gives

$$P = Fv = (ma)(at) = ma^2 t$$

This shows that the sprinter's power output increases linearly with time, reaching a maximum output at $t = 7$ s of $P_{\text{max}} = 1460$ W ≈ 2 hp. Figure 9-25 shows a graph of the sprinter's power output as a function of time. Her peak output power is substantially higher than the 0.5 hp quoted above—but that value was for a *sustained* output. Sprinters cannot sustain this output, which is why the average speed in races begins to drop rapidly after about 100 m. Sprinters, in fact, cannot even maintain constant acceleration for a full 100 meters. They reach their maximum power output earlier in the race, as in this example, and after that their acceleration quickly declines.

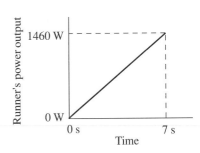

FIGURE 9-25 The power output as a function of time for a sprinter.

This last example has really taken us beyond the realm of single-particle mechanics. We needed to introduce the idea of stored chemical energy and its transformation into mechanical energy. While this is an important physical process, and the same ideas will apply to any system that "burns fuel" to acquire kinetic energy, a more complete understanding will require that we broaden our ideas of energy and energy transfer. That is the goal of the next two chapters.

Summary

◢ Important Concepts and Terms

energy

kinetic energy

potential energy

energy transformation

mechanical energy

work

state

joule

scalar product

dot product

net work

work-kinetic energy theorem

restoring force

elastic

displacement from equilibrium

Hooke's law

spring constant

power

watt

The concept of energy provides a new perspective on motion. The basic energy model that we are developing considers kinetic energy K, which is the energy of motion, and potential energy U, which is a "stored" form of energy. The sum of the kinetic and the potential energy is the mechanical energy

$$E = K + U.$$

In this chapter we have focused mainly on the ideas of *kinetic energy* and *work*. These are the first steps toward our broader goal of understanding energy and energy transfer. The relationships found in this chapter apply, strictly speaking, only to particles. The kinetic energy of a particle is

$$K = \tfrac{1}{2}mv^2.$$

Kinetic energy is one property, among many, that characterize the *state* of a system.

A force acting on a particle as it moves from $\vec{r}_i$ to $\vec{r}_f$ does an amount of work

$$W = \begin{cases} \vec{F} \cdot \Delta \vec{r} & \text{constant force} \\ \int_{x_i}^{x_f} F_x(x)dx & \text{variable force} \end{cases}$$

where $F_x(x)$ is the x-component of the force when the particle is at position x. Conceptually, work is the mechanical transfer of energy between a system (the particle) and the environment. Work *changes the state* of the system by changing its kinetic energy.

The work done on a particle was found to be related to the *change* in the particle's kinetic energy in a simple way:

$$\Delta K = W.$$

This relationship is called the *work-kinetic energy theorem*. In simple language, it tells us that if you push a particle in the direction of motion (positive work) the particle will go faster (increase in kinetic energy). If, on the other hand, you push against a particle opposite the direction in which it is moving (negative work) the particle will slow down (loss of kinetic energy). The work-kinetic energy theorem quantifies this common sense experience and allows us to make specific calculations.

Do not make the common error of memorizing "work equals force times distance." This is *not* the definition of work—it is merely a special case of work that happens to apply under special circumstances—namely, when the force is constant *and* when the force vector is parallel to the displacement vector. Learning this as the definition of work is guaranteed to lead to serious mistakes and confusion when you use work in the future.

Work is a method of energy transfer from the environment to or from a particle. The rate of energy transfer is called *power*. If work is the only form of energy transfer, then

$$P = \frac{dW}{dt} = \vec{F} \cdot \vec{v} \quad \text{(rate of energy transfer due to work)}.$$

Keep in mind, however, that this definition can, and will, be broadened to include the rate at which energy is transferred by other means.

Exercises and Problems

The ✐ icon indicates that these problems can be done on a Conservation Worksheet. Do not use the concept of potential energy for problems in this chapter.

Exercises

1. For the following vectors $\vec{A}$ and $\vec{B}$, show the vectors on a small grid and then evaluate the dot product $\vec{A} \cdot \vec{B}$.

   a. $\vec{A} = 2\hat{i} + 3\hat{j} \quad \vec{B} = -\hat{i} + 2\hat{j}$

   b. $\vec{A} = -3\hat{i} \qquad \vec{B} = 3\hat{i} + \hat{j}$

   c. $\vec{A} = 2\hat{i} - \hat{j} \quad \vec{B} = \hat{i} + 2\hat{j}$

2. Use the dot product to find the angle between the vectors $\vec{E} = 4\hat{i} + 2\hat{j}$ and $\vec{F} = 3\hat{i} + 4\hat{j}$.

3. Prove that for any vector $\vec{A}$, $|\vec{A}| = \sqrt{\vec{A} \cdot \vec{A}}$.

4. The law of cosines says that side C of the triangle in Fig. 9-26, with opposite angle θ, is given by
   $$C^2 = A^2 + B^2 - 2AB\cos\theta.$$
   Prove this law, using vectors and the dot product.

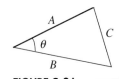

FIGURE 9-26

5. The graph in Fig. 9-27 shows the velocity-versus-time graph for a 2 kg object moving in one dimension under the influence of a single force. Determine the work done by the force during each of the five intervals AB, BC, CD, DE, and EF.

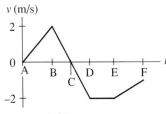

FIGURE 9-27

6. a. What is the kinetic energy of a 1500 kg car traveling at a speed of 30 m/s (≈65 mph)?

   b. From what height would the car have to be dropped to have this same amount of kinetic energy just before impact?

   c. Does your answer to b) depend on the car's mass?

7. A spring 10 cm long is hung from the ceiling. When a 2 kg mass is attached to the spring, the mass stretches the spring to a length of 15 cm.
   a.   What is the spring constant k?
   b.   How long is the spring when a 3 kg mass is suspended from it?
   c.   How much work is done by the spring on the 3 kg mass as the spring stretches from its initial 10 cm length to the length you found in b)?

8. A 5 kg mass, suspended from a spring scale, is slowly lowered onto a vertically oriented spring as shown in Fig. 9-28.
   a.   What does the scale read just before the mass contacts the spring?
   b.   The scale reads 20 N when the spring has been compressed by 2 cm. What is the value of the spring constant k?
   c.   At what compression will the scale read zero?

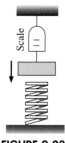

FIGURE 9-28

9. a.   How much power must be delivered by an elevator motor to lift a 1000 kg elevator a height of 100 m in 50 s at constant speed?
   b.   How much power must be delivered by you to push a 10 kg block of steel across a steel table at a steady speed of 1 m/s?

▲ Problems

✎ 10. Susan pulls a 10 kg wooden crate, initially at rest, across a wooden floor with a rope held at a 30° angle above the floor. The tension is kept constant at 30 N. What is the crate's speed after being pulled 3 m? Use the work-kinetic energy theorem.

✎ 11. A 1000 kg elevator accelerates upward at 1 m/s² for 10 m, starting from rest.
   a.   How much work is done on the elevator by gravity?
   b.   How much work is done on the elevator by the tension in the cable?
   c.   What is the kinetic energy of the elevator as it reaches 10 m? Use the work-kinetic energy theorem.
   d.   What is the speed of the elevator as it reaches 10 m?

✎ 12. A 50 kg ice skater is gliding along the ice, heading due north at 4 m/s. The ice has a small coefficient of static friction, but $\mu_k = 0$. Suddenly, the wind starts blowing *from* the northeast and exerts a force of 4 N on the skater.
   a.   What is the skater's speed after gliding 100 m in this wind? Use work and energy.
   b.   What is the minimum value of μ_s that is necessary for the skater to continue moving straight north?

✎ 13. Bob can throw a 500 g rock with a speed of 30 m/s. The distance through which his hand moves as he accelerates the rock forward, from rest until he releases it, is 1 m.
   a.   How much work does Bob do on the rock? Use the work-kinetic energy theorem.
   b.   How much force does Bob apply to the rock? (Assume the force is constant.)
   c.   What is Bob's maximum output power as he throws the rock?

14. Doug pushes a 5 kg crate up a 2 m high 20° frictionless slope by pushing it with a constant *horizontal* force of 25 N, as shown in Fig. 9-29. What is the speed of the crate as it reaches the top of the slope?

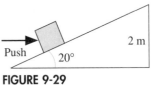

a. Solve this problem using the work-kinetic energy theorem.

b. Solve this problem using Newton's laws.

FIGURE 9-29

15. A 50 g rock is placed in a slingshot and the rubber band is stretched straight *down*, so that the rock can be launched straight up. The force of the rubber band on the rock is shown by the graph in Fig. 9-30. The coordinate is negative because the rock is being pulled down.

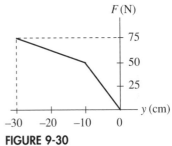

a. Does this rubber band obey Hooke's law? Explain, basing your answer on the graph.

b. What is the value of the spring constant k for *small* displacements of the rubber band?

c. The rubber band is stretched 30 cm and then released. How high does the rock go? (**Hint:** This is a two-part problem.)

FIGURE 9-30

16. An object of mass m, having initial conditions $x_i = 0$ and $v_i = v_0 > 0$, is subjected to the variable force $F(x) = F_0 \sin(cx)$ as it moves to the right along the x-axis. F_0 and c are constants.

a. What are the units of F_0?

b. What are the units of c?

c. What is the value of the force when the particle starts from position x_i?

d. At what position $x = x_{max}$ does the force first reach a maximum value? Your answer should be given in terms of the constant F_0 and c and other numerical constants.

e. Sketch a graph of F-versus-x from $x = 0$ to $x = x_{max}$.

f. What is the object's velocity upon reaching x_{max}? Your answer should be an algebraic expression involving m, v_0, F_0, and c.

17. Dan (80 kg) tries his first bungee jump from a bridge 100 m high. He attaches a 30 m long bungee cord to himself and to the bridge, then steps off the edge. A bungee cord, for practical purposes, is just a long spring, and this cord has a spring constant of 40 N/m.

a. Draw a pictorial model that shows Dan as he starts his jump, at the point where the cord first begins to stretch, and at the very bottom.

b. How far is Dan above the ground when the bungee cord begins to stretch?

c. How fast is Dan going at that point?

d. Draw a free-body diagram of Dan as the cord is stretching.

e. Use the work-kinetic energy theorem to find out how far Dan is above the ground when the cord reaches its maximum elongation. You will need to be careful with signs (y is decreasing) and with your formulation of Hooke's law. Don't forget to include *all* the forces doing work on Dan.

18. Figure 9-31 shows a velocity-versus-time graph of a 500 g object that starts at $x = 0$ and moves along the x-axis under the influence of a single force. Draw graphs of the following by calculating and plotting numerical values at $t = 0, 1, 2, 3,$ and 4 s. Then sketch lines or curves of the appropriate shape between the points. Make sure you supply an appropriate scale on both axes of each graph.

    a. Acceleration versus time.
    b. Position versus time.
    c. Kinetic energy versus time.
    d. Force versus time.
    e. Now draw a graph of force versus *position*. (**Hint:** This requires no calculations—just thinking carefully about what you have learned in parts a)–d).)
    f. Use your F-versus-t graph to determine the *impulse* delivered to the object during the time intervals: i) 0 s – 2 s and ii) 2 s – 4 s.

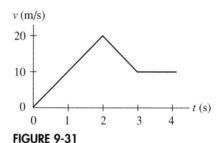

FIGURE 9-31

    g. Use your results from f) and the impulse-momentum theorem to determine the object's velocity at $t = 2$ s and at $t = 4$ s. Do your results agree with the velocity graph given?
    h. Use your F-versus-x graph to determine the *work* done on the object during the time intervals: i) 0 s – 2 s and ii) 2 s – 4 s.
    i. Use your results from h) and the work-kinetic energy theorem to determine the object's velocity at $t = 2$ s and at $t = 4$ s. Do your results agree with the velocity graph given?
    j. Draw a graph of kinetic energy versus position.

19. A Porsche 944 Turbo has a rated power of 217 hp (162 kW) and a mass, including driver, of 1480 kg.

    a. What is the maximum acceleration of the Porsche on a concrete surface ($\mu_s = 1.0$) if two-thirds of the car's weight is over the drive wheels? (**Hint:** What is the propulsion force?)
    b. If the Porsche accelerates at a_{max}, how long does it take until it reaches the maximum power output?
    c. What is its speed at that instant of time?
    d. What power would be required for the Porsche to still be accelerating at a_{max} at a speed of 60 mph?

20. A spring-loaded gun is used to launch a 10 g plastic ball. Pushing the ball into the barrel compresses the spring, of spring constant $k = 10$ N/m, by 10 cm. When the trigger is pulled, the spring is released and shoots the ball back out. What is the maximum power delivered by the spring to the ball? At what point in the motion does this occur? Use the following steps to solve this problem.

    a. Draw a pictorial model. Place the origin at the point where the ball rests against the compressed spring. (Note: This is a different coordinate system than the one used in Example 9-9.)
    b. Write Hooke's law for the spring force $F_{sp}(x)$ on the ball when the ball is at position x. (It is *not* $-kx$ in this problem.)

c. Calculate the work done by the spring on the ball as the ball moves from its starting position to position x.
d. Use your result from c) to determine the ball's velocity $v(x)$ as a function of its position x. Verify that your result agrees with Example 9-9 when $x = x_1 = 10$ cm.
e. Calculate and draw a graph of $v(x)$ for x from 0 to 10 cm.
f. Now use your results from b) and d) to find an expression $P(x)$ for the instantaneous power delivered to the ball at position x. This is the *rate* at which the ball's energy is increasing at x.
g. Calculate and draw a graph of $P(x)$ for x from 0 to 10 cm.
h. Give an explanation of why the power has the values it does at $x = 0$ and $x = 10$ cm.
i. Determine *precisely* (not just from the graph) the maximum power and the position at which it occurs. (**Hint:** This is a calculus problem.)

[**Estimated 10 additional problems for the final edition.**]

Chapter **10**

Concepts of Energy II: Potential Energy and Conservation

L O O K I N G B A C K | Sections 7.1–7.3; 8.4; 9.7–9.9

10.1 Interacting Particles

[**Photo suggestion: A compressed spring.**]

The work-kinetic energy theorem of Chapter 9 is an important statement about the motion of a *single* particle. The real world, however, is comprised of many interacting particles. If the concept of energy is to live up to its potential, we need to bring some ideas about interacting particles into the picture. That is what we will do in this chapter. For the most part, we will focus on just two interacting "particles," although we will use the term broadly. We will extend the ideas introduced here to many particles in the next chapter.

We have looked at interacting particles both from the standpoint of Newton's third law, with its action/reaction pairs, and with the concept of momentum. Each of these approaches required us to think carefully about which particles to include within "the system" and which to include as part of the environment. Recall, from Chapter 8, that a wise choice of the system allows us to use the law of conservation of momentum. Many problems that looked hopeless from a force and acceleration viewpoint turned out to be easy to understand when seen from a conservation of momentum perspective. Now we will be looking for a similar conclusion with regard to energy—namely, that an appropriately chosen system will allow us to solve motion problems by using a conservation law for energy.

Consider the simple example, of a falling rock, shown in Fig. 10-1a. If we select just the rock as a single-particle system, then the weight force is an external force exerted on the system by the environment of the earth. This force does work on the rock, causing the rock's kinetic energy to

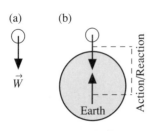

FIGURE 10-1 a) A falling rock considered as a single particle. b) The rock+earth as a combined system of two particles.

change. Newton's third law, however, suggests that we consider the rock and the earth *together* as a single system of interacting particles. From this perspective, shown in Fig. 10-1b, the gravitational forces are seen as an action/reaction pair of interaction forces. There are no external forces, so *no* work is done on the system *as a whole*.

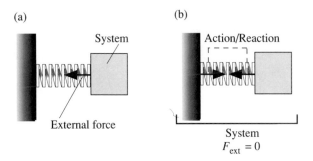

(a)

(b)

System

Action/Reaction

External force

System
$F_{ext} = 0$

FIGURE 10-2 a) The object as a single-particle system experiencing an external force. b) The object+wall as a combined system with $F_{ext} = 0$.

Figure 10-2 shows a similar situation. If an object attached to a spring is, by itself, considered the system, then the spring force is an external force that does work on the object and changes its kinetic energy. An alternative perspective, however, is to see the object and the wall together as a single system in which the "particles" interact with each other through the spring force. There is *no* external work done on this larger system.

In both these examples no work is done on the system as a whole. Yet the kinetic energy of the system definitely changes. The work-kinetic theorem $\Delta K = W$, which was derived for single-particle motion, clearly fails to apply to systems of particles. Our goal in this chapter is to expand the concept of energy to systems of interacting particles.

10.2 Where Did the Kinetic Energy Go? Stored Energy

Toss a ball straight up into the air, then let it return to its initial height—as shown in Fig. 10-3. (The "up" and "down" are displaced for clarity in the figure, but the motion is actually straight up and down along the same line.) If the ball alone is chosen as the system, then the force of gravity $\vec{W}$ does work on it. As the ball rises, the displacement Δy is positive while W_y, the y-component of the weight $\vec{W}$, is negative. Thus the work done by gravity is negative ($W_{grav} = W_y \Delta y < 0$) and the ball loses kinetic energy ($\Delta K < 0$). (Be careful to distinguish work W from the component W_y of the weight vector $\vec{W}$.) As the ball falls, W_y and Δy are both negative, so $W_{grav} > 0$ and the ball gains kinetic energy. Because the work done on the ball going up is equal in magnitude but opposite in sign from the work done on it as it falls, the *net* work done during the round trip is zero. We can conclude that $\Delta K = 0$ for the round trip and that $K_2 = K_0$. All of this is a straightforward application of the work-kinetic energy theorem of the last chapter.

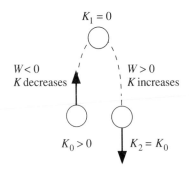

$K_1 = 0$

$W < 0$
K decreases

$W > 0$
K increases

$K_0 > 0$

$K_2 = K_0$

FIGURE 10-3 A ball moving under the influence of gravity, which does work to change the ball's kinetic energy.

Suppose, however, that we consider the ball and the earth *together* as being the system. Now the kinetic energy has to include both the ball and the earth: $K = K_{ball} + K_{earth}$. The ball+earth system is isolated, so its total momentum is conserved. The momentum is zero before you toss the ball. If you toss the ball with velocity v_{ball}, conservation of momentum requires

$$m_{ball} v_{ball} + m_{earth} v_{earth} = 0$$

$$\Rightarrow v_{earth} = -\frac{m_{ball}}{m_{earth}} v_{ball}.$$

The earth recoils in the $-y$-direction, but *very* slowly because the ratio m_{ball}/m_{earth} is extremely tiny. The earth's kinetic energy due to its recoil speed is

$$K_{earth} = \tfrac{1}{2} m_{earth} v_{earth}^2 = \tfrac{1}{2} m_{earth} \left(-\frac{m_{ball}}{m_{earth}} v_{ball} \right)^2$$

$$= \frac{m_{ball}}{m_{earth}} \left(\tfrac{1}{2} m_{ball} v_{ball}^2 \right) \tag{10-1}$$

$$= \frac{m_{ball}}{m_{earth}} K_{ball}$$

$$\ll K_{ball} \text{ because } \frac{m_{ball}}{m_{earth}} \ll 1.$$

In other words, the earth's kinetic energy is completely negligible in comparison with the ball's kinetic energy, so $K = K_{ball}$ is an extremely good approximation. Thus we are justified in considering only the ball's motion when we talk about the kinetic energy of the system. But keep in mind that you will need to include the kinetic energy of *all* the particles in the system if their masses are similar.

There are no external forces acting on the ball+earth system, so the work done on the system is zero. But the system certainly had kinetic energy at the start of the ball's motion, zero kinetic energy at the top of its trajectory, and kinetic energy again when it returned. Where did the kinetic energy "go" as the ball ascended, and from where was it recovered as the ball fell? Because $K_2 = K_0$, all of the kinetic energy that appears to have been "lost" on the way up is fully recovered on the way back down. This suggests that the energy was not lost at all, but was somehow "saved" or "stored" for later use.

Recall Jose's model of the monetary system from Chapter 9. Jose found that money comes in two basic forms: liquid assets and saved assets. These two types of assets can be converted back and forth. Saved assets, in particular, have the *potential* to become liquid assets if the conditions are right. Our basic energy model of Fig. 9-2 in Chapter 9 (which you might want to review) is exactly analogous. It says that energy comes in two basic forms: motional energy and stored energy. The motional energy we have named *kinetic energy*. The stored energy has the *potential* to be converted into kinetic energy. This stored energy is called **potential energy**. The symbol for potential energy is U, and its units—like those of kinetic energy and work—are joules.

It is important to keep in mind that the basic energy model is a *hypothesis*. We have suggested—with little evidence—that there is a physical quantity called *energy* that behaves very much like money. Now our goal is to provide support for this energy hypothesis. There are two essential aspects to the basic energy model. First, the sum of the kinetic and potential energy remains constant in the absence of external influences: $K + U$ = constant. Second, kinetic and potential energy can be converted back and forth into each other. When dealing with energy, we usually say that kinetic energy is *transformed* into potential energy ($K \rightarrow U$) or that potential energy is transformed into kinetic energy ($U \rightarrow K$). But the total quantity $K + U$ does not change as these conversions, or transformations, occur.

Figure 10-4 once again shows a ball being tossed straight up, but this time from the perspective of the ball+earth system. Because the net work on the system is zero, we need to analyze the motion with the tools of kinetic and potential energy. The ball+earth system starts with all kinetic energy. The energy is transformed ("saved") to potential energy ($K \rightarrow U$) as the ball rises. At the top of its trajectory, the conversion is complete and the system has only potential energy. The transformation goes the other direction as the ball falls ($U \rightarrow K$), with the energy of motion being recovered from the stored energy. Because the conversions are 100% efficient, the final kinetic energy K_2 equals the initial kinetic energy K_0. The value of the *total* quantity $K + U$ is the same at *all* points along the trajectory.

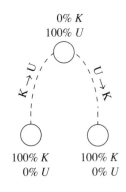

Notice that the potential energy depends on the separation of the two "particles." A ball 20 m above the ground has more *potential* for gaining kinetic energy than does a ball only 2 m above the ground. In more formal terms, we can say that the potential energy depends on the *configuration* of the two interacting particles. This is a general property of potential energy, one you will see each time a new potential energy is introduced. As a consequence, we can think of potential energy as an "energy of position," an energy that depends on "where things are." This is in contrast the "energy of motion" that depends on "how fast things move."

FIGURE 10-4 Ball toss seen as an energy transformation within the ball+earth system.

We can use the idea of energy transformation to understand the pendulum shown in Fig. 10-5. There is an additional force on the pendulum, namely the tension in the string, but that force does *no* work because it is always perpendicular to the direction of motion. So even though the pendulum moves along the arc of a circle, rather than straight up and down, the only force doing work on the pendulum is gravity. The combined pendulum+earth system is simply transforming potential energy to kinetic energy as the pendulum falls, then kinetic energy back to potential energy as it rises again. This process is repeated over and over.

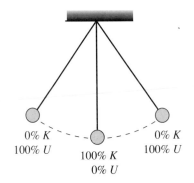

Figure 10-6 shows another example of motional energy being stored and later recovered. In this case, a moving rock strikes the elastic rubber band of a slingshot and stretches the rubber (elastic restoring forces) until the rock momentarily comes to rest ($K = 0$). But the kinetic energy is not lost. The energy is stored as potential energy in the *system* of rock+slingshot. The rock does not remain at

FIGURE 10-5 A pendulum transforms potential energy to kinetic energy and back to potential energy.

rest, of course, but is launched again as the stored energy is reconverted into kinetic energy: $U \rightarrow K$. The total value of $K + U$ stays constant *throughout* the interaction. Notice that the potential energy depends on the configuration of the system—how far the rubber band is stretched. The potential energy of elasticity, like the potential energy of gravity, is an energy of position. This example suggests that there are various kinds, or forms, of stored energy: gravitational potential energy, elastic potential energy, and perhaps others.

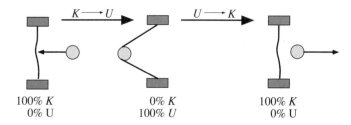

FIGURE 10-6 Energy transformation in a rock+slingshot system.

The sum of the kinetic energy and the potential energy of a system is called the **mechanical energy** $E_{\text{mech}} = K + U$. Here K is the total kinetic energy of all the particles in the system and U is the potential energy stored in the system. Our examples thus far suggest that the mechanical energy of an isolated system does not change. The kinetic energy and the potential energy can change, as they are transformed back and forth into each other, but their sum remains constant. We can express the unchanging value of E_{mech} as

$$E_{\text{mech}} = K + U = \text{constant}$$
$$\Rightarrow \Delta E_{\text{mech}} = \Delta K + \Delta U = 0.$$

(10-2)

Is this relationship always true?

Consider a box that is given a shove and then slides along the floor until it stops, as shown in Fig. 10-7. The two-particle box+earth system starts with kinetic energy that is subsequently lost. But in this case the energy is *not* saved or stored for future use. Unlike the ball at the top of its arc, the box does *not* start moving again. When the box stops moving, it does *not* have the potential to start again by transforming potential energy into kinetic energy. So the relationship $\Delta E_{\text{mech}} = 0$ is clearly not true in every situation. The box+earth system does not gain potential energy as its kinetic energy decreases.

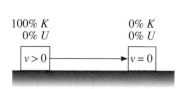

FIGURE 10-7 The kinetic energy is *not* saved when a box slides across a floor.

You know, of course, that after the box slides across the floor, both the box and the floor are slightly warmer than before. The kinetic energy has not been transformed into potential energy, but *something* has happened. We've already noted that there are different kinds of energy. Perhaps in the case of friction the kinetic energy is transformed into some form of energy other than potential energy.

The basic energy model hypothesis that $\Delta E_{mech} = \Delta K + \Delta U = 0$ for an isolated system seems valid for some situations but not for others. Our task, if the model is to be useful, is to discover:

1. Under what conditions is E_{mech} conserved?
2. What happens to the energy when E_{mech} isn't conserved?
3. How do you calculate the potential energy U?

There are many parts to the energy puzzle, and we must put them together piece by piece. We will answer the first and last of these questions in this chapter but will defer the second question until the next chapter.

10.3 Gravitational Potential Energy

[**Photo suggestion: A roller coaster car starting down a steep hill.**]

A ball tossed into the air seems to convert its kinetic energy into potential energy, then back to kinetic energy as it falls. Perhaps a closer look at the gravitational force will give us some insight for answering the questions at the end of the last section.

Consider the case of an object of mass m interacting with the earth (or other large planet) via an interaction force $|\vec{F}_{earth\ on\ mass}| = mg$. Our system will consist of two "particles," the object and the earth. It may seem a little strange to treat the earth as a particle, but actually we have been doing so all along. As long as interior processes of the earth do not enter into the problem, and as long as the object stays above the surface, the earth acts *as if* it were a single particle of very large mass exerting a gravitational force on the smaller object.

Where should we begin? Figures 10-3 and 10-4 give a good hint. If we choose the object alone as a single-particle system, then its kinetic energy changes as a consequence of the work done by the gravitational force. Alternatively, we can choose the two-particle object+earth system. In this case, potential energy is stored as the object rises $(K \rightarrow U)$ and is then converted back to kinetic energy as the object falls $(U \rightarrow K)$. This suggests that we start by calculating the work done on the object by the earth's gravity—a single-particle viewpoint—and that we then step back to see if we can view that work as a stored energy. So that is what we will be doing in the next few paragraphs.

Consider an object of mass m at the top of a frictionless hill of angle θ and height h, as shown in Fig. 10-8. How much will its kinetic energy *change* as it slides from the top to the bottom of the hill? (Note that we are *not* assuming $v_0 = 0$. The object may have some initial velocity.) Considering the object as a single particle, its change of kinetic energy is governed by the

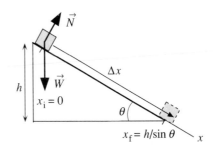

FIGURE 10-8 The work-kinetic energy theorem applied to an object sliding down a frictionless hill.

work-kinetic energy theorem: $\Delta K = W$. The normal force of the hill is perpendicular to the direction of motion so it does no work, as we saw in the last chapter. Because the object moves along the x-axis, we only need the x-component of the gravitational force: $W_x = mg\sin\theta$. The object's displacement Δx is related to the hill's height h by $\Delta x = h/\sin\theta$. Thus the work done by gravity as the object slides to the bottom is

$$W_{\text{gravity}} = \vec{W} \cdot \Delta \vec{r} = W_x \Delta x = (mg\sin\theta)\left(\frac{h}{\sin\theta}\right) = mgh. \qquad (10\text{-}3)$$

According to the work-kinetic energy theorem, the object's change in kinetic energy is

$$\Delta K = W_{\text{gravity}} = mgh. \qquad (10\text{-}4)$$

We find, surprisingly, that the change in kinetic energy does *not* depend on the slope of the hill—that is, it does not depend on the angle θ. The object could slide down either a short, steep hill or a long, not-so-steep hill. ΔK will be the same in both cases *if* both hills have the same height h.

Now suppose we simply *dropped* the object from height h. Then the y-component of the weight force is $W_y = -mg$ and the displacement is $\Delta y = -h$. The work-kinetic energy theorem gives $\Delta K = W_{\text{gravity}} = (-mg)(-h) = mgh$. In other words, ΔK for a falling object is the same as ΔK for an object sliding down a frictionless incline of the same height. The object travels farther on the incline, but it is going just as fast at the bottom.

The object gains kinetic energy $\Delta K = mgh$ as it descends height h, regardless of whether it falls straight down or slides along a slope of any angle. Is this only true for motion along a straight line? To decide this, let the object descend from height h along an arbitrary, but still frictionless, curved surface, as shown in Fig. 10-9. The normal force of the surface is perpendicular *at every point* to the motion, despite the fact that the surface curves, so it does *no* work on the object. Therefore, the object's kinetic energy changes because of the work done by the force of gravity alone.

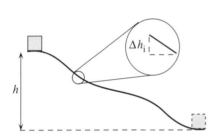

FIGURE 10-9 An object sliding down a frictionless curved surface. The inset shows one of many small "straight" segments of height Δh_i.

Let's subdivide the curved trajectory into tiny segments of height Δh_i, such that each segment is a very good approximation to a straight line. According to Eq. 10-3, the work done on the object by gravity as it moves through segment i is

$$W(\text{segment i}) = mg\,\Delta h_i.$$

The total work done by gravity as the object descends, which is equal to ΔK, is just the sum of the work done in each segment:

$$\Delta K = W_{\text{gravity}} = \sum_i W(\text{segment i})$$
$$= \sum_i mg\Delta h_i = mg\sum_i \Delta h_i = mgh. \qquad (10\text{-}5)$$

It seems—quite against common sense!—that the work done by the force of gravity is *independent of the path* by which the object descends. The kinetic energy change $\Delta K = mgh$ is the same along *any* trajectory that descends from the same height h—regardless of whether the object falls, slides down, or even slides along a curved, bumpy (but frictionless) surface.

Having examined the situation from the perspective of a single particle and the work-kinetic energy theorem, let us now step back and view the object+earth as a single system with *no* external forces. The system starts with $K = 0$ but ends up with $K = mgh$, regardless of the path followed by the object as it falls. This analysis strongly suggests that the object+earth system starts with a *stored energy* in the amount $U = mgh$ and that this stored energy is converted into kinetic energy as the object moves to the lower elevation. Thus the object loses potential energy $\Delta U = -mgh$ but gains kinetic energy $\Delta K = mgh$, where $h = y_i - y_f$. (Notice the order of the y's, because h is positive.) Mathematically we can write

$$\Delta K = \tfrac{1}{2} mv_f^2 - \tfrac{1}{2} mv_i^2 = mgh = mg(y_i - y_f)$$
$$\Rightarrow \tfrac{1}{2} mv_f^2 + mgy_f = \tfrac{1}{2} mv_i^2 + mgy_i \tag{10-6}$$
$$\Rightarrow K_f + mgy_f = K_i + mgy_i.$$

That is, the final value of the quantity $K + mgy$ is exactly equal to the initial value of $K + mgy$. This result is in perfect agreement with the basic energy model if we define the **gravitational potential energy** as

$$U_{grav} = mgy. \tag{10-7}$$

With this definition, we have

$$K_f + U_f = K_i + U_i$$
$$\Rightarrow K + U = E_{mech} = \text{constant}. \tag{10-8}$$

We have, as we had hoped, found an expression for the amount of energy stored by the object+earth system. If the object has vertical position y, the stored energy is $U_{grav} = mgy$. Notice that U_{grav} is an energy of position, not an energy of motion.

EXAMPLE 10-1   Vicky runs forward with her sled at 2 m/s. She hops onto the sled at the top of a 5 m high, very slippery slope. What is her speed at the bottom?

SOLUTION   Figure 10-10 shows a simple pictorial model in which we see a "before" of Vicky at the top of the slope, moving at an initial 2 m/s, and an "after" as she reaches the bottom with speed v_2. We are not told the angle of the slope, or even if it is a straight slope. The

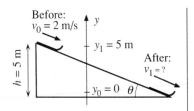

FIGURE 10-10 Pictorial model of Vicky's sled.

potential energy change depends only on the height Vicky descends and *not* on the trajectory she follows. We know from Eq. 10-8 that the total quantity $K + U$ is the same at the bottom of the hill as it was at the top. Thus

$$K_2 + U_2 = K_1 + U_1$$
$$\Rightarrow \tfrac{1}{2} mv_2^2 + mgy_2 = \tfrac{1}{2} mv_1^2 + mgy_1$$
$$\Rightarrow v_2 = \sqrt{v_1^2 + 2g(y_1 - y_2)} = \sqrt{v_1^2 + 2gh} = 10.1 \text{ m/s}.$$

Notice in this example that we did not need to know the mass of either Vicky or the sled. The mass appears in all terms of the energy equation and thus cancels out. This is an important observation.

The Zero of Potential Energy

Our expression for the gravitational potential energy $U_{grav} = mgy$ seems straightforward. But you might notice that the value of U depends on where you choose to put the origin of your coordinate system. Consider Fig. 10-11, where Betty and Billy are attempting to

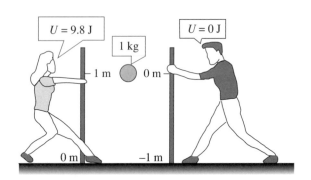

FIGURE 10-11 Betty and Billy attempting to determine the potential energy of a rock, using coordinate systems with different origins.

determine the potential energy of a 1 kg rock that is 1 meter above the ground. Betty chooses to put the origin of her coordinate system on the ground, measures $y_{rock} = 1$ m, and quickly computes $U = mgy = 9.8$ J. Billy, on the other hand, read Chapter 1 very carefully and recalls that it is entirely up to him where to locate the origin of his coordinate system. So he places his origin next to the rock, measures $y_{rock} = 0$ m and declares that $U = mgy = 0$ J!

How can the potential energy of one rock at one position in space have two different values? The source of this apparent difficulty comes from our interpretation of Eqs. 10-4 and 10-5. We learned that $\Delta K = mgh$, and this led us to propose that $U_{grav} = mgy$. But all we are *really* justified in concluding from $\Delta K = mgh$ is that the potential energy *changes* by $\Delta U = -mgh = mgy_f - mgy_i$. To go beyond this and claim $U = mgy$ is certainly consistent with a definition of $\Delta U = mgh$, we could just as easily claim that $U = mgy + C$ where C is a constant. Suppose, for example, that Allan measures the potential energy at the initial and final positions of an object's motion to be U_i and U_f. Anne, however, decides to add a constant to all of Allan's measurements and declares $U'_i = U_i + C$ and $U'_f = U_f + C$. Is there any reason that we should prefer one of these sets of potential energies over the other? The answer is "no," because *both* Anne's and Allan's potential energies give the same value for the *change* in potential energy: $\Delta U' = \Delta U$. Because it is only ΔU that is defined, both Anne's and Allan's potential energies are equally valid.

That is what happens with Betty and Billy in the example of Fig. 10-11. No matter where the rock is located, Betty's value of y will always equal Billy's value plus 1 meter. Consequently, her value of the potential energy will always equal Billy's value plus 9.8 J. While they disagree about the value of U, *both* will calculate exactly the *same* value for ΔU if the rock changes position.

This issue highlights an important point about energy. Whenever we use energy to analyze a physical situation, we are only interested in how the energy *changes*. The precise quantity of energy at one instant of time is not of interest and, as we have just seen, not even well defined. But the *change* of energy, $\Delta(K + U)$ *is* well-defined, and only the change will ever enter into a problem. A somewhat more formal way of saying this is that only ΔU is "physically meaningful," not U itself. As a consequence, you can place the origin of your coordinate system—and thus the "zero of potential energy"—wherever you choose and be assured of getting the correct answer. Your roommate can choose a different origin and have different values of U at each point, but he or she will arrive at the same answer. The following example illustrates this.

EXAMPLE 10-2 The rock shown in Fig. 10-11 is released from rest. Use both Betty's and Billy's perspective to calculate its speed upon impact with the ground.

SOLUTION If we choose the combined system of rock+earth there are no external forces, so

$$\Delta(K + U) = \Delta K + \Delta U = 0$$

$$\Rightarrow \Delta K = \tfrac{1}{2}mv_f{}^2 - \tfrac{1}{2}mv_i{}^2 = \tfrac{1}{2}mv_f{}^2 = -\Delta U.$$

Billy and Betty both agree that $K_i = 0$, because the rock was released from rest, and we have made use of that fact. According to Betty,

$$U_i = mgy_i = 9.8 \text{ J and } U_f = mgy_f = 0 \text{ J}$$

$$\Rightarrow \Delta U = U_f - U_i = -9.8 \text{ J}.$$

Thus the speed of the rock upon impact is

$$\left| v_f \right| = \sqrt{\frac{-2\Delta U}{m}} = \sqrt{\frac{-2(-9.8 \text{ J})}{1 \text{ kg}}} = 4.43 \text{ m / s}.$$

Billy, however, claims that his measurements of y give

$$U_i = mgy_i = 0 \text{ J and } U_f = mgy_f = -9.8 \text{ J}.$$

These are clearly different than Betty's values of U. Nonetheless, Billy finds

$$\Delta U = U_f - U_i = -9.8 \text{ J}$$

$$\Rightarrow \left| v_f \right| = \sqrt{\frac{-2\Delta U}{m}} = \sqrt{\frac{-2(-9.8 \text{ J})}{1 \text{ kg}}} = 4.43 \text{ m / s}.$$

Despite their disagreement over the value of U itself, both Betty and Billy arrive at the same answer for $|v_f|$. The reason, which we repeat for emphasis, is that only ΔU has physical significance, not U itself, and both Betty and Billy had the same value for ΔU. (Remember, the expression ΔQ *always* means $Q_{\text{final}} - Q_{\text{initial}}$, for any quantity Q.) The negative value of ΔU in this example was necessary to cancel the minus sign inside the square root, giving a proper value for v. One of the most common errors in energy problems is being careless with what is being subtracted from what. Sign errors introduced this way can wreak havoc later in the problem.

Before we move on it is worth noting that U can be negative, as U_f was for Billy in this example. All a negative value for U means is that the system has *less* potential for motion in that configuration than it does in the configuration defined to be $U = 0$. Contrast this with kinetic energy, which *cannot* be negative. The difference is because we directly defined K to be $\tfrac{1}{2}mv^2$, which cannot be negative, whereas for the potential energy we only defined ΔU.

10.4   Conservative and Nonconservative Forces

For an object that moves due to the earth's gravitational force, the *combined* object+earth system obeys $E_{\text{mech}} = K + U = $ constant, where $U = mgy$ is a stored or "potential" energy. Our *interpretation* of this result is that kinetic and potential energy are converted back and forth, with no loss, as the object moves. This is a useful interpretation because we can connect "before" with "after" by the relationship $(K + U)_{\text{after}} = (K + U)_{\text{before}}$. It is important to recognize that U is the potential energy of the object+earth *system*, not of the object alone. This is a consequence of treating gravity as an *interaction force* internal to the system.

We made a big deal when developing the gravitational potential energy about the fact that the work done by the force of gravity was *independent of the path* followed by the object. This path independence of the work is an essential ingredient of any interaction force for which there is a potential energy. To see this, consider an object that can move from point A to point B along two possible paths, 1 and 2. Potential energy is an energy of position, so the system has one value of the potential energy when the object is at A and a different value when the object is at B. The change ΔU has nothing to do with path 1 and path 2. Suppose, though, that the work differs along these two paths so that the object's final kinetic energy $K_{\text{B}1}$ when it follows path 1 is different than its final kinetic energy $K_{\text{B}2}$ when it follows path 2. But if this is the case, then it's not possible to have $\Delta K + \Delta U = 0$ along both paths. For potential energy to be a meaningful concept, we must require the work W done by the interaction force be independent of the path.

If the work done on a particle as it moves from an initial to a final position is independent of the path it follows, we define the force doing the work to be a **conservative force**. It is possible, although we will skip it, to give a mathematical proof that a potential energy can be associated with any conservative force. Gravity is a conservative force, which is why we could establish a gravitational potential energy, and you will meet others.

For the gravitational force we found that $\Delta U = mg(y_{\text{f}} - y_{\text{i}}) = -W_{\text{grav}}(\text{i} \rightarrow \text{f})$. That is, the change in potential energy as the object moved from the initial to the final position was defined as the *negative* of the work done by the interaction force. This turns out to be the general procedure to follow with any force. For a *system* of particles that interact via *conservative* forces, we can define the *change* in the potential energy, between configuration i and configuration f, to be

$$\Delta U = U_{\text{f}} - U_{\text{i}} = -W_{\text{c}}(\text{i} \rightarrow \text{f}) \qquad (10\text{-}9)$$

where $W_{\text{c}}(\text{i} \rightarrow \text{f})$ is the work done by the conservative forces.

The minus sign in the definition of Eq. 10-9 is an important part of our interpretation of energy. According to the work-kinetic energy theorem, a positive value of $W_{\text{c}}(\text{i} \rightarrow \text{f})$ corresponds to an increase in kinetic energy. But if the total quantity $K + U$ is to remain unchanged, ΔK can be positive (increasing K) only if ΔU is negative (decreasing U).

Not all forces are conservative forces. Consider, for example, a crate that slides along the floor under the influence of friction. If we choose crate+floor as the system,

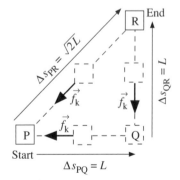

FIGURE 10-12 Top view of a crate sliding from P to R. The work done by friction *does* depend on the path that is followed.

then friction is an interaction force between them. Can we find a potential energy to describe this interaction?

Figure 10-12 shows a *top view* of a crate moving across a horizontal floor from P to R. The friction force, we know from experience, has magnitude $\mu_k|\vec{N}| = \mu_k mg$ and points opposite the direction of motion. It thus does *negative* work on the crate. Specifically, the work done by the friction force as the crate moves a distance Δs is $W = F_s\Delta s = -\mu_k mg\Delta s$. Is the work done by friction in moving directly from P to R, along the diagonal of the square, equal to the work done by following the indirect path PQR along the sides? In moving directly from P to R, a distance $(\Delta s)_{PR} = \sqrt{2}L$, the work is

$$W(P \rightarrow R) = -\mu_k mg(\Delta s)_{PR} = -\sqrt{2}\mu_k mgL. \tag{10-10}$$

In moving $P \rightarrow Q \rightarrow R$, the work is

$$\begin{aligned} W(P \rightarrow Q \rightarrow R) &= W(P \rightarrow Q) + W(Q \rightarrow R) \\ &= -\mu_k mg(\Delta s)_{PQ} + -\mu_k mg(\Delta s)_{QR} \\ &= -\mu_k mgL - \mu_k mgL \\ &= -2\mu_k mgL \\ &\neq W(P \rightarrow R). \end{aligned} \tag{10-11}$$

So here we have a case where the work done is *not* independent of the path followed. The work done by friction is directly proportional to the distance traveled, so more work (negative) is done on the longer path around the outside of the square.

An interaction force for which the work is *not* independent of the path is called a **nonconservative force**. It is not possible to define a potential energy for a nonconservative force. Friction is a nonconservative force, so friction is *not* an interaction force for which we can define a potential energy. This makes sense if you think about the fact that you cannot recover the kinetic energy lost to friction as the crate slides to a halt.

One thing that we notice when friction is present is that the sliding surfaces get warm. This suggests that *something* happens as the kinetic energy decreases, but that something cannot be converted back to kinetic energy. This is an idea we will look at in the next chapter, and again in more detail in thermodynamics, where we will see that the kinetic energy has been transformed to yet another type of energy called *thermal energy*. For now, we simply want to note that nonconservative forces are generally **dissipative forces**, meaning that the kinetic energy is "dissipated" into thermal energy rather than being stored as potential energy. All types of friction—sliding friction, air drag, or anything else—are nonconservative forces.

Consider a system of particles that interact via both conservative forces and nonconservative forces. As the system moves from configuration i to configuration f, the conservative forces do work W_c while the nonconservative forces do work W_{nc}. The total work done by *all* the interaction forces is $W_{int} = W_c + W_{nc}$. Therefore the change in the system's kinetic energy ΔK, as determined by the work-kinetic energy theorem, is

$$\Delta K = W_{int} = W_c(i \rightarrow f) + W_{nc}(i \rightarrow f). \tag{10-12}$$

But the work done by the conservative forces can be associated with a potential energy, as defined by Eq. 10-9: $W_c(i \rightarrow f) = -\Delta U$. With this definition of potential energy Eq. 10-12 becomes

$$\Delta K + \Delta U = \Delta E_{mech} = W_{nc}. \tag{10-13}$$

In the absence of dissipative, nonconservative forces the mechanical energy $E_{mech} = K + U$ is conserved: $\Delta E_{mech} = 0$. But nonconservative forces lead to a loss of mechanical energy that cannot be recovered. The ball tossed into the air is transforming kinetic energy into potential energy, but the sum $K + U$ isn't changing. The box that slides to a halt, however, is acted upon by nonconservative forces and really is losing mechanical energy. You can, though, still use energy concepts to analyze its motion—with Eq. 10-13—if you compute the work done by the nonconservative forces.

EXAMPLE 10-3   Vicky couldn't wait until winter to use her sled. In July she runs forward with the sled at 2 m/s toward her favorite hill. She hops onto the sled at the top of a 5 m high hill that has a 15° slope. The coefficient of kinetic friction between the sled and grass is 0.2. What is Vicky's speed at the bottom of the hill?

SOLUTION This is Example 10-1 with the addition of a nonconservative friction force. We solved Example 10-1 by using $\Delta K + \Delta U = 0$, but now we need to use Eq. 10-13: $\Delta K + \Delta U = W_{nc}$. In our analysis of the crate in Fig. 10-12 we found that the work done by friction in a distance Δs is $W = F_s \Delta s = -\mu_k mg \Delta s$. The pictorial model of Vicky's sled in Fig. 10-10 is still valid, but we are now also given the information that $\theta = 15°$. The length of the slope is thus

$$\Delta s = \frac{h}{\sin\theta} = \frac{5 \text{ m}}{\sin 15°} = 19.32 \text{ m}.$$

Energy Eq. 10-13 is

$$\Delta K + \Delta U = (K_2 - K_1) + (U_2 - U_1) = W_{nc}$$

$$\Rightarrow \tfrac{1}{2}mv_2^2 - \tfrac{1}{2}mv_1^2 + mgy_2 - mgy_1 = -\mu_k mg\Delta s.$$

Notice that the mass again cancels. Solving for v_2 gives

$$v_2 = \sqrt{v_1^2 + 2g(y_1 - y_2) - 2\mu_k g\Delta s} = \sqrt{v_1^2 + 2gh - 2\mu_k g\Delta s} = 5.12 \text{ m/s}.$$

Vicky's final speed in July is significantly less than the 10.1 m/s she achieved on a snow-covered slope in January. She still lost potential energy $\Delta U = -mgh$, but this time not all of her potential energy was transformed into kinetic energy. A portion of it was lost because of the work done by the nonconservative friction force.

10.5 Conservation of Mechanical Energy

Suppose you pick up a box from the floor and place it on a table. The box gains gravitational potential energy, but it doesn't lose kinetic energy. Both K_i and K_f are zero, so $\Delta K = 0$. Where did the energy come from? Or consider pushing the box across the table. The box gains kinetic energy, but not by transforming potential energy. There is no change in the gravitational potential energy as the box slides across a level surface.

These are two examples where an *external* force acts on the system. Our analysis up to this point has considered only interaction forces *within* the system, and we specifically noted that the ball+earth system and the rock+slingshot system had no external forces. As the examples of the preceding paragraph illustrate, external forces can change the mechanical energy of a system.

Figure 10-13 shows a general system of two particles with both interaction forces (internal forces) and external forces from the environment. Both the interaction forces and the external forces can do work on the particles. The change in the kinetic energy is given by the work-kinetic energy theorem as

$$\Delta K = W_{\text{total}} = W_{\text{int}} + W_{\text{ext}}. \qquad (10\text{-}14)$$

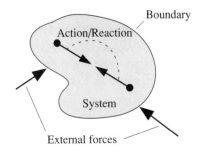

FIGURE 10-13 A system with both internal interaction forces and external forces.

The work done by interaction forces, as we've already noted in Eq. 10-12, can be divided into the work W_{c} done by conservative interaction forces and the work W_{nc} done by nonconservative interaction forces. But the work done by the conservative forces can be used to define a potential energy, as given by Eq. 10-9. Altogether, the change in kinetic energy as the system moves from configuration i to configuration f is

$$\begin{aligned} \Delta K = W_{\text{tot}} &= W_{\text{c}}(\text{i} \rightarrow \text{f}) + W_{\text{nc}}(\text{i} \rightarrow \text{f}) + W_{\text{ext}}(\text{i} \rightarrow \text{f}) \\ &= -\Delta U + W_{\text{nc}} + W_{\text{ext}}. \end{aligned} \qquad (10\text{-}15)$$

If we group the kinetic energy and potential energy terms, we end up with

$$\Delta E_{\text{mech}} = \Delta K + \Delta U = W_{\text{nc}} + W_{\text{ext}}. \qquad (10\text{-}16)$$

This is our most general conclusion about how the mechanical energy of a system changes.

In Chapter 8 we defined an *isolated system* as being a system on which the net external force is zero: $\vec{F}_{\text{tot ext}} = 0$. It follows that no external work is done on an isolated system: $W_{\text{ext}} = 0$. What we have found, with Eq. 10-16, is that *the mechanical energy of an isolated, non-dissipative system is conserved.* That is, $\Delta E_{\text{mech}} = \Delta K + \Delta U = 0$ for a system that is isolated ($W_{\text{ext}} = 0$) and has no dissipative forces ($W_{\text{nc}} = 0$). This is a significant result, one that we could not easily have anticipated in advance, and it has a special name:

> **Law of conservation of mechanical energy**: The mechanical energy of an isolated, non-dissipative system is a constant.

This statement is sometimes referred to as "the law of energy conservation," without the qualifier "mechanical," but that is not quite correct. We are going to extend the concept of energy further in the next chapter to encompass the work done by nonconservative forces. We will wait until we have a more complete definition of energy to see what is required for its conservation. Nonetheless, the law of conservation of mechanical energy is sufficiently powerful that we can use it to analyze many kinds of motion. It will be convenient, for the remainder of this chapter, to drop the subscript "mech" and write just E for the mechanical energy.

This is also a good point at which to summarize the strategy we have been developing for the using the concept of energy:

Strategy for Energy Problems

1. Identify which objects are part of the system and which are in the environment.
2. Identify any external forces or nonconservative forces.
3. Prepare a pictorial model showing the "before" and "after" of the problem. As always, the pictorial model should establish a coordinate system, define the symbols to be used, and identify known and unknown information.
4. Calculate the work W_{ext} done by any external forces and the work W_{nc} done by any nonconservative forces.
5. Invoke the energy equation $\Delta K + \Delta U = W_{nc} + W_{ext}$. If the system is isolated and non-dissipative, then the law of conservation of mechanical energy is $\Delta K + \Delta U = 0$. This can also be written $K_f + U_f = K_i + U_i$.
6. Determine appropriate expressions for each energy term.
7. Solve for the required unknown.

Much experience has shown that using the conservation law in the form $K_f + U_f = K_i + U_i$, when mechanical energy is conserved, greatly reduces the chances for careless errors. In addition, this form emphasizes that the mechanical energy is *really* constant and conserved. The upcoming examples will show this strategy in action.

A special word of caution: The law of conservation of mechanical energy most emphatically does *not* say either $K = U$ or $\Delta K = \Delta U$. These are two common errors made by students. Make sure that you understand *why* these are not correct. Now for some examples.

EXAMPLE 10-4   Renegade Prince Harry the Horrible made a little known attempt to launch the first satellite in the year 1547. He aimed his best cannon straight up and fired a 2 kg projectile with a muzzle velocity of 200 m/s. a) What was the highest point the projectile reached? b) What was the protectile's speed when it struck a poor, hapless bird halfway back to the ground? Ignore air resistance.

SOLUTION   This is certainly a problem that we could work using free-fall kinematics from Chapter 3. Now, however, we will try a different approach. First, let us choose projectile+earth as the system. This is an isolated system. Furthermore, there are no dissipative forces because we are ignoring air resistance. Thus the mechanical energy will be conserved. Figure 10-14 shows the pictorial model, which establishes the coordinate system origin as being the point at which the projectile is launched. The top of the trajectory is characterized, as in kinematics, by $v_1 = 0$.

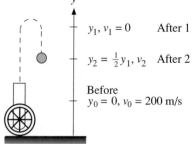

FIGURE 10-14 Pictorial model for using conservation of mechanical energy to analyze the motion of a projectile.

a) For this part the initial position is 0 and the final position is 1. Invoking the law of conservation of mechanical energy means first writing it generally, then filling in appropriate expressions for each term. Thus:

$$K_f + U_f = K_i + U_i$$

$$\Rightarrow \tfrac{1}{2}mv_1{}^2 + mgy_1 = \tfrac{1}{2}mv_0{}^2 + mgy_0$$

$$\Rightarrow y_1 = y_0 + \frac{v_0{}^2 - v_1{}^2}{2g} = \frac{v_0{}^2}{2g} = 2040 \text{ m.}$$

b) For this part the final position is 2, but where is the initial position? Either position 0 or position 1! Because E is conserved it has the same value at both position 0 and position 1, so you can use whichever seems most convenient. Let us use position 0 again as the initial position:

$$K_f + U_f = K_i + U_i$$

$$\Rightarrow \tfrac{1}{2}mv_2{}^2 + mgy_2 = \tfrac{1}{2}mv_0{}^2 + mgy_0$$

$$\Rightarrow |v_2| = \sqrt{v_0{}^2 + 2g(y_0 - y_2)} = \sqrt{v_0{}^2 - 2gy_2} = 141 \text{ m / s.}$$

Because we were asked for speed, the positive root is appropriate. The final *velocity*, of course, would be $v_2 = -141$ m/s.

It is worth exploring the implications of Example 10-4 graphically. If we go back to kinematics, we can easily find

$$y = v_0 t - \tfrac{1}{2}gt^2 \quad \Rightarrow \quad U(t) = mg(v_0 t - \tfrac{1}{2}gt^2)$$

$$v = v_0 - gt \quad \Rightarrow \quad K(t) = \tfrac{1}{2}m(v_0 - gt)^2 . \tag{10-17}$$

Let's look more carefully at the energy transfer. Figure 10-15 shows a graph of K and U as functions of time, as given by Eqs. 10-17. Here you can see clearly that the mechanical energy is completely kinetic at launch but that the potential energy starts increasing quickly as the kinetic energy decreases. By $t = 10$ s, the kinetic energy has decreased from 40,000 J to 10,400 J—a loss of 29,600 J—while U has increased from 0 J to 29,600 J. Most importantly, the *sum* of K and U, which gives the mechanical energy, does not change. The ascent is complete when all of the kinetic energy has been transformed to potential energy. Then the transformation of energy goes in the opposite direction—potential to kinetic—as the projectile falls.

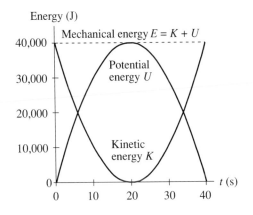

FIGURE 10-15 The kinetic and potential energies as a function of time for a projectile shot straight up into the air. Their sum, the mechanical energy, remains constant.

EXAMPLE 10-5 A roller-coaster car on a frictionless track starts 30 m above the track's low point and follows the curved track shown in the Fig. 10-16. a) What is the car's speed at point A, the low point? b) What is its speed at point B, 10 m above the low point?

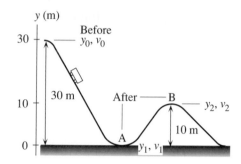

FIGURE 10-16 Pictorial model for using conservation of mechanical energy to analyze a roller-coaster car.

SOLUTION You may be wondering how we can solve this without knowing the exact shape of the track. In fact, we do not need the exact shape because the work done by gravity, and hence the change in gravitational potential energy, is *independent* of the path between two points. Furthermore, the normal force of the track does no work because it is perpendicular to the direction of motion. The car+earth system is an isolated system with no dissipative forces, so its mechanical energy is conserved. All we need to know is the change in y.

a) Figure 10-16 establishes a y-axis. The x-coordinates are not needed. Position 0 is clearly the initial position and position 1, at point A, is the final position. To find the speed at point A we can use energy conservation:

$$K_f + U_f = K_i + U_i$$

$$\Rightarrow \tfrac{1}{2}m|\vec{v}_1|^2 + mgy_1 = \tfrac{1}{2}m|\vec{v}_0|^2 + mgy_0$$

Solving this for $|\vec{v}|$ gives

$$|\vec{v}_1| = \sqrt{|\vec{v}_0|^2 + 2g(y_0 - y_1)} = \sqrt{2gy_0} = 24.2 \text{ m / s.}$$

We have used the more general definition of K, from Eq. 9-5, because the motion in this example is not one-dimensional.

b) Exactly the same reasoning now applies to point B. We simply have to rewrite the equation of a) with the 1 subscripts replaced by 2 subscripts. Doing so, we find

$$|\vec{v}_2| = \sqrt{|\vec{v}_0|^2 + 2g(y_0 - y_2)} = \sqrt{2g(y_0 - y_2)} = 19.8 \text{ m / s.}$$

Although this problem might have appeared at first to be impossible, because we were given so little specific information, it actually turns out to be fairly easy from an energy perspective. Once again, we are seeing energy transformation taking place. The system started with all potential energy, converted 100% of that into the car's kinetic energy by point A, then converted a portion of the kinetic energy back to potential energy by point B. It is also worth noting that we were not given, nor did we need, the car's mass.

EXAMPLE 10-6 From what minimum height h must a roller-coaster car start in order to successfully complete a loop-the-loop of radius R, as shown in Fig. 10-17a?

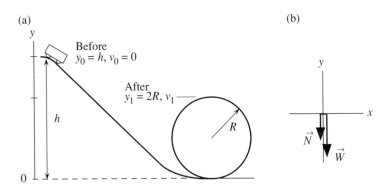

FIGURE 10-17 a) Pictorial model for a roller coaster with a loop-the-loop. b) Free-body diagram at the top of the loop.

SOLUTION Mechanical energy is conserved in this situation, assuming a frictionless track, so we can equate the mechanical energy at y_0, the start, to the mechanical energy at y_1, the top point of the loop. In Chapter 6 you saw that the car must maintain a certain speed at the top if it is to avoid coming off the tracks. Thus we can interpret the question as asking what minimum height is needed for the car to have enough speed not to come off the tracks at the top of the loop. The difficulty is that the conservation equation is going to contain *two* unknowns: the height h and the speed v_1 at the top of the loop. We need some other information.

First let's find the value for v_1. Be careful not to make the error of setting $v_1 = 0$ at the top of the loop. If the car were to come to rest at that point, even momentarily, it would fall off the track! The car must maintain sufficient speed for the track to continue exerting a normal force on the wheels; if $\vec{N} = 0$, the car is no longer in contact with the track. Figure 10-17b shows a free-body diagram at the top of the loop (reproduced from Fig. 6-28b). Because the car is in circular motion *if it is on the track*, $\vec{N}$ and $\vec{W}$ must combine to give the centripetal acceleration $\vec{a}_c = -(mv_1^2/R)\,\hat{\jmath}$. Newton's second law along the y-axis gives

$$\left[(F_{\text{net}})_y = -|\vec{N}| - |\vec{W}| = -|\vec{N}| - mg\right] = -\frac{mv_1^2}{R}.$$

The minimum speed for successfully completing the loop will be when $|\vec{N}| = 0$ and the car just starts to leave the track at the very top. Setting $|\vec{N}| = 0$ in the previous equation gives

$$(v_1)_{\text{min}} = \sqrt{gR}.$$

This is the value we want to use for v_1 in the conservation law to find the minimum starting height. The conservation of mechanical energy statement is then

$$K_f + U_f = K_i + U_i$$

$$\Rightarrow \tfrac{1}{2}mv_1^2 + mgy_1 = \tfrac{1}{2}mv_0^2 + mgy_0$$

$$\Rightarrow h = y_0 = y_1 + \frac{v_1^2 - v_0^2}{2g} = 2R + \frac{(gR) - 0}{2g} = \tfrac{5}{2}R.$$

Example 10-6 was a two-part problem where, to find the initial height, we had to combine some Newtonian analysis about the speed we wanted with a subsequent energy analysis. Notice that h is only $R/2$ above the $2R$ height of the top of the loop—probably a lower height than you might have guessed! A real loop-the-loop would start higher than this, of course, partly due to friction and partly due to the fact that you (and your insurance company) would like the car to cross the top of the loop *well* above the minimum speed!

EXAMPLE 10-7 A 150,000 kg rocket is launched straight up. The rocket motor generates a thrust of 5×10^6 N. What is the rocket's speed at a height of 1 km? Ignore air resistance and the slight mass loss due to burned fuel.

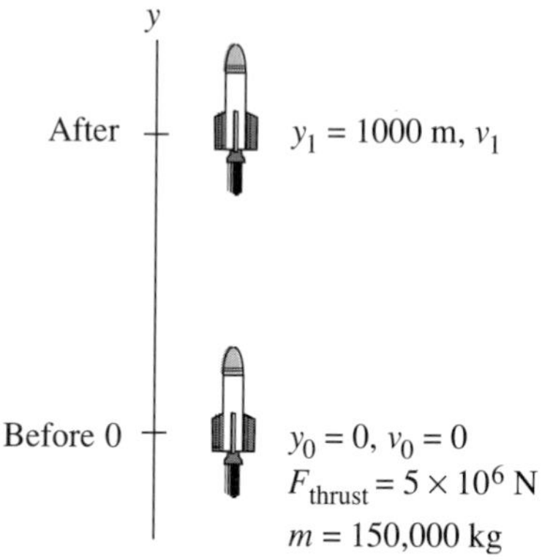

SOLUTION The gravitational force between the rocket and the earth is an interaction force within the rocket+earth system, but the thrust force is an *external* force that does work W_{ext}. Thus we need to use $\Delta K + \Delta U = W_{\text{ext}}$. Figure 10-18 shows a pictorial model. The external work done by the thrust force is

$$W_{\text{ext}} = (F_{\text{thrust}})_y \, \Delta y = |\vec{F}_{\text{thrust}}|(y_1 - y_0) = |\vec{F}_{\text{thrust}}|y_1.$$

The energy equation is then

$$\left[\Delta K + \Delta U = (\tfrac{1}{2}mv_1^2 - \tfrac{1}{2}mv_2^2) + (mgy_1 - mgy_2)\right] = \left[W_{\text{ext}} = |\vec{F}_{\text{thrust}}|y_1\right]$$

$$\Rightarrow \tfrac{1}{2}mv_1^2 + mgy_1 = |\vec{F}_{\text{thrust}}|y_1.$$

This is easily solved for the speed:

$$v_1 = \sqrt{\frac{2y_1}{m}|\vec{F}_{\text{thrust}}| - 2gy_1} = 217 \text{ m / s}.$$

FIGURE 10-18 Pictorial model of a rocket launch.

(figure labels: y; After, $y_1 = 1000$ m, v_1; Before 0, $y_0 = 0$, $v_0 = 0$, $F_{\text{thrust}} = 5 \times 10^6$ N, $m = 150{,}000$ kg)

10.6 Elastic Potential Energy

The law of conservation of mechanical energy would be of little consequence if it only worked for gravity. Fortunately there are many other conservative forces for which we can define an appropriate potential energy. An important example is the elastic force described by Hooke's law.

Figure 10-19a shows an object attached to a spring while the spring itself is attached to a wall. If we consider the object as a single particle, then the force of the spring $\vec{F}_{\text{sp}}$ does work on the particle and changes its kinetic energy. However, our experience with the force of gravity suggests that we would do well to consider object+wall as a single system consisting of two "particles," the object and the wall, interacting via the spring. There are two other forces acting on this system, namely $\vec{N}$ and $\vec{W}$, but they are perpendicular

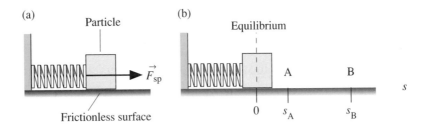

FIGURE 10-19 a) Work is done on a particle by a spring force exerted on it. b) The work done in moving from $s = 0$ to $s = s_A$ is independent of the path.

to the displacement and do zero work. If the surface is frictionless, which we will assume for now, then our larger, two-particle system is an isolated system with no dissipative forces. Thus $E = K + U$ will be conserved *if* the spring force is a conservative force for which we can define a potential energy U.

As you saw Section 10.4, the work done by the interaction force must be independent of the path followed by the object if a potential energy is to exist. That is, an elastic potential energy will exist only if the elastic force is a conservative force. We need to see if this is true. As an example, let's calculate the work done by the spring in Fig. 10-19b as the object moves from the equilibrium position to point A. We will first let it move directly from the origin to A, then we will let it move to A by way of point B.

Let the axis be a general position variable s and assume that the spring obeys Hooke's law. The force is $F_{sp} = -k(s - s_0)$, where s_0 is the object's position when the spring is unstretched and k is the spring constant. As Fig. 10-19b shows, we've chosen a coordinate system with the origin at the equilibrium position, so $s_0 = 0$ and Hooke's law is $F_{sp} = -ks$.

First let's calculate the work done in moving directly from the origin to A: $0 \rightarrow A$. Using the definition of work from Eq. 9-29, we find

$$W(0 \rightarrow A) = \int_0^{s_A} F(s)\,ds = -k\int_0^{s_A} s\,ds$$

$$= -\tfrac{1}{2}ks^2\Big|_0^{s_A} \tag{10-18}$$

$$= -\tfrac{1}{2}ks_A^2.$$

The work is negative because the spring force points opposite the displacement. Next, let's calculate the work done in moving along the path $0 \rightarrow B \rightarrow A$:

$$W(0 \rightarrow B \rightarrow A) = W(0 \rightarrow B) + W(B \rightarrow A)$$

$$= \int_0^{s_B} F(s)\,ds + \int_{s_B}^{s_A} F(s)\,ds$$

$$= -k\int_0^{s_B} s\,ds - k\int_{s_B}^{s_A} s\,ds \tag{10-19}$$

$$= -\tfrac{1}{2}ks^2\Big|_0^{s_B} - \tfrac{1}{2}ks^2\Big|_{s_B}^{s_A}$$

$$= -\tfrac{1}{2}ks_B^2 - \left(\tfrac{1}{2}ks_A^2 - \tfrac{1}{2}ks_B^2\right)$$

$$= -\tfrac{1}{2}ks_A^2.$$

The result is that $W(0 \rightarrow B \rightarrow A) = W(0 \rightarrow A)$. As we had hoped, the work done by the spring is, indeed, independent of the path followed and the elastic force is a conservative force. Knowing that, we can use the definition of Eq. 10-9 to find the change of elastic potential energy as the system moves from configuration i to configuration f:

$$\Delta U = U_f - U_i = -W_{\text{spring}}(s_i \rightarrow s_f)$$

$$= -k \int_{s_i}^{s_f} s\, ds \qquad (10\text{-}20)$$

$$= \tfrac{1}{2} k s_f^2 - \tfrac{1}{2} k s_i^2 .$$

Equation 10-20 suggests that $U_{\text{sp}} = \tfrac{1}{2} k s^2$. However, note that we made a particular choice of coordinate system with the equilibrium position s_0 at the origin. In the more general case, when we don't specify s_0 in advance, the **elastic potential energy** is

$$U_{\text{sp}} = \tfrac{1}{2} k (s - s_0)^2 = \tfrac{1}{2} k (\Delta s)^2 . \qquad (10\text{-}21)$$

As was the case with the gravitational potential energy, this is an energy of position that depends on how far the spring is stretched or compressed. Because Δs is squared, the energy stored in the spring will be positive for either stretching *or* compression.

In many problems you will find it convenient to use $s_0 = 0$ and the simpler $U_{\text{sp}} = \tfrac{1}{2} k s^2$. But don't blindly apply "$U = \tfrac{1}{2} k s^2$" every time you see a spring problem. There are some problems where $s_0 \neq 0$. In these cases it is crucial to remember that the potential energy depends on the spring's *displacement* from equilibrium Δs, *not* simply on the position s.

EXAMPLE 10-8 A rubber-band-powered slingshot has a spring constant $k = 40$ N/m. The rubber band is stretched by 15 cm and used to launch a 5 g paper wad. What is the paper wad's launch speed?

SOLUTION Figure 10-20 shows the initial position, in which the rubber band is stretched, and the final position in which the projectile is flying away. If we choose projectile+slingshot as our system, and if we launch horizontally so that gravity does no work, then we have an isolated system with no dissipative forces. The mechanical energy $E = K + U_{\text{sp}}$ is conserved. We can choose a coordinate system in which the rubber band's equilibrium position is $x_0 = 0$. Then

$$K_f + U_f = K_i + U_i$$

$$\Rightarrow \tfrac{1}{2} m v_2^2 + \tfrac{1}{2} k (x_2 - x_0)^2 = \tfrac{1}{2} m v_1^2 + \tfrac{1}{2} k (x_1 - x_0)^2$$

$$\Rightarrow |v_2| = \sqrt{\frac{k}{m}} x_1 = 13.4 \ \text{m / s}.$$

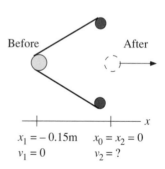

$x_1 = -0.15\text{m}$ $x_0 = x_2 = 0$
$v_1 = 0$ $v_2 = ?$

FIGURE 10-20 Pictorial model of paper wad launched from slingshot.

Once again, conservation of mechanical energy proves to be a very easy way to analyze the motion of a system. This is another example of an energy transformation—in this case potential energy U, stored in the stretched rubber band, is transformed to kinetic energy K. It is worth noting that the stored potential energy increases *quadratically* with the stretching distance. This explains why it is easy to pull a rubber band the first centimeter—you are not storing much energy—but very hard to pull it the last centimeter, when the energy you are storing during that centimeter is much larger.

Springs that expand and contract horizontally are only one possible arrangement. Springs can also be hung vertically or oriented at some angle above the horizontal. In one of these situations, the object+wall system is *not* an isolated system because the external force of gravity does work as the object moves vertically. The solution to this dilemma is to expand our system yet further to be "object+wall+earth." In doing so, we have brought all the interaction forces inside the system, and now $\vec{F}_{ext} = 0$. This larger system *is* isolated, so its mechanical energy will be conserved if there are no dissipative forces. However, the mechanical energy now contains *two* potential energy terms: $E_{mech} = K + U_{grav} + U_{sp}$. In other words, there are now two distinct ways of storing energy inside the system. We will demonstrate this in the following example.

EXAMPLE 10-9   Prince Harry the Horrible, frustrated by his first effort to launch a satellite, tried again. On his second attempt, he placed a 2 kg projectile on top of a very stiff spring 2 meters long that had a spring constant 50,000 N/m. Then the prince had his strongest men use a winch to crank the spring down to length of 80 cm. When released, the spring expanded and launched the projectile straight up. a) What was the launch speed? b) How high did this projectile go?

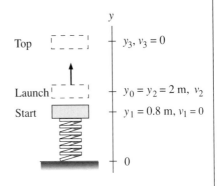

FIGURE 10-21 Pictorial model of Prince Harry's spring-launched projectile.

SOLUTION Figure 10-21 shows the projectile ready for launch. We have chosen to place the origin of the coordinate system on the ground, which means that the position of the end of the unstretched spring is *not* $y_0 = 0$, but is instead $y_0 = 2$ m.

a) We will consider "launch" to occur when the spring has expanded back to its unstretched position, so $y_2 = y_0$. The projectile+ground+earth system is isolated, so we can use conservation of mechanical energy to equate the energy at position 2 with that at position 1:

$$K_f + U_{sp\ f} + U_{grav\ f} = K_i + U_{sp\ i} + U_{grav\ i}$$

$$\Rightarrow \tfrac{1}{2}mv_2^2 + \tfrac{1}{2}k(y_2 - y_0)^2 + mgy_2 = \tfrac{1}{2}mv_1^2 + \tfrac{1}{2}k(y_1 - y_0)^2 + mgy_1$$

Solving this for v_2 gives:

$$v_2 = \sqrt{\frac{k}{m}(y_1 - y_0)^2 - 2g(y_2 - y_1)} = 190 \text{ m/s}.$$

This example was a bit more complex than the last example because there was more "bookkeeping" to do to keep track of all the different forms of energy. Nonetheless, the method is still fairly straightforward. You just have to be very careful in energy conservation problems to keep track of all the terms.

b) Position 3, at the highest point where $v_3 = 0$, is the final position. We can use either position 1 or position 2 as the initial position. If we use position 1, both K_i and K_f will equal zero. One word of caution, and this is *very* important: The projectile moves to position y_3, *but the spring does not!* The end of the spring is still at y_2. Consequently, energy conservation gives us:

$$K_f + U_{sp\,f} + U_{grav\,f} = K_i + U_{sp\,i} + U_{grav\,i}$$

$$\Rightarrow \tfrac{1}{2}mv_3^2 + \tfrac{1}{2}k(y_2 - y_0)^2 + mgy_3 = \tfrac{1}{2}mv_1^2 + \tfrac{1}{2}k(y_1 - y_0)^2 + mgy_1$$

$$\Rightarrow y_3 = \frac{k(y_1 - y_0)^2}{2mg} + y_1 = 1840 \text{ m}.$$

Notice that we used $U_{sp} = \tfrac{1}{2}k(y_2 - y_0)^2 = 0$ as the final potential energy of the spring, *not* $\tfrac{1}{2}k(y_3 - y_0)^2$. The spring's stored energy depends on Δy of the spring, not Δy of the projectile.

•

In Example 10-8 the energy was initially 100% potential energy due to the spring. This energy was converted *mostly* to kinetic energy at the launch point, but it was not completely kinetic because there was also a small increase in gravitational potential energy. This increase in both K and U_{grav} was at the expense of U_{sp}, which decreased to zero. Finally, at the very top, the energy was again 100% potential energy, only this time due to gravity alone. It is interesting to note that the energy is purely potential energy at both the bottom and the top. The initial potential energy, however, was in the spring and provided the potential to launch the projectile. The final potential energy is gravitational, giving the projectile the potential to gain kinetic energy as it falls back to earth.

•

EXAMPLE 10-10 A wooden crate of mass 10 kg sliding along a wooden floor collides with a spring of spring constant 100 N/m. If the crate's speed at the time of impact is 3 m/s, what is the maximum compression of the spring?

SOLUTION You may recognize this as Example 9-10 of Chapter 9, which we solved with the work-kinetic energy theorem. Now let's solve it using the basic energy model. Figure 10-22 shows that we have placed the origin at the spring's equilibrium position. Position x_2, where $v_2 = 0$, will be the point of maximum compression. The interaction with the spring can be represented via the elastic potential energy, but we will need to compute explicitly the work done by the nonconservative force of friction. As we found in Eqs.

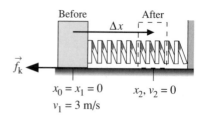

FIGURE 10-22 Pictorial model of a crate colliding with a spring.

10-11 and 10-12, the nonconservative work is $W_{\text{nc}} = -\mu_k mg\Delta x$. Thus

$$W_{\text{nc}} = \Delta K + \Delta U$$

$$\Rightarrow (-\mu_k mg)(x_2 - x_1) = \left(\tfrac{1}{2}mv_2^2 - \tfrac{1}{2}mv_1^2\right) + \left(\tfrac{1}{2}k(x_2 - x_0)^2 - \tfrac{1}{2}k(x_1 - x_0)^2\right)$$

$$\Rightarrow \tfrac{1}{2}kx_2^2 + \mu_k mgx_2 - \tfrac{1}{2}mv_1^2 = 0.$$

This is exactly the same quadratic equation we arrived at in Chapter 9, where we found the solution to be $x_2 = 0.773$ m. If you compare this to Example 9-10, you will see that much less effort is required when using potential energy instead of calculating explicitly the work done by the spring.

EXAMPLE 10-11 A spring with spring constant 2000 N/m is placed between a 1 kg block and a 2 kg block on a frictionless table. The blocks are pushed together so as to compress the spring by 10 cm, then released. What are their velocities as they fly apart?

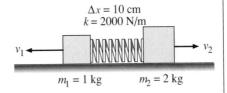

FIGURE 10-23 Pictorial model of blocks pushed apart by a spring.

SOLUTION Figure 10-23 shows the configuration of the blocks and spring. The two blocks form an isolated system with no dissipative forces, so $E = K + U$ is conserved. The initial energy is entirely potential, as the spring is compressed, while the final energy is entirely kinetic. Conservation of mechanical energy gives

$$K_f + U_f = K_i + U_i$$

$$\Rightarrow \tfrac{1}{2}m_1v_1^2 + \tfrac{1}{2}m_2v_2^2 + 0 = 0 + \frac{1}{2}k(\Delta x)^2.$$

Notice that *both* blocks contribute to the kinetic energy K_f. However, we now have an equation with two unknowns: v_1 and v_2. Fortunately, we also have another piece of information. Because the blocks are an isolated system, their momentum is also conserved. The initial momentum is zero, because both blocks are at rest, so we have

$$P_f = P_i$$

$$\Rightarrow m_1v_1 + m_2v_2 = 0$$

$$\Rightarrow v_1 = -\frac{m_2}{m_1}v_2.$$

If we substitute this expression for v_1 into the energy equation we find

$$\tfrac{1}{2}m_1\left(-\frac{m_2}{m_1}v_2\right)^2 + \tfrac{1}{2}m_2v_2^2 = \tfrac{1}{2}k(\Delta x)^2$$

$$\Rightarrow m_2\left(1 + \frac{m_2}{m_1}\right)v_2^2 = k(\Delta x)^2.$$

Solving for v_2 gives

$$v_2 = \sqrt{\frac{k(\Delta x)^2}{m_2(1 + m_2/m_1)}} = 1.826 \text{ m/s}.$$

Finally, we can go back to find v_1:

$$v_1 = -\frac{m_2}{m_1} v_2 = -3.65 \text{ m / s.}$$

•

Example 10-11, which uses the conservation of momentum and the conservation of mechanical energy together, shows just how powerful a problem-solving tool we have developed.

10.7 Energy Diagrams

Potential energy is an energy of position. The gravitational potential energy depends on the height of an object and the elastic potential energy depends on a spring's displacement. Other potential energies you will meet in the future will also depend in some way on position. Functions of position are easy to represent as graphs. For example, Fig. 10-24a is a graph of the gravitational potential energy $U_{grav} = mgy$ as a function of y. This graph is a straight line. The graph of a spring's potential energy $U_{sp} = \frac{1}{2}k(x - x_0)^2$ is shown in Fig. 10-24b. It is a parabola centered on x_0 because it depends on $(x - x_0)^2$.

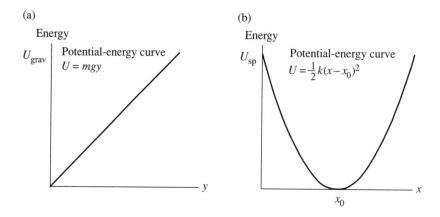

(a)

Energy

U_{grav} Potential-energy curve
$U = mgy$

y

(b)

Energy

U_{sp} Potential-energy curve
$U = \frac{1}{2}k(x - x_0)^2$

x_0 x

FIGURE 10-24 a) Potential-energy diagram of the gravitational potential energy U_{grav}. b) Potential-energy diagram of a spring's potential energy U_{sp}.

A graph of potential energy as a function of position is called a **potential-energy diagram**. These diagrams are used frequently in science and engineering to provide a graphical display of energy. Our goal in this section is to describe how information about the interaction forces and information about the kinetic energy can be obtained from a potential-energy diagram.

We already know how to find the potential energy of a conservative force, via Eq. 10-9. Can we go in reverse? That is, if we know the potential energy of two interacting particles can we then find the interaction force between them? Consider an interaction force $\vec{F}$—a gravitational force, an elastic force, or any other conservative interaction force—that acts as a particle moves through a *small* displacement Δs, where s could be either x or y. To keep things simple, we will assume that $\vec{F}$ is a one-dimensional force along the s-axis with s-component F. Let the interval Δs be sufficiently small so that the force F is essentially constant during the displacement. The work done by the interaction force to move the particle from s to $s + \Delta s$ is

$$W(s \rightarrow s + \Delta s) = F(s)\Delta s, \tag{10-22}$$

where $F(s)$ is the value of the force on the particle when it is at position s. The change in potential energy over this interval, according to the definition of Eq. 10-9, is

$$\Delta U = -W(s \rightarrow s + \Delta s) = -F(s)\Delta s,$$

which we can rewrite as

$$F(s) = -\frac{\Delta U}{\Delta s}. \tag{10-23}$$

Now let Δs become extremely small. In the limit $\Delta s \rightarrow 0$ we find:

$$F(s) = \lim_{\Delta s \rightarrow 0} \left(-\frac{\Delta U}{\Delta s} \right) = -\frac{dU}{ds}. \tag{10-24}$$

That is, the force on the particle is the *negative* of the derivative of the potential energy with respect to position. We can interpret this result graphically by saying

$$F(s) = \text{force on the particle at position } s$$
$$= \text{the negative of the slope of the } U\text{-versus-}s \text{ graph at } s. \tag{10-25}$$

In practice, of course, we will use either $F(x) = -dU/dx$ or $F(y) = -dU/dy$.

Consider first the case of gravitational potential energy: $U_{\text{grav}}(y) = mgy$. Figure 10-25a shows the potential-energy diagram U_{grav}-versus-y for a particle interacting with the earth via the gravitational force. It is, of course, simply a straight line passing through the origin. At a particular point y, the slope of this graph is $dU/dy = mg = $ constant. The force on the particle at position y, according to Eqs. 10-24 and 10-25, is simply $F_{\text{grav}} = -(\text{slope}) = -mg$. The negative sign here, as always, indicates that the force points in the $-y$-direction. Figure 10-25b shows the corresponding F-versus-y graph. At each point, the *value* of F is equal to the negative of the *slope* of the U-versus-y graph. Notice the similarity to position and velocity graphs, where the value of v at any time t is equal to the slope of the x-versus-t graph.

We already knew that $F_{\text{grav}} = -mg$, of course, but the point of this particular example is to illustrate the meaning of Eq. 10-24 rather than to find out anything new. Had we *not* known the force of gravity, we see is that it is possible to find it directly from the potential-energy diagram. Furthermore, we can deduce from the potential-energy diagram

(a)

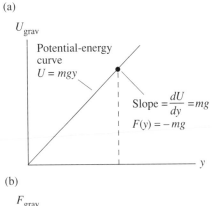

(b)

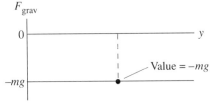

FIGURE 10-25 a) Potential-energy diagram of a particle's gravitational energy. b) The corresponding force diagram.

that the force of gravity is the same at all points, independent of the particle's height, because the slope of a straight-line graph has the same value at all points. Lastly, we can note that adding a constant to U shifts the entire potential-energy diagram up or down but does *not* change its slope. This is relevant to our discussion about the zero of potential energy in Section 10-3, where we found that nothing is changed by adding a constant to U. Equation 10-24, and our graphical interpretation of it, are telling us the same thing. The constant does not affect the slope and, hence, does not change the force.

Figure 10-26 shows a more interesting example, the potential-energy diagram for a particle on a horizontal spring. The graph of the function $U_{sp}(x) = \frac{1}{2}k(x - x_0)^2$ is a parabola centered at the spring's equilibrium position x_0. At a particular position x, the slope of this parabola is

$$\frac{dU}{dx} = k(x - x_0).$$

The force on the particle is thus

$$F = -k(x - x_0).$$

This is just Hooke's law! The mathematical interpretation of Hooke's law, that the force increases linearly with the displacement from equilibrium, we can now interpret graphically as the increasing steepness of the potential-energy diagram as we move away from x_0. The farther the particle is from x_0, the steeper the slope and, thus, the larger the restoring force. Because the slope is positive to the right of x_0, the force will be negative and pointing toward the left—exactly what a restoring force should do. Likewise, the negative slope everywhere to the left of x_0 means a positive restoring force to the right.

The slope of the potential-energy diagram is zero at x_0 because this point is the minimum of the curve. Thus the force is zero at x_0. If an object attached to a spring is released from rest at $x = x_0$, it will remain there in *static equilibrium*—no force,

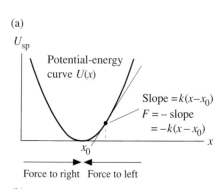

(a)

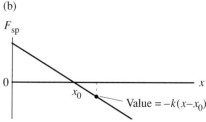

(b)

FIGURE 10-26 a) The potential-energy diagram for a particle on a spring. The slope at position x gives the magnitude of the force on the particle at that point. b) The corresponding force diagram.

no motion. This is, of course, just the position in which the spring is neither stretched nor compressed—its equilibrium length—so we didn't expect an object at rest to move from this position. But our conclusion is not restricted just to springs. For *any* kind of interaction force, a minimum in the potential-energy diagram represents an equilibrium position of the object because the slope—and thus the force—is zero at that point. As you use potential-energy diagrams you should look for minima in the curve to locate equilibrium points. Newton's first law can be phrased, "An object at rest at a point where U is a minimum will remain at rest at that point."

Figure 10-27a shows a more general potential-energy diagram. Although we do not know what kind of conservative interaction force is responsible for this potential energy, we can deduce how the force changes with position. The slope of this graph is negative for $x < x_1$, positive for $x_1 < x < x_2$, negative again for $x_2 < x < x_3$, and finally positive for $x > x_3$. This is enough information to sketch the force diagram of Fig. 10-27b. At each point, the *value* of the force is the negative of the *slope* of the potential energy. We aren't given an exact function $U(x)$, so the force diagram is only an approximation. Even so, we can easily recognize points at which the force is zero and points at which the force is maximum and minimum. Because the potential-energy diagram looks roughly like sections of parabolas, it is reasonable to expect that the force diagram will be comprised of sections that are mostly linear.

Notice that points x_1 and x_3 are minima of the potential-energy diagram, with zero slope. Consequently, these are points at which the force is zero—as indicated in the force diagram. An object at rest at x_1 or x_3 will be in static equilibrium. But what about point x_2? This is a local *maximum* of the potential energy, and the slope and the force at that point are zero. An object at rest at x_2 will also be in static equilibrium.

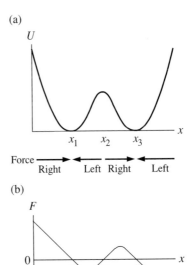

FIGURE 10-27 a) A more general potential-energy diagram. b) The corresponding force diagram.

But there is a big difference between point x_2 and points x_1 and x_3. The interaction force at x_1 and x_3 is a *restoring force*. If the object moves a little to the right of x_1, the force is negative—toward the left—and will push the object back toward x_1. The same is true if the object moves a little to the left of x_1. There the force is positive—toward the right—so again it pushes the object back toward x_1. This is like the restoring force of a spring. The object can oscillate back and forth about x_1 just like a mass oscillating on a spring. An equilibrium point with a restoring force is called a point of **stable equilibrium**.

Now look at point x_2. If the object moves a little to the right of x_2, the force is positive—to the right—and will push the object *away from x_2*. Similarly, the force is negative to the left of x_2. If the object moves a little to the left of x_2, the force will push it even further to the left. The force in the vicinity of x_2 is, in essence, an "anti-restoring force." Rather than returning the object to the equilibrium point, the force pushes the object away from equilibrium. This is called a point of **unstable equilibrium**. It is rather like trying to balance a pencil on its point. If you can get it perfectly balanced it will stay balanced, but with even the slightest wobble the pencil will fall over. A particle at rest at *exactly x_2* is, indeed, in static equilibrium. But with even the slightest wobble the force will push the particle away from this point. Its equilibrium is not stable.

If a system is isolated and has no dissipative forces, its mechanical energy $E = K + U$ is conserved. That is, the value of E is the same at all points. Figure 10-28 shows a potential-energy diagram to which a horizontal line has been added. This line indicates the mechanical energy of the system. The *energy line* has to be horizontal because E has the same value at all points.

Because $K = E - U$, we can interpret the distance *between* the potential energy curve $U(x)$ and the energy line E as the system's kinetic energy $K(x)$ at position x. This

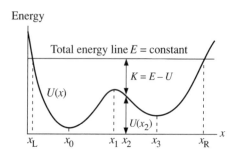

FIGURE 10-28 An energy diagram showing both the potential energy curve and the energy line E.

is illustrated at position x_2 in Fig. 10-28. As the particle changes position, we can see graphically that kinetic energy and potential energy are converted back and forth into each other but that $K + U$ remains constant. A diagram such as Fig. 10-28 that shows both the potential energy curve *and* an energy line is called an **energy diagram**. We can learn a lot about the motion from an energy diagram.

For example, suppose the particle represented in Fig. 10-28, which has an energy E, is at x_2 and moving toward the right (toward $+x$). As it moves, the potential energy is decreasing. Consequently, the particle's kinetic energy—the space between the $U(x)$ curve and the energy line E—is increasing. It is speeding up! It will continue to speed up until it reaches point x_3, where the potential energy is minimum. In other words, potential energy is being converted into kinetic energy from x_2 all the way to x_3. After passing point x_3, the potential energy begins to increase and the kinetic energy to fall. Now the particle is slowing down.

The particle will continue to slow down until it reaches the point x_R, where the energy line E crosses the potential energy curve. At that point $K = 0$, so the particle is instantaneously at rest. The particle cannot proceed forward because to do so would cause the kinetic energy to become negative, and that is *physically* impossible! There is a force on the particle at x_R. Because the slope of the potential energy curve is positive, the force is negative—to the left. Thus the particle will reverse direction at x_R and start back toward the left. Point x_R is called the *right turning point*. We introduced the idea of a turning point back in Chapter 3 where we discussed it in terms of velocity and acceleration. Just to remind you, with a physical example, it is the highest point in the motion of a ball thrown straight up, where the motion reverses from up to down, or the end point in the motion of a mass oscillating on a spring. Here we see a turning point characterized in terms of the energy as a point where the energy line crosses the potential energy curve.

As the particle heads back to the left it will gain speed (lose potential energy) until it reaches x_3, slow down on the way to x_1, and reach a minimum speed at x_1. Notice, though, that the speed at x_1 is not zero, because the kinetic energy is not zero, so the particle keeps moving to the left. It will speed up again after passing x_1 and reach a maximum speed at x_0. After x_0 the particle will slow down and will come to rest instantaneously at x_L. This is the *left turning point*, where the particle will again reverse direction.

You can see that the particle is confined to the region between x_L and x_R and will *oscillate* between those points. It will speed up and slow down, transforming energy between potential and kinetic, as it passes over the valleys and hills of the potential energy curve. Its total energy, however, remains constant.

What happens if we decrease the particle's energy E until the energy line is tangent to the potential energy curve at x_0, as shown in Fig. 10-29? Now the particle cannot be at any point *other* than x_0 because to do so would require a negative kinetic energy—not physically possible. It can exist at x_0 itself, but only if it is at rest ($K = 0$ because

$E = U(x_0)$). So a particle placed at x_0 at rest will remain there at rest. But we can say more. Suppose the energy is increased ever-so-slightly from that shown in Fig. 10-29, so that the energy line E slices off a small sliver of the potential energy curve around x_0. Now the particle has some speed at x_0 ($K \neq 0$), but it also has two turning points *very* close to where the energy line and the potential energy curve intersect. The particle's motion in this case will be a very small oscillation back and forth about x_0, rather like a marble rolling back and forth in the bottom of a bowl. As we have already noted, the point x_0 is a point of *stable equilibrium*. A particle displaced slightly from a point of stable equilibrium will simply oscillate about the equilibrium point.

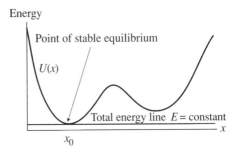

FIGURE 10-29 If $E = U(x_0)$, the particle is in static equilibrium at x_0. This is a point of stable equilibrium.

Figure 10-30 shows the particle with the energy line E tangent to the potential energy curve at x_1, the top of the potential energy hill. If the particle is placed at *precisely* x_1, it will be at rest ($K = 0$) and will remain there. Point x_1, like x_0, is a point of equilibrium. But what happens if we increase the energy ever-so-slightly, so that the particle has a very small but non-zero speed at x_1? Will it simply oscillate closely about the x_1 position? Not in this case. A very slight energy increase will now place the turning points far away at x_L and x_R. The particle, if displaced slightly from x_1, will go off on a large excursion. That is why we previously identified a maximum in the potential energy, such as x_1, as a point of *unstable equilibrium*. It is like balancing a marble on an upside-down bowl—the marble will remain stationary if the balance is perfect, but the slightest displacement will cause the marble to roll away.

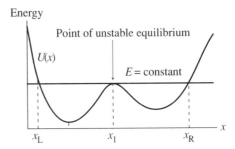

FIGURE 10-30 Point x_1 is also a point of equilibrium, with $K = 0$ if $E = U(x_1)$. However, x_1 is a point of unstable equilibrium.

A marble in a bowl is a pretty good analogy for all aspects of energy diagrams. Imagine a marble in a "bowl" shaped like the potential energy curve of Figs. 10-28, 10-29, and 10-30. You can give the marble energy E with an initial push. If you give it the energy shown in Fig. 10-28, the marble will roll back and forth between x_L and x_R, speeding up and slowing down as it goes through the valleys and over the hills. The force on the marble, which speeds it up or slows it down, is represented by the *steepness* of the walls. Steeper walls mean larger forces and thus larger accelerations. With somewhat less energy the marble may become "trapped" in either the left or right potential energy valley. Figure 10-29 shows the least possible energy the marble can have, a situation in which it is at rest at the lowest point in the bowl.

EXAMPLE 10-12 Take a spring of length L and spring constant k. Orient the spring vertically, as shown in Fig. 10-31a, and then place an object of mass m on it. What is the equilibrium length of the compressed spring?

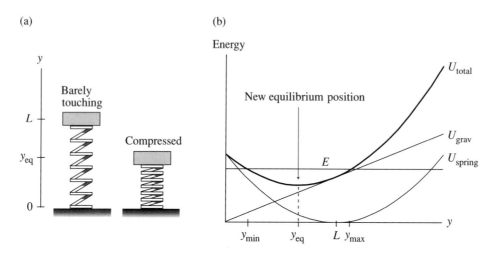

FIGURE 10-31 a) Object compressing a vertically-oriented spring. b) Energy diagram showing the equilibrium position and the turning points.

SOLUTION We can solve the problem by examining the potential-energy diagram. Because this is a vertical problem, the system has both gravitational potential energy $U_{grav} = mgy$ and elastic potential energy $U_{sp} = \frac{1}{2}k(y-L)^2$. The pictorial model of Fig. 10-31a has established a coordinate system with the origin at ground level, so the end of the spring is originally at $y_0 = L$. This is why U_{sp} is not simply $\frac{1}{2}ky^2$. Figure 10-31b shows the two potential energies separately and also shows the total potential energy $U_{total} = U_{grav} + U_{sp}$. The equilibrium position—the minimum of U_{total}—has shifted from $y = L$ to a smaller value of y, closer to the ground. To find the equilibrium position, we need to find the point of zero slope:

$$U_{total} = mgy + \tfrac{1}{2}k(y-L)^2$$

$$\Rightarrow \frac{dU}{dy} = mg + k(y-L).$$

The slope is zero when

$$mg + k(y_{eq} - L) = 0$$

$$\Rightarrow y_{eq} = L - \frac{mg}{k}.$$

Equilibrium occurs where the spring is compressed a distance mg/k from its original length L, giving it a new equilibrium length $L - mg/k$.

Example 10-12 illustrates a point of *stable* equilibrium, so the object will oscillate about y_{eq} if it is displaced slightly up or down. For example, Fig. 10-31b shows an energy line slightly above the potential energy minimum. If the object is given this much energy—by giving it a push—it will oscillate between points y_{min} and y_{max}.

Molecular Bonds

Let's end this chapter by seeing how these ideas can allow us to understand something about molecular bonds. A *molecular bond* consists of two atoms held together by electrical inter-action forces between the charged electrons and nuclei. As you will see later, the electric force is a conservative force and has an electric potential energy. Figure 10-32 shows the electric potential-energy diagram for the diatomic molecule HCl (hydrogen chloride), as it as been experimentally determined. The curve shows how the potential energy varies with the separation between the atoms. Note the very tiny distances: 1 nm = 10^{-9} m. This potential-energy diagram has some similarities to a spring, with a deep potential-energy valley, but also some significant differences. How should we interpret this?

First note that the molecule has a stable equilibrium position at an atomic separation of 0.13 nm = 1.3×10^{-10} m.

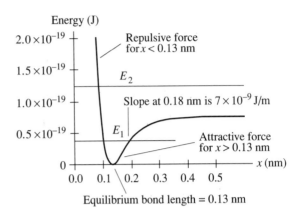

FIGURE 10-32 The potential-energy diagram of the diatomic molecule HCl. The equilibrium length of the molecular bond is seen to be 0.13 nm, or 1.3×10^{-10} m.

This is called the *bond length* of HCl and is the value you can find listed in chemistry books. If we try to push the atoms closer together (smaller x), the potential energy curve quickly becomes very steep with a negative slope. This indicates a large *repulsive* force between the atoms. Physically, this is the repulsive electrical force between the electrons orbiting each atom, preventing the atoms from getting too close. There are also attractive forces between atoms, called *polarization forces*. These are harder to understand, and we will defer discussion of such forces until our study of electricity, but they are similar to the static electricity forces by which a comb that has been brushed through your hair can attract small pieces of paper. If you try to pull the atoms apart (larger x), the attractive polarization force resists and is responsible for the increasing potential energy for $x > x_0$. The equilibrium position is where the repulsive electron force and the attractive polarization force are exactly balanced.

The repulsive electron force keeps getting stronger as you push the atoms together, and thus the potential energy curve keeps getting steeper on the left. The attractive polarization force, however, gets *weaker* as the atoms get farther apart. This is why the potential energy curve becomes *less* steep as the atomic separation increases. Ultimately, at very large x, the slope becomes zero. This is not surprising—two well-separated atoms do not exert forces on each other. Now, however, we can relate this information about forces to the shape of the potential-energy diagram and see that it corresponds to an *asymmetric* curve.

It turns out, for quantum physics reasons, that a molecule cannot have $E = 0$ and thus cannot simply rest at the equilibrium position. By requiring the molecule to have some energy, such as E_1, we see that the atoms oscillate back and forth between two turning points. This is a *molecular vibration*, and all molecular bonds are constantly vibrating. For a molecule having an energy $E_1 = 3.5 \times 10^{-19}$ J, as illustrated in Fig. 10-32, the bond oscillates in length between roughly 0.10 nm and 0.18 nm. At the right turning point we can draw the tangent to the curve and measure its slope to be $dU/dx \approx 7 \times 10^{-9}$ J/m = 7×10^{-9} N. This is the magnitude of the force pulling the atoms back together at that point!

Suppose we increase the molecule's energy to $E_2 = 12.5 \times 10^{-19}$ J. This can occur if the molecule absorbs some light or has a collision with a neighboring molecule. However the energy increase happens, the atoms are no longer bound together at large values of x. No matter how large x gets, the atoms will still have kinetic energy $K = E_2 - U > 0$ and will keep on moving apart. By raising the molecule's energy to E_2 we have *broken the molecular bond*. If the atoms are moving together at the time the energy changes, they will "bounce" once last time (there is still a *left* turning point at a bond length of ≈ 0.08 nm), then move away from each other and not return. The breaking of molecular bonds through the absorption of light is called *photodissociation*. It is an important process in many applications of chemistry.

Summary

▲ Important Concepts and Terms

potential energy	law of conservation of mechanical energy
mechanical energy	elastic potential energy
gravitational potential energy	potential-energy diagram
conservative force	stable equilibrium
nonconservative force	unstable equilibrium
dissipative force	energy diagram

In this chapter we have introduced a second form of energy—a stored or *potential* energy U. The potential energy arises due to interaction forces between the particles in a system, and U is the potential energy *of the system* rather than of an individual particle.

The interaction forces must be *conservative forces* for a potential energy to exist. A conservative force is one for which the work done in moving from configuration i to configuration f is independent of the paths followed by the particles in the system. If so, the *change* in potential energy of the system is

$$\Delta U = U_f - U_i = -W_c(i \rightarrow f),$$

where W_c is the work done by the conservative force. This defines ΔU in terms of the force. It is also possible to go the other direction and find the force from the potential energy via

$$F(s) = -\frac{dU}{ds}.$$

Potential energy has several important characteristics:

1. As a particle or system of particles moves, potential energy is converted into kinetic energy and vice versa. The system's energy flows back and forth between the kinetic energy of motion and the "stored" potential energy.

2. Only the *change* in potential energy ΔU is defined, not U itself. This implies that the zero of potential energy can be placed wherever convenient. Only changes in potential energy are used in problem solving.

3. Potential energy can only be defined for a conservative force—one for which the work done on a particle in moving from i to f is independent of the path followed.

4. Potential energy is an energy of position. It depends on the configuration, or positions, of the particles in the system. This allows us to draw potential-energy diagrams showing how U changes as a function of position.

Two specific potential energies introduced in this chapter are the gravitational potential energy

$$U_{\text{grav}} = mgy$$

and the elastic potential energy of a spring

$$U_{\text{sp}} = \tfrac{1}{2} k (\Delta s)^2 .$$

The mechanical energy of a system is defined to be $E_{\text{mech}} = K + U$. The mechanical energy can change if there are nonconservative forces, also called dissipative forces, or if work is done on the system by external forces. The change in the mechanical energy is

$$\Delta E_{\text{mech}} = \Delta K + \Delta U = W_{\text{nc}} + W_{\text{ext}} ,$$

where W_{nc} is the work done by any nonconservative forces and W_{ext} is the work done by any external forces. If a system is isolated ($W_{\text{ext}} = 0$) and not subject to dissipative forces ($W_{\text{nc}} = 0$), then

$$\Delta E_{\text{mech}} = \Delta K + \Delta U = 0$$

$$\Rightarrow E_{\text{mech}} = K + U = \text{constant}.$$

This is called the law of conservation of mechanical energy. The statement $E_f = E_i$ allows us to connect the "before" and "after" of an interaction without needing to know anything about the details of the interaction.

Chapters 9 and 10 have provided a framework for using the concept of energy to analyze the motion of a single particle and of systems of two or three particles. The language is rather different than that used up until this point in the text, with discussions of "before" and "after" and "work" replacing our earlier discussions of "acceleration" and "time interval." Nonetheless, the common feature that is still with us is the importance of the concept of force. Force was used in dynamics to determine the acceleration of an object; now, with energy, it is used to find the work and the potential energy. It remains the cornerstone of Newtonian mechanics.

In fact, this chapter has given us yet another way of thinking about the concept of force. We began, back in Chapter 4, with thinking of force as simply a push or a pull. That was an adequate viewpoint for using Newton's second law, but the limitations of this understanding of force became clear when we considered interacting systems. In Chapter 7 we found it necessary to broaden our view of force by considering it to be an interaction between two systems, with the interaction consisting of an action/reaction pair of forces. Now we can see, in Eq. 10-24, an even more sophisticated perspective where force is the slope of the potential-energy function (with a minus sign for technical reasons). The potential energy is an interaction energy, and the force of that interaction is now seen as the rate with which the interaction energy changes as the separation of the particles is varied. A small force indicates a potential energy that changes slowly with distance, while a large force occurs when the potential energy varies rapidly.

This brings us, finally, to the modern concept of force. Many of our future topics will be approached using energy ideas, and the forces will not be explicitly mentioned. Nonetheless, it will be important to keep in mind that it is forces of interaction that *cause* there to be a potential energy and that we can, if we desire, find those forces directly from the potential-energy function.

Exercises and Problems

Exercises

1. A boy reaches out of a window and tosses a ball straight up with a speed of 10 m/s. The ball is 20 m above the ground at the point he releases it. Use energy conservation to find:
   a. The ball's maximum height above the ground.
   b. The ball's speed as it passes the window on its way down.
   c. The speed of impact on the ground.

2. A 1000 kg safe is dropped 2 m onto a heavy-duty spring, which it compresses 50 cm. What is the spring constant of the spring?

3. A 50 g ball is released from rest 1 m above the bottom of a frictionless track, as shown in Fig. 10-33. The ball rolls down a straight 30° segment, then back up a parabolic segment whose shape is given by $y = x^2/4$. How high up will the ball go on the right before reversing direction and rolling back down?

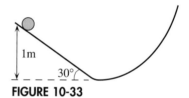

FIGURE 10-33

4. A pendulum is made by tying a 500 g ball to a string 75 cm long. The pendulum is pulled 30° to one side and then released.
   a. What is the ball's speed at the lowest point of its trajectory?
   b. To what angle does the pendulum swing on the other side?

5. A 500 g block on a frictionless surface is held against a horizontal spring, as was shown in Fig. 10-19a. The spring has a spring constant of 1250 N/m and is compressed by 4 cm. Then the block is released. What is the block's speed as it slides away?

Problems

6. As shown in Fig. 10-34, Sammy Skier (75 kg) starts down a 20° frictionless slope that is 50 m high. A strong headwind exerts a *horizontal* force of 200 N on him as he skies.
   a. Find Sammy's speed at the bottom using a work and energy analysis.
   b. Find Sammy's speed at the bottom using a Newton's second law analysis.

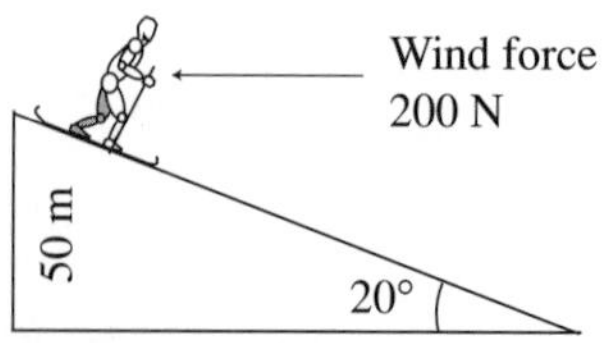

FIGURE 10-34

7. a. A 50 g ice cube on a 30° wooden slope is held against a spring, of spring constant 25 N/m. The spring has been compressed 10 cm (Fig. 10-35). Consider the coefficient of friction for ice on wood to be zero. When the ice cube is released, what total distance will it travel up the slope before reversing direction?

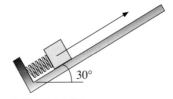

FIGURE 10-35

   b. If the ice cube is replaced by a 50 g wooden cube, how far will it travel up the slope?

8. Dan (80 kg) attaches a 30 m long bungee cord to himself and to a bridge that is 100 m above the river. Then Dan steps off the edge. A bungee cord, for practical purposes, is just a long spring, and this cord has a spring constant of 40 N/m.

   a. What should you choose as the system in this problem so as to have an isolated system whose mechanical energy is conserved?

   b. What interaction forces exist within this system? Are they all conservative forces?

   c. Use conservation of mechanical energy to find out how far Dan is above the water when the cord reaches its maximum elongation. Do this as a *single* step; you do not need to find his speed as the cord begins to stretch. Don't forget to include all forms of potential energy.

9. As shown in Fig. 10-36, a marble is spun in a *vertical* plane around the inside of a very smooth pipe. The marble's speed at the bottom of the pipe is 3 m/s.

   a. What is the marble's speed at the top of the pipe?

   b. Is this speed sufficient to keep the marble from falling off? (**Hint:** Think about apparent weight.)

   c. Find an expression for the marble's speed when it forms an angle θ above the bottom.

   d. Make a graph of v-versus-θ for one complete revolution.

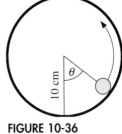

FIGURE 10-36

10. A roller-coaster car on the frictionless track shown in Fig. 10-37 starts from rest at height h. The track is straight until point A, then between points A and D the track consists of circle-shaped segments of radius R, as shown.

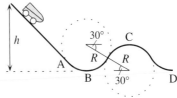

FIGURE 10-37

   a. What is the *maximum* height h_{max} from which the car can start so as not to fly off the track when going over the hill at point C? Your answer should be in terms of the radius R. (**Hint:** This is a two-part problem. First find v_{max} at C.)

   b. Evaluate h_{max} for a roller coaster having $R = 10$ m.

11. A sled starts from rest at the top of a frictionless, hemi-spherical, snow-covered hill of radius R (Fig. 10-38).

   a. What is the sled's speed when it is at angle ϕ ?

   b. Use a Newton's laws analysis to find the maximum speed the sled can have at angle ϕ without leaving the surface.

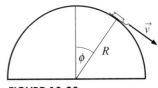

FIGURE 10-38

   c. At what angle ϕ_{max} does the sled "fly off" the hill?

12. The speed of a bullet can be measured with a device called a *ballistic pendulum*. A 10 g bullet is fired horizontally into a 1 kg block of wood that hangs from a string 1.5 m long. The bullet lodges in the block, which then swings out until the string is at a 45° angle. What was the initial speed of the bullet?

13. A 2 kg block has a massless spring attached to its front. The spring has a spring constant of 5000 N/m. The block slides with a speed of 4 m/s across a frictionless surface toward a stationary 1 kg block, as shown in Fig. 10-39.

    a. What is the maximum compression of the spring during the collision? (**Hint:** How are the velocities of the blocks related at the instant of maximum compression?)

    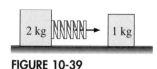

    FIGURE 10-39

    b. What is the speed of each block after the collision?

14. Figure 10-40 shows the potential-energy diagram for a 500 g particle as it moves along the x-axis. Consider a situation in which the particle's mechanical energy is 12 J.

    a. Where are the particle's turning points?

    b. What is the particle's speed at $x = 2$ m?

    c. What is the particle's maximum speed? At what position or positions does this occur?

    d. Describe the motion of the particle as it moves from the left turning point to the right turning point.

    e. Suppose the particle's energy is lowered to 4 J. Describe the possible motions.

    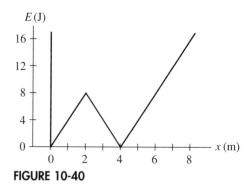

    FIGURE 10-40

15. A 100 g particle moving along the x-axis is subjected to the one-dimensional, conservative force F shown in Fig. 10-41.

    a. Draw a graph of the potential energy $U(x)$ function for this force from $x = 0$ to $x = 4$ m. Let the zero of the potential energy be at $x = 0$. (**Hint:** Think about the definition of potential energy *and* the geometric interpretation of the work done by a varying force.)

    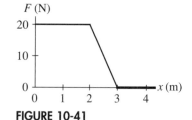

    FIGURE 10-41

    b. The particle is shot toward the left from position $x = 4$ m with a speed of 25 m/s. What is the particle's mechanical energy?

    c. Does the fact that E is negative cause a problem? Why or why not?

    d. Draw the energy line for this particle on your graph of part a).

    e. Describe, by interpreting your graph, how this particle will subsequently move.

    f. Find a numerical value for the turning point of the particle.

16. A 10 g particle experiences a conservative force as it moves along the x-axis. The potential-energy diagram for the particle is shown in Fig. 10-42.

    a. Determine and list the intervals in which the force vector points to the right, points to the left, and is zero. Explain how you determined these.

b. Draw a force-versus-position graph $F(x)$ from $x = 0$ to $x = 8$ cm.

c. What speed does the particle need at $x = 2$ cm to arrive at $x = 6$ cm with a speed of 10 m/s?

d. How much work is done on the particle as it moves from $x = 2$ cm to $x = 6$ cm? Answer this *without* doing any calculations, but explain how you know the answer.

e. What is the impulse on the particle as it moves from $x = 2$ cm to $x = 6$ cm?

f. How long does it take the particle to move from $x = 3$ cm to $x = 5$ cm?

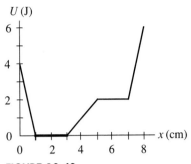

FIGURE 10-42

17. a. In Fig. 10-43, graph A shows the force $F(x)$ exerted on a particle that moves in one dimension along the x-axis. Draw a graph of $U(x)$, the particle's potential energy, as a function of position x. Let $U = 0$ be at $x = 0$. (**Hint:** Think about the definition of potential energy *and* the geometric interpretation of the work done by a varying force.)

b. Graph B shows the potential energy $U(x)$ of a particle that moves in one dimension along the x-axis. Draw a graph of $F(x)$, the force exerted on the particle at position x.

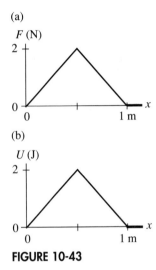

FIGURE 10-43

18. A vertical spring with spring constant 4.9 N/m hangs such that its free end is located at $y = 0$. A 100 g mass is then tied to the end of the spring.

a. Calculate and graph the potential-energy diagram from $y = -50$ cm to $y = +10$ cm. Draw the graph so it is approximately a half-page in size. Use graphing software if you know how; otherwise calculate and graph U every 5 cm.

b. Determine from your graph the equilibrium position for the 100 g mass.

c. Determine from your graph the total energy of the mass if it has turning points at $y = -5$ cm and $y = -35$ cm.

d. Determine from your graph the net force on the mass at $y = -35$ cm.

19. Protons and neutrons (together called *nucleons*) are held together in the nucleus of an atom by a force named the *strong force* (Fig. 10-44). The strong force is a very attractive force at very short distances—strong enough to overcome the repulsive electrical force between the protons—but the force quickly weakens as the distance increases. A well-established model for the potential energy of two nucleons interacting via the strong force is

$$U(x) = U_0\left[1 - e^{-x/x_0}\right],$$

where x is the separation between the centers of the two nucleons, x_0 is a constant having the value $x_0 = 2\times10^{-15}$ m, and $U_0 = 6\times10^{-11}$ J.

a. Calculate and draw an accurate potential-energy diagram from $x = 0$ to $x = 10\times10^{-15}$ m. Either calculate about 10 points by hand or use appropriate computer software.

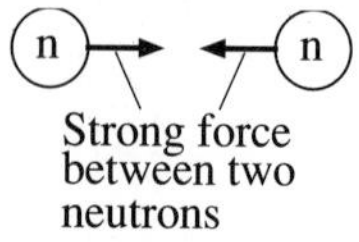

Strong force between two neutrons

FIGURE 10-44

b. What happens to the force between the nucleons as the separation increases? Explain how you can tell this from the potential-energy diagram.

c. Calculate the interaction force $F(x)$ between the nucleons as a function of the separation x.

d. Calculate and draw an accurate graph of $F(x)$ from $x = 0$ to $x = 10\times10^{-15}$ m.

e. Is $F(x)$ an attractive force or a repulsive force? Explain, making use of your graphs.

f. Quantum effects are essential for a proper understanding of how nucleons behave. Nonetheless, let us consider two neutrons *as if* they were small, hard, electrically-neutral spheres of mass 1.67×10^{-27} kg and diameter 1.0×10^{-15} m. (We will consider neutrons rather than protons so as to avoid complications from the electrical forces between protons. Both protons and neutrons experience the strong force in the same way.) Release two neutrons from rest at a separation $x_0 = 5\times10^{-15}$ m. Draw the total energy line for this situation on your diagram of part a).

g. What speed do *each* of these neutrons have as they crash together? Be sure to keep in mind that *both* neutrons are moving.

[**Estimated 10 additional problems for the final edition.**]

Chapter **11**

Expanding the Concept of Energy

LOOKING BACK | Sections 7.1; 8.3; 9.3–9.8; 10.4–10.5

11.1 Curious Examples of Forces That Do No Work

Jumping straight up into the air is a simple physical exercise. Figure 11-1 shows a jumper, initially crouched down with legs bent, then springing straight up and leaving the floor with velocity v_1. Once he leaves the surface, his motion becomes that of a projectile with initial velocity v_1—a problem we understand. But, now we want to consider the more difficult issue of how the motion starts. A careful look at this process will help to consolidate the energy concepts developed in the last two chapters.

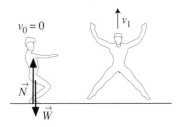

FIGURE 11-1 Forces on a person jumping straight up.

As Fig. 11-1 shows, there are two forces on the jumper as he pushes up from the crouched position: his weight downward and the upward normal force of the floor on his feet. The normal force forms an action/reaction pair with the force of the jumper pushing down against the floor. If the jumper were standing still, in static equilibrium, then the forces would be balanced: $|\vec{N}| = |\vec{W}|$ and $F_{\text{net}} = 0$. But this is *not* an equilibrium situation. The jumper is *accelerating* upward, so it must be the case that $|\vec{N}| > |\vec{W}|$. From a force/acceleration perspective we can say that the jumper presses down against the floor as he straightens his legs, causing the floor, in reaction, to exert an upward normal force that is larger than the weight force. This provides the net upward force that accelerates him upward until he has velocity v_1 as his feet lose contact with the floor.

From an energy perspective it appears that both the normal force and the weight force do work on the jumper, causing his kinetic energy to change. The total work should be

$$W_{\text{total}} = W_{\text{normal}} + W_{\text{gravity}} = |\vec{N}|\Delta y - mg\Delta y \,, \tag{11-1}$$

367

where Δy is the jumper's vertical displacement as he pushes up from the crouched position until his feet leave the floor. Using the work-kinetic energy theorem, his liftoff velocity should be given by

$$\Delta K = \tfrac{1}{2} m v_1{}^2 = W_{\text{total}} = \left(|\vec{N}| - mg \right) \Delta y . \tag{11-2}$$

This seems straightforward.

But wait! The normal force is applied only to the jumper's *feet*, and his feet *do not move* during the time that the normal force is applied to them. If there's no displacement, we should conclude that the normal force does *no* work: $W_{\text{normal}} = 0$. We seem to have a paradox. Is $W_{\text{normal}} = |\vec{N}| \Delta y$, as used in Eq. 11-1, or is $W_{\text{normal}} = 0$?

This problem has arisen because Eqs. 11-1 and 11-2 make an inappropriate application of the ideas of work and kinetic energy. The definition of work, and our subsequent discovery of the work-kinetic energy theorem, was for a *single particle*. We found that the work done by a constant force $\vec{F}$ on a particle displaced through Δs is $W = F_s \Delta s$. This was generalized to an integral for a variable force, but the idea is the same—namely, work is done *only* if the force is applied to a particle that is *displaced* by Δs. The atoms in the sole of the jumper's foot, the particles to which the normal force is applied, have no displacement and hence cannot have any work done on them. So W_{normal} *really* is zero.

A particle occupies only a single point in space; it has no extension and no internal structure. The particle model is often adequate for understanding the motion of an object, *but not always*! Sometimes we need to consider *systems* of particles. In contrast to a single particle, systems of particles have extension; have internal structure, motion, and forces; and can be deformed. As a consequence, systems of particles have additional modes of energy and additional mechanisms of energy transfer that are not available to a single particle.

The jumper is a deformable, extended object—changing from the crouched position to the upright position. He *cannot* be modeled as a single particle. So Eq. 11-1, which treats the jumper as a single particle, with the point of application of the normal force undergoing displacement Δy, is an invalid application of the definition of work. But the jumper most certainly does accelerate and gain kinetic energy as he jumps. If the normal force does no work, where does the kinetic energy come from?

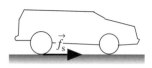

FIGURE 11-2 The friction force accelerating a car does no work.

As another example, Fig. 11-2 shows a car accelerating from rest. In Chapter 7 you learned that the forward propulsion force on the drive wheels is the *static* friction force. This force causes the forward acceleration of the car. But the atoms of the tire that are in contact with the ground, to which the static friction force is applied, undergo *no* displacement—they are momentarily at rest. (If the bottom of the tire were not momentarily at rest, it would be skidding or slipping.) Because the tire atoms have no displacement during the interval in which the force is applied to them, the propulsion force does *zero* work on the car! It is the cause of the car's acceleration, but it is *not* the source of the car's subsequent kinetic energy.

In Fig. 11-3 we see a car colliding with a solid wall and coming to rest. The wall does exert a force on the car to stop it, but that force does zero work. The point of application of the force—namely, the car's front bumper—undergoes *no* displacement while the force is being applied. So the force of the wall decelerates the car, but it is *not* the cause of the car's loss of kinetic energy.

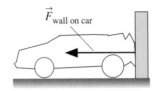

FIGURE 11-3 The force of the wall on the car does no work!

The car, like the jumper, is an extended, deformable object. To resolve the puzzle of forces that cause acceleration but do no work we need to expand the concepts of energy to *systems* of particles. That is the goal of this chapter. We will find that there is an important distinction between the *internal* motions of the particles and the *external* motion of the system as a whole. The ideas we develop in this chapter will take us beyond the particle model and will help form a bridge between dynamics and thermodynamics.

11.2 Microphysics and Macrophysics

All of the objects we handle and use every day are extraordinarily large compared to the size of atoms. These objects are all systems of very large numbers of particles (atoms). We want to distinguish between the motion of and forces on the object "as a whole" and the motion of and forces between the atoms inside. It is convenient to use the terms **macrophysics** to refer to the motion of the object as a whole and **microphysics** to refer to the motion of atoms. You recognize the prefix *micro*, meaning "small." You may not be familiar with *macro*, which simply means "large" or "large-scale."

The macrophysics of an object is very often the direct consequence of the underlying microphysics of its atoms. A good example that you have already studied is friction. Friction arises from complicated electrical forces between the atoms at the interface between two surfaces. This is the microphysics. Yet the net effect of very large numbers of atoms experiencing these forces is a very simple *macrophysics* model of friction that can be characterized by a single parameter: the coefficient of friction. In our later discussion of gases you will see that the motion and collisions of many atoms in the gas (the microphysics) all add up to give a model of gases in terms of macroscopic parameters, such as pressure and temperature.

The connection between microscopic and macroscopic behavior will be a new theme for this text, one that will grow in importance as we proceed. In fact, much of our evidence for the existence of atoms comes from the micro/macro connection. It is a topic we will explore in depth when we reach thermodynamics, in order to understand the bulk properties of materials, and again in electricity, to explain the electrical properties of conductors and insulators. For now, let's take a closer look at the kinetic and potential energies in a system of particles.

Kinetic and Potential Energy at the Microscopic Level

Figure 11-4 shows two different perspectives of an object. In the macrophysics perspective of Fig. 11-4a we see an object of mass M moving as a whole with what we will shortly define as the center of mass velocity v_{cm}. Although we will now give it a special name and subscript, the center of mass velocity is just the familiar velocity that we have been using all

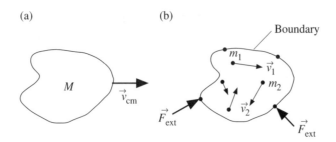

FIGURE 11-4 Two perspectives: a) The macroscopic motion of the system as a whole, and b) the microscopic motion of the atoms inside. These atoms each have kinetic and potential energy

along when we considered the object as a single particle. As a consequence of its motion, the object has macroscopic, or center of mass, kinetic energy given by $K_{cm} = \frac{1}{2}Mv_{cm}^2$. This motion of the object as a whole is often called the **translational motion**.

The microphysics is hidden from view in the Fig. 11-4a perspective, but it still exists. Figure 11-4b is a microphysics view of the same object, where now we see a *system of particles*. Many individual atoms of mass m_j are moving with velocities v_j as a consequence of both *internal* forces between the atoms and, perhaps, *external* forces. The combined microscopic masses of all the atoms, which is just the total mass of the object, is given by $M = \Sigma m_j$. We will use lowercase m to represent the mass of particles within the system and uppercase M for the mass of the system as a whole.

We see many kinds of atomic motions if we peer into the system with a microscope. The atoms in a solid vibrate back and forth around an equilibrium position, while the atoms in liquids and gases appear to move quite randomly. Regardless of the type of motion, each moving atom has a kinetic energy $K_j = \frac{1}{2}m_jv_j^2$. While the kinetic energy of one atom is exceedingly small, there are enormous numbers of atoms in a macroscopic object. The combined kinetic energy of all the atoms is what we call the *internal kinetic energy* $K_{int} = \Sigma K_j$. This is a mode of energy quite distinct from the macroscopic kinetic energy K_{cm}.

Because of the microscopic displacements of the atoms under the influence of internal forces, the atoms also have *internal potential energy* U_{int}. This potential energy comes from stretching or compressing the spring-like molecular bonds between the atoms. This is real stored energy, with the potential to be converted into the microscopic kinetic energy of atomic motion.

The combined internal kinetic and potential energy of the atoms is called the **thermal energy** of the system: $E_{therm} = K_{int} + U_{int}$. This is the energy, usually hidden from view in our macrophysics perspective, that is associated with the motions of the atoms. You will discover later, when we reach thermodynamics, that the thermal energy is proportional to the *temperature* of the system. In fact, temperature simply *measures* the amount of thermal energy in a system. A single particle, because it has no internal structure, cannot have a thermal energy. Thermal energy is a property of *systems* of particles.

A word of caution: The internal energy of the atoms in a system is *not* called "heat." The word *heat*, like the word *work*, has a very narrow and precise meaning in physics that is much more restricted than its use in everyday language. We will introduce the concept of heat later, when we need it. For the time being we want to use the correct term *thermal energy* to describe the random, thermal motion of the particles in a system. If the temperature of a system goes up (i.e., it gets hotter), it is because the system's thermal energy has increased, *not* because its heat has increased. Understanding the distinction between thermal energy and heat is of crucial importance, and we will return often to this discussion.

You can also see from Fig. 11-4b that the system has a well-defined boundary. External forces are exerted *only* on particles at the boundary, which means that external forces are exerted only on some particles in the system (often a very small fraction), not on all of them. Work is done on those particles only if they are displaced—that is, *work is done only if the boundary moves*. The work done by external forces is an energy transfer *across the boundary* of the system from the macroscopic environment to the kinetic energy of the particles. No work is done, and no energy transferred, by the external forces if the boundary on which they are exerted does not move. That is why the normal force exerted on the jumper and the wall force exerted on the car's bumper were forces that did no work.

11.3 The Center of Mass

[**Photo suggestion: Strobe photo of a wrench with its center of mass marked.**]

Thus far in this text we have been treating objects as if they were a mass located at a single point. But which specific point in the object is most appropriate to consider as *the* point for representing the object's motion? This is a question we have not had to deal with before. The point at which we assume all the mass of an object is concentrated in order to determine its motion in response to external forces is called the **center of mass**. For most practical purposes, this is the same as the *center of gravity*, which you may know how to find by balancing an object on a knife-edge. You will learn the precise definition of the center of mass in calculus (it involves integrals over the object's volume). Fortunately, we do not need a precise definition for our purposes. For the types of objects we will consider, the center of mass will be located at the object's center—which we can find without doing any calculations.

The work-kinetic energy theorem does not apply to the system as a whole, because it is defined only for single particles, but Newton's laws are still valid. The forces on the system cause its center of mass to accelerate in accordance with $\vec{F}_{net} = M\vec{a}_{cm}$. As an example, Fig. 11-5 shows a person jumping. The center of mass of the human body is located approximately at hip-level. As the jumper rises, his center of mass accelerates upward even though his feet remain fixed. The acceleration of the center of mass requires a net force. In this case $\vec{F}_{net} = \vec{N} + \vec{W} = (|\vec{N}| - mg)\,\hat{j}$, so the normal force *does* contribute to the acceleration. As the jumper leaves the ground, he has a center of mass velocity v_{cm} and an associated macroscopic kinetic energy K_{cm}.

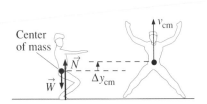

FIGURE 11-5 Motion of the center of mass of a jumper.

EXAMPLE 11-1   A 70 kg man crouches down and then jumps straight up into the air. His center of mass moves upward 20 cm before his feet leave the floor, and his liftoff velocity is 2 m/s. What is the normal force exerted by the floor?

SOLUTION   The situation is just that depicted in Fig. 11-5. The jumper's center of mass responds to the external forces, so the vertical component of Newton's second law is just

$$[(\vec{F}_{net})_y = |\vec{N}| - mg] = ma_{cm}.$$

If we make the reasonable assumption that the man's upward acceleration is constant as he jumps, then we can use kinematics to determine a_{cm}:

$$v_f^2 = v_i^2 + 2a_{cm}\Delta y_{cm}$$

$$\Rightarrow a_{cm} = \frac{v_f^2}{2\Delta y_{cm}} = \frac{(2 \text{ m/s})^2}{2 \cdot 0.2 \text{ m}} = 10 \text{ m/s}^2.$$

The only difference between this and the many problems we worked in Chapter 5 is that we are now applying the second law explicitly to the *center of mass* of a *deformable* object. Now that the upward acceleration is known, the normal force must be

$$|\vec{N}| = mg + ma_{cm} = 1368 \text{ N}.$$

Because $mg = 686$ N, we can also write $|\vec{N}| = 2.02mg$. The normal force as the man jumps is slightly more than double its equilibrium value of $|\vec{N}| = mg$ when he stands motionless. This increased normal force is a third law *reaction* to the force the jumper exerts on the floor ($\vec{F}_{jumper\ on\ floor}$) by rapidly extending his legs. But because the normal force does zero work, we are still left with the puzzle of how the jumper gains kinetic energy as he jumps.

11.4 The Energy Equation

The motion of the center of mass is a macrophysics problem. Now let's look more closely at the microphysics. Figure 11-6 shows a system of particles with a well-defined boundary. We can designate particles by a subscript j that ranges from 1 to N, where N equals the number of particles. Particle j has mass m_j and velocity v_j. The particles interact with each other through internal action/reaction pairs of forces, and we will assume that these internal forces are conservative. (This is a good assumption if our internal particles are atoms, which interact via the conservative electric force.) External forces on the system are seen, from the microphysics perspective, to be applied directly to particles on the boundary of the system.

Our definition of work and the work-kinetic energy theorem are valid for forces acting on particles—that is the situation for which work was defined. For a single particle j, the work-kinetic

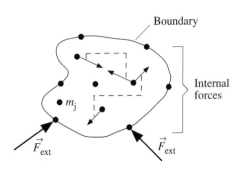

FIGURE 11-6 Internal and external forces acting on a system of particles. The dotted lines identify several action/reaction pairs.

energy theorem tells us

$$W_j = \int_i^f (F_{on\,j})_s\, ds_j = \Delta K_j, \tag{11-3}$$

where $K_j = \frac{1}{2}m_j v_j^2$ and ds_j is a small segment of the trajectory along which particle j moves. The force $\vec{F}_j$ is the *total* force on particle j, consisting of all internal and external forces applied to the particle. Summing Eq. 11-3 over all the particles gives

$$\left[W_{total} = \sum_j W_j = \sum_j \int_i^f (F_{on\,j})_s\, ds_j \right] = \sum_j \Delta K_j. \tag{11-4}$$

Here W_{total} is the total work done on all particles in the system.

Equation 11-4 relates a sum over works to a sum over kinetic energies. We need to examine each of these sums more carefully. The total force $\vec{F}_{on\,j}$ on particle j can be subdivided into external and internal forces as $\vec{F}_{on\,j} = \vec{F}_{ext\,on\,j} + \vec{F}_{int\,on\,j}$, with $\vec{F}_{ext\,on\,j}$ being non-zero only for those particles on the boundary that actually experience an external force. Then

$$\sum_j \int_i^f (F_{on\,j})_s\, ds_j = \sum_j \int_i^f (F_{ext\,on\,j})_s\, ds_j + \sum_j \int_i^f (F_{int\,on\,j})_s\, ds_j$$
$$= \sum_j W_{ext\,on\,j} + \sum_j (-\Delta U_j) \tag{11-5}$$
$$= W_{ext} - \Delta U_{int}.$$

In the second line of Eq. 11-5 we have made explicit use of our assumption that the internal forces are conservative. We used the Eq. 10-9 definition of potential energy to define ΔU_j for particle j in terms of the work done on the particle by these conservative internal forces. The potential energy U_j of particle j characterizes its interactions with other particles *inside* the system. For atomic systems, U_j is the electric potential energy stored in the molecular bonds between particle j and its neighbors—with some of those bonds stretched and others compressed. We have then defined the **internal potential energy** $U_{int} = \sum U_j$ as the *total* microscopic potential energy of the system.

The external work $W_{ext} = \sum W_{ext\,on\,j}$, which appears in Eq. 11-5, is the total work done by external forces on the particles at the system boundary. It is calculated as the force on particle j integrated over the displacement of j, then summed over all particles at the boundary.

Turning now to the right-hand side of Eq. 11-4, we will assert that a lengthy and somewhat tedious algebraic analysis leads to the following result:

$$\sum_j \Delta K_j = \sum_j \left[\frac{1}{2}m_j \left((v_j)_f^2 - (v_j)_i^2 \right) \right] = \Delta \left[\frac{1}{2}M v_{cm}^2 \right] + \Delta \sum_j \left[\frac{1}{2}m_j \tilde{v}_j^2 \right], \tag{11-6}$$

where $\tilde{v}_j = v_j - v_{cm}$ is the velocity of particle j *relative to* the center of mass. In other words, $\tilde{v}_j$ is the velocity that you would measure for particle j *if* the object as a whole were at rest. The relative velocities $\tilde{v}_j$ measure the *internal* motion of the system, which will exist whether or not the system as a whole is moving. Thus it makes sense to define the

internal kinetic energy K_{int} as

$$K_{int} = \sum_j \left[\tfrac{1}{2} m_j \tilde{v}_j^2 \right].$$ (11-7)

We have already noted that $\tfrac{1}{2} M v_{cm}^2$ is the center of mass kinetic energy K_{cm} of the system as a whole. Using this and Eq. 11-7 in Eq. 11-6 gives:

$$\sum_j \Delta K_j = \Delta K_{cm} + \Delta K_{int}.$$ (11-8)

The two equations Eq. 11-5 and Eq. 11-8 are expansions of the sums in the left-hand and right-hand sides of Eq. 11-4. If we now put them back into Eq. 11-4 we have

$$W_{ext} - \Delta U_{int} = \Delta K_{cm} + \Delta K_{int}$$
$$\Rightarrow W_{ext} = \Delta K_{cm} + (\Delta K_{int} + \Delta U_{int}).$$ (11-9)

We previously defined $K_{int} + U_{int} = E_{therm}$, the *thermal energy* of the system. Thus

$$W_{ext} = \Delta K_{cm} + \Delta E_{therm}.$$ (11-10)

Even though we have a *system* of particles, this is now a valid statement about work and energy. W_{ext} is work done on the system by the external forces.

For a few external forces, such as gravity or the force exerted by a spring, it makes sense to define a potential energy for the system as a whole. For example, we can consider the system+earth as a yet larger system in order to define a gravitational potential energy Mgy_{cm} for the mass of the entire system. A potential energy that applies to the system as a whole we will call U_{cm}, the potential energy of the center of mass. Any U_{cm} adds to the right side of Eq. 11-10, giving

$$W_{ext} = \Delta K_{cm} + \Delta U_{cm} + \Delta E_{therm}.$$ (11-11)

Here W_{ext}, by implication, is the work done by external forces for which no potential energy for the system as a whole has been defined.

Recalling the definition of mechanical energy as $E_{mech} = K + U$, we finally find the *energy equation*

$$W_{ext} = \Delta E_{mech} + \Delta E_{therm}.$$ (11-12)

This result is a very general statement about the energy in systems of particles. It explicitly recognizes two distinct types of energy: the mechanical energy of the system as a whole (center of mass energy) *and* the internal, microscopic energy of the particles within the system. These two types of energy can be transformed into each other under the proper circumstances, with one increasing and the other decreasing. External forces can transfer energy from the environment to the system (or vice versa) by doing work W_{ext} on the system. It is important to recognize work W_{ext} as a mechanism by which energy is *transferred* between the system and the environment.

Compare Equation 11-12 to Chapter 10's Eq. 10-16: $\Delta E_{mech} = W_{ext} + W_{nc}$, where W_{nc} is the work done by dissipative, nonconservative forces. There we noted that nonconservative forces, such as friction or air resistance, cause objects to become warmer. Now we see explicitly that the nonconservative forces are *internal* forces whose work increases the *thermal* energy of the system and raises its temperature.

As an example, we previously considered a box that slid to a stop across a level surface. The box has $\Delta E_{mech} = \Delta K_{cm}$ because its gravitational potential energy is unchanged. There are no external forces, so $W_{ext} = 0$. In our analysis of Chapter 10, where the box was treated as a particle, we found that $\Delta K_{cm} = W_{nc}$. That is, the box loses kinetic energy because of the work performed by the nonconservative friction force. (The work W_{nc} done by friction, you will recall, was a negative quantity.) With our new perspective of Eq. 11-12, we can now conclude that $\Delta E_{therm} = -\Delta K_{cm}$. Because $W_{ext} = 0$, the total energy of the box is not changed. It is simply transformed—via the internal friction forces of the box+floor system—from the kinetic energy of the center of mass into the thermal energy of the box's atoms!

Newton's second law describes the motion of the center of mass but does not characterize the internal motions of the system. The energy equation describes energy transfers but does not help us understand how the system as a whole can accelerate. Neither perspective alone gives a complete description of a system of particles, but when used *together* they allow us to make a complete and powerful analysis. The examples will help make this clear.

EXAMPLE 11-2   A 1000 kg car traveling at 5 m/s runs into a solid, unbending wall. The bumper and front of the car crumple upon impact, resulting in the car's center of mass moving forward 0.5 m before stopping. Analyze and interpret the forces and energy transfers of this collision.

SOLUTION Figure 11-7 shows a before and after pictorial model of the collision. This is clearly not a situation that we can analyze with the single particle model, so we must consider the car as a system of particles. The force of the wall on the car, as we have previously noted, does *zero* work because the front boundary of the car (the bumper) undergoes no displacement. So $W_{ext} = 0$. But the car's center of mass does decelerate during the collision, so the force is doing something. If we assume that $\vec{F}_{\text{wall on car}}$ remains constant during the time of the collision, then kinematics of the center of mass gives

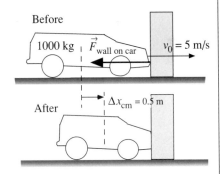

FIGURE 11-7 Pictorial model of car colliding with a solid wall.

$$a_{cm} = -\frac{v_0^2}{2\Delta x} = -25 \text{ m}/\text{s}^2.$$

Newton's second law allows us to determine the magnitude of the force:

$$|\vec{F}_{\text{wall on car}}| = m|a_{cm}| = 25{,}000 \text{ N}.$$

The energy equation gives us another perspective. We already noted that $W_{ext} = 0$. The car as a whole has no change in gravitational potential energy because it moves horizontally, so $\Delta E_{mech} = \Delta K_{cm}$. Thus

$$W_{ext} = 0 = \Delta K_{cm} + \Delta E_{therm}$$

$$\Rightarrow \Delta E_{therm} = -\Delta K_{cm} = -(-\tfrac{1}{2}Mv_0^2) = +12{,}500 \text{ J}.$$

It is worth noting that we can use the impulse-momentum theorem to determine the duration of the collision. Because $\vec{F}_{\text{wall on car}}$ is constant, the impulse of the wall is

$$\left[J = F_x \Delta t = -|\vec{F}_{\text{wall on car}}| \Delta t \right] = \left[\Delta p = -Mv_0 \right]$$

$$\Rightarrow \Delta t = \frac{Mv_0}{|\vec{F}_{\text{wall on car}}|} = 0.20 \text{ s}.$$

Make sure you agree with the signs.

How do we interpret this information? $\vec{F}_{\text{wall on car}}$ is a very real force of 25,000 N exerted on the car. It is this force that decelerates the car during a time of 0.20 s as the car's center of mass moves forward 0.50 m. This is the macrophysics. As we have noted, force $\vec{F}_{\text{wall on car}}$ does no work and cannot transfer any energy between the environment and the system. But the car *does* lose kinetic energy, so where does the kinetic energy go? The energy equation is telling us that the car's macroscopic kinetic energy is converted into the internal thermal energy of the atoms: $K_{\text{cm}} \rightarrow E_{\text{therm}}$. This energy *transformation* is accomplished not by the external force of the wall, but by *internal* forces. The increased thermal energy shows up as an increased temperature of the car—it gets hotter as a result of the impact!

•

An energy transformation from the macroscopic kinetic energy of the system as a whole into the microscopic, thermal energy of the particles is often called the *dissipation* of energy. This is not a completely accurate description because it seems to imply that the energy has been lost. However, the energy has been lost only to the macrophysics perspective—we can see from the microphysics perspective that, in fact, the energy has simply been transferred from one energy mode (K_{cm}) to another (E_{therm}). No energy was "lost."

A word of caution: If asked "Where did the energy go?" many students reply "It went into heat." But recall our earlier comment about the term *heat*. The kinetic energy went into the *thermal energy* of the particles in the system, thereby raising the temperature. Heat is something entirely different. The energy did *not* "go into heat."

As another example, consider a situation that we have looked at many times within the context of the single particle model: a force pushing a box along a horizontal floor with friction. We noted long ago that this procedure makes the surfaces warmer. Let us revisit the issue from our new model of considering the box to be a system of particles.

Figure 11-8a shows a box being accelerated from rest to velocity v_{cm} by force $\vec{F}$. The center of mass of the box is located at the box's geometric center. Figure 11-8b shows a free-body diagram of the box, with four forces being exerted on it. How many of these forces do work? The pushing force $\vec{F}$ does work W_{ext} because the *boundary* it is pushing

(a)

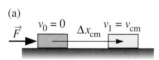

(b)

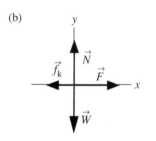

FIGURE 11-8 a) Box being accelerated by force $\vec{F}$. b) Free-body diagram.

on *does* undergo a displacement. Neither $\vec{N}$ nor $\vec{W}$ do any work because they are perpendicular to the displacement.

The difficulty is the kinetic friction force $\vec{f}_k$. The work due to $\vec{f}_k$ is not zero, but neither is it simply $-|\vec{f}_k|\Delta x_{cm}$. The work is done, in a very complicated fashion, on the particles at the boundary. At the boundary there are sections of the interface that stick and slip, others that bend and deform, and yet others that are entirely abraded away. Our simple model of friction, with $|\vec{f}_k| = \mu_k|\vec{N}|$, tells us the *average* force on the block as a whole *if* we treat the block as a single particle. That is fine if we want to use Newton's laws to find the acceleration of the center of mass, but we simply do not have the microphysics information that would be needed to calculate the work done.

This presents a problem for using the energy equation. If we choose the box alone as the system, then the friction force contributes to the work W_{ext} but, as we saw, we have no idea how to calculate that work. Suppose, on the other hand, that we cleverly choose our system to be the box+floor. Then the friction forces are *internal* forces. In this case, they transfer no energy between the surrounding environment and the system so they do not contribute to W_{ext}. The only work comes from force $\vec{F}$ pushing *on the boundary*, so the energy equation is:

$$\left[W_{ext} = |\vec{F}|\Delta x_{boundary} = |\vec{F}|\Delta x_{cm} = F\Delta x_{cm} \right] = \Delta K_{cm} + \Delta E_{therm}, \qquad (11\text{-}13)$$

where E_{therm} is the thermal energy of both the box *and* the floor. In calculating W_{ext} we have used the fact that $\Delta x_{boundary} = \Delta x_{cm}$.

Work is done by the pushing force—an energy transfer by mechanical means from the environment to the system. But unlike the situation for a single particle, not all of the energy transfer goes to the kinetic energy of the box! Some portion of it is transferred into the microscopic thermal energy of the atoms in the box and the floor—heating up the box and the floor. How much energy goes into each mode? We can use Newton's laws to find ΔK_{cm}, then whatever is left must be ΔE_{therm}.

The x-component of the second law gives

$$\left[(F_{net})_x = F - |\vec{f}_k| = F - \mu_k Mg \right] = Ma_{cm} \qquad (11\text{-}14)$$

$$\Rightarrow a_{cm} = \frac{F}{M} - \mu_k g.$$

Assuming $v_0 = 0$, the velocity after accelerating distance Δx_{cm} is

$$v_1^2 = v_{cm}^2 = 2a_{cm}\Delta x = \frac{2F\Delta x_{cm}}{M} - 2\mu_k g\Delta x_{cm} \qquad (11\text{-}15)$$

$$\Rightarrow \Delta K_{cm} = \tfrac{1}{2}Mv_{cm}^2 = F\Delta x_{cm} - \mu_k Mg\Delta x_{cm}.$$

Using this result for ΔK_{cm} in the energy equation, Eq. 11-13, gives a value for ΔE_{therm}:

$$F\Delta x_{cm} = \left[F\Delta x_{cm} - \mu_k Mg\Delta x_{cm} \right] + \Delta E_{therm}$$

$$\Rightarrow \Delta E_{therm} = \mu_k Mg\Delta x_{cm}. \qquad (11\text{-}16)$$

The increase in thermal energy is a direct consequence of the *internal* forces of friction. It depends, not surprisingly, on both the coefficient of friction (more friction, larger ΔE_{therm}) and on the distance the box slides (larger Δx, larger ΔE_{therm}).

Our analysis shows the following:

1. The force $\vec{F}$ is responsible for accelerating the box.
2. The force $\vec{F}$ does work $W_{ext} = F\Delta x_{cm}$.
3. *Unlike* the case for a single particle, the work W_{ext} is *not* all transferred to the box's kinetic energy. That is, $\Delta K \neq W_{ext}$ and the box does *not* obey the work-kinetic energy theorem. Instead, the work represents an energy transfer from the environment—whatever is doing the pushing—to both the box's kinetic energy *and* to the thermal energy of the box and the floor.
4. The increase in thermal energy is $\Delta E_{therm} = \mu_k Mg\Delta x_{cm}$. This increase of thermal energy raises the temperature of the box and the floor—exactly what we know really happens when friction is present. We emphasize, however, that nothing has been converted into "heat." The concept of heat does not enter into this problem at all.

To summarize the main idea of the last two sections, we have found that *two* relations are necessary to describe the changes in motion and energy of a system of particles. Newton's second law is a relationship among the *macro*scopic quantities that characterize the motion of the object as a whole—namely, forces and accelerations. The energy equation, on the other hand, describes the energy transformations that occur on both the macroscopic and the microscopic level. Proper use and interpretation of the energy equation requires a careful and explicit choice of the system to which it is applied.

11.5 The Law of Conservation of Energy

This chapter, you will have noticed, is not so much about doing calculations as it is about how we understand and think about the physics of energy. We have made good progress, but it is not yet enough. Consider, for example, a car accelerating from rest. In Section 11.1 we found that the static friction propulsion force $\vec{f}_s$ accelerates the car, but it does no work and hence transfers no energy to the car. Yet the car does acquire center of mass kinetic energy. Where does it come from? Could it be the case that the thermal energy of the car's atoms is transferred to the center of mass kinetic energy? Such an energy transfer would satisfy the energy equation $\Delta K_{cm} + \Delta E_{therm} = 0$, with $\Delta K_{cm} > 0$ and $\Delta E_{therm} < 0$. But decreasing the thermal energy would cause the car's temperature to decrease as it accelerates, and you know that doesn't happen. The car does *not* accelerate by converting thermal energy into center of mass kinetic energy!

Perhaps there are still other modes of energy that we have not yet included. What might they be? One possibility, suggested by the fact that a car burns fuel, is *chemical energy* E_{chem}. Chemical reactions occur when various molecules completely change their geometric structure in order to join together in new ways with other molecules. In the process the reaction products gain kinetic energy. Chemical energy, in some sense, is a form of potential energy because it is stored in the molecular configurations and can be converted into kinetic energy (and hence into increased thermal energy and temperature). But it is quite different than the molecular bond stretching and compressing that we considered earlier as U_{int}, so it is best to think of chemical energy as a distinct mode of energy storage.

Another form of energy is *nuclear energy* E_{nuc}. In this case, reactions within the nuclei of the atoms shoot out particles with exceptionally high kinetic energy. These particles collide with other particles in the system, and E_{therm} of the entire system can be substantially increased as a result of this energy transfer.

Yet other forms of energy include the energy associated with sound waves and with light. All of these additional forms of energy are *internal* to the system, rather than being associated with the motion of the system as a whole. This suggests that we need to broaden our concept of internal energy. Let us *postulate* that the conservation of energy equation really should be

$$W_{ext} = \Delta E_{mech} + \Delta E_{int} \qquad (11\text{-}17)$$

where the **internal energy** E_{int} consists of

$$E_{int} = E_{therm} + E_{chem} + E_{nuc} + E_{sound} + E_{light} + \cdots \qquad (11\text{-}18)$$

That is, the internal energy is the thermal energy *plus* chemical energy and other forms of energy *inside* the system. Notice that we have left open the possibility that there are yet other contributions to the internal energy.

With this postulate, we have now expanded the concept of energy well beyond Newtonian mechanics. As a postulate, of course, it must be experimentally verified before we can accept it. If verified, Eq. 11-17 is a statement of such depth and significance to be worthy of being called a law of nature. However, it is quite easy to show that Eq. 11-17 fails to account for very common situations. If you place a pan of water on a hot stove, the water gets hot. This means that $\Delta E_{therm} > 0$. Yet no work is being done on the water, there are no chemical or nuclear reactions taking place in the water, and so every other term in Eq. 11-17 is zero. Equation 11-17 fails to explain why the water's thermal energy and temperature increase.

The problem is that we have been focusing on finding all the possible modes of energy storage, but we have not thought to consider what other *energy transfer processes*, in addition to work, might exist for transferring energy between the system and the environment. Work is a *mechanical* transfer of energy across the boundary of a system that occurs when particles are pushed and pulled by external forces. It turns out that it is also possible to transfer energy across the boundary *non-mechanically* if there is a temperature difference between the system and the environment. This energy transfer due to a temperature difference is what we will call **heat**. The symbol for the energy transferred as heat is Q. If we now include both mechanical *and* non-mechanical forms of energy transfer between the system and the environment Eq. 11-17 becomes

$$\boxed{W_{ext} + Q = \Delta E_{mech} + \Delta E_{int}.} \qquad (11\text{-}19)$$

This statement is known as the **first law of thermodynamics**. It is the fullest and most complete statement there is about energy. Scientists have never found an experimental violation of the first law of thermodynamics, and it is one of the fundamental cornerstones of modern physics.

The quantities E_{mech} and E_{int} measure the energy of the system as a whole *and* the energy internal to the system. We have now included all the forms of energy that have been discovered, so it is useful to define the **total energy** E_{total} to be

$$E_{\text{total}} = E_{\text{mech}} + E_{\text{int}}. \tag{11-20}$$

With this definition, the first law of thermodynamics can written:

$$W_{\text{ext}} + Q = \Delta E_{\text{total}}. \tag{11-21}$$

In other words, the total energy of a system can be changed *only* if the environment does work on the system or if the environment transfers heat to the system. This is a powerful idea.

We can also extend our earlier definition of an *isolated system* to say that a system is isolated if the environment neither does work on it nor transfers heat to it. That is, a system is isolated if no energy is transferred to or from the environment. Thus for an isolated system:

$$\Delta E_{\text{tot}} = 0 \quad \text{(isolated system)}. \tag{11-22}$$

We had to discover and recognize many contributions to the total energy. But having done so, we have found a quantity that *never changes* for an isolated system. We can state our conclusion as a new law of nature:

Law of conservation of energy: The total energy of an isolated system does not change.

The law of conservation of energy takes its place with Newton's laws of motion as one of the fundamental principles of physics. But while it is important to learn that the total energy is conserved for an isolated system, it is equally important to recognize how interactions with the environment cause the total energy to change. The first law of thermodynamics describes how those energy *transfers* take place.

The point of our introducing the first law of thermodynamics at this time is to make explicit the connection between mechanics, which you already have been studying, and thermodynamics, which you will soon study. Newtonian mechanics establishes the concepts of energy and energy transfer, but (as we see with the example of heating water) it does not quite reach the level of providing a full explanation of phenomena that we can observe and measure. But with the introduction of a very few new concepts—namely the concept of heat and the realization of other modes of energy—we have the first law of thermodynamics. So there is a very close connection between the energy concepts of mechanics and the broader ideas of thermodynamics.

For the time being we will continue to deal only with situations where $Q = 0$, in which case Eq. 11-17 is a correct statement of the energy equation. Further, we will limit ourselves to situations in which $E_{\text{int}} = E_{\text{therm}} + E_{\text{chem}}$. Other forms of internal energy are negligible except under special circumstances, but the inclusion of chemical energy is essential for understanding how systems can accelerate. The following examples will show how this helps.

EXAMPLE 11-3 A 70 kg man crouches down and then jumps straight up into the air. His center of gravity moves upward 20 cm before his feet leave the floor, and his lift-off velocity is 2 m/s. Analyze and interpret the forces and energy transfers of his jump.

SOLUTION We did the force analysis for this problem in Example 11-1, finding that $|\vec{N}| = 1368$ N $= 2.02mg$. This provides a net upward force that accelerates the jumper. However, $\vec{N}$ does no work and transfers no energy to the jumper because the boundary to which $\vec{N}$ is applied, the jumper's feet, do not move until the jumper leaves the ground and becomes a projectile. Now we are ready to complete the analysis by considering energy. The conservation of energy equation is

$$W_{ext} = 0 = \Delta E_{mech} + \Delta E_{int}$$
$$= \Delta K_{cm} + \Delta U_{cm} + \Delta E_{therm} + \Delta E_{chem}.$$

Although jumping up and down repeatedly can raise your body temperature, it doesn't seem reasonable that one jump would make much change in the jumper's thermal energy. So we will assume $\Delta E_{therm} \approx 0$ and rearrange the energy equation to give

$$\Delta E_{chem} = -\Delta K_{cm} - \Delta U_{cm} = -\Delta K_{cm} - Mg\Delta y_{cm}.$$

This relationship is the essence of the energy transfer. The jumper gains translational kinetic energy and also gravitational potential energy as his center of mass rises. Where did that energy come from? Not from the normal force pushing up on him—that force does no work and transfers no energy. It comes, instead, from what he ate for breakfast! The food you eat is processed by your body into various chemical "fuels" that are stored in your cells. Those fuels are "burned" to enable you to exert muscular forces, such as pressing your feet down against the floor. This is a conversion of chemical energy (a *loss* of E_{chem}, hence the negative signs) to kinetic and potential energy. The energy equation tells us that exactly the right amount of chemical energy is released to lift the jumper. The energy conversion occurs via *internal* forces—there is no transfer of energy between the jumper and his environment. Using the given data we can calculate the chemical energy release to be $\Delta E_{chem} = -277$ J.

To reiterate, the force that accelerates the jumper upward is $(F_{net})_y = |\vec{N}| - Mg$. This force is greater than zero because the jumper exerts a large downward force against the ground, causing the reaction force $\vec{N}$ on his feet to be larger than Mg. But the *reason* he can exert a large downward force on the ground is that he has an internal source of stored chemical energy, enabling his muscles to perform the necessary motions. Thus the *source* of his increase in kinetic and potential energy is the decrease of stored chemical energy within his body.

We can do a similar analysis of an accelerating car. No energy is transferred to it from the environment. Instead, the chemical energy of the fuel decreases as it burns, and that energy is transferred both to the forward kinetic energy and to increased thermal energy. The full analysis will be left as a homework problem.

Before leaving this section we can make a final interesting observation about energy, one that will provide a glimpse of things to come. We have seen several examples—such as the car colliding with the wall—where macroscopic center of mass kinetic energy is "dissipated" into microscopic thermal energy: $K_{cm} \rightarrow E_{therm}$. You can certainly think of many other examples of this. They are situations where a system *decelerates* by converting its kinetic energy into thermal energy. But what about a system that *accelerates*? At the beginning of this section we noted that a car does *not* accelerate by converting thermal energy into its center of mass kinetic energy.

Apparently it is all right to decelerate by having $K_{cm} \rightarrow E_{therm}$, but it is *not* all right to accelerate by converting $E_{therm} \rightarrow K_{cm}$. Why not? The energy equation, Eq. 11-17, would not be violated by such a process. If $W_{ext} = 0$, then energy can be conserved either with $\Delta K_{cm} < 0$ and $\Delta E_{therm} > 0$ (deceleration) *or* with $\Delta K_{cm} > 0$ and $\Delta E_{therm} < 0$ (acceleration). In fact, nothing that we have introduced thus far would be violated if a car were to accelerate forward *without using fuel* simply by converting its thermal energy into macroscopic kinetic energy, thus lowering its temperature. But it does not happen!

Something else must be going on that allows energy to be converted or transferred in one direction, $K_{cm} \rightarrow E_{therm}$, but not in the other. That "something" is the *second* law of thermodynamics. The first law simply tells us that energy has to be conserved, but the second law will tell us that energy can flow only in certain directions but not others. There are many fascinating implications of the second law of thermodynamics, including such things as how we are able to tell the past from the future. Something to look forward to in Part IV!

11.6 Power Revisited

We introduced *power* in Chapter 9 as the rate of change of energy: $P = dE/dt$. If a force does work on a particle, then the small change of energy dE is caused by a small amount of work dW. In that case we could find power as the rate at which work is done: $P = dW/dt$. But what happens if the force that causes an object to accelerate happens to be a force that does no work? We need to reconsider the issue of power for a *system* of particles.

Recall Example 9-14 from Chapter 9, where we calculated the power output of a sprinter. We asserted that *her* power output—as distinct from the power of forces exerted on her—was a consequence of the conversion of chemical energy to kinetic energy. Now we have the tools to analyze this idea more thoroughly.

The force that accelerates the sprinter is the propulsion force $\vec{F}_{ground\ on\ foot}$. But we have shown that this force does no work because the point of its application—the sprinter's foot—does not move while the force is applied. So while the propulsion force accelerates the sprinter, it is *not* the cause of her increase in kinetic energy. No energy is transferred to her from the environment.

The energy equation, applied to the sprinter with $W_{ext} = 0$, is

$$\begin{aligned} W_{ext} = 0 &= \Delta E_{mech} + \Delta E_{int} \\ &= \Delta K_{cm} + \Delta U_{cm} + \Delta E_{therm} + \Delta E_{chem}. \end{aligned} \tag{11-23}$$

$\Delta U_{cm} = 0$ because she is moving horizontally, and her thermal energy is not likely to change significantly during the few seconds of a sprint. (This assumption would not be valid for a longer run in which the runner heats up and then transfers that energy to the environment as *heat* by evaporating perspiration.) Thus the sprinter obeys $\Delta K_{cm} = -\Delta E_{chem}$, and we conclude that her gain in kinetic energy results from a *loss* of the chemical energy stored in her muscles.

The sprinter's *power output* is $P_{out} = -dE_{chem}/dt$, the rate at which she is converting chemical energy into kinetic energy. Because she is losing chemical energy, we need the minus sign to make P_{out} a positive quantity. But because $-\Delta E_{chem} = \Delta K_{cm}$, we can also write the sprinter's power output as the rate at which she gains kinetic energy: $P_{out} = dK_{cm}/dt$. This quantity we can compute from kinematics, using $K_{cm} = \frac{1}{2}Mv_{cm}^2$ and $v_{cm} = at$:

$$K_{cm} = \frac{1}{2}Mv_{cm}^2 = \frac{1}{2}M(at)^2 = \frac{1}{2}Ma^2t^2$$

$$\Rightarrow P_{out} = \frac{dK_{cm}}{dt} = Ma^2t. \tag{11-24}$$

This is identical with our earlier result in Chapter 9, giving a peak power output of 1460 W as the sprinter reaches full speed at $t = 7$ s.

The major point, as with most examples in this chapter, is not the numerical answer—which, at least in this case, the single-particle model of Chapter 9 gave correctly. Instead our goal is understanding *how* energy is being utilized. The Chapter 9 exercise treated the sprinter as a particle and *assumed* that the propulsion force did work on her, causing her kinetic energy to increase. As we have now learned, the propulsion force does cause her to accelerate but it is *not* the source of her increasing kinetic energy. Instead, she is busy converting stored chemical energy into her kinetic energy of motion. So it really does make sense to talk about *her* power output. Likewise with an accelerating automobile—the propulsion force on the tires accelerates the car, but no energy is being transferred from the environment to the car. The energy conversion is from the fuel's chemical energy to the car's kinetic energy, so we are justified in referring to the *car's* power output as the *rate* at which chemical energy is being used.

EXAMPLE 11-4 A 40 kg boy climbs up a 10 m long hanging rope in 20 s. What is his power output?

SOLUTION Other than a brief interval to get started, rope climbers ascend at steady speed—they are not accelerating as they go up! At steady speed, the boy is in dynamic equilibrium with $\vec{F}_{net} = 0$. The only two forces acting on him, as seen in Fig. 11-9, are his weight and the upward force $\vec{F}_{rope}$ of the rope on his hands. The rope force is a *static* friction force because his hands do not slip. Thus, as with other propulsion forces, the rope force does *no* work on the boy as he climbs. The boy does gain potential energy as he climbs, but this energy is *not* transferred to him from the environment. It is another internal energy conversion.

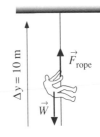

FIGURE 11-9
Chemical energy is converted to potential energy.

Of the two forces on the boy, $\vec{F}_{rope}$ does no work and the work of $\vec{W}$ has already been included as the gravitational potential energy $U_{cm} = U_{grav}$. So $W_{ext} = 0$ in the energy equation, and we are left with

$$W_{ext} = 0 = \Delta K_{cm} + \Delta U_{cm} + \Delta E_{therm} + \Delta E_{chem}.$$

$\Delta K_{cm} = 0$ because the boy climbs at steady speed, and, as with the sprinter, $\Delta E_{therm} \approx 0$ over the short time interval involved. Thus the boy's increase in potential energy occurs as a conversion from his stored chemical energy: $\Delta U_{cm} = -\Delta E_{chem}$. His power output is then

$$P_{out} = -\frac{dE_{chem}}{dt} = \frac{dU_{cm}}{dt} = \frac{d}{dt}(Mgy) = Mg\frac{dy}{dt} = Mgv = 196 \text{ W}.$$

where we used $v = \Delta y / \Delta t = 0.5$ m/s.

11.7 Collisions Revisited

[**Photo suggestion: Two billiard balls colliding.**]

We will close our initial study of energy with another look at two colliding objects. We introduced one-dimensional collisions in Chapter 8. We found that the *momentum* is conserved in a collision between two particles of masses m_1 and m_2:

$$m_1 v_{1i} + m_2 v_{2i} = m_1 v_{1f} + m_2 v_{2f}.$$

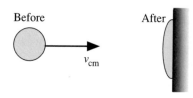

Before   After

v_{cm}

FIGURE 11-10 The force of the wall does no work on the ball of clay.

Now we can apply energy concepts to collisions and extend our analysis. Suppose you were to throw a ball of clay against a wall with velocity v_{cm}. The clay sticks to the wall, as shown in Fig. 11-10. This is analogous to a car hitting the wall. The wall does indeed exert a force on the clay, but it does no work because the point of application of the force does not move. The clay distorts under the influence of *internal* forces—it is a *system* of particles!—and those internal forces transfer the initial center of mass kinetic energy to thermal energy. The energy equation is

$$W_{ext} = 0 = \Delta E_{mech} + \Delta E_{int}$$
$$= \Delta K_{cm} + \Delta E_{therm}$$

Because the ball sticks, $\Delta K_{cm} = -K_i = -\frac{1}{2}mv_{cm}^2$. Solving for ΔE_{therm}:

$$\Delta E_{therm} = -\Delta K_{cm} = -(-\tfrac{1}{2}mv_{cm}^2) > 0, \tag{11-25}$$

where the fact that kinetic energy is *lost* means that thermal energy (and hence temperature) is *gained*. No other form of internal energy is relevant in this problem ($\Delta E_{chem} = 0$), and U_{cm} does not change because the clay's center of mass moves horizontally.

Now suppose, instead, we were to throw a springy ball at the wall, and that it bounces back. Again, the force of the wall does no work on the ball, but we now need to include the elastic potential energy of the ball's spring in the energy equation:

$$W_{ext} = 0 = \Delta K_{cm} + \Delta U_{sp} + \Delta E_{therm}. \tag{11-26}$$

Apply this to the *full* collision, from just before impact until just after the ball has rebounded from the wall with kinetic energy K_f. Then $\Delta U_{sp} = 0$ because the elastic ball has fully re-expanded. Thus

$$0 = \Delta K_{cm} + 0 + \Delta E_{therm} = (K_f - K_i) + \Delta E_{therm}$$
$$\Rightarrow \Delta E_{therm} = K_i - K_f. \tag{11-27}$$

That is, any macroscopic kinetic energy that appears to be "lost" in the bounce has simply been converted to the ball's thermal energy.

When the ball of clay stuck to the wall, it had no springiness at all. A collision in which the objects stick together and do not rebound ($K_f = 0$) is called a *perfectly inelastic collision*, and that is the situation we analyzed in Chapter 8.

At the other extreme, *all* of the ball's original kinetic energy could be stored as elastic potential energy. Once the ball is fully compressed, it expands and pushes off the wall with its kinetic energy fully restored. In this case $K_f = K_i$ and $\Delta E_{therm} = 0$. Such a collision, in which the ball's kinetic energy is conserved, is called a **perfectly elastic collision**.

Needless to say, most real collisions fall somewhere between perfectly elastic and perfectly inelastic. A rubber ball bouncing on the floor might "lose" 10% of its kinetic energy on each bounce, where it is converted to thermal energy. Therefore the ball only returns to 90% of the height of the previous bounce. Perfectly elastic and perfectly inelastic collisions are limiting cases rarely seen in the real world, but they are nonetheless instructive for demonstrating the major ideas without making the mathematics too complex.

Let us consider an important example of a perfectly elastic collision, namely the head-on collision of a ball of mass m_1, having velocity v_{1i}, with a ball of mass m_2 that is initially at rest. As shown in Fig. 11-11, the balls' velocities after the collision are v_{1f} and v_{2f}. These velocities, as always, are the x-components of velocity vectors and have signs. Mass m_1, in particular, could bounce backward and have a negative value for v_{1f}.

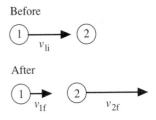

Before

After

FIGURE 11-11 A perfectly elastic collision.

We have two conditions that the collision must obey: conservation of momentum (obeyed in any collision) and conservation of kinetic energy (because the collision is perfectly elastic). Thus

momentum:
$$m_1 v_{1i} = m_1 v_{1f} + m_2 v_{2f} \tag{11-28}$$

energy:
$$\tfrac{1}{2} m_1 v_{1i}^2 = \tfrac{1}{2} m_1 v_{1f}^2 + \tfrac{1}{2} m_2 v_{2f}^2. \tag{11-29}$$

Momentum conservation alone is not sufficient to analyze the collision because there are two unknowns—the two final velocities. That is why we did not consider perfectly elastic collisions in Chapter 8. But energy conservation gives us another condition. Equations 11-28 and 11-29 are two equations in two unknowns, so we can solve them. Isolating v_{1f} in Eq. 11-28 gives

$$v_{1f} = v_{1i} - \frac{m_2}{m_1} v_{2f}. \tag{11-30}$$

Substituting this into Eq. 11-29:

$$\tfrac{1}{2} m_1 v_{1i}^2 = \tfrac{1}{2} m_1 \left(v_{1i} - \frac{m_2}{m_1} v_{2f} \right)^2 + \tfrac{1}{2} m_2 v_{2f}^2$$
$$= \tfrac{1}{2} \left(m_1 v_{1i}^2 - 2 m_2 v_{1i} v_{2f} + \frac{m_2^2}{m_1} v_{2f}^2 + m_2 v_{2f}^2 \right). \tag{11-31}$$

The first two terms on each side cancel. With a bit of algebra, the resulting equation can be rearranged to give

$$v_{2f} \cdot \left[\left(1 + \frac{m_2}{m_1} \right) v_{2f} - 2v_{1i} \right] = 0. \tag{11-32}$$

One possible solution to this equation is seen to be $v_{2f} = 0$. However, this solution is of no interest; it is the case where m_1 misses m_2. The other solution is

$$v_{2f} = \frac{2}{1 + m_2 / m_1} v_{1i} = \frac{2m_1}{m_1 + m_2} v_{1i},$$

which can be substituted back into to Eq. 11-30 to yield v_{1f}. The complete solution is then

$$\begin{cases} v_{1f} = \dfrac{m_1 - m_2}{m_1 + m_2} v_{1i} \\ v_{2f} = \dfrac{2m_1}{m_1 + m_2} v_{1i}. \end{cases} \tag{11-33}$$

These equations are a little difficult to interpret as they are, so let us look at three special cases.

Case 1: $m_1 = m_2$. This is the case of one billiard ball striking another of equal size. Eqs. 11-33 give, for this case,

$$\begin{cases} v_{1f} = 0 \\ v_{2f} = v_{1i}. \end{cases}$$

In other words, the incoming ball *stops*, while the second ball is knocked forward with the first ball's original velocity—exactly what billiard balls do!

Case 2: $m_1 \gg m_2$. This is the case of a bowling ball running into a pea. We do not want an exact solution here, but an approximate solution for the limiting case that $m_1 \to \infty$. Evaluating Eqs. 11-33 in this limit gives

$$\begin{cases} v_{1f} \approx v_{1i} \\ v_{2f} \approx 2v_{1i}. \end{cases}$$

In this case the heavy incoming ball hardly slows down—it barrels forward, unaware of the collision it had. The lightweight ball that was struck is knocked forward at twice this speed.

Case 3: $m_1 \ll m_2$. Now we have the reverse case of a pea colliding with a bowling ball. Here we are interested in the limit $m_1 \to 0$, in which case Eqs. 11-33 become

$$\begin{cases} v_{1f} \approx -v_{1i} \\ v_{2f} \approx 0. \end{cases}$$

Now the incoming pea simply bounces off, reversing direction without losing speed, while the bowling ball does not even respond.

These cases agree well with our expectations and give us confidence that Eqs. 11-33 accurately describe a perfectly elastic collision. A partially elastic collision, where $K_f \neq K_i$, can be analyzed in the same way except that the energy equation, Eq. 11-29, needs a slight modification to take into account the loss of kinetic energy. This will be left as a homework problem.

EXAMPLE 11-5 A 200 g steel ball is tied to a string 1 m long and hung as a pendulum. The ball is pulled sideways until the string is at a 45° angle, and is then released. At the very bottom of its swing the ball strikes a 500 g steel block that is resting on a frictionless table. What is the speed of the block after the collision, and to what angle does the ball rebound?

SOLUTION This is a three-part problem. First the ball swings down as a pendulum. Second, the ball and block undergo a collision. Steel balls bounce off each other very well, so we will assume that the collision is perfectly elastic. Third, the ball will swing back up as a pendulum after it bounces off the block. Figure 11-12 shows four distinct moments of time: as the ball is released, an instant before the collision, an instant after the collision but before the ball and block have had time to move, and as the ball reaches its highest point on the rebound. Call the ball A and the block B, so $m_A = 0.2$ kg and $m_B = 0.5$ kg.

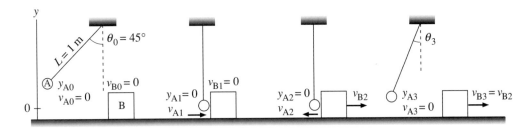

FIGURE 11-12 Four moments in the collision of a pendulum with a block.

First part: The first part involves the ball only. Its initial height is
$$y_{A0} = L - L\cos\theta_0 = L(1 - \cos\theta_0) = 0.293 \text{ m}.$$
The tension in the string does no work, because the force at each point is perpendicular to the direction of motion, so $K + U_{grav}$ is conserved. We can use conservation of mechanical energy to find the ball's velocity at the bottom, just before impact on the block:
$$K_f + U_f = K_i + U_i$$
$$\Rightarrow \left[\tfrac{1}{2} m_A v_{A1}^2 + m_A g y_{A1} = \tfrac{1}{2} m_A v_{A1}^2 \right] = \left[\tfrac{1}{2} m_A v_{A0}^2 + m_A g y_{A0} = m_A g y_{A0} \right]$$
$$\Rightarrow v_{A1} = \sqrt{2 g y_{A0}} = 2.40 \text{ m/s}.$$

Second part: The ball and block undergo an elastic collision in which the block was initially at rest. These are the conditions for which Eq. 11-33 was derived. Using Eq. 11-33, the velocities *immediately* after the collision, prior to any motion, are
$$v_{A2} = \frac{m_A - m_B}{m_A + m_B} v_{A1} = -1.03 \text{ m/s}$$
$$v_{B2} = \frac{2 m_A}{m_A + m_B} v_{A1} = +1.37 \text{ m/s}.$$

The ball rebounds toward the left with a speed of 1.03 m/s while the block moves to the right at 1.37 m/s. The block will continue with this speed because the table is frictionless.

Third part: Now the ball is a pendulum with an initial speed of 1.03 m/s. Mechanical energy is again conserved, so we can find its maximum height at the point where $v_{A3} = 0$:

$$K_f + U_f = K_i + U_i$$

$$\Rightarrow \left[\tfrac{1}{2} m_A v_{A3}^2 + m_A g y_{A3} = m_A g y_{A3} \right] = \left[\tfrac{1}{2} m_A v_{A2}^2 + m_A g y_{A2} = \tfrac{1}{2} m_A v_{A2}^2 \right]$$

$$\Rightarrow y_{A3} = \frac{v_{A2}^2}{2g} = 0.0541 \text{ m.}$$

The height y_{A3} is related to angle θ_3 by $y_{A3} = L(1 - \cos\theta_3)$. This can be solved to find the angle of rebound:

$$\theta_3 = \cos^{-1}\left(1 - \frac{y_{A3}}{L} \right) = 18.9°.$$

The block speeds away at 1.37 m/s and the ball rebounds to an angle of 18.9°.

Summary

Important Concepts and Terms

macrophysics	internal energy
microphysics	heat
translational motion	first law of thermodynamics
thermal energy	total energy
center of mass	law of conservation of energy
internal potential energy	perfectly elastic collision
internal kinetic energy	

This chapter has focused on how to think about the concept of energy when dealing with a system of particles. Our original definitions of work and energy were for single particles only, so some modification has been necessary to apply them to systems of particles. We found, in particular, that the boundary of the system must be displaced in order for work to be done on the system. Many external forces that are responsible for the acceleration of the system do *no* work on the system. That is, they do not transfer any energy from the environment to the system. Instead, the increase (or decrease) of the system's center of mass energy comes about via *conversion* of energy from one form to another.

The internal energy of a system of particles is

$$E_{int} = E_{therm} + E_{chem} + E_{other}$$

where E_{therm} is the thermal energy due to the microscopic motions of the atoms; E_{chem} is a particular form of potential energy stored in molecular bonds and available for release through chemical reactions; and E_{other} is other forms of energy, such as nuclear energy, sound energy, and so on.

As long as no heat flow takes place, the energy equation for a system of particles is

$$W_{ext} = \Delta E_{mech} + \Delta E_{int}$$
$$= \Delta K_{cm} + \Delta U_{cm} + \Delta E_{therm} + \Delta E_{chem} + \ldots$$

Energy can be transferred to the system from the environment (or vice versa) via the work of external forces that displace the boundary. Equally important in many situations, in the absence of work being done, are *conversions* of energy between macroscopic and microscopic forms. Single particles have no internal structure, so these new forms of energy and new ways of thinking about energy transfer appear only when we begin to consider systems of particles.

The combination of mechanical energy and internal energy is the *total energy* of a system of particles:

$$E_{\text{total}} = E_{\text{mech}} + E_{\text{int}}.$$

The law of conservation of energy says that the total energy of an *isolated system* is conserved. An isolated system is one on which the environment does no work and to which it transfers no heat.

We have now spent three chapters developing the concepts of energy and energy transfer. We began with the relationship between work and kinetic energy for a single particle. We expanded this idea in Chapter 10 with the addition of potential energy, and we expanded it yet again in this chapter with the inclusion of thermal and chemical energy. Now we have even gone so far as to suggest another energy transfer process called *heat*. Now is a good time to step back and see the big picture of how these ideas are related.

Figure 11-13 provides a "Knowledge Structure of Energy." The first law of thermodynamics is the basic principle about energy transfer. Work and heat are the two means by which energy is transferred between the system and its environment. Work and heat *change* the system's energy according to $\Delta E_{\text{total}} = W + Q$. The total energy is subdivided into mechanical energy and internal energy, and each of these is further subdivided. Energy conversions *within* the system can transform one kind of energy to another, but these conversions do not change the total energy. Only work or heat from the environment can change the total energy.

It is not always obvious when to use energy conservation and when to use Newton's laws for solving a problem. The major "clue" is whether or not the problem requires knowing the details of the interaction. If the problem asks you to relate "before" and "after," without asking about the time interval or the forces of interaction, then energy

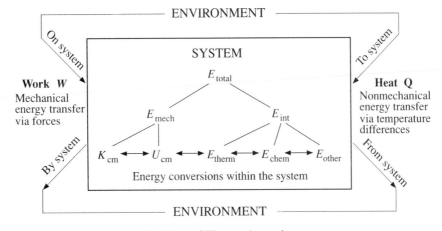

First Law of Thermodynamics

$$W + Q = \Delta E_{\text{total}} = \Delta K_{\text{cm}} + \Delta U_{\text{cm}} + \Delta E_{\text{therm}} + \Delta E_{\text{chem}} + \dots$$
Net energy transferred = Total change in system energy

FIGURE 11-13 The Knowledge Structure of Energy.

conservation (and maybe momentum conservation) is the appropriate strategy. But if you need to know times, forces, or other details of the interaction itself, then Newton's laws are required. Many problems can be worked either way, but energy conservation is almost always easier and shorter than using the full apparatus of Newton's laws.

Exercises and Problems

Exercises

1. A 1 kg rubber ball is dropped from a height of 10 m and undergoes a perfectly elastic collision with the earth. What is the earth's velocity after the collision? (Assume it was initially at rest.) How many years would it take the earth to move 1 mm at this speed?

2. A 50 g marble traveling at 2 m/s strikes a 20 g marble at rest. What is the speed of each marble immediately after the collision?

3. A 100 kg weight lifter crouches, picks up a 100 kg barbell, and holds it against his chest. He then rises to a standing position, raising his center of mass by 60 cm.
   a. What is the work W_{ext} on the weight lifter as he stands?
   b. What is ΔK_{cm}?
   c. What is ΔU_{cm}?
   d. What is ΔE_{therm}?
   e. What is ΔE_{chem}?
   f. What is the lifter's power output if he takes 0.5 s to rise to the standing position?

Problems

Problems 4–7 are open-ended. That is, they not asking for a specific answer but, instead, for an analysis and interpretation analogous to that of Examples 11-1 to 11-3. You should, among other things, clearly identify the system, identify which forces do work and which do no work, determine the numerical values of important quantities, discuss the relevance of Newton's third law, and discuss where the energy for the motion originates and where it ends up (i.e., energy transfers and conversions).

4. A 1500 kg automobile accelerates from rest up a 20° slope. It achieves a speed of 10 m/s after traveling a distance of 50 m. Analyze and interpret the forces and the energy transfers involved in this acceleration.

5. A 50 g ball of clay is dropped from a height of 1 m and sticks to the floor. Analyze and interpret the forces and the energy transfers involved in this situation.

6. A 50 g wood ball is dropped from a height of 1 m and bounces back to a height of 40 cm. Analyze and interpret the forces and the energy transfers involved from the ball's initial release until it reaches the maximum height on the rebound.

7. Bob, who has a mass of 80 kg, puts on a pair of rollerblades and stands facing a wall with his arms bent. He then pushes off from the wall by straightening his arms, causing him to roll backwards away from the wall. The coefficient of rolling friction between the rollerblades and the ground is 0.10. Bob rolls 50 cm before his fingers lose contact with the wall. His speed at that point is 5 m/s. He then coasts

until coming to a halt. Analyze and interpret the forces and the energy transfers involved in Bob's motion from the time he starts his push until he is again at rest.

8. It takes Cindy (50 kg) 10 s to run up three flights of stairs at steady speed, gaining 10 m of elevation. a) Analyze and interpret the forces and energy transfers involved in her motion. b) What is Cindy's power output?

9. a. How much power output would a car need to travel at a steady 60 mph on a frictionless road if there were no air resistance?
   b. While rolling friction on the road is fairly negligible, the air resistance on a car traveling 60 mph is 900 N. What is the car's power output when driving a steady 60 mph?
   c. The magnitude of the air resistance force is proportional to v^2. If the car's maximum power output is 100 kW (134 hp), what is its maximum speed (in mph)?

✍ 10. A 20g ball is fired with a speed of 5 m/s toward a 100 g block attached to a spring of constant 20 N/m (Fig. 11-14). The block slides on a frictionless surface and, in doing so, compresses the spring.
   a. If the collision is perfectly elastic, what are the velocities of each object immediately after the collision?
   b. What is the maximum compression of the spring?
   c. Repeat parts a) and b) for the case of a perfectly inelastic collision.

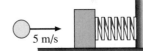

5 m/s

FIGURE 11-14

✍ 11. A 20 g ball is fired with initial speed v_0 toward a 100 g ball that is hanging motionless from a 1 m long string, as shown in Fig. 11-15. The balls undergo a head-on, perfectly elastic collision, after which the 100 g ball swings out to make a maximum angle $\theta_{max} = 50°$. What was v_0?

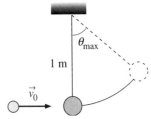

θ_{max}

1 m

$\vec{v}_0$

FIGURE 11-15

✍ 12. Ball 1, having a mass of 100 g and traveling at 10 m/s, collides head on with ball 2, which has a mass of 300 g and is initially at rest.
   a. If the collision is perfectly elastic, what are the final velocities of each ball?
   b. If the collision is perfectly *in*elastic, what are the final velocities of each ball?

13. A perfectly elastic collision is defined as one in which the kinetic energy is conserved: $K_f = K_i$. In general, some fraction ε of the initial kinetic energy is "lost" in the collision, and is converted to thermal energy. Thus $K_f = (1 - \varepsilon)K_i$. A perfectly elastic collision has $\varepsilon = 0$.
   a. The maximum possible energy loss, and thus the maximum value for ε, occurs for a perfectly inelastic collision where the objects stick together. What is ε_{max} for the two balls in Problem 12?
   b. What value of ε causes ball 1 to be at rest after the collision ($v_{1f} = 0$)?
   c. What are the final velocities of each ball if $\varepsilon = 0.4$? (Hint: You will need to derive expressions for v_{1f} and v_{2f} for the case where $\varepsilon \neq 0$.)

[**Estimated 10 additional problems for the final edition.**]

Chapter **12**

Newton's Theory of Gravity

LOOKING
BACK

Sections 6.4–6.6; 9.8; 10.3–10.5

12.1 A Little History

[**Photo suggestion: Total eclipse of the sun.**]

Every ancient culture was fascinated with the motion of the heavens above. Without city lights or urban haze, the nighttime sky and the daytime sun were ever-present, powerful experiences. The unknown people that built Stonehenge clearly used it as a solar observatory, and the ancient Babylonians learned to predict the occurrence of solar eclipses. In fact, much of the early development of mathematics was stimulated by a desire to predict future events in the heavens.

The study of the structure of the universe is called **cosmology**. The ancient Greeks developed a cosmology, illustrated in Fig. 12-1, in which the earth was located at the center of the universe while the moon, the sun, the planets, and the stars were points of light that turned about the earth on large "celestial spheres." Their ideas were based on the common-sense observation that the earth seems to be at rest. Even today, when we regularly see pictures taken from space, our senses

FIGURE 12-1 The earth-centered cosmology of the ancient Greek and medieval periods.

392

give us no indication at all that the ground under our feet is in motion, and it appears very much that the heavenly objects revolve around the earth. Greek cosmology started with these everyday observations, but it evolved into a sophisticated, mathematical theory.

The second century Egyptian astronomer Ptolemy (the P is silent) further expanded Greek cosmology. He developed an elaborate mathematical model of the solar system that was quite accurate for predicting the complex planetary motions. Ptolemy's earth-centered cosmology was accepted without question for over a thousand years in medieval Europe, playing a prominent role in the writings of Dante and others at the peak of late-medieval culture. These ideas also became closely intertwined with Roman Catholic theology. Islamic astronomers made additional advances in astronomy during this time period, although those were not known to western Europe until much later.

Then, in the year 1543, the medieval world was turned on its head with the publication of Nicholas Copernicus' *De Revolutionibus*. Copernicus argued that it is not the earth at rest in the center of the universe—it is the sun! Furthermore, Copernicus asserted that all of the planets, including the earth, revolve about the sun in circular orbits. What better proof of God's existence and perfection could be found, Copernicus reasoned, than a universe designed of perfect circles?

Copernicus was suspicious that his ideas would not sit well with the authorities, and he tactfully waited until on his deathbed to have his book published. The Church did, indeed, angrily denounce his views and for over one hundred years persecuted astronomers who used Copernican ideas. After all, Copernicus' attempt to remove the earth, and with it humans, from the center of the universe undermined the very foundation of Catholic faith and theology. It wasn't until a century later, when Galileo used a telescope to study the heavens, that the Copernican view became widely accepted.

Galileo could *see* moons orbiting Jupiter, just as Copernicus suggested. He could *see* that Venus has phases, like the moon, which implied its orbital motion about the sun. And he could *see* spots moving across the face of the Sun, indicating that it was not a perfect and unchanging symbol of God. Galileo was tried and convicted of heresy (this was still the time of the Inquisition), forced to publicly recant his views, and left a broken man. But his evidence was too persuasive, and by the time of his death in 1642 (the year of Newton's birth) the Copernican revolution was complete.

The greatest astronomer of this period was Tycho Brahe, a Dane born in 1546, just three years after Copernicus' death. Tycho began astronomical observations as a teenager, and for roughly thirty years—from 1570 to 1600—compiled the most voluminous and most accurate astronomical observations the world had known. This was still prior to the invention of the telescope, so Tycho's observations were all with the naked eye. However, Tycho did develop various mechanical sighting devices that allowed him to determine the positions of stars and planets in the sky with unprecedented accuracy. At his peak, when he was supported by the Danish king, Tycho employed a large staff of astronomers and craftsmen—the first research institute! Despite his careful observations, Tycho never accepted Copernicus' assertion that the earth was in motion.

Tycho willed his records to a young mathematical assistant named Johannes Kepler. Kepler had wanted to prepare for the ministry, but his family was poor and could not afford his studies. So he turned to teaching mathematics and—for an income supplement—preparing astrological calendars. It was the need for an accurate location of the planets on certain dates, both in the past and the future, that brought him to Tycho.

Kepler spent ten years trying to find patterns in Tycho's data. To appreciate the difficulty of his task, keep in mind that Kepler was working before the development of graphs or of calculus, and certainly before calculators! His mathematical tools were algebra, geometry, and trigonometry, and he was faced with thousands upon thousands of individual observations of planetary positions measured as angles above the horizon. Kepler had, early on, become one of the first outspoken defenders of Copernicus, yet he could not find evidence for circular orbits in Tycho's records.

Kepler was motivated to continue by a rather mystical philosophy in which symbolic meanings of the various planets were entwined with the geometry of variously shaped polyhedrons. Eventually his "method" led him to think that Copernicus' insistence on the circularity of orbits was unnecessary, and that perhaps orbits could have other shapes. This was the mental barrier that had to be broken. Once beyond it, Kepler was quickly able to deduce that the orbit of Mars is, indeed, not a circle but an *ellipse*! Furthermore—and this was an essential finding for him to discover—the speed of the planet is not constant but varies as it moves about the ellipse.

Kepler published his results in 1609, announcing the discovery of his first two laws in a paper entitled *On the Motion of Mars*. He continued to work, applying his ideas to the other planets, and his so-called third law was announced in 1619 in his *Harmonies of the World*.

Stated in modern terms, Kepler's laws of planetary orbits are:

1. The planets move in elliptical orbits, with the sun at one focus of the ellipse.

2. A line drawn between the sun and any planet sweeps out equal areas during equal intervals of time.

3. The square of the orbital period of any planet is proportional to the cube of the semimajor axis of its orbit.

Figure 12-2 will clarify these laws. Figure 12-2a shows that an ellipse has two *foci* (plural of *focus*), and the sun occupies one of these. The long axis is called the semimajor axis, while the short axis is the semiminor axis. Figure 12-2b shows several lines drawn from the sun to a planet orbiting the sun. During a time interval Δt these lines "sweep out" an area. If the two areas shown are required to be equal, for equal Δt, then the planet must move faster when near the sun and slower when further away.

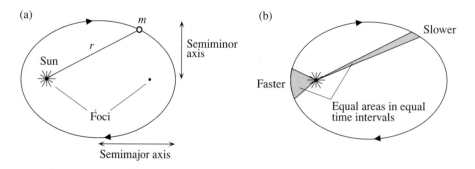

FIGURE 12-2 The elliptical orbit of a planet about the sun. a) The sun at one focus of the ellipse, with the semimajor and semiminor axes defined. b) Illustration of Kepler's second law, showing the equal areas swept out during equal time intervals.

For most of the planets, excepting Mercury and Pluto, the elliptical orbits are only very slightly distorted circles. A circle is the limiting case of an ellipse in which the two foci move together and the semimajor and semiminor axes become equal to the radius of the circle. Because the mathematics of ellipses is difficult, we will consider only circular orbits in this chapter. In that case, the second law implies that the planet's speed will be *constant* and the third law says that the square of the period will be proportional to the cube of the *radius* of the orbit.

Kepler made an additional contribution that is less widely recognized but that was essential to prepare the way for Newton. The role of the sun in Ptolemy's system was merely to light and warm the earth. This did not change with Copernicus. The sun now had to light all the planets, but it had no role in their movement. Those few astronomers who thought about the issue at all accepted that each planet moved in a circle, because that is the "natural" motion for planets, at a speed determined by God. Kepler was the first to suggest that the sun was a center of force that somehow *caused* the planetary motions. Now Kepler, keep in mind, was working well before Galileo and Newton, so he did not speak in terms of forces and centripetal accelerations. He thought that some type of rays or spirit emanated from the sun and pushed the planets around their orbits. The value of his contribution was not the specific mechanism he proposed but his introduction of the idea that the sun somehow exerts forces on the planets to determine their motion. His third law is a consequence of this line of thought, because he was able to show that the planets' periods are not just arbitrary choices of God but that they follow a mathematical law.

In hindsight, we can see that Kepler's analysis and Galileo's observations had set the stage for a major theoretical leap. All that was needed was a great intellect to recognize and pull together these ideas. Enter Isaac Newton, perhaps one of the most brilliant people ever to live.

12.2 Isaac Newton

Isaac Newton was born to a poor farming family in 1642, the year of Galileo's death. His father, who died three months before Newton was born, could not even sign his name, and his mother abandoned him at an early age to the care of his grandmother. He entered Trinity College at Cambridge University at age nineteen as a "subsizar"—a poor student who had to work his way through school. These circumstances may explain the extreme feelings of insecurity that dominated Newton's adult life.

[**Photo suggestion: Portrait of Newton.**]

Newton received a Bachelor of Arts degree in 1665, at age 23, just as the universities were closing for a two-year period due to an outbreak of the plague in England. He returned to his family farm for that period, during which he made major experimental discoveries in optics, laid the foundations for his theories of mechanics and gravitation, and made major progress toward his invention of calculus as a whole new branch of mathematics. At age twenty-six, upon the reopening of the universities, Newton was appointed to the prestigious position of Lucasian Professor of Mathematics at Cambridge University.

A popular image has Newton thinking of the idea of gravity after an apple fell on his head. It is an amusing story, and it is at least close to the truth. Newton himself said that the "notion of gravitation" came to him as he "sat in a contemplative mood" and "was occasioned by the fall of an apple." It occurred to him that, perhaps, the apple was attracted to the *center* of the earth, but was prevented from getting there by the earth's surface. And if the apple was so attracted, why not the moon?

Robert Hooke, discoverer of Hooke's law, had already suggested that the planets might be attracted to the sun with a strength proportional to the inverse square of the distance between the sun and the planet. This seems to have been a hunch, rather than based on any particular evidence, and Hooke failed to follow up on the idea. Newton's genius was not just his successful application of Hooke's suggestion, but his sudden realization that the force of the sun on the planets was *identical* to the force of the earth on the apple. In other words, gravitation is a *universal* force between all objects in the universe! No one before had ever thought that the mundane motion of objects on earth had any connection at all with the stately motion of the planets around the sun.

Newton's analysis was exactly what we presented in Section 6.6, with the critical step being his comparison of the ratio $g_{\text{on earth}}/g_{\text{at moon}}$ (actually, he used forces rather than g's) to the ratio $(R_{\text{m}}/R_{\text{e}})^2$—exactly as we did in Eq. 6-32. In Newton's words:

> I deduced that the forces which keep the planets in their orbs must be reciprocally as the squares of their distances from the centers about which they revolve; and thereby compared the force requisite to keep the Moon in her orb with the force of gravity at the surface of the Earth; and found them answer pretty nearly.

This is worth reading a couple of times, because the language is a bit antiquated. Note that his two ratios were not identical—the distances were not known as well then as now, and, as we saw in Eq. 6-32, even modern values give a discrepancy of nearly 1%. But he found them "pretty nearly" equal, and he knew that he had to be on the right track.

With this flash of insight, our conception of the universe was changed forever. Copernicus displaced humans from the center of the universe, and now Newton had shown that the laws of the heavens were no different than earthly laws. Nonetheless, Newton did not publish his results for a long twenty-two years. The issue that troubled him was treating the sun, the earth, and the other planets as if they were single particles with all the mass at the center. If his idea about a universal force was correct, then *every* atom in the earth exerts a force on *every* atom in the moon. Newton had to show that all of these forces add up to give a result that is identical with treating the bodies as single particles. This is a difficult problem in integral calculus, and Newton had first to develop the necessary mathematics. He did eventually succeed, and his theory of gravitation was published in 1687 along with his theory of mechanics (Newton's laws) in his great work *Philosophia Naturalis Principia Mathematica* (*Mathematical Principles of Natural Philosophy*). This work, now referred to as simply Newton's *Principia*, was written in Latin, as was customary, and not translated into English until 1729. The rest is history.

12.3 Newton's Law of Gravity

Newton proposed that *every* object in the universe attracts *every other* object with a force that has the following properties:

1. The force is inversely proportional to the square of the distance between the objects.

2. The force is directly proportional to the product of the masses of the two objects.

3. The force is an attractive force, with the force vectors directed along a line joining the two objects.

Property 3 indicates that Newton already had a good understanding of what were to become his laws of motion. The force on a planet does *not* need to be tangential to the orbit, which earlier scientists thought it must to keep the planet moving—Newton recognized that inertia will do that. Instead, the force needs to be directed *toward* the sun to create a centripetal acceleration and change the planet's velocity vector. So you can see that the theory of gravity proposed by Newton was an inspired blend of earlier astronomical ideas with his own new theory of dynamics.

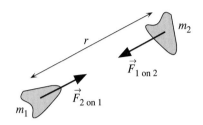

FIGURE 12-3 The gravitational forces on masses m_1 and m_2.

Figure 12-3 shows two masses, separated by a distance r, exerting attractive gravitational forces on each other. In accordance with Newton's third law, the magnitude of $\vec{F}_{2 \text{ on } 2}$ must be equal to the magnitude of $\vec{F}_{2 \text{ on } 1}$. **Newton's law of gravity** states that the magnitude of these two forces is given by

$$|\vec{F}_{1 \text{ on } 2}| = |\vec{F}_{2 \text{ on } 1}| = \frac{Gm_1 m_2}{r^2} \quad \text{(Newton's law of gravity).} \quad (12\text{-}1)$$

The constant G, called the **gravitational constant**, is a proportionality constant necessary to relate the masses, measured in kilograms, to the force, measured in newtons. In the SI system, G has the value

$$G = 6.67 \times 10^{-11} \text{ N m}^2/\text{kg}^2.$$

It is important to note that *two* forces are involved in Eq. 12-1, forming an action/reaction pair with one acting on each object. The direction of each force is directly toward the other object.

Strictly speaking, Eq. 12-1 is valid only for particles. However, as we noted, Newton was able to show that this equation also applies to spherical objects, such as planets, if r is measured as the distance between the *centers* of the spheres. Even though our intuition and common sense suggest this to us, as they did to Newton, the proof is difficult and we will omit it.

Knowing G, we can calculate the size of the gravitational force. Consider two objects of mass 1 kg each spaced a distance 1 m apart. According to Newton's theory, these two masses exert an attractive gravitational force on each other. The magnitude of that attractive force is given by Eq. 12-1:

$$F_{\text{grav}} = |\vec{F}_{1 \text{ on } 2}| = |\vec{F}_{2 \text{ on } 1}| = \frac{(6.67 \times 10^{-11} \text{ N m}^2/\text{kg}^2)(1 \text{ kg})(1 \text{ kg})}{(1 \text{ m})^2} = 6.67 \times 10^{-11} \text{ N.}$$

This is an exceptionally tiny force, especially when you compare it to the weight of one of the masses: $|\vec{W}| = mg = 9.8$ N.

The fact that the gravitational force between two ordinary-size objects is so small is the reason that we—as well as everyone before Newton—are not aware of it. As you sit there reading, you are being attracted to this book, to the person sitting next to you, and to every object around you, but the forces are so tiny in comparison to your weight and to the normal forces and friction forces acting on you that they are completely undetectable. Only when one (or both) of the masses gets to be exceptionally large—planet-size—does the force of gravity become important.

If we calculate the force *of the earth* on one of the 1 kg masses from the previous example, we find a more respectable result:

$$|\vec{F}_{\text{earth on 1 kg}}| = \frac{Gm_{\text{earth}}m_{1\text{ kg}}}{r^2}$$

$$= \frac{(6.67 \times 10^{-11} \text{ N m}^2 / \text{kg}^2)(5.98 \times 10^{24} \text{ kg})(1 \text{ kg})}{(6.37 \times 10^6 \text{ m})^2} = 9.8 \text{ N},$$

where the distance r between the 1 kg mass and the center of the earth is just the earth's radius. (Table 12-2 in Section 12.6, also reprinted inside the front cover of the book, contains astronomical data that will be used for some of the examples and homework in this chapter.)

Notice that $|\vec{F}_{\text{earth on 1 kg}}| = 9.8$ N. This is exactly the weight of a 1 kg mass: $mg = 9.8$ N. Is this a coincidence? Of course not—the weight $|\vec{W}| = mg$ of an object *is* the "force of gravity" acting on that object. Newton's law of gravity has now given us a way to determine explicitly the gravitational force that the earth exerts on an object, based on the masses and the radius of the earth. But there is only one force, being calculated two different ways, so they have to agree. This suggests that we should be able to use Newton's law of gravity to *predict* the value of g. (We will return to this idea in Section 12.4.)

Gravity, despite its weakness, is a *long-range* force. No matter how far away two objects may be, there is a gravitational attraction between them given by Eq. 12-1. Because no other force in nature has this range, gravity is the most ubiquitous force in the universe. It is responsible not only for keeping your feet on the earth, but for the earth orbiting the sun, the solar system orbiting the center of the Milky Way galaxy, and the entire Milky Way galaxy performing an intricate orbital dance with other galaxies making up what is called the "local cluster" of galaxies. Figure 12-4 shows a photograph of a galaxy, containing perhaps 10^{11} stars. The dynamics of the stellar motions, covering distances of hundreds of thousands of light years, is governed by Eq. 12-1.

It is extremely important to notice that the gravitational force between two objects, and hence their acceleration, will *vary* with the distance between them. Thus the kinematics of

FIGURE 12-4 A galaxy of 10^{11} stars, all of whose motions are governed by Newton's law of gravity.

motion due to gravitational forces is *not* constant acceleration. None of the Chapter 3 formulas for constant acceleration will have any validity in this chapter. Incorrect application of constant acceleration kinematics to gravitational motion is a common student error; do not let it happen to you!

Newton's *theory* of gravity, it is worth noting, is more than just Eq. 12-1. The *theory of gravity* consists of: 1) A specific force law for gravity, as given by Eq. 12-1, *and* 2) an assertion that Newton's three laws of motion are *universally* applicable—as valid for the distant stars as they are for earthly objects. As a consequence, everything we have learned about forces, motion, and energy is relevant to the dynamics of satellites, planets, and galaxies.

The Principle of Equivalence

Newton's law of gravity contains a rather curious assumption. The concept of *mass* was introduced in Chapter 4 by considering the relationship between the force and the acceleration of an object. In fact, Section 4.3 defined the slope of the force-versus-acceleration graph as something called the *inertial mass*. This is the mass that appears in Newton's second law. The inertial mass of an object can be determined by measuring the object's acceleration a in response to a force F:

$$m_{\text{inert}} = \text{inertial mass} = \frac{F}{a}.$$

Notice that gravitational forces played no role in this definition of mass.

The quantities m_1 and m_2 in Newton's law of gravity, Eq. 12-1, are being used very differently than the way the quantity m, representing the inertial mass, is used in Newton's second law. Masses m_1 and m_2 govern the strength of the gravitational attraction between two objects. The mass used in Newton's law of gravity is called the **gravitational mass**. The gravitational mass of an object can be determined by measuring the attractive force exerted on it by another mass M a distance r away:

$$m_{\text{grav}} = \text{gravitational mass} = \frac{r^2 F_{\text{grav}}}{GM}.$$

The concept of acceleration does not enter into the definition of the gravitational mass.

These are two very different concepts of mass. Yet Newton, in his theory of gravity, asserts that the inertial mass in his second law is the very same mass that governs the strength of the gravitational attraction between two objects. The assertion that $m_{\text{grav}} = m_{\text{inert}}$ is called the **principle of equivalence**. It says that inertial mass is *equivalent to* gravitational mass. As a hypothesis about nature, the principle of equivalence is subject to experimental verification or disproof. There have been exceptionally clever experiments designed to look for any difference between the gravitational mass and the inertial mass, and they have shown that any difference, if it exists at all, is less than ten parts in a trillion! As far as we know today, the gravitational mass and the inertial mass are exactly the same thing. But why should a quantity associated with the dynamics of motion, relating force to acceleration, have anything at all to do with the gravitational attraction?

Einstein was especially intrigued with the equivalence between the gravitational mass and the inertial mass. The principle of equivalence forms one of the cornerstones of his general theory of relativity—the theory about curved space-time and black holes. Perhaps some future Newton or Einstein can shed some more light on this puzzle. Despite the impression you might get from an introductory book, such as this, physics is *not* a completed body of knowledge—and probably never will be!

12.4 Little *g* and Big *G*

We noted in the previous section that using Newton's law of gravity to calculate the force exerted by the earth on a 1 kg mass gives the same result as using $|\vec{W}| = mg$. Why then, you might wonder, bother with Newton's more complicated law if *mg* works well? The reason is that *mg* is the gravitational force on an object of mass *m only* on the surface of the earth. The value of 9.8 m/s^2 that we know for *g* is specifically an earthly value. The formula *mg* will not help us find the force exerted on the same object if it were on the moon's surface, nor can we use it to find the force of attraction between the earth and the moon. Newton's law of gravity provides a more fundamental starting point because it describes a *universal* force that exists between all objects.

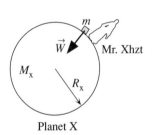

Figure 12-5 shows an object of mass *m* on the surface of Planet X. Mass, you will recall, is a characteristic of the *object*, so it has the same value anywhere in the universe. Planet X inhabitant Mr. Xhzt places the object on a scale to weigh it and exclaims (translated), "Aha! It has a weight $|\vec{W}| = mg_X$, which is the magnitude of the gravitational force between the object and X. Constant g_X is the acceleration due to gravity on my planet." We, taking a more cosmic perspective, reply, "Yes, it has that weight *because* of a universal force of attraction between X and the object. The size of the force is determined by Newton's law of gravity."

FIGURE 12-5 Weighing an object on Planet X.

This is a case where we are both correct. Because there is only one force on the object, we and Mr. Xhzt must arrive at the *same* numerical value for the magnitude of that force. Thus

$$\left[\text{Mr. Xhzt' s value } = |\vec{W}| = mg_X \right] = \left[\text{our value } = |\vec{F}_{X \text{ on } m}| = \frac{GM_X m}{R_X{}^2} \right] \tag{12-2}$$

$$\Rightarrow g_X = \frac{GM_X}{R_X{}^2}.$$

By equating the two different expressions for the force, we have arrived at a *prediction* of the "local value" of *g*, the acceleration due to gravity on Planet X. It depends on the mass and radius of the planet as well as on the value of *G*, which establishes the overall strength of the gravitational force. (Note that we are using uppercase letters for the planet and lowercase for objects on the planet. This is a fairly traditional notation.)

The expression for *g* in Eq. 12-2 is valid for any planet or star, not just Planet X. Using the values for the mass and radius of the earth, from Table 12-2, we can use Eq. 12-2 to *predict* the earthly value of *g*:

$$g_{earth} = \frac{GM_e}{R_e^2} = \frac{(6.67 \times 10^{-11} \text{ N} \cdot \text{m}^2 / \text{kg}^2)(5.98 \times 10^{24} \text{ kg})}{(6.37 \times 10^6 \text{ m})^2} = 9.83 \text{ m} / \text{s}^2. \quad (12\text{-}3)$$

Galileo could measure *g*, but he did not know *why* it was 9.8 m/s² rather than some other number. Now, with a more fundamental theory, we can see that value for *g* is a direct consequence of the size and mass of the earth. Similarly,

$$g_{moon} = \frac{(6.67 \times 10^{-11} \text{ N} \cdot \text{m}^2 / \text{kg}^2)(7.36 \times 10^{22} \text{ kg})}{(1.74 \times 10^6 \text{ m})^2} = 1.62 \text{ m} / \text{s}^2$$

$$g_{jupiter} = \frac{(6.67 \times 10^{-11} \text{ N} \cdot \text{m}^2 / \text{kg}^2)(1.90 \times 10^{27} \text{ kg})}{(6.99 \times 10^7 \text{ m})^2} = 25.9 \text{ m} / \text{s}^2.$$

A falling object on the moon would accelerate only about one-sixth as rapidly as on earth, while on Jupiter it would accelerate about 2.6 times more rapidly.

You may have noticed that the value of *g* calculated in Eq. 12-3 is slightly larger than the "accepted" value of 9.80 m/s². Is there something wrong with the theory? No, the difference arises because we have not considered what effects there might be from the earth's rotation. Figure 12-6 shows an object on the earth's equator as seen by an observer above the north pole. There are two forces on the object—the gravitational force $\vec{F}_{grav}$ and the normal force $\vec{N}$. You will recall that the *effective weight* of the object, what a set of scales would read, is the magnitude $|\vec{N}|$ of the normal force. If the earth were not rotating, the object would be in equilibrium and the two forces would exactly balance. That is the situation described by the Eq. 12-2 value of *g*.

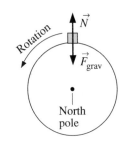

FIGURE 12-6 The earth's rotation affects the value of *g*.

But the earth *is* rotating and the object moves in a circle. The object must have a *net* force acting on it, directed toward the center of the earth, to give it the necessary centripetal acceleration. Thus $|\vec{N}| < |\vec{F}_{grav}|$ for an object on a rotating planet. (Notice how similar this is to our earlier analysis of the effective weight of roller coasters doing loop-the-loops and buckets of water swung overhead.) But $|\vec{N}|$ is just the effective weight W_{eff}. Thus

$$\left[W_{eff} = mg_{eff}\right] < \left[|\vec{F}_{grav}| = mg_{true}\right]$$

$$\Rightarrow g_{eff} < g_{true}.$$

The details of this calculation will be left as a homework problem, but the final result is that the measured value of *g*, which is g_{eff}, is slightly *less* than the value calculated for a non-rotating earth. At mid-latitudes, where most of us live, the reduction is just about 0.03 m/s², and hence the value for *g* on a rotating earth is 9.80 m/s².

A gravitational problem of particular practical importance is that of orbiting spacecraft and satellites. We will consider this more carefully in Section 12.6, but now is a good time to think about the value of *g* for a satellite. Consider a satellite orbiting a height *h* above the earth. Its distance from the center of the earth is then $r = R_e + h$. Most people have a mental image that satellites orbit "far" from the earth, but in reality h is typically 200 miles $\approx 3 \times 10^5$ m, while $R_e = 6.37 \times 10^6$ m. Thus $h/R_e \approx 0.05$—the satellite is barely "skimming" the earth at a height only about 5% of the earth's radius!

The value of g at height h will be, according to Eq. 12-2,

$$g(h) = \frac{GM_e}{(R_e + h)^2} = \frac{GM_e}{R_e^2 (1 + h/R_e)^2} = \frac{g_0}{(1 + h/R_e)^2}, \tag{12-4}$$

where $g_0 = GM_e/R_e^2 = 9.83$ m/s^2 is the value calculated in Eq. 12-3 for $h = 0$ on a non-rotating earth. Table 12-1 shows the value of g evaluated at several values of h.

TABLE 12-1   Variation of the acceleration due to gravity g with height above the ground.

Height h	Example	g (m/s^2)
0 m	ground	9.83
4500 m	Mt. Whitney	9.82
10,000 m	jet airplane	9.80
300,000 m	Space Shuttle	8.90
35,900,000 m	communications satellite	0.22

The significance of this calculation is that the acceleration due to gravity on a satellite such as the Space Shuttle is only slightly less than the ground-level value. The acceleration *has* to be this large to keep the satellite in its orbit. The "weightlessness" of objects in orbit is *not* because there is no gravity or because the force of gravity has become exceptionally weak—it is only about 10% less than at the surface. Weightlessness, as you learned in Sections 5.5 and 6.6, is a consequence of an object being in free fall.

Weighing the Earth

You just saw how to predict g on the basis of a more fundamental law of gravity—but how do we know the value of M_e? We cannot place the earth on a giant pan balance, so how is its mass known? Furthermore, how do we know what the value of G is? These are interesting and important questions.

Newton himself did not know the value of G! He could say that the gravitational force is proportional to the product $m_1 m_2$ and inversely proportional to r^2, but he had no means of knowing the value of the proportionality constant. He worked, instead, with ratios in which the factor of G cancels out.

To determine a numerical value for G it is necessary to make a *direct* measurement of the gravitational force between two known masses at a known separation. This is not easy, due to the small size of the gravitational force between ordinary size objects. Yet the English scientist Henry Cavendish came up with an ingenious way of doing so with a device called a *torsion balance*. Two fairly small masses m, typically about 10 g, are placed on the ends of a lightweight rod. The rod is hung from a thin fiber, as shown in Fig. 12-7, and allowed to reach equilibrium. If the rod is then twisted slightly and released, a *restoring force* will return it to equilibrium. This is analogous to displacing a spring from equilibrium, and, in fact, the restoring force and the angle of displacement obey a version of Hooke's law: $F_{\text{restore}} = k\Delta\theta$. The "torsion constant" k can be determined, like the spring constant, by timing the period of oscillations—as you will learn shortly when we get to oscillations. Once k is known, a force that twists the rod slightly away from equilibrium can be measured by the product $k\Delta\theta$. It is possible to measure

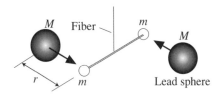

FIGURE 12-7 Cavendish's experiment to measure *G*.

very small angular deflections, so this device can be used to determine very small forces.

Two larger masses M (typically lead spheres with $M \approx 10$ kg) are then brought close to the torsion balance. The gravitational attraction that they exert on the smaller hanging masses causes a small but measurable twisting of the balance—enough to measure F_{grav}! Because m, M, and r are all known, Cavendish was able to determine G from

$$G = \frac{F_{\text{grav}} r^2}{Mm}.$$

(12-5)

His first results were not highly accurate, but improvements over the years in this and similar experiments have produced the value of G accepted today.

With an independently determined value of G, we can return to Eq. 12-2 to find

$$M_{\text{e}} = \frac{g_{\text{e}} R_{\text{e}}^2}{G}.$$

(12-6)

We have weighed the earth! The value of g is known with great accuracy from kinematics experiments, while the earth's radius R_{e} is determined by surveying techniques. Combining our knowledge from these very different measurements has given us a way to determine the mass of the earth. Equation 12-6 is exactly how the mass of the earth is calculated. The most precise values take into account small effects due to the earth's rotation and due to the fact that the earth is not perfectly spherical, but these are minor corrections to Eq. 12-6.

The constant G is a *universal constant*. We have certainly had constants in problems before; the mass is usually a constant within a single problem, but there is no "universal mass"—each object is different. The acceleration due to gravity g is constant for any given planet, but each planet is different. G, however, is a constant of a different nature. Its value establishes the basic strength of one of the fundamental forces of nature. As far as we know today, the gravitational force between two masses would be the same anywhere in the universe at any time, past or future. Universal constants tell us something about the most basic and fundamental properties of nature. You will meet other universal constants before long.

12.5 Gravitational Potential Energy

We have, thus far, been looking at gravity from a force/acceleration perspective. Not surprisingly, after what we have learned in the last three chapters, energy and conservation laws can give us a different perspective and a different set of tools. In order to employ conservation of energy, we will need to determine an appropriate form for the gravitational potential energy for two particles interacting via Newton's law of gravity.

Recall from Chapter 10 that the definition of potential energy is:

$$\Delta U = U_f - U_i = -W_{\text{interaction force}}(\text{position i} \rightarrow \text{position f}). \qquad (12\text{-}7)$$

Strictly speaking, of course, this defines only ΔU, which is the *change* in potential energy. To find an explicit expression for U, we must choose a zero-point of the potential energy. For a flat earth, we used $F = -mg$ and chose the zero point so that $U = 0$ at the surface ($y = 0$). This gave us the expression $U_{\text{grav}} = mgy$. This result for U_{grav} is valid only for $y \ll R_e$, when the earth's curvature and size are not apparent. Now we need to generalize this earlier result. We want to find an expression for the gravitational potential energy of masses that interact over *large* distances.

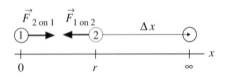

FIGURE 12-8 Calculating the work done by the gravitational force between two particles of mass m_1 and m_2.

Figure 12-8 shows two particles of mass m_1 and m_2 exerting gravitational forces on each other, as described by Eq. 12-1. Let mass m_1 be at the origin. Now let's calculate the work done on mass m_2 by force $\vec{F}_{1 \text{ on } 2}$ as m_2 moves from an initial position $x_i = r$ to a final position very far away: $x_f \rightarrow \infty$. Because the force on m_2 points to the left while the displacement of m_2 is to the right, this force will do a *negative* amount of work. Thus the change ΔU in potential energy, due to the minus sign in Eq. 12-7, will be *positive*. The pair of particles will *gain* potential energy as they move further apart—just as a particle near the earth's surface gains potential energy as it moves to higher elevations.

Because this is a variable force, we need the full definition of work from Eq. 9-29:

$$W = \int_{x_i}^{x_f} F_x(x)dx, \qquad (12\text{-}8)$$

where F_x is the x-component of the force vector. Here $\vec{F}_{1 \text{ on } 2}$ points toward the left, so it is important that we correctly use $(\vec{F}_{1 \text{ on } 2})_x = -Gm_1m_2/x^2$. We can then proceed to use Eqs. 12-7 and 12-8 to evaluate ΔU:

$$\Delta U = U(\infty) - U(r) = -\int_r^\infty (\vec{F}_{1 \text{ on } 2})_x dx = -\int_r^\infty \left(\frac{-Gm_1m_2}{x^2}\right)dx$$

$$= +Gm_1m_2 \int_r^\infty \frac{dx}{x^2} = -\frac{Gm_1m_2}{x}\bigg|_r^\infty \qquad (12\text{-}9)$$

$$= \frac{Gm_1m_2}{r}.$$

Because gravity is a conservative force, ΔU is independent of the path followed by m_2. So while we chose to integrate along the x-axis, if m_2 moves from an initial distance r to infinity along *any* path, the potential energy will increase by Gm_1m_2/r. Although Eq. 12-9 looks rather similar to the force law of Eq. 12-1, notice that it depends only on $1/r$, *not* on $1/r^2$.

As expected, the change in potential energy was found to be positive. To proceed further and find an explicit expression for $U(r)$, we need to choose the point where $U = 0$. We would like our choice to be valid for any star or planet, regardless of its mass and radius. This will be true if we set $U = 0$ at the point where the interaction between the masses vanishes. According to Eq. 12-1 the strength of the interaction is zero only when $r = \infty$. Two masses infinitely far apart will have no tendency, or potential, to move together, so we will *choose* to place the zero of potential energy at $r = \infty$. That is, $U(r = \infty) = 0$.

Having made this choice, Eq. 12-9 immediately gives us the gravitational potential energy

$$U_{\text{grav}}(r) = -\frac{Gm_1m_2}{r}. \tag{12-10}$$

This is the potential energy of masses m_1 and m_2 when their *centers* are separated by distance r. It is important to keep in mind the fact that U_{grav} is the potential energy of the *system* of masses $m_1 + m_2$. It is not the energy of either mass alone.

It may seem disturbing that the potential energy is *negative*, but we encountered similar situations in Chapter 10. All a negative potential means is that the potential energy of the two masses at separation r is *less* than their potential energy at infinite separation, where $U = 0$. It is only the *change* in U that has physical significance, and the change will be the same no matter where we place the zero of potential energy.

Figure 12-9 shows a graph of U_{grav} as a function of the separation r between the masses. Notice that it asymptotically approaches 0 as $r \to \infty$. You might also note that U is infinite at $r = 0$, but that is not a physically meaningful situation because two masses cannot occupy the same point in space—which is what $r = 0$ would imply. Suppose two masses initially separated by distance r_1 are released from rest. How would they move? From a force perspective, you would note that each experiences an attractive force that accelerates it toward the other mass. The energy perspective of Fig. 12-9 tells us the same thing: by moving toward a smaller r (that is, $r_1 \to r_2$), the system *loses* potential energy and *gains* kinetic energy while conserving E_{total}. The mass is falling "downhill," although in a more general sense than we think about on a flat earth. Remember that the kinetic energy K is the distance between the potential energy curve and the energy line, just as you learned in Chapter 10. Similarly, if the masses start with initial separation r_2 but moving away from each other, they will move "uphill" as their kinetic energy is converted into potential energy, coming to rest (momentarily) at r_1 where $K = 0$.

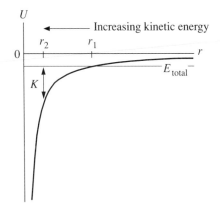

FIGURE 12-9 The gravitational potential energy curve. The two masses gain kinetic energy as they move closer together from r_1 to r_2.

EXAMPLE 12-1   Suppose the earth suddenly stopped revolving around the sun. The gravitational force would then pull it directly into the sun. What would the earth's speed be as it crashed?

SOLUTION Figure 12-10 provides a pictorial model for this gruesome cosmic event, showing the "before" and "after" that we need for a conservation law problem. It is essential to realize that the "crash" occurs as the earth touches the sun, at which point the separation between their centers is $r_2 = R_s + R_e$. Initial separation r_1 is the radius of the earth's *orbit* about the sun, *not* the radius of the earth. Because there are no friction forces, the mechanical energy is conserved. Strictly speaking, the kinetic energy is the *sum* $K = K_{earth} + K_{sun}$. However, the sun is so much more massive than the earth that the light-

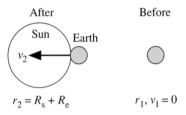

FIGURE 12-10 Pictorial model of the earth crashing into the sun.

weight earth does almost all of the moving and $K_{sun} \ll K_{earth}$. (We justified this claim in Chapter 9 for a light ball moving on the heavy earth.) It is a reasonable approximation to consider the sun as remaining at rest. In that case:

$$K_f + U_f = K_i + U_i$$

$$\Rightarrow \tfrac{1}{2} M_e v_2^2 - \frac{GM_s M_e}{(R_s + R_e)} = 0 - \frac{GM_s M_e}{r_1}$$

$$\Rightarrow v_2 = \sqrt{2GM_s \left(\frac{1}{R_s + R_e} - \frac{1}{r_1} \right)} = 6.13 \times 10^5 \text{ m/s}.$$

In other words, the earth is really flying along at over 1 million miles per hour! It is worth noting that we do not have the mathematical tools to solve this problem using Newton's second law because the acceleration is not constant. But the solution is straightforward using energy conservation.

EXAMPLE 12-2   A 1000 kg rocket is fired straight up from the surface of the earth. What speed does the rocket need to "escape" from the gravitational pull of the earth and never return? Assume a non-rotating earth.

SOLUTION This is an important example that illustrates what is called the **escape speed**. In a simple model of the universe, consisting of only the earth and the rocket, an insufficient launch speed will result in the rocket eventually falling back to earth. It might reach a distance of a million light years, decelerating the whole way, before reaching $v = 0$, but when it does there will be an attractive force that ever-so-slowly pulls it back to the earth. How, then, can the rocket ever escape? It will escape forever if either a) it never reaches $v = 0$ and thus never has a turning point, or b) when it reaches $v = 0$ there

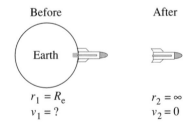

FIGURE 12-11 The pictorial model for a rocket launched with sufficient speed to escape the earth.

is no force on it. Condition b) will be realized if the rocket reaches $v = 0$ at $r = \infty$! Now infinity, of course, is not a "place," so a statement like this means that we want to rocket's speed to approach $v = 0$ asymptotically as $r \to \infty$, so that there is never a value of $v = 0$ for any finite value of r. Condition a) calls for the rocket to asymptotically approach a final velocity $v_f > 0$ as $r \to \infty$. Because this requires more kinetic energy at $r = \infty$ than condition b), the rocket must have started with more energy. We want the *minimum* launch speed for escape, so we want the rocket to satisfy condition b). Figure 12-11 shows a pictorial model with this information. By applying the law of energy conservation we can find v_1.

$$K_f + U_f = K_i + U_i$$

$$\Rightarrow 0 + 0 = \tfrac{1}{2}mv_1^2 - \frac{GM_em}{R_e}$$

$$\Rightarrow v_1 = \sqrt{\frac{2GM_e}{R_e}} = 11,200 \text{ m / s} \approx 25,000 \text{ mph.}$$

The answer does *not* depend on the rocket's mass, so this is the escape speed for any object.

Example 12-2 was mathematically easy; the difficulty was deciding how to interpret it. That is why the "physics" of a problem consists of thinking, interpreting, and modeling—quite distinct from the mathematics of obtaining a specific number at the very end. We will see variations on this situation again in the future, both with gravitation and electricity, so you might want to review the reasoning involved.

Equation 12-10 is the general form of the gravitational potential energy, but how is it related to our previous use of $U_{grav} = mgy$ on a flat earth? Consider an object of mass m located a height y above the surface of the earth. The object's distance from the earth's center is $r = R_e + y$. The potential energy of its interaction with the earth is, by Eq. 12-10,

$$U_{grav}(y) = -\frac{GM_em}{r} = -\frac{GM_em}{R_e + y} = -\frac{GM_em}{R_e(1 + y / R_e)} \tag{12-11}$$

where, in the last step, we have factored R_e out of the denominator.

Our earlier analysis of the gravitational potential energy was restricted to objects very close to the earth's surface, $y \ll R_e$, in which case the ratio $y/R_e \ll 1$. There is an approximation you will learn about in calculus, called the *binomial approximation*, that says

$$(1 + x)^{-1} \approx 1 - x \text{ if } x \ll 1.$$

If we call $y/R_e = x$ and use the binomial approximation in Eq. 12-11, we find:

$$U_{grav}(y \ll R_e) \approx -\frac{GM_em}{R_e}\left(1 - \frac{y}{R_e}\right) = -\frac{GM_em}{R_e} + m\left(\frac{GM_e}{R_e^2}\right)y. \tag{12-12}$$

Now the first term, $-GM_em/R_e$, is just the gravitational potential energy $U_{grav}(y = 0)$ when the object is at ground level ($y = 0$). In the second term, we can recognize $GM_e/R_e^2 = g$ from the definition of g in Eq. 12-2. Using this information, we can write Eq. 12-12 as

$$U_{grav}(y \ll R_e) = U_{grav}(y = 0) + mgy. \tag{12-13}$$

Although we chose $U(r = \infty) = 0$ in Eq. 12-10, we are always free to change our minds. If we change the zero-point of potential energy to $U(y = 0) = 0$, which is the choice we made in Chapter 10, then Eq. 12-13 is

$$U_{\text{grav}}(y \ll R_e) = mgy. \qquad (12\text{-}14)$$

Thus, Eq. 12-10 for the gravitational potential energy is consistent with our earlier "flat earth" expression for the potential energy when $y \ll R_e$.

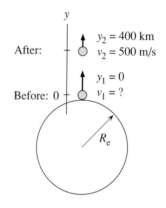

EXAMPLE 12-3 A projectile is launched straight up from the earth. a) With what speed should you launch it if it is to have a speed of 500 m/s at a height of 400 km? Ignore air resistance. b) By what percentage would your answer be in error if you used a flat earth approximation?

SOLUTION a) Figure 12-12 shows a pictorial model of the situation. Although the height is exaggerated in the picture, 400 km = 400,000 m is high enough that we cannot ignore the earth's spherical shape. Mechanical energy is conserved, so we can use the law of energy conservation:

$$K_f + U_f = K_i + U_i$$

$$\Rightarrow \tfrac{1}{2}mv_2^2 - \frac{GM_e m}{R_e + y_2} = \tfrac{1}{2}mv_1^2 - \frac{GM_e m}{R_e + y_1}.$$

FIGURE 12-12 Pictorial model of a projectile launched straight up.

We've used the distance $r = R_e + y$ between the projectile and the earth's center to find the potential energy. Notice that the projectile mass m cancels and is not needed. Solving for v_1, the launch speed, gives

$$v_1 = \sqrt{v_2^2 + 2GM_e\left(\frac{1}{R_e} - \frac{1}{R_e + y_2}\right)} = 2770 \text{ m / s}.$$

This is about 6000 miles per hour, significantly less than the escape speed.

b) Calculating the initial speed in the flat-earth approximation proceeds the same as in the previous part except that we need to use $U_{\text{grav}} = mgy$. Thus

$$\tfrac{1}{2}mv_2^2 + mgy_2 = \tfrac{1}{2}mv_1^2 + mgy_1$$

$$\Rightarrow v_1 = \sqrt{v_2^2 + 2gy_2} = 2840 \text{ m / s}.$$

The actual speed is somewhat less than that calculated in the flat-earth approximation because the force of gravity decreases with height. It takes less effort with a decreasing force than it would with the flat-earth force of mg at all heights.

The flat-earth value of 2840 m/s is 70 m/s too big. The error, as a percentage of the correct 2770 m/s, is

$$\text{error} = \frac{70}{2770} \times 100 = 2.5\%.$$

12.6 Satellite Orbits and Energies

Kepler's first law tells us that the "solution" to the problem of motion under the influence of the gravitational force is a set of elliptical trajectories. These trajectories are called *orbits*. Because the mathematics of ellipses is rather difficult, we will restrict our analysis to the limiting case in which an ellipse becomes a circle. Most planetary orbits differ only very slightly from circular, so our simpler purely-Copernican analysis will give good, but not perfect, results. (The earth's orbit has a ratio of (semiminor axis/semimajor axis) = 0.99986—awfully close to a true circle!)

The basic physics of orbits was presented in Section 6.6. We found that a force that always points toward a central point can cause a centripetal acceleration and thus a circular motion. Such a central-pointing force is generally called a **centripetal force**. We did not introduce this term earlier because we wanted to focus on how forces *act* rather than on what they are named, but now it is reasonable to begin using the term. A circular motion of speed $|\vec{v}|$ at radius r is caused by a centripetal force of magnitude

$$|\vec{F}_{cent}| = m|\vec{a}_c| = \frac{m|\vec{v}|^2}{r}.$$

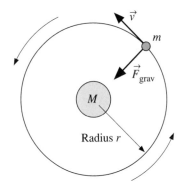

FIGURE 12-13 Orbital motion of a satellite due to the force of gravity.

Consider a massive body M, such as the earth or the sun, with a lighter body m—called a **satellite**—orbiting it as shown in Fig. 12-13. The gravitational force on the satellite is always directed toward the center of M. Thus the gravitational force *acts* as a centripetal force. If the satellite is to move in a circular orbit, it *must* have a speed $|\vec{v}|$ such that the required centripetal force $m|\vec{v}|^2/r$ is exactly provided by the gravitational attraction toward M. That is, circular motion requires:

$$\left[|\vec{F}_{cent}| = \frac{m|\vec{v}|^2}{r}\right] = \left[|\vec{F}_{grav}| = \frac{GMm}{r^2}\right]$$

$$\Rightarrow \quad |\vec{v}|^2 = \frac{GM}{r}.$$

(12-15)

Equation 12-15 tells us that a satellite must have a specific speed in order to have a circular orbit of radius r about the larger mass M. If the velocity differs from this value, the orbit will become elliptical rather than circular. Notice that there is no dependence on the satellite's mass m. This is consistent with everything you have learned previously about gravity—that motion due to gravity is independent of the mass.

EXAMPLE 12-4   If the Space Shuttle in a 200-mile-high orbit wants to capture a smaller satellite for repairs, what speeds do both the Shuttle and the satellite need?

SOLUTION Despite their different masses, the Shuttle, the smaller satellite, and the astronaut working in space to make the repairs all travel side-by-side with the same speed.

They are simply in free-fall together. Using $r = R_e + h$ and $h = 200$ miles $= 3.22 \times 10^5$ m, Eq. 12-15 can be evaluated to give

$$|\vec{v}| = 7720 \text{ m/s} \approx 16,000 \text{ mph.}$$

The answer depends on the mass of the earth but *not* on the mass of the satellite. ●

An important concept for circular motion is the *period*, which we introduced in Section 6.4. Period T, you will recall, is the time to complete one full orbit. We also discovered, in Eq. 6-19, the relationship between speed, radius, and period:

$$|\vec{v}| = \frac{1 \text{ circumference}}{1 \text{ period}} = \frac{2\pi r}{T}. \tag{12-16}$$

By using Eq. 12-16 in Eq. 12-15 we can find a relationship between the period and the radius of an orbit:

$$|\vec{v}|^2 = \frac{4\pi^2 r^2}{T^2} = \frac{GM}{r}$$

$$\Rightarrow T^2 = \left(\frac{4\pi^2}{GM}\right) r^3. \tag{12-17}$$

In other words, the *square* of the period is proportional to the *cube* of the radius. But that is Kepler's third law! We have shown that Kepler's third law is a direct consequence of Newton's law of gravity. We have also shown that circular orbits exist, as long as they satisfy Eq. 12-15. It can be shown, although the mathematics gets complex, that elliptical orbits exist as well. Thus Kepler's first law also follows from Newton's theory. Even Kepler's second law—that equal areas are swept out during equal time intervals—can be shown to follow from the fact that Newton's law of gravity is a *central force*, one where the force is always directed toward the central focus occupied by mass M.

Kepler's laws simply summarize observational data about the motions of the planets; they do not form a theory. Newton, however, put forward a *theory*: a specific set of relationships between force and motion that allow *any* motion to be understood and calculated. These general laws of motion, along with a specific proposal for how to determine the force of gravity, have allowed us to *deduce* Kepler's laws and, thus, to understand them at a more fundamental level. Furthermore, Kepler's laws are not perfectly accurate. The planets, in addition to their attraction to the sun, are also attracted toward each other and toward their orbiting moons. The consequences of these additional forces are small, but over time they provide measurable effects not contained in Kepler's laws. With the Newtonian theory, however, it is possible to use Eq. 12-1 to calculate *all* of the forces, add them to get the net force acting on each planet, and then proceed to solve Newton's second law to determine the dynamics. The mathematics of the solution can be exceedingly difficult, and today is all done with computers, but even with the hand calculations used in the mid-nineteenth century this procedure predicted the existence of an undiscovered planet that was having minor effects on the orbital motion of Uranus. The planet Neptune was discovered in 1846, just where the calculations predicted.

TABLE 12-2 Useful astronomical data.

Planetary body	Mean distance from Sun (m)	Period (years)	Mass (kg)	Mean radius (m)
Sun	–	–	1.99×10^{30}	6.96×10^{8}
Moon	$3.84 \times 10^{8\dagger}$	27.3 days	7.36×10^{22}	1.74×10^{6}
Venus	1.08×10^{11}	0.615	4.88×10^{24}	6.06×10^{6}
Earth	1.50×10^{11}	1.00	5.98×10^{24}	6.37×10^{6}
Mars	2.28×10^{11}	1.88	6.42×10^{23}	3.37×10^{6}
Jupiter	7.78×10^{11}	11.9	1.90×10^{27}	6.99×10^{7}
Saturn	1.43×10^{12}	29.5	5.68×10^{26}	5.85×10^{7}
Uranus	2.87×10^{12}	84.0	8.68×10^{25}	2.33×10^{7}
Neptune	4.50×10^{12}	165	1.03×10^{26}	2.21×10^{7}

† distance from earth

Table 12-2 contains astronomical information about the sun, the earth, the moon, and the planets. We can use this data to check the validity of Eq. 12-17. Figure 12-14 shows a graph of $\log T$ versus $\log r$ for all the planets in Table 12-2. Notice that the scales on each axis are increasing logarithmically—by *factors* of 10—rather than linearly. As you can see, the graph is a very straight line with a statistical "best fit" equation of $\log T = 1.500 \log r - 9.264$. You will be asked, as a homework problem, to show that the slope of 1.500 for this "log-log graph" does, indeed, confirm the prediction of Eq. 12-17. The y-intercept value of this line also contains useful information, and you will use it to use it to "measure" the mass of the sun.

A particularly interesting application of Eq. 12-17 is to communication satellites that are in **geosynchronous orbits** above the earth. A satellite in such an orbit appears to

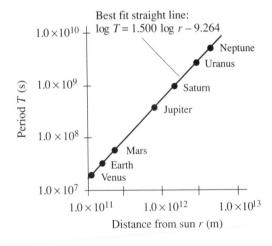

FIGURE 12-14 Graph of $\log T$ versus $\log r$ for the planetary data of Table 12-2. The statistical "best fit" line through the data is shown.

remain stationary over the earth's equator. For this to happen, the period of the satellite has to be the same 24 hours as the period of the point on the ground beneath it. The satellite's orbital motion is synchronous with the earth's rotation, and hence the term *geosynchronous*. Eq. 12-17 allows us to compute the radius of an orbit with this period. When using Eq. 12-17, note that the period *must* be in SI units of seconds! Thus

$$r = R_e + h = \left[\left(\frac{GM}{4\pi^2}\right)T^2\right]^{1/3}$$

$$= \left[\left(\frac{(6.67 \times 10^{-11} \text{ N m}^2 / \text{kg}^2)(5.98 \times 10^{24} \text{ kg})}{4\pi^2}\right)(86,400 \text{ s})^2\right]^{1/3}$$

$$= 4.22 \times 10^7 \text{ m} \tag{12-18}$$

$$\Rightarrow h = r - R_e = 3.59 \times 10^7 \text{ m} = 35,900 \text{ km} \approx 22,300 \text{ miles}.$$

This is much higher than the so-called "low earth orbit" used by the Space Shuttle and remote sensing satellites, where $h \approx 200$ miles. Communications satellites in geosynchronous orbits were first proposed in 1948 by science fiction writer Arthur C. Clarke, ten years before the first artificial satellite of any type!

EXAMPLE 12-5 Astronomers using the most advanced telescopes have only recently seen evidence of planets orbiting nearby stars. Suppose a planet is observed to have a 1200 day period as it orbits a star. The orbital radius of the planet is measured to be the same as that of Jupiter. What is the mass of the star in solar masses? (1 *solar mass* is defined to be the mass of the sun.)

SOLUTION Here "day" means earth days, as used by astronomers to measure the period. Thus the planet's period in SI units is $T = 1200$ days $= 1.037 \times 10^8$ s. The orbital radius is that of Jupiter, which is $r = 7.78 \times 10^{11}$ m. Solving Eq. 12-17 for the mass of the star gives

$$M = \frac{4\pi^2 r^3}{GT^2} = 2.59 \times 10^{31} \text{ kg}.$$

Using the conversion factor 1 solar mass $= 1.99 \times 10^{30}$ kg gives the final result:

$$M = (6.64 \times 10^{31} \text{ kg}) \times \frac{1 \text{ solar mass}}{1.99 \times 10^{30} \text{ kg}} = 13 \text{ solar masses}.$$

This is a large, but not extraordinary, star.

Orbital Energetics

Let us conclude this chapter by thinking about the energetics of orbital motion. As you saw in Section 12.5, the potential energy of two masses M and m separated by distance r is $U_{grav} = -GMm/r$. You also learned, in Eq. 12-15, that a satellite in a circular orbit is required to have $GM/r = |\vec{v}|^2$. Combining these two pieces of information gives the potential energy of a satellite in a circular orbit of radius r:

$$U_{grav} = -\frac{GMm}{r} = -m\left(\frac{GM}{r}\right) = -m|\vec{v}|^2 = -2K. \tag{12-19}$$

This is an interesting result. In all our earlier examples, the kinetic and potential energy were two independent variables. But Eq. 12-19 tells us that a satellite can move in a circular orbit *only* if there is a very specific relationship between K and U. It is not that K and U *have* to have this relationship, but if they do not the trajectory will be something other than a circular orbit. The mechanical energy of a satellite in a circular orbit is thus:

$$E_{mech} = K + U = -K = \tfrac{1}{2}U. \qquad (12\text{-}20)$$

Because the gravitational potential energy is negative, Eq. 12-20 says that the *total* mechanical energy is also negative. Negative total energy is characteristic of a **bound system**, which is what we have when the satellite is *bound* to the central mass by the gravitational force. If E were zero, then the satellite could just barely manage to escape, reaching infinity (where $U = 0$) with $K = 0$. And if E were positive, the satellite could easily escape, reaching infinity with enough remaining kinetic energy to keep it going. A negative value of E is the necessary condition for the satellite to be unable to escape.

Figure 12-15 shows a satellite's kinetic energy, potential energy, and total energy as a function of the radius of its orbit. If the satellite moves from a smaller orbit of radius r_1 to a larger orbit of radius r_2, we can see that its kinetic energy decreases—it moves more slowly in the larger orbit. But its potential energy *increases* more than its kinetic energy decreases, so moving from r_1 to r_2 actually requires a net energy *gain* $\Delta E > 0$, as the arrow shows. Where does this increase of energy come from?

Artificial satellites are raised to higher orbits by firing their rocket motors and experiencing a thrust force. This force does work, because the point of application of the force moves (unlike the propulsion force on car tires). The energy equation of Chapter 11 says that this work on the satellite increases its energy: $W_{ext} = \Delta E_{mech}$. The ultimate source of this energy, of course, resides in the chemical energy of the rocket fuel.

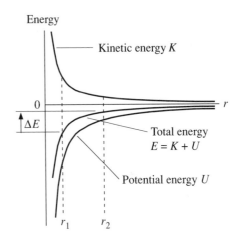

FIGURE 12-15 The kinetic, potential and total energy of a satellite as a function of the radius of its orbit.

EXAMPLE 12-6 How much work must be done to boost a 1000 kg communications satellite from a low earth orbit with $h = 300$ km, where it is released by the Space Shuttle, to a geosynchronous orbit?

SOLUTION The required work is $W_{ext} = \Delta E_{mech}$, and from Eq. 12-20 we see that $\Delta E_{mech} = \tfrac{1}{2}\Delta U$. The radius of a geosynchronous orbit was calculated in Eq. 12-18, so we can find

$$W_{ext} = \Delta E_{mech} = \tfrac{1}{2}\Delta U = \tfrac{1}{2}(-GM_e m)\left(\frac{1}{r_{geosync}} - \frac{1}{r_{Shuttle}}\right) = 2.52 \times 10^{10}\ \text{J}.$$

It takes a lot of energy to boost satellites to high orbits!

You might naively think that moving a satellite to a higher orbit would require pointing its thrusters toward the earth and blasting outward. You would be wrong. That would work fine *if* the satellite were initially at rest and moved outward along a straight trajectory, but an orbiting satellite is already moving and has significant inertia. A force directed straight outward would *change* the satellite's velocity vector in that direction but would not cause it to *move* along that line (recall all those earlier motion diagrams for motion along curved trajectories). In addition, a force directed outward would be almost at right angles to the motion and would do essentially zero work. Navigating in space is not as easy as it appears on Star Trek!

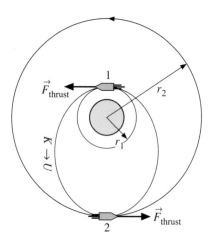

FIGURE 12-16 A satellite changing orbits. Firing the motor at point 1 moves the satellite onto the elliptical orbit. A second firing at point 2 transfers it to the larger circular orbit of radius r_2.

Suppose that a satellite in a circular orbit of radius r_1 needs to be moved to a larger circular orbit of radius r_2, as shown in Fig. 12-16. At point 1, the satellite is oriented so that its thrusters apply a brief thrust force in the direction the satellite is moving—that is, *tangent* to the circle. This force does a significant amount of work because the displacement Δx during the force is large. Thus the satellite quickly gains kinetic energy ($\Delta K > 0$). If the thrust is of short duration, the satellite does not have time for any significant change of distance from the earth, so $\Delta U \approx 0$. With the kinetic energy changed, but not the potential energy, the satellite no longer meets the requirements for a circular orbit and it goes into an elliptical orbit.

In the elliptical orbit, the satellite moves "uphill" toward point 2 by converting kinetic energy into potential energy. At point 2, the satellite has arrived at the desired distance from earth and has the "right" value of the potential energy. But it turns out (from an analysis more detailed than we want to pursue) that its kinetic energy is *less* than what is needed for a circular orbit. If no action is taken, the satellite will continue on its elliptical orbit and "fall" back to point 1. But another forward thrust at point 2 increases its kinetic energy, without changing U, until the kinetic energy reaches the value $K = -\frac{1}{2}U$ required for a circular orbit. Presto! The second burn kicks the satellite into the desired circular orbit of radius r_2. The work $W = \Delta E$ is the *total* work done in both burns. It would take a more extended analysis to see how the work has to be divided between the two burns, but even without those details you are now in possession of enough knowledge about orbits and energy to understand the ideas that are involved.

Summary

Important Concepts and Terms

cosmology	escape speed
Newton's law of gravity	centripetal force
gravitational constant	satellite
gravitational mass	geosynchronous orbit
principle of equivalence	bound system

This chapter has pulled together many ideas about force, motion, and energy in order to analyze and understand a very fundamental problem about motion in response to the gravitational force. If you look at all the knowledge that had to go into this analysis, you may be surprised at how much physics you have now learned!

Newton's theory of gravity consists of two distinct parts: first, a specific force law for the force of attraction between two masses m_1 and m_2:

$$|\vec{F}_{1 \text{ on } 2}| = |\vec{F}_{2 \text{ on } 1}| = \frac{Gm_1m_2}{r^2} \, ;$$

and second, the assertion that Newton's three laws of motion are universally applicable.

While Kepler's laws had provided a *description* of orbital motion, Newton's theory led to a detailed *understanding* of the dynamics of the solar system. In particular, a small mass m can orbit a large body of mass M in a circular orbit of radius r if and only if $|\vec{v}| = (GM/r)^{1/2}$. This leads to Kepler's third law for the orbital period T:

$$T^2 = \left(\frac{4\pi^2}{GM}\right)r^3 .$$

Further, the force *very near* the surface of the larger body can be written as $F_{\text{grav}} = mg$ where

$$g = \frac{GM}{R^2} .$$

Because gravity is a conservative force, we can define a potential energy of m_1 and m_2 as

$$U_{\text{grav}} = -\frac{Gm_1m_2}{r} .$$

It is important to understand the negative sign, which is a consequence of choosing the zero point of potential energy to be $U = 0$ at $r = \infty$. It was shown that this expression for potential energy is equivalent to the more familiar $U = mgy$ when the mass m is very near the surface of a planet ($y \ll R$). The mechanical energy $E_{\text{mech}} = K + U_{\text{grav}}$ of a mass is conserved unless there is work done by an external force, such as a rocket engine.

Exercises and Problems

Exercises

1. What is the period, in minutes, of the Space Shuttle in a typical 200-mile-high orbit? What is the Shuttle's speed in this orbit?

2. The *asteroid belt* circles the sun between the orbits of Mars and Jupiter. What is the orbital radius for an asteroid that has a period of 5 earth years? What is the asteroid's speed?

3. Two Jupiter-size planets are released from rest 1×10^{11} m apart. What are their speeds as they crash together? (**Hint:** *Both* planets are moving as they collide.)

▲ Problems

✍ 4. a. A 1 kg object is released from rest 500 km (300 miles) above the earth. What is its impact speed on the ground?
   b. What would be the impact speed if the earth were flat?
   c. By what percent is the flat-earth calculation in error?
   d. Suppose a Space Shuttle astronaut, while out for a stroll, "drops" a 1 kg hammer. Would it fall to earth? Explain why or why not.

✍ 5. There is strong evidence that the "age of the dinosaurs" ended 65 million years ago when a large asteroid struck the earth. Consider a simplified solar system consisting of just the earth and an asteroid, very far away, that is initially at rest. The asteroid is attracted to the earth and collides with it, causing the earth to "recoil" with a speed of 100 m/s.
   a. What was the asteroid's mass?
   b. Is this an elastic or an inelastic collision? Explain.
   c. How much energy is released in this collision?

6. Consider an object of mass m on the equator of a planet having mass M, radius R, and period of rotation T. Figure 12-17 shows the view from above the planet's north pole, looking down on the object as it rotates. If the planet did not rotate, the object would be in static equilibrium and its effective weight would be

$$W_{eff} = |\vec{N}| = |\vec{F}_{grav}| = mg_{true} = W_{true},$$

where $g_{true} = MG/R^2$.

Because of the planet's rotation, however, the object is rather like a roller-coaster car going over the top of a hill. It has an effective weight $W_{eff} = |\vec{N}| = mg_{eff} \neq mg_{true}$. Any experiments performed by a physics student to measure the acceleration of gravity will determine the value of g_{eff}, not g_{true}.
   a. Find an algebraic expression for g_{eff} in terms of M, R, T, and G.
   b. Evaluate both g_{true} and g_{eff} for the earth.
   c. Which of these accelerations due to gravity have we been using in this text?
   d. How will the value of g measured at the planet's north pole compare to the value measured at the equator? Will it be larger, smaller, or the same? Explain.

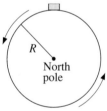

FIGURE 12-17

7. Figure 12-14 in the text shows a graph of $\log T$-versus-$\log r$ for the planetary data given in Table 12-2. Such a graph is called a log-log graph. A log-log graph is often useful when graphing data having the functional relationship $v^p = Cu^q$, where C is a constant and p and q are exponents (not necessarily integers).

Notice that the scale in Fig. 12-14 is labeled with the values of T and r, but the scale itself is logarithmic, not linear. That is, each division along the y-axis corresponds to a *factor* of ten increase in the value of T. The "correct" labels would be 7, 8, 9, and 10 because these are the logarithms of 10^7, ... , 10^{10}. It is customary, however, to label the tick marks with values of T rather than values of $\log T$. It is not how the scale is labeled that determines if the graph is linear or logarithmic, but how the scale is laid out.

a. Define $x = \log u$ and $y = \log v$. If u and v are related by the expression $v^p = Cu^q$, find the algebraic relationship for y in terms of x.
b. What *shape* will a graph of $\log v$-versus-$\log u$ have? Explain.
c. What *slope* will a graph of $\log v$-versus-$\log u$ have? Explain.
d. Fig. 12-14 shows that the "best fit" line passing through all the planetary data points has the equation $\log T = 1.500 \log r - 9.264$. This is an *experimentally* determined relationship between $\log T$ and $\log r$, using measured data; it is not a theory. Is this experimental result consistent with what you would expect from Newton's *theory* of gravity? How can you tell? Explain.
e. Use the experimentally determined "best fit" line to find the mass of the sun.

8. Large stars, somewhat more massive than the sun, can explode as they finish burning their nuclear fuel, causing a *supernova*. The explosion blows away the outer layers of the star, but the forces that push the outer layers away have *reaction forces* that are inwardly directed on the core of the star. These forces compress the core and can, if conditions are right, cause the core to undergo a *gravitational collapse*. The gravitational forces keep pulling all the matter together tighter and tighter, crushing atoms out of existence, until the nuclei come into contact with each other. Under conditions of extreme temperature and pressure a proton and an electron can be squeezed together to form a neutron. This begins to happen during the gravitational collapse, ultimately converting all the core's matter into neutrons. If the collapse is halted when the neutrons all come into contact with each other, the result is an astronomical object called a *neutron star*—an entire star consisting of solid nuclear matter.

A neutron star has a mass about that of our sun, but a radius of only ≈ 10 km! Furthermore, it rotates about its axis with a period of ≈ 1 s and, as it does so, sends out a pulse of electromagnetic waves once a second. Neutron stars were discovered by astronomers in the 1960s and they are now called *pulsars*.

a. Consider a neutron star with a mass equal to the sun, a radius of 10 km, and a rotation period of 1 s. What is the speed of a point on the equator of the star?
b. What is g at the surface of this neutron star?
c. What would be the weight of an object on this star that weighs 10 N on earth?
d. What is the radius of a geosynchronous orbit about the neutron star?
e. What would be the speed of a satellite orbiting 1 km above the surface?
f. How many revolutions per second would such a satellite make?

9. a. How much energy must a 50,000 kg Space Shuttle lose to descend from an orbit 400 miles high to one that is 200 miles high?
   b. Give a *qualitative* description, including a sketch, of how the Shuttle would do this.

[**Estimated 10 additional problems for the final edition.**]

PART **II** Scenic Vista

The Newtonian Synthesis

Now I a fourfold vision see,
And a fourfold vision is given to me;
'Tis fourfold in my supreme delight
And threefold in soft Beulah's night
And twofold Always. May God us keep
From Single vision & Newton's sleep!

William Blake

This is a good point to take stock of where we have been and what we have learned. So far, our single vision has been the science of motion—mechanics—as formulated by Newton. We have developed two parallel perspectives, each with its own ideas and techniques. The first of these, which we focused on in Part I, dealt with the relationship between force and motion. Newton's second law is the principle most central to the force/motion perspective, although other concepts are important ingredients (such as kinematics, the description of motion, and Newton's third law, a relationship between forces).

Our second perspective, as developed in Part II, looked at the relationship between "before" and "after." The central principle in this perspective is the idea of a conservation law. Newton's laws are essential in the development of conservation laws, but they remain hidden in the background when the conservation laws are applied. We also found some hints, to be explored further, that conservation laws have wider validity than just Newtonian mechanics and, in that regard, are more fundamental to our understanding of nature than are Newton's laws.

But why two perspectives? Why are Newton's laws not enough? There are certainly problems that can be worked equally well from either the force/acceleration perspective or the conservation law perspective, and we gave examples of those. This indicates—correctly—that there is no "hard barrier" between the two, nor could there be because they are not two different theories but simply two different ways of formulating a single theory. Nonetheless, you have likely seen a pattern emerge. If the situation you want to analyze involves a single object that can legitimately be treated as a particle, then the force/acceleration perspective is often more straightforward. Furthermore, if you need specific *dynamical* information, such as the position or velocity at certain instants of time, then *only* the force/acceleration perspective will work. Conservation laws, keep in mind, provide *no* information about the time at which things occur.

On the other hand, situations that involve two or more interacting particles suggest that some quantities might be conserved as they interact. This is particularly true if there is a clear "before" and "after" situation and if knowing the time interval is not a concern. Even if the momentum and energy are not strictly conserved (not an isolated system), it may be the case that the impulse or the work due to external forces can be readily determined. One of the trickier aspects of using conservation laws is judging what to include in the system. It is always worth starting an interacting-particles problem with conservation laws in mind, resorting to a force/acceleration analysis only if it seems absolutely necessary. You may recall, from Chapter 7, how complex action/reaction problems can be! While direct application of Newton's laws is always an option, it is generally best to try steering around those complexities by using the conservation law perspective.

The knowledge structure table on the next page summarizes our knowledge of Newtonian mechanics. You will recognize parts of it from the smaller table in the Scenic Vista from Part I. The knowledge structure is a pyramid consisting of a very few general principles at the top and a broad base of specific applications at the bottom. In approaching a new problem or situation, you want to start at the top by considering which general principles are relevant. With that decision made, you can work down the table, layer by layer, bringing in at each level the specific definitions and knowledge appropriate to the particular problem at hand. We have tried to illustrate this approach to problem solving in the examples of this book.

Our knowledge structure of mechanics is now essentially complete. We will add new ideas as we need them, making minor modifications to the table, but much of the next two sections of this text will consist of applying this knowledge to new situations. So this is an opportune moment to step back a bit to see the "big picture." Just what was the significance of Newton's theories? How was he influenced by earlier ideas and discoveries, and what was his effect on those that followed? While it may seem all very factual and straightforward to us now, it is important to keep in mind that these ideas are all human inventions. There was a time when they did not exist and when our concepts of nature were quite different than they are today.

The Newtonian Synthesis

Newton's achievements are often called the *Newtonian synthesis*. *Synthesis* means "the combining of separate elements to form a coherent whole." It is often said that Newton "united the heavens and the earth." Newton was an unparalleled genius, but he did not bring about this synthesis single-handedly. As Newton remarked, "If I have been fortunate enough to see further than others, it is only because I have stood on the shoulders of giants." Three of those giants were Galileo Galilei, Francis Bacon, and René Descartes. These were the primary advocates for three lines of thought that had been evolving during the 75 years preceding Newton. Newton synthesized their ideas into a single, coherent theory.

In earlier chapters you saw that Galileo developed of the concepts of kinematics. His work was based on a belief that science requires measurements and evidence, not just "pure

TABLE SVII-1 The knowledge structure of Newtonian mechanics.

General Principle: Force Laws

Second Law $\vec{F}_{net} = m\vec{a}$

Third Law $\vec{F}_{1\,on\,2} = -\vec{F}_{2\,on\,1}$

Force Laws: Relate forces to acceleration for single particles.

Linear Motion	*Trajectory Motion*	*Circular Motion*
$\sum F_x = ma_x$	$\sum F_x = ma_x$	$\sum F_r = m\lvert\vec{v}\rvert^2/r$
$\sum F_y = 0$	$\sum F_y = ma_y$	$\sum F_z = 0$

Kinematics: Use to find position and velocity

Linear and trajectory motions (applies to x and y) | Circular motion

Uniform motion:

$a = 0$

$v = $ constant

$x_f = x_i + v\Delta t$

Uniform acceleration:

$a = $ constant

$v_f = v_i + a\Delta t$

$x_f = x_i + v_i\Delta t + \frac{1}{2}a(\Delta t)^2$

$v_f^2 = v_i^2 + 2a\Delta x$

General

$a \neq $ constant

$v_f = v_i + \int_{t_i}^{t_f} a(t)dt$

$x_f = x_i + \int_{t_i}^{t_f} v(t)dt$

Circular motion

$a_c = \dfrac{\lvert\vec{v}\rvert^2}{r}$

$\lvert\vec{v}\rvert = \dfrac{2\pi r}{T} = 2\pi rf$

$s = \lvert\vec{v}\rvert\Delta t$

General Principle: Conservation Laws

Momentum p

Energy E

Conservation Laws: Relate "before" to "after" for systems of interacting particles.

Impulse and Momentum	*Work and Energy*
$\left[J = \int_{t_i}^{t_f} F_x(t)dt \right] = \Delta p$	$\left[W = \int_{x_i}^{x_f} F_x(x)dx \right] = \Delta K$

Conservation: Use for isolated system

Momentum | Energy

Momentum

Total momentum:

$P = \sum_i p_i$

Isolated system ($J = 0$):

$P_f = P_i$

Energy

$\Delta U = -W(i \rightarrow f)$

$U_{grav} = mgy$ or $-\dfrac{Gm_1m_2}{r}$

$U_{spring} = \frac{1}{2}k(\Delta x)^2$

Mechanical energy:

$E_{mech} = K_{cm} + U_{cm}$

Total energy:

$E_{total} = E_{mech} + E_{int}$

Conservation of energy:

$W_{ext} = \Delta E_{mech} + \Delta E_{int}$

Isolated system ($W_{ext} = 0$):

$E_f = E_i$

thought." In the last chapter we looked at how astronomy and cosmology changed from the mid-sixteenth-century ideas of Copernicus to the mid-seventeenth-century observations of Galileo. Galileo's observations and methods had a profound influence on Newton.

An English contemporary of Galileo was Francis Bacon, who is today generally recognized as the pioneer of the modern scientific method. His writings were the first to promote the idea of the scientific method, in which experiments lead to hypotheses that in turn lead to new predictions that can be tested experimentally. Bacon hoped that a *method* for science would, with step-by-step progress, gradually unlock the secrets of nature. His goal was purely pragmatic: to give humans control over nature by means of inventions and discoveries. He is regarded as the originator of the saying "Knowledge is power."

The idea of a scientific method was picked up and amplified by the French mathematician, scientist, and philosopher René Descartes. We remember Descartes most directly for his invention of analytical geometry and graphs—Cartesian coordinates! He was, in addition, an immensely influential philosopher, immortalized by his saying "I think, therefore I am." But Descartes was also very interested in physics, especially mechanics. He developed the idea of a mechanical model of the world in which material particles are influenced by their interactions with each other, and by gravity, and move about in accordance with mathematical laws. By using models of nature, Descartes emphasized that complex problems can be broken down into simpler problems that can be analyzed. This reduction of a complex situation to a collection of simple situations is called *reductionism*. Descartes also strongly emphasized the idea that nature follows *mathematical* laws, and that physics is about the mathematical relationships between measurable quantities.

If none of this sounds shocking, it is because Cartesian ideas have been so thoroughly assimilated into twentieth-century thought. Nonetheless, Bacon's and, especially, Descartes' ideas were quite radical at the time. This was a completely new way to try understanding the world around us. The point of interest to us is Descartes' *concept* of mechanics as a set of mathematical laws governing the interactions among material particles. He was not able to develop a successful theory, but he put the ideas into place.

All of these different lines of thought were brought together by Newton's genius into a single, coherent theory. His achievement was praised by no less than Einstein as "perhaps the greatest advance in thought that a single individual was ever privileged to make." Newton provided a consistent, wide-ranging mathematical theory of the world that was the unquestioned foundation of science until the twentieth century, and Newtonian mechanics remains the cornerstone of most engineering and applied science to this day.

But different thoughts and ideas are not all that Newton joined together. He also united the heavens and the earth, and in doing so he changed forever the way we view ourselves and our relationship to the universe. Greek and medieval cosmology, as we noted in Chapter 12, considered the heavenly bodies to be perfect, unchanging objects quite unrelated to the imperfect, changeable earthly matter. Their perfection and immortality reflected the perfection of God above, while the material bodies of humans below were imperfect and mortal. The same structure was evident in medieval feudal society: the king—perfect and ordained by God—surrounded by a small circle of nobles and then a larger circle of serfs and peasants. These ideas and institutions—science, religion, and society—together form what we call the medieval *worldview*. Their worldview, in all its many facets, was hierarchical and authoritarian, reflecting what they thought was the "natural order" of things.

Newton's mechanics, and especially his theory of gravity, was the final blow that caused the collapse of the cosmology that formed the underpinnings of the medieval worldview. That is why the Catholic Church, whose fundamental authority was being questioned, opposed Copernicus and Galileo so fiercely. Galileo's telescope had said "Look! The heavens are not perfect and unchanging." Then the success of Newton's theories said that the sun and the planets were just ordinary matter, obeying the same ordinary laws as earthly matter. The Newtonian synthesis overturned the medieval worldview, but Newton alone could not put a new worldview in place.

Newton's Influence

Newton's success changed the way we see and think about the universe. Rather than whirling celestial spheres, people began to think of the universe as being reduced to the motion of material particles following rigid laws. Many thinkers of the seventeenth and eighteenth centuries concluded that God had created the world by placing all the particles in their original positions, then giving them all a push to get them going. God, in this role, was called the "prime mover." But once the universe was started, it went along perfectly well just by obeying Newton's laws. No divine intervention or guidance was needed. This was certainly a very different picture of people's relationship to God and the universe than was contained in the medieval worldview.

This Newtonian conception of the cosmos is often called a "clockwork" universe. The technology of clocks and watchmaking was progressing rapidly in the eighteenth century, and people everywhere admired the consistency and predictability of these little machines. The Newtonian universe is just a very large machine, but one that is consistent, predictable, and law-abiding—a perfect clock. This idea of a perfectly predictable universe intrigued the French mathematician and philosopher Laplace, who wrote:

> An intelligence which at a given instant knew all the forces acting in nature and the positions of every object in the universe—if endowed with a brain sufficiently vast to make all the necessary calculations—could describe with a single formula the motions of the largest astronomical bodies and those of the lightest atoms. To such an intelligence, *nothing* would be uncertain; the future, like the past, would be an open book.

The implications are troubling. Even if no one *knows* all the forces and positions, they do exist. So is not the entire future of the universe—including your behavior and mine—already determined by the forces and the particles existing at this present instant of time? What happens to free will in a Newtonian universe? Maybe it *seems* free to us, but it would be entirely predictable to Laplace's "great intelligence." These implications of *determinism* have troubled philosophers since the time of Newton. Fortunately, our twentieth-century understanding of science erases this issue—but, perhaps, only to replace it with others equally troubling.

The Newton's ideas influenced not only philosophy and theology but also the realm of technology. The early stirrings of the industrial revolution had their origin in social and economic conditions more than in technology, and the precursors to the industrial revolution were already underway at the time of Newton. Nonetheless, Newton's theories powered great advances in the understanding and technology of machinery, and his influence was strongly felt by the time the industrial revolution reached its peak in the nineteenth century. Modern engineering disciplines are an outgrowth of these efforts to control and use machines and, later, other technologies. It is in the applied science of engineering that the incredible predictive power of Newtonian mechanics is most apparent.

Newton also helped to change the way people think about themselves and their society. The seeming perfection of physics in describing and controlling material objects, and the success of natural laws, set others to wondering about the implications for human nature, human behavior, and human institutions. The main protagonist in this school of thought was the English philosopher and political scientist John Locke, a contemporary of Newton. Locke developed a theory of human behavior based on such scientific ideas as natural laws and empirical evidence. We will not go into his theories in this text, but the success of Newton helped to propel Locke's ideas into the mainstream of eighteenth century political thought.

Locke's writings had a great influence on a young American thinker named Thomas Jefferson. The concept of natural laws, as they apply to individuals, is very much behind Jefferson's enunciation of "unalienable rights" in the Declaration of Independence. In fact, the very first sentence of the Declaration refers explicitly to "Laws of Nature and of Nature's God." The idea of *checks and balances*, built into the American Constitution, are very much a *mechanical* and clock-like model of how political institutions function.

So just as medieval feudalism mirrored the medieval understanding of the universe, contemporary constitutional democracy mirrors, in many ways, our contemporary Newtonian cosmology. Figure SVII-1 shows a timeline for the development of our modern worldview. Hierarchy and authority have been replaced by equality and law because that is what now seems to us the "natural order" of things. Having grown up with this modern worldview, it is difficult for us to imagine any other. Nonetheless, it is important to realize that other vastly different worldviews have existed at different times and in different cultures—and who can tell what today's scientific breakthroughs may imply for our future worldview.

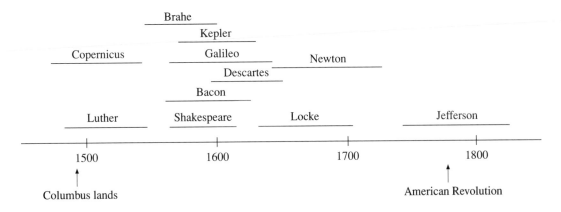

FIGURE SVII-1 Timeline for the development of the modern worldview from 1500 to 1800.

The Darker Side

Newton's genius and influence are undeniable, but there is another side to the story. The medieval worldview, as well as those of earlier cultures, regarded nature as a living thing, an intricate organism. Our ideas of "Mother Earth" stem from these earlier times. In addition, God or other spirits were present in the world, never far removed from the day-to-day appearances of things. Humans generally considered themselves a part of a living, spiritual world rather than set apart from it.

This began to change with Bacon. His goal for science, as we noted above, was to gain *control* over nature. Historian Theodore Roszak summarizes Bacon well.

> With Bacon we arrive at the pregnant idea that there is a kind of knowledge that grows incrementally and systematically over time, not as the result of hit-and-miss discovery or lucky accident, but as the product of a deliberate activity of the mind ... This knowledge will not be simply contemplative. It will be capable of asserting power over the world.

> True, we find no "hard science" in Bacon's work that will compare with the accomplishments of Kepler or Galileo. But we find what is far more important to the future of science in our society; we find the moral, aesthetic, and psychic raw materials of the scientific worldview. They are all there in his writing—the bright hopes and humanitarian intentions, obscurely mingled with hidden forces of dehumanization. More than any other figure in the western tradition, it was Bacon ... who foreshadowed the bleakest aspects of scientized culture: the malaise of spirit [and] the nightmare of environmental collapse.

Bacon was not shy about stating his opinions as to the means and methods by which nature's secrets were to be discovered. In describing the interrogation of nature, Bacon made analogies to torture chambers wherein "...nature exhibits herself more clearly under the *trials* and *vexations* of art [mechanical devices for experimentation] than when left to herself." He wrote that men of science must not think that the "inquisition of nature is in any part interdicted or forbidden." Nature must be "bound into service" and made a "slave," put in "constraint" and "molded" by the experimenter.

Bacon was quite clear that the purpose of science, as he saw it, is the control and exploitation of nature. Nature is a resource to be used, manipulated, and consumed. Quite a different relationship between humans and nature than the medieval worldview!

While Bacon set humans apart from nature, it was Descartes that reduced nature to mere dead, inert matter. As an outgrowth of his interest in the mechanical view of nature as nothing more than interacting material particles, Descartes developed a entire philosophy—called *dualism*—that divided the world into "mind" and "matter." The mind is the realm of conscious thought, feelings, sensations, emotions, and the soul. Only humans enter the world of the mind, because only humans have a soul. Animals, according to Descartes, have no soul and are completely devoid of feelings, emotions, and consciousness. They are no more than machines. Nature, other than humankind, is simply inert matter blindly following the laws of physics.

Bacon and Descartes had immense influence, not just on Newton but on other leaders of the new "scientific revolution." Their ideas were incorporated into the Newtonian worldview as it developed. By the time the industrial revolution reached its peak, our views of nature had shifted dramatically. Few even questioned that nature was a resource to be exploited, that the natural world was intended to be manipulated, and that other species were to be dominated and subjugated.

Newton, we should emphasize, did not create this worldview single-handedly. And it is likely that neither Bacon nor Descartes envisioned what their suggestions would mean with the power of nineteenth- and twentieth-century technology behind them. But their thoughts and ideas were propelled into the intellectual mainstream by Newton's success, and the negative exploitative aspects were inseparably bound up with the positive liberating aspects.

A few artists and poets have always remained troubled by this. The Newtonian worldview values only those things that can be observed, repeated, measured, and calculated. It is a single, sharply-focused vision of reality, and that is the basis for Blake's plea, "May God us keep from Single vision & Newton's sleep," that was quoted at the beginning of this Scenic Vista. There is little room, in the Newtonian world, for the vision of the artist, the poet, or the mystic.

The Future

If the medieval worldview was overly irrational and superstitious, then the Newtonian worldview has been overly rational, calculating, and impersonal. Perhaps the pendulum is swinging again toward a more balanced understanding of nature and our place in it. The twentieth century has seen another revolution in science that has changed our conceptions of reality. Einstein's relativity has dismissed many of the ideas of "absolutes" that are implicit in Newton's mechanics. Quantum physics has overthrown the predictable, deterministic, clockwork model of the universe and replaced it with a universe full of uncertainty. There are also implications in quantum physics that humans are not separable from the world they observe.

New sciences, unknown in Newton's time, have also changed our knowledge of the world. Ecology and evolution have established that humans are a part of nature, not separate from it, and psychology and neurology have made Descartes' mind/body dualism untenable. These scientific ideas are slowly working their way into other areas of thought and human activity, and bit by bit they are changing the ways in which we see ourselves, our society, and our relationship to nature.

Will these ideas foster an entirely new worldview, or merely modify and soften the edges of the Newtonian world? The future, as always, is unpredictable. (See! A non-Newtonian idea right there.) But one thing is certain. Change—the very first idea back in Chapter 1—change will continue to occur.

P A R T III

Oscillations and Waves

PART **III** Overview

Oscillations and Waves

Parts I and II of this text have been about the physics of particles and systems of particles. The concept of a *particle* is one of the two basic models of physics. The other, to which we now turn our attention, is a *wave*.

Waves are ubiquitous in nature. Two types of waves that might immediately come to mind are water waves, such as those on the surface of a lake or the ocean, and sound waves. A rather different type of wave is that created by a vibrating string, such as a guitar string or a violin string, or a vibrating membrane, such as a drumhead or a loudspeaker cone. Light is another type of wave—an *electromagnetic wave*. You will see, in this part of the text, the evidence for the wave nature of light.

The physics of waves is the subject of Part III, the next stage of our journey into physics. Despite the great diversity of types and sources of waves, there is a single, elegant physical theory that is capable of describing them all. Our exploration of wave phenomena will call upon water waves, sound waves, and light waves for examples, but we want to emphasize the unity and coherence of our understanding of *all* types of waves.

We will start with a study of the oscillations and vibrations of particles, such as a mass oscillating on a spring. This will be the connection between the Newtonian mechanics of particles, with which you are now quite familiar, and the new ideas of waves. We will then examine the oscillations that travel outward through some medium—like the ripples that spread out from a pebble hitting a pool of water or the sound waves that spread out from a loudspeaker. These are called *traveling waves*, and we will investigate their properties carefully. After arriving at an understanding of a single traveling wave, we will demonstrate what happens when several traveling waves are combined. In one case, this will lead to the very important idea of *standing waves*. In another, the combination of waves will lead to a discussion of the phenomena called *interference*, one of the most important defining characteristics of waves. We will then go on to examine how waves interact with their environment. For example, we will look at what happens when waves are forced to pass through an opening in an obstacle or a barrier.

Waves will also bring us into closer contact with atoms and their properties. Many of the sources of light and color in our environment are due to the emission and absorption of light by atoms and molecules. Each element in the periodic table emits a unique

"fingerprint" of light that reveals clues about the structure of the atoms. The classical physics of the nineteenth century was unable to explain these observations, and this "crisis" ultimately led to the development of quantum physics at the beginning of the twentieth century. Part III will conclude with an initial look at the connection between atoms and light. We will then return to this important topic in Part VI.

Our journey into the atomic world will also lead us to a closer examination of the relationship between waves and particles. The classical physics of the eighteenth and nineteenth centuries made a fundamental distinction between waves and particles. These were the two basic *models* of physics, and each of the objects in nature could be characterized as being either a particle or a wave. This was a well-defined wave-particle dichotomy, an either-or situation. Planets, projectiles, and atoms were clearly particles, or systems of particles, while sound and light were waves. That is not to say that particles (air molecules) are not relevant to sound, but sound itself is a collective, wave-like behavior of the air; sound does not follow a parabolic trajectory as a particle would.

Table OVIII-1 lists of some of the basic, defining characteristics of the wave and particle models.

TABLE OVIII-1 Basic characteristics of particles and waves.

Particles	Waves
discrete	continuous
localized (here)	non-localized (everywhere)
individual	collective

Particularly important is the idea that a particle exists at a specific location—it has a position coordinate x, you can put your finger on it, it is *here*—but a wave is diffuse, spread out, not found at a single point in space. For instance, if you consider a cork bobbing up and down on an ocean swell, there is no doubt about which is a particle and which a wave. Yet as we enter the atomic realm, we will be faced with evidence that its inhabitants are not so easily classified. Electrons, protons, and even light itself will be found to have characteristics of both particles *and* waves. In the strange world of quantum physics, "either-or" is replaced with "both-and." Rather than the wave-particle dichotomy of classical physics, we will come to recognize a "wave-particle duality" in quantum physics. This breakdown of the distinction between waves and particles is one of the discoveries of the twentieth century that has undermined the Newtonian worldview. So our study of waves, in addition to its inherent interest and its many important applications, will advance us along our journey toward understanding atoms and the atomic structure of matter.

Chapter **13**

Oscillations

LOOKING BACK | Sections 9.9; 10.6–10.7

13.1 Periodic Motion

[**Photo suggestion: Child on a swing.**]

In Parts I and II you studied linear motion, projectile motion, and circular motion. There is another important type of motion that we haven't spent much time on—oscillatory, or periodic, motion. An object that repeats the same motion over and over, taking the same amount of time for each repetition, is called **periodic motion**.

Uniform circular motion is one example of a periodic motion. Objects that *oscillate* or vibrate also exhibit periodic motion. You are already familiar with many examples of oscillating motion, such as a car bouncing up and down on its springs, or a child swinging back and forth on the playground "swing." Other examples include vibrating pieces of machinery, the "ringing" of a bell or a crystal goblet that has been lightly tapped, the twisting back-and-forth motion of a mass hanging on a rope, and the oscillation of current in an electrical circuit used to drive an antenna. Objects or systems of objects that undergo an oscillatory periodic motion are called **oscillators**.

Our goal in this chapter is to study the physics of oscillators. We will start with the kinematics of oscillation—the mathematical description of the motion. The kinematics will lead into a force/acceleration analysis of oscillators based on Newton's laws. We will also examine oscillators from an energy perspective. Finally, we will look at how oscillations are built up by *driving forces* and how they decay away over time. Most of our analysis will focus on the most basic form of periodic motion, called *simple harmonic motion*. Qualitative and graphical descriptions of the motion will be emphasized and are just as important as the mathematical description.

Although there are many types of oscillators, all have certain properties in common. Figure 13-1 shows some examples of displacement-versus-time graphs for several different oscillating systems. Each of these graphs shows an oscillatory, periodic motion. The time interval required to complete one full oscillation is called the **period** of the motion, and it is given the symbol T.

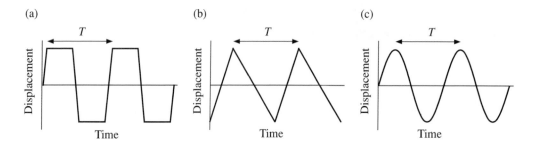

FIGURE 13-1 Several examples of displacement-versus-time graphs for oscillating systems. The period T is shown for each. Only sinusoidal graph (c) represents simple harmonic motion.

The period T tells us the number of seconds required per oscillation. A related piece of information is the number of oscillations completed per second. If the period is $1/10$ s, then the oscillator can perform 10 oscillations in one second. Conversely, an oscillation period of 10 s allows only a fraction of the oscillation—$1/10$, to be precise—to be completed per second. In general, T seconds per oscillation implies that $1/T$ oscillations will be completed each second. The number of oscillations per second is called the **frequency** f of the oscillation. The relationship between frequency and period is

$$f = \frac{1}{T} \quad \text{or} \quad T = \frac{1}{f}. \tag{13-1}$$

The units of frequency are *hertz*, abbreviated "Hz," named in honor of the German physicist Heinrich Hertz who produced the first artificially-generated radio waves in 1887. By definition, 1 Hz = 1 oscillation per second = 1 s^{-1}. We will frequently deal with very rapid oscillations and make use of the units 1 kilohertz = 1 kHz = 10^3 Hz; 1 megahertz = 1 MHz = 10^6 Hz; and 1 gigahertz = 1 GHz = 10^9 Hz. (Note: uppercase and lowercase letters *are* important. 1 MHz is 1 megahertz = 10^6 Hz, but 1 mHz is 1 millihertz = 10^{-3} Hz.)

EXAMPLE 13-1 What is the oscillation period for the broadcast of a 100 MHz FM radio station?

SOLUTION The frequency of current oscillations in the radio transmitter is 100 MHz = 10^8 Hz. These current oscillations are sent to the antenna and transmitted as a wave of this frequency. The period is

$$T = \frac{1}{f} = \frac{1}{1 \times 10^8 \text{ Hz}} = 1 \times 10^{-8} \text{ s} = 10 \text{ ns}.$$

13.2 Simple Harmonic Motion

Although there are lots of ways in which a system can oscillate, we are especially interested in the smooth *sinusoidal* oscillation shown in Fig. 13-1c. This sinusoidal oscillation turns out to be the most basic of all oscillatory motions. It is called **simple harmonic motion**, which is sometimes abbreviated as SHM. Before we dive into the mathematics of simple harmonic motion, let's look at a qualitative and graphical description.

[**Photo suggestion: Strobe photo of a mass oscillating on a spring.**]

Imagine an experiment in which you place a spring on a support, attach an object to its lower end, then pull the object down a few centimeters and release it. The object will oscillate up and down about the *equilibrium position*. You can use a meter stick to measure the object's position y at different instants of time. You can also use a stop watch to measure the time interval Δt during which the object undergoes a displacement Δy. With these measurements you can calculate the object's velocity at each point as $v = \Delta y / \Delta t$.

Figure 13-2 shows the results of an actual experiment like the one described except that a computer was used—rather than measurements by hand—to measure the position quite accurately 20 times per second. It is important to notice several things about this graph. First, it is the *displacement* from equilibrium Δy, not y itself, that we are graphing on the vertical axis. The $\Delta y = 0$ value corresponds to the object being at the **equilibrium position**, which is where it would hang at rest if you stopped the motion. The location of the origin, where $y = 0$, is not relevant. Second,

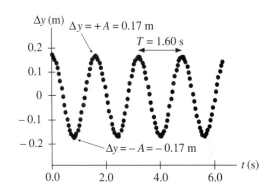

FIGURE 13-2 Measured displacement-versus-time graph of a real oscillating spring.

the motion is very obviously sinusoidal, indicating that an object oscillating on a spring undergoes simple harmonic motion. Third, the motion is *symmetrical* about the equilibrium position. That is, the maximum distance above equilibrium is equal to the maximum distance below equilibrium.

The maximum value of the displacement, Δy_{max}, is called the **amplitude** A of the motion. The displacement oscillates between a minimum of $-A$ and a maximum of $+A$. Note that the amplitude is the distance from the *axis* to the maximum, *not* the distance from the minimum to the maximum. You can see in this experiment that the amplitude is 0.17 m. You can also, from the graph, measure the period to be 1.60 s. The oscillation frequency is thus $f = 1/T = 0.625$ Hz.

What about the velocity of the object as it oscillates? At the very top and the very bottom—Δy_{max} and Δy_{min}—the object is at a *turning point*. The instantaneous velocity at these points is zero, just as it is at the turning point when a vertically-tossed ball reaches the top of its trajectory. The maximum *speed*, which occurs at the points of maximum slope, appears to occur as the object passes through the middle—at $\Delta y = 0$. The *velocity* is positive as the object passes $\Delta y = 0$ heading upward but *negative* as the object passes $\Delta y = 0$ heading downward. The velocity, as measured by the slope, appears to change smoothly, rather than abruptly, as the object moves from the center to an end point and back again. Because this behavior of the velocity occurs repetitively, it would seem that the velocity is also undergoing a sinusoidal oscillation.

Figure 13-3 confirms that this is so. The top graph is the displacement-versus-time graph from Fig. 13-2. Directly below it is a velocity-versus-time graph that the computer produced by taking the derivative of the displacement graph: $v = \Delta y / \Delta t$. You should convince yourself that the *value* of the velocity graph at each instant of time is equal to the *slope* of the displacement graph at that time. You can see that the velocity graph is, indeed,

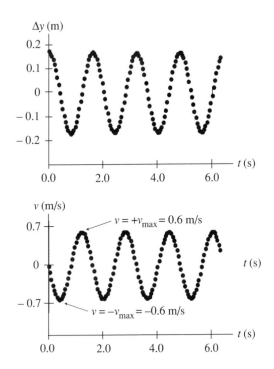

FIGURE 13-3 Displacement and velocity graphs for an oscillating spring.

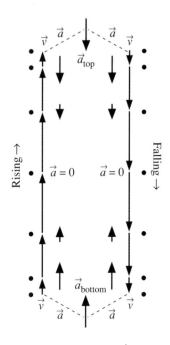

FIGURE 13-4 Motion diagram for an oscillating spring.

sinusoidal, oscillating between $-v_{max}$ and $+v_{max}$. As expected, $v = 0$ when $\Delta y = \pm A$. Similarly, $v = \pm v_{max}$ when $\Delta y = 0$. These observations should come as no big surprise if you remember that the derivative of a sinusoidal function (sine or cosine) is just another sinusoidal function (cosine or sine).

Now let's look at the object's acceleration. Acceleration is harder to visualize, but a motion diagram can help. Figure 13-4 shows a motion diagram for one cycle of the motion, with the upward motion and downward motion separated horizontally simply to make the diagram clear. You can see several things from the diagram. First, the velocity is large as the object passes through the equilibrium point, but it is *not changing* at that point. Because acceleration measures the *change* of the velocity, it is zero at $\Delta y = 0$. On the other hand, the velocity is changing rapidly at the turning points: at the upper turning point it is changing from an upward-pointing vector to a downward-pointing vector. It takes a large downward (negative) acceleration $\vec{a}_{top}$ to turn the velocity vector around. A similarly large acceleration is required at the bottom turning point, only this time the acceleration needs to be positive to change the downward-pointing velocity into an upward-pointing velocity.

Our motion diagram analysis suggests that the acceleration will be a maximum when the displacement is most negative, a minimum (most negative) when the displacement is a maximum, and zero when $\Delta y = 0$. We can confirm this by taking the derivative of the velocity curve, as shown in Fig. 13-5. We again see a sinusoidal oscillation, this time between $-a_{max}$ and $+a_{max}$. Now that we have displacement, velocity, and acceleration all graphed together, it is worth noting that all three kinematic variables have the *same* period T.

It is especially worth noticing that the bottom acceleration graph is found simply by flipping the top displacement graph over. At any instant of time, the acceleration is

$a(t) = -(\text{constant}) \times \Delta y(t)$. (The constant is necessary because displacement and acceleration have different units and different scales.) The basic observation is that the acceleration is proportional to the *negative* of the displacement Δy. This is an important point to which we will return.

Radians and Small Angle Approximations

The mathematics of SHM is somewhat different than we have been using up until now, and it will be useful to review some geometry and trigonometry. Figure 13-6 shows an angle θ. If you draw a circle of radius r with the vertex of the angle at the center, then the angle encloses a piece of the circle of length s called the *arc length*. Regardless of the size of the circle, the ratio s/r of arc length to radius is a constant. This ratio is a natural way to measure the size of the angle. We define the angular measure of *radians* to be:

$$\theta(\text{radians}) = \frac{s}{r}. \qquad (13\text{-}2)$$

Recall from geometry that the arc length of a full circle is its circumference $2\pi r$. This implies that the angular measure of a full circle is

$$\theta(\text{full circle}) = \frac{2\pi r}{r} = 2\pi.$$

This relationship is the basis for the well-known conversion factor: 2π radians = 360°.

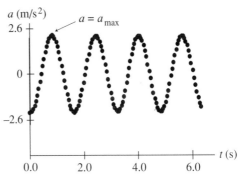

FIGURE 13-5 Displacement, velocity, and acceleration graphs for an oscillating spring.

Similarly, a 90° angle encloses one-quarter of a circle, having an arc length $2\pi r/4 = \pi r/2$, so a 90° angle has a radian measure of $\pi/2$.

An important consequence of Eq. 13-2 is that the arc length spanning an angle θ, measured in radians, is

$$s = r\theta.$$

This is a result that we will use often.

When we calculate angular quantities in physics, they will nearly always be in radians. This will be especially true during our study of oscillations and waves. A common error is to perform calculations with your calculator set for degrees! It's important to get into the habit of always setting your calculator for radians before doing physics problems. Even if you forget, you will likely catch your error *if* you begin building some intuition for radians and if you *assess* the results of your calculations to see if they make sense.

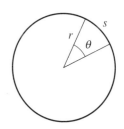

FIGURE 13-6
Measuring an
angle in radians.

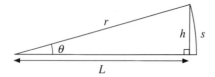

FIGURE 13-7 The geometrical basis of the "small angle approximation."

While an exact conversion gives 1 radian = 57.3°, a good approximation to remember for assessment purposes is that 1 radian ≈ 60°.

There is an additional piece of geometry that we will find useful. Figure 13-7 shows a circular arc of length s and angle θ, along with a right triangle that has been constructed by dropping a perpendicular from the top of the arc to the axis. The height of the triangle is $h = r\sin\theta$, and we have already seen that $s = r\theta$. Suppose that the angle θ is "small," as we have drawn it, in which case there is very little difference between h and s. If $h \approx s$, then

$$r\sin\theta \approx r\theta$$

$$\Rightarrow \sin\theta \approx \theta.$$

The result, that $\sin\theta \approx \theta$ for small angles, is called the **small angle approximation**. It is valid *only* if the angle θ is in radians! We can similarly note that $L \approx r$ for small angles. Because $L = r\cos\theta$ we have

$$r\cos\theta \approx r$$

$$\Rightarrow \cos\theta \approx 1.$$

Finally, we can ratio the sine and cosine to find $\tan\theta$:

$$\tan\theta = \frac{\sin\theta}{\cos\theta} \approx \sin\theta \approx \theta.$$

Table 13-1 summarizes the results of the small angle approximation. You will learn in calculus to justify these approximations, but the geometrical approach used here is easy to visualize and understand.

An obvious question is "How small does θ have to be to justify using the small angle approximations?" Table 13-2 shows several angles in radians, the corresponding angle in degrees (which are easier to think about), $\sin\theta$, and the percentage difference between $\sin\theta$ and θ. This table will help you judge when it makes sense to use the small angle approximation.

The small angle approximation is good to three significant figures—an error of $\leq 0.1\%$—up to angles of about 5°. In practice, you

TABLE 13-1 Small angle approximations.

$$\sin\theta \approx \theta$$
$$\cos\theta \approx 1$$
$$\tan\theta \approx \sin\theta \approx \theta$$

TABLE 13-2 Testing the $\sin\theta \approx \theta$ small angle approximation.

θ (radians)	θ (degrees)	$\sin\theta$	% error of $\sin\theta = \theta$	comments
0.0174533	1°	0.0174524	0.005%	excellent
0.087266	5°	0.087155	0.1%	good
0.17453	10°	0.17364	0.5%	marginal
0.3490	20°	0.3420	2%	poor

can use the approximation up to about $10°$, but you can see that it rapidly looses validity and would produce unacceptable results for angles larger than $10°$.

Simple Harmonic Motion and Circular Motion

Now let's consider a particle undergoing uniform circular motion of radius r with speed $v = |\vec{v}|$, as shown in Fig. 13-8. At any point in time, we can locate the particle by the angle ϕ (Greek "phi"), which we will define as measured *counterclockwise* from the x-axis. The particle's position measured along the arc is $s = r\phi$, where we have used the definition of radians from Eq. 13-2. Because the particle's speed is $v = ds/dt$, and because r is a constant, we must have

$$v = \frac{ds}{dt} = r\frac{d\phi}{dt} = r\omega, \qquad (13\text{-}3)$$

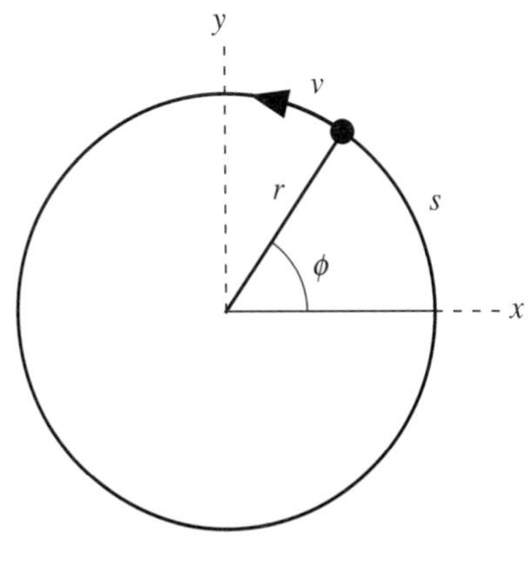

where

$$\omega = \frac{d\phi}{dt} \text{ radians per second} \qquad (13\text{-}4)$$

is defined as the **angular frequency**. (The symbol ω is a lowercase Greek "omega." It is *not* a w.)

FIGURE 13-8 A particle moving in a circle of radius r at speed v.

The reason for calling ω a "frequency" is that it is directly related to the number of revolutions per second, which is the particle's actual frequency f. As the particle moves through an angle ϕ, it will complete N revolutions, given by $N = \phi/2\pi$. (N may be less than or greater than 1, depending on whether ϕ is less than or greater than 2π.) The frequency f is the number of revolutions per second or, equivalently, the rate of increase of revolutions: $f = dN/dt$. Thus

$$\phi = 2\pi N$$

$$\Rightarrow \left[\frac{d\phi}{dt} = \omega\right] = \left[2\pi\frac{dN}{dt} = 2\pi f\right] \qquad (13\text{-}5)$$

$$\Rightarrow \omega = 2\pi f = \frac{2\pi}{T}.$$

Stated in words, if the particle undergoes f revolutions per second, then the angle ϕ that locates the particle increases by $2\pi f$ radians per second. Thus $\omega = 2\pi f$ is an *angular frequency*.

The terms "oscillations" and "revolutions" are not true units—they are simply counted events—so the frequency f has units of hertz, where 1 Hz $= 1$ s^{-1}. Similarly, radians are not true units because the definition of θ in Eq. 13-2 is a dimensionless ratio of two lengths. Thus the actual units of ω are also simply s^{-1}. We often say "oscillations per second" or "radians per second" as a way to indicate whether we are talking about f or ω, but in terms of checking units both are simply s^{-1}. Note that the unit hertz is *specifically* "oscillations per second" and is used for f but *not* for ω.

If the particle shown in Fig. 13-8 is moving with *constant* speed, then $\omega = d\phi/dt = $ constant. We can integrate this to find the particle's angle $\phi(t)$ as a function of time:

$$d\phi = \omega dt$$

$$\Rightarrow \int d\phi = \int \omega dt = \omega \int dt \qquad (13\text{-}6)$$

$$\Rightarrow \phi(t) = \omega t + \text{integration constant} = \omega t + \phi_0 .$$

As the particle moves about the circle at constant speed, the angle ϕ increases linearly with time. Now the particle did not necessarily start on the x-axis ($\phi = 0$) at $t = 0$, so the integration constant ϕ_0 serves the purpose of identifying the *initial condition*: $\phi(t = 0) = \phi_0$. The angle ϕ is called the **phase angle** of the particle or, often, just the **phase**. The *phase* is simply an *angle*—in radians—that locates the particle at an instant of time t. The constant ϕ_0 is called the **phase constant**, and it merely specifies the initial condition of the system. The initial angle ϕ_0 is related to the initial position s_0 by $\phi_0 = s_0/r$.

As the particle moves around the circle at constant angular frequency ω, its x-coordinate is given by

$$x(t) = r \cos \phi(t) = r \cos(\omega t + \phi_0) . \qquad (13\text{-}7)$$

This is illustrated in Fig. 13-9a. You can see that the particle's x-coordinate starts at $x_0 = x(t = 0) = r \cos \phi_0$ and then *oscillates* as the phase ϕ of the particle increases with time. Figure 13-9b shows a graph of x-versus-t. It is simple harmonic motion! Notice that this is neither a pure sine nor a pure cosine function because the graph does not start at zero or at the maximum. Instead, we see how the phase constant ϕ_0 determines the initial condition for x. The *period* of the oscillation is given by $T = 1/f = 2\pi/\omega$.

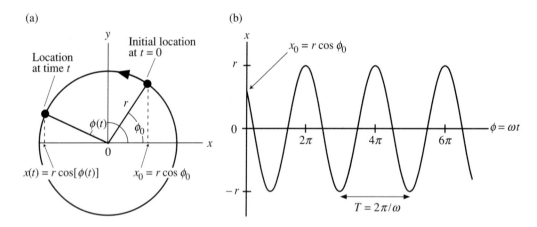

FIGURE 13-9   a) A particle undergoing circular motion at constant speed. b) The x-coordinate of the particle oscillates with simple harmonic motion.

Suppose you placed a small block on the edge of a turntable and then positioned your eye in the *plane* of the turntable so that you looked directly across its surface. As Fig. 13-10 shows, you would see the block oscillate back and forth as the turntable moved. In fact, you would be "seeing" just the x-coordinate of the block's motion, which would be perpendicular to your line of sight, and not the y-component, which would be along your line of sight. This motion of the block is simple harmonic motion—identical to the motion of a mass oscillating back and forth on a horizontal spring! It is this very close connection between circular motion and simple harmonic motion that has prompted us to develop the mathematics of this section—and we will see very shortly how this mathematical description is connected to the SHM equation of motion.

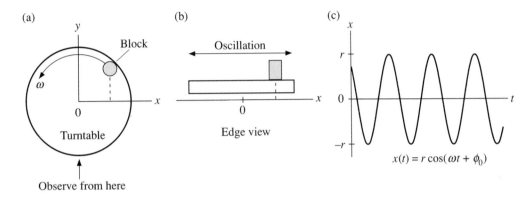

FIGURE 13-10 a) Top view of a block revolving on a turntable. b) Edge view of the block moving back and forth along x-axis. c) Graph of the x-motion, which is simple harmonic motion

EXAMPLE 13-2 A small block is placed on the edge of turntable that is 20 cm in diameter. The block is viewed from the edge as the turntable rotates at 45 rpm. At $t = 0$, the block is observed to be 5 cm to the right of the center and moving to the right. Draw a position-versus-time graph for the block.

SOLUTION From the edge we can see only the x-component of the block's motion. The position is described by Eq. 13-7: $x(t) = r\cos(\omega t + \phi_0)$. The turntable is 20 cm in diameter, so its radius is $r = 10$ cm. The rotation frequency and the angular frequency are

$$f = \frac{45 \text{ revolutions}}{\text{minute}} \times \frac{1 \text{ minute}}{60 \text{ seconds}} = 0.75 \text{ s}^{-1} = 0.75 \text{ Hz}$$

$$\Rightarrow \omega = 2\pi f = 4.71 \text{ s}^{-1}.$$

We can find the phase constant ϕ_0 from the initial condition $x_0 = 5$ cm $= r\cos\phi_0$. This condition gives:

$$\phi_0 = \cos^{-1}\left(\frac{x_0}{r}\right) = \cos^{-1}\left(\frac{5 \text{ cm}}{10 \text{ cm}}\right) = \pm\frac{\pi}{3} = \pm 60°.$$

There are two possible angles, $\pm 60°$, with a cosine of $\frac{1}{2}$. These angles correspond to the particle in Fig. 13-9a being above the x-axis ($+60°$) or below the x-axis ($-60°$). Because the block is moving to the *right* at $t = 0$, it must be *below* the x-axis. Thus ϕ_0 is $-\pi/3$. The block's position at time t is thus:

$$x(t) = (10 \text{ cm})\cos\left((4.71 \text{ s}^{-1})t - \frac{\pi}{3}\right).$$

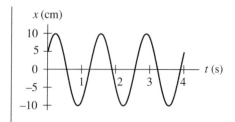

FIGURE 13-11 The position-versus-time graph for Example 13-2.

The position-versus-time graph for the first four seconds is shown in Fig. 13-11. Notice how a correct choice of the phase constant causes the graph to start at $x_0 = 5$ cm and then to *increase* as the block moves to the right. A phase constant of $+\pi/3$ would have given a graph that started at 5 cm but then decreased with time.

13.3 The Equation of Motion

Now that we've described the basics of simple harmonic motion, we can being to analyze the motion in more detail. The prototype of all simple-harmonic-motion is a mass oscil-

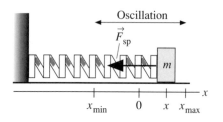

FIGURE 13-12 The prototype of simple-harmonic-motion: a mass oscillating on a horizontal spring without friction.

lating on a spring. Consider the situation shown in Fig. 13-12 in which a block of mass m is attached to a spring of spring constant k and allowed to oscillate on a horizontal frictionless surface. This is the simplest possible oscillation, with no distractions due to frictional or gravitational forces. We will consider the spring itself to be massless, as we have done previously. From Hooke's law, we know the restoring force on the block to be $F = -k\Delta x$. For mathematical simplicity, let us place the origin of the coordinate system at the equilibrium position of the block. In that case $\Delta x = x$ and thus $F = -kx$. The oscillation is one-dimensional motion along the x-axis. Using Newton's second law we find:

$$\left[\sum F_x = -kx\right] = \left[ma = m\frac{d^2x}{dt^2}\right],$$
(13-8)

where we have used $a = dv/dt = d^2x/dt^2$. We can rearrange Eq. 13-8 to be

$$\frac{d^2x}{dt^2} + \frac{k}{m}x = 0 \qquad \text{(equation of motion for a spring).}$$
(13-9)

Equation 13-9 is called the **equation of motion**. Unfortunately for us, this equation is a second-order differential equation, so the solution for $x(t)$ is not trivial. The solution *cannot* be found by direct integration.

One interesting property of differential equations is that their solutions are *unique*—there is only *one* solution to Eq. 13-9. If we were able to *guess* a solution, the uniqueness property would tell us that we had found the *only* solution. This might seem a rather strange way to solve equations, but in fact differential equations are frequently solved by using your knowledge of what the solution needs to "look like" to guess an appropriate functional form. Let's give it a try.

We know from experimental evidence that the oscillatory motion of a spring is sinusoidal. (This was shown in Fig. 13-2.) You may also recall our discovery that the acceleration is proportional to the negative of the displacement for an oscillating spring: $a \propto -\Delta y$. That is exactly what Eq. 13-8 is telling us (with Δy replaced by x): $a = -(k/m)x$. All of this strongly suggests that the function seen in Fig. 13-2 must be a solution of the differential equation of Eq. 13-9. Let us *guess* that the solution should have the functional form

$$x(t) = A\cos(\omega t + \phi_0). \tag{13-10}$$

In writing $x(t)$ this way we have used what you learned in Section 13.2 about the connection between SHM and circular motion. At this time, A, ω, and ϕ_0 are unspecified constants that we can adjust to any values that might be necessary to satisfy Eq. 13-9.

If you were going to guess that a solution to the algebraic equation $x^2 - 4 = 0$ is $x = 2$, you would verify your guess by substituting it into the original equation to see if it works. We need to do the same thing here: substitute Eq. 13-10 into Eq. 13-9 to see if, for an appropriate choice of the three constants, it "works." To do so, we are going to need the second derivative of $x(t)$. Finding this is straightforward:

$$x(t) = A\cos(\omega t + \phi_0)$$

$$\Rightarrow \frac{dx}{dt} = -\omega A\sin(\omega t + \phi_0) \tag{13-11}$$

$$\Rightarrow \frac{d^2x}{dt^2} = -\omega^2 A\cos(\omega t + \phi_0).$$

If we now substitute the first and third of Eqs. 13-11 into Eq. 13-9 we find

$$\left[-\omega^2 A\cos(\omega t + \phi_0)\right] + \frac{k}{m}\left[A\cos(\omega t + \phi_0)\right] = 0$$

$$\Rightarrow \left(\frac{k}{m} - \omega^2\right)A\cos(\omega t + \phi_0) = 0. \tag{13-12}$$

The only way Eq. 13-12 can be true at *all* instants of time is if $\omega^2 = k/m$. There do not seem to be any restrictions on the two constants A and ϕ_0, and, in fact, the solution to a second-order differential equation always has two "integration constants" that specify the initial conditions. (We will see later in this section how A and ϕ_0 are related to the initial conditions x_0 and v_0.)

So we have found—by guessing!—a solution to Eq. 13-9. Because solutions are unique, we have found *the* solution to the equation of motion for an oscillating spring:

$$\begin{cases} x(t) = A\cos(\omega t + \phi_0) \\ \text{with } \omega = 2\pi f = \sqrt{\dfrac{k}{m}} \end{cases} \quad \text{(SHM of an oscillating spring).} \tag{13-13}$$

Interpreting the Equation of Motion

Finding the solution to the equation of motion is one thing, but interpreting and understanding that solution is something else—and that is where the challenge lies!

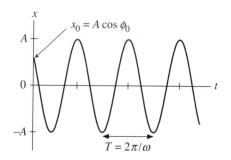

FIGURE 13-13 Displacement-versus-time graph for the solution of the SHM equation of motion. Note the initial condition: $x_0 = A\cos\phi_0$.

A graph of x-versus-t for Eq. 13-13, shown in Fig. 13-13, is identical with Fig. 13-9b. Now, however, we can specifically identify the constant A as being the *amplitude* of the motion, with x oscillating between $x_{min} = -A$ and $x_{max} = +A$. The *phase constant* ϕ_0 is the initial value of the *phase angle* $\phi = \omega t + \phi_0$, and it establishes the initial position to be $x_0 = A\cos\phi_0$. There are no restrictions on the constants A or ϕ_0—we can set them to any value we wish.

On the other hand, the constant ω can *only* take the value $\omega = \sqrt{k/m}$. We see, by the role it plays in $x(t)$, that ω is the *angular frequency* and that it is uniquely determined by the spring constant k and the mass m of the block. Selecting a specific spring and a specific block establishes the period and frequency as:

$$\text{frequency:} \quad f = \frac{\omega}{2\pi} = \frac{1}{2\pi}\sqrt{\frac{k}{m}}$$

$$\text{period:} \quad T = \frac{1}{f} = \frac{2\pi}{\omega} = 2\pi\sqrt{\frac{m}{k}}. \tag{13-14}$$

Equations 13-14 have an important implication: The period and frequency of an oscillator depend on the spring constant k and on the mass m but *not* on the amplitude of oscillation. A small amplitude oscillation and a large amplitude oscillation have exactly the same period and frequency.

We can find the velocity $v(t)$ and the acceleration $a(t)$ of the oscillating mass by taking the time derivatives of $x(t)$, as given in Eq. 13-13. First,

$$v(t) = \frac{dx}{dt} = -\omega A\sin(\omega t + \phi_0) = -v_{max}\sin(\omega t + \phi_0) \tag{13-15}$$

where $v_{max} = \omega A$ is the "amplitude" of the velocity. Note that the amplitude is inherently positive, which is why we defined $v_{max} = \omega A$ rather than $v_{max} = -\omega A$. Similarly

$$a(t) = \frac{dv}{dt} = -\omega^2 A\cos(\omega t + \phi_0) = -a_{max}\cos(\omega t + \phi_0) \tag{13-16}$$

where $a_{max} = \omega^2 A$ is the maximum acceleration.

The three equations 13-13, 13-15, and 13-16 together describe the kinematics of simple harmonic motion. They bear no resemblance to the kinematics of constant acceleration motion! Figure 13-5 has already shown how the graphs of x, v, and a are related to each other. It would be good to look back at Fig. 13-5 now that you have seen the mathematical expressions for x, v, and a.

In Section 13.1 we discovered from experiment that $a \propto -x$. Now we see from Eq. 13-16 that $a(t) = -(\omega^2)\cdot x(t)$. This result confirms our observation and tells us that the proportionality constant is ω^2. Our *theory* of simple harmonic motion, which is based on Hooke's law and Newton's second law, is in good agreement with the evidence.

An issue that we need to look at more carefully is just what the *phase constant* ϕ_0 represents. Many students become confused by ϕ_0 and try to omit it from equations (effectively setting $\phi_0 = 0$), but you cannot do that! It contains important information about the initial conditions of the oscillator. The initial position x_0 and velocity v_0, which we can find by setting $t = 0$ in Eqs. 13-13 and 13-15, are

$$x(t = 0) = x_0 = A\cos\phi_0$$

$$v(t = 0) = v_0 = -\omega A\sin\phi_0.$$

(13-17)

You can see from Eq. 13-17 that the phase constant ϕ_0, along with the amplitude A, describes the initial conditions x_0 and v_0. But problems typically give values of x_0 and v_0, rather than of A and ϕ_0, so we want to "solve" Eqs. 13-17 for the constants A and ϕ_0 in terms of x_0 and v_0. First divide the second of Eqs. 13-17 by ω, then square both sides of each equation:

$$x_0^2 = A^2\cos^2\phi_0$$

$$\left(\frac{v_0}{\omega}\right)^2 = A^2\sin^2\phi_0.$$

Now add these two equations, giving

$$x_0^2 + \left(\frac{v_0}{\omega}\right)^2 = A^2\cos^2\phi_0 + A^2\sin^2\phi_0 = A^2$$

$$\Rightarrow A = \sqrt{x_0^2 + \left(\frac{v_0}{\omega}\right)^2}.$$

(13-18)

Here we used the trigonometric identity $\cos^2\alpha + \sin^2\alpha = 1$. Equation 13-18 tells us that the amplitude of the motion is determined by *both* the initial position *and* the initial velocity. Do not make the error of assuming $A = x_0$. That assumption is valid only if $v_0 = 0$, which describes the situation of releasing the block from one of its turning points.

Now return to Eqs. 13-17, again divide the second equation by ω, then divide the initial velocity equation by the initial position equation:

$$\frac{v_0/\omega}{x_0} = \frac{-A\sin\phi_0}{A\cos\phi_0} = -\tan\phi_0$$

$$\Rightarrow \phi_0 = \begin{cases} -\tan^{-1}\left(\dfrac{v_0}{\omega x_0}\right) & x_0 > 0 \\[2ex] -\tan^{-1}\left(\dfrac{v_0}{\omega x_0}\right) + \pi & x_0 < 0. \end{cases}$$

(13-19)

You may recall, from trigonometry, that there is always some ambiguity when taking inverse tangents because $\tan(\theta + \pi) = \tan\theta$. Consequently, the calculation of ϕ_0 needs an additional π added when $x_0 < 0$. Equations 13-18 and 13-19 meet our goal of being able to determine the amplitude A and the phase constant ϕ_0 from the initial conditions x_0 and v_0.

Let us look at how these ideas work in two common situations. First, consider a block that is released from the right turning point at $t = 0$. This situation is shown in the picture of Fig. 13-14a. Here we have $x_0 = A$ and $v_0 = 0$, from which we find (using Eq. 13-19) the phase constant to be $\phi_0 = 0$. Eq. 13-18 gives the amplitude as $A = x_0$. Thus

$$x(t) = x_0\cos\omega t$$

$$v(t) = -\omega x_0\sin\omega t.$$

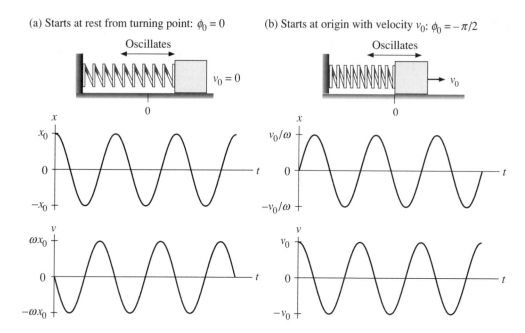

FIGURE 13-14 Graphical illustration of the SHM solution for $x(t)$ and $v(t)$. Parts a) and b) have different initial conditions and thus different values of the phase constant.

This is the simplest possible oscillator solution, but it is a specific solution that applies *only* to one very specific set of initial conditions. The graphs in Fig. 13-14 show x and v as functions of time. Notice how they agree with the physical situation of the picture.

But you're free to start your clock whenever you wish. Suppose that you choose to start the clock, thus defining $t = 0$, as the block is passing the origin and moving toward the *right*. This situation is shown in Fig. 13-14b. In this case $x_0 = 0$ and $v_0 > 0$, giving $\phi_0 = -\tan^{-1}(+\infty) = -\pi/2$. The amplitude is $A = \sqrt{(v_0 / \omega)^2} = v_0/\omega$. The position and velocity for this situation are given by the equations

$$x(t) = \frac{v_0}{\omega} \cos(\omega t - \frac{\pi}{2}) = \frac{v_0}{\omega} \sin \omega t$$

$$v(t) = -\omega \cdot \frac{v_0}{\omega} \sin(\omega t - \frac{\pi}{2}) = v_0 \cos \omega t.$$

The graphs of x and v are shown in Fig. 13-14b. Notice that this is the *same motion* as Fig. 13-14a. The same block is oscillating on the same spring with the same period. All that is different is the initial conditions—where the oscillation begins. This physical difference is represented mathematically by a different value of the phase constant ϕ_0.

What we learn from this example is that the *single* equation Eq. 13-13 can represent *any* sinusoidal curve by properly choosing the phase constant ϕ_0 to represent the initial conditions. If you are working a problem in which an oscillation is taking place but you do not know the initial conditions, then you can omit ϕ_0. But if the initial conditions are known, you must use them to determine the specific value of ϕ_0.

EXAMPLE 13-3   A 1 kg block oscillates on a spring having a spring constant of 20 N/m. At $t = 0$, the block is observed to be 20 cm to the right of the equilibrium position and moving toward the left with a speed of 100 cm/s. Determine the period of oscillation and draw graphs of position and velocity versus time.

SOLUTION We can begin by determining the angular frequency

$$\omega = \sqrt{\frac{k}{m}} = \sqrt{\frac{20 \text{ N/m}}{1 \text{ kg}}} = 4.47 \text{ radians/s},$$

from which we find the period to be $T = 2\pi/\omega = 1.40$ s. The period is independent of the initial conditions. The position and velocity are given by Eqs. 13-13 and 13-15, but we need to determine specific values for the amplitude A and the phase constant ϕ_0. These do depend on the initial conditions, as given by Eqs. 13-18 and 13-19. We can calculate that

$$A = \sqrt{x_0^2 + (v_0/\omega)^2}$$
$$= \sqrt{(0.20 \text{ m})^2 + ((-1.00 \text{ m/s})/(4.47 \text{ radian/s}))^2} = 0.300 \text{ m}$$

and

$$\phi_0 = -\tan^{-1}\left(\frac{v_0}{\omega x_0}\right)$$
$$= -\tan^{-1}\left(\frac{-1.00 \text{ m/s}}{(4.47 \text{ radians/s})(0.20 \text{ m})}\right)$$
$$= +0.841 \text{ radians} = +48.2°.$$

Notice that the we used a *negative* initial velocity because the block is moving to the left. The position and velocity equations for this oscillator are thus

$$x(t) = 0.300 \text{ m} \cdot \cos(4.47t + 0.841)$$
$$v(t) = -1.34 \text{ m/s} \cdot \sin(4.47t + 0.841).$$

These two equations are graphed in Fig. 13-15. Note how the proper choice for ϕ_0 starts both graphs at the correct initial values, that the block is moving to the left (x decreasing), that the position graph does reach an amplitude of 0.30 m, and that the period is 1.40 s.

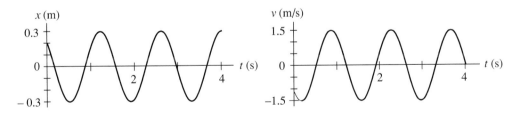

FIGURE 13-15 Position and velocity graphs for the solution of Example 13-3.

EXAMPLE 13-4 A 500 g block oscillating on a spring is first observed at $x = 15$ cm and moving to the right. Then 0.30 s later the block reaches a maximum displacement of 25 cm. What is the spring constant?

SOLUTION The only way we can find the spring constant is by first finding the angular frequency, then using $\omega^2 = k/m$. The position equation of the block is $x(t) = A\cos(\omega t + \phi_0)$, which we can use to find ω if we know where the block is at a specific instant of time. Let's start our clock at the instant of the first observation, so $x_0 = 0.15$ m. We can interpret the "maximum displacement" in the problem statement as the amplitude: $A = 0.25$ m. With these two pieces of information, the initial condition can be written

$$x_0 = A\cos\phi_0$$

$$\Rightarrow \phi_0 = \cos^{-1}\left(\frac{x_0}{A}\right) = \cos^{-1}(.6) = \pm 0.927 \text{ radians.}$$

Either the positive or negative solution is mathematically acceptable (remember that $\cos(-\alpha) = \cos(\alpha)$), but we need to select the *physically* meaningful solution. Because the phase constant affects both x_0 and v_0, we need to use the fact that v_0 is *positive*. Going back to Eq. 13-17, which was $v_0 = -\omega A\sin\phi_0$, we see that only the *negative* value of the phase constant will give a positive initial velocity. So $\phi_0 = -0.927$ radians.

The block reaches its maximum displacement $x_{\max} = A$ at time $t_{\max} = 0.30$ s. At that instant of time

$$x_{\max} = A = A\cos(\omega t_{\max} + \phi_0)$$

$$\Rightarrow \cos(\omega t_{\max} + \phi_0) = 1.$$

This can be true only if $\omega t_{\max} + \phi_0 = 0$, in which case we have

$$\omega = \frac{-\phi_0}{t_{\max}} = \frac{-(-0.927 \text{ radians})}{(0.30 \text{ s})} = 3.09 \text{ radians / s.}$$

Now that we know ω it is straightforward to compute

$$k = m\omega^2 = (0.500 \text{ kg})(3.09 \text{ radians/s})^2 = 4.78 \text{ N/m.}$$

13.4 Vertical Oscillations

The analysis we just completed was for a horizontally oscillating spring. But the experimental data we looked at in Section 13.2 was for a vertically oscillating spring. Is it safe to assume that a vertical oscillation is the same as a horizontal oscillation? Or does the additional force of gravity change the motion? Let us look at this more carefully.

Figure 13-16 shows a block of mass m on a vertical spring of spring constant k. The difficulty with this problem is that the *equilibrium* position of the block is *not* where the spring is at its unstretched length. We first see the empty spring, at its unstretched length,

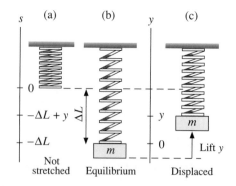

FIGURE 13-16 Vertical oscillations. a) An unstretched spring. b) The equilibrium position. c) A displacement from equilibrium.

in Fig. 13-16a. The s-axis measures the displacement of the spring from this point. Figure 13-16b then shows the block attached and hanging motionless. This is the equilibrium position of the block, with the spring stretched by ΔL.

Determining the stretching ΔL is a static equilibrium problem. There are two forces on the block in the equilibrium situation of Fig. 13-16b: the upward spring force and a downward weight force. The y-component of the spring force is given by Hooke's law: $F_{sp} = -ks = +k\Delta L$. Here we have noted ΔL is simply a *distance*, which is a positive number, and that the block's position is $s = -\Delta L$. Newton's first law for a block in equilibrium is:

$$\sum F_y = 0 = k\Delta L - mg$$

$$\Rightarrow k\Delta L = mg.$$

We can then find that

$$\Delta L = \frac{mg}{k}. \tag{13-20}$$

This is the distance the spring stretches to reach equilibrium when the block is hung on it.

Let's establish a y-axis so that its origin is at the block's equilibrium position of Fig. 13-16b. (This is to be consistent with our analysis of the horizontal spring, where we chose the origin to be at the equilibrium position.) Then displace the block to position y, as shown in Fig. 13-16c. In this position, the spring is stretched by an amount $\Delta L - y$ and the block's position relative to the spring is $s = -(\Delta L - y)$. The block is not in equilibrium at this point, but Newton's second law can be used to find:

$$\left[\sum F_y = ma = m\frac{d^2y}{dt^2}\right] = F_{sp} - |\vec{W}|$$
$$= -ks - mg \tag{13-21}$$
$$= k\Delta L - ky - mg.$$

But from Eq. 13-20 we have $k\Delta L - mg = 0$, so the right side of Eq. 13-21 is simply $-ky$. Thus Eq. 13-21 can be rearranged to read:

$$\frac{d^2y}{dt^2} + \frac{k}{m}y = 0. \tag{13-22}$$

Equation 13-22 for the motion of a vertical spring about its *equilibrium* position (*not* about its unstretched length) is identical to Eq. 13-9 for the motion of a horizontal spring. It must, therefore, have exactly the same oscillatory solution:

$$\begin{cases} y(t) = A\cos(\omega t + \phi_0) \\ \text{with } \omega = 2\pi f = \sqrt{\dfrac{k}{m}} \end{cases} \tag{13-23}$$

where the amplitude A and phase constant ϕ_0 are again given by Eqs. 13-18 and 13-19.

The vertical oscillatory motion of block about its equilibrium position is the same simple harmonic motion as that of a block on a horizontal spring. This is an important finding. It was not obvious that the motion would still be simple harmonic motion when gravity was included. Because the motions are the same, everything we have said about horizontal oscillations is equally valid for vertical oscillations.

EXAMPLE 13-5 A 200 g block is attached to a spring that has a spring constant of 10 N/m. The block is pulled down to the point where the spring is stretched 30 cm, then released. Where is the block and what is its speed 3 seconds later?

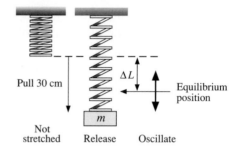

SOLUTION Figure 13-17 shows the situation. If the block simply hangs at rest, the spring's stretch will be given by Eq. 13-20 as $\Delta L = mg/k = 0.196$ m = 19.6 cm. Because the spring was stretched by 30 cm, the block is released 10.4 cm below the equilibrium point. Oscillations are symmetrical about the equilibrium point, *not* about the spring's

FIGURE 13-17 A block on a vertical spring is pulled down 30 cm and released. It then oscillates about its equilibrium position.

original end point, so the block oscillates with an amplitude $A = 10.4$ cm about a point that is 19.6 cm beneath the spring's original end point. The block's position as a function of time, as measured from the equilibrium position, is

$$y(t) = (10.4 \text{ cm})\cos(\omega t + \phi_0)$$

where $\omega = \sqrt{k/m} = 7.07$ s^{-1}. The phase constant is determined by the initial conditions, which are $y_0 = -A$ and $v_0 = 0$. From Eq. 13-19, $\phi_0 = \pi$. This is the appropriate phase constant to give $y(t = 0) = A\cos(\pi) = -A$. At $t = 3$ s, the block is at position

$$y(t = 3 \text{ s}) = (10.4 \text{ cm})\cos\left((7.07 \text{ s}^{-1})(3 \text{ s}) + \pi\right) = 7.4 \text{ cm}.$$

The block is 7.4 cm above the equilibrium position, or 12.2 cm below the original end of the spring. Its velocity at this point is

$$v(t) = -\omega A\sin(\omega t + \phi_0)$$

$$\Rightarrow v(t = 3 \text{ s}) = -(7.07 \text{ s}^{-1})(10.4 \text{ cm})\sin\left((7.07 \text{ s}^{-1})(3 \text{ s}) + \pi\right) = 52 \text{ cm/s}.$$

13.5 The Pendulum

[**Photo suggestion: A grandfather clock with an ornate pendulum.**]

Now let's look at another very common oscillator: a pendulum. Figure 13-18a shows a mass m attached to a string of length L. The mass is free to swing back and forth. There are two forces acting on the mass: the string tension $\vec{T}$ and the weight $\vec{W}$. Because the mass moves tangent to the circle, along the arc of length s, it is convenient to divide the forces into components parallel to the motion (i.e., tangent to the circle) and perpendicular to the motion (i.e., parallel to the string). As the free-body diagram of Fig. 13-18b shows, the tension has only a perpendicular component $T_\perp = |\vec{T}|$. The weight

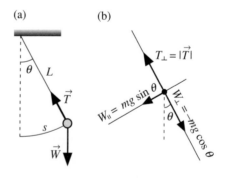

FIGURE 13-18 a) The motion of a pendulum. b) Free-body diagram.

has components $W_\parallel = -mg\sin\theta$ and $W_\perp = -mg\cos\theta$. Notice the minus sign of $W_\parallel$, implying that this component of $\vec{W}$ acts as a restoring force.

Newton's second law can be applied to the component parallel to the motion (along s). Note that the angle θ is related to the arc length s through $\theta = s/L$. We find

$$\left[\sum F_\parallel = -mg\sin\theta = -mg\sin\left(\frac{s}{L}\right)\right] = \left[ma = m\frac{d^2s}{dt^2}\right] \qquad (13\text{-}24)$$

$$\Rightarrow \frac{d^2s}{dt^2} + g\sin\left(\frac{s}{L}\right) = 0.$$

The second of Eqs. 13-24 is the equation of motion for an oscillating pendulum. It is even more complicated, with the sine function, than Eq. 13-9 for an oscillating spring! (Note: Both θ and s are negative when the pendulum is to the left of center.)

Suppose, however, that we restrict the pendulum's oscillations to *small angles* of less than about 10°. This is an excellent description for many practical applications of pendulums. With that restriction, we can use the small angle approximation: $\sin\theta \approx \theta = s/L$. Eq. 13-24 then becomes

$$\frac{d^2s}{dt^2} + \frac{g}{L}s = 0 \qquad \text{(equation of motion for a pendulum).} \qquad (13\text{-}25)$$

This is *exactly* the same equation as Eq. 13-9 for the oscillating spring. The variable names are different, with x replaced by s and k/m by g/L, but that does not make it a different equation. Because we know the solution to the spring problem, we can immediately write the solution to the pendulum problem just by changing variables and constants:

$$\begin{cases} s(t) = A\cos(\omega t + \phi_0) \\[2mm] \text{with } \omega = 2\pi f = \sqrt{\dfrac{g}{L}} \end{cases} \qquad \text{(SHM of a pendulum),} \qquad (13\text{-}26)$$

where the amplitude A and the phase constant ϕ_0 are again determined by the initial conditions, as given in Eqs. 13-18 and 13-19. You begin to see how it is that we have solved *all* simple harmonic motion problems once we have solved the case of the horizontal spring!

The pendulum is interesting in that the frequency, and hence the period, is *independent* of the mass. It depends only on the length of the pendulum.

EXAMPLE 13-6   What length pendulum has a period of exactly 1 s?

SOLUTION From Eq. 13-29 we can find

$$T = \frac{1}{f} = 2\pi\sqrt{\frac{L}{g}}$$

$$\Rightarrow L = g\cdot\left(\frac{T}{2\pi}\right)^2 = 0.248 \text{ m.}$$

EXAMPLE 13-7   A pendulum with a 300 g mass on a 30 cm string has a velocity of 25 m/s as it passes through the lowest point. a) What maximum angle does the pendulum reach? b) What is its velocity when the angle is half of the maximum?

SOLUTION a) The angular frequency of the pendulum is

$$\omega = \sqrt{\frac{g}{L}} = \sqrt{\frac{9.8 \text{ m/s}^2}{0.30 \text{ m}}} = 5.72 \text{ radians/s}.$$

The initial conditions are $s_0 = 0$ at the lowest point and $v_0 = v_{max} = \omega A$, so the amplitude is given by Eq. 13-18 as

$$A = s_{max} = \frac{v_0}{\omega} = \frac{0.25 \text{ m/s}}{5.72 \text{ /s}} = 0.0437 \text{ m}.$$

We can change the arc length s_{max} to an angle θ_{max} by $\theta_{max} = s_{max}/L = 0.146$ radians = 8.35°. Because the maximum angle is less than 10°, our analysis based on the small angle approximation is valid.

b) To find the velocity at the point in the motion where $\theta = \frac{1}{2}\theta_{max}$, we first need to find out *when* it reaches that point. The time interval to move from $\theta = 0$ to $\theta = \theta_{max}$ is $T/4$, a quarter of a period. However, the time to move to $\frac{1}{2}\theta_{max}$ is *not* $T/8$ because the velocity is not constant. The pendulum's position is given by $s = A\cos(\omega t + \phi_0)$. We need to use this to find the time at which $s = A/2$ (which corresponds to $\theta = \frac{1}{2}\theta_{max}$):

$$s = \frac{A}{2} = A\cos(\omega t + \phi_0)$$

$$\Rightarrow \cos(\omega t + \phi_0) = \frac{1}{2}.$$

The phase constant is determined by the initial conditions to be $\phi_0 = -\tan^{-1}(+\infty) = -\pi/2$. Thus the instant of time when $\theta = \frac{1}{2}\theta_{max}$ is found (using the trig identity $\cos(\alpha - \pi/2) = \sin\alpha$) to be:

$$\cos(\omega t - \frac{\pi}{2}) = \sin \omega t = \frac{1}{2}$$

$$\Rightarrow t = \frac{1}{\omega}\sin^{-1}\left(\frac{1}{2}\right) = 0.0915 \text{ s}.$$

Now that we know the instant of time when $\theta = \frac{1}{2}\theta_{max}$, we can calculate the velocity to be

$$v = -\omega A\sin(\omega t - \frac{\pi}{2}) = +\omega A\cos \omega t = 0.187 \text{ m/s} = 18.7 \text{ cm/s}.$$

The restoring force of a spring, given by Hooke's law $F = -kx$, is proportional to the displacement x from equilibrium. The pendulum's restoring force is proportional to the displacement s when the small angle approximation is valid. A restoring force that is directly proportional to the displacement is called a **linear restoring force**. It turns out that *any* linear restoring force gives an equation of motion that is identical—other than maybe using different variables—to the spring equation. Consequently, the motion of any system that oscillates in response to a linear restoring force—and there are many in science and engineering—will have *exactly* the same simple harmonic motion as a block on a spring. So you can see why we consider an oscillating spring as the prototype of SHM: everything that you learn about it can be immediately applied to the oscillations of any other linear restoring force. Vibrations of machinery, the motion of electrons in electrical circuits, and many other diverse applications all turn out to have an equation of motion identical to Eqs. 13-9. It is one of the most important equations in science and engineering.

13.6 Energy in Oscillating Systems

Our analysis of oscillatory motion thus far has been in terms of force and acceleration—a Newton's laws analysis. Another tool that we have available, however, is an energy analysis. We already have all of the information we need for such an analysis because we derived the elastic potential energy back in Section 10.6.

Consider, once again, the simple harmonic oscillator shown in Fig. 13-5, which has a horizontally oscillating block of mass m attached to a spring of constant k. The block's kinetic energy at time t is given by

$$K = \tfrac{1}{2}mv^2 = \tfrac{1}{2}m\left[-\omega A \sin(\omega t + \phi_0)\right]^2 = \tfrac{1}{2}m\omega^2 A^2 \sin^2(\omega t + \phi_0). \qquad (13\text{-}27)$$

But we can also write $m\omega^2 = k$, using the definition of ω from Eq. 13-13, which gives

$$K = \tfrac{1}{2}kA^2 \sin^2(\omega t + \phi_0). \qquad (13\text{-}28)$$

The kinetic energy, not surprisingly, oscillates with time as the block's velocity oscillates.

Similarly, we can find the elastic potential energy stored in the spring due to the block-wall interaction. We have $\Delta x = x$ because we have chosen to use the equilibrium position of the block as the origin. Thus

$$U = \tfrac{1}{2}k(\Delta x)^2 = \tfrac{1}{2}kx^2 = \tfrac{1}{2}k\left[A\cos(\omega t + \phi_0)\right]^2$$
$$= \tfrac{1}{2}kA^2 \cos^2(\omega t + \phi_0). \qquad (13\text{-}29)$$

The total mechanical energy is the sum of Eqs. 13-28 and 13-29:

$$E = K + U = \tfrac{1}{2}kA^2 \sin^2(\omega t + \phi_0) + \tfrac{1}{2}kA^2 \cos^2(\omega t + \phi_0)$$
$$= \tfrac{1}{2}kA^2\left[\sin^2(\omega t + \phi_0) + \cos^2(\omega t + \phi_0)\right] \qquad (13\text{-}30)$$
$$= \tfrac{1}{2}kA^2$$

where we used $\sin^2\alpha + \cos^2\alpha = 1$. We find a very simple result! Despite the fact that K and U are oscillating in time, their *sum* is a constant—exactly as expected if energy is conserved.

Figure 13-19 shows graphically how the kinetic and potential energy change with time (oscillating but *positive*, because x and v are squared). There is a continuous transfer of energy back and forth between the kinetic energy of the moving block and the stored potential energy of the spring, but notice that their sum is constant. Also notice that K and U both oscillate *twice* each period; make sure you understand why.

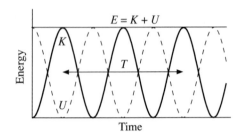

FIGURE 13-19 Kinetic energy(solid line), potential energy (dashed line), and the total energy (horizontal line) for simple harmonic motion.

We can interpret Eq. 13-30 better by writing it in several different forms. We know that $\omega^2 = k/m$, that $x_{max} = A$, and that $v_{max} = \omega A$. With this information we can write

$$E = \tfrac{1}{2} kA^2 = \tfrac{1}{2} kx_{max}^2 = \tfrac{1}{2} m\omega^2 A^2 = \tfrac{1}{2} mv_{max}^2 . \tag{13-31}$$

These are all alternative ways of expressing the total energy, and all can be useful in different contexts. Note especially the two forms involving x_{max} and v_{max}. As we have seen, the block reaches its maximum displacement at the turning points, where $v = 0$. At those points, the energy is *purely* potential energy of magnitude $\tfrac{1}{2} kx_{max}^2$. Similarly, the block reaches its maximum speed as it passes through the $x = 0$ equilibrium point. At that instant, the energy is *purely* kinetic energy of magnitude $\tfrac{1}{2} mv_{max}^2$.

EXAMPLE 13-8   A 500 g block on a spring is pulled a distance of 20 cm and released. The subsequent oscillations are measured to have a period of 0.80 s. What is the block's speed when it is 12 cm from equilibrium?

SOLUTION We can analyze this problem from either the force/acceleration perspective or the energy perspective. It will be instructive to do it both ways, so that you can see what is involved. Let us start with force/acceleration. The block's velocity is given by

$$v(t) = v_{max} \sin(\omega t + \phi_0) = -\omega A \sin(\omega t + \phi_0) .$$

We can deduce the amplitude from the problem statement as being $A = 0.20$ m. Because we are given the period, we can determine the angular frequency $\omega = 2\pi/T = 7.85$ radians/s. The phase constant depends on the initial conditions, but we are lucky in this case: Eq. 13-19 gives $\phi_0 = 0$ because $v_0 = 0$. (These are the same conditions as the situation shown in Fig. 13-14a.) So all we need in order to find the velocity is to know the instant of time when $x = 0.12$ m. For that, we go to the position equation:

$$x(t) = A\cos(\omega t + \phi_0)$$

$$\Rightarrow t = \frac{1}{\omega}\left[\cos^{-1}\left(\frac{x}{A}\right) - \phi_0 \right] = \frac{1}{7.85 \text{ s}^{-1}}\left[\cos^{-1}\left(\frac{12 \text{ cm}}{20 \text{ cm}}\right) - 0 \right] = 0.118 \text{ s}.$$

Remember to set your calculator to radians before taking the inverse cosine. Knowing the time when the block reaches $x = 12$ cm, we can find the velocity at that point to be:

$$v = -(7.85 \text{ s}^{-1})(0.20 \text{ m})\sin\left((7.85 \text{ s}^{-1})(0.118 \text{ s}) + 0\right) = -1.26 \text{ m / s}.$$

The negative sign implies that the motion is to the left. But the question asked for the *speed*, so the final answer is $|v| = 1.26$ m/s.

Now let us try an energy analysis. Let the time at which the block is released be the "before" and the time when it reaches 12 cm displacement be the "after." The block starts from the point of maximum displacement where $K_i = 0$ and $U_i = \frac{1}{2}kA^2$. Conservation of energy is thus

$$K_f + U_f = K_i + U_i$$

$$\Rightarrow \tfrac{1}{2}mv^2 + \tfrac{1}{2}kx^2 = 0 + \tfrac{1}{2}kA^2$$

$$\Rightarrow v = \sqrt{\frac{k}{m}(A^2 - x^2)} = \omega\sqrt{(A^2 - x^2)}$$

$$= 7.85 \text{ s}^{-1}\sqrt{(0.20 \text{ m})^2 - (0.12 \text{ m})^2} = 1.26 \text{ m}/\text{s}.$$

Which is the easier approach? You can be the judge!

Figure 13-20 is an energy diagram based on Fig. 10-26 in Chapter 10. The potential-energy curve of a spring, $U = \frac{1}{2}k(\Delta x)^2$, is a parabola—indicative of a linear restoring force. A particle with energy E is constrained to oscillate between the left turning point at x_L and the right turning points x_R. To go beyond these points would require a negative kinetic energy, which is physically impossible. The particle has purely potential energy at the two turning points and purely kinetic energy as it passes through the equilibrium point. Note that the motion is SHM if and only if the potential energy curve is parabolic because that is the "signature" of a linear restoring force.

You might also want to review the potential energy diagram of a diatomic molecule, which we showed in Fig. 10-32. If the molecular bond has a very small energy, such that E is very near the minimum of the curve, the potential energy curve looks "approximately" like the parabola of a linear restoring force. In that case we expect that the molecular vibration to be approximately simple harmonic motion. If, on the other hand, the bond energy

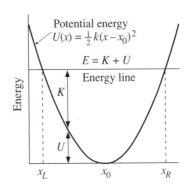

FIGURE 13-20 The energy diagram of a simple harmonic oscillator. A particle with energy E oscillates between x_L and x_R.

is well above the minimum—such as the E_1 actually shown in the figure—then the potential-energy curve is *not* the simple parabola of a spring obeying Hooke's law. The molecular bond will still oscillate at a regular frequency, but the motion will *not* be SHM. The vibration frequencies of molecules are very high—typically 5×10^{13} Hz!

13.7 Damped Oscillations

One obvious flaw in our analysis of oscillators is that we have not included any friction or dissipative forces. All real oscillators do run down—some very slowly but others quite quickly. As they do, their mechanical energy is transformed into the thermal energy of the oscillator and its environment, and the amplitude of oscillation decreases with time. Motion that runs down and stops is said to be *damped*, and when this idea is applied to an oscillator we refer to **damped oscillations**.

There are many possible reasons for the dissipation of energy: air resistance, friction at the point of support of a spring or pendulum, internal forces within the metal of the spring as it expands and contracts, and so on. While we could never account for all of these, a reasonable model is to consider only air resistance because, in many practical cases, it will be the predominant dissipative force.

We did not study air resistance or drag back in Part I. We had enough to worry about at the time without introducing the mathematical complexities of drag forces! Now, however, is a good time to see the effects of drag forces. Like friction, drag forces can only be modeled. These is no "law of air resistance" to tell us exactly how big air resistance forces are. Nonetheless, it is known from experimental evidence that the air-resistance drag force on *slowly* moving objects is proportional to their velocity: $\vec{F}_{drag} \propto \vec{v}$. Because air resistance slows particles down, rather than speeds them up, the *direction* of the drag force vector must always be *opposite* to the direction of the velocity vector. For one-dimensional motion, this gives us a model of the air resistance force:

$$F_{drag} = -bv \qquad \text{(model of the air resistance force).} \qquad (13\text{-}32)$$

Here F_{drag} and v are vector components and have signs. The minus sign is the mathematical statement that the force direction is opposite the velocity direction. The **damping constant** b depends in a complicated way on the shape of the object (long, narrow objects have less air resistance than wide, flat ones) *and* on the viscosity of the air or other medium in which the particle moves. It plays the same role in our model of air resistance that the coefficient of friction does in our model of friction.

The units of b need to be such that they will give units of force when multiplied by units of velocity. As you can confirm, these units are kg/s. A value of $b = 0$ kg/s corresponds to the limiting case of no resistance, in which case the mechanical energy will be conserved. A typical value of b for a spring or a pendulum in air is ≤ 0.10 kg/s. Poorly-shaped objects or objects moving in a liquid (which is much more viscous than air) can have significantly larger values of b.

We need to modify Eq. 13-8 for a horizontal spring to include the effects of air resistance. When the drag force of Eq. 13-32 is included, Newton's second law becomes

$$\left[\sum F_x = -kx - bv \right] = \left[ma = m\frac{d^2x}{dt^2} \right]$$

$$\Rightarrow \frac{d^2x}{dt^2} + \frac{b}{m}\frac{dx}{dt} + \frac{k}{m}x = 0, \qquad (13\text{-}33)$$

where we have used $v = dx/dt$. Equation 13-33 is the equation of motion of a damped oscillator. As was the case with Eq. 13-9, this is a second-order differential equation.

Equation 13-33 is *really* hard to solve, so we are not even going to try. We will simply assert that the solution is

$$\begin{cases} x(t) = Ae^{-bt/2m}\cos(\omega t + \phi_0) \\ \text{with } \omega = \sqrt{\dfrac{k}{m} - \dfrac{b^2}{4m^2}} = \sqrt{\omega_0^2 - \dfrac{b^2}{4m^2}} \end{cases} \qquad \text{(damped SHM),} \qquad (13\text{-}34)$$

where $\omega_0 = \sqrt{k/m}$ is the angular frequency of an undamped oscillator ($b = 0$). The constant e is the base of natural logarithms, so $e^{-bt/2m}$ is an *exponential function*. If you wish, you can confirm that this is, indeed, the solution by taking the first and second derivatives and substituting them into Eq. 13-33.

This solution makes sense, if we look at it carefully. First notice that Eq. 13-34 reduces to our previous solution, Eq. 13-13, when $b = 0$. In that case, $\omega = \omega_0$ and $e^0 = 1$. Second, for *lightly-damped* oscillations, where the system oscillates many times before stopping, $b/2m << \omega_0$. In that case, $\omega \approx \omega_0$ is a good approximation—that is, the presence of light damping does not affect the oscillation frequency. (*Heavy-damping*, which stops the motion within a couple of oscillations, causes the frequency to be significantly lowered.) We will focus on lightly-damped systems for the rest of this section.

Figure 13-21 shows a graph of the solution $x(t)$, from Eq. 13-34, for a lightly-damped oscillator. What we see is that the term $Ae^{-bt/2m}$, which is shown by the dashed line, acts like a slowly-varying time-dependent amplitude: $x_{max}(t) = Ae^{-bt/2m}$. The oscillation keeps bumping up against this line, slowly dying out with time. Note that $x_{max}(t = 0) = A$, so A is now the *initial* amplitude. A slowly-changing line that provides a border to a rapid oscillation is called the **envelope** of the oscillations. In this case, we can say that the oscillations have an *exponentially-decaying envelope*. Figure 13-21 shows how real oscillations decay with time, so Eq. 13-34 seems an adequate mathematical description. (Make sure you study Fig. 13-21 long enough to be completely convinced that it describes what you "see" when watching a damped oscillation.)

Changing the amount of damping, by changing the value of b, will affect how quickly the oscillations decay. Figure 13-22 shows just the envelope $x_{max}(t)$ for several oscillators that are identical except for the value of the damping constant b. (You need to imagine a rapid oscillation within each envelope, as in Fig. 13-21.) One of the envelopes (with $b = 0.10$ kg/s) is identical to Fig. 13-21. Increasing b causes the oscillations to damp more quickly, while decreasing b makes them last longer. A value of $b = 0$—no damping at all—produces a horizontal line indicative of an unchanging amplitude and conserved energy.

Because b has units of kg/s, the units of m/b must be seconds. It will be convenient to define what we will call the **time constant** τ

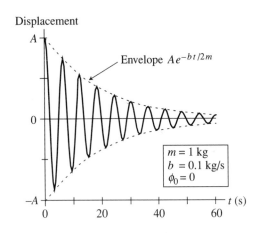

FIGURE 13-21 Displacement-versus-time graph for a damped oscillator.

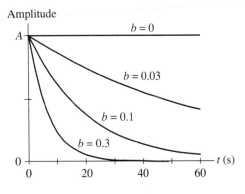

FIGURE 13-22 Several oscillation envelopes, which correspond to different values of the damping constant b.

(Greek "tau," pronounced to rhyme with "raw") to be

$$\tau = time \ constant = \frac{m}{b}. \tag{13-35}$$

With this definition, we can write the oscillation amplitude as $x_{max}(t) = Ae^{-t/2\tau}$.

The introduction of the idea of a time constant is particularly relevant for considering the oscillator's mechanical energy. The mechanical energy is no longer conserved, but at any particular time we can compute the mechanical energy from

$$E(t) = \tfrac{1}{2} k x_{max}^2 = \tfrac{1}{2} k \left(Ae^{-t/2\tau} \right)^2 = \left(\tfrac{1}{2} kA^2 \right) e^{-t/\tau} = E_0 e^{-t/\tau}, \tag{13-36}$$

where $E_0 = \tfrac{1}{2} kA^2$ is the initial energy at $t = 0$ and where we have used $(z^m)^2 = z^{2m}$. In other words, the oscillator's energy decays exponentially with time constant τ.

The graph of Fig. 13-23 clarifies what this means. The oscillator starts with an initial energy E_0, but at time $t = \tau$ the energy has decreased to $E(\tau) = E_0 e^{-1} = 0.37 E_0$. That is, the energy at time τ is 37% of its initial value. By $t = 2\tau$, the energy has decreased to $E(2\tau) = E_0 e^{-2} = 0.13 E_0$. This is 13% of its initial value. The significance of the time constant τ is that it measures the "characteristic time" during which the energy of the oscillation is dissipated. Roughly two-thirds of the initial energy is gone after an interval of one time constant has elapsed, while nearly 90% has dissipated after two time constants have gone by. For practical purposes, we can speak of the time constant as the *lifetime* of the oscillation—about how long it lasts.

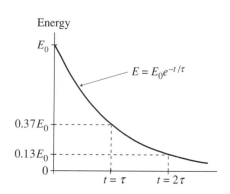

FIGURE 13-23 Exponential decay of the energy of an oscillator.

An oscillator with $\tau = 3$ s will oscillate for roughly 3 s, while one with $\tau = 30$ s will continue for roughly 30 s. Mathematically, there is never a time when the oscillation is "over." (Note that Eq. 13-34 approaches zero asymptotically, but it never gets there in any finite time.) So there is not any specific number X where we can say, "The oscillation is over at time X." The best we can do is define a characteristic time when the motion is "almost over," and that is what the time constant τ does.

There are many, many examples of exponential decay in science and engineering—not just for oscillations, but for many situations that change with time. You will see several more in this text, and you will find many others in more advanced courses. Wherever exponential decay occurs, the idea of a time constant is used to characterize how long the decay lasts. No matter what is decaying, it will have only 37% of its original value (roughly one-third) after a time interval of one time constant. This is an exceedingly important concept. If this is the first time you have seen the idea of a time constant, you will want to study this section carefully.

EXAMPLE 13-9   The amplitude of a 500 g pendulum on a 60 cm string is observed to decay to half its initial value after 35 s. a) What is the time constant for this oscillator? b) At what time will the *energy* have decayed to half its initial value?

SOLUTION a) The amplitude of oscillation is given by

$$x_{max}(t) = Ae^{-t/2\tau}.$$

The initial amplitude is $x_{max}(t = 0) = A$, and at $t = 35$ s the amplitude is $x_{max}(t = 35 \text{ s}) = 0.50A$. Notice that we do not need to know A itself—the *ratio* of the amplitudes at the two different times is enough information to solve for τ. First write the ratio:

$$\frac{1}{2} = \frac{x_{max}(t)}{x_{max}(t = 0)} = \frac{Ae^{-t/2\tau}}{A} = e^{-t/2\tau}.$$

Next, take the natural logarithm of both sides of the equation:

$$\left[\ln\left(\frac{1}{2}\right) = -\ln 2\right] = \left[\ln e^{-t/2\tau} = -\frac{t}{2\tau}\right]$$

$$\Rightarrow \tau = \frac{t}{2\ln 2} = \frac{35 \text{ s}}{2\ln 2} = 25.2 \text{ s}.$$

We could, if we desired, then determine the damping constant to be $b = m/\tau = 0.020$ kg/s.

b) The energy at time t is given by

$$E(t) = E_0 e^{-t/\tau}.$$

The time at which an exponential decay is reduced to half its initial value has a special name—it is called the **half-life** and given the symbol $t_{1/2}$. The concept of the half-life is widely used in the study of radioactive decay, as well as other applications. To connect it to τ,

$$E(t = t_{1/2}) = E_0 e^{-t_{1/2}/\tau} = \frac{E_0}{2}$$

$$\Rightarrow \frac{1}{2} = e^{-t_{1/2}/\tau}.$$

Again, take the natural logarithm of both sides:

$$\left[\ln\left(\frac{1}{2}\right) = -\ln 2\right] = \left[\ln e^{-t_{1/2}/\tau} = -t_{1/2}/\tau\right]$$

$$\Rightarrow t_{1/2} = \tau \ln 2 = 0.693\tau.$$

This relationship—that $t_{1/2}$ is 69% of τ—is valid for any exponential decay. In this particular problem, we find that half the energy is gone at $t_{1/2} = (.693)(25.2 \text{ s}) = 17.5$ s. The oscillator loses energy faster than it loses amplitude, which is what we should expect because the energy depends on the *square* of the amplitude.

13.8 Driven Oscillations and Resonance

So far, we have looked only at the free oscillations of an isolated system. Some initial disturbance displaces the system from equilibrium, but then the system oscillates freely until its energy is dissipated. These are very important situations, but they do not exhaust the possibilities of oscillating systems. One significant class of oscillations are those "driven" by a *periodic* external force.

A simple example of a **driven oscillation** is pushing a child on a swing, in which case you supply a periodic external force to the system. A more complex example would be a car driving over a series of equally-spaced bumps that cause a periodic upward force on the car's shock absorbers—big heavily-damped springs. The electromagnetic coil on the back of a loudspeaker cone provides a periodic magnetic force that drives the cone back and forth, causing it to send out sound waves. The electromagnetic waves picked up by an antenna force the electrons in the receiver circuit of a radio or TV to oscillate back and forth. These oscillations are then amplified and decoded to provide the sound and picture. Air turbulence moving across the wings of an aircraft can exert periodic forces on the wings and other aerodynamic surfaces, causing them to vibrate if they are not properly designed.

As these examples suggest, the topic of driven oscillations has many important applications. It is, however, a mathematically complex subject. We will simply hint at some of the results, saving the details for more advanced classes.

Consider an oscillating system that, when not subjected to external forces, oscillates at a frequency f_0. We will call this the **natural frequency** of the oscillator. It would be $\frac{1}{2\pi}\sqrt{k/m}$ for a spring or $\frac{1}{2\pi}\sqrt{g/L}$ for a pendulum, but it might be some other expression for another type of oscillator. Regardless of the expression, f_0 is simply the frequency of the system if it is displaced from equilibrium and released.

Now suppose that this system is subjected to a *periodic* external force of frequency f_{ext}. If we assume, for simplicity, that this force is sinusoidal, we can write it as $F_{ext} = F_0\sin\omega_{ext}t$, where $\omega_{ext} = 2\pi f_{ext}$ is the angular frequency. It is important to note that the external frequency f_{ext}—or the **driving frequency,** as it is often called—is completely independent of the oscillator's natural frequency f_0. Somebody or something in the environment selects the frequency f_{ext} of the driving force.

Including this external force in Newton's second law adds yet another term to Eq. 13-33 for the damped oscillator and ends up giving the differential equation:

$$\frac{d^2x}{dt^2} + \frac{b}{m}\frac{dx}{dt} + \omega_0^2 x = \frac{F_0}{m}\sin\omega_{ext}t \,. \tag{13-37}$$

We have replaced k/m, which is specific to a spring, with ω_0^2 so that Eq. 13-37 describes *any* damped, driven simple harmonic oscillator.

This frightening looking equation can be solved, but for our purposes we do not even need to write down the solution. We will be better off by looking at a graphical interpretation of the solution. We will make two assumptions, both of which are valid for most applications of driven oscillations. First, we will assume that the driving force has been applied for a long time. In this case, the system's motion will have settled into a steady pattern. Second, we will assume that the system is lightly damped.

With these assumptions, there are two important characteristics of the solution to Eq. 13-37. First, the system oscillates at the driving frequency f_{ext}, *not* at its natural frequency. This seems strange at first, but think about a situation where you put your hand underneath a block suspended from a spring and then move your hand up and down, which provides a periodic driving force to the block. The block might bounce around a little at first, but after a few seconds it moves *with your hand*—that is, with frequency f_{ext}. (The initial bouncing around is at the natural frequency f_0, but that motion is damped after a couple of time constants have passed and is no longer a factor when the driving force has been going for a long time.) The driving force at frequency f_{ext} "overrides" the system's tendency to oscillate at its natural frequency f_0.

The second characteristic—and the most important for our purposes—is that the oscillation *amplitude* depends very sensitively on the frequency f_{ext} of the driving force. This response to the driving frequency is shown graphically in Fig. 13-24 for a system of 1 kg mass and natural frequency $f_0 = 2$ Hz. Three different values of the damping constant b are considered. Figure 13-24 is called the *response curve* of the oscillator. This response-curve diagram occurs in so many different applications that it is worth careful study so that you understand the information being conveyed.

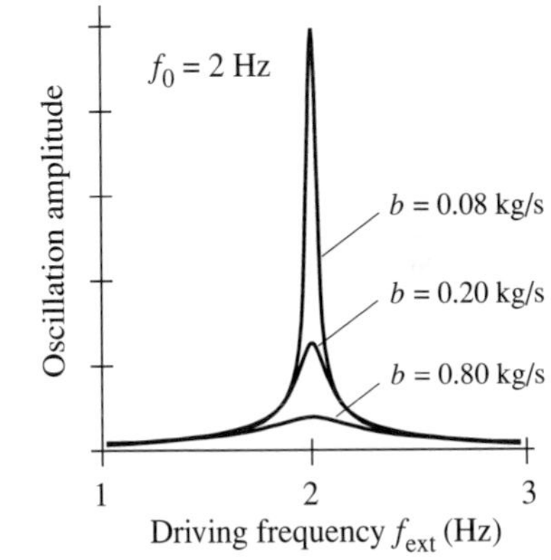

When the driving frequency is substantially different than the oscillator's natural frequency, at the far right and far left edges of Fig. 13-24, the system may oscillate at the driving frequency but its amplitude is very small. The system simply does not respond well to a driving frequency that differs much from f_0. As the

FIGURE 13-24 The amplitude of a driven oscillator at frequencies near its resonance frequency of 2 Hz.

driving frequency gets closer and closer to the natural frequency, the amplitude of the oscillation starts rising dramatically. After all, f_0 is the frequency with which the system *wants* to oscillate, so it is quite happy to respond to a driving frequency near f_0. It is not surprising, then, that the amplitude reaches a maximum exactly when $f_{ext} = f_0$—that is, when the driving frequency matches the system's natural frequency. The amplitude can become exceedingly large when the frequencies match, especially if the damping constant is very small.

This large-amplitude response to a driving force whose frequency matches the natural frequency of the system is a phenomenon called **resonance**. Mathematically the condition for resonance is $f_{ext} = f_0$, and this is often called the **resonance frequency**.

There are many interesting examples of resonance. One that you have likely experienced is our earlier example of a driven oscillation—pushing a swing. To push a child on a swing, you match the frequency of your pushing to the frequency of the swing so that $f_{ext} = f_0$. In response, the swing gains a large amplitude. Imagine trying to push the swing at some other frequency—for instance, pushing it sometimes as it goes forward and thus accelerating it, but pushing it at other times as it comes back and decelerating it! The net result will be a very small amplitude. Only the frequency-matching condition builds up the amplitude—a resonance.

You may have heard that soldiers "break step" when crossing a footbridge, rather than marching in unison. The bridge has a natural frequency with which it bounces up and down. If a group of soldiers came marching across at exactly the bridge's natural frequency they would cause a resonance—and potentially a disaster. A similar situation occurs when a singer is able to shatter a crystal goblet by singing the right note. The goblet has natural frequency—the musical sound you hear if you lightly tap a wine glass. The singer causes a sound wave to impinge on the goblet, exerting a small driving force at the frequency of the note she is singing. If the singer's frequency matches the natural frequency of the goblet—resonance! The vibration amplitude of the glass may grow so large that the glass has a structural failure and shatters.

Other interesting examples are electrical. The air is full of electromagnetic waves of many frequencies—all the various radio and TV stations in your neighborhood. How does your radio "know" which one to play? The antenna of a radio or TV goes to the *input circuit,* which is an electrical oscillator. The oscillator can be tuned to different natural frequencies by changing the values of various electrical components, which is what happens when you change channels or stations. If the receiver's natural frequency is tuned to match the broadcast frequency of a local station, then a resonance occurs and the very weak electromagnetic wave causes a large amplitude current to build up in the oscillator. The amplitude due to all the other stations is very tiny, because they are not resonant, so those signals are not detected. The resonance phenomena in this case provides a "filter" by which just one out of many different signals is selected.

A last example is the microwave oven in your kitchen. Water molecules have a certain natural frequency with which they like to rotate. Normally, these rotations are random and of small amplitude. However, an electromagnetic wave (that is what microwaves are) tuned exactly to that frequency causes a resonance. The amplitude and speed of the molecules' rotation increase dramatically as they experience a driving force from the microwaves, which means that their *energy* also increases. As the molecules collide with each other, their mechanical energy of rotation is transformed into thermal energy and your food gets hot! It would not work if your microwave oven had any other frequency.

One thing you will notice from Fig. 13-24 is that the amplitude of the resonance depends on the size of the damping constant. A heavily-damped system responds fairly little, even at resonance, which is why your car has shock absorbers. As shock absorbers wear out, their damping decreases and the amplitude of the car's bounces goes up! Very lightly-damped systems can reach exceptionally high amplitudes, but notice that the range of frequencies to which the system responds becomes less and less as b decreases. In other words, a heavily-damped system may not respond much, but it responds to a wide range of driving frequencies. A very lightly-damped system responds vigorously, but only if the resonance condition $f_{ext} = f_0$ is very tightly met.

This allows us to understand why singers can break crystal goblets but not inexpensive, everyday glasses. An inexpensive glass gives a "thud" when tapped, but a fine crystal goblet "rings" for several seconds. In physics terms, the goblet has a much longer time constant than the glass! That, in turn, implies that the goblet is very lightly damped while the ordinary glass is heavily damped (due to internal forces within the glass, because it is not a high-quality crystalline structure). Only the lightly-damped goblet—like the top curve in Fig. 13-24—can reach amplitudes large enough to shatter. The restriction, though, is that its frequency has to be matched very precisely, which is why such demonstrations are done by professional singers!

It is worth noting that there are many mechanical systems for which it is essential that all resonances be avoided, such as the wings on airplanes! It is important to understand the conditions of resonance in order to design structures that avoid them.

Summary
● Important Concepts and Terms

periodic motion	equation of motion
oscillator	linear restoring force
period	damped oscillation
frequency	damping constant
simple harmonic motion	envelope
equilibrium position	time constant
amplitude	half-life
small angle approximation	driven oscillation
angular frequency	natural frequency
phase angle	driving frequency
phase	resonance
phase constant	resonance frequency

A system can exhibit periodic motion, or oscillations, when there is a restoring force that tries to return the system to equilibrium. An oscillating system is characterized by its amplitude A (its maximum displacement from equilibrium), its period T (how long a cycle lasts), and its phase ϕ (where the system is in its cycle). The oscillation frequency f is the inverse of the period: $f = 1/T$.

A particularly important type of oscillation, called simple harmonic motion, occurs for a linear restoring force such as that exerted by an ideal spring. A mass m attached to a spring of spring constant k oscillates as

$$x(t) = A\cos(\omega t + \phi_0)$$

$$v(t) = -v_{max}\sin(\omega t + \phi_0) = -\omega A\sin(\omega t + \phi_0),$$

where the angular frequency is $\omega = 2\pi f = 2\pi/T = \sqrt{k/m}$. The amplitude $A = x_{max}$ is the maximum displacement, while the maximum speed is $v_{max} = \omega A$. Both the amplitude A and the phase constant ϕ_0 are determined by the initial conditions $x_0 = x(t = 0)$ and $v_0 = v(t = 0)$ through the relationships

$$x_0 = A\cos\phi_0$$

$$v_0 = -\omega A\sin\phi_0.$$

The motion of a pendulum of length L is described by the same mathematics, at least for small-angle motion, except that the angular frequency is given by $\omega = \sqrt{g/L}$.

An object oscillating on a spring has both kinetic and potential energy. These are continuously converted back and forth into each other. The total energy is conserved in the absence of damping and is given by

$$E = \tfrac{1}{2}kA^2 = \tfrac{1}{2}kx_{max}^2 = \tfrac{1}{2}m\omega^2 A^2 = \tfrac{1}{2}mv_{max}^2 .$$

An oscillator subject to a drag force $F_{drag} = -bv$, such as an air resistance force, is a damped oscillator, where b is called the damping constant. The motion is described by

$$x(t) = Ae^{-t/2\tau}\cos(\omega t + \phi_0),$$

where $\tau = m/b$ is the time constant—the interval of time during which the energy of oscillation decays to $e^{-1} \approx 37\%$ of its initial value.

An oscillator subject to an external driving force will respond by oscillating at the driving frequency, although usually with a very small amplitude. If, however, the driving frequency matches the system's natural frequency ($f_{ext} = f_0$), the result is a very large amplitude response called resonance. A lightly-damped oscillator has a larger resonance response to a driving force than does a heavily-damped oscillator.

Exercises and Problems

Exercises

1. A block on a spring with an unknown spring constant oscillates with a period of 2 s. What is the period if
   a. The mass is doubled?
   b. The mass is halved?
   c. The amplitude is doubled?
   d. The spring constant is doubled?

2. A 200 g mass on a horizontal spring oscillates about $x = 0$ with a frequency of 2 Hz. At $t = 0$, the mass has a displacement $x = 5$ cm and a velocity $v = -30$ cm/s. Determine:
   a. The period.
   b. The angular frequency.
   c. The amplitude.
   d. The phase constant.
   e. The maximum velocity
   f. The maximum acceleration.
   g. The total energy.
   h. The $x(t)$ equation of oscillation.
   i. The displacement at $t = 0.4$ s.
   j. The velocity at $t = 0.4$ s.
   k. The potential energy at $t = 0.4$ s.
   l. The kinetic energy at $t = 0.4$ s.
   m. Sketch and carefully label a graph of $x(t)$ from $t = 0$ to $t = 2$ s.

3. A 50 g oscillating mass is described by $x(t) = (2\text{ cm}) \cdot \cos(10t - \pi/4)$. Determine:
   a. The period.
   b. The frequency.
   c. The amplitude.
   d. The phase constant.
   e. The maximum velocity.
   f. The maximum acceleration.
   g. The total energy.
   h. The spring constant.
   i. The initial conditions.
   j. The velocity at $t = 0.4$ s.
   k. Sketch and carefully label a graph of $x(t)$ from $t = 0$ to $t = 2$ s.

4. An object oscillates with SHM with a period of 4 s and an amplitude A on a spring that is 10 cm long. How long does it take the object to move from $x = A$ to $x = A/2$?

▲ Problems

5. Calculate and draw an accurate displacement graph from $t = 0$ to $t = 10$ s of an oscillator with a frequency of 2 Hz and a damping time constant of 5 s.

6. A 100 g mass on a spring with $k = 2.5$ N/m oscillates horizontally on a frictionless table. It is observed to have $v = 20$ cm/s when $x = -5$ cm.
   a. What is the amplitude of oscillation?
   b. What is the maximum acceleration of the mass?
   c. What is the position of the mass when the acceleration is maximum?
   d. What is the speed of the mass when $x = 3$ cm? Find the answer first using the oscillatory equations for x and v, and then using conservation of energy.

7. A spring of spring constant k is suspended vertically from a support and a mass m is attached. The mass is initially held in a position where the spring is *not stretched*. Then the mass is gently released and begins to oscillate up and down with its lowest point measured 20 cm below the point where it was released. What is the oscillation frequency?

8. A block on a frictionless table is connected to two springs having spring constants k_1 and k_2, as shown in Fig. 13-25. Show that the block's oscillation frequency is given by

   FIGURE 13-25

$$f = \sqrt{f_1^2 + f_2^2},$$

   where f_1 and f_2 are the frequencies at which the block would oscillate if attached to spring 1 or spring 2 alone. (**Hint:** There is no formula to give the answer. You must analyze the restoring forces on the mass to find its oscillation frequency.)

9. An oscillator with a mass of 500 g and a period of 0.5 s has an amplitude that decreases by 2% during each complete oscillation.
   a. If the initial amplitude is 10 cm, what will be the amplitude after 25 oscillations?
   b. At what time will the energy be reduced to 60% of its initial value?

10. A 200 g undamped oscillator placed in a vacuum is observed to have a frequency of 2 Hz. When air is admitted, the oscillation decreases to 60% of its initial amplitude in 50 s. How many oscillations will have been completed when the amplitude is 30% of its initial value?

11. A 1000 kg car carrying two 100 kg occupants travels over a bumpy "washboard" road, which has speed bumps every 3 m. The driver finds that the car bounces up and down with maximum amplitude when he drives at a speed of 5 m/s ($\approx$11 mph). The car then stops and picks up three more 100 kg passengers. By how much does the car body "sag" on its suspension when these three additional passengers get in?

12. A molecular bond can be modeled as a spring between two atoms that vibrate with simple harmonic motion. Figure 13-26 shows a SHM approximation for the potential energy of an HCl molecule. For $E < 4 \times 10^{-19}$ J, the figure is a good representation of

the more accurate HCl potential-energy curve that was shown in Fig. 10-32. Because the chlorine atom is so much more massive than the hydrogen atom, it is reasonable to assume that the hydrogen atom ($m = 1.67\times10^{-27}$ kg) vibrates back and forth while the chlorine atom remains at rest. Use the data given in Fig. 13-26 to determine the vibrational frequency of the HCl molecule. (**Hint:** Think about how you can use the graph to determine the spring constant of the molecular bond.)

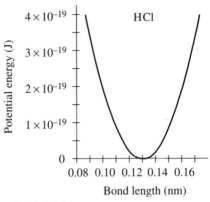

FIGURE 13-26

13. Invent a device, based on the physics principles from this chapter, with which a "weightless" astronaut in space could measure her mass. To recognize your invention, the patent office will need:
    a. A sketch.
    b. A description of the operating principles.
    c. Values for all major components.
    d. Sample calculations showing that these component values are reasonable.

[**Estimated 12 additional problems for the final edition.**]

Chapter **14**

Traveling Waves

LOOKING BACK | Sections 7.4; 13.2–13.3

14.1 Wave Physics

[**Photo suggestion: Surfers on a large ocean wave.**]

You can find waves all around you. The ripples that occur when you throw a rock into a pool of water, the crashing waves rolling into an ocean beach, musical sounds, and even light are all examples of waves. The "waviness" of water waves is readily apparent as you look at them, whereas careful observations and experiments are necessary to demonstrate that sound and light are waves. Regardless of whether the waves are big or small, easy to detect or difficult, there is a single physical theory that describes the motions and interactions of waves. This is the first of four chapters devoted to the physics of waves. Our goal is to understand the common characteristics of waves of all types.

In this chapter we are going focus on *traveling waves*. A **traveling wave** is an organized disturbance that travels through a medium, starting from a source, with a well-defined wave speed. There are several key words in this

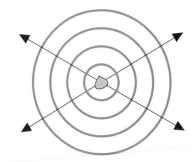

FIGURE 14-1 Ripples on a pond are a traveling wave.

definition to which you need to pay close attention: source, organized, disturbance, medium, and wave speed.

The **medium** of a wave is the substance through or along which the wave moves. For example, the medium of a water wave is the water, the medium of a sound wave is the air, and the medium of a wave on a stretched string is the string. A basic characteristic of any medium is that the medium is *elastic*. That is, there is some restoring force that attempts to bring the medium back to equilibrium if it is displaced or disturbed. A stretched string has a spring-like tension that pulls it back straight if you pull and release it. Air is compressible—if you squeeze and release, the air expands back to its original shape and volume.

As a wave passes through a medium, the particles that make up the medium undergo a displacement. The way in which the medium is displaced from equilibrium is called the **disturbance**. A disturbance might be the ripples on a pond, a pulse traveling down a string after you give the end of the string a quick shake, or the shock wave of a sonic boom created by a jet traveling faster than the speed of sound. The disturbance of a wave is *organized*, like the ripples on a pond or a compression traveling through the air. This organized motion is distinct from the *random* motions of the water molecules or air molecules that are constantly taking place.

A wave disturbance is created by a *source*. The source of a wave could be a rock thrown into water, your hand plucking a stretched string, or a loudspeaker cone pushing the air back and forth. Once created, the disturbance travels, or *propagates*, outward through the medium at the **wave speed**. This is the speed with which a ripple moves across the water or a pulse travels down a string.

There are two basic types of wave motion that we can identify. For some waves, such as a wave moving along a string, the particles in the medium move *perpendicular* to the direction in which the wave is traveling. If the string is stretched horizontally, the wave travels in a horizontal direction while successive sections of the string itself oscillate up and down vertically, as shown in Fig. 14-2a. This type of wave is called a **transverse wave**. String waves and electromagnetic waves (which we will discuss shortly) are both examples of transverse waves.

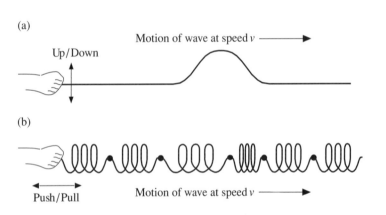

For other waves, the particles in the medium move *parallel* to the direction in which the wave travels. Consider a chain of masses connected by springs—like a Slinky—as shown in Fig. 14-2b. If you give the end of the chain a sharp push, you create a disturbance that compresses and expands the springs as it travels down the chain. The masses oscillate back and forth parallel to the direction in which the wave travels. This type of wave is called a **longitudinal wave**. Sound waves are the primary example of longitudinal waves. A loudspeaker cone pushes and pulls on the air, compressing and expanding the air much like the coils of the Slinky.

FIGURE 14-2 a) In a transverse wave the particles in the medium move perpendicular to the direction of the wave. b) In a longitudinal wave the particles of the medium move parallel to the direction of the wave. In both, the disturbance moves to the right with speed *v*.

Water waves are somewhat more complex, with characteristics of both transverse and longitudinal waves. Although the surface of the water moves up and down vertically, individual water molecules actually move both perpendicular *and* parallel to the direction of the wave. The same is true for waves that travel through the earth from the epicenter of an earthquake. We will not analyze these more complex waves in this text.

One especially important characteristic of a traveling wave is that the disturbance propagates through the medium, but *the medium itself does not move forward*! That is, the ripples on the pond (the disturbance) move outward from the source, but the water molecules themselves (the medium) simply bob up and down. The molecules do *not* move outward from the source. Likewise with a pulse traveling down a string, where the sections of the string oscillate up and down but do *not* move in the direction of the pulse. No material or substance is transferred from the source to any person or instrument that detects the wave. Nonetheless, waves *do* transmit energy. It takes energy to wiggle your eardrum (or the diaphragm of a microphone) or to activate the photosensors in your eye. That energy *is* carried forward by the wave, even though the particles of the medium do not move forward.

This observation that the motion of particles *in* the medium is separate and distinct from the motion of the wave *through* the medium is quite important. It is one of the features that distinguishes the concept of a wave from the concept of a particle. You can transmit energy and information with particles—say by throwing pebbles at a window to attract someone's attention—but the particle itself travels from the source to the receiver. A wave, by contrast, can transmit the same energy and information but does so as an organized disturbance moving *through* the medium rather than by any actual motion of the medium between the source and receiver.

Regardless of what the medium is, or whether the waves are transverse or longitudinal, there is a *single* set of physical ideas that characterize and describe waves. The ideas developed in this chapter, and the following three, apply just as much to light waves as they do to sound waves or water waves. We will draw upon many types of waves for examples and homework, but keep in mind that our topic for study is *wave physics* in general.

14.2 Traveling Waves

How does a wave travel? This is not an easy question because, as you have seen, the wave has a motion that is distinct from the motion of the particles that comprise the medium. We cannot apply Newton's laws to the wave itself—it is not a particle, nor does it have a mass. We can, however, apply Newton's laws to the medium because the motion of the wave is a consequence of the elastic restoring forces within the medium. It is important to see how these *dynamical properties* of the medium cause a disturbance to travel.

One of the simplest types of waves to examine is a *wave pulse* traveling down a stretched string. A wave pulse is generated by quickly shaking the end of the string up and down or by "plucking" a taut string. Figure 14-3 shows a pulse—a transverse disturbance—that is traveling to the right along a stretched string. The very top of the wave pulse is called the *crest* of the wave. Our goal is to understand how the pulse keeps moving.

The tension in a string, you will recall, is a pulling force directed *parallel* to the string. This

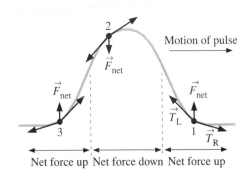

FIGURE 14-3 Tension forces on a segment of the string are at different angles because of the string's curvature.

means that the direction of the tension force vector varies as the string curves in the vicinity of the pulse. Consider a small segment of the string at point 1 near the right edge of the wave pulse. Because the string is curved, the tension force $\vec{T}_L$ directed toward the left is tilted upward at a slightly larger angle than the tension force $\vec{T}_R$ directed toward the right. As a consequence, there is a net *upward* force being exerted on this segment of string. Thus this segment of the string must have an upward acceleration.

At point 2, near the top of the pulse, the string's curvature is such that there is a net *downward* force on a segment of the string. This segment has a downward acceleration. Then at point 3, on the left edge, the net force and the acceleration are again upward. At some point between 1 and 2, and again between 2 and 3, the net force must be zero as the force changes from upward to downward.

Imagine painting a little dot at a point on the string so that you can watch its motion as a wave passes by. As the pulse approaches the dot from the left, as shown in Fig. 14-4,

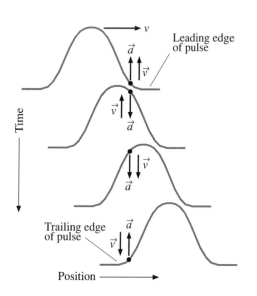

there is an upward force on this segment of the string. The dot accelerates upward from rest, gaining upward velocity. This is the **leading edge** of the wave pulse as it moves to the right. A short time later, as the peak gets closer, the force on and acceleration of the dot change to point downward. The dot's segment of string, however, still has an *upward* velocity. Thus the downward force *decelerates* the dot, slowing it until it reaches $v = 0$ exactly at the crest of the wave. This is the turning point in the dot's motion, exactly as if it were oscillating on a spring.

Upon reaching the crest, the net force on the dot's segment is still downward and so the dot begins to accelerate downward. The acceleration continues until the pulse has moved sufficiently far to the right that the force again turns upward. The upward force causes the dot

FIGURE 14-4 The acceleration and velocity of a point on a string as a wave passes by.

to decelerate until it comes to rest when the pulse has passed completely by. This is the **trailing edge** of the pulse.

You should compare the sequence of Fig. 14-4 to the accelerations and velocities of a block on a spring (see Fig. 13-4) that is pulled down and released from rest. This will help you to understand the important connection between oscillations and waves.

What we have shown with this analysis is that no new physical principles are required to understand how a wave moves. The motion of a pulse along a string is a direct consequence of the restoring forces acting on the segments of the string—as described by Newton's laws—even though the motion of the string itself is perpendicular to the travel direction of the wave. The origin of the pulse requires an external force—the source—to make the initial disturbance. But once it has started, the pulse continues to move because of the *internal dynamics* of the medium. A similar analysis applies to longitudinal waves, such as sound. It is not terribly difficult to use Fig. 14-3 to find a mathematical relationship between the acceleration of a segment of the string and the curvature (second derivative) of the string at that point. The resulting equation is called the *wave equation*. Because the wave equation is a partial differential equation, which would take us into mathematical realms beyond what is appropriate for this text, we will defer it to more advanced courses. But even though we will forgo the mathematical analysis, the *physical* analysis of how the wave moves, as shown in Fig. 14-4, is quite important and worth careful thought.

14.3 One-Dimensional Waves

Some waves travel in one dimension (waves on a string), some in two dimensions (ripples on a pond), and yet others in three dimensions (light from the sun). We will begin our study with an analysis of one-dimensional waves, which are the easiest to visualize. The ideas we develop in this section will carry over to three-dimensional sound waves and light waves.

Waves on a String

The most common example of a one-dimensional wave is a wave on a string. A string wave is a transverse wave that propagates down a string that is stretched with tension $T_s = |\vec{T}_{string}|$. (We've added the subscript s to the tension symbol T to distinguish it from the symbol for the *period* of oscillation.) String waves are interesting in their own right, and they are also the prototype for waves that travel along rigid beams and rods. As such, they are extremely important to engineering applications.

One important parameter that characterizes a string is the tension. But the tension alone is not sufficient to determine how a wave moves on a string. Some strings are fat while others are skinny, and some strings are made of dense materials while others are of less dense materials. These features also affect the wave motion.

Consider a string of total length L and total mass m. The ratio of mass to length is called the **linear density** of the string and is given the symbol μ:

$$\mu = \frac{m}{L} \text{ (linear density).} \tag{14-1}$$

For example, a string 2 m in length with a mass of 4 g has a linear density of

$$\mu = \frac{0.004 \text{ kg}}{2 \text{ m}} = 0.002 \text{ kg / m.}$$

The units of "kilograms per meter" tells us that the *numerical* value of μ is the mass that a 1-meter-long section of string would have. That is, the linear density is the mass *per meter* of length of the string. But because μ is a ratio, we can apply it to a segment of string of any length. If a string has linear density μ, a length L of this string (even if L is not the entire string) has a mass

$$m = \mu L. \tag{14-2}$$

Thus a 30 cm section of the 2-meter-long string we just considered has a mass

$$m = \mu L = (0.002 \text{ kg / m})(0.3 \text{ m}) = 0.0006 \text{ kg} = 0.6 \text{ g.}$$

The linear density allows us to characterize the *type* of string we are using. A fat string will have a larger value of μ than a skinny string, even when both are under the same tension. Similarly, a steel wire will have a larger value of μ than a plastic string of the same diameter. This is our first time—but certainly not our last—to meet the idea of a *density*. Density is an important concept, and we will return for a more complete discussion of density in later chapters.

The wave speed on a string depends on both the tension in the string and the string's linear density. A mathematical analysis of Fig. 14-3, which we said leads to the wave equation, shows that the wave speed of a traveling wave on a string is given by

$$v_{\text{string}} = \sqrt{\frac{T_s}{\mu}} \qquad \text{(wave speed on a stretched string)}. \qquad (14\text{-}3)$$

The crest of a wave pulse will travel with this speed. You can increase the wave speed either by *increasing* the string's tension (make it tighter) or by *decreasing* the string's linear density (make it skinnier). This has implications for stringed musical instruments, which we will explore in Chapter 15. It is important to note that Eq. 14-3 gives the wave *speed*, not the wave velocity, so v_{string} always has a positive value. It is *not* the component of a vector.

EXAMPLE 14-1 A 2 m long, 4 g string is tied to a wall at one end, stretched horizontally to a pulley 1.5 m away, then attached to a mass M that hangs over the pulley. What hanging mass M will cause the speed of a wave traveling along the string to be 30 m/s?

SOLUTION Figure 14-5a shows a pictorial model of the situation, with a pulse on the string traveling to the right at 30 m/s. Finding the string's tension is an equilibrium problem, which we have done many times before. Figure 14-5b shows a free-body diagram of mass M from which we conclude, because the mass is in static equilibrium, that $T_s = |\vec{T}| = |\vec{W}| = Mg$. Recall that the tension for a string in equilibrium is constant throughout the string, so $T_s = Mg$ is also the tension in the horizontal section of the string where the wave travels.

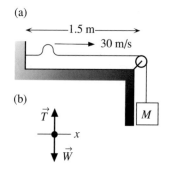

(a)

(b)

FIGURE 14-5 a) Pictorial model of wave pulse on string. b) The free-body diagram.

You might think you need to find the mass of the upper 1.5 m segment to determine the string's linear density. That, however, is unnecessary. The linear density is the same for *any* piece of the string, so μ for the 1.5 m segment is exactly the same as μ for the entire string, which we have already computed in the calculation following Eq. 14-1 to be 0.002 kg/m. The length of the string between the wall and the pulley is irrelevant. We can proceed to find M by squaring Eq. 14-3 and doing some minor algebra:

$$v^2 = \frac{T_s}{\mu} = \frac{Mg}{\mu}$$

$$\Rightarrow M = \frac{\mu v^2}{g} = \frac{(0.002 \text{ kg/m})(30 \text{ m/s})^2}{9.8 \text{ m/s}^2} = 0.184 \text{ kg} = 184 \text{ g}.$$

In principle, we should include the weight of the piece of string hanging over the pulley as a contributor to the tension. In practice, however, we know that a string is very light in comparison to a mass that we would hang on it to create tension, so we have used the simplifying assumption that the additional mass of the string itself (1 g in this case) is negligible.   •

Snapshot Graphs and History Graphs

The difficulty we face with an analysis of waves is that, for the first time, we will be dealing with functions of *two* variables. Until now, we have been concerned with quantities that depend only on time, such as $x(t)$ or $v(t)$. These are functions of the one variable t. This is all right for a particle, because a particle is only in one place at a time. But a wave is not localized—it is everywhere at each instant of time. The mathematical function that describes a wave must specify not only *when*, but also *where*. Consequently, it will be helpful to think about waves graphically before we get into a mathematical analysis.

Consider a wave pulse moving along a stretched string, as shown in Fig. 14-6. (We will consider somewhat artificial triangular and square-shaped pulses in this section so that it is clear where the edges of the pulse are.) The upper graph shows the string's displacement Δy as a function of position x along the string at an instant of time t_1. This is a "snapshot" of the wave, much like you might make with a camera whose shutter is opened at t_1. A short time later, at t_2, the wave has moved to the right and is described by the lower graph. This is another snapshot. A graph that shows the wave's displacement as a function of position at a single instant of time is called a **snapshot graph**.

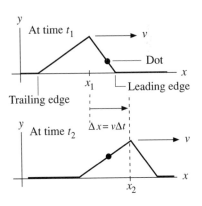

FIGURE 14-6 A snapshot graph of a wave pulse on a string. The crest moves forward a distance $\Delta x = x_2 - x_1 = v \, \Delta t$ during the time interval $\Delta t = t_2 - t_1$.

The crest of the wave moves from position x_1 at time t_1 to position x_2 at time t_2. The wave speed is thus

$$v = \frac{\Delta x}{\Delta t} = \frac{x_2 - x_1}{t_2 - t_1}. \tag{14-4}$$

During the time interval Δt the wave crest moves forward a distance $\Delta x = v\Delta t$.

A snapshot graph tells only half the story. It tells us *where* the wave is and how it varies with position, but only at one instant of time. It gives us no information about how the wave *changes* with time. As a different way of portraying information about the wave, suppose we follow the dot marked on the string in Fig. 14-6 and produce a graph of how the displacement of this dot changes with time. The result, shown in Fig. 14-7, is a displacement-versus-time graph at a single position in space. A graph that shows the wave's displacement as a function of time at a single position in space is called a **history graph**. It tells the history of that particular point in the medium.

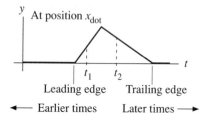

FIGURE 14-7 A history graph for the dot seen in Fig. 14-6.

You might think we have made a mistake—the graph of Fig. 14-7 is reversed compared to Fig. 14-6. It is not a mistake, but it requires careful thought to see why. As the wave moves toward the dot, the steep leading edge will cause the dot to rise quickly. On the displacement-versus-time graph, however, *earlier* times (smaller values of t) are to the *left* and later times (larger t) to the right. Thus the leading edge of the wave is on the *left* side of the Fig. 14-7 graph. As you move to the right on Fig. 14-7, you are going toward later times and you see the slowly falling trailing edge of the wave as it moves past the dot.

The snapshot graph of Fig. 14-6 and the history graph of Fig. 14-7 portray complementary information. The snapshot graph tells us how things look throughout all of space, but at only one instant of time. The history graph tells us how things look at all times, but at only one position in space. We need them both to have the full story of the wave. An alternative representation of the wave is the three-dimensional series of graphs of Fig. 14-8, where we can get a clearer picture of the wave moving forward. But such graphs are essentially impossible to draw by hand, so it is necessary to move back and forth between the snapshot graph of Fig. 14-6 and the history graph of Fig. 14-7.

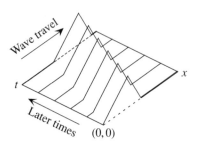

FIGURE 14-8 A 3-D look at a traveling wave.

EXAMPLE 14-2 Figure 14-9a shows a snapshot graph at $t = 0$ of a wave moving to the right at a speed of 2 m/s. Draw a history graph for the position $x = 8$ m.

SOLUTION The snapshot graph of Fig. 14-9a shows the wave at all points on the x-axis at $t = 0$. You can see that nothing is happening at $x = 8$ m at this instant of time because the wave hasn't yet reached $x = 8$ m. In fact, at $t = 0$ the leading edge of the wave is still 4 m away from $x = 8$ m. Because the wave is traveling at 2 m/s, it will take 2 s for the leading edge to reach $x = 8$ m. Thus the history graph for $x = 8$ m will be zero until $t = 2$ s. The first part of the wave causes a *downward* displacement of the medium, so immediately after $t = 2$ s the displacement at $x = 8$ m will be negative. The negative portion of the wave pulse is 2 m wide and takes 1 s to pass $x = 8$ m, so the midpoint of the pulse reaches $x = 8$ m at $t = 3$ s. The positive portion takes another 1 s to go past, so the trailing edge of the pulse arrives at $t = 4$ s. You could also note that the trailing edge was initially 8 m away from $x = 8$ m and

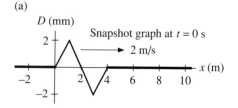

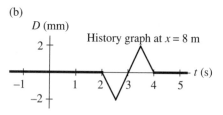

FIGURE 14-9 a) The $t = 0$ snapshot graph of Example 14-2. b) The corresponding history graph at position $x = 8$ m.

needed 4 s to travel that distance at 2 m/s. The displacement at $x = 8$ m returns to zero at $t = 4$ s and remains zero for all later times. This information is all portrayed on the history graph of Fig. 14-2b.

The Displacement Function

Our goal is to describe many different waves with a single wave physics. As a step in this direction, let's adopt a single variable D to represent the *displacement* of the medium. For a string wave, D represents the perpendicular displacement Δy of a point on the string. For a sound wave, D represents the parallel displacement Δx of an air molecule. For any other wave, D represents the appropriate displacement. Because the displacement depends both on *where* (x) and on *when* (t), D is described mathematically as

$$\text{wave displacement } = D = f(x,t), \tag{14-5}$$

where $f(x, t)$ is a function of the two variables x and t. For example, perhaps $f(x, t) = x^2 - 3t$. The values of *both* variables must be specified before you can evaluate the displacement D.

Not every function of two variables describes a traveling wave. A wave traveling in the positive x-direction with wave speed v must be a function of the form

$$D = f(x - vt) \quad \text{(wave traveling toward } +x \text{ with speed } v). \tag{14-6}$$

Examples of such functions include $D = (x - vt)^2$, $D = e^{x - vt}$, and $D = \sin(x - vt)$. Each and every wave has its own function, depending on the shape and speed of the wave, so there is not a single "wave function" that describes all waves. But *all* waves, regardless of their shape, are described by *some* function of the form $f(x - vt)$.

The key to understanding Eq. 14-6 is that the *crest* of the wave travels with *constant* speed v. The crest of the wave is the point of maximum displacement. In other words,

$$D(\text{crest}) = D_{max} = \text{maximum of the function } f(x, t).$$

The crest of the wave shown in Fig. 14-6 is located at position x_1 at time t_1. Later, at time t_2, the crest is at position x_2. Suppose that $f(x, t)$ is the function that describes the wave pulse of Fig. 14-6. Then

$$D_1 = \text{displacement at position } x_1 \text{ at time } t_1 = D_{max} = f(x_1,t_1)$$
$$D_2 = \text{displacement at position } x_2 \text{ at time } t_2 = D_{max} = f(x_2,t_2). \tag{14-7}$$

Because $D_1 = D_2$ we can conclude that

$$f(x_1,t_1) = f(x_2,t_2). \tag{14-8}$$

We also know that the wave crest is traveling with the speed v given by Eq. 14-4. From this we can conclude that

$$v = \frac{\Delta x}{\Delta t} = \frac{x_2 - x_1}{t_2 - t_1} \tag{14-9}$$

$$\Rightarrow x_2 - vt_2 = x_1 - vt_1.$$

If the function $f(x, t)$ is of the form $f(x - vt)$, then the speed relationship of Eq. 14-9 will guarantee that Eq. 14-8 is satisfied. That is, if $x_2 - vt_2 = x_1 - vt_1$ then $f(x_2 - vt_2) = f(x_1 - vt_1)$. This says that the placement at x_2 at time t_2 is the same as the displacement had been at x_1 at the earlier time t_1.

So a function of the form $f(x - vt)$ describes a disturbance that travels, or propagates, from (x_1, t_1) to (x_2, t_2) at speed v. The specific function depends on the shape of the specific wave. The logic here is a bit tricky, but an example can help make it clear.

EXAMPLE 14-3   Show, with a series of graphs, that the displacement function

$$D = f(x,t) = \begin{cases} 1 & \text{if } |x - 3t| \leq 1 \\ 0 & \text{if } |x - 3t| > 1 \end{cases}$$

describes a square-pulse wave traveling to the right with a speed of 3 m/s.

SOLUTION   This is a function of the form $f(x - vt)$ because it depends only on the value of $x - vt$. According to Eq. 14-6, this function represents a traveling wave moving to the right at 3 m/s. Let us compute and draw a series of graphs at time increments of one second. At $t = 0$, the non-zero portion of the wave is located in the range $|x| \leq 1$, or $-1 \leq x \leq +1$. The displacement is zero for all other values of x. This is the top graph in Fig. 14-10. One second later, at $t = 1$ s, the non-zero portion occurs where $|x - 3| \leq 1$, or for $2 \leq x \leq 4$. This is the second graph of Fig. 14-10. Continuing this

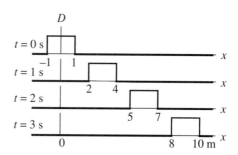

FIGURE 14-10   The square-pulse wave of Example 14-3 at four instants of time.

process produces the third and fourth graphs at times $t = 2$ s and $t = 3$ s. This series of graphs depicts a square-pulse wave moving to the right. How fast? The front edge of the pulse is located at $x = 1$ m at $t = 0$ s, at $x = 4$ m at $t = 1$ s, at $x = 7$ m at $t = 2$ s, and so on. In other words, the pulse is moving at a steady 3 m/s—exactly what Eq. 14-6 predicts.

●

It should come as no surprise that a wave moving in the *negative x*-direction with wave speed v will have the form

$$D = f(x + vt) \quad \text{(wave traveling toward } -x \text{ with speed } v\text{).} \quad (14\text{-}10)$$

A homework problem, similar to Example 14-3, will allow you to verify that functions of this form do move to the left. It is also true that not every wave in the universe chooses to move along the x-axis. Clearly a change of variables will allow us to describe a wave traveling in the positive y-direction as $D = f(y - vt)$ and a wave traveling in the negative z-direction as $D = f(z + vt)$. We will usually stick to waves along the x-axis, just for convenience, but do not become fixated on specific variables and lose sight of the fact that we can equally well describe waves moving in any direction.

14.4 Sinusoidal Waves

Some waves are created by sources that oscillate with simple harmonic motion. Such a source generates a **sinusoidal wave** that travels outward through the medium as a sinusoidal disturbance. Examples include a paddle that oscillates up and down on a pond, sending out a sinusoidal water wave; a loudspeaker cone that oscillates with SHM, radiating a sinusoidal sound wave; and the oscillating charges in an antenna, transmitting a sinusoidal electromagnetic wave.

Figure 14-11 shows, in a series of three-dimensional graphs, a sinusoidal wave moving forward. The source of the wave is located at $x = 0$ and it is undergoing simple harmonic motion. Notice how the crest moves, at a steady speed, to larger values of x at later times t.

If we monitor the displacement of the medium at a point in space, we can generate the displacement-versus-time history graph of Fig. 14-12a, which shows how the medium oscillates with time. The medium at this point oscillates with SHM at *exactly* the same frequency f as the source of the wave. For a string wave, the oscillation of a segment of the string is perpendicular to the travel direction. For a sound wave, the oscillation of the air molecules is parallel to the travel direction. The graph shows the period T, which is related to the frequency by $T = 1/f$.

Displacement-versus-time is only half the story, so Fig. 14-12b shows a displacement-versus-position snapshot graph for the same wave at an instant in time. This is a snapshot of the wave stretched out in space, moving to the right with speed v. You can see the *crests* of the wave, at the maximum *positive* displacements, and the *troughs*, at maximum *negative* displacements. The *amplitude* A of the wave is defined as the maximum value of the displacement. Thus $D_{crest} = A$ and $D_{trough} = -A$.

The most important characteristic of a sinusoidal wave is that it is periodic *in space*, as well as in time. That is, as you look at the wave "frozen" at an instant of time, the disturbance in space repeats itself over and over. The distance in which the wave repeats itself is called the **wavelength** of the wave. Wavelength is symbolized by λ (Greek "lambda") and it has units of meters. The wavelength is shown in Fig. 14-12b as the distance between two crests, although it could equally well be the distance between two troughs. Note the analogy to the wave's period. The period T is the *time* in which the disturbance at a single point in space repeats itself. The wavelength λ is the *distance* in which the disturbance at a single instant of time repeats itself.

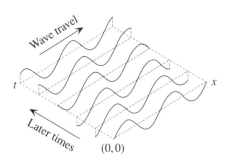

FIGURE 14-11 Sinusoidal wave moving forward along the x-axis.

(a)

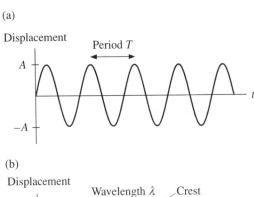

(b)

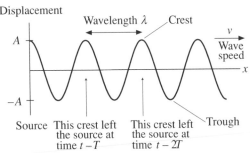

FIGURE 14-12 a) Displacement-versus-time history graph of a wave at a position x in space. b) Displacement-versus-position snapshot graph of the same wave at an instant t in time.

Is there a relationship between the wavelength and the period of a wave? We see that the snapshot graph of the wave in Fig. 14-12b was taken at a time t when the source, located at $x = 0$, happened to be exactly at maximum displacement—creating a wave crest at $x = 0$. The wave is moving to the right, so crests to the right of $x = 0$ were emitted *before* time t. The first crest to the right—one wavelength away—must have left the source exactly one period ago, at time $t - T$, when the source last reached maximum displacement. The crest beyond that, two wavelengths from the source, was emitted exactly two periods ago, at time $t - 2T$, and has been traveling to the right at speed v ever since.

In other words, *during a time interval of exactly one period T, each crest of the wave travels forward a distance of exactly one wavelength λ.* For the wave to travel distance λ in a time interval T at a constant speed v requires that

$$\lambda = vT = v \cdot \frac{1}{f}. \tag{14-11}$$

This is the relationship between the period and the wavelength of a sinusoidal wave.

Notice something important: the frequency f (and period T) is a property of the *source* of the waves, while the wave speed v is a property of the *medium* through which the waves are traveling. The wavelength λ then, according to Eq. 14-11, depends on *both* the source *and* the medium. If the same source were to be placed in a different medium with a different wave speed, the waves would travel out with the same frequency and period as before but with a *different* wavelength. This is not surprising, if you stop to think about it. If the medium oscillates with the same frequency but the waves travel at a slower speed, then the crests will get "bunched up," like cars on a freeway during heavy traffic. The spacing between crests will decrease.

It is customary to write Eq. 14-11 in the form

$$\lambda f = v. \tag{14-12}$$

Although this equation has no special name, it is *the* fundamental relationship for sinusoidal waves. However, it is vital to keep in mind that the *physical* meaning of this statement is that the wave moves forward a distance of one wavelength during a time interval of one period. Make sure you understand how that physical idea is embodied in Eq. 14-12.

There is another useful way to give a graphical representation of sinusoidal waves. Suppose you were to take a photograph of ripples spreading out on a pond that are created by a source, such as a paddle, that oscillates up and down with SHM. If we mark the location of the crests on the photo, we arrive at Fig. 14-13. The lines that locate the crests are called the **wave fronts**, and they are spaced precisely one wavelength apart. The diagram shows only a single instant of time, but you can imagine a movie in which you would see the wave fronts moving outward from the source at speed v. A wave like this is called a *circular wave*, because that is the shape of the wave fronts.

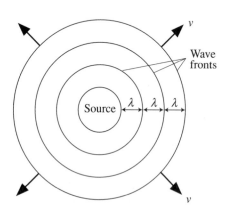

FIGURE 14-13 The moving wave crests are called *wave fronts*. They spread out from the source as circles or spheres.

Imagine investigating a circular wave very, very far away from the source. Although the wave fronts are still circles, you would hardly notice the curvature if the radius of the circle is much, much larger than the piece of the wave front you are able to observe. In that case, the wave fronts would appear to be parallel lines, still spaced one wavelength apart and traveling at speed v. A good example of this is an ocean wave reaching a beach. Ocean waves are generated by storms and wind far out at sea, hundreds or thousands of miles away. By the time they reach the beach the small segments of the crests that you can see appear to be straight lines. An aerial view of the ocean would show a wave diagram like Fig. 14-14, where the wave fronts are parallel lines. We call this type of wave a *line wave*. While perfect line waves may not exist in nature, we will consider them to be the limiting case of a source far away.

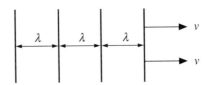

FIGURE 14-14 Far from the source, the wave fronts appear as a *line wave*.

Water waves are unusual in that they move only in two dimensions, along the surface. Most waves of interest, such as sound waves or light waves, move in a full three dimensions. For example, loudspeakers and light bulbs send out **spherical waves**. Their waves, in essence, are three-dimensional ripples. We will still find it advantageous to draw wave diagrams such as Fig. 14-13, but now you must keep in mind that the wave fronts on the diagram, still separated by λ, are slices through what are really spherical shells locating the wave crests.

Now imagine investigating a spherical wave very, very far from its source. The small piece of the wave front that you can observe is a little patch on the surface of a very large sphere. If the radius of the sphere is sufficiently large, you will not notice the curvature and this little patch—locating the wave crest—appears to be a plane. You will see a series of these planes, separated by distance λ, moving toward you at speed v. We will call this a **plane wave**. Figure 14-15 provides an illustration.

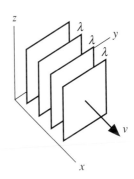

FIGURE 14-15 Plane waves are seen far from the source.

This is a little tricky to visualize, so imagine a sound wave traveling toward you from a loudspeaker very far away. This would be like standing on the x-axis of Fig. 14-15, facing the origin as the waves move toward you. The air molecules oscillate back and forth parallel to the direction in which the wave travels, so in this case they will oscillate toward you and away from you. If you were to locate all of the air molecules that, at one instant of time, were at their maximum displacement toward you, they would all be located in a plane perpendicular to the travel direction. To your right and your left, above you and below you, all the air molecules in this plane are doing exactly the same thing at that instant of time. This plane, which locates those air molecules at maximum displacement, is moving forward at speed v. There will be another plane one wavelength behind it where the molecules are also at maximum displacement, and another two wavelengths behind, and so on. These are the wave fronts moving toward you that will arrive one period from now, two periods from now, and so on. This is what a plane wave is. While there are no perfect plane waves in nature, it is a good approximation for many waves of practical interest.

When we write a wave in the form $D = f(x - vt)$ we are, in fact, describing a plane wave. The reason is that the wave's displacement depends on x but not on y or z. Once you specify a value for x, the displacement is the same at every point in the yz-plane that slices the x-axis at that value (i.e., one of the planes shown in Fig. 14-15). To describe a spherical wave, which we will need to do later, we need to change the mathematical description to read $D = f(r - vt)$, where r is the radius measured outward from the source. Then at a single instant of time, the displacement will be the same at every point having the same radius r—namely, at every point on a sphere.

14.5 Mathematics of Sinusoidal Waves

In Section 14.3 we investigated the displacement function $D(x, t)$ for one-dimensional waves. Now let's look at the specific displacement function for a sinusoidal wave. We know the displacement must be of the form $D = f(x - vt)$, and from the shape of the wave we can guess that it should involve the sine or cosine of something. It cannot be simply $\sin(x - vt)$ because the quantity $x - vt$ is a distance, in meters, and not an angle. However, we can convert $x - vt$ to an angle if we multiple it by a constant k that has units of radians per meter. So let's guess that a sinusoidal plane wave traveling in the positive x-direction with speed v is described by the function

$$D(x,t) = A\sin\left[k(x - vt) + \phi_0\right], \tag{14-13}$$

where A is the amplitude of the displacement of the medium and k and ϕ_0 are constants.

If Eq. 14-13 really describes a sinusoidal wave, then it *must* be true that $D(x + \lambda, t) = D(x, t)$—that is, the medium's displacement at position $x + \lambda$ must be exactly the same as its displacement at position x. This requirement of the wave can be expressed, using Eq. 14-13, as

$$\left[D(x + \lambda, t) = A\sin\left(k(x + \lambda - vt) + \phi_0\right) = A\sin\left(k(x - vt) + \phi_0 + k\lambda\right)\right]$$
$$= \left[D(x,t) = A\sin\left(k(x - vt) + \phi_0\right)\right]. \tag{14-14}$$

Because $\sin(\alpha + 2\pi) = \sin\alpha$, the two sides of Eq. 14-14 will be equal *if* the product $k\lambda = 2\pi$. Thus the constant k is required to be

$$k = \frac{2\pi}{\lambda}. \tag{14-15}$$

Stated differently, Eq. 14-13 represents a sinusoidal traveling wave, having wavelength λ, if and only if the constant k is specifically $k = 2\pi/\lambda$. No other value of k will do.

The constant k is called the **wave number**. It is, most emphatically, *not* a spring constant, even though it uses the same symbol. This is a most unfortunate use of symbols, but every major textbook and professional tradition uses the same symbol k for these two very different meanings, so we have little choice but to follow along. The wave number k is directly analogous to the angular frequency ω. Recall the definition of ω as

$$\omega = \frac{2\pi}{T}.$$

You see that ω is 2π times the reciprocal of the period in time while k is 2π times the reciprocal of the period in space.

From the fundamental relationship $\lambda f = v$, we can find an analogous relationship involving ω and k:

$$v = \lambda f = \frac{2\pi}{k} \cdot \frac{\omega}{2\pi} = \frac{\omega}{k} \tag{14-16}$$
$$\Rightarrow \omega = vk.$$

This result contains no new information—it is another version of Eq. 14-12—but it is often convenient when working with k and ω.

The term kvt, which appears in Eq. 14-13 if k is multiplied through, can be written as

$$kvt = \frac{2\pi}{\lambda} vt = 2\pi \frac{v}{\lambda} t = 2\pi f t = \omega t , \tag{14-17}$$

where we have used $\lambda f = v$ and the definition of ω. With the information of Eqs. 14-15 and 14-17, Eq. 14-13 can be written as

$$\begin{cases} D(x,t) = A\sin(kx - \omega t + \phi_0) \\ \text{with } k = \dfrac{2\pi}{\lambda} \text{ and } \omega = 2\pi f \end{cases} \quad \text{(sinusoidal plane wave).} \tag{14-18}$$

Notice that the displacement still is a function of $x - vt$, just rewritten a bit. It is also a wave traveling to the right; a leftward traveling wave, as we noted earlier, would be $A\sin(kx + \omega t + \phi_0)$. Eq. 14-18 was graphed versus both x and t in Fig. 14-12, so you might want to refer back to it.

It is not difficult to show, from Eq. 14-18, that $D(x, t + T) = D(x, t)$. The displacement is periodic in time, with period T, just as it is periodic in space, with period λ. We will leave this demonstration as a homework problem.

The constant ϕ_0 is again called the *phase constant*, and it plays the same role here that it did for SHM: it characterizes the initial conditions. At $(x, t) = (0, 0)$ Eq. 14-18 becomes

$$D_0 = D(0,0) = A\sin\phi_0 . \tag{14-19}$$

So different values of ϕ_0 describe different initial conditions for the wave.

EXAMPLE 14-4   A sinusoidal wave of amplitude 1 cm and frequency 100 Hz travels with a speed of 200 m/s in the +y-direction. At $t = 0$, a wave crest is passing the point $y = 1$ m.

a) Determine the values of A, v, λ, k, f, ω, T, and ϕ_0 for this wave.

b) Write the equation for the wave's displacement as it travels.

c) On a single set of axes, draw snapshot graphs of the wave at the two instants of time $t = 0$ and $t = 5$ ms.

SOLUTION a) There are quite a few numerical values associated with a traveling wave, but they are not all independent. Equations 14-12 and 14-15 are two important relationships between the wave quantities, and our Chapter 13 relationships between f, T, and ω remain valid for sinusoidal waves. From the problem statement itself we learn that

$$A = 1 \text{ cm} \qquad v = 200 \text{ m/s} \qquad f = 100 \text{ Hz.}$$

We can then find:

$$\lambda = v/f = 2.00 \text{ m}$$
$$k = 2\pi/\lambda = 3.14 \text{ m}^{-1}$$
$$\omega = 2\pi f = 628 \text{ s}^{-1}$$
$$T = 1/f = 0.010 \text{ s} = 10 \text{ ms}.$$

The phase constant ϕ_0 determines the initial conditions, which are that a wave crest $D = A$ is passing $y_0 = 1$ at $t_0 = 0$. For a wave traveling toward the $+y$-direction, the displacement is $D(y, t) = A\sin(ky - \omega t + \phi_0)$. (Note the use of y rather than x.) The initial condition is thus

$$\left[D(y_0 = 1 \text{ m}, t_0 = 0 \text{ s}) = A \right] = A\sin(ky_0 + \phi_0)$$
$$\Rightarrow \sin(ky_0 + \phi_0) = 1$$
$$\Rightarrow ky_0 + \phi_0 = \frac{\pi}{2}$$
$$\Rightarrow \phi_0 = \frac{\pi}{2} - ky_0 = \frac{\pi}{2} - \frac{2\pi}{2 \text{ m}} \cdot 1 \text{ m} = -\frac{\pi}{2}.$$

b) With the information gleaned from part a), we can write the displacement function as

$$D(y,t) = 1.0 \cdot \sin\left(3.14y - 628t - \frac{\pi}{2} \right) \text{ cm}.$$

You should note that $D(1 \text{ m}, 0 \text{ s}) = 1.0\sin(\pi/2) = 1.0$ cm, a wave crest as expected. Notice also that we *have* included units for D.

c) A snapshot graph shows the wave in space—$D(y)$—at a specific instant of time. To generate such a graph, you do *not* need to calculate a whole series of values (although you could). The task is much easier if we utilize what we know about the properties of waves, in particular the fact that this wave repeats itself after every interval $\Delta y = \lambda = 2$ m. At $t = 0$, we already know that $y = 1$ m is a wave crest. Because this is $y = \lambda/2$, we expect to find a wave trough at $y = 0$, which we can confirm from $D(0,0) = 1.0\sin(-\pi/2) = -1.0$ cm. We can then draw the $t = 0$ snapshot wave seen in Fig. 14-16.

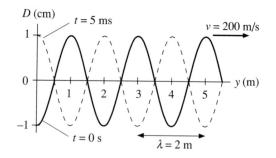

FIGURE 14-16 The traveling wave of Example 14-4 at $t = 0$ s and $t = 5$ ms.

At $t = 5$ ms $= 0.005$ s $= T/2$, we find that $\omega t = 3.14 = \pi$. So $D(y, t = 5 \text{ ms}) = 1.0\sin(3.14y - 3\pi/2)$ cm. At $y = 0$, we find $D(0 \text{ m}, 5 \text{ ms}) = 1.0\sin(-3\pi/2) = 1.0$ cm, which is once again a wave crest. This makes sense if you recall that a wave moves forward a distance λ during a time interval of T. There was a trough at $y = 0$ at $t = 0$, so 5 ms later (half a period) the wave has moved a distance $\lambda/2$ and the trough has been replaced by a crest. The $t = 5$ ms wave is also shown on Fig. 14-16, where we can see visually that the wave has traveled 1 m $= \lambda/2$ during this 5 ms interval.

It is important in Example 14-4 to notice that *each point* of the medium undergoes *simple harmonic motion*, oscillating about its equilibrium position as the wave passes through. Pick a specific point $x = x_1$ in the medium. At that point,

$$D(x = x_1, t) = A\sin(kx_1 - \omega t + \phi_0) = A\sin(-\omega t + kx_1 + \phi_0)$$

$$= A\sin(\omega t - (kx_1 + \phi_0 + \pi)) \qquad (14\text{-}20)$$

$$= A\sin(\omega t + \text{constant}),$$

where we have used the identity $\sin(-x) = \sin(x - \pi)$ and where, at position x_1, the term $kx_1 + \phi_0 + \pi$ is simply a constant. The medium at x_1 undergoes a sinusoidal oscillation—namely SHM! This was displayed in Fig. 14-12a.

EXAMPLE 14-5   A very long string having a linear density of 2 g/m is tied at one end to an oscillator that vibrates perpendicularly to the string with a frequency of 100 Hz and an amplitude of 2 mm. The string is stretched to have a tension of 5 N. a) Write the equation for the traveling wave on this string, stating clearly any assumptions you make. b) What is the displacement of a point on the string 2.7 m away from the oscillator at a time when the oscillator reaches its maximum displacement?

SOLUTION a) Figure 14-17 is a pictorial model, showing our assumptions that the oscillator is located at $x = 0$ and that the wave travels to the right along the positive x-axis. Let us also assume that we will start our clock, defining $t = 0$, at an instant when the oscillator's displacement is at its maximum value: $\Delta y_{\text{osc}} = A = 2$ mm. The equation for the wave will be

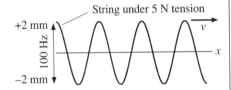

FIGURE 14-17 Pictorial model of an oscillator creating a traveling wave on a string under 5 N of tension.

$$D = A\sin(kx - \omega t + \phi_0),$$

with A, k, ω, and ϕ_0 to be determined. The wave amplitude is the same as the amplitude of the oscillator that generates the wave, so $A = 2$ mm $= 0.002$ m. The initial condition for the wave, by our choice of $x = 0$ and $t = 0$, is $D(0, 0) = A = A\sin(\phi_0)$, which can only be true if $\phi_0 = \pi/2$. The angular frequency is $\omega = 2\pi f = 2\pi \cdot 100$ radians/s because the wave's frequency is set by the 100 Hz frequency of the source.

We still need $k = 2\pi/\lambda$, but we do not know the wavelength. We do, however, have enough information to determine the wave speed, and we can then use either $\lambda = v/f$ or $k = \omega/v$. The speed is given by

$$v = \sqrt{T_s/\mu} = \sqrt{5 \text{ N} / 0.002 \text{ kg/m}} = 50 \text{ m/s},$$

from which we find $\lambda = 0.5$ m and $k = 2\pi \cdot 2$ radians/m. Thus the wave's equation is

$$D(x, t) = (0.002 \text{ m})\sin[2\pi(2x - 100t) + \pi/2].$$

Notice that we have separated out the 2π, rather than multiplying it through. This is not essential, but for some problems it makes subsequent steps easier to have the 2π separate.

b) One instant of time when the oscillator is at its maximum displacement is $t = 0$. At that instant, the wave's spatial shape is given by $D(x, t = 0) = (0.002 \text{ m})\sin(4\pi x + \pi/2)$. If we evaluate this at $x = 2.7$ m,

- $$D(2.7 \text{ m}, 0 \text{ s}) = -0.00162 \text{ m} = -1.62 \text{ mm}.$$

The total quantity $(kx - \omega t + \phi_0)$ is, as was the case for SHM, called the *phase* ϕ. The phase of a wave will be an important idea in Chapters 16 and 17, where we will explore the consequences of adding various waves together. For now, we can note that the wave fronts seen in the wave diagram for a plane wave are "surfaces of constant phase." The wave fronts, you recall, locate the *crests* of the wave. The crests are found by requiring the displacement to be equal to the amplitude: $D_{\text{crest}} = A$. This requirement is met when and where $\sin(kx - \omega t + \phi_0) = 1$, which in return requires the phase of the wave to satisfy

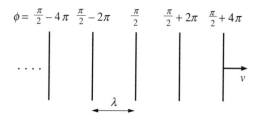

FIGURE 14-18 A plane wave seen from the side, showing the edges of the planes. Each plane is a surface of constant phase, which differs from the phase of neighboring planes by 2π.

$$\phi = kx - \omega t + \phi_0 = \dots, \frac{\pi}{2} - 4\pi, \frac{\pi}{2} - 2\pi, \frac{\pi}{2}, \frac{\pi}{2} + 2\pi, \frac{\pi}{2} + 4\pi, \dots$$

$$= \frac{\pi}{2} \pm n \cdot 2\pi \text{ where } n = 0, 1, 2, 3, \dots$$

(14-21)

Each wave front in the plane wave corresponds to one of these values of the phase, as illustrated in Fig. 14-18. Note that moving from one crest of the wave to a neighboring crest is accomplished by changing the wave's phase by $\pm 2\pi$. You can see this directly in Fig. 14-18, or we can show it mathematically by using the fact that crests are separated by a distance λ. Thus

$$\Delta\phi = \phi(x + \lambda) - \phi(x) = (k(x + \lambda) - \omega t + \phi_0) - (kx - \omega t + \phi_0)$$

$$= k\lambda = \frac{2\pi}{\lambda} \cdot \lambda = 2\pi.$$

(14-22)

This is an important idea: moving from one crest of the wave to the next corresponds to changing the *distance* by λ and to changing the *phase* by 2π.

- **EXAMPLE 14-6** A 100 Hz sound wave travels with a wave speed of 343 m/s. a) What is the phase difference between two points 60 cm apart along the direction the wave is traveling? b) How far apart are two points whose phase differs by 90°?

SOLUTION a) The phase of a wave at a point x and time t is $\phi = kx - \omega t + \phi_0$. The phase difference between two points of different x, but at the same time t, is

$$\Delta\phi = \phi_2 - \phi_1 = (kx_2 - \omega t + \phi_0) - (kx_1 - \omega t + \phi_0) = k(x_2 - x_1) = k\Delta x = 2\pi \cdot \frac{\Delta x}{\lambda}.$$

In this case, $\Delta x = 60$ cm $= 0.60$ m. We need to find the wavelength:

$$\lambda = \frac{v}{f} = \frac{343 \text{ m / s}}{100 \text{ Hz}} = 3.43 \text{ m.}$$

Knowing the wavelength, we can then find

$$\Delta\phi = 2\pi \cdot \frac{0.60 \text{ m}}{3.43 \text{ m}} = 0.350\pi = 63.0°.$$

b) A phase difference of 90° is $\pi/2$ radians. This will be the phase difference between two points when $\Delta x/\lambda = 1/4$, or when $\Delta x = \lambda/4 = 85.8$ cm in this case.   •

Notice that two points separated by $\lambda/4$ have a phase difference of $\pi/2 = (2\pi)/4$. The idea of a *phase difference* allows us describe the distance between two points as a *fraction* of the wavelength without having to know a specific value of λ. A separation of $\frac{1}{4}$ of a wavelength corresponds to a phase difference of $\frac{1}{4}$ of 2π. Make sure you understand how this relates to Fig. 14-18. We will use this idea to understand the interference of waves in Chapter 16.

14.6 Sound and Light

While there are many kinds of waves in nature, two are especially significant for us as humans. These are sound waves—the basis of hearing—and light waves—the basis of seeing. A light wave, as you will see in Part V of this text, is an **electromagnetic wave**, an oscillation of the electromagnetic field. Other electromagnetic waves, such as radio waves, microwaves, and ultraviolet light, have the same physical characteristics as light waves even though we cannot sense them with our eyes. In this section we will look more closely at sound waves and electromagnetic waves.

Sound Waves

A **sound wave** is a longitudinal disturbance that travels though a medium. Although we usually think of sound waves traveling in air, they can travel through any gas, through liquids, and even through solids. Figure 14-19 shows how longitudinal waves are created in a long tube of gas with a piston in one end. If you give the piston a quick push, it collides with the air molecules and compresses the gas at the left end of the tube. It also gives those molecules an increased velocity toward the right, so a *compression pulse* begins to move in that direction. The specific air molecules that were against the piston do not travel far before colliding with other molecules, giving them a velocity to the right while stopping the original

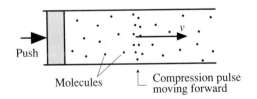

FIGURE 14-19 A sound wave is a compression pulse moving through a gas.

molecules. Through these ongoing collisions, the compression pulse is transmitted down the tube—a sound wave! Much the same happens if you hit the end of a metal rod with a hammer, sending a compression wave through the metal.

There is a difficulty in representing sound waves graphically. For a transverse wave on a string, the displacement Δy really is perpendicular to the travel direction x. A standard xy-graph is thus a literal picture, or snapshot, of the wave. But what about a sound wave, where the displacement Δx of the air molecules is *parallel* to the travel direction? We cannot draw a "x versus x graph." We can, and will, continue to draw displacement-versus-position graphs, but the displacement that is *shown* on the y-axis is *actually* occurring along the x-axis. In other words, the graph is a mathematical statement about how the longitudinal displacement varies with position, but it is *not* a literal picture of the wave.

The speed of sound waves, like the speed of waves on a string, depends on the properties of the medium. A gas or liquid compresses and expands as a sound wave disturbance passes through. The compression and expansion of gases and liquids is a topic studied in thermodynamics. A thermodynamical analysis finds that the wave speed in a gas depends both on the temperature and on the molecular weight of the gas. For air at room temperature (20° C), we find

$$v_{\text{sound}} = 343 \text{ m / s} \quad \text{(sound speed in air at 20° C).}$$

The speed of sound is a little bit slower at colder temperatures and a little faster at warmer temperatures. Liquids and solids are less compressible than air, and that causes the speed of sound in those media to be higher than in air. Table 14-1 gives the speed of sound in several different substances.

TABLE 14–1 Speed of sound in several different substances.

Medium	Speed (m/s)
air (0°C)	331
air (20°C)	343
methyl alcohol	1190
water (20°C)	1480
granite	6000
aluminum	6420

A speed of 343 m/s may seem fast, but it really is not. A distance of 1 mile is 1610 m, so the time required for sound to travel 1 mile is $t = d/v = (1610 \text{ m})/(343 \text{ m/s}) = 4.7$ s. You may have learned to estimate the distance to a bolt of lightning by timing the number of seconds between when you see the flash and when you hear the thunder, then dividing by 5 to get the number of miles. The reason this works is that the speed of sound is $\approx 1/5$ mile/s. Even a distance as small as 100 yards is enough to notice a slight delay between when you see something, such as a person hammering a nail, and when you hear it.

Your ears are able to detect sinusoidal sound waves having a frequency between about 20 Hz and about 20,000 Hz, or 20 kHz. Low frequencies are perceived as a "low pitch"— a bass note—while high frequencies are heard as a "high pitch"—a treble note. The high-frequency range of hearing can deteriorate either with age or as a result of exposure to loud sounds, which damage the ear.

We can use Eq. 14-12 to find the wavelengths for sounds of various frequencies. A typical midrange sound has a frequency of about 400 Hz. Spoken and sung notes are near this frequency, as are notes played by striking keys near the center of a piano keyboard. The wavelength of this sound in air at room temperature, where $v = 343$ m/s, is

$$\lambda = \frac{v}{f} = \frac{343 \text{ m / s}}{400 \text{ Hz}} = 0.86 \text{ m.}$$

This is a very convenient "human scale" size for the wavelength. If you listen to a 400 Hz source, the wave fronts moving toward you at 343 m/s are separated by 0.86 m, or about 34 inches. You might note that many musical instruments—flute, clarinet, saxophone, violin, guitar—are a meter or a little less in size. This is not a coincidence! We will see in the next chapter how the size of musical instruments is related to the wavelength of the notes they produce.

At the high-frequency end of hearing the wavelength of a 20 kHz note is a small 1.7 cm, while at the other extreme a 20 Hz note has a huge wavelength of 17 m ($\approx$ 55 feet)! This is because the wave moves forward one wavelength during a time interval of one period, and a wave traveling at 343 m/s can move 17 m during the 1/20 s period of a 20 Hz note.

Longitudinal sound waves can exist at frequencies well above 20 kHz, even though we no longer hear them. These high frequencies are called *ultrasonic* frequencies. Special types of oscillators can vibrate at frequencies of many MHz, and these generate the ultrasonic waves used in ultrasound medical imaging. A 3 MHz frequency traveling through water (which is basically what your body is) at a sound speed of 1480 m/s has a wavelength of just about 0.5 mm. The very small wavelength is what allows ultrasound to image very small objects.

[**Photo suggestion: Ultrasound image.**]

Electromagnetic Waves

Electromagnetic waves, which include light as well as radio waves and microwaves, are considerably more complex than string waves or sound waves. It is easy to demonstrate that light will pass unaffected through a container from which all the air is removed. And light gets here from distant stars through the vacuum of interstellar space. Such observations raise interesting but difficult questions: If light can travel through a region in which there is no matter, then what is the *medium* of a light wave? What is it that is waving?

It took scientists 50 years, pretty much the second half of the nineteenth century, to answer these questions. We will examine the answers in more detail in Part V after we introduce the ideas of electric and magnetic fields. For now we can say that light waves are a "self-sustaining oscillation of the electromagnetic field." Being self-sustaining means that light waves require *no medium* in order to travel. Fortunately, we can learn about the wave properties of light without having to understand electromagnetic fields. In fact, the discovery that light is a wave was made 60 years before it was realized that light is an electromagnetic wave. We, too, will be able to discuss much about the wave nature of light without having to know just what it is that is waving.

It was predicted theoretically in the 1860s, and has been subsequently confirmed experimentally with outstanding precision, that all electromagnetic waves travel through vacuum with the same speed, called the **speed of light**. The value of the speed of light is

$$v_{\text{light}} = c = 3.00 \times 10^8 \text{ m/s} \quad \text{(electromagnetic wave speed in vacuum)},$$

where we have noted that the special symbol c is used to designate the speed of light. (This is the c in Einstein's famous formula $E = mc^2$.) Now *this* is really moving—about one million times faster than the speed of sound in air! At this speed, light could circle the earth 7.5 times in a mere one second—*if* there were some way to make it go in circles.

Light has wavelengths that are unimaginably small. You will learn in Chapter 16 how the wavelengths of light are determined, but for now we will note that visible light is an electromagnetic wave with a wavelength in the range of roughly 400 nm ($400{\times}10^{-9}$ m) to 700 nm ($700{\times}10^{-9}$ m). Each wavelength is perceived as a different color, with the longer wavelengths seen as orange or red light and the shorter wavelengths seen as blue or violet light. A prism is able to spread the different wavelengths apart, from which we learn that "white light" is all the different colors, or wavelengths, combined. The spread of colors seen with a prism, or seen in a rainbow, is called the *visible spectrum*.

If the wavelengths of light are unbelievably small, the oscillation frequencies are unbelievably large. The frequency for a 550 nm wavelength of light is

$$f = \frac{v}{\lambda} = \frac{3.00 \times 10^8 \text{ m/s}}{550 \times 10^{-9} \text{ m}} = 5.5 \times 10^{14} \text{ Hz}.$$

The frequencies of light waves are roughly a factor of a trillion (10^{12}) higher than sound frequencies!

Electromagnetic waves can also exist at many frequencies other than the rather limited range that our eyes detect. One of the major technological gains of the twentieth century has been learning how to generate and detect electromagnetic waves at many frequencies, ranging from low-frequency radio waves to the extraordinarily high frequencies of x rays. Figure 14-20 displays the full *electromagnetic spectrum*, showing the visible spectrum as being simply a small slice out of the wide range of electromagnetic waves. Notice that the frequencies and wavelengths span an incredible 12 *orders of magnitude*—that is, a factor of 10^{12}.

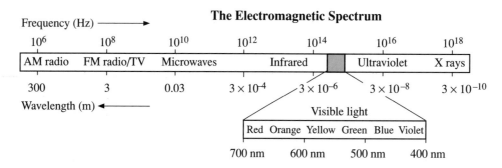

FIGURE 14-20 The electromagnetic spectrum from 10^6 Hz TO 10^{18} Hz.

EXAMPLE 14-7 AM radio stations broadcast with a frequency of about 1 MHz. a) What is the wavelength of an AM radio wave? b) How long does it take the signal to travel from the antenna to your home, 10 km away?

SOLUTION a) Radio waves are electromagnetic waves that travel with speed c. Thus

$$\lambda = \frac{c}{f} = \frac{3.00 \times 10^8 \text{ m/s}}{1.00 \times 10^6 \text{ Hz}} = 300 \text{ m}.$$

b) All electromagnetic waves travel at speed c, regardless of their frequency. The time to travel 10 km (≈ 6 miles) is

$$\Delta t = \frac{\Delta x}{c} = \frac{10,000 \text{ m}}{3.00 \times 10^8 \text{ m / s}} = 3.3 \times 10^{-5} \text{ s} = 33 \text{ } \mu\text{s}.$$

Light waves travel with speed c in vacuum, but they slow down as they pass through transparent materials such as water or glass or even, to a very slight extent, air. The slow-down is a consequence of interactions between the electromagnetic field of the wave and the electrons in the substance. The speed of light in a material is characterized by the material's **index of refraction** n, which is defined as

$$n = \frac{\text{speed of light in vacuum}}{\text{speed of light in the material}} = \frac{c}{v} \qquad (14\text{-}23)$$

where v is the speed of light in the material. The index of refraction of a material is always greater than 1 because $v < c$. A vacuum has $n = 1$ exactly.

Table 14-2 shows the index of refraction for several materials. You can see that liquids and solids have larger indices of refraction than gases. The index of refraction of air is relevant only in very precise measurements, and we will assume $n_{\text{air}} = 1$ in this text.

An interesting issue is what happens to the frequency and wavelength of light as it enters into a transparent material, such as glass. Because $\lambda f = v$, either λ or f or both have to change when v changes. As an analogy, think of a sound wave in the air as it impinges on the surface of a pool of water. As the air molecules oscillate back and forth they push and pull on the surface of the water. Their pushing and pulling on the surface generates the compressions and expansions of the sound wave that continues on into the water. But because each push/pull of the air molecules causes one compression/expansion of the water, the frequency of the sound wave in the water is exactly the same as the frequency of the sound wave in the air. In other words, there is no change in the frequency as the wave moves from one medium to another.

The same is true for electromagnetic waves, although the pushes and pulls are a bit more complex as the electric and magnetic fields of the wave interact with the atoms at the surface of the material. Nonetheless, the frequency does not change as the wave moves from one material to another that has a different index of refraction.

Figure 14-21 shows a light wave in air entering a transparent material with index of refraction n. As the wave travels through the air it has wavelength λ_{air} and

TABLE 14–2 Typical indices of refraction at $\lambda = 600$ nm.

Material	Index of Refraction
Vacuum	1 exactly
Air	1.0003
Water	1.33
Glass	1.50
Diamond	2.42

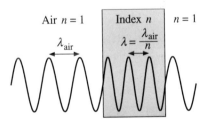

FIGURE 14-21 As light enters into a transparent material with index of refraction n, the wavelength decreases but the frequency remains unchanged.

frequency f_{air} such that $\lambda_{air} f_{air} = c$. (We're assuming $n_{air} = 1$.) The frequency does not change as the wave enters the material, so $f_{mat} = f_{air}$. But if the frequency does not change, the wavelength must. The wavelength in the material is given by

$$\lambda_{mat} = \frac{v}{f_{mat}} = \frac{c}{nf_{mat}},$$

where we've used $v = c/n$ from Eq. 14-23. But $f_{mat} = f_{air}$ and $c/f_{air} = \lambda_{air}$, so

$$\lambda_{mat} = \frac{c}{nf_{air}} = \frac{\lambda_{air}}{n}. \tag{14-24}$$

The wavelength in the material is shorter than in air. This makes sense. Suppose a marching band is marching at 1 step per second at a speed of 3 mph. Suddenly they slow their speed to 2 mph but maintain their march at 1 step per second. The only way they can go slower while marching at the same pace is to take *smaller steps*. When a light wave enters a material, the only way it can go slower while oscillating at the same frequency is to have a *smaller wavelength*.

EXAMPLE 14-8   Orange light with a wavelength of 600 nm is incident upon a glass microscope slide that is 1 mm thick. How many wavelengths of the light are inside the slide?

SOLUTION From Table 14-2 we see that the index of refraction of glass is $n = 1.50$. The wavelength of the light in air is $\lambda_{air} = 600$ nm, so the wavelength inside the glass is

$$\lambda_{glass} = \frac{\lambda_{air}}{n} = \frac{600 \text{ nm}}{1.50} = 400 \text{ nm} = 4.00 \times 10^{-7} \text{ m}.$$

N wavelengths span a distance $d = N\lambda$, so the number of wavelengths in $d = 1$ mm is

$$N = \frac{d}{\lambda} = \frac{1.00 \times 10^{-3} \text{ m}}{4.00 \times 10^{-7} \text{ m}} = 2500.$$

14.7 Power and Intensity

A traveling wave can transfer energy from one point to another. A sound wave from a loudspeaker, for example, can set your eardrum into motion while electromagnetic waves from the sun warm the earth with the form of energy transfer called heat. We are often more interested in the *rate*, in joules per second, at which a wave transfers energy. As you learned in Chapter 9, the rate of energy transfer is what we called *power*, measured in watts. A loudspeaker might emit 2 W of power, meaning that energy in the form of sound waves is radiated at the rate of 2 joules per second. A light bulb might emit 2 W, or 2 joules per second, of visible light wave power. (This is about right for a so-called "100-Watt bulb," with the other 98 W of power being emitted as heat, or infrared radiation, rather than as visible light waves.)

Imagine two experiments with a light bulb emitting 2 W of visible light. In the first, you hang the bulb in the center of a room and allow the light to fall on the walls. In the second experiment, you use mirrors and lenses to "capture" the bulb's light and focus it onto a small screen—like what happens in a slide projector. Is there a difference in effect? The bulb emits light energy at the rate of 2 J/s in both cases, but focusing the light onto a small area makes the screen much *brighter* than the walls of the first experiment. We would say that the light on the screen is more *intense* than the light on the walls, even though both receive the same power. Similarly, a loud-speaker that beams its sound forward into a small area produces a *louder* sound than a speaker of equal power that radiates the sound in all directions. It appears that easily measurable quantities—brightness, loudness—depend not only on the rate of energy transfer, or power, but also on the size of the *area* that receives that power.

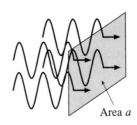

FIGURE 14-22 Plane waves of power P impinge on area a with intensity $I = P/a$.

Consider the plane wave of Fig. 14-22 impinging on a surface, perpendicular to the wave's travel direction, of area a. This might be a real, physical surface, such as your eardrum or a photovoltaic cell, but it could equally well be a mathematical surface in space that the wave passes through. If the wave has power P, we will define the **intensity** I of the wave to be

$$I = \frac{P}{a} = \text{power per unit area.} \qquad (14\text{-}25)$$

The SI units of intensity are W/m². Because intensity is a power-to-area ratio, a wave focused into a small area will have a larger intensity than a wave of equal power that is spread out over a large area.

EXAMPLE 14-9   Just above the earth's atmosphere, the intensity of electromagnetic waves from the sun is 1.4 kW/m². Roughly 80% of this intensity reaches the surface at noon on a clear summer day.

a) Joe is lying in the midday sun, working on his tan. His back is approximately a 30 cm × 50 cm rectangle. How many joules of solar energy are received by his back in 1 hour?

b) A helium-neon laser, which emits the familiar red light used in classroom demonstrations and supermarket scanners, typically emits 1 mW of power into a laser beam that is 1 mm in diameter. Compare the intensity of the laser beam to the intensity of the sun.

SOLUTION   a) The intensity of the light waves as they reach Joe's back is $I \approx 0.8 \times 1400$ W/m² = 1120 W/m². The light power, or joules per second of energy, on his back is given by

$$P = aI = (0.3 \text{ m} \times 0.5 \text{ m}) \times 1120 \text{ W/m}^2 = 168 \text{ W} = 168 \text{ J/s.}$$

Because power is $P = \Delta E/\Delta t$, during $\Delta t = 1$ hour = 3600 s, Joe's back receives energy

$$\Delta E = P\Delta t = (168 \text{ J/s}) \times (3600 \text{ s}) = 605,000 \text{ J.}$$

b) The light waves of the laser beam pass through a mathematical surface that is a circle of diameter 1 mm. The intensity of the laser beam is thus

$$I = \frac{P}{a} = \frac{P}{\pi r^2} = \frac{0.001 \text{ W}}{\pi \cdot (0.0005 \text{ m})^2} = 1270 \text{ W / m}^2.$$

So we see that $I_{laser} \approx I_{sun}$. The difference between the sun and a small laser is not their intensities, which are about the same, but their *powers*. The laser has a small power of 1 mW. It can produce a very intense wave *only* because the area through which the wave passes is very small. The sun, by contrast, radiates a total power $P_{sun} \approx 4 \times 10^{26}$ W. This immense power is spread through *all* of space, producing an intensity at a distance of 1.5×10^{11} m (the earth's distance) of 1400 W/m^2.

Many wave "detectors" respond to the *intensity* of the wave rather than to its amplitude. Examples include your ears, your eyes, film, and video cameras. It seems likely that there is some relationship between the wave's amplitude and its intensity because, after all, increasing the maximum displacement of the medium somehow ought to transfer more energy through the medium. Each little piece of the medium—an atom in a string or a molecule in the air—oscillates back and forth in simple harmonic motion (assuming a sinusoidal wave). The medium itself is elastic, which is why it can oscillate, so the energy of that little piece of the medium is $E = \frac{1}{2}kA^2$, where k is the spring constant of the medium (*not* the wave number). It is this oscillatory energy of the medium that is being transferred, atom to atom, as the wave moves through the medium.

The details of how this transfer occurs depend on the nature of the medium—string, air, water, etc.—and for our purposes we do not need an exact expression for the resultant wave intensity. But because the oscillatory energy in the medium is proportional to the *square* of the amplitude, we can infer that for *any* wave

$$I \propto A^2. \tag{14-26}$$

That is, the intensity of a wave is proportional to the *square* of its amplitude, with a proportionality constant that depends on the nature of the wave's medium. (Note: Electromagnetic waves are more complex because they do not require a medium. Nonetheless, Eq. 14-26 is a valid relationship for electromagnetic waves.) If we double the amplitude of a wave, its intensity will increase by a factor of four. Because it is the amplitude that enters into our mathematical description of a wave but it is the intensity that we detect, this relationship between intensity and amplitude will turn out to be very important.

14.8 The Principle of Superposition

Waves have yet another characteristic that distinguishes them from particles—they can pass directly through each other! Figure 14-23 shows a stereo system with two loudspeakers and two listeners, A and B. Both listeners hear the music quite well and without distortions, so it seems that the sound wave from the left loudspeaker going to listener B must pass right through the wave going from the right speaker to A. If sound consisted of little particles, they would crash together in the middle of the room and fall to the floor! But if waves are passing through each other, what is happening to the air molecules at that point?

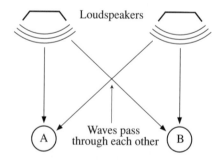

FIGURE 14-23 The waves from two sources pass through each other.

As a wave passes through a medium, it exerts forces on the particles in the medium that cause those particles to be displaced. In a medium that has a *linear* restoring force and obeys Hooke's law $F = -kD$, where D is the displacement and k is the spring constant, the displacement is directly proportional to the force being exerted. If several waves arrive simultaneously at a point in the medium, they will exert a *net* force given by the vector sum of the individual forces and cause a *net* deflection given by $F_{net} = -kD_{net}$. We will make the simplifying assumption, to keep the problem manageable, that all the forces and displacements are in the same direction and can thus be represented by scalar components rather than full vectors. In that case, the net displacement is

$$D_{net} = -\frac{F_{net}}{k} = -\frac{\sum_i F_i}{k} = -\frac{\sum_i (-kD_i)}{k} = \sum_i D_i. \tag{14-27}$$

In other words, the net displacement of a point in the medium due to several waves is just the *sum* of the displacements due to each individual wave (assuming a linear restoring force, which is a very good assumption in most cases).

This is a very important idea because it tells us how to combine waves. This idea is known as the **principle of superposition**, which states that when more than one wave is *simultaneously* present at a single point in space, the net displacement of the medium at that point is given by the sum of the displacements of each individual wave: $D_{net} = \Sigma D_i$.

When different objects are laid on top of each other, they are said to be *superimposed*. But through some quirk in the English language, the process of superimposing objects is called a *superposition*—without the syllable "im" in the middle. When one wave is "placed" on top of another wave, we have a "superposition of waves." Hence the name of the principle that tells us what happens as a result.

It is most important to notice that we are adding the *displacements* of the waves—a sum that will vary at each and every point in the medium because the displacements are different at each point. We are *not* adding the amplitudes of the waves. A common error is to add amplitudes rather than displacements. Make sure you understand the difference.

We will be using the principle of superposition extensively in the next few chapters. As a warm-up, let us use the principle of superposition to investigate a phenomenon easily demonstrated and well known to musicians. If you listen to two sound sources that are widely different in frequency, such as a "high note" and a "low note," you hear two distinct tones. But if the frequency difference is very small—say just one or two hertz—then you hear a single tone whose loudness is *modulated* once or twice every second. That is, the "note" you hear goes up and down in volume—loud, soft, loud, soft, and so on—making a very distinctive sound called **beats**. Can we understand beats on the basis of our theory of traveling waves?

Consider two sinusoidal waves of different frequencies moving along the *x*-axis toward some kind of detector—such as your ear, if the waves happen to be sound waves. The two waves are described by $D_1 = A\sin(k_1 x - \omega_1 t + \phi_1)$. and $D_2 = A\sin(k_2 x - \omega_2 t + \phi_2)$, where the subscripts 1 and 2 indicate that the frequencies, wave numbers, and phase constants of the two waves are different. To simplify the analysis, let us make several assumptions:

1. The two waves are moving in the same direction (the x-axis).
2. The two waves have the same amplitudes A.
3. Your ear, or some other detector, is located at the origin ($x = 0$).
4. We have chosen $t = 0$ such that the initial conditions give $\phi_1 = \pi$.
5. By happy coincidence, $\phi_2 = \pi$ as well.

The first two assumptions had already been made in the way we wrote D_1 and D_2. None of these assumptions are essential to the outcome. All could be different and we would still come to basically the same conclusion, but the mathematics would be far more messy. Making these assumptions allows us to emphasize the physics with the least amount of mathematics.

With these assumptions, the two waves as they reach the detector at $x = 0$ are

$$D_1(x = 0, t) = A\sin(-\omega_1 t + \pi) = A\sin\omega_1 t$$
$$D_2(x = 0, t) = A\sin(-\omega_2 t + \pi) = A\sin\omega_2 t. \tag{14-28}$$

The principle of superposition tells us that the *net* displacement of the medium at the detector is the sum of the displacements of the individual waves. Thus

$$D_{\text{net}}(x = 0, t) = D_1 + D_2 = A(\sin\omega_1 t + \sin\omega_2 t). \tag{14-29}$$

There is a trigonometric identity, which you may have proven in math, that states

$$\sin\alpha + \sin\beta = 2\cos\left[\tfrac{1}{2}(\alpha - \beta)\right]\sin\left[\tfrac{1}{2}(\alpha + \beta)\right]. \tag{14-30}$$

We can use this identity to re-write Eq. 14-29 in the form

$$D_{\text{net}}(x = 0, t) = 2A\cos\left[\tfrac{1}{2}(\omega_1 - \omega_2)t\right]\sin\left[\tfrac{1}{2}(\omega_1 + \omega_2)t\right]$$
$$= \left[2A\cos(\omega_{\text{mod}}t)\right]\sin(\omega_{\text{avg}}t), \tag{14-31}$$

where $\omega_{\text{avg}} = (\omega_1 + \omega_2)/2$ is the *average* angular frequency, and $\omega_{\text{mod}} = (\omega_1 - \omega_2)/2$ is what we will call the *modulation frequency*. D_{net} tells us how the detector will respond.

Although Eq. 14-31 is quite general, the particular situation in which we are interested is when the two frequencies are very nearly equal: $\omega_1 \approx \omega_2$. In that case, ω_{avg} hardly differs from either ω_1 or ω_2, but ω_{mod} is going to be very near to—but not exactly—zero. With ω_{mod} very small, the term $\cos(\omega_{\text{mod}}t)$ is going to oscillate *very* slowly. We have grouped it with the $2A$ term because, together, they provide a slowly changing "amplitude" for the much more rapid oscillation at frequency ω_{avg}.

When you listen to sound waves, your ear detects the oscillations of the air molecules at a single position in space. We need to look at a displacement-versus-time graph of Eq. 14-31—a history graph at $x = 0$—in order to know what you would hear. Figure 14-24 is such a graph, showing the displacement-versus-time of

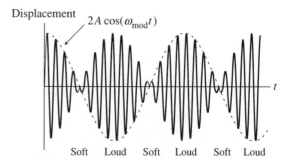

FIGURE 14-24 The superposition of two waves of nearly identical frequency results in a modulation of the amplitude called beats.

the air molecules against your eardrum. How do we interpret it? First, notice that the displacement is oscillating rapidly. This is an oscillation at frequency $f_{avg} = \omega_{avg}/2\pi = (f_1 + f_2)/2$, due to the $\sin(\omega_{avg}t)$ term. The frequency of this oscillation determines the note you hear. It differs little from the two notes at frequencies f_1 and f_2. Second, and particularly striking, notice that the amplitude of this rapid oscillation keeps rising and falling. That is the time-dependent amplitude given by the term $2A\cos(\omega_{mod}t)$, which is shown as a dotted line. It provides an envelope to the rapid oscillation. This is called a **modulation** of the frequency f_{avg} oscillation, which is where ω_{mod} gets its name.

As the amplitude rises and falls, the sound alternates loud, soft, loud, soft, etc. But that is exactly what you hear when you listen to beats! The phenomenon of beats results from the superposition of two waves of slightly different frequency.

Physically, the alternating loud and soft sounds arise from the waves being alternately "in step" and "out of step." Imagine two people walking side-by-side at just slightly different paces. Initially both of their right feet hit the ground together, but after awhile they get out of step so that one person's right foot strikes simultaneously with the other's left. A little bit later they are back in step again and the process alternates. The sound waves are doing the same. Initially the crests of each wave, of amplitude A, arrive together at your ear and the net displacement is doubled to $2A$. But after awhile the two waves, being of slightly different frequency, get out of step so that the crest of one arrives with a trough of the other. When this happens, the two waves try to displace the air in the opposite directions, canceling each other and giving a net displacement of zero. This process alternates over and over, loud and soft.

Notice, from the figure, that the sound intensity rises and falls *twice* during one cycle of the modulation envelope. Each "loud-soft-loud" is one beat that you hear, so the **beat frequency** f_{beat}, which is the number of beats per second that you hear, is *twice* the modulation frequency $f_{mod} = \omega_{mod}/2\pi$. From the definition of ω_{mod}, the beat frequency is

$$f_{beat} = 2f_{mod} = 2\frac{\omega_{mod}}{2\pi} = 2\frac{(\omega_1/2\pi - \omega_2/2\pi)}{2} = f_1 - f_2, \qquad (14\text{-}32)$$

where, to keep f_{beat} from being negative, we will always let f_1 be the larger of the two frequencies. The beat frequency is simply the *difference* between the two individual frequencies.

If one flutist plays a note of 510 Hz while a second plays a note of 512 Hz, you (and they) would hear a note of $f_{avg} = 511$ being modulated in intensity with a beat frequency of $f_{beat} = 2$ beats per second. Your ear and brain are not able to tell the difference between a 510 Hz note and a 512 Hz note played separately—they are perceived as the same pitch—but when played together the quite obvious beats tell you that the frequencies are slightly different. Musicians learn to make constant minor adjustments in their tuning as they play in order to eliminate beats between themselves and other players. It is through this "feedback" process—listen for beats, then make corrections to eliminate them—that musicians can play in tune.

Figure 14-25 shows a graphical example of beats. Two "fences" of slightly different frequencies—25 lines per inch and 27 lines per inch—are superimposed on each other. As they alternate in step and out of step the density of lines varies, giving a visual "loud/soft" alternation. The difference in the two frequencies is 2 lines per inch. You can confirm, with a ruler, that the figure has two "beats" per inch—in agreement with Eq. 14-32.

27 lines per inch

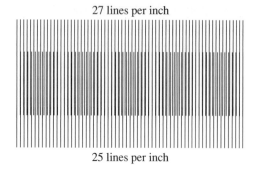

25 lines per inch

FIGURE 14-25 A graphical example of beats.

Understanding beats allows you to understand other phenomena as well. For example, you have probably seen movies where rotating wheels "seem" to turn slowly backwards. Why is this? Suppose the movie camera is shooting at 30 frames per second but the wheel is rotating 31 times per second. The combination of the two produces a "beat" of 1 Hz, meaning that the wheel "appears" to rotate once per second—very slowly. The same is true if the wheel frequency is rotating 29 times per second, but in this case the wheel frequency slightly lags the camera frequency, so it appears to rotate *backwards* once per second!

These ideas are also widely used in radio and television communications for the process of *modulating* and *demodulating* electromagnetic waves to carry signals. The term "AM," used in "AM radio," means *amplitude modulation*—which is exactly what is happening in Fig. 14-24.

Summary

Important Concepts and Terms

traveling wave	wave front
medium	spherical wave
disturbance	plane wave
wave speed	wave number
transverse wave	electromagnetic wave
longitudinal wave	sound wave
leading edge	speed of light
trailing edge	index of refraction
linear density	intensity
snapshot graph	principle of superposition
history graph	beats
sinusoidal wave	modulation
wavelength	beat frequency

A traveling wave is an organized disturbance that travels through a medium. In a longitudinal wave the medium is displaced parallel to the travel direction, while in a transverse wave the medium is displaced perpendicular to the travel direction. The wave speed depends on the medium. The speed of light in vacuum is a fixed 3.00×10^8 m/s. The speed of sound in air at room temperature is 343 m/s, but the value will differ in different media or at other temperatures. The speed of waves along a stretched string depends on both the tension and the linear density of the string and is given by $v = \sqrt{T / \mu}$.

Any displacement function of the form $D(x, t) = f(x - vt)$ represents a wave traveling in the $+x$-direction at speed v. Because the displacement is a function of two variables, we introduced both snapshot graphs (D-versus-x at an instant of time t) and history graphs (D-versus-t at a position x) to convey information about waves. Sinusoidal waves are a special class of traveling waves that are characterized by a frequency f and wavelength λ, related by $\lambda f = v$. A sinusoidal wave is represented mathematically by the displacement

$$D(x,t) = A\sin(kx - \omega t + \phi_0),$$

where $k = 2\pi/\lambda$ is the wave number, $\omega = 2\pi f$ is the angular frequency, and ϕ_0 is the phase constant. A sinusoidal wave can be represented pictorially as a series of wave fronts, surfaces of constant phase spaced one wavelength apart.

A wave transfers energy at a rate given by its power P. If the wave is incident on an area a, the intensity of the wave is $I = P/a$. The wave intensity is proportional to the square of the wave amplitude: $I \propto A^2$.

When more than one wave is present at a point in space, the net displacement of the medium is the sum of the displacements of the individual waves. This is called the principle of superposition, and it is an extremely important property of waves. One application of superposition is to the phenomena of beats, in which two waves of nearly equal frequency $f_1 \approx f_2$ are combined to produce a slow intensity modulation at the beat frequency $f_{\text{beat}} = f_1 - f_2$.

Exercises and Problems

Exercises

1. a. What is the wavelength of a "mid-range" sound of 800 Hz in room temperature air?
   b. What is the wavelength of a 2 MHz ultrasound pulse traveling through aluminum?
   c. What frequency of electromagnetic wave would have the same wavelength as the ultrasound pulse of part b)?

2. A hammer taps on the end of a 4 m long metal bar at room temperature, as shown in Fig. 14-26. A microphone at the other end of the bar picks up two "pulses" of sound separated in time by 11.0 ms. What is the speed of sound in this metal?

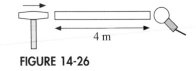

FIGURE 14-26

3. Figure 14-27 shows a snapshot graph at $t = 0$ of a 5 Hz wave traveling to the left along a string.
   a. What is the wave speed?
   b. What is the phase constant of the wave?
   c. Write the equation for this traveling wave.

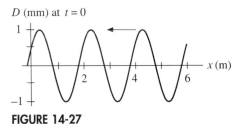

FIGURE 14-27

4. Consider the displacement function $D(x,t) = \begin{cases} 1 & \text{if } |x + 2t| \leq \frac{1}{2} \\ 0 & \text{if } |x + 2t| > \frac{1}{2} \end{cases}$.

a. Draw a series of five graphs of D-versus-x at 1 second intervals from $t = -2$ s to $t = +2$ s. Use a horizontal axis from –6 to 6 m. Stack the five graphs vertically, as in Fig. 14-10.
b. Is this a traveling wave? If so, in which direction is it traveling?
c. Determine the wave velocity from the graphs. Explain how you did so.
d. Determine the wave velocity from the equation for $D(x, t)$. Explain how you did so.

Problems

5. A plane wave traveling along the x-axis is described by $D(x, t) = A\sin(kx - \omega t + \phi_0)$. Prove that $D(x, t + T) = D(x, t)$—that is, that the wave has a temporal period of T.

6. Andy (80 kg) pulls Bob (60 kg) across the floor ($\mu_k = 0.20$) at a constant speed of 1 m/s with a 3 m long rope. Bob signals to Andy to stop by "plucking" the rope, sending a wave pulse forward along the rope. The pulse reaches Andy 150 ms later. What is the mass of the rope? How long would it take the signal to reach Andy if Bob used a laser beam to zap Andy in the back?

7. A sound wave in air is given by $D(y, t) = (0.02 \text{ mm})\sin[(8.96 \text{ m}^{-1})y + (3140 \text{ s}^{-1})t + \pi/4]$.
a. In what direction is this wave traveling?
b. Along which axis are the air molecules oscillating?
c. What is the wavelength?
d. What is the wave speed?
e. What is the period of oscillation?
f. Draw a displacement-versus-time graph $D(y = 1 \text{ m}, t)$ at $y = 1$ m from $t = 0$ to $t = 4$ ms.
g. Is the air temperature less than, greater than, or equal to 20°C? Explain.

8. A string having a linear density of 2 g/m is tied at one end to a vibrator that oscillates up and down in SHM with a maximum displacement of 1 mm at a frequency of 100 Hz. The string is stretched over a pulley that is very far away and then tied to a hanging mass of 2040 grams (Fig. 14-28). At $t = 0$, the vibrator is at its lowest point.

FIGURE 14-28

a. What is the angular frequency of the oscillation?
b. What is the phase constant of the oscillator?
c. Write the oscillator's equation $y(t)$.
d. What is the wave speed on the string?
e. What is the wavelength of the traveling wave?
f. What is wave number of the traveling wave?
g. What is the amplitude of the traveling wave?
h. What is the phase constant of the wave?
i. Write the wave equation for the displacement $D(x, t)$ of the traveling wave.
j. What is the string's displacement at $x = 0.50$ m and $t = 15$ ms?
k. Draw a graph of $D(x, t = 15 \text{ ms})$ as a function of x.

    l.   Draw a graph of $D(x = 0.50$ m, $t)$ as a function of t.

    m. What are the first *four* times after $t = 0$ when the displacement at $x = 0.75$ m is 0.50 mm?

9. Ann plays a note on the flute with a frequency of 440 Hz. When she plays simultaneously with Betty, they hear the sound "pulsing" in intensity twice every second. When Ann plays simultaneously with Cindy, the pulsing occurs only once every second. Then Betty and Cindy play notes simultaneously. Can you predict how many pulses per second will be heard? If so, do so and explain how. If not, explain the difficulty and list the possible options.

10. A sound wave of frequency 686 Hz is traveling as a plane wave in the $+y$–direction. At time $t = 0$, one crest of the wave is located in the plane $y = 12.5$ cm.

    a. Draw a graph of $D(y, t = 0)$ as a function of y from $y = -150$ cm to $y = +150$ cm.

    b. Draw a graph of $D(y = 0, t)$ as a function of t from $t = -5$ ms to $t = +5$ ms.

    c. What is the phase constant for this wave? There are *two* possible values that satisfy the mathematics, so justify your answer.

    d. Write the wave equation $D(y, t)$ for the wave, leaving the amplitude A unspecified.

    e. Draw a set of xy-axes with y ranging from -150 cm to 150 cm. On the axes draw the wave fronts of the wave (seen from the edge, as in Fig. 14-18) at time $t = 0$. Number each of the wave fronts (1, 2, ...) at the side of the graph.

    f. Repeat e) at times $t = 0.729$ ms and $t = 1.458$ ms. Keep the same number attached to each wave front for each of these three graphs so that you can "see" the wave fronts "moving."

    g. At $t = 0.729$ ms, what is the phase of the wave at $y = +12.5$ cm?

    h. At $t = 0.729$ ms, what is the phase of the wave at $y = -12.5$ cm?

    i. What *fraction* of 2π is the *phase difference* between $y = +12.5$ cm and $y = -12.5$ cm?

    j. What *fraction* of λ is the *spatial difference* between $y = +12.5$ cm and $y = -12.5$ cm?

11. a. A standard 100 W light bulb actually produces about 2 W of visible light. What is the light intensity on a wall 2 m away from the light bulb?

    b. A krypton laser generates a 2 mm diameter red laser beam with about 2 W of visible light. What is the light intensity on a wall 2 m away from the laser?

12. Lasers are used in manufacturing processes to drill or cut material. One such laser generates a series of high-intensity "pulses," rather than a continuous beam of light. Each pulse contains 100 mJ of energy and lasts 10 ns. The laser fires 10 such pulses per second.

    a. What is the "peak power" of the laser light—the power output during the 10 ns that the laser is "on?"

    b. What is the average light power output of the laser? (Hint: The *average* power is the total energy delivered in 1 s.)

    c. In actual use, a lens focuses the laser beam to a 10 μm diameter circle on the target. During a pulse, what is the light intensity on the target?

    d. The text noted that the ground intensity of sunlight at midday is about 80% of 1400 W/m^2, or roughly 1100 W/m^2. How many times intense as sunlight is the laser intensity on a target?

[**Estimated 12 additional problems for the final edition.**]

Chapter **15**

Standing Waves

LOOKING BACK

Sections 13.2–13.3; 14.3–14.6

15.1 Waves with Boundaries

Why do you hear an echo when you shout in front of a large building? What allows you to see your reflection in a mirror? Why do you hear a musical note when you pluck a guitar string or blow into a flute? All of these phenomena occur when a traveling wave strikes the boundary, or edge, of its medium. When this happens, the wave "bounces off" or is *reflected* from the boundary and travels back toward the original source. The interaction of a wave with a boundary produces phenomena such as echoes and musical notes.

A good example of a wave with boundaries occurs if you pluck a string that is stretched between two *fixed points*—such as a rubber band stretched between your fingers or a string on a guitar. When you pull the string to the side and release it, it vibrates back and forth. In many cases the vibration is so rapid that you only see a blur. A diagram of a "strobe photo" that freezes the string's motion at several different times is shown in Fig. 15-1. As you can see, the string oscillates back and forth between the positions of maximum displacement.

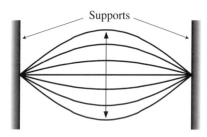

FIGURE 15-1 A "strobe photo" showing a string at several instants of time as it vibrates between two points of support.

In Chapter 14 you learned that plucking a long string creates a wave pulse that travels down the string. But there is no evidence, either in Fig. 15-1 or in your experience of watching strings vibrate, of any wave "traveling" up and down the string. The string simply oscillates back and forth.

Is this motion a wave? We certainly cannot describe the string's motion as that of a particle, and it is a collective, non-localized motion of the string as a whole. Such a motion certainly seems to fit within the wave model, but it is clearly distinct from the motion of a traveling wave. A wave in which the medium oscillates but no disturbance propagates, such as that of Fig. 15-1, is called a **standing wave**.

Our goal in this chapter is to understand the properties of standing waves. We will need to expand our description of waves by looking at what happens when a wave reaches a boundary. In particular, we will find that a standing wave such as the one shown in Fig. 15-1 is produced by a traveling wave as it reflects back and forth between two fixed boundaries.

15.2 Reflections and Superposition

As you stand in front of the mirror to brush your hair, the light from an overhead light bulb reflects from your head, travels toward the mirror, reflects from the mirror, then travels to and enters your eye, as shown in Fig. 15-2. The cornea and lens of your eye focus this light onto your retina, and you "see" the top of your head in the mirror. A mirror works because it is a *boundary* that causes a traveling light wave incident upon it to *reflect* and, as a result, travel back away from the boundary.

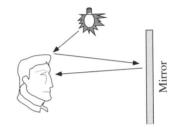

FIGURE 15-2 A mirror works because light waves *reflect* at the *boundary*.

Echoes occur for similar reasons. Echoes are simply sound waves that reflect from hard surfaces. If the surface is far enough away, such as happens outdoors with cliffs or large buildings, then the delay between when you shout and when you hear the reflected sound is long enough to be perceptible. You do not hear echoes when you speak indoors because the delay time is too small to be perceived. Nonetheless, the reflections of sound waves from interior walls do play an important role in establishing the acoustical quality of a room.

Something you can notice with either light waves reflected from a mirror or sound waves reflected as an echo is that the wavelength and frequency of the reflected waves are the same as for the incident waves. How do you know this? The reflected light has the same color, and the reflected sound has the same pitch! This property of reflected waves is important.

Whenever a traveling wave reaches a boundary, part or all of the wave is reflected. To understand what happens, let's look at a transverse wave on a stretched string. If you take a fairly long string or rope and tie one end to a firm support, such as a wall, you can observe that a pulse generated at the free end—by plucking it or by giving it a quick shake—travels down the rope. This pulse is called the **incident wave**. When the pulse reaches the boundary, the end of the rope cannot move. At this point, the pulse exerts a force on the support, which in turn exerts a force back on the rope. This reaction force sets up a **reflected wave** pulse that travels in the reverse direction. Figure 15-3 shows a sketch of the situation. Figure 15-4 shows a strobe photograph of a pulse as it travels up a spring-like rope, reflects at the top, and travels back down the rope. Notice that the reflected wave travels along the string at the *same speed* as the incident wave.

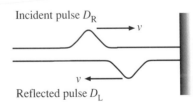

FIGURE 15-3 A traveling string wave reflecting from a boundary. The lower wave has been shifted vertically for clarity, but it actually reflects directly back along the string.

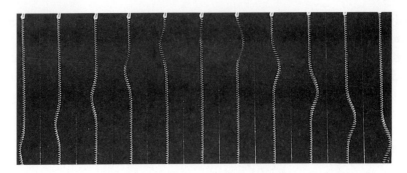

FIGURE 15-4 Strobe photo of a pulse traveling along a spring-like rope. The direction of the displacement reverses when it reflects at the top.

An interesting observation about the reflection is seen in the middle photo of Fig. 15-4. At that instant of time the string is perfectly straight and has *no displacement* at any point. The wave seems to have vanished! Our first task will be to understand what is happening at this instant of time.

To begin, notice a curious feature of the reflected pulse in Figs. 15-3 and 15-4. The displacement of the reflected pulse is *reversed*. The pulse shape is the same, but the upward (or leftward) displacement becomes a downward (or rightward) displacement after the pulse reflects from the boundary.

We can use our understanding of traveling waves from Chapter 14 to form a mathematical statement of this reversal. For a string of length L, let the free end be at $x = 0$ and

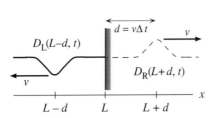

FIGURE 15-5 The reflected wave and the wave that "would-have-been" had the boundary not been there.

the boundary, where the string is tied, be at $x = L$. A time interval Δt *after* the reflection, the pulse's center is located at position at $x = L - d$ where $d = v\Delta t$ is the distance the pulse has traveled since the reflection. Had the boundary not been there, the incident wave would have continued and would be located, at this same instant of time, at $x = L + v\Delta t$. As Fig. 15-5 shows, the reflected pulse and the "would-have-been" pulse are symmetrically spaced about the boundary, but the reflected pulse has had its displacement reversed.

Let the incident wave pulse moving to the right in the Fig. 15-3 have a displacement $D_R(x, t)$ and the reflected wave pulse moving to the left have a displacement $D_L(x, t)$. Because the reflected pulse at $x = L - d$ has the same shape as the "would have been" pulse at $x = L + d$, but differs in sign, it must be the case that

$$D_L(L - d, t) = -D_R(L + d, t). \qquad (15\text{-}1)$$

This mathematical statement says that the reflected wave looks like the incident wave, except the reflected is traveling the opposite direction and is reversed in displacement. Because there is nothing special about pulses, other than that they are easy to visualize, Eq. 15-1 must apply to *any* reflected wave, including sinusoidal waves.

Suppose that a medium has a source generating a *continuous* wave that travels to the right and reaches a boundary. The wave will reflect from the boundary, as we have seen, and travel back toward the left. As a consequence, there will *simultaneously* be *two* traveling waves in the medium, as shown in Fig. 15-6, with one traveling to the right and the other, the reflected wave, to the left. This is exactly the type of situation to which the principle of superposition applies. That principle, you will recall, says that when more than one wave is simultaneously present at a single point in space, the net displacement at that point is the *sum* of the displacements of the individual waves.

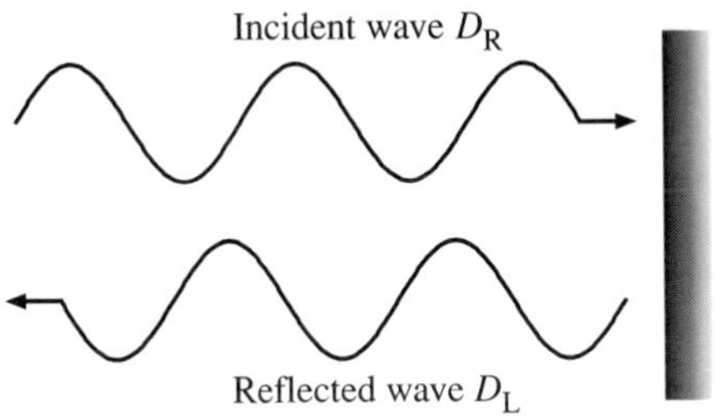

If the displacement of the wave traveling to the right is given by $D_R(x, t)$ and that of the wave traveling to the left is $D_L(x, t)$, then the net displacement of the string at position x—which is what we actually see—is

Incident wave D_R

Reflected wave D_L

FIGURE 15-6 A continuous wave reflecting from a boundary. The displacement of the reflected wave is reversed from that of the incident wave.

$$D_{net}(x,t) = D_R(x,t) + D_L(x,t). \qquad (15\text{-}2)$$

Make sure you notice that we are adding the *displacements* (*not* the amplitudes!) of the two waves at the *same* point in space x at the *same* instant of time t.

The principle of superposition is what we need to understand the observation that there is an instant of time in the reflection of a pulse when the string as no displacement at all. Figure 15-7 shows five instants as a wave pulse reflects from a boundary. The first column shows the displacement D_R of the incident pulse and the second column shows the displacement D_L of the reflected pulse as it "emerges" from the boundary. Notice that the

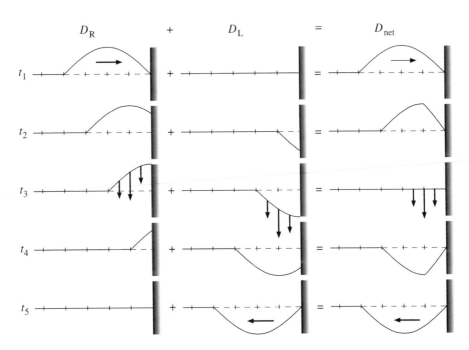

FIGURE 15-7 Five instants in the reflection of a wave pulse from a boundary. The incident and reflected pulses are shown separately, but it is the superposition D_{net} that you actually observe.

leading edge of the pulse reflects first, so the left edge of the reflected pulse corresponds to what had been the right edge of the incident pulse. However, the displacement of the reflected pulse is reversed.

You cannot observe either D_R or D_L alone. The actual displacement of the string, which is what you see, is the superposition $D_{net} = D_R + D_L$ at *each point* on the string. The right column of Fig. 15-7 shows the net displacement of the string during the reflection of the pulse. Notice two things. First, because $D_L(x = L) = -D_R(x = L)$ at the boundary, according to Eq. 15-1, the net displacement at the boundary is *always zero*. This has to be the case because the string is tied at that point. Second, there really is an instant of time—at t_3—when the string is straight and has no displacement.

But how can there be a wave if the string has no displacement? The key to understanding this is to look at the velocity of points on the string. The incident wave at time t_3 is the *trailing* edge of the pulse. You learned in Chapter 14 that the points on the trailing edge are moving downward, back toward equilibrium. This is indicated in Fig. 15-7 by the downward arrows on the t_3 graph. The reflected wave at t_3 is the *leading* edge of the reflected pulse. Points on the leading edge are moving *away* from equilibrium. Because the displacement has been reversed, the points on the reflected wave pulse are also moving downward.

The superposition of D_R and D_L at time t_3 produces a string that has no displacement but on which *every point*, within the width of the pulse, is moving downward. Thus the wave still exists in the *kinetic energy* of the medium. The wave at this instant is analogous to a block oscillating on a spring as the block passes through the midpoint. The spring is not stretched, but the block is moving rapidly and its inertia carries it forward. Similarly the inertia of the moving string will carry it downward until the entire pulse is "reborn" at time t_5.

15.3 Standing Waves on a String

As we observed in Section 15.1, a standing wave occurs when a wave reflects back and forth between two boundaries. Suppose that a medium has two boundaries, one at $x = 0$ and another at $x = L$. If a disturbance occurs at some point in the medium and starts moving to the right, it will be incident on the boundary at $x = L$, as seen in Fig. 15-8, and will reflect and then move toward the left. After traveling a ways, the disturbance will be a wave incident on the boundary at $x = 0$, where it will reflect and head back toward the right. But this is how the wave started. It has made a full round trip and will be able to continue this back-and-forth motion indefinitely if no energy is lost (no damping occurs).

When you pluck a string, by pulling it out and releasing it, you create a disturbance. That disturbance moves outward in *both* directions as two traveling waves. But the waves quickly reach the boundaries at $x = 0$ and $x = L$ and reflect at those points. Within a very short interval of time, the entire length of the string

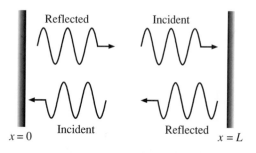

FIGURE 15-8 A wave reflecting back and forth between boundaries at $x = 0$ and $x = L$.

will have two traveling waves simultaneously moving in both directions! The net displacement of *any* point on the string will then be given by Eq. 15-2 as the sum of the displacements of the two oppositely-directed waves.

Let us make the hypothesis that this combination of two traveling waves moving in opposite directions will cause a standing wave. If our hypothesis is correct, then a standing wave is not a new type of wave. It is, instead, the net result of traveling waves moving both directions through the medium because of reflections at the boundaries.

First, let's focus on transverse standing waves on a string. (We will consider other types of waves later.) We want to look at the consequences of two waves traveling simultaneously in both directions along the string. Because we are particularly interested in waves of a fixed frequency, let us consider the two traveling waves both to be sinusoidal.

A sinusoidal wave traveling to the right along the x-axis with angular frequency $\omega = 2\pi f$, wave number $k = 2\pi/\lambda = \omega/v$, and amplitude a is written as

$$D_R = a\sin(kx - \omega t), \tag{15-3}$$

while an equivalent wave traveling to the left is

$$D_L = a\sin(kx + \omega t). \tag{15-4}$$

We previously used the symbol A for the wave amplitude, but here we will let a lowercase a represent the amplitude of each individual wave and reserve A for the amplitude of the net wave.

The net displacement of the string is given, according to the principle of superposition and Eq. 15-2, as the sum of D_R and D_L. We will drop the subscript "net" and simply call the string's displacement $D(x, t)$, which is given by

$$D(x,t) = D_R + D_L = a\sin(kx - \omega t) + a\sin(kx + \omega t). \tag{15-5}$$

We can use the trigonometric identity $\sin(\alpha \pm \beta) = \sin\alpha\cos\beta \pm \cos\alpha\sin\beta$ to simplify this expression. Doing so gives

$$D(x,t) = a(\sin kx \cos \omega t - \cos kx \sin \omega t) + a(\sin kx \cos \omega t + \cos kx \sin \omega t)$$

$$= (2a\sin kx)\cos\omega t \tag{15-6}$$

$$= A(x)\cos\omega t,$$

where we have defined the **amplitude function** $A(x)$ to be

$$A(x) = 2a\sin kx. \tag{15-7}$$

The net displacement $D(x,t)$, as given by Eq. 15-6, is neither a function of $x - vt$ nor a function of $x + vt$, so it is *not* a traveling wave. Instead, Eq. 15-6 describes a string on which each point oscillates up and down with frequency $f = \omega/2\pi$, as given by the $\cos\omega t$ term, with a position-dependent amplitude given by the amplitude function $A(x) = 2a\sin kx$.

Figure 15-9a shows a graph of Eq. 15-6 every one-eighth of a period from $t = 0$ to $t = T/2$. Only the left boundary at $x = 0$ is seen in this picture; the wave continues oscillating as shown until it reaches the right boundary at $x = L$. At $t = 0$, when $\cos\omega t = 1$, the displacement *is* the amplitude function: $D(x, t = 0) = A(x) = 2a\sin kx$. As time passes, the segment of the string at position x oscillates up and down between its maximum displacement $A(x)$ and its minimum displacement $-A(x)$.

(a)

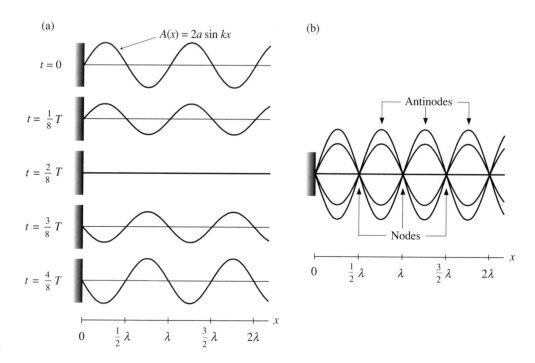

(b)

$A(x) = 2a \sin kx$

$t = 0$

$t = \frac{1}{8}T$

$t = \frac{2}{8}T$

$t = \frac{3}{8}T$

$t = \frac{4}{8}T$

Antinodes

Nodes

$0 \quad \frac{1}{2}\lambda \quad \lambda \quad \frac{3}{2}\lambda \quad 2\lambda$

FIGURE 15-9 a) The superposition of two oppositely-directed traveling waves, seen every one-eighth of a period. b) A "strobe photo" of the string as it oscillates.

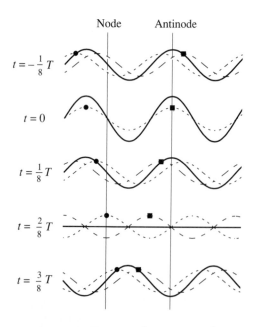

Node   Antinode

$t = -\frac{1}{8}T$

$t = 0$

$t = \frac{1}{8}T$

$t = \frac{2}{8}T$

$t = \frac{3}{8}T$

FIGURE 15-10 Two traveling waves (dotted lines) and their sum shown at five instants of time. The circle and square locate the crests of the waves so that you can see how they move.

The graphs of Fig. 15-9a are collapsed into a single graph in Fig. 15-9b. This is a "strobe photo" diagram similar to Fig. 15-1. Here you can see that the amplitude function $A(x)$ provides the outer edge, or the *envelope*, of the oscillation. Although the oscillation is composed of two traveling waves, you cannot "see" either of them. The net displacement of the string is a disturbance that oscillates in place rather than a disturbance that moves to the right or left. Thus the superposition of two equal-frequency waves traveling in opposite directions is a standing wave.

Figure 15-10 shows that the standing wave really is a superposition of two oppositely-traveling waves. Each of the individual waves (dotted lines) is a traveling wave; their crests move to the right or left from one frame to the next. (The two waves are exactly overlapped in the second frame.) At each point on the string the displacements of the individual waves are added to give the net displacement. Notice, in particular, that the string is completely straight—no displacement—at $t = T/4$.

Figure 15-11 looks at the velocities on just one segment of the string, having a length of one wavelength. Both at $t = 0$ and $t = T/2$, *every* point on the string is at its maximum displacement and instantaneously *at rest*. You know, from Chapter 13, that the very top and very bottom points in motion of an oscillator are turning points, where $v = 0$. This is what is happening at these instants to every point on the string. At $t = T/8$, the points on the string are accelerating back toward their equilibrium positions and each point has a velocity directed toward equilibrium.

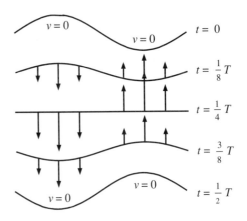

FIGURE 15-11 Velocity vectors at different points on the string for several instants of time. The string is moving most rapidly at the instant when it is straight.

The string is instantaneously straight at $t = T/4$, but it is *not* at rest. This is, in fact, the instant when every point on the string has maximum velocity—again, just like an oscillating spring passing through equilibrium. The string keeps moving, because of its inertia, past the equilibrium position. At $t = 3T/8$ the velocity vectors are still in the same directions as the last two pictures, but now each point on the string is *decelerating*. Finally, at $t = T/2$, all points come instantaneously to rest and are ready to reverse direction. Not surprisingly, based on what you have learned previously about simple harmonic motion, the string has exclusively potential energy at time $t = 0$ and $t = T/2$ when it is stretched but motionless, and exclusively kinetic energy at $t = T/4$ when it is not stretched but is moving rapidly.

A particularly striking feature of the standing wave pattern is the existence of points that *never move*! These points, shown in Figure 15-9b, are called **nodes**. They are located at the points where the amplitude function equals 0, so in a sense the nodes are oscillating like every other point on the string, only with zero amplitude. The nodes are located at positions x for which $A(x) = 0$:

$$A(x) = 2a \sin kx = 0$$

$$\Rightarrow kx_n = \frac{2\pi x_n}{\lambda} = n\pi, \qquad n = 0, 1, 2, 3, 4, \ldots \tag{15-8}$$

Thus the position x_n of the nth node is

$$x_n = \frac{n\lambda}{2}, \tag{15-9}$$

where n is an integer.

Equation 15-9 tells us, and Fig. 15-9b shows, that a node is located along the axis at every integer multiple of $\lambda/2$, starting at $x = 0$. The spacing between one node and the next is thus $\lambda/2$, *not* λ as you might suspect. If you look at the $A(x)$ curve in Fig. 15-9a, you can see that the spatial interval in which it repeats itself—one wavelength, by definition—is *twice* the spacing between the nodes.

Halfway between the nodes are the points where the string reaches its maximum amplitude $A_{\max} = 2a$, twice the amplitude of the two traveling waves that are being superimposed. These are the points at which $\sin kx = 1$. These points of maximum amplitude are called the **antinodes**, and you can see that they are also spaced $\lambda/2$ apart.

How is it possible that there can be points on the string that do not move at all? After all, there are two waves traveling on the string, causing the points of the string to be displaced in an oscillatory manner. It seems counterintuitive that some points can end up with no motion at all.

If you study Fig. 15-10, showing the two traveling waves and their superposition, you will see that the nodes occur at the points where $D_R = -D_L$ at *every* instant of time. (This statement does *not* mean that D_R has a negative value at those points; rather, it is a relational statement saying that the value of D_R at those points is always the negative of the value of D_L.) We can relate this observation to the *phases* of the two traveling waves.

The nodes, according to Eq. 15-8, occur at points where $kx_n = n\pi$, where n is an integer. At these points, the leftward traveling wave has the displacement

$$D_L(x_n, t) = a\sin(n\pi + \omega t) = a(\sin n\pi \cos \omega t + \cos n\pi \sin \omega t)$$
$$= a(-1)^n \sin(\omega t), \tag{15-10}$$

where we have used $\sin n\pi = 0$ and $\cos n\pi = (-1)^n$. At these same points, the rightward traveling wave has displacement

$$D_R(x_n, t) = a\sin(n\pi - \omega t) = a[\sin n\pi \cos \omega t - \cos n\pi \sin \omega t]$$
$$= -a(-1)^n \sin(\omega t). \tag{15-11}$$

By comparing Eqs. 15-10 and 15-11 you can see that $D_R = -D_L$ at the nodes. This result is in agreement with the graphs of Fig. 15-10. Thus

$$D(x_n, t) = D_R(x_n, t) + D_L(x_n, t) = 0 \text{ at the nodes for all } t. \tag{15-12}$$

But we can also write the second line of Eq. 15-11 as

$$D_R(x_n, t) = a(-1)^n \sin(\omega t + \pi). \tag{15-13}$$

Compare this to D_L in Eq. 15-10. Because the coefficients in front are the same, the *only* difference between D_L and D_R at the nodes is that their *phase differs by* π. It is the extra π in the phase of D_R, at the positions of the nodes, that makes $D_R = -D_L$ (because $\sin(\alpha + \pi) = -\sin\alpha$). Two oscillations that differ in phase by π are said to be "180° out of phase." The superposition of two oscillations that are 180° out of phase will always result in a perfect cancellation and no net oscillation at all. This is our first example of what, in the next chapter, we will call *destructive interference*.

By comparison, you can see from Fig. 15-10 that the positions of the antinodes correspond to points where $D_R = D_L$ at all times. If the two oscillations at those points are identical, they must have the *same phase*. Two oscillations with no difference in phase are said to be "completely in phase." Their superposition results in an oscillation with twice the amplitude of each individual oscillation, which is what occurs at the antinodes. We will call this behavior *constructive interference*. As a homework problem you will be asked to demonstrate mathematically that the two traveling waves are always completely in phase at the positions of the antinodes.

We will explore the phenomenon of interference more thoroughly in the next chapter. Some of the ideas encountered there will be easier to understand if you realize that the nodes and antinodes of a standing wave are examples of destructive and constructive interference.

EXAMPLE 15-1   A very long string has a linear density of 5 g/m and is stretched with a tension of 8 N. Sinusoidal waves with a frequency of 100 Hz and an amplitude of 2 mm are generated at the right end of the string. The waves travel to the left and reflect from the left end of the string at $x = 0$. a) Where are the first two antinodes located? b) What is the maximum displacement of the string at the point $x = 25$ cm?

SOLUTION a) The situation is the same as that shown in Fig. 15-9. The speed of waves on the string is given by

$$v = \sqrt{\frac{T_s}{\mu}} = \sqrt{\frac{8 \text{ N}}{0.005 \text{ kg}/\text{m}}} = 40 \text{ m}/\text{s}.$$

The wavelength is thus

$$\lambda = \frac{v}{f} = \frac{40 \text{ m}/\text{s}}{100 \text{ Hz}} = 0.40 \text{ m} = 40 \text{ cm}.$$

Nodes are spaced $\lambda/2 = 20$ cm apart, so nodes are located at $x = 0$ cm, 20 cm, 40 cm, and so forth. The antinodes are midway between the nodes, so antinodes are located at $x = 10$ cm, 30 cm, 50 cm, and so on. The first two antinodes, for which the question asked, are at $x = 10$ cm and $x = 30$ cm.

b) The point $x = 25$ cm is neither a node or an antinode. The maximum displacement at this point is given by the amplitude function

$$A(x = 25 \text{ cm}) = 2a \sin kx = 2a \sin\left(2\pi \frac{x}{\lambda}\right)$$

$$= (4 \text{ mm}) \sin\left(2\pi \frac{25 \text{ cm}}{40 \text{ cm}}\right)$$

$$= -2.83 \text{ mm}.$$

The point $x = 25$ cm is just beyond the first node, so it has its most negative displacement at $t = 0$. This is why $A(x = 25 \text{ cm})$ is a negative number. The maximum displacement at this point is the absolute value of A, or $D_{max} = 2.83$ mm. The string at this point oscillates up and down between +2.83 mm and −2.83 mm.

15.4 The Wavelengths and Frequencies of Standing Waves

Our hypothesis was that standing waves occur as a consequence of the superposition of two waves traveling in opposite directions due to reflections at the boundaries. We have now shown that two oppositely-directed traveling waves do generate a standing wave, but we made no reference to the boundaries. In fact, the results of Section 15.3 would apply to a long string with SHM oscillators at each end generating the two sinusoidal traveling waves. To connect the mathematical analysis of Section 15.3 with the physical reality of a string tied down at the ends, we need to "impose" the **boundary condition** that $D(x = 0, t) = 0$ and $D(x = L, t) = 0$. That is, the displacements at the ends of the string are zero at *all* times t. To state it another way, we must require that both ends of the strings be nodes.

The "solution" to the standing wave problem is Eq. 15-6: $D(x, t) = (2a\sin kx)\cos\omega t$. We see that this solution already satisfies $D(x = 0, t) = 0$; that is, the origin has already been located at a node. The condition we must impose is the presence of a node at $x = L$, which requires

$$A(x = L) = 2a\sin kL = 0 \quad \text{(boundary condition at } x = L). \quad (15\text{-}14)$$

This requirement of the string will be satisfied if $\sin kL = 0$, or if

$$kL = \frac{2\pi L}{\lambda} = n\pi \quad n = 1,2,3,4,\ldots$$

$$\Rightarrow L = n\frac{\lambda}{2}. \quad (15\text{-}15)$$

We can give this result a physical interpretation. The boundary condition requires the end of the string to be a node. The nodes are spaced every $\lambda/2$ along the string, as we showed in Section 15.3. If the end of the string is the nth node, then the string's length must be $n(\lambda/2)$.

It will be useful to determine the possible wavelengths that can exist on a string of length L. Solving Eq. 15-15 for λ gives

$$\lambda_n = \frac{2L}{n} \quad n = 1,2,3,4,\ldots \quad (15\text{-}16)$$

In other words, a standing wave can exist on the string *only* if its wavelength is given by Eq. 15-16. The nth possible wavelength $\lambda_n = 2L/n$ is just the right size so that its nth node is located at the end of the string (at $x = L$). Other wavelengths, which would be perfectly acceptable for a traveling wave, cannot exist as a *standing* wave of length L because they cannot meet the boundary conditions requiring a node at each end of the string.

If standing waves are possible only for certain wavelengths, then only a few specific oscillation frequencies are allowed. Because $\lambda f = v$ for any wave, the oscillation frequency corresponding to wavelength λ_n is given by

$$f_n = \frac{v}{\lambda_n} = \frac{v}{2L/n} = n\frac{v}{2L} \quad n = 1,2,3,4,\ldots \quad (15\text{-}17)$$

The allowed frequencies are integer multiples of the lowest allowed frequency f_1,

$$f_1 = \frac{v}{2L} \quad \text{(fundamental frequency).} \quad (15\text{-}18)$$

This lowest frequency, which corresponds to wavelength $\lambda_1 = 2L$, is called the **fundamental frequency** of the string. The allowed frequencies of Eq. 15-17 can be written in terms of the fundamental frequency as

$$f_n = nf_1 \quad n = 1,2,3,4,\ldots \quad (15\text{-}19)$$

Figure 15-12 shows graphs of the first four possible standing waves on a string of fixed length L. These possible standing waves are called the **normal modes** of the string. Each mode, numbered by the integer n, has a unique wavelength and frequency. Keep in mind that these drawings—which are standard ways of picturing standing waves—simply show the *envelope*, or outer edge, of the oscillations. The string is continuously oscillating at all positions between these edges, as we showed in more detail in Figure 15-9.

There are several important things to note about the normal modes of a string. First, n gives the number of *antinodes* on the standing wave, *not* the number of nodes (the number of nodes is $n + 1$, due to the $n = 0$ node at $x = 0$). You can tell which mode of oscillation a string is performing by counting the number of antinodes. Second, the *fundamental mode*, where $n = 1$, has $\lambda = 2L$, *not* $\lambda = L$. Only half of a wavelength is contained between the boundaries, and this is a direct consequence of the fact that the spacing between nodes is $\lambda/2$. For $n = 1$, the two boundaries are two adjacent nodes. Third, the frequencies of the normal modes form an arithmetic series:

$$f_1 = 1f_1$$
$$f_2 = 2f_1$$
$$f_3 = 3f_1$$
$$f_4 = 4f_1$$
$$\vdots$$

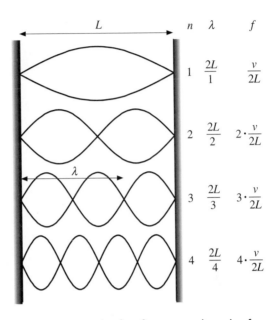

FIGURE 15-12 The first four normal modes for standing waves on a string of fixed length L.

The frequency of higher modes can be many times higher than the fundamental frequency.

Figure 15-13 shows a time-exposure photograph of a person shaking one end of a string to generate standing waves modes. The nodes and antinodes are quite distinct, and you can see that the wavelength is different in each case. If this were a movie, you would be able to see that the person has to shake the end three times as fast to generate the $n = 3$ mode than to generate the fundamental $n = 1$ mode. You might object—and correctly so—that the person's hand is not a node because he is wiggling the string at that point, so our analysis based on $D(x = 0, t) = 0$ does not apply to this situation. Strictly speaking, that is true. Nonetheless, notice that the string is blurred, due to its large-amplitude motion, while the person's hand is quite distinct. The amplitude of his hand is *very small*, so it is *approximately* a node. Thus Eqs. 15-16 and 15-17, while not exact in this case, remain very good approximations.

If the person seen in Fig. 15-13 were to shake the string at a frequency $f = 2.5f_1$, *nothing would happen!* The string would quiver a little in a random and chaotic manner, but no standing wave would appear. This is not a frequency whose corresponding wavelength satisfies the boundary conditions, so it cannot exist as a standing wave. The string can oscillate *only* if it satisfies the boundary conditions. The most common standing wave oscillation, such as you produce by plucking a rubber band stretched between your fingers, is the fundamental mode with $n = 1$. There is only one possible frequency and wavelength that satisfy the boundary conditions for $n = 1$, so that is the frequency with which the string will oscillate.

You saw, in Fig. 15-5 and Eq. 15-1, that the displacement of a reflected wave is reversed from that of the incident wave. The mathematical statement of this observation

$n = 1$ $n = 2$ $n = 3$ $n = 4$

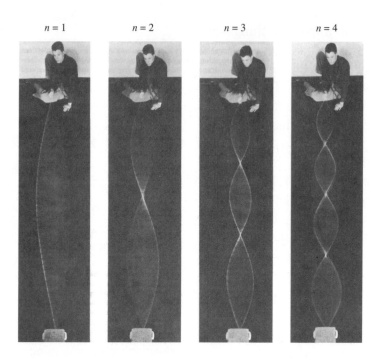

FIGURE 15-13   Time-exposure photographs of the first four standing wave modes on a stretched string.

was the requirement that $D_L(L - d,t) = -D_R(L + d,t)$. We will leave it as a homework problem for you to show that this requirement, when used with the sinusoidal functions of Eqs. 15-3 and 15-4 for D_R and D_L, leads to the requirement that $\lambda = 2L/n$—exactly Eq. 15-16. In other words, the Eq. 15-1 requirement on the reflection at the boundary is really equivalent to the requirement that there be a node at the boundary—they are two different ways of stating the same boundary condition. If there is a node at the boundary, then it must be the case that the displacement is reversed upon reflection. Conversely, if the displacement is reversed upon reflection, it must be the case that there is a node at the boundary.

EXAMPLE 15-2   A 10 g string of length 2.5 m is stretched between two fixed points. When excited with a frequency of 100 Hz, the string is observed to have a standing wave that has nodes at 1.0 m and 1.5 m from one end and at no points in between. a) What is the mode number? b) What is the tension in the string?

SOLUTION a) If there are no nodes between the two at 1.0 m and 1.5 m, then they must be adjacent nodes and so are separated by 0.5 m = $\lambda/2$. The number of 0.5 m half-wavelength segments that fit into a 2.5 m length is 5, so this must be the $n = 5$ mode.
b) The wavelength is 1.0 m and the frequency is 100 Hz, so the wave speed on the string must be $v = \lambda f = 100$ m/s. The wave speed is determined from the string's tension and linear density by

$$v = \sqrt{\frac{T_s}{\mu}}.$$

Solving for the tension gives

$$T_{\mathrm{s}} = \mu v^2 = \frac{m}{L} v^2 = \frac{.010 \text{ kg}}{2.5 \text{ m}} (100 \text{ m/s})^2 = 40 \text{ N}.$$

Harmonics

[**Photo suggestion: A harp.**]

An important application of standing waves on strings is to stringed musical instruments, such as the guitar, the piano, and the violin. These instruments all have strings that are fixed at the ends and placed under tension. A disturbance is generated on the string by plucking, striking, or bowing it. Regardless of how it is generated, the disturbance causes standing waves to be established on the string. The fundamental frequency of the vibrating string is

$$f_1 = \frac{v}{2L} = \frac{1}{2L} \sqrt{\frac{T_{\mathrm{s}}}{\mu}}, \tag{15-20}$$

where T_{s} is the tension in the string and μ is its linear density. The fundamental frequency is the "note" you hear when the string is sounded. Notice that v is the wave speed on the string, *not* the speed of sound in air.

While the fundamental frequency of the standing wave determines the pitch of the note you hear, the strings of real musical instruments actually vibrate simultaneously in *several* normal modes—not just the $n = 1$ mode. That is, the standing wave is actually a *superposition* of standing waves having frequencies f_1, f_2, f_3, and so on. In musical terminology, the higher frequency modes are called **harmonics**, with the $n = 2$ mode at frequency f_2 called the *second harmonic*, the $n = 3$ mode called the *third harmonic*, and so on. The vibration amplitudes of the harmonics are less than the amplitude of the fundamental frequency, but their presence is what accounts for the "tone" of the sound. A guitar, a violin, and a piano can all play the same note, having the same value for f_1, but each has a very different sound quality or tone. The difference is because each—due to its design and the way the string is excited—has a very different combination of harmonics that are vibrating simultaneously with the fundamental frequency.

For stringed instruments like the guitar or the violin, the strings are all the same length and under approximately the same tension. Were that not the case, the neck of the instrument would tend to twist toward the side of higher tension. The strings have different frequencies because they differ in linear density—the lower-pitched strings are "fat" while the higher-pitched strings are "skinny." This changes the frequency by changing the wave speed. *Small* adjustments are then made in the tension, by turning the tuning screws, to bring each string to the exact desired frequency. Once it is tuned, you play the instrument by using your fingertips to alter the effective length of the string—your finger is then the boundary which forces the string to have a node at that point. As you shorten the length, the frequency and pitch go up.

A piano covers a much wider range of frequencies than a guitar or violin. This cannot be done by changing only the linear densities of the strings—the high end would have strings too thin to use without breaking, and the low end would have solid rods rather than flexible wires! So a piano is tuned through a combination of changing the linear density *and* the length of the strings. The bass note strings are not only fatter, they are also longer, which is why a grand piano has the characteristic curved outline.

In all stringed instruments, one end of the string is the point where it crosses over the *bridge*. As the string vibrates, it causes small vibrations of the bridge. The bridge transfers these vibrations to the *sounding board*, which is the upper body of an acoustic guitar or a violin and the back of an upright piano. (An electric guitar does not have a sounding board. The string vibrations are sensed electronically to produce a current that is amplified and sent to loudspeakers.) The sounding board has a much larger surface area than the string, allowing it to push easily on the air, and its vibrations are the source of a *traveling* sound wave that leaves the instrument and travels to your ear. The sound wave has the same frequency as the vibrating standing wave on the string, but a very different wavelength determined by the speed of sound in air. Make sure you do not confuse the speed and wavelength of the standing wave on the string with the speed and wavelength of the sound wave radiated by the sounding board.

15.5 Standing Electromagnetic Waves

There are countless examples of standing waves other than on stretched strings. Another transverse wave, as we have seen, is an electromagnetic wave. Standing electromagnetic waves can be established between reflective boundaries such as mirrors. In fact, a laser operates by establishing a standing light wave between two mirrors facing each other, as shown in Fig. 15-14, to form what is called a *laser cavity*. The mirrors act exactly like the points to which a string is tied, and the light wave has a node at the surface of each mirror. (One of the mirrors is not perfectly reflective, allowing some of the light to escape the laser cavity and form the *laser beam*.)

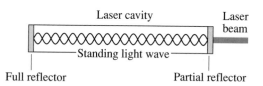

FIGURE 15-14 A laser works by establishing a standing light wave between two mirrors.

Because the boundary conditions are the same, Eqs. 15-16 and 15-17 apply just as much to a laser as to a guitar string. The primary difference is the size of the wavelength. A typical laser cavity has a length $L \approx 30$ cm, and visible light has a wavelength $\lambda \approx 600$ nm. The standing light wave in a laser cavity must then have a mode number n of approximately

$$n = \frac{2L}{\lambda} \approx \frac{2 \times 0.30 \text{ m}}{6 \times 10^{-7} \text{ m}} = 1 \times 10^6 .$$

In other words, there are approximately one million antinodes in the standing light wave inside a laser cavity! This is a consequence of the very small wavelength of light.

EXAMPLE 15-3   Helium-neon lasers emit the red laser light commonly used in classroom demonstrations and supermarket scanners. A helium-neon laser emits light with a wavelength of precisely 633.110 nm. Careful measurements show that the spacing between the mirrors is 310.43 mm. a) In which mode does this laser operate? b) What is the next longest wavelength that could form a standing wave in this laser cavity?

SOLUTION a) From Eq. 15-16, $\lambda_n = 2L/n$, we can find the value of n (the mode) to be

$$n = \frac{2L}{\lambda_n} = \frac{2 \times 0.31043 \text{ m}}{6.3311 \times 10^{-7} \text{ m}} = 980,651 .$$

There are 980,651 antinodes in the standing light wave between the mirrors.

b) The next longer wavelength to "fit" in this laser cavity will have one less node. It will be the $n = 980,650$ mode and its wavelength will be

$$\lambda = \frac{2L}{n} = \frac{2(0.31043 \text{ m})}{980,650} = 633.111 \text{ nm} .$$

The wavelength increases by a mere 0.001 nm, or by approximately 1 part in a million. •

Many lasers actually "lase" on three or four different modes simultaneously and are called *multimode* lasers. They emit three or four slightly different wavelengths. But as Example 15-3 shows, the wavelengths differ by so little that your eye, and even most instruments, perceive the laser light as a single wavelength.

Microwave radiation, which has a wavelength of a few centimeters, can also set up standing waves between reflective surfaces. This is not always good. If the microwaves in a microwave oven were to form a standing wave, there would be nodes where the electromagnetic field intensity was always zero. These nodes would cause "cold spots" where the food would not heat! Designers of microwave ovens use special precautions to try to prevent standing waves. They are not completely successful, and ovens usually do have cold spots—spaced a distance $\lambda/2$ apart—representing nodes in the microwave field. Turntables are widely used in microwave oven to keep the food moving so that no part of the food remains at a node.

EXAMPLE 15-4   Cold spots in a microwave oven are found to be 1.25 cm apart. What is the frequency of the microwaves?

SOLUTION The cold spots are nodes in a microwave standing wave. Nodes are spaced $\lambda/2$ apart, so the wavelength of the microwave radiation must be $\lambda = 2.5$ cm $= 0.025$ m. The speed of microwaves is the speed of light, $v = c$, so the frequency is

$$f = \frac{c}{\lambda} = \frac{3.00 \times 10^8 \text{ m / s}}{0.025 \text{ m}} = 1.2 \times 10^{10} \text{ Hz} = 12 \text{ GHz} .$$

•

15.6 Standing Sound Waves

A long, narrow column of air, such as in a tube or a pipe, can support a *longitudinal* standing sound wave inside. As a sound wave travels down a tube, the air molecules oscillate *parallel* to the tube. If the tube is closed at the end, with a cap, the molecules cannot oscillate. Therefore there must be a node at the closed end of a column of air.

From a wave perspective, a column of air of length L closed at *both* ends is no different than a string of length L tied at both ends. Each allows standing waves for only certain wavelengths and frequencies. Because there must be a node at both ends, the possible wavelengths and frequencies of a closed air column are the same as for a string:

$$\begin{cases} \lambda_n = \dfrac{2L}{n} \\ f_n = n \cdot \dfrac{v}{2L} = nf_1 \end{cases} \qquad n = 1, 2, 3, 4, \ldots \qquad (15\text{-}21)$$

EXAMPLE 15-5 A shower stall is 8 feet tall. For what frequencies less than 500 Hz are there standing sound waves in the shower stall?

SOLUTION The shower stall, at least to a first approximation, is a column of air 8 feet = 2.44 m long. It is closed at the ends by the ceiling and floor. The fundamental frequency for a standing sound wave in this air column is

$$f_1 = \frac{v}{2L} = \frac{343 \text{ m/s}}{2(2.44 \text{ m})} = 70 \text{ Hz}.$$

The possible standing wave frequencies are integer multiples of the fundamental frequency. These are 70 Hz, 140 Hz, 210 Hz, 280 Hz, 350 Hz, 420 Hz, and 490 Hz.

The many possible standing waves in a shower cause the sound to "resonate," which is why some people like to sing in the shower! However, air columns that are closed at both ends are of limited interest except for situations, such as Example 15-5, where you are inside the column.

Columns of air that *emit* sound are open at one or both ends. Many musical instruments fit this description. For example, a flute is a tube of air that is open at both ends. The flutist blows across one end to create a standing wave inside the tube, and a note of that frequency is emitted from both ends of the flute. (The blown end of a flute is open on the side, rather than across the tube. This is necessary for practical reasons of how flutes are played, but from a physics perspective this is the "end" of the tube because it opens the tube to the atmospheric pressure of the surrounding air.) A trumpet, however, is open at the bell end but is *closed* by the player's lips at the other end.

When a traveling sound wave reaches the open end of a tube of air, some of the wave's energy is transmitted out of the tube. This becomes the sound wave that you hear. But some portion of the wave is also reflected back into the tube. The air in the room acts like a spring. As a sound wave exits the tube it compresses the air in the room, then as the air expands again it pushes a reflected wave back into the tube. Because there is a wave reflection from the open end of a tube of air, standing sound waves can exist in a tube that is open at either one or both ends.

Not surprisingly, the *boundary condition* at the open end of a column of air is not the same as the boundary condition at a closed end. A closed end must be a node because the air molecules have no room to vibrate. An open end, by contrast, is required to be an antinode. (A careful analysis shows that the antinode is actually just outside the open end, but for our purposes we'll assume the antinode is exactly at the open end.)

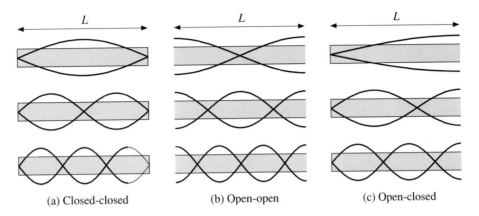

(a) Closed-closed   (b) Open-open   (c) Open-closed

FIGURE 15-15 The first three standing sound wave modes in a column of air.
a) A tube closed at both ends. b) A tube open at both ends. c) A tube open at
one end, closed at the other.

Figure 15-15 shows the first three standing wave modes of a closed-closed tube, an open-open tube, and an open-closed tube, all with the same length L. The open-open tube looks like the closed-closed tube except that the positions of the nodes and antinodes are interchanged. The wavelengths and frequencies of an open-open tube are the same as a closed-closed tube, given by Eq. 15-21, because there are still n half-wavelength segments between the ends.

The open-closed tube is different. As you can see, the fundamental mode has $L = \lambda/4$ and thus the $n = 1$ wavelength is $\lambda_1 = 4L$. This is twice the λ_1 wavelength of an open-open or a closed-closed tube. Consequently, the fundamental frequency will be half that of an open-open or a closed-closed tube of the same length. It will be left as a homework problem for you to show that the possible wavelengths and frequencies of an open-closed tube of length L are

$$\begin{cases} \lambda_n = \dfrac{4L}{n} \\ f_n = n \cdot \dfrac{v}{4L} = nf_1 \end{cases} \quad n = 1, 3, 5, 7, \ldots \quad (15\text{-}22)$$

Notice that n in Eq. 15-22 takes on only *odd* values.

Because sound is a longitudinal wave, the graphs of Fig. 15-15 are *not* "pictures" of the wave as they are for a string wave. The graphs show the displacement Δx, *parallel* to the axis, versus position x. The tube itself is shown merely to indicate the location of the open and closed ends, but the diameter of tube is *not* related to the amplitude of the displacement.

The frequencies of woodwind instruments, and hence the note they play, are determined by Eq. 15-21 and Eq. 15-22. The player's fingers, which cover holes in the side, determine the length of the tube and thus its frequency. As we noted for a flute, the fact that the holes are on the side makes very little difference. A hole becomes an antinode because the air molecules are free to oscillate in and out of the opening.

An interesting implication of Eqs. 15-21 and 15-22 is that an instrument's frequency depends on the speed of sound *inside* the instrument. The speed of sound depends on the temperature of the air. When a wind player first blows into the instrument, the air inside starts to rise in temperature. This increases the sound speed, which in turn raises the instrument's frequency for each note. Wind players must "warm up" their instruments before tuning them. This is not a difficulty for string instruments. For strings, the speed appearing in Eq. 15-17 is the wave speed on the string as determined by the tension, not the sound speed in air.

EXAMPLE 15-6   An organ pipe open at both ends sounds its second harmonic at a frequency of 524 Hz. (Musically, this is the note one octave above middle C.) What is the length of the pipe from the "sounding hole" to the end?

SOLUTION   An organ pipe, similar to a flute, has a "sounding hole" where compressed air, rather than a player's breath, is blown across the edge of the pipe. This is one end of an open-open tube, with the other end at the true "end" of the pipe. The second harmonic is the $n = 2$ mode, which for an open-open tube has frequency

$$f_2 = 2\frac{v}{2L}.$$

Thus the length of the organ pipe is

$$L = \frac{v}{f} = \frac{343 \text{ m / s}}{524 \text{ Hz}} = 0.655 \text{ m} = 65.5 \text{ cm}.$$

EXAMPLE 15-7   A clarinet is 66 cm long. The speed of sound in warm air is 350 m/s. a) What is the frequency of the lowest note on a clarinet? b) What is the frequency of the next highest harmonic?

SOLUTION   a) The lowest frequency is the fundamental frequency. A clarinet is an open-closed tube because the player's lips and the reed seal the tube at the upper end. Thus

$$f_1 = \frac{v}{4L} = \frac{350 \text{ m / s}}{4(0.66 \text{ m})} = 133 \text{ Hz}.$$

b) Equation 15-22 says that an open-closed tube has only *odd* harmonics. The next highest harmonic is $n = 3$, with frequency $f_3 = 3f_1 = 399$ Hz.

15.7 Standing Wave Resonances

In the absence of energy loss—damping—a standing wave would sustain itself indefinitely. Real standing waves, of course, die away unless there is a continuous input of energy to keep them going. The input energy represents a *driving force*. Recall from Chapter 13 that a driving force can cause a resonance—a large-amplitude response—when the driving frequency matches a natural frequency of the system.

An oscillator such as a spring or a pendulum has only a single natural frequency, but a system that can support standing waves has a whole collection of natural frequencies at which it would like to oscillate—the f_n given by Eqs. 15-17 for a string and by Eqs. 15-21 or 15-22 for a column of air. A driving force at any of these frequencies will generate a large-amplitude standing wave. That is exactly what is happening in the photographs of Fig. 15-13. The person shaking the end of the rope is hardly moving his hand at all, but because his driving frequency matches the standing wave frequencies of the rope he has created large-amplitude standing waves. Our earlier example of shattering a crystal goblet by singing the right note is similar. The goblet is really not a simple oscillating particle, like a pendulum or a mass on a spring; instead, it is a collective oscillation of the curved surface of the glass—a standing wave on the glass surface!

We noted, for a cylinder of air, that the reflection at the boundary is not 100% perfect. It is a *partial reflection*, where most of the sound wave is reflected, but some small fraction is *transmitted* through the boundary. That is essential—otherwise the sound could not get out of a musical instrument to be heard!—but it also allows the converse: a sound wave impinging from the outside will mostly reflect from the end of the pipe, but a small fraction will be transmitted *into* the pipe where it can act as an external driving force.

Suppose we arrange a long pipe with a loudspeaker at one end and your ear at the other, as shown in Fig. 15-16. The loudspeaker plays a note of frequency f_{ext}, which can be adjusted, with a very low intensity such that without the pipe you would barely hear it. If f_{ext} is varied, you will occasionally hear the sound become very loud. This happens when f_{ext} coming from the loudspeaker matches one of the standing wave frequencies f_n of the pipe, in which case it creates a *standing wave resonance* inside the pipe—just like the string resonances seen in Fig. 15-13. With a large-amplitude wave inside, even the small fraction transmitted out to your ear makes for a loud sound.

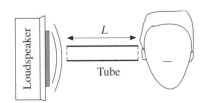

FIGURE 15-16 Listening to standing wave resonances in a tube of air.

Many wind instruments have a "buzzer" at one end of the pipe—a vibrating reed on a saxophone or clarinet, vibrating lips on a trumpet or trombone. Buzzers such as these generate a continuous range of frequencies rather than single notes, which is why they sound like a "squawk" if you play on just the mouthpiece without the rest of the instrument. When connected to the body of the instrument, most of those frequencies cause no response of the air molecules. But the frequency from the buzzer that matches the fundamental frequency of the instrument causes the build-up of a large-amplitude response at just that frequency—a standing wave resonance. It is this energy input that generates and sustains the musical note.

A laser, as we noted, has one mirror that is only a "partial reflector." Like the open end of an air cylinder, this mirror reflects enough of the wave to create a standing wave inside, but it also allows some energy to escape and form the laser beam. But what is the energy *input* into the standing wave? That comes from atoms inside the laser cavity that are forced to emit light, which occurs if you pass an electrical current through a gas and create a discharge, such as a neon light. The atoms emit light of many different frequencies, and these create the distinctive color you see when looking at a gas discharge. When placed inside a laser cavity, however, some of these frequencies may exactly match one or more of the standing light wave frequencies of the laser cavity. If that occurs—and lasers are designed so that it will—then a large-amplitude standing light wave builds up inside the laser cavity. Even the small part of the wave that is transmitted through the partial reflector can be a very intense beam of light. So a laser is a standing wave resonance, with the vibrating electrons of atoms providing the energy input while the laser beam itself is the energy output. This is analogous to a wind instrument, where the energy input is from a vibrating buzzer and the energy output is the sound radiated from the end of the instrument.

Summary

Important Concepts and Terms

boundary	antinode
standing wave	boundary condition
incident wave	fundamental frequency
reflected wave	normal modes
amplitude function	harmonics
node	

Standing waves are an essential part of wave physics, with important applications throughout science and engineering. A standing wave is the superposition of traveling waves moving in both directions through a medium as a result of reflections at the boundaries. This causes the particles of the medium to oscillate back and forth in a pattern that remains fixed in space—in contrast with a traveling wave. The amplitude of these oscillations varies with position, as described by an amplitude function. The points of maximum amplitude are called antinodes, while points at which the oscillation amplitude is zero are called nodes.

A standing wave is described by the displacement

$$D(x,t) = (2a\sin kx)\cos\omega t = A(x)\cos\omega t,$$

where a is the amplitude of each traveling wave and $A(x)$ is the amplitude function. This oscillation is a series of nodes and antinodes, with the nodes spaced $\lambda/2$ apart.

A standing wave can exist on a string only when the wavelength and frequency satisfy

$$\lambda_n = \frac{2L}{n} \qquad f_n = n\cdot\frac{v}{2L} = nf_1 \qquad n = 1,2,3,\dots$$

where f_1 is the fundamental frequency. No other wavelengths or frequencies are permitted. This result for the allowed wavelengths and frequencies is also valid for standing electromagnetic waves reflected between two mirrors and for standing sound waves reflected from the end of a column of air that is open at both ends or closed at both ends. The boundary conditions, however, do depend on the type of wave. Strings and electromagnetic waves require nodes at the boundaries, while a standing sound wave in a tube must have a node at a closed end but an antinode at an open end.

Exercises and Problems

▲ Exercises

1. A 120 cm string of mass 4 g oscillates with a maximum amplitude of 5 mm at its third harmonic frequency of 180 Hz.
   a. What is the wavelength on the string?
   b. What is the tension in the string?

2. A violin string is 30 cm long. It sounds the musical note A (440 Hz) when played without fingering. How far from the end of the string should you place your finger to play the note C (524 Hz)?

3. A vertical tube 1 m long is filled with water. A tuning fork vibrating at 580 Hz is held just over the top of the tube as the water is slowly drained from the bottom. At what heights of the water will there be a standing wave resonance in the tube?

4. A guitar string has a linear density of 7 g/m. It is stretched between supports that are 60 cm apart. The string is observed to form a standing wave with three antinodes when driven at a frequency of 420 Hz.
   a. What is the frequency for the fifth harmonic of this string?
   b. What is the tension in the string?

5. A microwave generator can be tuned to any frequency between 10 GHz and 20 GHz. The microwaves are aimed, through a hole, into a "microwave cavity" consisting of a 10 cm long cylinder with reflective ends, as shown in Fig. 15-17. Which frequencies will create standing wave resonances in the microwave cavity? For which of these frequencies is the cavity midpoint an antinode?

Microwaves

FIGURE 15-17

▲ Problems

6. Equation 15-1 is the mathematical statement that the wave reflected from a boundary is like the incident wave except that it is traveling in the opposite direction and reversed in displacement. This is a condition that a standing wave must obey. Use this condition to prove that a standing wave must satisfy the condition $\lambda_n = 2L/n$ where $n = 1, 2, 3, \ldots$ (Hint: Use the sinusoidal Eqs. 15-3 and 15-4 for D_R and D_L, then explore the conditions under which Eq. 15-1 is true.)

7. In Section 15.3 we showed that the two traveling waves forming a standing wave on a string are always completely out of phase at the positions of the nodes. Prove that the two traveling waves forming a standing wave on a string are always completely *in* phase at the positions of the antinodes. (Hint: Show that $D_R = D_L$ at the positions of the antinodes.)

8. Use a graphical analysis of standing sound waves in an open-closed tube to show that the possible wavelengths and frequencies are given by Eq. 15-22.

9. A violinist places her finger so as to make the vibrating length of a 1 g/m string have a length of 30 cm, then she draws her bow across the string. A listener nearby hears a note of wavelength 40 cm. What is the tension in the string?

10. A narrow cylinder of air is found to have standing wave resonances at frequencies of 390 Hz, 520 Hz, and 650 Hz and at no frequencies in between. The behavior of the tube at frequencies less than 390 Hz or greater than 650 Hz is not known.
    a. How long is the tube?
    b. Draw a graph of the 520 Hz standing wave in the tube.
    c. The air in the tube is replaced with carbon dioxide, which has a sound speed of 280 m/s. What are the new frequencies of these three modes?

11. Astronauts visiting Planet X have a 2.5 m string of mass 5 g. They tie the string to a support, stretch it horizontally over a pulley 2.0 m away, then hang a 1.0 kg mass on the free end. Then the astronauts begin to excite standing waves on the string. Their data show that standing waves exist at frequencies of 64 Hz and 80 Hz, but at no frequencies in between. What is the value of g, the acceleration due to gravity, on Planet X?

12. A sound wave of 280 Hz is directed into one end of a trombone slide and a microphone is placed at the other end to record the intensity of sound waves that are transmitted through the tube (Fig. 15-18). The straight sides of the slide are 80 cm in length and 10 cm apart with a semicircular bend at the end. For what slide extensions s will the microphone detect a maximum of sound intensity? (Hint: For what values of s will the tube have a standing wave resonance at 280 Hz?)

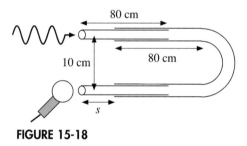

FIGURE 15-18

[**Estimated 12 additional problems for the final edition.**]

Chapter **16**

Interference

L O O K I N G
B A C K | Sections 14.3–14.5; 14.8

16.1 Wave Interactions

[**Photo suggestion: Interference pattern of two water waves.**]

One of the most basic characteristics of waves is their ability to pass through each other and combine into a single wave whose displacement is given by the principle of superposition. In the previous chapter, you saw how two waves traveling in opposite directions can combine to produce a standing wave. In some places the individual waves combined to produce an increased wave amplitude, while in other places the combined waves had zero amplitude—the antinodes and nodes of the standing wave. This type of combination, or superposition, of waves is called **interference**. Interference can be seen in many types of wave patterns, not just standing waves.

Interference is an *interaction* between two different waves. The interactions of waves are very different than the interactions of particles: Waves interact by combining rather than colliding with each other. Understanding how waves interact will complement what you have learned about how waves are generated and how waves propagate.

In this chapter, we will look at what happens when two waves with the same wavelength and frequency, but traveling along different paths, meet and combine. Although most examples of wave interference occur in two or three dimensions, we will first consider waves traveling in one dimension. This will allow us to focus on the physics and the mathematics in a relatively simple situation. Then we will expand our discussion to include interference in two and three dimensions.

16.2 Interference in One Dimension

Figure 16-1a shows two light waves impinging on a partially-silvered mirror that is oriented at a 45° angle. Such a mirror partially transmits and partially reflects each wave. As a result, two *overlapped* light waves travel along the x-axis to the right of the mirror. Or

(a)

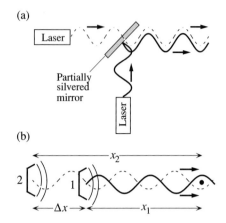

(b)

FIGURE 16-1 a) Two overlapped light waves traveling along the x-axis. b) Two loudspeakers sending sound waves along the x-axis.

consider two loudspeakers that are oriented as shown in Fig. 16-1b. The sound wave from the left speaker passes just barely to the side of the right speaker, so there are two overlapped sound waves traveling to the right along the x-axis. We want to find out what happens when two waves, as in these examples, travel together along the same axis.

Figure 16-1b shows a point on the x-axis where the overlapped waves are passing by. This point is distance x_1 from loudspeaker 1 and distance x_2 from loudspeaker 2. (We will use loudspeakers and sound waves for many of our examples, but our analysis will be valid for any one-dimension wave.) Keep in mind what Fig. 16-1 is showing. Each sinusoidal graph represents the displacement of the medium at that point on the x-axis. The displacement of wave 2 at the point indicated in Fig. 16-1b is equal in magnitude but opposite in direction to the displacement of wave 1.

Let's consider two sinusoidal waves that have the same frequency, the same wavelength, the same amplitude, and that travel to the right along the x-axis. We can represent these two waves as the displacements

$$D_1 = a\sin(kx_1 - \omega t + \phi_{10})$$
$$D_2 = a\sin(kx_2 - \omega t + \phi_{20}), \tag{16-1}$$

where x_1 and x_2 are measured from the two sources, as shown in Fig. 16-1b. Both waves have the same values for the amplitude a, the wave number $k = 2\pi/\lambda$, and the angular frequency $\omega = 2\pi f$.

Recall, from Chapter 14, that the *phases* of the waves are

$$\phi_1 = kx_1 - \omega t + \phi_{10}$$
$$\phi_2 = kx_2 - \omega t + \phi_{20}. \tag{16-2}$$

The *phase constants* ϕ_{10} and ϕ_{20} describe the initial conditions of the sources. The source of wave 1 is located at $x_1 = 0$ and that of wave 2 is at $x_2 = 0$. At $t = 0$, the displacements *at the sources* are

$$D_1(x_1 = 0, t = 0) = a\sin\phi_{10}$$
$$D_2(x_2 = 0, t = 0) = a\sin\phi_{20}. \tag{16-3}$$

You can see that the phase constants ϕ_{10} and ϕ_{20} determine what the sources are doing at $t = 0$. We will often consider *identical sources*, by which we mean that $\phi_{20} = \phi_{10}$.

Figure 16-2 shows snapshot graphs at $t = 0$ of waves being emitted by sources having phase constants of 0, $\pi/2$, and π. Each is simply a graph of $D = a\sin(kx + \phi_0)$. Notice that speaker 1 in Fig. 16-1b has phase constant $\phi_{10} = 0$ while speaker 2 has $\phi_{20} = \pi$.

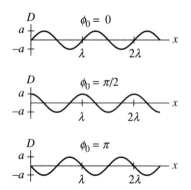

FIGURE 16-2 Snapshot graphs of the waves from sources having phase constants of 0, $\pi/2$, and π.

Before delving into a mathematical analysis, let's first examine the situation graphically. There are two important cases to consider. The first case, shown in Fig. 16-3, occurs when the crests of the two waves overlap as they travel outward along the x-axis. Figure 16-2a shows graphs of the two displacements D_1 and D_2, while Fig. 16-2b shows the wave fronts of the two waves moving along the x-axis. Recall from Chapter 14 that the wave fronts locate the crests of the waves. The graphs and the wave fronts in this figures are slightly displaced from each other so that you can see what each wave is doing, but the *physical situation* we are describing is one in which the waves are traveling *on top of* each other.

The displacement of the medium by a wave is $D = a\sin\phi$, where $\phi = kx - \omega t + \phi_0$ is the phase at that point in the medium. Because the two waves of Fig. 16-3 have the same displacement at *every* point ($D_1 = D_2$), they must have the same phase ($\phi_1 = \phi_2$ or, more precisely, $\phi_1 = \phi_2 \pm m \cdot 2\pi$ where m is an integer). Two waves that are aligned crest-to-crest and trough-to-trough are said to be **in phase**. Waves that are in phase march along "in step" with each other.

When we combine two in-phase waves, using the principle of superposition, we find that the net displacement at each point is twice the displacement of each individual wave. That is, the superposition of two in-phase waves of amplitude a is a wave with amplitude $2a$. Figure 16-3c shows the superposition of the two waves of Fig. 16-3a. This type of wave interaction is called **constructive interference**.

The second important case, shown in Fig. 16-4, occurs when the crests of one wave align with the troughs of the other wave. As these waves move forward, they are marching along "out of step." In this case $D_1 = -D_2$ at every point. Because $\sin(\phi + \pi) = -\sin\phi$, the relationship between the phases of these two waves is seen to be $\phi_1 = \phi_2 + \pi$. Two waves that are aligned crest-to-trough are said to be *180° out of phase* or, more generally, just **out of phase**.

The displacement of wave 2 is opposite that of wave 1 at every point along the x-axis. Consequently, the net displacement is *zero* at *every* point along the axis. If we combine two waves that are aligned crest-to-trough, the waves *cancel* and give no wave at all! This type of wave interaction is called **destructive interference**.

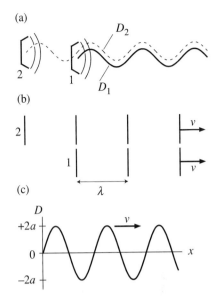

FIGURE 16-3 a) Two waves moving along the x-axis in phase with each other. b) The wave fronts for these waves are aligned. c) The superposition of two in-phase waves gives constructive interference.

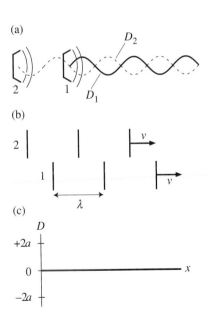

FIGURE 16-4 a) Two out-of-phase waves moving along the x-axis. b) The wave fronts show that the waves are out of phase. c) The superposition of two out-of-phase waves gives destructive interference.

Let's look more closely at the **phase difference** between the two waves. The condition of being in phase is stated mathematically as

$$\Delta\phi = \phi_2 - \phi_1 = 0, 2\pi, 4\pi, 6\pi, \ldots = m \cdot 2\pi \qquad m = 0, 1, 2, 3, \ldots \qquad (16\text{-}4)$$

The phase difference $\Delta\phi$ can be written, using Eq. 16-2 and $k = 2\pi/\lambda$, as

$$\phi_2 - \phi_1 = (kx_2 - \omega t + \phi_{20}) - (kx_1 - \omega t + \phi_{10})$$
$$= k(x_2 - x_1) + (\phi_{20} - \phi_{10}) \qquad (16\text{-}5)$$
$$= 2\pi\frac{\Delta x}{\lambda} - \Delta\phi_0$$

where $\Delta x = x_2 - x_1$ is the distance between the sources and $\Delta\phi_0 = \phi_{20} - \phi_{10}$ is the phase difference between the sources. Both Δx, the *path length difference*, and $\Delta\phi_0$, any inherent phase difference between the sources, contribute to the phase difference $\Delta\phi$ between the waves.

Thus the condition for constructive interference is:

$$\Delta\phi = 2\pi\frac{\Delta x}{\lambda} + \Delta\phi_0 = m \cdot 2\pi \qquad m = 0, 1, 2, 3, \ldots \text{(constructive)}. \qquad (16\text{-}6)$$

For identical sources, which have $\Delta\phi_0 = 0$, constructive interference occurs when $\Delta x = m\lambda$. That is, identical sources cause constructive interference when they are spaced an integer number of wavelengths apart.

Figure 16-5 shows two identical sources spaced one wavelength apart. Because a wave moves forward exactly one wavelength during a time interval of one period, a crest emitted from speaker 2 at time t will pass speaker 1 at time $t + T$. Speaker 1 is identical to speaker 2, so speaker 1 also emits a crest at time t and emits its next crest at time $t + T$—exactly as the crest of wave 2 passes by. The two waves are "in step," with their crests and troughs aligned, so the two waves interfere constructively and produce a wave of amplitude $2a$.

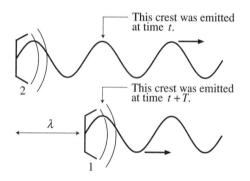

FIGURE 16-5 Two identical sources cause constructive interference if they are exactly one wavelength apart.

Destructive interference occurs when two waves are out of phase:

$$\Delta\phi = \phi_2 - \phi_1 = \pi, 3\pi, 5\pi, 7\pi, \ldots = (m + \tfrac{1}{2})2\pi \qquad m = 0, 1, 2, 3, \ldots \qquad (16\text{-}7)$$

Using Eq. 16-5 for $\Delta\phi_0$, the condition for destructive interference is

$$\Delta\phi = 2\pi\frac{\Delta x}{\lambda} + \Delta\phi_0 = (m + \tfrac{1}{2})2\pi \qquad m = 0, 1, 2, 3, \ldots \text{(destructive)}. \qquad (16\text{-}8)$$

Destructive interference between two waves occurs when the phases of the waves differ by an odd multiple of π. The phases of the waves can differ because the sources are located at different positions, because the sources themselves are not in phase ($\phi_{20} \neq \phi_{10}$), or because of a combination of both of these situations.

Figure 16-6 illustrates these ideas by showing three different ways in which two waves can undergo destructive interference. In Fig. 16-6a, the two sources are side-by-side

($\Delta x = 0$) but are oscillating out of phase ($\phi_{10} = 0$, but $\phi_{20} = \pi$ so $\Delta\phi_0 = \pi$). Figure 16-6b shows two identical sources ($\Delta\phi_0 = 0$) that are spaced $\Delta x = \lambda/2$ apart. Finally, Fig. 16-6c shows two sources that have a phase difference $\Delta\phi_0 = \pi/2$ and that are spaced $\Delta x = \lambda/4$ apart. Each of these three arrangements of the sources creates *waves* having a phase difference $\Delta\phi = \pi$ and that interfere destructively. Note that the phase difference of the waves ($\Delta\phi$) is not the same thing as the phase difference of the sources ($\Delta\phi_0$).

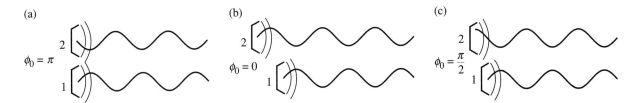

FIGURE 16-6 Destructive interference three ways. a) The sources are out of phase. b) Identical sources are separated by half a wavelength. c) The sources are both separated and partially out of phase.

EXAMPLE 16-1   Two loudspeakers are placed side by side and play continuous notes of the same frequency while you stand in front of them and listen to their combined sound. Initially there is almost no sound at all. Then one of the speakers is moved slowly away from you. As the separation between the speakers is increased you hear the sound intensity increasing. The intensity reaches a maximum when the speakers are 0.75 m apart, then decreases. What is the next separation distance at which you will hear a minimum intensity?

SOLUTION The minimum sound intensity that was heard initially implies that the two sound waves are undergoing destructive interference. Because the loudspeakers are side-by-side, the situation must be the same as in Fig. 16-6a with $\Delta x = 0$ and $\Delta\phi_0 = \pi$. That is, the speakers themselves are out of phase. Moving one speaker does not change $\Delta\phi_0$. The first point of maximum intensity, a constructive interference, is reached when

$$\Delta\phi = 2\pi\frac{\Delta x}{\lambda} + \Delta\phi_0 = 2\pi\frac{\Delta x}{\lambda} + \pi = 2\pi$$

$$\Rightarrow \Delta x = \frac{\lambda}{2}.$$

A speaker separation of $\lambda/2$ creates a phase difference of π. When combined with the phase difference π of the speakers themselves, the total phase difference of the waves is 2π and the interference is constructive. This situation is illustrated in Fig. 16-7.

Because $\Delta x = 0.75$ m is $\lambda/2$, the sound's wavelength is $\lambda = 1.50$ m. The next point of destructive interference occurs when

$$\Delta\phi = 2\pi\frac{\Delta x}{\lambda} + \Delta\phi_0 = 2\pi\frac{\Delta x}{\lambda} + \pi = 3\pi$$

$$\Rightarrow \Delta x = \lambda = 1.50 \text{ m}.$$

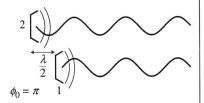

FIGURE 16-7 Two sources that are out of phase will generate waves that are in phase if the sources are one-half wavelength apart.

A separation of λ would give constructive interference for two *identical* speakers ($\Delta\phi_0 = 0$), as was shown in Fig. 16-5. Here the phase difference between the speakers (one is pushing forward as the other pulls back) gives destructive interference at this separation.

•

The Mathematics of Interference

Let's look at the superposition of the two waves more closely. As the two waves travel together along the x-axis, the net displacement of the medium at each point is described by

$$D_{\text{net}} = D_1 + D_2 = a\sin(kx_1 - \omega t + \phi_{10}) + a\sin(kx_2 - \omega t + \phi_{20})$$
$$= a\sin\phi_1 + a\sin\phi_2, \tag{16-9}$$

where the phases ϕ_1 and ϕ_2 were given in Eq. 16-2.

Recall the trigonometric identity we used to analyze the phenomenon of beats in Chapter 14:

$$\sin\alpha + \sin\beta = 2\cos\left[\tfrac{1}{2}(\alpha - \beta)\right]\sin\left[\tfrac{1}{2}(\alpha + \beta)\right]. \tag{16-10}$$

This identity is directly applicable to Eq. 16-9, allowing us to rewrite it as

$$D_{\text{net}} = \left[2a\cos\frac{\Delta\phi}{2}\right]\sin\left(kx - \omega t + (\phi_0)_{\text{avg}}\right), \tag{16-11}$$

where $\Delta\phi = \phi_2 - \phi_1$ is the phase difference between the two waves, as given in Eq. 16-5, and $(\phi_0)_{\text{avg}} = (\phi_{10} + \phi_{20})/2$ is the average phase constant of the two sources.

We need to examine Eq. 16-11 carefully to interpret its meaning. The sine term indicates that the combination of waves D_1 and D_2 is still a traveling wave, which differs from the two individual waves only by a slightly different phase constant—the average of the phase constants of the two waves being combined. So if you are an observer, you still see a wave coming toward you—a *single* combined wave!—with the same wavelength and frequency as the original waves.

The more interesting question is, How *big* is this wave compared to the size of the two original waves? They each had amplitude a, but the amplitude of their superposition is given by Eq. 16-11 as

$$A = \left|2a\cos\left(\frac{\Delta\phi}{2}\right)\right|, \tag{16-12}$$

where we have used an absolute value sign because amplitudes must be positive. Depending upon the phase difference of the two waves, the amplitude of their superposition can be anywhere from zero to $2a$.

The amplitude will reach its maximum value $A = 2a$ if $\cos(\Delta\phi/2) = \pm 1$. This occurs when

$$\text{Maximum amplitude:} \quad \Delta\phi = m \cdot 2\pi. \tag{16-13}$$

where m is any integer. Similarly, the amplitude will be $A = 0$ if $\cos(\Delta\phi/2) = 0$, which occurs when

$$\text{Minimum amplitude:} \quad \Delta\phi = (m + \tfrac{1}{2})2\pi. \tag{16-14}$$

But Eqs. 16-13 and 16-14 are precisely the conditions of Eqs. 16-6 and 16-8 for constructive and destructive interference. We initially found these conditions graphically by considering the alignment of the crests and troughs; now we have confirmed them with an algebraic addition of the waves.

It is entirely possible, of course, that the crests of one wave align with neither the crests nor the troughs of the other wave. That is, the two waves are neither exactly in phase nor exactly out of phase. Even so, Eqs. 16-11 and 16-12 allow us to calculate the superposition. For example, Figure 16-8 shows the calculated interference—using Eq. 16-11—of two waves that differ in phase by 40°, by 90°, and by 160°. These situations occur as the physical separation between two identical sources is slowly increased from zero to $\lambda/2$. The interference is basically constructive for $\Delta\phi = 40°$ in Fig. 16-8a, but the amplitude is only $2a\cos20° = 1.88a$ rather than the full $2a$ of true constructive interference. Likewise Fig. 16-8c shows the interference of two waves having a phase difference of 160°. They nearly cancel—having an amplitude of only $0.35a$—but not quite. We will tend to focus on those situations where the waves are either exactly in phase or exactly out of phase because those situations locate for us the points of maximum and minimum amplitude.

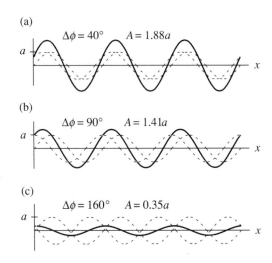

FIGURE 16-8 The interference of two waves (dotted lines) for cases where the phase difference is a) 40°, b) 90°, and c) 160.° The net wave is shown as a solid line.

Nonetheless, it is important to keep in mind that most interference falls somewhere in between. Being *exactly* in or out of phase is *not* a requirement for interference.

EXAMPLE 16-2   A 500 Hz signal is sent to two loudspeakers. Each loudspeaker alone broadcasts a sound wave of 0.1 mm maximum displacement. Speaker 2 is 1.0 m behind speaker 1, and the signal to speaker 2 is delayed by 0.5 ms. What is the amplitude of the sound wave at a point 2.0 m in front of speaker 1?

SOLUTION   The amplitude is given by Eq. 16-12 as $A = |2a\cos(\Delta\phi/2)|$, where $a = 0.1$ mm and

$$\Delta\phi = \phi_2 - \phi_1 = 2\pi\frac{\Delta x}{\lambda} + (\phi_{20} - \phi_{10}).$$

The wavelength is $\lambda = v_{sound}/f = 0.686$ m. The distances $x_1 = 2.0$ m and $x_2 = 3.0$ m are measured from the speakers, so $\Delta x = 1.0$ m. We are free to start our clock any time we choose, so let's select $t = 0$ such that $\phi_{10} = 0$. Then what is ϕ_{20}?

The signal takes 0.5 ms longer to reach speaker 2 than to reach speaker 1. A 500 Hz signal has a period $T = 2.0$ ms, so the delay is $T/4$, or one-quarter of a period. The wave broadcast by speaker 2 at $t = 0$ will be the same as the wave broadcast by speaker 1 at the earlier time $t = -0.5$ ms $= -T/4$. Speaker 1, at $x_1 = 0$ with $\phi_{10} = 0$, oscillates as

$$D_{\text{speaker 1}} = a\sin(-\omega t + \phi_{10}) = a\sin\left(-\frac{2\pi t}{T}\right).$$

At $t = -T/4$, speaker 1 has the displacement

$$D_{\text{speaker 1}}(t = -T/4) = a\sin(-(-\pi/2)) = a.$$

Speaker 2 has this displacement one-quarter of a period later, at $t = 0$:

$$D_{\text{speaker 2}}(t = 0) = a\sin(\phi_{20}) = a,$$

from which we can conclude $\phi_{20} = \pi/2$. We can now determine the phase difference to be

$$\Delta\phi = 2\pi\frac{\Delta x}{\lambda} + (\phi_{20} - \phi_{10})$$

$$= 2\pi\frac{1\text{ m}}{0.686\text{ m}} + \frac{\pi}{2}$$

$$= 10.73 \text{ radians}.$$

Finally, from Eq. 16-12, we can calculate the amplitude of the wave at this point to be

$$A = \left|2a\cos\left(\frac{\Delta\phi}{2}\right)\right| = (0.2\text{ mm})\cos\left(\frac{10.73}{2}\right) = 0.121\text{ mm}.$$

16.3 Interference in Two and Three Dimensions

Many sources of waves emit circular waves in two dimensions (ripples on a lake) or spherical waves in three dimensions (sound from a loudspeaker or light from a light bulb). We noted, at the end of Section 14.4, that the mathematical description of a circular or spherical wave needs to change from $D = f(x - vt)$ to $D = f(r - vt)$, where r is the radius measured outward from the source. A sinusoidal circular or spherical wave is represented by

$$D(r,t) = a\sin(kr - \omega t + \phi_0), \tag{16-15}$$

where the one-dimensional coordinate x has been replaced with a more general radial coordinate r. Figure 16-9 shows the wave fronts of a circular or spherical wave.

Now let's look at what happens when two of these waves overlap. Consider two circular waves emitted by two paddles that are oscillating up and down on the surface of water. We will assume that the two paddles are identical sources emitting identical waves. Figure 16-10a shows the wave fronts for the two waves. The ripples begin to overlap and cause interference as they travel outward from the two sources. We would like to know what the net effect will be.

We showed in the last section that constructive interference occurs where two crests or two troughs align. The net displacement at these points is twice the size of each individual displacement. Several locations of constructive interference are marked in Fig. 16-10a at points where the wave fronts intersect. Several points of destructive interference, where a crest of one wave aligns with a trough of the other wave are also shown.

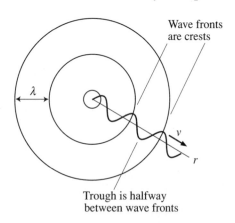

Wave fronts are crests

Trough is halfway between wave fronts

FIGURE 16-9 A circular or spherical wave, showing both the displacement and the wave fronts.

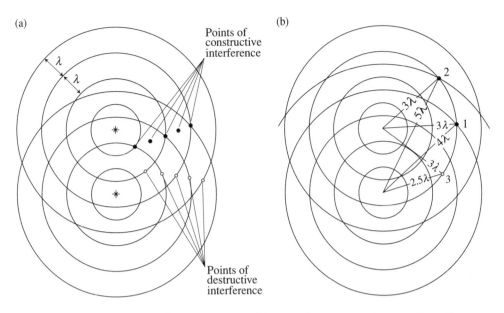

FIGURE 16-10 a) Two circular waves emitted by nearby sources. Some points of constructive and destructive interference are noted. b) A closer look at the interference. Point 1 has $\Delta r = 1\lambda$, point 2 has $\Delta r = 2\lambda$, and point 3 has $\Delta r = \lambda/2$.

Wave fronts, remember, are spaced exactly one wavelength apart. We can tell the distances from the sources to a point on the water's surface simply by counting the number of "rings" in the wave front pattern. Point 1 in Fig. 16-10b, a point of constructive interference, is seen to be a distance of 3λ from one source and 4λ from the other. The *difference* in the distances is thus $\Delta r = 1\lambda$, which is exactly the condition for constructive interference of identical sources. Point 2 is similar, with $\Delta r = 2\lambda$. Point 3, on the other hand, has $\Delta r = \lambda/2$, which you can see is a point of destructive interference (crest aligned with trough). So our *physical* understanding from one dimension carries over into two (and three) dimensions: constructive interference due to two identical sources occurs at points such that the difference in the path lengths to the sources is $\Delta r = m\lambda$. Similarly, destructive interference occurs at points such that $\Delta r = (m + \frac{1}{2})\lambda$.

It is important to keep in mind that the wave front picture of Fig. 16-10 is a snapshot at a single instant of time. A movie would show that the wave fronts are not fixed in space but are moving outward from the sources, with new wave fronts being created at the centers. But as the wave fronts move, the points of constructive and destructive interference remain fixed! If $\Delta r = m\lambda$, then two crests at time t will be replaced by two troughs at $t + T/2$ and by the next two crests at $t + T$. In other words, the displacements of the two traveling waves as they arrive at a point where $\Delta r = m\lambda$ are *always* identical (in phase). Therefore the net displacement at this point is always twice the displacement of each individual wave, and the medium at that point oscillates with maximum amplitude $2a$. Similarly at a point of destructive interference. The point cannot have a crest and a trough present at every instant of time. But the displacement of one wave is guaranteed to be exactly the opposite (out of phase) the displacement of the other wave so that their *sum* will be zero at all times.

With this in mind, we can draw lines connecting all the points where $\Delta r = 0$, $\Delta r = 1\lambda$, $\Delta r = 2\lambda$, and so on. These lines, shown in Fig. 16-11, locate all the points of constructive interference on the surface of the water—the points where the water is oscillating up and down with maximum amplitude. Note particularly that the line bisecting the sources ($\Delta r = 0$) is a line of constructive interference. Any point on this line is equidistant from both sources, so the two waves travel equal distances, in equal times, and arrive in phase. Halfway between these lines of constructive interference are the lines of destructive interference, where the water surface oscillates with minimum (ideally zero) amplitude.

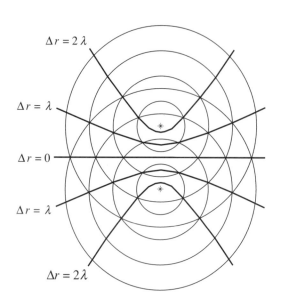

FIGURE 16-11 Lines showing all the points of constructive interference, where $\Delta r = m\lambda$.

Note the similarity to standing waves, where in one dimension we had *points* of maximum amplitude (antinodes) and minimum amplitude (nodes). We can now see those as the points of constructive and destructive interference of two waves traveling in opposite directions. In the two-dimensional case, we have *lines* of maximum amplitude, corresponding to antinodes, and *lines* of minimum amplitude, corresponding to nodes. Most points on the surface are neither purely constructive nor purely destructive interference, but somewhere in between—just as most points on a standing wave are neither nodes nor antinodes.

The mathematical description of interference in two (or three) dimensions is very similar to that of one-dimensional interference. The displacement at a point on the surface, or in space, is given by

$$D_{\text{net}} = D_1 + D_2 = a\sin(kr_1 - \omega t + \phi_{10}) + a\sin(kr_2 - \omega t + \phi_{20})$$
$$= a\sin\phi_1 + a\sin\phi_2, \tag{16-16}$$

where we have again assumed equal amplitudes and wavelengths. The only difference between Eq. 16-16 and the one-dimensional Eq. 16-9 is that the linear coordinates x_1 and x_2 have been changed to radial coordinates r_1 and r_2. Applying the trigonometric identity Eq. 16-10 leads to

$$D_{\text{net}} = \left[2a\cos\left(\frac{\Delta\phi}{2}\right)\right]\sin\left(kr_{\text{avg}} - \omega t + \phi_{\text{avg}}\right), \tag{16-17}$$

where $r_{\text{avg}} = (r_1 + r_2)/2$. The sine term indicates that the superposition is still a traveling wave. However, our interest is in the amplitude, because that will tell us where the constructive and destructive interference occurs.

The amplitude of the combined waves is

$$A = \left|2a\cos\left(\frac{\Delta\phi}{2}\right)\right|. \tag{16-18}$$

This result is identical with Eq. 16-12 for one-dimensional waves, except that now

$$\Delta\phi = 2\pi\frac{\Delta r}{\lambda} + \Delta\phi_0. \tag{16-19}$$

The term $2\pi(\Delta r/\lambda)$ is the phase difference that arises when the waves travel different distances from the sources to the point at which they combine. The other term, $\Delta\phi_0$, is the inherent phase difference of the sources themselves.

Constructive interference occurs (just as in one dimension) at those points on the surface where $\cos(\Delta\phi/2) = \pm1$. Similarly, destructive interference occurs at points where $\cos(\Delta\phi/2) = 0$. Using Eq. 16-19 for $\Delta\phi$, we can summarize the conditions for constructive and destructive interference as

$$
\begin{aligned}
\text{Constructive:} \quad &\Delta\phi = 2\pi\frac{\Delta r}{\lambda} + \Delta\phi_0 = m\cdot 2\pi \\
\text{Destructive:} \quad &\Delta\phi = 2\pi\frac{\Delta r}{\lambda} + \Delta\phi_0 = (m+\tfrac{1}{2})2\pi
\end{aligned}
\qquad m = 0, 1, 2, \ldots \tag{16-20}
$$

Although we derived Eq. 16-20 by considering two-dimensional water waves, which are easy to visualize, the result applies equally well in three dimensions to sound or electromagnetic waves.

For identical sources that oscillate in phase ($\Delta\phi_0 = 0$) the conditions for constructive and destructive interference are very simple:

$$
\begin{aligned}
\text{Constructive:} \quad &\Delta r = m\lambda \\
\text{Destructive:} \quad &\Delta r = (m+\tfrac{1}{2})\lambda
\end{aligned}
\qquad m = 0, 1, 2, \ldots \quad \text{(identical sources).} \tag{16-21}
$$

This was the situation shown in Fig. 16-10.

EXAMPLE 16-3   Two loudspeakers are 6 m apart and are emitting sound waves of the same frequency. You hear a maximum sound intensity if you stand 10 m in front of the speakers along the line that bisects them. As you move sideways, parallel to the speakers, the sound first gets softer than again gets louder, reaching another maximum of intensity after you have moved 2 m. a) What is the phase relationship of the two speakers? b) What frequency sound waves are they emitting?

SOLUTION a) Originally you are equidistant from both speakers, so $\Delta r = 0$. The phase difference is $\Delta\phi = \Delta\phi_0$, the phase difference between the speakers themselves. Because the sound intensity is maximum, you must be at a point of constructive interference. Thus $\Delta\phi_0 = 0$—the two speakers are in phase. (You might think about what you would hear at this point if the two speakers were out of phase.)

b) As you move sideways, Δr is no longer zero. Figure 16-12 shows the geometry of the

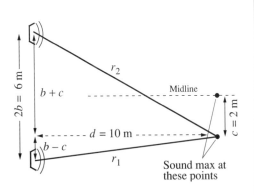

FIGURE 16-12 Geometry of Example 16-3.

situation. Because your original position corresponded to $\Delta r = 0\lambda$ and because the position two meters away is the *next* point of constructive interference, it must be a point where $\Delta r = 1\lambda$. (See Fig. 16-11 where the lines of constructive interference were shown.) The figure defines the quantities $b = 3$ m, $c = 2$ m, and $d = 10$ m. From the right triangles we have

$$\lambda = \Delta r = r_2 - r_1$$
$$= \sqrt{d^2 + (b+c)^2} - \sqrt{d^2 + (b-c)^2}$$
$$= 0.720 \text{ m}.$$

Because this is a sound wave with $v = 343$ m/s (assuming room temperature), the frequency must be

$$f = \frac{v}{\lambda} = \frac{343 \text{ m/s}}{0.720 \text{ m}} = 476 \text{ Hz}.$$

16.4 Interference of Light Waves

Let's consider an experiment in which a light beam from a laser is aimed at an opaque screen containing two long, narrow slits spaced very close together. We will assume that the light beam illuminates both slits equally. Any light that passes through the slits impinges on a viewing screen. Figure 16-13a shows the view of the experiment from above, looking down on the top ends of the slits and on the top edge of the viewing screen. Figure 16-13b shows the two slits as they are seen from the laser. The pair of slits is called a **double slit**. In a typical experiment the slits are ≈0.1 mm wide and spaced ≈0.5 mm apart—barely discernible by eye.

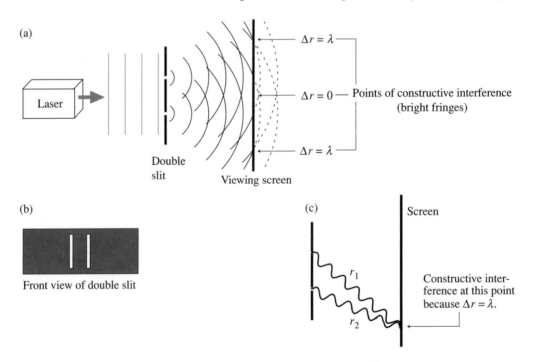

FIGURE 16-13 A double-slit interference experiment. a) Top view of the apparatus, showing wave fronts and points of constructive interference on the screen. b) The two slits as seen by the approaching light wave. c) Distance r_1 is one wavelength more than distance r_2, causing constructive interference at this point on the viewing screen.

A light wave impinging on these slits will squeeze through and then spread out on the other side—like water waves squeezing through a small opening in a jetty and then spreading out on the other side. (This "spreading out" is called *diffraction*, and is the subject of the next chapter.) The two slits act as wave sources, similar to two loudspeakers or two antennas. The waves from these two sources overlap and thus interfere, exactly as shown in Fig. 16-10.

Because the light is coming through slits, rather than circular holes, the spreading waves are cylinders rather than spheres. The wave fronts seen in Fig. 16-13a are thus the tops of cylinders extending above and below the page. When two cylindrical wave fronts intersect at the viewing screen they cause constructive interference along a *line* extending above and below the page. This is a line along which the light intensity is maximum, and you would see it as a bright band of light on the screen. The waves will cancel where a crest and a trough intersect at the screen, creating a dark band with no light at all. The interference pattern, as seen on the screen, should be a series of parallel bright and dark bands.

Figure 16-14 is a photograph of the viewing screen in a double-slit interference experiment. The bands of light and dark are called **interference fringes**. The bright fringes are points of constructive interference where the waves from the two slits arrive in phase. They occur at points where the lines of constructive interference shown in Fig. 16-11 intersect the viewing screen. The dark fringes are points of destructive interference where the light intensity is zero.

FIGURE 16-14 Photograph of the interference fringes in a double-slit interference experiment. Bright and dark fringes are areas of constructive and destructive interference, respectively.

We have already done the analysis needed to understand the double-slit experiment, starting with Eq. 16-16 and leading up to Eq. 16-21. The two sources of this experiment (the two slits) are identical and in phase because both are illuminated by the *same* wave front from the laser. Thus $\Delta\phi_0 = 0$. According to Eq. 16-21, constructive interference will occur at points having $\Delta r = m\lambda$, where $\Delta r = r_2 - r_1$ is the *difference* in the distances from the slits to the screen. The midpoint on the screen is equally distant from both slits, so at this point $\Delta r = 0$ and a bright fringe is produced. This bright fringe is called the **central maximum**. Figure 16-13c shows a point on the screen where $r_1 = 6.25\lambda$ and $r_2 = 5.25\lambda$. This is also the location of a bright fringe because $\Delta r = \lambda$ even though the distances r_1 and r_2 themselves are not integer multiples of the wavelength.

Figure 16-15 shows the geometry of the double-slit experiment. We will call the spacing between the two slits d and the distance between the slits and the viewing screen L. Waves leaving the slits travel distances r_1 and r_2 to a point C on the screen where they meet and recombine. Point C is at an angle θ from the midpoint between the slits. It will be more convenient to locate point C by measuring its distance y from the midpoint on the

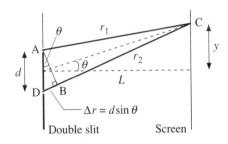

FIGURE 16-15 Geometry of the double-slit experiment.

viewing screen. Angle θ and distance y are related by

$$y = L \tan \theta. \tag{16-22}$$

The line AB in Fig. 16-15, drawn from the upper slit to the r_2 line, is a circular arc with its center at point C on the screen where the two waves recombine. Thus AB divides the r_2 line into a segment BC of length r_1 and a much smaller segment BD of length $\Delta r = r_2 - r_1$. Segment BD is the *difference* in the travel distances, which determines if point C is a point of constructive interference.

In practice, the angle θ in a double-slit experiment is very small ($<1°$). For small angles, the arc AB is indistinguishable from a straight line forming the side of right triangle ABD. You can then see from the geometry that

$$\Delta r = BD = d \sin \theta. \tag{16-23}$$

Bright fringes (constructive interference) occur at angles θ_m such that

$$\Delta r = d \sin \theta_m = m\lambda \qquad m = 0, 1, 2, 3, \dots \tag{16-24}$$

We have added the subscript m to make clear that θ_m is the angle of the mth bright fringe, starting with $m = 0$ at the center.

Equation 16-22 allows us to predict the *position* of the mth bright fringe, as measured from the center of the pattern. Because θ_m is a small angle, we can use the small angle approximation

$$\tan \theta_m \approx \sin \theta_m = \frac{m\lambda}{d}. \tag{16-25}$$

Using Eq. 16-25 in Eq. 16-22, the mth bright fringe occurs at position

$$y_m = \frac{m\lambda L}{d} \qquad m = 0, 1, 2, 3, \dots \qquad \text{(bright fringes)}. \tag{16-26}$$

Keep in mind that the interference pattern is symmetrical about the center, so there is an mth bright fringe at the same distance on either side of the center. (Note that Eq. 16-26 does *not* apply to the interference of sound waves from two loudspeakers because the small angle approximation is not valid in that situation.)

The interference pattern described by Eq. 16-26 is a series of *equally-spaced* bright lines on the screen, exactly as shown in Fig. 16-14. How do we know they are equally spaced? The spacing between the m fringe and the $m + 1$ fringe is

$$\Delta y = y_{m+1} - y_m = \frac{(m+1)\lambda L}{d} - \frac{m\lambda L}{d} = \frac{\lambda L}{d}. \tag{16-27}$$

Because Δy is independent of m, *any* two bright fringes will have the same spacing.

The dark fringes in the photograph are bands of destructive interference. Destructive interference occurs where $\Delta r = (m + \frac{1}{2})\lambda$. Using Eq. 16-23 for Δr and the small angle approximation, the dark fringes are located at positions

$$y'_m = (m + \tfrac{1}{2})\frac{\lambda L}{d} \qquad m = 0, 1, 2, 3, \dots \qquad \text{(dark fringes)}. \tag{16-28}$$

We have used y'_m, with a prime, to distinguish the location of the mth minimum from the mth maximum at y_m. You can see from Eq. 16-28 that the dark fringes are located exactly halfway between the bright fringes.

Equations 16-26 and 16-28 locate the positions of *maximum* and *minimum* intensity. Other points on the screen, at in-between values of y, have an intensity somewhere between maximum and minimum. You can see in the photograph of Fig. 16-14 that the intensity varies smoothly from minimum to maximum. Figure 16-16 shows the results of a more detailed calculation for the light intensity at each point on the screen. Note that this graph is rotated 90° counterclockwise from "normal," so that the position is graphed vertically, to match the geometry of the experiment, while the light intensity increases toward the left.

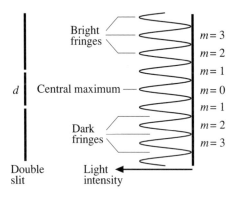

FIGURE 16-16 Graph of the intensity of the interference fringes in a double-slit experiment.

EXAMPLE 16-4   Light from a helium-neon laser ($\lambda = 633$ nm) illuminates two slits spaced 0.4 mm apart. A viewing screen is placed 2 m behind the slits. a) What is the position of the $m = 2$ bright fringe? b) What is the position of the $m = 2$ dark fringe?

SOLUTION a) Equation 16-26 gives the positions of the bright fringes. For $m = 2$:

$$y_m = \frac{m \lambda L}{d} = \frac{2(633 \times 10^{-9} \text{ m})(2 \text{ m})}{4 \times 10^{-4} \text{ m}} = 6.33 \times 10^{-3} \text{ m} = 6.33 \text{ mm}.$$

The $m = 2$ bright fringe is 6.33 mm from the central maximum.

b) The positions of the dark fringes are given by Eq. 16-28. For $m = 2$:

$$y'_m = (m + \tfrac{1}{2}) \frac{\lambda L}{d} = 7.91 \text{ mm}.$$

Notice that as the fringes are counted outward from the center, the $m = 2$ bright fringe occurs *before* the $m = 2$ dark fringe.

EXAMPLE 16-5   A two-slit interference pattern is produced with two slits 0.30 mm apart and is observed on a screen 1.0 m away. Ten bright fringes are easily seen, spanning a distance of 1.65 cm. What is the wavelength of the light?

SOLUTION When you observe interference fringes, it is not always obvious which fringe is the $m = 0$ central maximum. Slight imperfections in the slits can make the interference fringe pattern less than ideal. However, you do not need to identify the $m = 0$ fringe because you can make use of the fact that the spacing Δy between fringes is uniform. If you observed ten bright fringes, there are *nine* spaces between them (not ten—be careful!) that span a total distance of 1.65 cm. Thus the spacing between adjacent fringes is

$$\Delta y = \frac{1.65 \text{ cm}}{9} = 1.833 \times 10^{-3} \text{ m}.$$

Using the fringe spacing in Eq. 16-27 gives

$$\lambda = \frac{d}{L} \cdot \Delta y = 5.50 \times 10^{-7} \text{ m} = 550 \text{ nm}.$$

●

The Nature of Light

The interference patterns of light have helped resolve an age-old debate about the nature of light. The first Greek scientists and philosophers did not make a distinction between light and vision. Light, to them, was not something that existed apart from seeing. But gradually there arose a view that light actually "exists," that light is some sort of physical entity that is present regardless of whether or not someone is looking.

If light is a physical entity, what is it? What are its characteristics? Is it a wave, like sound, or a collection of small particles that blows by like the wind? Newton was one of the early investigators of the nature of light—he dispersed sunlight into colors with a prism, recombined them to show that all colors combine to produce white, and looked carefully at the properties of lenses and mirrors. (It is worth keeping in mind that artificial light sources were not invented until the late nineteenth century. Newton and his contemporaries had only the sun and candles with which to experiment.)

Newton knew that water waves, after passing through an opening, did not make sharp "shadows." Instead, the water waves immediately spread out to fill the space behind the opening. (This is the problem of diffraction, which we will investigate in the next chapter.) But a beam of sunlight passing through an opening, such as a window or door, makes a very sharp-edged shadow if it then falls upon a viewing screen. This is exactly what you would expect from particles traveling in straight lines—some pass through the opening and making a bright area on the screen while others do not get through the opening and cause a shadow. Newton concluded that light consisted of some type of very small, light, fast particles that he called "corpuscles."

His conclusion was vigorously opposed by Robert Hooke (of Hooke's law) and the Dutch scientist Christian Huygens, who argued that light was some sort of wave. Although the debate was lively, and sometimes acrimonious, the fame and stature of Newton eventually allowed him to prevail. The belief that light consisted of corpuscles was not seriously questioned for the next one hundred years.

The first evidence that light is a wave was provided in 1801 when the English scientist Thomas Young announced that he had produced interference between two waves of light. Young's experiment was a version of the double-slit experiment we have just analyzed. This was a painstakingly difficult experiment with the technology of his era. Nonetheless, Young quickly settled the debate in favor of a wave theory of light because interference is a distinctly wave-like phenomenon.

Young's discovery of the interference of light overturned the scientific thinking of his day and set the stage for the nineteenth century to be a period devoted to new, and more sophisticated, investigations as to the nature of light. Young's measurements of the wavelengths of light—vastly smaller than anyone had previously considered—helped to explain how light waves, under ordinary conditions, cast sharp-edged shadows. We will see this, and understand Newton's error, in the next chapter.

But if light is a wave, what is it that is waving? This was the question posed to the nineteenth century, and the history of responses is a fascinating story. We have neither the time nor the background yet to pursue this topic, so we will fast-forward toward the end of the century. It was ultimately established, through brilliant theoretical and experimental efforts by numerous scientists, that light is an *electromagnetic wave*—an oscillation of the electromagnetic field that requires no material medium in which to travel. Further, as we have already seen, light is just one small slice out of a vastly broader *electromagnetic spectrum* that includes phenomena ranging from radio waves to x rays.

It is so easy today to demonstrate the wave-nature of light with a laser that we often forget the extreme technical difficulties faced by Newton, Young, and other early investigators of the nature of light. Indeed, it would seem that devices such as bar-code scanners and technologies such as holography, both of which make direct use of the interference of light, would have established the wave-nature of light beyond any question. Yet, as always seems to happen in science, the unexpected is lurking nearby. Investigations of the twentieth century have shown that light is a wave—*and* a particle. Our story is not over, and we will return to take a closer look at the nature of light in Chapter 18.

16.5 Interferometers

Scientists and engineers have devised many ingenious methods for using interference to make very precise measurements. For example, constructive interference can be changed to destructive interference—an easy change to detect and record—if the distance traveled by a wave is changed by a mere half wavelength. If the distance is measured precisely, perhaps with a pair of calipers, a precise wavelength can be determined. A device that uses interference to make very accurate measurements of wavelengths or distances is called an **interferometer**. In this section, we will look at acoustical and optical examples of interferometers.

Interference requires the combination of waves of *exactly* the same wavelength. This can be difficult to achieve with two independent sources. One way of guaranteeing that two waves have exactly equal wavelengths is to divide a *single* wave into two parts and then, at some later time and a different point in space, to recombine the parts. The recombination process produces an interference pattern that is exactly the same as if the waves had come from different sources. The interference will be constructive or destructive depending upon whether the waves are in phase or out of phase as they are recombined. Interferometers are based on the division and recombination of a single wave.

Figure 16-17 shows an acoustical interferometer that uses wave division and recombination. Sound waves from the loudspeaker are sent into the left end of the tube. The wave splits into two parts at the junction and waves of smaller amplitude travel down each side. After traveling distances r_1 and r_2, the waves are recombined and a superposition of the two travels out of the right end to the microphone. Distance L_2 can be changed by sliding the upper tube in and out

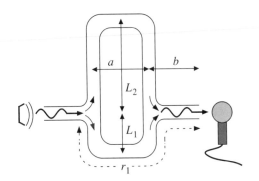

FIGURE 16-17 An acoustical interferometer. Sound waves enter the tube from left, follow two different paths, then are recombined at the right after traveling path lengths r_1 and r_2. Lengths L_1 and L_2 can be varied.

like a trombone. The phase relationship between the two waves as they are recombined will determine if the sound emerging from the right end is a maximum, a minimum, or somewhere in between.

Let's analyze the acoustical interferometer of Fig. 16-17. The two waves that take the two paths through the interferometer started from the *same* source—the loudspeaker. Thus the phase difference $\Delta\phi_0$ between the wave sources is automatically zero. The recombined waves have a phase difference $\Delta\phi$ only because they travel different distances r_1 and r_2. The analysis that led to Eq. 16-21 for identical sources is valid here, so the condition for constructive and destructive interference is again

$$\text{Constructive:}\quad \Delta r = m\lambda$$
$$\text{Destructive:}\quad \Delta r = (m+\tfrac{1}{2})\lambda \qquad m = 0, 1, 2, \ldots$$

The difference Δr depends on the geometry of the interferometer. You can see from Fig. 16-17 that

$$r_1 = L_1 + a + L_1 + b = 2L_1 + a + b$$
$$r_2 = L_2 + a + L_2 + b = 2L_2 + a + b,$$

from which we find

$$\Delta r = r_2 - r_1 = 2(L_2 - L_1). \qquad (16\text{-}29)$$

With this value for Δr the conditions for constructive and destructive interference are:

$$\text{Constructive:}\quad L_2 - L_1 = m\frac{\lambda}{2}$$
$$\text{Destructive:}\quad L_2 - L_1 = (m+\tfrac{1}{2})\frac{\lambda}{2} \qquad m = 0, 1, 2, \ldots \qquad (16\text{-}30)$$

These conditions involve $\lambda/2$ rather than just λ because the waves travel distance L_1 and L_2 *twice*—once up and once down. If $L_2 - L_1 = \lambda/2$, then the upper wave travels a full wavelength λ more than the lower wave before they recombine.

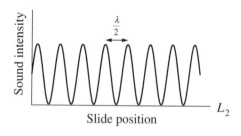

FIGURE 16-18 Interference maxima and minima alternate as the top slide on an acoustical interferometer is pulled out.

The interferometer is used by recording the alternating maxima and minima in the sound as the top is pulled out and L_2 changes. The interference changes from a maximum to a minimum and back to a maximum every time L_2 increases by half a wavelength. Figure 16-18 shows a graph of the sound intensity at the microphone as distance L_2 is increased. It is easy to count the number of maxima that occur as the slide is pulled out. N new maxima will appear if the slide is moved distance $N\lambda/2$.

EXAMPLE 16-6   A loudspeaker broadcasts a 5440. Hz sound wave into an acoustical interferometer that is at a temperature of 20°C. The bottom tube is adjusted to make the recorded sound intensity a maximum. How far must the top slide be pulled out to cause 24 new maxima to appear?

SOLUTION A new maximum appears each time L_2 increases by $\lambda/2$, which causes the path difference Δr to increase by λ. The wavelength is

$$\lambda = \frac{v}{f} = \frac{343 \text{ m/s}}{5440 \text{ Hz}} = 0.06305 \text{ m} = 6.305 \text{ cm}.$$

Notice that we've calculated the wavelength to *four* significant figures because the frequency was given to four significant figures. A total of 24 new maxima will appear if L_2 is increased by

$$\Delta L_2 = 24\frac{\lambda}{2} = 75.66 \text{ cm}.$$

Distances can be measured very accurately simply by counting maxima if the wavelength is accurately known. That is one important use of interferometers.

The Michelson Interferometer

Young's experiment demonstrated that light is a wave and allowed an initial determination of the wavelengths. But despite its historical importance, double-slit interference is not a practical means of making measurements. Following Young's discovery, scientists throughout the nineteenth century developed ever better techniques for turning the interference of light into a useful measuring tool.

The first optical interferometer was invented by Albert Michelson. Michelson was one of the first American scientists of international stature, and he was the first to receive a Nobel Prize—in 1907 for his development of precision optical techniques and interferometry. The interferometer he invented has been widely used for over a century to make extremely precise measurements. One of its first uses was in the famous Michelson-Morely experiment, which provided conclusive evidence that light does not need a medium in which to travel. This experiment prepared the way for Einstein's theory of relativity just a few years later, although historians still debate the extent to which Einstein knew of Michelson's work. A contemporary version of Michelson's interferometer several kilometers in length is being built in old mine shafts in Colorado to detect "gravity waves" arriving at the earth from distant supernovae. These are just a couple of examples of how interferometry has been, and continues to be, used to explore the universe.

Figure 16-19 shows a Michelson interferometer. A light wave enters from the left and strikes a *beamsplitter*, a partially-silvered mirror that reflects half the light but transmits the other half. This is the point of wave division. Separate waves travel at right angles to each other toward mirrors M_1 and M_2, where they reflect. As the waves reach the beamsplitter on their return trip, half of the wave returning from M_1 is transmitted through the beamsplitter, where it recombines with the reflected half of the wave returning from M_2. The combined wave travels on to a light detector— originally a human observer but now more likely an electronic photodetector.

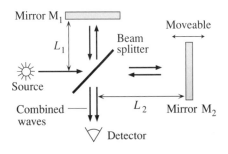

FIGURE 16-19 A Michelson interferometer.

Mirrors M_1 and M_2 are distances L_1 and L_2 from the beamsplitter. The waves travel distances $r_1 = 2L_1$ and $r_2 = 2L_2$, with the factors of two appearing because the waves travel both to the mirrors and back again. The path difference between the two waves is thus $\Delta r = 2L_2 - 2L_1$. This is all we need to determine if the light wave seen by the detector exhibits constructive interference—seen as bright light—or destructive interference—seen as no light. The condition for constructive interference is $\Delta r = m\lambda$, which in this geometry is

$$\text{Constructive:} \quad L_2 - L_1 = m\frac{\lambda}{2} \quad m = 0, 1, 2, \ldots. \tag{16-31}$$

You should recognize this result as being identical to Eq. 16-30 for an acoustical interferometer. Although the physical construction is different, the *physics* of the two interferometers is the same. In both a wave is divided, sent along two paths that differ in length by $\Delta r = 2(L_2 - L_1)$, and then is recombined.

Our analysis has been for light waves that strike the beamsplitter at exactly 45° and then impinge on the mirrors exactly perpendicular to the surface. Some of the light waves in an actual experiment travel at slightly different angles and, as a result, have slightly different path lengths. The observer looking into the interferometer sees a pattern of circular interference fringes, as shown in Fig. 16-20. Equation 16-31 is the condition for constructive interference at the *center* of the circle (a bright central spot). The surrounding circular fringes are from the light waves that traveled at slightly different angles before recombining. Their analysis is more complicated and will be left to more advanced courses in optics.

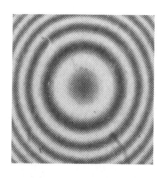

FIGURE 16-20
Photograph of the interference fringes seen in a Michelson interferometer.

The Michelson interferometer can be, and is, used to make very precise measurements of the wavelength of light. To do so, mirror M_2 is placed on a stand that can be slowly moved by turning a screw. As M_2 moves, and Δr changes, the central spot in the fringe pattern alternates between bright and dark. This is exactly analogous to the alternating loud and soft sounds shown in Fig. 16-18 as the slide on the acoustical interferometer was withdrawn. Suppose the interferometer is adjusted initially to produce a bright central spot. The next bright spot will appear when M_2 is moved by half a wavelength. The appearance of N new maxima implies that distance L_2 changes by

$$\Delta L_2 = \frac{N\lambda}{2}$$

$$\Rightarrow \lambda = \frac{2\Delta L_2}{N}. \tag{16-32}$$

Very precise wavelength measurements can be made by moving the mirror a distance ΔL_2 while counting the number of new bright spots that appear at the center of the pattern. The number N is counted and known exactly. The only limitation on how precisely λ can be measured this way is the precision with which distance ΔL_2 can be measured. Unlike λ, which is microscopic, ΔL_2 is typically a few millimeters—a macroscopic distance that can be measured extremely precisely using precision screws, micrometers, precision spacing blocks, and other techniques. Michelson's invention provided a way to transfer the precision of macroscopic distance measurements to an equal precision of wavelength measurements of light.

EXAMPLE 16-7 An experimenter uses a Michelson interferometer to measure one of the wavelengths of light emitted by a neon atom. She slowly moves mirror M_2 until 10,000 new bright central spots have appeared. (In a modern experiment, this would be done with a photodetector and a computer, eliminating the possibility of "experimenter error" during counting.) She then measures that the mirror has moved a distance of 3.1641 mm. What is the wavelength of the light?

SOLUTION This is not intended to be a difficult problem; instead, it is a straightforward application of Eq. 16-32. Using $\Delta L_2 = 3.1641$ mm $= 3.1641 \times 10^{-3}$ m, we can compute λ to be

$$\lambda = \frac{2\Delta L_2}{N} = 6.3282 \times 10^{-7} \text{ m} = 632.82 \text{ nm}.$$

Notice that a measurement of ΔL_2 accurate to five significant figures allowed us to determine λ to five significant figures. This happens to be the neon wavelength that is emitted as the laser beam in a helium-neon laser.

The technology of wavelength measurements using a Michelson interferometer became so good that in 1960 an international scientific committee decided to use it as the *definition* of the meter. They defined the meter to be exactly 1,650,763.73 wavelengths of a particular orange color of light emitted by the krypton isotope ^{86}Kr. Scientists would aim this light at a Michelson interferometer, then move mirror M_2 while counting out a predetermined number of new bright spots. The total distance traveled by M_2 was then, *by definition*, exactly 1 meter (or, in actual practice, some fraction of 1 meter). This was a vast improvement over the previous definition of the meter as the distance between two scratches on a special bar kept in a safe in Paris, and it stood until 1983 when the meter was again refined in terms of the speed of light.

Modern versions of interferometers are used to study and measure small deflections and vibrations of objects. Figure 16-21 shows an *interferogram* of a vibrating violin body. The bright and dark fringes can be analyzed to provide information about the displacement at different points of the violin. These applications of interferometry are widely used in engineering to study the vibrations of machinery and the deflections of structures under the influence of applied forces.

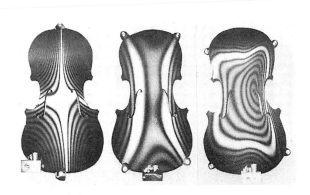

FIGURE 16-21 An interferogram of a violin as it vibrates.

Interference also forms the basis of *holography*, which is used as an engineering tool as well as a medium for artistic expression. Both interferograms and holograms are created by dividing a light wave at one point, letting one wave interact with the object, then recombining the waves on a piece of film where the interference pattern is recorded. The physical principles are just the constructive and destructive interference of light that you have seen at work in the double-slit experiment and the Michelson interferometer.

Summary

▲ Important Concepts and Terms

interference	phase difference
in phase	double slit
constructive interference	interference fringes
out of phase	central maximum
destructive interference	interferometer

As you review this chapter you should see that all the examples, no matter how different they appear, involve *exactly the same physics*. Interference is an *interaction* between two waves of the same wavelength when they overlap in the same region in space. The net displacement at each point, according to the principle of superposition, is the sum of the individual displacements. The starting point for our analysis of interference, Eqs. 16-9 and 16-16, was simply the addition of the displacements of the two waves.

It is the phase difference $\Delta\phi$ between the two waves that determines what happens when two waves interfere. Constructive interference, with maximum amplitude, occurs at points where the two waves are in phase. Destructive interference, with zero amplitude, occurs at points where the waves are out of phase. For waves in two and three dimensions, both constructive *and* destructive interference occur at the same time but at different points in space. This is the reason that Young's double-slit experiment gives both bright and dark interference fringes.

Whether two waves are in phase, out of phase, or somewhere in between depends on two factors. First, the two waves may have *started out* with a phase difference between them due to differences between the two sources. This initial phase difference is described by the term $\Delta\phi_0$. Second, the two waves may *travel different distances* to the point at which they are combined. The path length difference Δr contributes the term $2\pi(\Delta r/\lambda)$. Taken together, these provide the conditions for constructive and destructive interference:

Constructive interference: $\Delta\phi = 2\pi\dfrac{\Delta r}{\lambda} + \Delta\phi_0 = m \cdot 2\pi$

$\qquad\qquad\qquad\qquad\qquad\qquad\qquad\qquad\qquad\qquad m = 0, 1, 2, \ldots$

Destructive interference: $\Delta\phi = 2\pi\dfrac{\Delta r}{\lambda} + \Delta\phi_0 = (m + \frac{1}{2})2\pi$

Finding phase differences thus becomes a *geometrical* problem of finding the difference Δr in the two path lengths. The phase difference $\Delta\phi$ increases by 2π each time Δr increases by λ. As you review the examples in this chapter, you will see that they focus on the geometry of the physical arrangement to find Δr.

Young's double-slit experiment demonstrates the interference between light waves spreading out from two closely-spaced slits. A series of equally-spaced bright and dark interference fringes are seen on a viewing screen behind the slits. Bright fringes are found at positions

$$y_m = \frac{m \lambda L}{d} \qquad m = 0, 1, 2, 3, \ldots$$

where y is measured from the central maximum at the midpoint of the screen.

Interferometers make accurate measurements of wavelengths and distances by dividing a wave into two pieces, and then recombining the pieces after they have traveled along different paths. The condition for constructive equilibrium in both the acoustical interferometer and the Michelson interferometer is

$$\text{Constructive}: \quad L_2 - L_1 = m \frac{\lambda}{2} \qquad m = 0, 1, 2, \ldots$$

where L_1 and L_2 are the lengths of the two sides of the interferometer.

Exercises and Problems

Exercises

1. A double-slit interference pattern is set up using two slits spaced 0.20 mm apart. The distance between the first minimum and the fifth minimum, as seen on a screen 60 cm from the slits, is 6.0 mm. What is the wavelength of the light used in this experiment?

2. Light from a helium-neon laser ($\lambda = 633$ nm) is used to illuminate two slits. The interference pattern is observed on a screen 3 m behind the slits. Twelve bright fringes are seen, spanning a distance of 52 mm. What is the spacing between the slits?

3. A double-slit experiment is performed with light having a wavelength of 600 nm. The bright interference fringes are spaced 1.8 mm apart on the viewing screen. What will the fringe spacing be if the light is changed to a wavelength of 400 nm?

4. Two loudspeakers broadcast identical 1000 Hz sound waves. What distance should one speaker be placed behind the other for the sound to have an amplitude 1.5 times that of each speaker alone?

Problems

5. You are standing 2.5 m directly in front of one of two loudspeakers, as shown in Fig. 16-22. They are 3 m apart and both are playing exactly the same 686 Hz tone.
   a. As you begin to walk directly away from the speaker, at what positions (what distances from the speaker) do you hear a *minimum* sound intensity?
   b. Draw a wave front diagram, similar to Fig. 16-10, with a scale of either 1 cm = 1 m or 0.5 in = 1 m. Indicate on this diagram the *closest* position you found in a).
   c. Explain how you can see *from the diagram* that the position you indicated in b) is a point of minimum sound intensity.

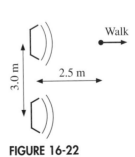

FIGURE 16-22

6. Two radio antennas separated by 2.0 m broadcast identical 750 MHz waves. If you walk around the antennas in a circle of radius 10 m, how many maxima will you detect?

7. Light of wavelength 600 nm passes though a double slit having a separation of 0.20 mm and is then observed on a screen 1.0 m away. The location of the central maximum is marked on the screen and labeled $y = 0$.
    a. At what distance, on either side of $y = 0$, are the $m = 1$ bright fringes?
    b. A very thin piece of glass is then placed in one slit. Because light travels slower in glass than in air, the wave passing through the glass is "delayed" by 1.25×10^{-14} s in comparison to the wave going through the other slit. What fraction of the oscillatory period of the light wave is this delay?
    c. With the glass in place, what is the phase difference $\Delta\phi_0$ between the two waves as they leave the slits?
    d. The glass causes the interference fringe pattern on the screen to shift sideways. Which way does the central maximum move (toward or away from the slit with the glass) and by how much?

8. Modern broadcast antennas are often designed to use multiple transmitters. The waves radiated by these transmitters interfere, just like the waves in Young's experiment. The locations of the maxima and minima can be controlled by adjusting the phases between the transmitters. Thus the broadcast energy can be "steered" to selected locations and removed from others.

    The basic idea can be illustrated with a broadcast technique used by AM radio stations. If the antenna is located at the edge of town, the station would like to beam all of the radiated energy into town and none into the countryside. A single antenna would radiate energy equally in all directions. Consider two antenna separated by a distance L, as shown in Fig. 16-23. Both antennas broadcast a sinusoidal radio signal at wavelength λ. Antenna 2 can delay its broadcast relative to antenna 1 by a time interval Δt in order to create a phase difference $\Delta\phi_0$ between the sources. As the designer of the transmitter, your task is to find values of L and Δt such that the waves interfere constructively on the town side and destructively on the country side. Let antenna 1 be at $x = 0$ and antenna 2 be at $x = L$. Choose $t = 0$ to make $\phi_{10} = 0$. To the right of both antennas, which is the town side, the two traveling waves are given by Eq. 16-1 with $\phi_{10} = 0$ and $x_1 = x_2 + L$.
    a. Write the equations for the two waves on the *left* side (country side) of both antennas.
    b. What is the relationship between x_1 and x_2 on the left side?

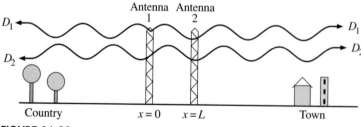

FIGURE 16-23

c. What is the smallest value of L, in terms of λ, for which you can create perfect constructive interference on the town side and perfect destructive interference on the country side? (Hint: Use the phase difference $\Delta\phi = \phi_2 - \phi_1$ for the waves on the right and on the left. This will give you two simultaneous equations with unknowns L and ϕ_{20}.)

d. What phase constant ϕ_{20} of antenna 2 is needed?

e. The phase constant can be generated by delaying the broadcast from antenna 2 for a time interval Δt. What fraction of the oscillation period T must Δt be?

f. Evaluate both L and Δt for the realistic AM radio frequency of 1000 KHz.

9. You may have noticed that lenses on cameras and binoculars have a slight purple color. This comes from an "antireflection coating" on the glass. Ordinary glass surfaces only transmit about 96% of incident light and reflect about 4%. These reflections can seriously degrade the performance of optical instruments. However, reflections can be eliminated if the glass is coated with a very thin layer of transparent material, causing there to be *two* reflections. The first reflection is from the front surface of the coating, and the second reflection is from the glass itself at the coating/glass interface. If the coating has the correct thickness, the two reflections will be out of phase and will interfere destructively, thereby "canceling" the reflection. (In practice, the interference is perfectly destructive only for wavelengths near the center of the visible spectrum. The purplish color of the coating is due to incomplete cancellation of reflections for wavelengths at the edges of the visible spectrum.)

A piece of glass is coated with a transparent material having an index of refraction 1.33. Consider the surfaces to be straight and parallel. Figure 16-24 shows one reflected wave "above" the other so that the situation can be clearly seen, but in reality the two reflected waves overlap each other. What is the minimum thickness of the coating that will give perfect destructive interference for a wavelength of 550 nm?

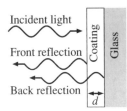

FIGURE 16-24

10. Consider light of wavelength 500 nm in vacuum.

a. What are the frequency and wavelength of this light as it travels through water, which has $n = 1.33$?

b. Suppose that a 1 mm thick layer of water could somehow be inserted into one arm of a Michelson interferometer, as shown in Fig. 16-25. How many "extra" wavelengths does the light now have to travel?

c. By how many fringes will this water layer shift the interference pattern?

d. The water layer is replaced by a 0.10 mm thick piece of glass. This causes the fringe pattern to shift by 200 fringes. What is the index of refraction of this piece of glass?

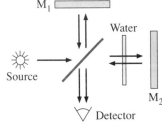

FIGURE 16-25

[**Estimated 12 additional problems for the final edition.**]

Chapter **17**

Diffraction

L O O K I N G B A C K | Sections 14.4; 14.6; 16.3–16.4

17.1 Light in the Shadows

One of the most important characteristics of waves, which we discussed briefly in the last chapter, is that they spread out as they travel. By contrast, a beam of particles will travel in a perfectly straight line if not acted upon by external forces. In this chapter we will take a closer look at this characteristic of waves.

Figure 17-1 shows a water wave passing through a small opening in a barrier. We easily see the wave spreading out from the opening in circular arcs. This is similar to Fig. 16-13 in Chapter 16, where we showed the light waves in a double-slit interference experiment spreading out after passing through the narrow openings of the slits. A beam of particles, by contrast, would emerge from the barrier as a smaller beam, with the width of the opening, and would continue forward without spreading.

Sound waves offer another example of spreading waves. You can *hear* a source of sound from around a corner, even though you cannot *see* the source. The sound waves somehow spread out to fill in the area behind the corner—much like the water waves in the Fig. 17-1. The corner does not block the sound or create a sound shadow. A sound wave, like a water wave, does not have a well-defined edge.

But light waves can be blocked by obstacles, and when they are blocked they cast shadows. Sunlight passing through a window makes a bright, rectangular patch of light on the wall surrounded by the shadow of the window frame. If you look at the shadow, it appears to have a well-defined edge. It was observations such as this, where light

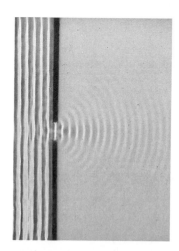

FIGURE 17-1 Water waves spreading out behind a small opening in a barrier.

seems to behave quite differently than sound or water waves, that led Newton to the conclusion that light must consist of a beam of particles, or corpuscles.

Suppose, however, that Newton could have seen the photograph shown in Fig. 17-2. This is an enlargement of the "edge" of the shadow cast by an exceptionally sharp-edged obstacle. We see now that the edge of the shadow is not at all sharp. Instead, it exhibits a transition region of alternating light and dark, with some of the light areas even extending slightly into the *geometric shadow*—that region that would be dark if light consisted of particles following straight lines. There is light in the shadows! Newton, very likely, would have reached a quite different conclusion than the one he did.

The inexorable spreading of waves that you see in Fig. 17-1 is the wave phenomenon called **diffraction**. Our goal in this chapter is to understand how and why diffraction occurs and what the implications are. Not only is this very basic wave physics, but we will find that diffraction can be used as a tool of tremendous practical importance for investigating atoms and the atomic properties of materials.

Diffraction is an interference effect that occurs when a wave *interacts* with a barrier of some sort. There is no clearly defined distinction between what is called "diffraction" and "interference." The usage of the two terms is mostly a historical tradition rather than a logical division. As a general rule of thumb, the interference of a small number of waves, such as in the last chapter, is called "interference," while the interference of a large number of sources along a wave front, which is what we will examine in this chapter, is called diffraction. The mathematics of diffraction can be quite complex, so we will take a somewhat more qualitative approach and leave the mathematics to more advanced courses. However, the basic idea will still be based on the principle of superposition and the comparison of phases.

FIGURE 17-2 Expected shadow cast by a sharp edge if light is a particle (top) compared with a photograph of the observed light pattern (bottom). The light area extends slightly into the geometric shadow.

17.2 The Diffraction Grating

In Chapter 16 we considered the interference that occurs when light passes through two closely-spaced slits—Young's experiment. We found that alternating bright and dark fringes are seen on a viewing screen and that the intensity changes slowly from bright to dark and back to bright. (It would be good to review Figs. 16-13 through 16-16.)

Suppose we were to replace the double slit with an opaque screen having a very large number N of closely-spaced slits. When illuminated from one side, each of these many slits becomes a source of light waves. The net displacement at a point on the viewing screen is the superposition of the N light waves arriving there. The light intensity pattern that we observe is the result of the interference of N discrete sources. Such a multi-slit device is called a **diffraction grating**.

Figure 17-3 shows a section from such a diffraction grating in which N slits are equally-spaced a distance d apart. This is a top view of the grating, and the slits extend above and below the page. Although only ten slits are shown, a practical grating will have many hundreds or even thousands of slits. We can illuminate the grating with a *plane wave* of light of wavelength λ that approaches from the left. A plane wave arrives *simultaneously* at each of the slits. This causes the wave emerging from each slit to be *in phase* with the wave emerging from every other slit. Each of these emerging waves spreads out, just like the water wave you saw in Fig. 17-1, and after a short distance they all overlap each other.

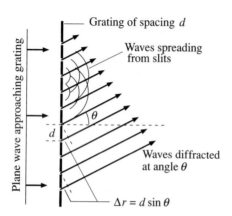

FIGURE 17-3 Top view of a diffraction grating having spacing d between the slits.

We are interested in how the pattern of light will appear on a screen located a large distance $L \gg d$ behind the grating. The light wave at the screen will be a superposition of all N waves, from N slits, as they spread and overlap. Because the distance to the screen is very large in comparison with the small section of the grating shown in the figure, the path followed by the light from one slit to a point on the screen is *very nearly* parallel to the path followed by the light from any other slit. (The paths cannot be perfectly parallel, of course, or they would never meet to interfere, but their slight deviation from perfect parallelism is too small to notice.) We can see, from the geometry, that each slit is going to be a distance $\Delta r = d\sin\theta$ farther from the screen than the slit above it, and a distance $\Delta r = d\sin\theta$ less than the slit below it. This is exactly the same reasoning we used to analyze the double-slit experiment in Fig. 16-13.

If the angle θ is such that $\Delta r = d\sin\theta = m\lambda$, where m is an integer, then the light wave arriving at the screen from each slit will be *exactly in phase* with the light waves arriving from the two slits next to it. But each of those is then in phase with the slits next to them, and so on until we reach the end of the grating. We can conclude that the N light waves, from N different slits, arriving at a point on the screen located at angle θ_m will *all* be in phase with each other if

$$d\sin\theta_m = m\lambda \quad \text{(constructive interference).} \tag{17-1}$$

If each wave has amplitude a, then adding N in-phase waves will cause constructive interference to occur at that point on the screen with a net amplitude of Na. The screen will show a bright constructive interference fringe at the values of θ_m given by Eq. 17-1. When this happens, we say that the light is "diffracted at angle θ." The distance y_m from the center to the mth maximum is

$$y_m = L\tan\theta_m. \tag{17-2}$$

Notice that the condition Eq. 17-1 for constructive interference in a grating of N slits is exactly the same as the condition for just two slits. Equation 17-1 is simply the requirement that the path difference from adjacent slits is $m\lambda$, regardless of the number of slits. But unlike the situation with double-slit interference, the angles of constructive interference from a diffraction grating are generally *not* small angles. The value of d used in a diffraction grating is so small that $m\lambda/d$ is not a small number, even for $m = 1$. Thus we cannot use the small angle approximation to simplify Eq. 17-2 as we did for double-slit interference.

While it was straightforward to locate the angles for constructive interference, by requiring all N slits to be in phase, it is much more difficult to find angles for which there is perfect destructive interference. With two slits, we only had to find a point where the two waves were exactly out of phase. But with N slits all contributing, finding the angles where all N terms add to zero is a challenge. There are complex mathematical techniques for adding N sinusoidal terms that each differ in phase from the proceeding term, allowing one to find the amplitude at any point on the screen, but we will save those techniques for advanced courses and be content with a qualitative description.

The results of a detailed calculation are shown graphically in Fig. 17-4a. You should compare this with the double-slit interference pattern of Fig. 16-16. There is still a *central maximum* corresponding to $m = 0$, just as with two slits, and bright fringes for $m = 1, 2, 3$, and so on are located symmetrically on either side. But rather than the slowly changing light intensity that we found for two slits, the interference pattern for a diffraction grating with $N \gg 2$ slits consists of a small number of *very* bright and *very* narrow fringes with most of the screen dark. You might imagine taking a two-slit interference pattern and "squeezing" the bright fringes so that all of the light is concentrated into bright, narrow lines and the rest of the screen goes dark. Figure 17-4b is a simulation of how the viewing screen behind a diffraction grating appears.

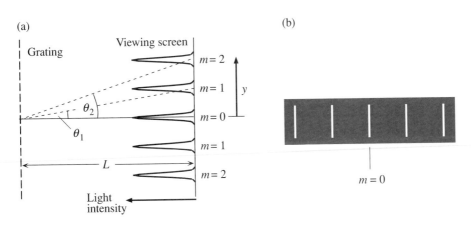

FIGURE 17-4 a) The light intensity pattern created by a diffraction grating that is illuminated with light of wavelength λ. b) A simulation of how the viewing screen appears.

The width of each bright fringe can be shown to be proportional to $1/N$. The more slits the grating has, the sharper and narrower the fringes become. This result makes diffraction gratings useful for the measurement of wavelengths. For example, suppose that the light impinging on the slits consists of *two* different wavelengths λ_1 and λ_2. The first of

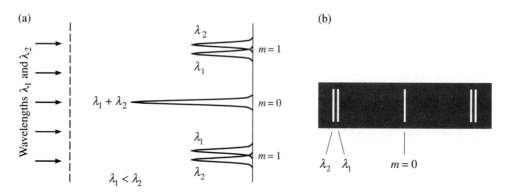

FIGURE 17-5 a) The intensity pattern seen when two slightly different wavelengths illuminate the grating. b) A simulation of the viewing screen.

these will cause a bright interference fringe to occur on the screen at an angle $\theta_1 = \sin^{-1}(\lambda_1/d)$ (assuming $m = 1$). The second will create a bright fringe at $\theta_2 = \sin^{-1}(\lambda_2/d)$. If the width of the fringe for λ_1 is sufficiently narrow, then the intensity maximum for λ_2 will be located at a "dark" area in the λ_1 interference pattern and both fringes will be easily visible—even if the difference between the two wavelengths is very small. Figure 17-5 illustrates this idea. Contrast this with a two-slit experiment, where the fringes are so broad and overlapped that it would not be possible to distinguish the fringes of one wavelength from those of another.

EXAMPLE 17-1 Light from a sodium lamp passes through a diffraction grating having 1000 slits per millimeter and then falls on a screen located 1 m away from the grating. Two bright yellow lines are seen at distances of 72.88 cm and 73.00 cm from the central maximum, with no yellow lines in between. What are the wavelengths of these two lines?

SOLUTION Figure 17-5a shows the geometry for this situation. The two fringes observed must both be $m = 1$ because there are no other bright yellow fringes closer to the central maximum. The diffraction angles θ_1 of these two wavelengths can be found from their positions y_1 as

$$\theta_1 = \tan^{-1}\left(\frac{y_1}{L}\right) = \begin{cases} 36.08° & \text{fringe at 72.88 cm} \\ 36.13° & \text{fringe at 73.00 cm.} \end{cases}$$

Keep in mind that the small angle approximation is not valid here. These angles must satisfy $d\sin\theta_1 = \lambda$. So the wavelengths are

$$\lambda = d\sin\theta_1.$$

What is d? If a 1 mm length of the grating has 1000 slits, then the spacing from one slit to the next must be 1/1000 mm, or $d = 1\times10^{-6}$ m. We can now determine the wavelengths creating the two bright fringes to be

$$\lambda = d\sin\theta_1 = \begin{cases} 589.0 \text{ nm} & \text{fringe at 72.88 cm} \\ 589.6 \text{ nm} & \text{fringe at 72.88 cm.} \end{cases}$$

Notice that we had data accurate to four significant figures, and that all four were necessary to distinguish the two wavelengths which differ by only 0.1%.

The science of measuring the wavelengths of atomic and molecular emissions is called **spectroscopy**, and we will examine it more closely in the next chapter. The diffraction grating is the primary tool of spectroscopy because it provides for the accurate measurement of wavelengths and allows the fringes of two closely-spaced wavelengths to be separated. The two sodium wavelengths of the example are called the *sodium doublet*, a name given to two closely-spaced wavelengths emitted by the atoms of one element. This doublet is an identifying characteristic of sodium. Because no other element emits these two wavelengths, the doublet can be used to identify the presence of sodium in an unknown sample, even if the sodium is only a very minor constituent. This procedure is called *spectral analysis*, and it is widely used in chemistry and materials engineering.

Although we have analyzed what is called a *transmission grating*, with many parallel transparent slits, most practical gratings are manufactured as *reflection gratings*. These are, in essence, mirrors with thousands of narrow, parallel scratches. When light hits the mirror, only the "good" strips of the mirror between the scratches reflect the light. The result, though, is exactly like that shown in Fig. 17-3 and the same condition locates the angles for constructive interference.

In recent years it has become possible to manufacture low-quality plastic reflection gratings at very low cost, and these are now widely sold as toys and novelty items. When illuminated with white light, from the sun or ordinary light bulbs, all the different wavelengths in the visible spectrum are diffracted at different angles and rainbows appear on the walls! It also happens that reflection gratings are found in nature. The *iridescence* of some bird feathers and insect shells occurs as a consequence of biological structures that have many very small, parallel ridges. These ridges act as a reflection grating and diffract the different wavelengths of sunlight at different angles. As the animal moves, the angle between the grating and your eye is constantly changing and, as a result, the wavelength of diffracted light reaching your eye changes. This creates the color-changing aspect of iridescence.

17.3 Single-Slit Diffraction

We opened this chapter with a photograph of a water wave passing through an opening in a barrier and then spreading out on the other side. It is time to look more carefully at just what happens to a wave passing through a narrow opening. This phenomena is called **single-slit diffraction.**

Figure 17-6 shows the experimental arrangement for observing the diffraction of light from a single slit of width a. A viewing screen is placed a distance $L \gg a$ behind the slit. We want to know what the light intensity pattern will be as a function of the angle θ away from the centerline. As before, the figure is a top view of the experiment and the opening in the screen is a long, narrow slit that extends above and below the page.

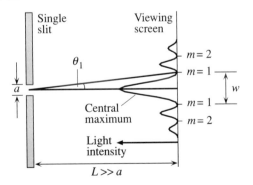

FIGURE 17-6 The diffraction from a single slit as light approaches from the left.

FIGURE 17-7 Photograph, greatly enlarged, of a single-slit diffraction pattern.

A photograph of the light pattern seen on the viewing screen is shown in Fig. 17-7. If light consisted of particles, the photo would be a uniformly bright rectangle in the center with sharp edges and well-defined shadows. That is clearly not what is observed. The observed pattern consists of a central maximum flanked by a series of weaker **secondary maxima** and dark fringes. Notice that the central maximum is significantly broader than the secondary maxima; it is also significantly brighter than the secondary maxima, although that is hard to tell here because this photograph has been over-exposed to make the secondary maxima show up better.

To analyze single-slit diffraction, imagine dividing the slit into a large number N of parallel "microslits," each of width a/N. Each of these microslits can be thought of as a source of waves—rather like the diffraction grating of Fig. 17-3 but with no boundaries between the sources. This is illustrated in Fig. 17-8 for $N = 6$. Such a division is clearly an artificial construct—the microslits are not real and there is no specific value for N other than "large"—but this approach can be justified in a more rigorous theory of waves. Because N has no specific value, we are safe in assuming it to be an even number. The diffraction of a slit results from the interference of these N continuous sources.

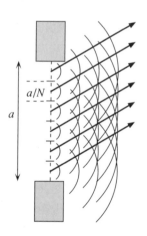

FIGURE 17-8 Dividing a single slit into N microslits, each a source of waves.

Because the viewing screen is very distant compared to the width of the slit, we are again justified in considering the paths from each microslit to the screen to be parallel to each other, all at angle θ. The situation where $\theta = 0°$ is shown in Fig. 17-9a. In this case, the light travels straight ahead and the path from each microslit to the screen is the same length as every other path. Therefore, the waves from *all* N microslits will arrive at the screen *in phase*. The line on the screen (extending above and below the plane of the figure) with $\theta = 0°$ experiences constructive interference from all N microslits and is thus bright. This line of constructive interference at $\theta = 0°$ is the central maximum of the diffraction pattern. (It is good to keep in mind, even though the figure does not show them explicitly, that we are still talking about sinusoidal waves moving along each of these paths.)

Now consider light traveling at an angle θ, as shown in Fig. 17-9b. Path 4, from the fourth microslit down from the top, is longer than path 1 from the first microslit by a distance Δr_{14}, so these two waves will arrive at the screen with *different* phases. If Δr_{14} happens to be $\lambda/2$, these two waves will arrive exactly out of phase and cause destructive interference. But if Δr_{14} is $\lambda/2$, then we can see from the geometry that the difference Δr_{25}

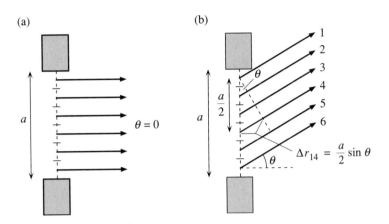

FIGURE 17-9 a) Waves diffracted at angle $\theta = 0$. b) Waves diffracted at angle $\theta > 0$, creating *phase differences* due to the different lengths traveled by each wave.

(between paths 2 and 5) as well as the difference Δr_{36} (between paths 3 and 6) are also $\lambda/2$. Therefore those pairs of waves will also cause destructive interference. In other words, $\Delta r_{14} = \lambda/2$ is the condition such that *every* microslit in the opening will be paired with another microslit, a distance $a/2$ away, that is exactly out of phase. When we sum all N displacements at the screen, they will—pair by pair—add to zero. This is the condition for complete destructive interference, and the viewing screen at this position will be dark.

From Fig. 17-9b it is not hard to see that $\Delta r_{14} = (a/2)\sin\theta$, so the condition for destructive interference is that the light travels at an angle θ_1 such that

$$\frac{a}{2}\sin\theta_1 = \frac{\lambda}{2},$$

or, equivalently, that $a\sin\theta_1 = \lambda$. The number N of microslits does not enter the final result—which is proper, because the microslits are not real but simply provide a way to think about what is going on. Regardless of the value of N, the condition $a\sin\theta_1 = \lambda$ guarantees that every microslit will be paired with another microslit a distance $a/2$ away that is exactly out of phase. This is the main idea of our analysis, and one worth thinking about carefully.

Once we have determined θ_1, we can extend the idea to find other angles of complete destructive interference. Suppose each microslit is paired with another microslit a distance $a/4$ away that is exactly out of phase. When all N displacements are summed they will again cancel in pairs to give zero. The angle θ_2 at which this occurs is found by replacing $a/2$ in our first analysis with $a/4$, leading to the condition $a\sin\theta_2 = 2\lambda$. This process can be continued, and we find that the condition for complete destructive interference is

$$a\sin\theta_m = m\lambda \qquad m = 1, 2, 3, \ldots \qquad \text{(\textit{m}th minimum of single slit).} \qquad (17\text{-}3)$$

Equation 17-3 gives the angles of the dark minima in the intensity pattern of Fig. 17-7.

A couple of things are worth noticing. First, this result looks *mathematically* the same as the condition for the *m*th *maximum* of the double-slit interference pattern. But the *physical* meaning is here quite different. Equation 17-3 locates the *minima* (dark fringes) of the single-slit diffraction pattern. Second, the value $m = 0$ is explicitly *excluded* in Eq. 17-3.

The $m = 0$ condition corresponds to the straight-ahead position at $\theta = 0$, but you saw in Fig. 17-9a that $\theta = 0$ is the central *maximum*, not a minimum.

It is possible, although difficult, to compute the entire light intensity pattern. The results of such a calculation were shown graphically in Fig. 17-6. There you can see the destructive interference minima as well as the bright central maximum. Compare this graph to the photograph of Fig. 17-7 and make sure you see the agreement between the two.

The Width of Single-Slit Diffraction Patterns

It is useful to talk about the *width* of the diffraction pattern of a single slit. We will define the width w of the pattern to be the distance between the two $m = 1$ minima on either side of the central maximum (see Fig. 17-6). The central maximum is much brighter than the secondary maxima, so its width is the primary visual pattern that you observe on the screen. If y_1 is the position of the $m = 1$ minimum, then the width is twice this or $w = 2y_1$.

From the geometry of the figure:

$$w = 2y_1 = 2L \tan \theta_1 .$$

The small angle approximation is very good for the single-slit diffraction of light, so $\tan \theta_1 \approx \sin \theta_1 = \lambda/a$. Thus the width of the single-slit diffraction pattern, as viewed on the screen, is

$$w = \frac{2\lambda L}{a} \quad \text{(width of single-slit diffraction pattern).} \qquad (17\text{-}4)$$

This has some interesting implications. First, the farther back you move the screen (increasing L), the wider the pattern of light on it. In other words, the light waves are *spreading out* behind the slit, and they fill a wider and wider region as they travel farther. Second, the narrower the slit width a, the *wider* the diffraction pattern. For waves, it seems, the smaller the opening you squeeze them through, the *more* they spread out on the other side. Suppose, for example, that $\lambda = 500$ nm and $L = 2$ m. A 1 mm slit produces a diffraction pattern with central maximum width of 2 mm—almost no spreading at all—but a 0.1 mm slit causes the central maximum to expand to a 20 mm width.

This suggests, correctly, that the diffraction of light waves is not easily observed unless the opening through which the light passes is <<1 mm in size. The underlying reason is the incredibly small wavelength of light. Because visible light has a wavelength near 500 nm, the factor 2λ in Eq. 17-4 is roughly 1000 nm = 10^{-3} mm = 10^{-6} m. If we estimate that the width w of a diffraction pattern needs to be at least 1 mm in size before we are likely to notice it, then Eq. 17-4 tells us that we need a ratio $L/a \geq 1000$. This implies either observing the pattern at very large distances, which is usually not practical, or using very small openings.

The situation is quite different for sound or water waves. Although such waves have a wide range of possible wavelengths, we can take 1 m to be a typical value of λ. This value is also a fairly typical size for an opening through which a sound or water wave might pass. This tells us that sound and water waves typically diffract with the ratio $\lambda/a \approx 1$, so the first minimum in their diffraction pattern will be located at $\theta = \sin^{-1}(\lambda/a) \approx 90°$. In other words, diffraction causes a sound or water wave to spread so as to fill *completely* the space behind the opening! (Note that the small angle approximation is *not* valid for the diffraction of sound or water waves.)

Figure 17-10 illustrates the difference between the diffraction of a short wavelength and a long wavelength if both are diffracted through the same opening. The dotted lines are at the angles θ_1, which locate the edges of the central maximum as it spreads out. This is the region containing most of the intensity, but keep in mind that these are not sharp edges. The secondary maxima beyond the $m = 1$ minimum spill over into the region outside the dotted lines. The shorter wavelength moves forward as a "beam" of light, spreading slowly, while the longer wavelength quickly spreads to fill the region behind the opening. A typical sound or water wave is like the long-wavelength pattern but a light wave has the short-wavelength pattern.

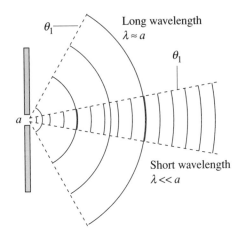

FIGURE 17-10 The difference between diffraction of a long wavelength and a short wavelength.

Now we can better appreciate Newton's dilemma. With everyday-sized openings, which are in the range of 1 m, familiar sound and water waves diffract at such large angles as to fill completely the space behind the opening. This is what we come to expect for the behavior of waves. However, Newton saw no evidence of this for light passing through openings. We see now that light really does spread out behind the opening, but the very small value of the wavelength greatly reduces the width of the diffraction pattern. Only if the opening is reduced to $a < 1$ mm does a diffraction pattern begin to be discernible, and in practice $a \approx 0.1$ mm is usually necessary for the pattern to become clear. If we wanted the light wave to *fill* the space behind the opening ($\theta \approx 90°$), we would need to reduce the size of the opening to $a \approx \lambda < 0.001$ mm! These are not the kinds of openings with which we have any experience! Although such small openings can be made today, with the processes used to make integrated circuits, the *amount* of light able to pass through such a small opening is so small that its intensity, after it spreads to fill the space behind, is too small to see by eye.

EXAMPLE 17-2 Light from a helium-neon laser ($\lambda = 633$ nm) passes through a single slit of width 0.1 mm and then travels to a screen 1 m away. a) What is the angle of the first minimum in the diffraction pattern? b) What is the width of the central maximum?

SOLUTION a) This is a straightforward calculation problem. Using the small angle approximation,
$$\theta_1 \approx \sin\theta_1 = \lambda/a = 6.33\times10^{-3} \text{ radians} = 0.363\,°.$$
This illustrates nicely how small the angles are in the diffraction of light.

b) The width of the central maximum a distance L away is
$$w = 2\lambda L/a = 12.7 \text{ mm}.$$

This width is easily seen and measured.

We are not going to treat the problem of diffraction past a straight edge—it is mathematically very complicated. But our understanding of the diffraction of a slit gives us some insight into what a single straight edge should do, because a slit, after all, is simply two straight edges placed close together. We then might expect that the diffraction pattern of a single straight edge would look rather like half of a single-slit diffraction, which is what we observed in Fig. 17-2.

17.4 Circular-Aperture Diffraction

Diffraction will occur if a wave passes through an opening of any shape. Diffraction by a single slit establishes the basic ideas of diffraction, but the most common situation of practical importance is diffraction of a wave by a **circular aperture**. Consider some examples. A loudspeaker cone generates sound by the rapid oscillation of a diaphragm, but the sound wave must pass through the circular aperture defined by the outer edge of the speaker cone before it travels into the room beyond. This is diffraction by a circular aperture. Or consider a radar dish. A small electronic tube generates the microwaves, beams them downward at the large circular radar dish, and the dish (acting as a curved mirror) reflects them into the sky as a radar beam. But the dish itself acts like a circular opening to the sky, and the radar waves diffract "through" this circular opening. A telescope, microscope, or other optical instrument is just the reverse. Waves originate somewhere outside and need to enter the instrument. To do so, they must pass through a circular aperture—the input lens. This causes diffraction.

The limits of performance of these and many other modern instruments is determined by the diffraction of the openings through which the waves must pass. Circular diffraction is mathematically more complex than diffraction from a slit, and we will present results without derivation. However, the *ideas* behind the results are the same as for a single slit.

Figure 17-11a is a photograph of the circular diffraction pattern seen on a screen after light has passed through a small circular hole. You should compare this to Fig. 17-7 to note the similarities and differences. The circular diffraction pattern still has a *central maximum*, now circular in shape, that is surrounded by a series of secondary bright fringes. Most of the intensity is contained within the central maximum, so it will once again be

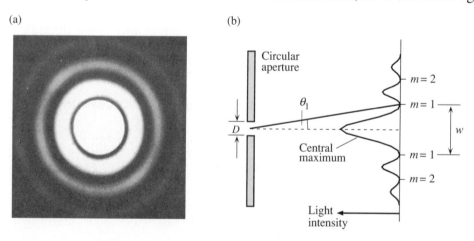

(a)   (b)

FIGURE 17-11 a) Photograph and b) figure drawn to the same scale illustrating the diffraction of light by a circular opening of diameter D.

convenient to refer to the width of the central maximum as *w*. In this case, *w* is the *diameter* of the central maximum.

A diagram of the circular diffraction is shown in Fig. 17-11b. We can still define an angle θ_1 to the first minimum in the intensity, where there is complete destructive interference. The mathematical analysis of circular diffraction leads to the result

$$\theta_1 = \frac{1.22\lambda}{D} \quad \text{(first minimum of a circular aperture)}, \tag{17-5}$$

where *D* is the *diameter* of the circular opening. Note that this is very similar to the $m = 1$ minimum of a single slit, but not quite the same. Equation 17-5 has assumed the small angle approximation. This is always valid for the diffraction of light, where λ is very small, but usually is not valid for the diffraction of sound. Within the small angle approximation, the width of the central maximum is

$$w = 2y_1 = 2L\tan\theta_1 \approx \frac{2.44\lambda L}{D} \quad \text{(width of circular diffraction pattern)}. \tag{17-6}$$

As was the case for diffraction through a slit, the light spreads out behind the circular aperture. The width increases with distance *L* but decreases if the size *D* of the aperture is increased.

EXAMPLE 17-3   A radar or microwave transmitter (Fig. 17-12) consists of a microwave source and a curved circular reflector that collects waves from the source and beams them in the forward direction. An ideal transmitter, in the absence of diffraction, would generate a perfectly parallel microwave beam that would propagate without spreading. The reflector dish, however, acts as a circular "opening" through which the microwaves must pass before making it into the transmitted beam. The diffraction of this opening causes the beam to spread out in space. A typical microwave frequency is 10 GHz. The transmitters used for communications (the towers you often see on mountain tops) have reflectors about 3 m in diameter. a) What is the angular spread of the transmitted beam? b) Microwave repeater stations are spaced about every 50 miles. What is the diameter of the central maximum of the beam when it reaches the repeater?

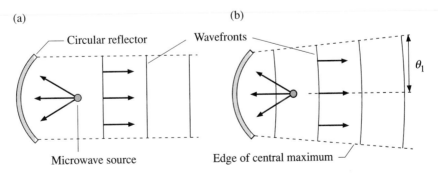

(a)   (b)

Circular reflector   Wavefronts

θ_1

Microwave source   Edge of central maximum

FIGURE 17-12 a) An ideal microwave transmitter generates a parallel beam. b) The beam of a real transmitter spreads because of diffraction.

SOLUTION a) A frequency of 10 GHz = 10×10^9 Hz gives a microwave wavelength $\lambda = c/f$ = 0.030 m. The angular spread of the transmitted beam is determined by θ_1, the angle that defines the edges of the central maximum. From Eq. 17-5,

$$\theta_1 = \frac{1.22\lambda}{D} = 0.0122 \text{ radians} = 0.70°.$$

b) The width of the central maximum at a distance L is $w = 2.44\lambda L/D$. A distance of 50 miles converts to $L = 8.0 \times 10^4$ m, from which we can calculate $w = 2000$ m. A good-sized circle! The receiving antenna on the repeater station intercepts only a very tiny fraction of the microwave power in this 2000 m diameter beam. The repeater then amplifies the received signal back up to its original power and transmits it on to the next station.

The circular reflectors of microwave transmitters grow in proportion to the distance the signal needs to travel because making D larger reduces θ_1 and minimizes the spread of the beam. Communications needing to go only a few kilometers use reflectors a meter or so in diameter. Radar for aircraft and missile tracking, which need to "see" hundreds of kilometers, often have reflector dishes 10 m or more in diameter. The deep-space network that communicates with satellites on missions to other planets uses giant reflectors of roughly 100 m diameter—and would make them even larger if the technology were feasible. It is all a matter of needing to operate within the physical constraints set by diffraction of the microwaves.

17.5 Diffraction by a Lens

A particularly important application of circular-aperture diffraction is the focusing of light by a lens. The purpose of a lens is to focus incoming light waves to a point. You are familiar with this if you have used a magnifying lens to focus the sun's light rays to a point, or if you think about a telescope lens focusing starlight to the "points" you see as stars in astronomy photographs. We have not examined the properties of lenses in this text, but it is reasonable to think that waves converging *to* a point will look like just the opposite of waves emerging *from* a source—hence the wave fronts are inward-converging circles, as seen in Fig. 17-13a.

A lens has a finite diameter, or *aperture*. Only those waves passing *through* the aperture of the lens can be focused, so the lens acts like a circular opening of diameter D in an opaque barrier, as shown in Fig. 17-13b. In Fig. 17-13c, we have separated the focusing properties of the lens from the diffraction properties of the lens aperture by placing an "ideal" diffractionless lens next to a circular aperture having the diameter D. This is a *model* of how the lens functions—a combination of diffraction properties and focusing properties.

For the purposes of this discussion, you only need to know three things about a lens:

1. Light waves traveling along parallel paths—called **rays**—are bent by the lens to converge at a single point called the **image**.

2. The one ray that passes through the *center* of the lens is not bent. Its angle of travel upon leaving the lens is exactly the same as the angle with which it entered.

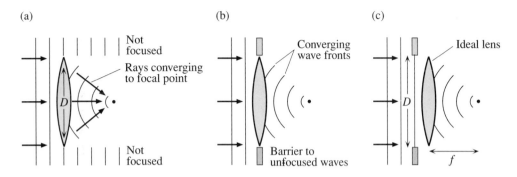

FIGURE 17-13 a) Plane waves being focused by a lens. b) Only those waves entering the lens are focused, so the lens acts as an opening in a barrier. c) A *model* of a lens in which the diffraction properties of the opening are separated from the focusing properties of the lens.

3. The convergence point is located a distance f behind the lens, where f is called the **focal length** of the lens. The focal length is a fixed property of the lens, determined by how it is designed.

Now imagine that a bundle of parallel rays—waves traveling along parallel paths—enters a lens. Take the one ray passing through the center and extend it in a straight line. Then draw a plane (the *focal plane*) a distance f behind the lens. The point where this ray intersects the focal plane is the image point, and all of the other rays will be bent by the lens so as to converge to this point. This is what happens when you hold a magnifying lens in the sun and see the light focused to a small point. Rays from the sun are striking the lens over its entire surface, but they are then bent by the lens so as to meet at a single point.

Figure 17-14 shows an example of this. One group of rays (solid lines) enters perpendicular to the plane of the lens ($\theta = 0$). The ray through the center continues without bending until it intersects the focal plane a distance f behind the lens. All of the other rays are bent so as to meet at this point. The dotted lines show a second bundle of rays traveling at an angle $\theta \neq 0$. Again, the ray through the center is not bent, and the point where it intersects the focal plane is the point where all of the rays are focused. Keep in mind that the focal plane is not a physical object but simply the mathematical plane a distance f from the lens. If you were to place a screen at that position, you would see a bright dot where the light is focused. But the light waves still converge to that point in space whether the screen is there or not.

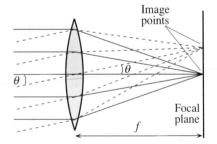

FIGURE 17-14 A lens focuses light waves to an image point at the intersection of the focal plane with an undeviated ray through the center of the lens.

If light consisted of particles, a perfect lens would be able to focus the light to an arbitrarily small point in the focal plane. The only limitation would be how well the lens was designed and manufactured. But light is a wave, not a particle, and the diffraction of the wave through the lens opening has very real effects. Let's consider light approaching the model lens of

Fig. 17-13c from the left. A viewing screen is placed in the focal place of the lens at a distance f on the right side of the lens, as shown in Fig. 17-15.

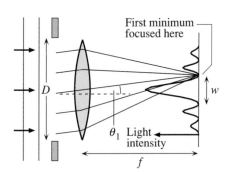

FIGURE 17-15 Diffraction by the opening of a lens causes the light to be focused to a diffraction pattern of width w.

The opening of diameter D diffracts the light, causing it to spread at different angles rather than to continue ahead as a perfectly parallel beam. Light waves traveling at angle $\theta = 0$ are all in phase. The lens will focus those waves at the center of the screen where they will undergo constructive interference to create a bright spot—the central maximum. Some waves, however, will be diffracted at angle $\theta_1 = 1.22\lambda/D$. These will enter the lens as a parallel bundle and will also be focused onto the screen. But these waves undergo *destructive* interference when they meet, so the viewing screen will be *dark* at that point.

When all of the different diffraction angles are considered, and all of the image points mapped on the screen, we find that the screen shows the standard circular diffraction pattern of Fig. 17-11a. A lens, therefore, focuses light waves *not* to a perfect point but to a circular diffraction pattern having a central maximum of finite width w as well as secondary bright fringes.

A lens, no matter how well made, *cannot* focus light to an arbitrarily small point. The reason it cannot is the diffraction of the light waves by the lens itself. The width of the central maximum of the focused light is (using the small angle approximation)

$$w = 2f \tan \theta_1 = 2f\theta_1 = \frac{2.44 \lambda f}{D} \qquad \text{(minimum spot size).} \qquad (17\text{-}7)$$

This is called the *minimum spot size* to which a lens can focus light. Many lenses cannot focus light to a spot this small because they are limited by factors in the design of the lens. But a well-crafted lens, for which this is the minimum spot size, is called a *diffraction-limited lens*. This truly is a limit. The wave nature of light is such that no optical design can overcome the spreading of light due to diffraction, and it is this spreading that is responsible for the image having a minimum spot size.

For various optical reasons, it is extremely difficult (essentially impossible) to produce a diffraction-limited lens having a focal length less than its diameter. That is, $f \geq D$ for any realistic lens. This implies, based on Eq. 17-7, that the smallest diameter to which you can focus a spot of light—no matter how hard you try—is

$$w_{\min} \approx 2.5\lambda. \qquad (17\text{-}8)$$

This is a fundamental limit on the performance of optical equipment—you simply cannot focus light to a spot smaller than about two-and-a-half wavelengths in diameter. Diffraction has very real consequences!

Are these consequences really important? One example of the limits caused by the diffraction of light can be found in the manufacturing of integrated circuits. Integrated circuits are manufactured by creating a "mask" of all the components and their connections. This mask is focused, using a lens, onto the surface of a semiconductor wafer. The wafer is coated with a photographic-like substance that is exposed by the bright areas in the mask. Subsequent processing steps use chemicals to etch away areas that had been exposed while leaving behind areas that were in the dark areas of the mask and not exposed. This whole process is called *photolithography.*

The density of elements on a semiconductor chip—which determines how powerful a microprocessor is or how much memory can be placed on a chip—is directly dependent on how small the elements in the circuit can be made. Because the mask is optically projected onto the chip, diffraction dictates that a circuit element can be no smaller than the smallest spot to which light can be focused. That is, no feature on the chip can be smaller than $w_{min} \approx 2.5\lambda$. If the mask is projected with the bluest light in the visible spectrum, having $\lambda \approx 400$ nm = 0.4 μm, then the smallest elements that can be built on a chip are $w_{min} \approx 1$ μm. This is, in fact, just about the current limits of technology. Further improvements are not likely unless either projection light sources of much shorter wavelength are developed or entirely different non-optical means are developed for writing the mask on the wafers.

EXAMPLE 17-4   A telescope has a lens 12 cm in diameter and a focal length of 1 m. What is the diameter of the image of a star in the focal plane if the lens is diffraction limited *and* if the earth's atmosphere is not a limitation?

SOLUTION Stars are so far away that they appear to even the very best telescopes as simply points in space. An ideal diffractionless lens would be able to focus the images of the stars to arbitrarily small points. But the real lens of this problem, according to Eq. 17-7, will produce a minimum spot size in the focal plane of $w = 2.44$ $\lambda f/D$, where D is the lens diameter. What is λ? Because stars emit white light, the *longest* wavelengths are going to spread the most and will determine the size of the image that is seen. If we use $\lambda = 700$ nm as the approximate upper limit of visible wavelengths, we find $w = 1.4\times10^{-5}$ m = 14 μm. This is certainly small, and it would appear as a perfect point to your unaided eye, but the finite size would be easily noticed if it were recorded on film and enlarged. In fact, turbulence and temperature effects in the atmosphere—the causes of the "twinkling" of stars—generally prevent images with ground-based telescopes from being this good. Space-based telescopes, however, really are diffraction limited.

EXAMPLE 17-5   High-power lasers are now routinely used to cut and weld materials. They do this by using a lens to focus the laser beam to a very small spot, thus concentrating the laser beam's energy into a very tiny area. This is no different than using a magnifying lens to focus the sun's light to a small spot that can burn things. As an engineer, you have designed a laser-cutting device in which the material to be cut is placed 5 cm behind the lens. You have selected a certain high-power laser having a wavelength of 1.06 μm. (These are both very typical values.) Your calculations indicate that the laser must be focused to a spot size of 5 μm (i.e., 5 μm in diameter) to have sufficient power to make the cut. What is the minimum diameter of the lens you must install?

SOLUTION A lens of diameter D will create a minimum spot size, when focused, of $w = 2.44$ $\lambda f/D$. The focus occurs a distance f behind the lens, so we will need a lens of focal length $f = 5$ cm. The diameter of a diffraction-limited lens giving a spot size of 10 μm is then

$$D = \frac{2.44\lambda f}{w} = 2.6 \times 10^{-2} \text{ m} = 2.6 \text{ cm} \approx 1 \text{ inch.}$$

This is the *minimum* usable size because a smaller lens cannot focus light to a spot as small as 5 μm.

The Resolution of Optical Instruments

Suppose you point a telescope at two nearby stars in a galaxy far, far away. If you use the best possible detector on your telescope, will you be able to distinguish separate images for the two stars, or will they blur into a single blob of light? A similar question could be asked of a microscope. Can two microscopic objects, no matter how small and close together, always be distinguished if sufficient magnification is used, or is there some size limit at which they will blur together so they cannot be seen separately? These are questions about the **resolution** of optical instruments, and they are clearly questions of great importance for the practical application of optics.

If light consisted of particles, the images of the two objects could be made arbitrarily sharp if enough effort were made to produce lenses and other components of sufficient precision. But, as you have seen, the wave nature of light precludes focusing an image with perfect sharpness. The image of a point of light, such as a distant star, is not a point but, because of diffraction, a circle of light of some width w. If two objects are viewed simultaneously, the image of each will be a small circle. Our question, then, really amounts to asking how close together two circles can be and still be distinguished as two separate circles.

Figure 17-16 shows two objects being imaged by a lens of diameter D. The angle between the two objects is α. Although there are many rays from each object striking the lens, this figure shows just the ray from each that passes through the center of the lens. Because these rays continue as straight lines to the image points, we see that the angle between the two images is also α.

The light waves from each object spread, due to diffraction, after passing through the lens opening. The angle θ_1 locates the edge of the central maximum of the image, determining the spot size w. Angle α is greater than angle θ_1 in Fig. 17-16, and the diffraction patterns of the two images are clearly separate. If it should happen that $\alpha \ll \theta_1$, the two diffraction patterns will fall almost exactly on top of each other and appear as a single circle of light. The two images, in that case, would not be resolvable.

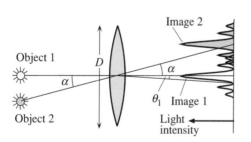

FIGURE 17-16 Two objects imaged simultaneously by a lens create two diffraction patterns separated by the same angle as the objects.

One of the major scientists of the nineteenth century, Lord Rayleigh, studied this problem and suggested a reasonable rule of thumb. If two objects are separated by an angle $\alpha > \theta_1$, where θ_1 (Eq. 17-5) is the edge of the central maximum of a circular diffraction pattern, then their images are resolvable. If their separation is $\alpha < \theta_1$, they are not resolvable because their diffraction patterns overlap too much. The case $\alpha = \theta_1$ is marginally resolvable. Notice that $\alpha = \theta_1$ means that the central maximum of one image falls exactly on top of the first dark minimum of the other image. This criteria for whether or not the images of two objects can be resolved is called **Rayleigh's criteria**.

Figure 17-17 shows enlarged photographs of the images in the focal plane for these three cases. The images are not points but, as expected, circular diffraction patterns. In a), where the objects are separated by $\alpha > \theta_1$, the two images are close but distinct. In c), by contrast, there really are two objects, but their separation is $\alpha < \theta_1$ and their images have blended together. The two images are not resolved. Image b) meets Rayleigh's criteria, with $\alpha = \theta_1$, and you can see that the two images are just barely resolved.

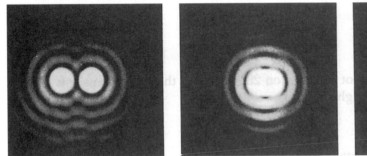

FIGURE 17-17 Enlarged photographs of the images of two closely spaced objects. The objects are separated by a) more than, b) equal to, and c) less than the Rayleigh criteria.

For telescopes, the angle $\theta_1 = 1.22\lambda/D$ is called the *angular resolution* of the telescope. Two stars, or other objects, with an angular separation α larger than θ_1 can be resolved whereas two objects with angular separation less than this cannot. Note that the angular resolution is a function of only the lens diameter and the wavelength; the magnification is not a factor. Two overlapped, unresolved images will remain overlapped and unresolved no matter what the magnification. Because λ is fairly well fixed by the need to use visible light, the only parameter over which the astronomer has any control is the diameter of the lens or mirror of the telescope. The urge to build ever larger telescopes is motivated, in part, by a desire to improve the angular resolution. (Another motivation is to increase the light-gathering power so as to see sources farther away.)

A microscope is rather like a telescope in reverse. In this case, the objects are located a distance f, the focal length, in *front* of the lens and the images are formed much farther away on the other side of the lens. (Think how close the objective of a microscope is placed to the object and how far away, by comparison, the eyepiece is.) Figure 17-18 shows two objects that are just marginally

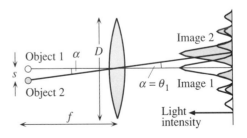

FIGURE 17-18 Two objects that are marginally resolved when imaged by a microscope.

resolved by the microscope, with the central maximum of one falling on the first dark minimum of the other—Rayleigh's criteria. (See the photo of Fig. 17-17b.) This marginal resolution occurs when the two objects are separated by angle α_{min} such that $\alpha_{min} = \theta_1 = 1.22\lambda/D$. The physical separation s between the two objects is related to α and the focal length by $s = f\tan\alpha \approx f\alpha$, because the small angle approximation is valid. The smallest separation of the objects that can be resolved occurs when $\alpha = \alpha_{min} = \theta_1$, in which case

$$s_{min} = f\alpha_{min} = f\theta_1 = \frac{1.22\lambda f}{D} \geq 1.22\lambda . \qquad (17\text{-}9)$$

In the last step we have used, as previously noted, the fact that $f/D \geq 1$ for any real lens.

Diffraction by the circular lens aperture once again limits the performance of an optical instrument. In this case, we find that a microscope can only resolve objects down to a certain size s_{min}. Objects smaller than this will be so blurred by diffraction as to be unresolved, no matter what the magnification. Consequently, the smallest object any optical microscope will ever be able to see is

$$s_{min} \approx \lambda.$$

In other words, objects smaller than about one wavelength of light *cannot* be resolved by a microscope. This is a conclusion of fundamental importance.

Only recently, with the invention of a non-optical technique called the scanning tunneling microscope, have atoms been imaged for the first time. Now we can understand why. Atoms are approximately 0.1 nm in diameter, vastly smaller than the wavelength of visible or even ultraviolet light. There is no hope of ever seeing atoms with an optical microscope. This is not simply a matter of needing a better design or more precise components—it is a fundamental limit set by the wave nature of the light with which we see.

17.6 X-Ray Diffraction

In 1895, the German physicist Wilhelm Röntgen made a remarkable discovery that has had profound implications for twentieth-century science and technology. The late nineteenth century was a period in which the technology of vacuum tubes was being perfected, and Röntgen was experimenting with the production and propagation of electrons through a vacuum (Fig. 17-19). He sealed an electron-producing filament—basically the same as the filament in an incandescent light bulb—and a metal electrode into a vacuum tube, Then he applied a high voltage between them to accelerate the electrons to very high velocity before they struck the electrode. Röntgen and others had done similar experiments many times before, but one day he happened—by chance—to have a sealed envelope containing film next to the vacuum tube. He discovered, to his surprise, that the film had been exposed even though it had never been removed from the envelope. This serendipitous discovery was the beginning of x rays.

Röntgen quickly found that the vacuum tube was the source of whatever was exposing the film, but he had no idea what was coming from the tube—so he

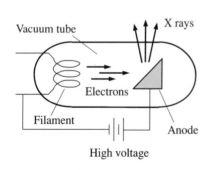

FIGURE 17-19 Röntgen's x-ray tube.

called the phenomena **x rays**, using the algebraic symbol *x* as meaning "unknown." x rays were unlike anything—particle or wave—ever discovered. Röntgen was not successful at reflecting the rays or at focusing them with a lens. It was possible to show that they traveled in straight lines, like particles, but they also passed right through most solid materials—something no known particle could do.

By the early 1900s it was suspected that x rays were an electromagnetic radiation with a incredibly short wavelength of $\lambda \approx 0.1$ nm. If so, x rays should exhibit diffraction. But while diffraction gratings were able to diffract visible light, the ratio λ/d for x rays was far too small to produce a measurable spreading.

In this same time period, scientists were first discovering that atoms have sizes ≈ 0.1 nm. Several individuals had suggested that solids might consist of regular arrays of atoms in crystalline lattices, with the spacing between atoms also being ≈ 0.1 nm. In 1912, the German scientist Max von Laue suggested that *if* x rays are waves with $\lambda \approx 0.1$ nm, and *if* solids are atomic crystals with spacing $d \approx 0.1$ nm, then x rays passing through such a crystal ought to undergo diffraction from the "three-dimensional grating" of the crystal. In fact, the diffraction angles ought to be fairly large because $\lambda/d \approx 1$.

Von Laue's prediction of x-ray diffraction by crystals was soon confirmed experimentally. Measurements showed that x rays are indeed electromagnetic waves—not fundamentally different than visible light—with wavelengths in the range 0.01 nm–10 nm.

However, for reasons you will see shortly, it was not easy to know how to interpret an x-ray diffraction pattern. Another physicist, W. L. Bragg, quickly suggested a simpler and easier way to measure and understand the diffraction caused by a crystal.

Figure 17-20 shows a simple cubic-lattice crystal, with each of the little circles representing the position of an atom in a solid. You need to imagine that the square array you see in the figure is extended into the page, forming a cubic array; we are looking at one *plane* of atoms. (Nearly all materials have crystalline structures more complex than this, but a cubic array will help you to understand the ideas.)

There are many ways in which this crystal can be "sliced" into parallel planes of atoms. The figure shows dotted horizontal lines spaced distance d_1 apart—these are the *edges* of planes of atoms extend-

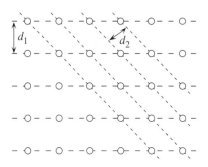

FIGURE 17-20 Atoms arranged in a cubic-lattice crystal. There are many ways in which this lattice can be viewed as parallel planes of atoms, each with a different spacing *d*.

ing above and below the page. We could also "slice" the crystal along the parallel planes of atoms seen tilted at a 45° angle, having a different spacing d_2, and there are yet other possible slices that have not been shown.

Suppose that a beam of x rays is incident on a *single* plane of atoms, as shown in Fig. 17-21a. Most of the beam will be transmitted through the plane, because we know that x rays penetrate solids, but a small portion may be reflected. The reflected wave obeys the law of reflection—that is, the angle of reflection equals the angle of incidence—and the figure has been drawn accordingly. (The law of reflection for x rays from a plane of atoms is actually a *consequence* of their diffraction—an issue you can explore further as a homework problem.)

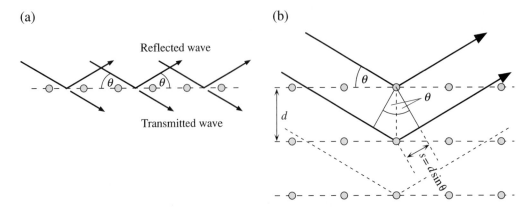

FIGURE 17-21 a) Waves diffract from a single plane of atoms so as to obey the law of reflection. b) The reflections from parallel planes interfere with each other to cause constructive interference for certain angles θ.

As x rays penetrate into a crystalline solid, a small fraction reflects from each of the *many* parallel planes of atoms. Each reflection is a wave traveling along a slightly different but *parallel* path, as shown in Fig. 17-21b. The reflections all interfere where they combine at a point outside the crystal. Notice how similar Fig. 17-21b is to Fig. 17-3 for a diffraction grating—it is exactly the same idea in both cases. The reflection from any specific plane has to travel an extra distance Δr before being combined with the reflection from the plane immediately above it. From the geometry of Fig. 17-21b you can see that the extra travel distance is $\Delta r = 2s$, where $s = d\sin\theta$. The angle θ is the angle between the incident x rays and the plane of atoms.

If $\Delta r = m\lambda$, these two reflected waves will be in phase when they recombine. But the same geometry applies to *all* the planes, so if these two reflections are in phase it must be the case that *all* the reflections from *all* the planes are in phase. This will cause constructive interference of the reflected waves. Atomic places with spacing d will diffract x rays at an angle θ if

$$\Delta r = 2d\sin\theta = m\lambda, \quad m = 1, 2, 3, \ldots. \tag{17-10}$$

This result is called the **Bragg condition** for x-ray diffraction.

To make practical use of this idea, a solid crystal is placed in an x-ray beam and mounted so that it can be rotated through a wide range of angles (Fig. 17-22a). An x-ray detector looks for a diffracted x-ray beam at an angle equal to the angle of incidence. A graph of detected intensity versus angle θ is called the *x-ray diffraction spectrum,* and it contains valuable information about the structure of the crystal. The complicating factor is that *any* set of parallel planes of atoms, all of which have different values of d, will cause diffraction—possibly at several different angles corresponding to different values of m. Figure 17-22b shows a simulated x-ray diffraction spectrum for just the two sets of planes shown for the simple cubic crystal of Fig. 17-20, assuming typical values of $d_1 = 0.20$ nm and $\lambda = 0.12$ nm.

Real x-ray diffraction spectra are usually much more complicated than the one shown in Fig. 17-22b. Nonetheless, such spectra contain the information necessary to deduce the crystalline structure of the solid. It becomes a challenging puzzle to find a structure that corresponds to the observations.

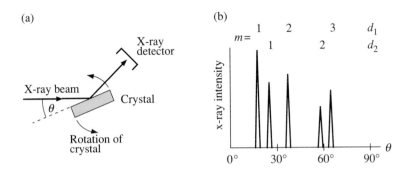

FIGURE 17-22 a) An experiment with a rotating crystal to measure the Bragg diffraction angles of a crystal. b) Simulated x-ray diffraction spectrum for the crystal of Figure 17-20. Two sets of parallel planes, with spacings d_1 and d_2, contribute to the diffraction.

Although the Bragg procedure is straightforward, most practical x-ray diffraction studies are done using the method that von Laue suggested—looking at the diffraction of a beam of x rays that is *transmitted* through a crystal. Figure 17-23a shows a typical experimental arrangement for measuring an x-ray diffraction pattern. An x-ray tube generates a variety of wavelengths, so Bragg diffraction is first used to select just one of these wavelengths by rotating a crystal with known spacing d to the angle meeting the Bragg condition. This part of the apparatus is called an *x-ray monochromator*—a device that selects one (mono) wavelength or color (chrome).

Once a single, known wavelength is selected, it passes through the sample and is diffracted by the three-dimensional grating of the crystal lattice. A piece of x-ray film behind the sample records the locations of constructive interference. Because of the three-dimensional nature of the grating, the diffraction pattern consists of bright points, rather than lines or fringes. Figure 17-23b shows a typical diffraction pattern recorded in this fashion—in this case for a crystal of niobium diboride. You can see that it is quite complicated, which is why we noted earlier that the interpretation of such patterns is not easy. However, crystallographers have developed many powerful analysis tools for deciphering such patterns. These techniques are computationally very intense, but modern supercomputers have made such analyses routine.

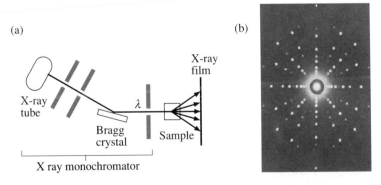

FIGURE 17-23 a) Modern technique for using x-ray diffraction to study the atomic structure of a sample. b) x-ray diffraction pattern for niobium diboride.

The technique of x-ray diffraction has revolutionized our ability to study the atomic and molecular structure of matter. Nearly everything we know about the structure of solids has come from the analysis of diffraction patterns such as the one in Fig. 17-23b. The most important properties of solids—their strengths, their chemical properties, their abilities to be cut or welded, their optical properties—are consequences of their crystal structure. Modern engineering could not exist without the knowledge of materials gained through x-ray diffraction.

X-ray diffraction also is one of the most important tools in solid state physics, where the properties of semiconductor devices are studied; in materials engineering, where x-ray diffraction reveals the structure of alloys; and in biophysics, where x-ray diffraction was used to deduce the double-helix structure of DNA and continues to elucidate the structure of biological molecules such as proteins. The techniques of x-ray diffraction are likely to become even more important in the future as physicists develop new superconducting materials, molecular biologists produce "designer drugs," and engineers develop techniques for producing "nanostructures" at the atomic level.

Summary

◢ Important Concepts and Terms

diffraction	image
diffraction grating	focal length
spectroscopy	resolution
single-slit diffraction	Rayleigh's criteria
secondary maxima	x rays
circular aperture	Bragg condition
rays	

We have come a long way from the double-slit interference experiment to unraveling the structure of DNA with x-ray diffraction. But, amazingly, it is all the same physics! The wide range of examples and applications you have seen in the last two chapters have all just been exploiting the principle of superposition to show how waves interact with their surroundings and with other waves. You will have succeeded in your study of wave physics if you can see that all of these examples are just variations on a common theme.

Diffraction is the name given to the interference of a large number of sources along a wave front. These sources may be discrete, as in a diffraction grating, or continuous, as for a single slit. In either case, the underlying idea of diffraction is that the superposition of waves causes constructive and destructive interference due to phase differences.

A diffraction grating, either transmission or reflection, creates a large number of very closely spaced wave sources. These produce constructive interference at angles θ such that $d\sin\theta = m\lambda$. This condition for constructive interference is the same as for Young's double-slit experiment. A grating, however, produces interference fringes that are much narrower and much brighter than a double-slit apparatus.

Openings, such as a single slit or a circular aperture, produce diffraction patterns rather than sharp-edged shadows. The patterns are a bright central maximum surrounded by dark fringes and weaker secondary maxima. The edge of the central maximum is located at angle θ_1 and the central maximum, when viewed at distance L, has a width w. These are given by:

Single slit of width a: $\qquad\qquad \theta_1 = \dfrac{\lambda}{a} \qquad w = \dfrac{2\lambda L}{a}$

Circular aperture of diameter D: $\qquad \theta_1 = \dfrac{1.22\lambda}{D} \qquad w = \dfrac{2.44\lambda L}{D}$

Diffraction by circular apertures has important implications in the resolution of optical instruments. Rayleigh's criteria says that two objects separated by angle α can be resolved if $\alpha \geq \theta_1$. When applied to a microscope, this leads to the conclusion that the smallest resolvable object has a physical size $s_{min} \approx \lambda$.

Short wavelength x rays can be diffracted by the atomic planes in a crystal. This is analogous to a diffraction grating. X rays will be constructively diffracted from a crystal at angle θ if they obey the Bragg condition $2d\sin\theta = m\lambda$, where d is the distance between parallel planes of atoms. There are many such parallel planes in a real crystal, and thus many values of d. This causes x-ray diffraction spectra to be rather complex. Nonetheless, they are widely used to determine the atomic structure of materials.

Exercises and Problems

Exercises

1. A helium-neon laser ($\lambda = 633$ nm) illuminates a single slit, and the light pattern is observed on a screen 1.5 m away. The distance between the first and second minima of the diffraction pattern is 4.75 mm.
   a. What is the width of the slit?
   b. What is the width of the central maximum of the diffraction pattern?

2. a. A 0.5 mm wide slit is illuminated by 500 nm light. What is the width of the central maximum on a screen 2 m away?
   b. A 0.5 mm diameter hole is illuminated by 500 nm light. What is the width of the central maximum on a screen 2 m away?

3. A diffraction grating with 600 slits per millimeter is illuminated with 500 nm light. A viewing screen is 2 m behind the grating.
   a. What is the distance between the two $m = 1$ bright fringes?
   b. What is the maximum number of bright fringes that can be seen?

Problems

4. Light emitted by Element X passes through a transmission diffraction grating having 1200 slits per mm. The diffraction pattern is observed on a screen 75.0 cm behind the grating. Bright lines are *seen* on the screen at distances of 56.2 cm, 65.9 cm, and 93.5 cm from the central maximum.
   a. What is the value of m for these diffracted wavelengths? How can you tell? Explain why only one value is possible.
   b. What are the wavelengths of light emitted by Element X?

5. A helium-neon laser ($\lambda = 633$ nm) is constructed using a glass tube, having an inside diameter of 1.5 mm, placed between two mirrors (Fig. 17-24). One mirror is partially transmitting to allow the laser beam out. Electrodes at both ends create an electrical discharge down the "bore" of the tube, causing it to glow like a neon light. From an optical perspective, the laser beam is a light wave that is diffracted out through a 1.5 mm circular opening.

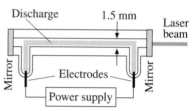

FIGURE 17-24

 a. Can a laser beam be *perfectly* parallel, with no spreading? Why or why not?

 b. The angle θ_1 to the first minimum is called the *divergence angle* of a laser beam. What is the divergence angle of this laser beam?

 c. What is the diameter of the laser beam at a distance of 10 m?

 d. What is the diameter of the laser beam at a distance of 1 km?

6. Scientists use *laser range-finding* to measure the distance to the moon with great accuracy. A very short laser pulse is fired at the moon and the time interval is measured until the "echo" is seen by a telescope. A special wide-beam laser must be used that has a very low divergence angle. The beam creates a "spot" on the moon that is 1 km in diameter. If $\lambda = 500$ nm, what is diameter of the circular opening out of which the laser beam emerges? (Use Table 12-2 for astronomical data.)

7. Alpha Centauri, the nearest star to earth, is 4.3 light years away, where 1 light year is the distance traveled by light in one year. Assume that Alpha Centauri has a planet with an advanced civilization. Professor Dhg, at the Astronomical Institute, wants to design a telescope with which he will be able to discover if there are planets orbiting the Sun.

 a. What minimum diameter mirror does he need, if all other conditions are perfect, to just barely resolve Jupiter and the Sun? Assume $\lambda = 600$ nm. (Use Table 12-2 for data.)

 b. Building a telescope of the necessary size does not appear to be a problem. What other factors, though, might make Prof. Dhg's experiment impossible with a telescope this size? (**Hint:** Think about the relative brightness of the objects being viewed.)

 c. What diameter mirror would Prof. Dhg need to see the Sun as a disk, rather than just a point of light?

8. Optical-disk data storage, such as used with compact-disk players and computer optical-disk memories, uses a small semiconductor laser with $\lambda \approx 800$ nm (infrared) to "read," via reflected light, small "pits" that are burned into a plastic surface (Fig. 17-25). The maximum storage density of such a media depends on the smallest spot to which the laser beam can be focused.

 a. What is the smallest spot size to which an 800 nm laser beam can be focused?

 b. Assume that the pits are spread on a two-dimensional grid with a spacing 50% larger than the laser focal spot size. (Any closer would risk reading errors.) Each pit records 1 bit of information, and it takes 8 bits to form 1 byte, the standard unit of data storage. What is the storage density, in "bits per square centimeter," of such an optical disk?

c. A standard optical disk has a usable surface with an inner diameter of 4 cm and an outer diameter of 11 cm. How many *bytes* of data can be stored on such a disk? (For comparison, a high-density floppy disk has a capacity of 2 MB or 2×10^6 bytes.)

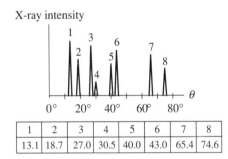

Magnified view of pits on an optical disk

FIGURE 17-25

d. There is a large effort in industrial research laboratories to develop a "blue" semiconductor laser. Suppose that a laser can be developed at a 400 nm wavelength. By what *factor* would this increase the storage density on optical disks?

9. Figure 17-26 shows an x-ray diffraction spectrum taken by rotating a crystal in a beam of x rays having a wavelength of 0.10 nm. Each observed diffraction angle is called a *line* in the spectrum. The lines have been numbered for convenience, and their diffraction angles are given in the table. It is suspected that the crystal under investigation has a cubic lattice.

a. What is the spacing of atoms in this lattice, assuming it to be cubic?

b. Can you *confirm* that the lattice is cubic by "assigning" all the lines in the spectrum? To assign the lines, you need to identify for each line both the set of parallel planes causing the diffraction and the value of *m*. If all lines can be assigned, then you have good reason to accept the hypothesis that this is a cubic lattice.

(**Hint:** There are three sets of parallel planes contributing to this spectrum. By comparing the intensities of the lines, pick out three "families" of lines diffracting from the same set of planes but with different values of *m*.)

X-ray intensity

1	2	3	4	5	6	7	8
13.1	18.7	27.0	30.5	40.0	43.0	65.4	74.6

FIGURE 17-26

10. Light of wavelength λ is incident at angle ϕ on a *reflection* grating of spacing d. Figure 17-27 shows both the "big picture" as well as a "close up" of two adjacent reflection spots. You need to find the angles θ at which constructive interference occurs. These will be the diffraction angles for light incident at angle ϕ.

a. First consider light shining straight onto the grating with $\phi = 0$. What is the path length difference Δr for the waves reflecting from two adjacent spots at angle θ? (Δr is the difference between the points at which the incident wave front strikes

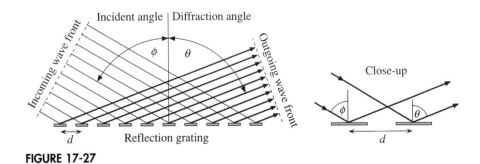

FIGURE 17-27

the grating and the points at which the two waves are recombined in the outgoing wave front.)

b. Show that the diffraction condition for $\phi = 0$ is $d\sin\theta_m = m\lambda$, which is exactly the same as for a transmission grating (Eq. 17-1).

c. Now let the incident angle ϕ be non-zero, as shown in the figure. Find an expression for the path length difference between the incoming wave front and the outgoing wave front for waves reflecting from two adjacent spots.

d. Using your result from c), find an expression (analogous to Eq. 17-1) for the angles θ at which diffraction occurs when the light is incident at angle ϕ. Your expression should agree with your answer to b) when $\phi = 0$.

e. In Section 17.5 we *asserted* that x rays incident at angle ϕ on a plane of atoms would "reflect" at angle $\theta = \phi$. That is, the angle of reflection equals the angle of incidence. Show that this behavior is simply the "zeroth order" diffraction from a single plane of atoms.

f. Light of $\lambda = 500$ nm is incident at an angle $\phi = 40°$ on a grating having 700 reflection lines per millimeter. At what angles θ is this light diffracted?

g. Draw a picture showing a *single* 500 nm light ray incident at $\phi = 40°$ and showing all the diffracted rays at the correct angles.

[**Estimated 12 additional problems for the final edition.**]

Chapter **18**

A Closer Look at Light and Matter

L O O K I N G
B A C K

Sections 14.8; 15.3–15.4; 16.4; 17.2–17.3; 17.6

18.1 Physics in Crisis

Our journey into physics has led us to the ideas and discoveries of the mid-nineteenth century, about 150 years ago. The physics of particles and the physics of waves were well understood by then, and it seemed that Newtonian physics would soon succeed in explaining all the phenomena of nature. But trouble was on the horizon. Discoveries made during the second half of the nineteenth century and the opening years of the twentieth century refused to yield to a Newtonian analysis. At first these unexplained discoveries were few enough to be dismissed as just a nuisance, but their numbers grew until physics found itself in "crisis" at the beginning of the twentieth century. It eventually took two radical new theories—relativity and quantum physics—to resolve the crisis.

We will return to study quantum physics more thoroughly in Part VI. However, it is fitting to conclude our study of waves with a preliminary look at the crisis that led to quantum physics. At the heart of this crisis was a breakdown of the basic particle and wave models of physics. Our goal in this chapter is to use our knowledge of particles and waves to examine some of the experimental evidence for this breakdown. As you will see, light sometimes refuses to act like a wave and seems more like a collection of particles. And matter, such as electrons or neutrons, sometimes seems to behave more like a wave than a particle.

Although we will look at the evidence, we will not present an explanation of these experiments at this time. The explanations must await the introduction of quantum physics in Part VI. Nonetheless, it is interesting and instructive to discover the limits of the wave and particle models that we've developed. In the process, we'll find important new information in our quest to understand atoms.

18.2 Spectroscopy: Unlocking the Structure of Atoms

Light is emitted and absorbed by atoms. In fact, much of what we know about atoms has been learned by studying the light they emit. The technique of studying light by separating and measuring its wavelengths is called **spectroscopy**, and it has been the major tool for learning about the properties of atoms for more than one hundred years.

The basic discoveries of the interference and diffraction of light were all made early in the nineteenth century. These phenomena were well understood by the end of the century, and the knowledge was used to design practical tools for measuring wavelengths with great accuracy. The primary device used to measure the wavelengths emitted by atoms is called a **spectrometer**. The heart of a spectrometer is a diffraction grating that diffracts different wavelengths of light at different angles. Figure 18-1 shows a simplified diagram of such an instrument. A lens is used to collect light from a light-emitting sample and to focus that light upon the *entrance slit*. As the light spreads out beyond the slit it impinges upon a reflection-type diffraction grating. Unlike the diffraction grating we analyzed in Chapter 17, this one is *curved*—formed by thousands of parallel lines on a curved mirror. It still diffracts light at different angles, depending on the wavelength, but it also—because of the curvature—*focuses* the light. The light comes to a focus on a *photographic plate*, a photographic emulsion coated onto a rigid glass plate.

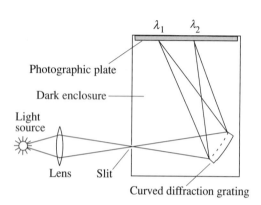

FIGURE 18-1 A diffraction spectrometer for the accurate measurement of wavelengths. Different wavelengths are focused to different lines on the photographic plate.

Each wavelength present in the light is diffracted at a different angle and focused to a different point on the plate. When the plate is developed, it shows a *line* for each wavelength in the light. After the spectrometer is calibrated, the wavelength of each line can be determined simply by measuring its position on the photographic plate. The pattern of wavelengths emitted by a source of light is called the **spectrum** of the light.

Spectroscopists discovered very early that there are two types of spectra: continuous spectra and discrete spectra. Hot, incandescent objects—such as the sun or an incandescent light bulb—emit a *continuous spectrum* in which a rainbow of light is emitted at every possible wavelength. Other sources of light, such as the light emitted when an electric current passes through a dilute gas (as in neon signs), were found to emit only certain discrete, individual wavelengths. Such a spectrum is called a **discrete spectrum**.

Figure 18-2 shows examples of discrete spectra as they would appear on the photographic plate of a spectrometer. Each bright line—called a **spectral line**—is the result of *one* specific wavelength of the light being diffracted by the grating and focused onto the photographic plate at that point. A discrete spectrum is sometimes called a **line spectrum** because of its appearance on the plate. While the spectra in Fig. 18-2 differ in their complexity and the number of lines, the important point is that each consists *only* of certain discrete wavelengths and not a continuum of wavelengths. You can see, for example, that a neon light has its familiar reddish-orange color because nearly all of the wavelengths emitted by neon atoms fall within the wavelength range 600–700 nm that we perceive as orange and red.

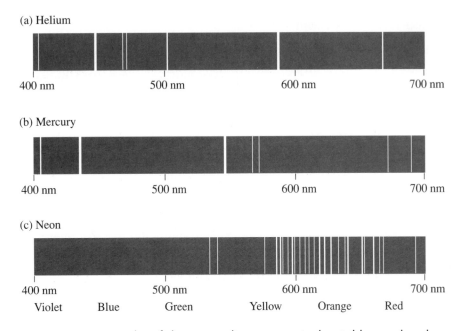

FIGURE 18-2 Examples of discrete, or line, spectra in the visible wavelength of range 400–700 nm. a) Helium spectrum. b) Mercury spectrum. c) Neon spectrum.

It was established by the end of the nineteenth century that every element in the periodic table has a unique, reproducible spectrum that is entirely different than the spectrum of any other element. An atomic spectrum is thus a "fingerprint" that can be used to identify elements. This discovery was the basis for many contemporary technologies for analyzing the composition of unknown materials, monitoring air pollutants, and studying the atmospheres of the earth and other planets. The discovery also led scientists to the hypothesis that the specific wavelengths emitted by an atom are an indication of its internal structure. If scientists just knew how to "decode" these wavelengths, they would be able to determine the atomic structure for that element. A correct theory of why and how atoms emit light would allow scientists to unlock the atoms' structure and discern their inner workings.

Despite heroic attempts by some of the best scientists of the late nineteenth century, Newtonian mechanics and the (then) new theory of electromagnetism were completely unable to provide an explanation of atomic spectra. Not only did they fail to predict why one element's spectra should differ from another, these "classical" theories could not even explain why atomic spectra are discrete rather than continuous. In fact, nineteenth-century physics predicted that atoms should not even be stable and that the electrons should spiral into the nucleus—destroying the atoms and the universe—in a small fraction of a second! This prediction was obviously incorrect.

Physics from the time of Newton through the mid-nineteenth century had been spectacularly successful at explaining the motions of the heavens and of everything on earth, at forming a theory of electricity and magnetism, and at predicting—and then verifying—the existence of electromagnetic waves. But the physics of particles and waves was completely unable to explain how atoms emit light. Physicists' inability to understand spectra was the first sign that new ideas and theories were needed. The first hint of a new direction in which to turn was made in 1885 by a Swiss schoolteacher, Johann Balmer.

Balmer and the Hydrogen Atom

Balmer was interested in the line spectrum of hydrogen. Hydrogen is the simplest atom, with a single atom orbiting a proton, and it also has the simplest atomic spectrum. The *visible spectrum* (the spectrum between 400 nm and 700 nm) of hydrogen consists of a mere four lines, whose wavelengths are given in Table 18-1. Physicists felt certain that such a simple spectrum must have a simple and straightforward explanation, but they had no more success at understanding the spectrum of hydrogen than for more complicated spectra such as mercury or neon.

TABLE 18–1 Wavelengths of visible lines in the hydrogen spectrum*

656.46 nm
486.27 nm
434.17 nm
410.29 nm

*vacuum wavelengths

In 1885, Johann Balmer discovered by trial and error that the four wavelengths in the visible spectrum of hydrogen could be represented by the formula

$$\lambda = \frac{91.18 \text{ nm}}{\left(\dfrac{1}{2^2} - \dfrac{1}{n^2}\right)}, \qquad n = 3, 4, 5, 6. \tag{18-1}$$

Balmer's formula was able to reproduce the measured wavelengths with much better than 0.1% accuracy. Not only was his formula accurate, it was also *simple*—in keeping with expectations that there should be a simple explanation of the hydrogen spectrum.

Balmer only knew the four visible wavelengths shown in Table 18-1, so the n in his formula only included four values. But an obvious question to ask was whether Eq. 18-1 also predicts wavelengths for $n = 7$, 8, 9, and so on. The prediction for $n = 7$ is $\lambda = 397.1$ nm, an ultraviolet wavelength. Spectroscopists were just beginning to extend their craft into the ultraviolet and infrared regions of the spectrum, and it was soon confirmed that Balmer's formula does, indeed, work for *all* values of n.

Balmer's formula predicts a *series* of spectral lines of gradually decreasing wavelength, converging to the *series limit* wavelength of $\lambda(n = \infty) = 364.7$ nm. Although there are an infinite number of spectral lines in this series, their intensities rapidly get weaker as n increases until, for large values of n, they blur together and cannot be resolved. This series of spectral lines is now called the **Balmer series**. Figure 18-3 shows a photograph of the Balmer series of hydrogen, in which the series limit is quite obvious.

With the success of the Balmer series, it was natural to ask what happens if the 2^2 in Eq. 18-1 is changed to 1^2 or 3^2 or, in general, m^2 where m is an integer. It was easy to calculate that all spectral lines in the series with 1^2, if they existed, would have fairly extreme ultraviolet wavelengths while all those in the series with 3^2 would be in the infrared range. Spectroscopists accepted the challenge and went to work developing the techniques for infrared and ultraviolet spectroscopy. (The difficulty is not with the gratings, which work fine over a very wide wavelength range, but with finding appropriate detectors to replace the photographic plate.)

The $m = 1$ series was discovered by Lyman, and is now called the Lyman series, while the $m = 3$ series was found by Paschen and, not surprisingly, is called the Paschen

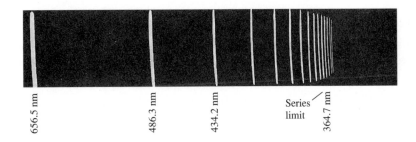

656.5 nm   486.3 nm   434.2 nm   Series limit   364.7 nm

FIGURE 18-3 The Balmer series of hydrogen, seen as a line spectrum on the photographic plate of a spectrometer. The spectral lines extend as a series to the limit at 364.7 nm

series. They provided confirmation, beyond any doubt, that Balmer's formula could be generalized to

$$\lambda = \frac{91.18 \text{ nm}}{\left(\dfrac{1}{m^2} - \dfrac{1}{n^2} \right)}, \quad \begin{cases} m = 1 & \text{Lyman series} \\ m = 2 & \text{Balmer series} \\ m = 3 & \text{Paschen series} \\ \vdots \end{cases} \quad n = m+1, m+2, \ldots \quad (18\text{-}2)$$

As the spectroscopists acquired ever more data, it became increasingly clear that Eq. 18-2 could predict *every* line in the hydrogen spectrum, from the extreme ultraviolet to the far infrared.

Surely Balmer's success was not a mere coincidence. There must be some underlying meaning to his formula—but what? Balmer did not present a *theory* but simply said "Here's a formula that accurately calculates the wavelengths in the hydrogen spectrum." This, then, was a challenge that any successful theory of atoms was going to have to meet: *derive* Balmer's formula (and the Eq. 18-2 extension of Balmer's formula) from the basic laws and principles of the theory. It was thirty years before the first theory was proposed that could meet this challenge—as you will learn in Part VI.

It is particularly striking that Eq. 18-2 depends on two *integers*. Hydrogen atoms simply do not emit wavelengths for $m = 1.6$ or for $n = 3.4$—n and m must be integers. This must tell us *something* important about the structure of the hydrogen atom. Newtonian mechanics does not deal in such "discrete" quantities. Masses, forces, velocities, energies—these quantities can take on any value at all; they are not restricted to having only some values but not others. We have, however, seen one exception to this: standing waves. Standing waves exist only for certain frequencies and wavelengths that are described by an *integer* called the mode number. Could there, somehow, be a connection between standing waves and the structure of atoms?

We are seeing here some of the first hints of quantum physics, where we will find that energies—at the atomic level—*cannot* be varied continuously but are restricted to certain discrete *energy levels*. Section 18.5 will provide a first look at the connection between standing waves and atoms and at the existence of discrete energy levels. First, however, we must probe yet deeper into the nature of light and matter.

18.3 Photons

If you take a picture with an incorrect shutter speed on your camera you may get a picture that is seriously underexposed. The entire picture is present, but it is dim and faded. If you keep decreasing the exposure, you could expect the picture to gradually fade away, rather like turning down the volume on your stereo until you can no longer hear the sound.

Figure 18-4 shows a series of photographs made with a special camera in which the film has been replaced by a special, highly-sensitive detector. A correct exposure shows a perfectly normal photograph of a woman. But with very faint illumination (upper left), the picture is *not* just a dim version of the properly-exposed photo. Instead, it is a collection of dots where some points on the detector have registered the presence of light but most have not. As the illumination increases, the density of these dots increases until the dots form a full picture.

FIGURE 18-4 Photographs made with an increasing level of light intensity. The very low light levels show individual points, as if light were a particle arriving at the detector. At higher light levels this particle-like behavior is not noticeable.

These results are completely unexpected. If light is just a wave, reducing its intensity should cause the detector output to drop slowly but *steadily* to zero. The picture would grow dimmer and dimmer until it disappeared, but the entire picture would remain present. But Fig. 18-4 looks more like someone randomly throwing "pieces" of light at the detector, causing full exposure at a few *discrete* points but no exposure at others. Nothing in waves physics allows us to explain the results of this experiment.

If we did not "know" that light is a wave, we would interpret the results of this experiment as evidence that light is a stream of some type of particle-like object. If these particles arrive frequently enough, they overwhelm the detector and it senses a steady "river" instead of the individual particles in the stream. But at sufficiently low intensities we become aware of the particles. How can we reconcile this disparity between the wave nature of light and this particle-like behavior?

Notice that the particle-like behavior of light becomes apparent in Fig. 18-4 only for light of very low intensity. Let's return to the experiment that showed most dramatically the wave nature of light—Young's double-slit interference experiment—and lower the light intensity by inserting filters between the light source and the slits. We cannot expect to see the interference fringes by eye for such a low intensity, so we will replace the viewing screen with the same detector used to make the photographs of Fig. 18-4.

What should we expect to see as the light intensity is decreased? If light is a wave, there is no reason to think that the nature of the interference fringes will change—they will just become dimmer and dimmer. The detector should continue to show alternating light and dark bands.

Figure 18-5 shows the outcome of such an experiment at three low light levels. The detector does not show bands at all but, instead, shows dots—just like those seen in Fig. 18-4. The detector is registering particle-like objects. They arrive one-by-one, and they are localized at a specific point on the detector. This is particle-like behavior, not wave-like behavior. (Waves, recall, are not localized at a specific point in space.) But these dots of light are not entirely random—they are grouped into bands at *exactly* the positions where we expected to see bright constructive interference fringes.

Figures 18-4 and 18-5 are our first evidence of the particle-like nature of light. The results are certainly not what wave-physics would lead us to expect, they are consistent with the idea that light arrives in particle-like pieces. These particle-like components of light are called **photons**.

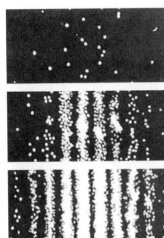

FIGURE 18-5 A simulation of a double-slit interference experiment with very low but increasing levels of light.

The photon concept was introduced by Einstein to explain an experiment called the *photoelectric effect*—an experiment we will investigate in detail in Part VI. Einstein made the hypothesis that light consists of discrete, massless units moving at the speed of light (3.00×10^8 m/s). Each unit, or photon, has the energy

$$E_{\text{photon}} = hf, \qquad (18\text{-}3)$$

where f is the frequency of the light and where h is a universal constant called **Planck's constant**. The value of Planck's constant is

$$h = 6.63 \times 10^{-34} \text{ J} \cdot \text{s}.$$

Most light sources with which you are familiar emit such vast numbers of photons that you are only aware of their wave-like superposition—just as you only notice the roar of a heavy rain on your roof and not the individual raindrops. But with extremely low intensities the light begins to appear as a stream of individual photons, like the random patter of raindrops when it is barely sprinkling. Each dot on the detector in Figs. 18-4 and 18-5 signifies a point where one individual photon delivered its energy and caused a measurable signal.

EXAMPLE 18-1   What is the energy of an ultraviolet photon having a wavelength of 200 nm?

SOLUTION   The frequency of the photon is

$$f = \frac{c}{\lambda} = \frac{3.00 \times 10^8 \text{ m/s}}{200 \times 10^{-9} \text{ m}} = 1.50 \times 10^{15} \text{ Hz}.$$

Using Eq. 18-3, the energy of this photon is

$$E_{\text{photon}} = hf = 9.95 \times 10^{-19} \text{ J}.$$

This is an extremely small energy. A typical light bulb emits about 1 joule of energy every second, which corresponds to an emission of $>10^{18}$ photons every second.

If photons are particles, they are certainly not "classical" particles because they do not travel in straight lines. Classical particles would travel in straight lines through the two slits and make just two bright areas on the detector. Instead, the *particle*-like photons seem to be landing at places where a *wave* undergoes constructive interference, thus forming the bands of dots.

Suppose that the detector in the double-slit interference experiment is 30 cm behind the slits and that the light intensity is so low that only 10^6 photons arrive per second. This is experimentally quite feasible. On average, a new photon passes through the slits every 1 μs (10^{-6} s). A photon moving at the speed of light travels a distance $d = ct = 300$ m during 1 μs. So while one photon is traveling the 30 cm between the slits and the detector, the next photon is an average of 300 m away. Or in the likely case that the light source is closer to the slits than 300 m, the next photon has not yet even been emitted by the light source! Under these conditions, only one photon at a time is passing through the double-slit apparatus.

If photons are arriving at the detector in a banded pattern as a consequence of interference, but if only one photon at a time is passing through the experiment, with what is it interfering? The only possible answer is that the photon is interfering *with itself*. There is nothing else present. But if each photon interferes with itself, rather than with other photons, then each photon, despite the fact that it is a particle-like object, must somehow go through *both* slits!

This all seems pretty crazy. We just spent several chapters accumulating evidence that light is a wave. It interferes, it diffracts, it was predicted and confirmed from electromagnetic theory. Now we come along and present evidence that light is a particle—but not a particle like any we've seen before. This light particle somehow seems to go through *both* slits of a double-slit experiment and then *interfere* with itself!

Crazy or not, this is the way light behaves. Sometimes it exhibits particle-like behavior and sometimes it exhibits wave-like behavior. Even when the light intensity is low enough for us to detect individual photons, the points at which the photons arrive are determined by the behavior of a wave. In some strange way, light is acting like *both* a particle *and* a wave. This completely defies our common sense, which says things are either a particle or a wave but cannot be both. But we need to keep in mind that our intuition and common sense were developed by our experience with everyday-sized, macroscopic objects. We have no direct experience with how microscopic objects such as atoms behave or how they emit and absorb light.

If you are expecting that we will now bring forth an "explanation" so that these observations will all "make sense"—sorry. This is simply how light really and truly behaves. The thing we call "light" is stranger and more complex than it first appears, and there just is not a way for these seemingly contradictory behaviors to make sense. We have to accept nature as it is rather than hoping that nature will conform to our expectations. Furthermore, it turns out that this half-wave/half-particle behavior is not restricted to light.

18.4 Matter Waves

The Davisson-Germer Experiment

A fascinating and historic experiment took place in 1927 at the Bell Telephone Laboratories in New York. Two physicists, Clinton Davisson and Lester Germer, were studying how electrons reflect from the surface of metals—a series of experiments they had begun several years before. This time, they used a well-crystallized piece of nickel as their target. When they rotated the sample, as shown in Fig. 18-6a, they discovered that the intensity of the reflected electron beam exhibited clear minima and maxima! Note that Davisson and Germer were using *exactly* the same experimental arrangement as that used to study x-ray diffraction (see Fig. 17-22), where minima and maxima are also observed.

Their results were certainly reminiscent of Bragg diffraction of x rays by a crystal. Davisson and Germer found a *scattering* angle $\phi = 50°$—the angle between the incident and the reflected electrons—when using electrons having a speed of 4.35×10^6 m/s. Although we "know" that electrons are material particles, completely unlike light waves, suppose we were to analyze this result *as if* electrons were a wave undergoing Bragg diffraction.

Figure 18-6b shows a crystal and a set of parallel atomic planes giving a scattering angle of 50°. Because the angle of reflection equals the angle of incidence, we can see

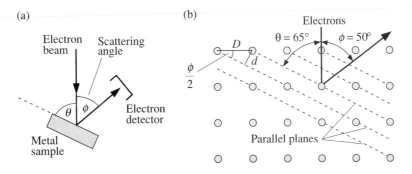

FIGURE 18-6 a) The Davisson-Germer experiment used to study electrons scattered from metal surfaces. b) Interpretation of the scattering as Bragg diffraction from the crystal lattice of the metal.

from the geometry that $\theta = 90° - (\phi/2) = 65°$. We can also see that the spacing d between the planes is related to the interatomic spacing D by

$$d = D\sin(\phi/2) = D\sin(90° - \theta) = D\cos\theta. \qquad (18\text{-}4)$$

The Bragg condition for the diffraction of a wave by the atomic planes was found in Chapter 17 to be (Eq. 17-10):

$$2d\sin\theta = m\lambda, \; m = 1, 2, 3, \ldots \qquad (18\text{-}5)$$

The geometrical relationship of Eq. 18-4 allows us to rewrite the Bragg condition in terms of the interatomic spacing D, rather then the plane spacing d, as

$$2(D\cos\theta)\sin\theta = D(2\sin\theta\cos\theta) = D\sin(2\theta) = m\lambda. \qquad (18\text{-}6)$$

The interatomic spacing of nickel was already known, from x-ray diffraction, to be $D = 0.215$ nm. Combining this with the measured angle $\theta = 65°$, and assuming $m = 1$, gives a "wavelength"

$$\lambda = D\sin(2\theta) = 0.165 \text{ nm}.$$

This seems like a pointless exercise. So what if, for some reason, electrons reflect from a nickel surface with a scattering angle of 50°? Electrons are particles of matter, so there must be some explanation in terms of the collision of the electron with the atoms at the surface of the crystal—right? But inspired by their initial results, Davisson and Germer searched for, and quickly found, *twenty* other reflections obeying the Bragg condition for *exactly* the same "wavelength" of 0.165 nm.

These results could not be a coincidence. Electrons—particles of matter, having a mass—were somehow, in some way, being *diffracted* by the grating of a crystal. Particles of matter, for the first time ever, were being observed to have wave-like properties!

The de Broglie Wavelength

Three years earlier, in 1924, a French graduate student named Louis-Victor de Broglie (Fig. 18-7) was puzzling over the growing evidence that light seemed to have both wave-like and particle-like properties. Sometimes light acted like a classical wave, exhibiting interference and diffraction. Other times, however, light seemed to come in small, localized pieces like a particle. Einstein had just won the Nobel Prize in 1921 for his explanation of the photoelectric effect in terms of particle-like photons of light.

If light—something that we generally think of as a wave—can act like a particle, then it occurred to de Broglie that perhaps some object we generally think of as a particle would, if conditions were right, act like a wave! What are the most "particle-like" entities we can think of? Very likely electrons and protons, the

FIGURE 18-7 Louis-Victor de Broglie.

basic building blocks of matter. Can an electron or a proton act like a wave? What would that mean? What behavior would they exhibit that is wave-like? And what is the "wavelength" of an electron, if it has one?

De Broglie postulated that a particle of mass m and momentum $p = mv$ has a wavelength

$$\lambda = \frac{h}{p}, \tag{18-7}$$

where h is Planck's constant. The Eq. 18-7 wavelength for material particles is now called the **de Broglie wavelength**. It depends *inversely* on the particle's momentum, so the largest wave effects will occur for particles having the smallest momentum.

What led de Broglie to this postulate? The constant that we now symbolize as h was first introduced into physics in the year 1900 by the German physicist Max Planck as part of an explanation of how incandescent objects emit the continuous spectra that are observed. Planck's explanation was not widely accepted until five years later, in 1905, when Einstein showed that the photoelectric effect could be understood if the energy E of a photon of light is related to its frequency f by Eq. 18-3: $E_{photon} = hf$.

This was the relationship that intrigued de Broglie. He reasoned that matter, if it has wave-like properties, should also obey Einstein's $E = hf$. But he also knew that the kinetic energy of a particle of mass m is related to its momentum by

$$E = \tfrac{1}{2}mv^2 = \tfrac{1}{2}m\left(\frac{p}{m}\right)^2 = \frac{p^2}{2m}. \tag{18-8}$$

What relationship between momentum and wavelength would allow these two statements about the particle's energy to be consistent with each other? The only possibility he could find was $\lambda = h/p$. The details of his reasoning, although not difficult, are not important to us. Our goal, instead, is to understand the experimental evidence for and some of the implications of de Broglie's bold and imaginative suggestion.

It is worth noting that there was absolutely *no* evidence for matter waves in 1924. But, as de Broglie must have reasoned, perhaps the lack of evidence was because no one had looked in the right places or used the right equipment and techniques. If Eq. 18-7 is right, what evidence would you expect to see? You now know that the most obvious characteristics of waves are their ability to exhibit interference and diffraction. But you have also learned that diffraction effects are not easily observable unless the opening through which the wave passes is comparable in size to the wavelength. If a wave passes through an opening of size $a \gg \lambda$, it will show no obvious spreading. So what wavelengths do material particles have?

EXAMPLE 18-2   Estimate the de Broglie wavelength for a very slow 1 μm diameter particle.

SOLUTION One of the smallest macroscopic particles we could imagine using for an experiment would be a very fine smoke particle. These are about 1 μm in diameter, too small to see with the naked eye and just barely at the limits of resolution of a microscope. Such a particle has a mass $m \approx 10^{-18}$ kg. Let it be moving at about the slowest speed we can perceive by eye, $v \approx 1$ mm/s $= 10^{-3}$ m/s. Its momentum is then $p \approx 10^{-21}$ kg m/s, giving it a de Broglie wavelength

$$\lambda = \frac{h}{p} \approx 6.6 \times 10^{-13} \text{ m}.$$

The wavelength in Example 18-2 is only $\approx 1\%$ the size of an atom! We should not expect to see the wave-like nature of a 1 μm particle unless we can force it through a hole about the size of an atom. If a smoke particle has a wavelength this small, a baseball must have one even smaller by many factors of ten. It is thus little wonder, if de Broglie's suggestion is correct, that we do not see macroscopic objects exhibiting wave-like behavior!

EXAMPLE 18-3   Find the de Broglie wavelength of an electron having a speed of 4.35×10^6 m/s, the speed used in the Davisson-Germer experiment.

SOLUTION   Using the known electron mass of 9.11×10^{-31} kg, the de Broglie wavelength is

$$\lambda = \frac{h}{p} = \frac{h}{mv} = 0.167 \text{ nm}.$$

This result is in near-perfect agreement with Davisson and Germer's measurement of 0.165 nm!

Electrons moving with speeds in the range found in Example 18-3 have de Broglie wavelengths very similar to those of x rays—exactly the right scale to be diffracted by atomic crystals. Davisson and Germer's subsequent measurements, as well as those of others, fully confirmed de Broglie's hypothesis that material particles have wave-like properties.

Davisson's Nobel Prize lecture puts the issue in perspective:

> It would be pleasant to tell you that no sooner had Elsasser's suggestion [that free electrons might show diffraction] appeared than the experiments were begun in New York which resulted in a demonstration of electron diffraction—pleasanter still to say that work was begun the day after copies of de Broglie's thesis reached America. The true story contains less of perspicacity and more of chance. The work actually began in 1919 ...

> Out of this grew an investigation of the distribution in angle of these elastically scattered electrons. And then chance again intervened; it was discovered, purely by accident, that the intensity of elastic scattering varies with the orientations of the scattering crystals. Out of this grew, quite naturally, an investigation of elastic scattering by a single crystal of predetermined orientation. The initiation of this phase of the work occurred in 1925, the year following publication of de Broglie's thesis ... Thus the New York experiment was not, at its inception, a test of the wave theory. Only in the summer of 1926, after I had discussed the investigation in England ... did it take on this character.

> The search for diffraction beams was begun in the autumn of 1926, but not until early in the following year were any found—first one and then twenty others in rapid succession. Nineteen of these could be used to check the relationship between wavelength and momentum, and in every case the correctness of the de Broglie formula was verified to within the limit of accuracy of the measurements.

This example illustrates a common occurrence in the history of science. Davisson and Germer had not set out to perform a breakthrough experiment; they were simply continuing research that had started years earlier. But as they proceeded they found unexpected and unexplainable results. They were open-minded enough to seek out the advice and opinions of others, where they learned of the possibility that they might be able to demonstrate electron diffraction. There was a large element of chance and luck involved—they just happened to be doing the "right" experiments at the right time. But they were also prepared, through their careful thought and study, to recognize a unique opportunity when it came along. It was their willingness to give a fair test to a really novel idea—that electrons might be waves—that allowed Davisson and Germer to be the first to witness directly the wave-nature of matter.

The Interference and Diffraction of Matter

Further evidence in support of de Broglie's hypothesis was soon forthcoming. The English physicist G. P. Thompson performed a diffraction experiment with an electron beam transmitted *through* a crystal—exactly equivalent to Fig. 17-23 for x-ray diffraction. Figure 18-8 shows photographs of both x-ray and electron diffraction patterns when passed through an aluminum foil target. (The foil is not a single crystal but, instead, thousands of tiny crystal grains at random orientations. As a consequence, the single-crystal diffraction spots of Fig. 17-23b get rotated about the axis of the x-ray or electron beam and form concentric diffraction circles.) The primary observation to make from Fig. 18-8 is that electrons act exactly like x rays—they really *are* waves and undergo diffraction!

FIGURE 18-8 The diffraction patterns produced by x rays (left) and electrons (right) due to an aluminum foil target.

G. P. Thompson shared the 1937 Nobel Prize with Davisson for their discovery of the wave-like properties of matter. G. P. Thompson was the son of J. J. Thompson, who discovered the electron in 1897—a discovery for which J. J. Thompson had been awarded the Nobel Prize in 1906. It has often been noted that the father won the Nobel Prize for showing that the electron is a particle while the son won the Nobel Prize for showing that the electron is a wave!

It later became possible to show that de Broglie's hypothesis applies to other material particles as well. Neutrons have a much larger mass than electrons—which tends to decrease their de Broglie wavelength—but it is also possible to generate very slow neutrons. The much smaller speeds compensate for the heavier mass, so neutron wavelengths are comparable to electron wavelengths. Figure 18-9 shows a neutron diffraction pattern. It is similar to the x-ray and electron diffraction patterns, although of lower quality because neutrons are harder to detect. A neutron, too, is a matter wave. In fact, in recent years it has become possible to observe the interference and diffraction of entire atoms!

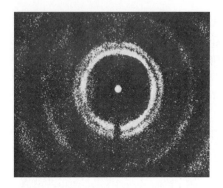

FIGURE 18-9 Neutron diffraction pattern.

The classic test of "waviness" is Young's double-slit interference experiment. If an electron, or other material object, has wave-like properties it should exhibit interference when passing through two slits. Does it? This is not an easy experiment to do because the spacing between the two slits has to be very tiny—comparable to the electron's wavelength. The technical challenges of such an experiment could not be met until around 1960, when it became possible to produce slits in a thin foil with a spacing of ≈2 μm = 2000 nm. Even so, the electron beam apparatus had to accelerate the electrons to much higher velocities than Davisson and Germer used (for various techni-cal reasons), reducing their de Broglie wavelength to ≈0.005 nm. This rather significant discrepancy between the wavelength and the slit spacing would be the equivalent of doing an optical double-slit experiment with a slit spacing of 20 cm! Nonetheless, the experi-ment was performed and Fig. 18-10a shows the highly-enlarged electron pattern that was detected. Interference fringes! Amazing as it seems, electrons—one of the basic building blocks of matter—exhibit interference when passing through a double slit.

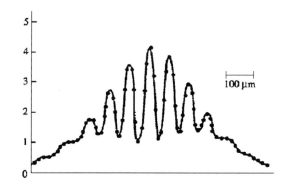

FIGURE 18-10 a) Photograph of a double-slit interference pattern of electrons. b) Recorded double-slit interference pattern of neutrons.

Later, during the 1970s and 1980s, techniques were developed for observing the two-slit interference of neutrons. Figure 18-10b shows the pattern recorded when neutrons were passed through a double slit having a separation of 104 μm. Once again, despite the much larger mass of the neutron, the characteristic interference fringes are easily observed.

Figure 18-11 shows an electron double-slit experiment in which the intensity of the electron beam was reduced to only a few electrons per second. You can see that each elec-tron is detected on the screen as a *particle*—a localized dot of light where the electron hits—but that the pattern of dots is the interference pattern of a *wave* of wavelength $\lambda = h/p$. Compare this picture to Fig. 18-5 for photons. They are the same! (Note that Fig. 18-5 was a simulation, but Fig. 18-11 is a photograph from a real experiment.) In both cases, electrons and photons, we see a combination of both wave-like and particle-like behaviors.

We noted, for photons, that each photon must in some sense interfere with itself. The same is true for electrons. If only a few electrons arrive per second, then only one electron at a time is in the region of the slits and the screen. Somehow each electron goes through both slits, has a wave-like interference with itself, but is finally detected at the screen as a particle-like dot.

We are *not* saying that photons and electrons are the same thing. We are saying that both light and electrons are found to share both wave-like and particle-like properties, so under similar experimental conditions we can expect to see similar behavior. Nonetheless, electrons are matter—particles with mass and charge and obeying $\lambda = h/p$—while photons have no mass, no charge, and obey $\lambda = c/f$. There are many other situations where the behavior of electrons and photons is quite distinct.

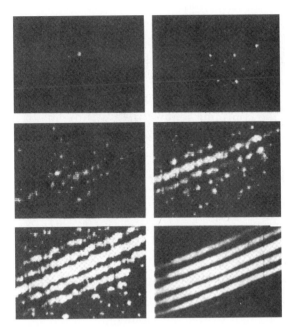

FIGURE 18-11 A double-slit interference pattern of electrons is seen to be built up electron by electron.

18.5 Energy Is Quantized!

An electron passing through two slits acts as a traveling wave as it moves from left to right. If de Broglie's hypothesis is correct, and material particles have wave-like properties, what are the implications for a particle that is confined to a specific region of space?

You have seen what happens to waves that are confined to a localized region—they form standing waves. That is what Chapter 15 was all about. A traveling wave reaches the boundary, reflects, then travels to the opposite boundary and reflects again. The region between the two boundaries is filled with waves traveling in both directions, and it is their superposition that gives us standing waves.

To see how this applies to particles, let's consider a situation that physicists call "a particle in a box." Figure 18-12 shows a particle of mass m confined in a box of length L. We will, for simplicity, consider only the one-dimensional motion parallel to the length of the box. Figure 18-12a shows a "classical" particle, such as a

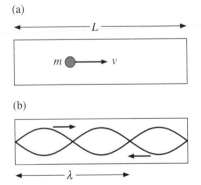

FIGURE 18-12 a) A classical particle confined in a box of length L. b) A de Broglie wave forms a standing wave.

macroscopic dust particle, in the box. According to Newton's laws it will simply bounce back and forth at constant speed. But de Broglie's hypothesis about the wave-like nature of the particle suggests that we need to consider a *wave* reflecting back and forth from the two ends of the box. This will set up the standing wave shown in Fig. 18-12b.

You already know how the wavelength of a standing wave is related to the length of the confining region—that was the major result of Section 15.4 (Eq. 15-16):

$$\lambda_n = \frac{2L}{n} \qquad n = 1, 2, 3, 4, \dots. \tag{18-9}$$

But the particle must also satisfy the de Broglie condition that $\lambda = h/p$. Equating these two statements about the wavelength leads to the conclusion, for de Broglie waves, that

$$\frac{h}{p} = \frac{2L}{n}$$

$$\Rightarrow p_n = \left(\frac{h}{2L}\right)n \qquad n = 1, 2, 3, 4, \dots. \tag{18-10}$$

In other words, the momentum of the particle described by a de Broglie wave cannot take on just any value but only those *discrete* values allowed by Eq. 18-10.

The momentum of the particle is directly related to the particle's energy by Eq. 18-8:

$$E = \tfrac{1}{2}mv^2 = \frac{p^2}{2m},$$

where we've used the fact that the potential energy in the box is $U = 0$. If we now use Eq. 18-10 for the momentum, we find that the *only* allowed energies for the particle in the box are given by

$$E_n = \left(\frac{h^2}{8mL^2}\right)n^2 \qquad n = 1, 2, 3, 4, \dots \tag{18-11}$$

This conclusion is one of the most profound discoveries of physics. Because of the wave-nature of matter—which, we have seen, has ample experimental confirmation—the energy of a confined particle can *only* take on certain discrete values. It is simply not possible for the particle to exist in the box with any energy other than one of the values given by Eq. 18-11. We call this the **quantization** of energy, or say that energy is *quantized*.

The number n is called the **quantum number,** and each value of n characterizes one **state,** or **energy level,** of the *system* of the particle in the box (both terms are used). The least energetic state, with $n = 1$, is called the **ground state** of the system.

Not only is the energy quantized, we can see from Eq. 18-11 that the energy of the particle in the box cannot be reduced below a minimum value $E_{\min} = E_1 = h^2/8mL^2$. But because $U = 0$ inside the box, E is the particle's kinetic energy. Thus E_1 is the *least* kinetic energy a particle can have. It must be that the particle is *always* in motion—it cannot be made to stay at rest! These properties of a wave-like particle in a box are in stark contrast to those of a classical Newtonian particle, for which the possible energies are continuous and the minimum energy is zero. If we think of an atom as an electron confined in a somewhat more complicated "box," then we can begin to see why Newtonian physics failed to provide an explanation of atomic properties.

Notice that the allowed energies are inversely proportional to both m and L^2. We know that the quantization of energy is not apparent with macroscopic objects, or else we would have known about it long ago. Therefore both m and L have to be exceedingly small before energy quantization has any significance. This is an issue we need to look into, because any new theory about matter and energy cannot be in conflict with our observations of macroscopic objects. Newtonian physics still works for baseballs.

EXAMPLE 18-4   What is the ground state energy of the very small 1 μm particle of Example 18-2 if it is confined to a box 10 μm in length?

SOLUTION This is about as small as we can easily imagine making macroscopic particles and boxes. In Example 18-2 we estimated that such a particle has $m \approx 10^{-18}$ kg. The ground state energy, $n = 1$, is then

$$E_1 = \frac{h^2}{8mL^2} \approx 5 \times 10^{-40} \text{ J}.$$

This is an unimaginably small amount of energy. By comparison, the kinetic energy of this particle moving at a barely perceptible speed of 1 mm/s is $K \approx 5 \times 10^{-25}$ J—a factor of 10^{15} larger. There is no way we could ever observe or measure discrete energies this small, so it is not surprising that we are unaware of energy quantization for macroscopic objects.   •

EXAMPLE 18-5   What is the ground state energy of an electron confined to a box of 0.1 nm?

SOLUTION Using the known mass of an electron (9.11×10^{-31} kg) we can calculate

$$E_1 = \frac{h^2}{8mL^2} = 6 \times 10^{-18} \text{ J}.$$

Such an energy corresponds to an electron speed $v = 3.6 \times 10^6$ m/s, roughly 1% of the speed of light. A box length of 0.1 nm is about the size of an atom. The very large speed of the *minimum* electron energy in an atomic-size box suggests that the wave-nature of electrons *is* important in the physics of atoms. The electron orbits must, in some sense, be standing waves. We will study the atom much more carefully in Part VI, but this example shows that the quantization of energy *is* important at the microscopic scale of the atom.   •

What is the implication of these two last examples? Even for a particle as small and well confined as the particle in Example 18-4, the size of one quantum of energy is so tiny that we stand no chance of being able to observe the effects of energy quantization. Energy quantization is simply not an issue for the physics of macroscopic objects—Newtonian physics works fine. But at the much smaller scale of atoms the physics is very new and different.

These examples raise more questions than they answer. If matter is some kind of wave, what is waving? What is the medium of a matter wave? What kind of displacement does it undergo? De Broglie's hypothesis is not a *theory*, and it provides no answers to important questions such as these. Nonetheless, you have seen strong experimental evidence confirming de Broglie's hypothesis about the wave-nature of matter.

De Broglie's suggestion came nearly forty years after Balmer's discovery—forty years during which the atom was being explored and the failures of classical physics were becoming ever more apparent. His suggestion was the final spark, igniting a whirlwind of activities and new ideas that led within a year to a complete and revolutionary new theory—quantum physics. We will revisit these issues later, for a more complete study, but for now we want to note just how far we have been able to come with our study of waves.

18.6 Wave Packets

The experimental evidence is overwhelming that light sometimes acts like a particle and that electrons, and other matter, sometimes act like waves. This is completely at odds with the models we have developed of particles and waves—the physics of water waves and baseballs. These "classical" ideas of particles and waves are mutually exclusive—an object can be one or the other, but not both. These classical ideas apparently fail to work at the atomic level. So with what should we replace them? A useful model is that of a *wave packet*.

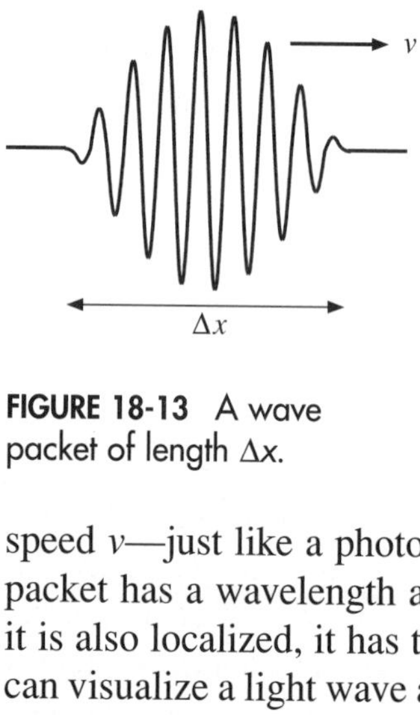

A wave oscillates and has a wavelength. A particle, on the other hand, can be localized within a small region of space. Consider a wave such as the one shown in Fig. 18-13. Unlike the sinusoidal waves we have considered previously, which stretch throughout space, this wave is bunched up, or localized, within a finite region of space of size Δx. The localization is particle-like while the oscillation is wave-like. Such a localized wave is called a **wave packet**.

FIGURE 18-13 A wave packet of length Δx.

A wave packet travels through space with constant speed v—just like a photon in a light wave or an electron in a force-free region. A wave packet has a wavelength and thus will undergo interference and diffraction. But because it is also localized, it has the possibility of making a "dot" when it strikes a detector. You can visualize a light wave as consisting of a very large number of these wave packets moving along together, while a beam of electrons is a series of wave packets spread out in a line. Wave packets are not a perfect model of photons or electrons—we need a full treatment of quantum physics to get a more accurate description—but they do provide a useful way of thinking about photons and electrons in many circumstances.

Notice that the wave packet of Fig. 18-13 looks very much like the earlier Fig. 14-24, where we illustrated the phenomenon of beats. Recall that beats occur if we superimpose two waves of frequencies f_1 and f_2 where the two frequencies are very similar: $f_1 \approx f_2$. One beat, or one wave packet, lasts for a duration Δt from when the beat begins until it ends. In the case of beats, another beat begins as soon as this one ends and the pattern repeats over and over.

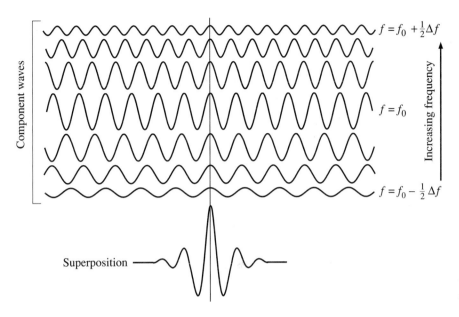

FIGURE 18-14 A wave packet is the superposition of many component waves of very similar wavelength and frequency.

A more advanced treatment of waves, called Fourier analysis, reveals that a single, *non*-repeating wave packet is a superposition of *many* waves of very similar frequency. Figure 18-14 shows that a wave packet is the sum, or superposition, of a whole series of waves with very similar frequencies. All of these waves are *in phase* along the vertical line, and their constructive interference at that instant of time produces the maximum amplitude of the wave packet. At increasing distance from this center line, the phases of the different waves become more and more different. Their superposition then has equal numbers of positive and negative contributions and the net displacement becomes zero.

When only two frequencies f_1 and f_2 are superimposed, creating beats, the wave packet duration Δt is the repetition interval between beats or, equivalently, the *period* of the beats: $\Delta t = T_{\text{beat}}$. The beat frequency was found to be

$$f_{\text{beat}} = f_1 - f_2 = \Delta f, \tag{18-12}$$

where Δf is the *range* of frequencies included in the superposition that forms the wave packet. Because the period and the frequency are inverses of each other, the duration Δt is

$$\Delta t = T_{\text{beat}} = \frac{1}{f_{\text{beat}}} = \frac{1}{\Delta f},$$

which we can rewrite in the form

$$\Delta f \, \Delta t = 1. \tag{18-13}$$

There is nothing new here—we are simply writing what we already knew in a somewhat different form. Equation 18-13 is just a combination of three things: the relationship $f = 1/T$ between period and frequency, rewriting T_{beat} as Δt, and the specific knowledge that the beat frequency f_{beat} is the difference Δf of the two waves contributing to the wave packet.

A more general wave packet, such as the one in Fig. 18-14, is formed by the superposition of *many* waves of similar frequency. While this makes the mathematics more complicated, it is still the same idea as forming beats with just two frequencies. The wave packet itself lasts for a certain length of time Δt, and it is formed by the superposition of many waves within some range of frequencies Δf. How, in this more general case, are Δt and Δf related to each other?

An advanced analysis of wave packets can prove the following result for any wave packet:

$$\Delta f \, \Delta t \approx 1 . \tag{18-14}$$

The relationship between Δf and Δt for a general wave packet is not as precise as Eq. 18-13 for the two frequencies contributing to beats. There are two reasons for this. First, wave packets come in a variety of shapes. The exact relationship between Δf and Δt depends somewhat on the shape of the wave packet. Second, we have not given a precise definition of Δt and Δf for a general wave packet. The quantity Δt is "about how long the wave packet lasts" while Δf is "about the range of frequencies needing to be superimposed to produce this wave packet." For our purposes, we will not need to be any more precise than this.

Equation 18-14 is a purely classical result for waves of any kind. It tells you what range of frequencies you need to superimpose to construct a wave packet of duration Δt. Alternatively, Eq. 18-14 tells you that a wave packet created as a superposition of various frequencies cannot be arbitrarily short but *must* last for a time interval $\Delta t \approx 1/\Delta f$.

EXAMPLE 18-6 A short-wave station has a frequency of 10.00 MHz. What range of frequencies must be superimposed to broadcast a short-wave pulse lasting 0.8 μs?

SOLUTION The period of a 10.00 MHz oscillation is 0.1 μs. A pulse 0.8 μs in length is 8 oscillations of the wave, exactly as shown in the wave packet of Fig. 18-13. This pulse is not a pure 10.00 MHz oscillation because a pure oscillation would last forever. Instead, this pulse has been created by the superposition of many waves over a range a frequencies

$$\Delta f \approx \frac{1}{\Delta t} = \frac{1}{0.8 \times 10^{-6} \text{ s}} = 1.25 \times 10^{6} \text{ Hz} = 1.25 \text{ MHz} .$$

This range of frequencies will be centered at the 10.00 MHz broadcast frequency, so the actual frequencies used to create this pulse span the range

$$f_0 - \tfrac{1}{2}\Delta f \le f \le f_0 + \tfrac{1}{2}\Delta f$$
$$\Rightarrow 9.375 \text{ MHz} \le f \le 10.625 \text{ MHz} .$$

Short pulses like this are used to transmit digital signals.

There is another way of looking at what Eq. 18-14 is telling us. Suppose you want to identify "when" a wave packet arrives at a specific point—perhaps using a detector of some sort. At what instant of time can you say that the wave packet arrives? When the front edge arrives? When the maximum amplitude arrives? When the back edge arrives?

Because the wave packet is spread out in time, there is not a unique and well-defined time *t* at which the packet arrives. Instead, all we can say is that it arrives within some *interval* of time Δ*t*. We are *uncertain* about the exact arrival time.

Similarly, suppose we would like to know the oscillation frequency of a wave packet. There is not a precise value for *f* because the wave packet is constructed from many waves within a range of frequencies Δ*f*. All that we can say is that the frequency is within this range—we are *uncertain* about the exact frequency.

Equation 18-14 tells us that the uncertainty in our knowledge about the arrival time of the wave packet is related to our uncertainty about the packet's frequency: Δ*f* Δ*t* ≈ 1. The more precisely and accurately we know one quantity, the less precisely we will be able to know the other. Figure 18-15 shows two different wave packets. The wave packet of Fig. 18-15a is very narrow—very *localized* in time. As it travels, our knowledge of when it will arrive at a specified point is pretty precise. That is, Δ*t* is pretty small. But a very wide range of frequencies Δ*f* is required to create a wave packet with a very small Δ*t*. So the price we pay for being fairly certain about the time is a very large uncertainty Δ*f* about the frequency of this wave packet.

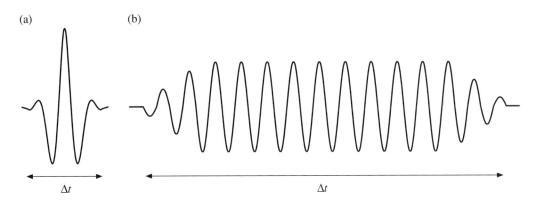

FIGURE 18-15 a) A wave packet well-localized in time (Δ*t* small) but with a large uncertainty Δ*f* in frequency. b) A packet with a well-defined frequency (Δ*f* small) but with a large uncertainty Δ*t* in its time of arrival.

Figure 18-15b shows the opposite situation, where the wave packet oscillates many times and the frequency of these oscillations is pretty clear. Our knowledge of the frequency is good, with minimal uncertainty Δ*f*. But such a wave packet is so spread out that there is a very large uncertainty Δ*t* as to its time of arrival.

Now it may be the case that technical limitations cause the uncertainties in our knowledge of *f* and *t* to be even larger than Eq. 18-14 implies. That is, the statement Δ*f* Δ*t* ≈ 1 is really a lower limit. In practice, the product Δ*f* Δ*t* may be much larger than 1. A better statement about our *knowledge* of the wave packet is

$$\Delta f \, \Delta t \gtrsim 1. \tag{18-15}$$

The fact that waves are spread out makes it impossible, even meaningless, to try to specify an exact frequency and an exact arrival time simultaneously. This is an inherent feature of "waviness" that applies to waves of any type.

18.7 The Heisenberg Uncertainty Principle

Equation 18-15, as we noted, applies to waves of any kind. If matter has wave-like aspects and a de Broglie wavelength, then Eq. 18-15 must somehow apply to matter. How? And what are the implications?

Consider a particle represented by a wave packet having de Broglie wavelength $\lambda = h/p$ and traveling with speed v along the x-axis. During the time interval Δt, which measures the duration of the wave packet, the packet moves forward by

$$\Delta x = v\Delta t = \frac{p}{m}\Delta t, \tag{18-16}$$

where $p = p_x = mv$ is the particle's x-momentum. The quantity Δx is the physical length of the wave packet. We can write the wave packet's duration in terms of its length as

$$\Delta t = \frac{m}{p}\Delta x. \tag{18-17}$$

We also know that any wave must satisfy the wave condition $\lambda f = v$. This is where we are going to introduce the idea of matter waves by using the de Broglie wavelength for λ. Isolating the frequency f gives

$$f = \frac{v}{\lambda} = \frac{(p/m)}{(h/p)} = \frac{p^2}{hm}. \tag{18-18}$$

Using Eq. 18-18, we can relate a small range of frequencies Δf to a small range of momenta Δp_x by

$$\Delta f = \frac{2p\Delta p_x}{hm}. \tag{18-19}$$

We have assumed that $\Delta f \ll f$ and $\Delta p_x \ll p$ (a reasonable assumption) and thus treated the ranges Δf and Δp_x as if they were differentials df and dp.

If we now multiply the Eqs. 18-17 and 18-19 expressions for Δt and Δf we find

$$\Delta f \Delta t = \frac{2p\Delta p_x}{hm} \cdot \frac{m\Delta x}{p} = \frac{2}{h}\Delta x \Delta p_x. \tag{18-20}$$

But we know, for any wave, that $\Delta f \Delta t \gtrsim 1$. So with one last rearrangement of Eq. 18-20 we can write, for a matter wave,

$$\Delta x \Delta p_x \gtrsim \frac{h}{2} \qquad \text{(Heisenberg uncertainty principle).} \tag{18-21}$$

This statement about the relationship between position and momentum of a particle was proposed by Werner Heisenberg, creator of the first successful complete quantum theory. Physicists often just call it the **uncertainty principle**. The uncertainty principle is sometimes stated with the right-hand side being $h/2$, as we have it, but other times with the right-hand side being just h or containing various factors of π. The specific number is not especially important because it depends on exactly how Δx and Δp are defined as well as what the exact wave packet shape is. The important idea is that the product of Δx and Δp_x for a particle cannot be significantly less than Planck's constant h. A similar relationship for $\Delta y \Delta p_y$ applies along the y-axis.

Now that we have stated Heisenberg's uncertainty principle, what does it mean? Our interpretation of the uncertainty principle is similar to our interpretation of Eq. 18-15: it is a statement about our *knowledge* of the properties of a particle. If we want to know *where* a particle is located, we measure its position x. That measurement is not absolutely perfect but has some uncertainty Δx. Likewise, if we want to know how fast the particle is going, we need to measure its velocity v or, equivalently, its momentum p. This measurement also has some uncertainty Δp_x.

There are always uncertainties associated with experimental measurements, but better procedures and techniques can reduce those uncertainties. Newtonian physics places no limits on how small the uncertainties can be. A Newtonian particle at any instant of time has an exact position x and an exact momentum p, and with sufficient care we can measure both x and p simultaneously with such precision that the product $\Delta x \, \Delta p_x \to 0$. There are no inherent limits to our knowledge about a classical, or Newtonian, particle.

Heisenberg, however, made the bold and original statement that our knowledge has real limitations. No matter how clever you are, and no matter how good your experiment, you *cannot* measure both x and p simultaneously with arbitrarily good precision. Any measurements you make are limited by the condition that $\Delta x \Delta p_x \gtrsim h/2$. Our knowledge about a particle is *inherently* uncertain.

Why? Because of the wave-like nature of matter. The "particle" is spread out in space and there simply is not a precise value of its position x. Similarly, the de Broglie relationship between momentum and wavelength implies that we cannot know the momentum of a wave packet any more exactly than we can know its wavelength or frequency. Our ideas that position and momentum have precise values are tied to our classical concept of what a particle is. That concept works well for Newtonian physics but, as we have seen, fails at the level of electrons and atoms. As we revise our ideas of what atomic particles are like, our old ideas about position and momentum are also going to have to be revised.

EXAMPLE 18-7   Return to our earlier example of a 1 µm particle ($m \approx 10^{-18}$ kg) confined within a 10 µm box. a) Can we know with certainty if the particle is at rest? b) If not, within what range is its velocity likely to be found?

SOLUTION a) If we know with *certainty* that the particle is at rest, then we would have $\Delta p_x = 0$. For that to be true, according to the uncertainty principle, the uncertainty in our knowledge of the particle's position would have to be $\Delta x \to \infty$. In other words, we would have no knowledge at all about the particle's position—it could be anywhere! But that is not the case. We know the particle is *somewhere* in the box, so the uncertainty in our knowledge of its position is $\Delta x = L = 10$ µm. With a finite Δx, the uncertainty Δp_x *cannot* be zero. We cannot know with certainty if the particle is at rest inside the box.

b) The best we can say is that the particle's momentum is somewhere within the range $\Delta p \approx h/(2\Delta x) = h/2L$. (We have assumed the most accurate measurements possible so that the "greater than or approximately equal to" in Heisenberg's uncertainty principle becomes simple an "approximately equal to.") The velocity will then be within a range of velocities

$$\Delta v = \frac{\Delta p}{m} \approx \frac{h}{2Lm} \approx 3.0 \times 10^{-11} \text{ m / s}.$$

This range of possible velocities will be centered on $v = 0$ if we have done our best to have the particle be at rest, but all that we can say with certainty is that the particle's actual velocity is somewhere within the interval -1.5×10^{-11} m/s $\leq v \leq 1.5 \times 10^{-11}$ m/s.

Now for all practical purposes you might consider the result from Example 18-7 to be a satisfactory definition of "at rest." After all, a particle moving with a speed of 1.5×10^{-11} m/s would need 6×10^7 s to move a mere 1 mm. That is about two years! Nonetheless, it is not perfectly at rest. We cannot even say whether it is moving to the right or to the left.

EXAMPLE 18-8   What range of velocities might an electron have if confined to a region of 0.1 nm (about the size of an atom)?

SOLUTION The analysis is the same as Example 18-7. If we know that the electron's position is located within an interval $\Delta x \approx 0.1$ nm, then the best we can know is that its velocity is within the range

$$\Delta v = \frac{\Delta p_x}{m} \approx \frac{h}{2Lm} \approx 4 \times 10^6 \text{ m / s}$$

Because the electron's *average* velocity is zero, the best we can say about the electron is that its velocity is somewhere in the interval -2×10^6 m/s $\leq v \leq 2 \times 10^6$ m/s. It is simply not possible to know the electron's velocity any more precisely than this. Unlike Example 18-7, where Δv was so small as to be of no practical consequence, our uncertainty about the electron's velocity is enormous—about 1% of the speed of light!

The point of these examples is that we can see, once again, that even the smallest of macroscopic objects behaves very much like a classical Newtonian particle. If a 1 μm particle is slightly fuzzy and has a slightly uncertain velocity, it is far beyond the measuring capabilities of even the very best instruments to detect this wave-like behavior. But the effects of the uncertainty principle at the atomic scale are stupendous. We are unable to determine the velocity of an electron in an atom to any better accuracy than about 1% of the speed of light!

What Does It Mean?

Let's take a closer look at the consequences of the uncertainty principle. Consider a single-slit diffraction experiment with a beam of electrons, as shown in Fig. 18-16. The source of electrons in this experiment is extremely far to the left, so as the electrons arrive at the slit they are traveling as perfectly as we desire along the x-axis and have no component of velocity along the y-axis. That is, $p_y = 0$. We will assume that the electron source is designed to emit all electrons with the same velocity, so they all have the same p_x as they reach the slit.

Is this motion possible without violating the Heisenberg uncertainty principle? If we *know* that p_y has the *specific* value $p_y = 0$, then $\Delta p_y = 0$. There is no uncertainty in our knowledge of p_y. According to the uncertainty principle, this can only happen if $\Delta y \rightarrow \infty$—that is, if we have no knowledge at all of the electron's y-coordinate. But that is exactly what is happening in this experiment. The de Broglie waves arriving at the slit from a *very* distant source are *plane waves*. We cannot localize a plane wave at all—it is spread out along the entire y-axis, so our uncertainty in the value of y for any particular electron does, indeed, approach infinity. Now it is true that there are no perfect plane waves, so we are considering

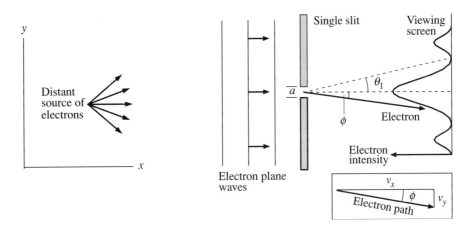

FIGURE 18-16 A single-slit experiment with electrons. Electrons reach the slit from a distant source with $p_y = 0$, then diffract at angle ϕ. The inset shows how an electron's diffraction angle ϕ is related to v_x and v_y, and thus to p_x and p_y.

an idealized experiment. Nonetheless, it *is* possible to make Δp_y extremely close to zero by allowing Δy to become extremely large. The uncertainty principle is satisfied.

As an electron reaches the slit it is forced to pass through a small hole of width a. We know the electron's y-coordinate with good accuracy as it *emerges* from the slit: it is within the small range of values $-a/2 \leq y \leq a/2$, so $\Delta y = a$. We have, in effect, *measured* the y-coordinate of those electrons that pass through the slit with an uncertainty $\Delta y = a$.

Have we beaten the uncertainty principle? We have made Δy very small, and we had $\Delta p_y \approx 0$ by arranging the experiment so that the electrons arrive with $p_y = 0$. Thus we seem to have $\Delta y \Delta p_y < h$ for the electron as it leaves the slit—right? No! While we knew p_y very accurately *before* the electron reached the slit, we do *not* know its value now that it has emerged from the slit on the other side. By measuring y we have *lost* information about p_y.

If p_y were still zero after passing through the slit, the electrons would all continue straight ahead and make a small dot of width a on the viewing screen. That is, they would act like classical particles. But that is not what happens! Instead, as we have seen, they *diffract* and spread out behind the slit. Some of the electrons that were traveling perfectly horizontally before reaching the slit are moving at an angle upon leaving it, so the electrons must have acquired a non-zero value of p_y as they went through the slit.

You learned, in your study of diffraction, that nearly all the intensity is located in the central maximum. This implies that nearly all the electrons will diffract at an angle ϕ somewhere within the range $-\theta_1 \leq \phi \leq \theta_1$, where $\theta_1 = \lambda/a$ is the angle of the $m = 1$ diffraction minimum (the edge of the central maximum). Now the angle at which an electron travels is related to its velocity and momentum components by

$$\tan \phi \approx \phi = \frac{v_y}{v_x} = \frac{p_y}{p_x}$$

$$\Rightarrow p_y \approx p_x \phi,$$

(18-22)

as shown in Fig. 18-16.

If ϕ is nearly always in the range $-\lambda/a \leq \phi \leq \lambda/a$, then the y-component of momentum is nearly always in the range $-p_x \lambda/a \leq p_y \leq p_x \lambda/a$. There would be no diffraction pattern if this range of y-momenta did not exist. Because of the diffraction, our *uncertainty* about the value of p_y after the electron passes through the slit is thus

$$\Delta p_y \approx \frac{2 p_x \lambda}{a} . \qquad (18\text{-}23)$$

Now the de Broglie wavelength of the electron is $\lambda = h/p$. But $p_y << p_x$ (because $\phi <<$ 1 rad), so to a very good approximation we can write $p \approx p_x$ and $\lambda \approx h/p_x$. In addition, as we just noted, our uncertainty in the electron's position upon leaving the slit is $a = \Delta y$. Combining these pieces, we can write the uncertainty Δp_y as

$$\Delta p_y \approx \frac{2 p_x \lambda}{a} = \frac{2 p_x (h / p_x)}{\Delta y} = \frac{2h}{\Delta y} . \qquad (18\text{-}24)$$

One last algebraic rearrangement gives

$$\Delta y \Delta p_y \approx 2h . \qquad (18\text{-}25)$$

Heisenberg's uncertainty principle! Not exactly Eq. 18-21, because $h/2$ has become $2h$, but keep in mind our earlier comments that factors of two or so are irrelevant because we have not given exact definitions of Δy and Δp.

The point of this analysis is that the diffraction and spreading of matter waves is a consequence of the uncertainty principle. If we try to improve our knowledge of y by localizing the electron with a narrower slit, the price we pay is an increased range of possible p_y and, thus, an increased spreading of the electrons behind the slit.

Now it is true that we knew p_y very accurately *before* the electron passed through the slit and y very accurately *after* it has passed through. That is not prevented by the uncertainty principle. It is concerned with what we can know about the values of x and p_x, or y and p_y, *simultaneously*, because that is the information needed to predict where the electron will be in the future. The fact that the electron went through the slit means that we *no longer* have accurate knowledge of p_y. Our measurement of the electron's position has *changed* its momentum!

The implications of this are profound. We did nothing to the electron that would change the value of p_y for a classical particle. After all, according to Newton, changing p_y requires a force component F_y, and there are no such forces here. All we did was to *measure* the value of y by selecting electrons that passed through the slit. It would appear that we cannot make measurements on a system without disturbing the system in some way. In this case we measured y and, in doing so, changed the value of p_y!

To state this more generally, an observer cannot study a phenomenon and make measurements without, in some way, changing the phenomenon which he or she is studying. This is a very non-classical idea. We have *assumed* all along that we could measure whatever information we need about a system—such as the velocity of a car—without changing the system. What we learn from quantum physics and the existence of matter waves is that our commonsense assumptions are at best of limited validity and, in some cases, totally wrong. True, the consequences of the uncertainty principle applied to a car are so far from being measurable as to be of no practical interest—but we *do* change a car's position ever so slightly by measuring its velocity! The implications at the atomic level are far more significant.

Summary

Important Concepts and Terms

spectroscopy	de Broglie wavelength
spectrometer	quantization
spectrum	quantum number
discrete spectrum	state
spectral line	energy level
line spectrum	ground state
Balmer series	wave packet
photon	uncertainty principle
Planck's constant	

A closer look at the nature of light and the nature of matter reveals surprising evidence that cannot be explained by either the particle model or the wave model of classical physics. In this chapter we have looked at experiments showing that light and atomic-level particles each exhibit both particle-like and wave-like properties.

Light consists of a stream of particle-like photons of energy

$$E_{\text{photon}} = hf,$$

where h is Planck's constant. The superposition of large numbers of photons has wave-like properties, but for very low light intensities we can detect the particle-like photons. However, these are not classical particles. The double-slit experiment implies that each photon somehow passes through both slits and then interferes with itself to create wave-like bands of particle-like dots on the detector.

Matter waves have a de Broglie wavelength

$$\lambda = \frac{h}{p} = \frac{h}{mv}.$$

As a consequence, atomic particles exhibit interference and diffraction just like a light wave of the same wavelength. Particles, like photons, seem to go through both slits of a double-slit apparatus and then interfere with themselves.

A particle confined in a region of space of length L must form a de Broglie standing wave. As a result, the particle can only have certain allowed energies given by

$$E_n = \left(\frac{h^2}{8mL^2} \right) n^2 \qquad n = 1, 2, 3, 4, \ldots.$$

The energy is quantized. Each allowed value of E is called an energy level, and the $n = 1$ energy level is called the ground state.

It is useful to think of photons and atomic particles as wave packets of length Δx and duration Δt. A wave packet is created by the superposition of many waves over a range of frequencies Δf such that

$$\Delta f \, \Delta t \approx 1.$$

This result is valid for any kind of wave. When applied to matter waves it leads to the Heisenberg uncertainty principle:

$$\Delta x \Delta p_x \gtrsim \frac{h}{2}.$$

The uncertainty principle is a statement about our knowledge of a particle. Because a particle has wave-like properties it is impossible to know both its position and its velocity with perfect accuracy.

The concepts of photons and matter waves raise more questions than they answer. Neither Einstein's hypothesis about photons nor de Broglie's postulate of matter waves is a *theory*. They provide no answers to questions such as "What is waving in a matter wave?" or "How do atoms emit and absorb photons of light?" We do not yet know what should replace Newton's physics as a theory about how matter behaves at the atomic level. All we have are hints that a correct theory must go beyond the wave and particle models of classical physics to include objects having both wave-like and particle-like properties. We will return to this topic in Part VI, where you will learn that the new theory of quantum physics is able to answer these questions and to *explain* such phenomena as the discrete spectra emitted by atoms.

Exercises and Problems

Data: $m_{elec} = 9.11 \times 10^{-31}$ kg $m_{proton} = m_{neutron} = 1.67 \times 10^{-27}$ kg

Exercises

1. Estimate your de Broglie wavelength while walking at a speed of 1 m/s.
2. a. What is the speed of an electron having a de Broglie wavelength of 0.20 nm?
   b. What is the speed of an proton having a de Broglie wavelength of 0.20 nm?
   c. What is the speed of a 100 g baseball having a de Broglie wavelength of 0.20 nm?
3. What is the smallest box in which you can confine an electron if you want to know for certain that the electron's speed is no more than 10 m/s?
4. What is the energy of a photon of light that has a wavelength of 500 nm?

Problems

5. a. Calculate the wavelengths of the first four members of the Lyman series in the spectrum of hydrogen.
   b. Calculate the wavelengths of the first four members of the Paschen series in the spectrum of hydrogen.
   c. What is the series limit for the Lyman series?
   d. What is the series limit for the Paschen series?
   e. Light from a hydrogen discharge is passed through a transmission diffraction grating and registered on a detector 1.5 m away. The $m = 1$ Lyman-alpha line is located 37.6 cm away from the central maximum. What is the position of the second member of the Lyman series?

6. A helium-neon laser emits a laser beam with a wavelength of 633 nm. The power of the laser beam is 1 mW.
   a. What is the energy of one photon of laser light?
   b. How many photons does the laser emit each second?

7. a. What are the energies of the first three quantum states of an electron confined in a one-dimensional box of length 0.5 nm?
   b. How much energy would the electron have to lose to change from the $n = 2$ state to the ground state?
   c. Suppose that the electron changes from the $n = 2$ state to the ground state by emitting a photon of light. If energy is conserved, what must the photon's wavelength be? Give your answer in nm.

8. Electrons with a speed of 2.0×10^6 m/s pass through a double-slit apparatus. A detector measures interference fringes with a fringe spacing of 1.5 mm.
   a. What will the fringe spacing be if the electrons are replaced by neutrons having the same speed?
   b. What speed must neutrons have to produce interference fringes with the same spacing of 1.5 mm?

9. The two-slit neutron diffraction pattern shown in Fig. 18-10b was measured 3 m behind two slits having a separation of 0.10 mm. Based on measurements that you can make on the figure (notice the scale on the figure), determine the speed of the neutrons.

10. You learned previously that sound waves are easily transmitted through a confined tube of air *only* if the sound's wavelength λ matches one of the possible standing wave resonances in the tube. This idea can help you understand a modern semiconductor device called a *quantum well device*. In a quantum well device, as shown in Fig. 18-17, a current of electrons is transmitted through a thin layer of material, of thickness L, that "confines" the electrons. That is, it is not easy for electrons to enter or leave this layer. A current will flow through this layer *only* if the electrons are able to excite de Broglie standing wave resonances in the confinement layer.

    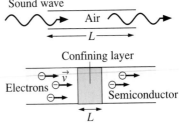

    FIGURE 18-17

    a. A typical semiconductor confinement layer has a thickness $L = 5$ nm. What are the energies of the first four energy levels of an electron in this "box?"
    b. What are the four slowest electron velocities for which a current will flow through this layer?

11. A wave packet of light is created by the superposition of many waves within a range of frequencies Δf. These frequencies span the range $f_0 - \frac{1}{2}\Delta f \leq f \leq f_0 + \frac{1}{2}\Delta f$, where $f_0 = c/\lambda$ is called the *center frequency* of the wave packet. Laser technology can generate a wave packet of light that has a wavelength of 600 nm and that lasts an incredibly small time interval of only 6 fs (1 fs = 1 femtosecond = 10^{-15} s).
    a. What is the center frequency of this wave packet of light?
    b. How many cycles, or oscillations, of the light wave are completed during the 6 fs interval that the light is "on?"
    c. What is the spatial length of the laser pulse as it travels through space?

    d. Draw a snapshot graph of this wave packet.

    e. What range of frequencies must be superimposed to create this pulse?

12. A proton is confined, by the strong force, into an atomic nucleus of diameter 4 fm (1 fm = 1 femtometer = 10^{-15} m). Use the uncertainty principle to estimate the range of speeds with which a proton in a nucleus might be found.

[**Estimated 10 additional problems for the final edition.**]

PART **III** Scenic Vista

Wave-Particle Duality

The common division of the world into subject and object, inner world and outer world, body and soul is no longer adequate.

Werner Heisenberg

We end our study of waves a long distance from where we started. Who would have guessed, as we examined our first oscillating spring, that we would end up with quantum numbers, the uncertainty principle, and a whole new non-Newtonian way of viewing the universe? But despite the wide disparity between oscillating springs and de Broglie standing waves, there have been a few key ideas that have stayed with us throughout this part of the book. These include simple harmonic motion, the principle of superposition, the phenomena of interference, and the properties of standing waves and resonances. You should, as part of your final study of waves, trace the influence of these ideas through the chapters of Part III.

A point we have tried to emphasize throughout Part III is the *unity* of wave physics. We did not need separate theories of string waves and sound waves and light waves, despite the great differences in their media and of their scales of times and lengths. Instead, a few basic ideas were found to provide us with a good explanation of both. By focusing on similarities, rather than differences, we have been able to analyze sound and light, as well as strings and electrons, in a single part of this book.

Unfortunately, wave physics is not as easy to summarize as particle physics. Newton's laws and the conservation laws are two very general sets of principles that can be used to analyze and solve most problems about particles. These general principles allowed us to develop the powerful problem-solving strategies of Part I and Part II. You have probably noticed that we have not found such general problem-solving strategies for wave problems. Wave problems, by their very nature, are much more varied and not as easily classified.

This is not to say that there is no structure to wave physics. Rather that the knowledge structure of waves is organized more around *phenomena* than around general principles. The three most general wave phenomena are wave sources, wave propagation, and wave interactions. Oscillations are the sources of waves (as well as interesting in their own right); traveling waves, standing waves, and characteristics of the medium are aspects of wave propagation; and interference and diffraction are wave interactions.

Table SVIII-1 summarizes the knowledge structure of wave physics. Unlike the knowledge structure of Newtonian mechanics, which was a "pyramid of ideas" radiating downward from the general principles at the top to a wide base of specific applications, the knowledge structure of waves is a logical grouping of the many topics you have studied. This is a different way of structuring knowledge, but it still provides you with a mental framework for analyzing and thinking about wave problems. As you review, you should think about where all the different topics of Part III appear in this table.

TABLE SVIII-1 The knowledge structure of wave physics.

Wave Motion

Source		Propagation	
periodic	*nonperiodic*	*wave type*	*propagation characteristics*
		electromagnetic	wave speed v
source characteristics:	pulse shape	sound	pulse waves: shape
frequency f		string	periodic waves:
amplitude A		matter	wavelength $\lambda = v/f$
phase constant ϕ_0		other	amplitude A
		traveling waves	*standing waves*
SHM sources:		history/snapshot graphs	boundary conditions
Hooke's Law		phase	modes
spring constant k		intensity	
$x/v/a$ relations			
energetics			

Wave Interactions

Principle of Superposition

single wave	*multiple waves*		
diffraction through apertures of various shapes	*different frequencies*	*same frequency*	
	beats: $f_{beat} = \Delta f$	constructive interference: $\Delta\phi = m \cdot 2\pi$ destructive interference: $\Delta\phi = (m + \frac{1}{2}) \cdot 2\pi$	
		two waves: Young's experiment Michelson interfer. loudspeakers	multiple waves: diffraction grating x-ray diffraction	other situations: light sound particles

The success of a theory or idea can often be measured by how widely it can be applied and by the number of phenomena it can explain. While wave physics is not a theory *per se*—not like Newton's mechanics—it is a set of concepts and ideas having a remarkable range of influence and validity. There is hardly any area of science or engineering that is free from the ideas of oscillations and waves. And waves, perhaps more than any other aspect of physics, bridge the chasm between classical and quantum physics. In the rest of this Scenic Vista, we will examine one aspect of this "bridge" more closely.

Wave-Particle Duality

Prior to the twentieth century, physicists classified phenomena as being *either* a particle *or* a wave. Planets, projectiles, and atoms were particles, or collections of particles, while sound and light were clearly waves. Particles followed trajectories given by Newton's laws; waves obeyed the principle of superposition and exhibited interference and diffraction. Everything seemed under control until physicists encountered evidence that light sometimes acts like a particle and, even stranger, that matter sometimes acts like a wave. This amazing behavior of matter was completely unexpected, and the wave-like aspects of matter continue even today to baffle physicist and non-physicist alike. But the experimental evidence for matter having wave-like properties is overwhelming and unrefutable. In fact, quantum physics—the full wave theory of matter that we will study in Part VI—is by far the most accurate and successful theory physicists have ever developed.

You might first think that light and matter are *both* a wave *and* a particle. But that idea does not work. The basic definitions of particleness and waviness are mutually exclusive. Something that is a wave—spread out, nonlocalized—cannot simultaneously be a hard, localized particle. It is more profitable to conclude that light and matter are *neither* a wave *nor* a particle. Waves and particles are concepts of classical physics. While all the macroscopic phenomena with which we are familiar seem to be easily classified as "waves" or "particles," we have little justification for thinking that phenomena outside the range of our experience have to fall into one or the other of these categories. So we should not be too surprised if we find phenomena somewhere in the universe where neither the wave nor the particle model seems to apply.

And, indeed, such phenomena do exist at the microscopic scale of atoms and their constituents—a physical scale completely inaccessible to our five senses. The classical concepts of wave and particle are simply too limited to explain the subtleties of nature that we find on the atomic scale. Rather than to force every phenomenon in nature into the molds of "wave" or "particle," we need to adopt a more open-minded approach. Let an electron be an electron, and a photon a photon, and we will let nature, rather than our assumptions, guide us as to just what these entities are.

For matter and light, the experimental evidence is that they have both wave-like aspects and particle-like aspects—but only one at a time. If we arrange an experiment to measure wave-like properties, such as interference, then we find photons and electrons acting like waves, not particles. An experiment to look for particles will find photons and electrons behaving as particles, not waves. They have two sides to their personalities, and we must be willing to accept both if we wish to understand them. This two-sided point of view is called *wave-particle duality*.

These two aspects of light and matter are *complementary* to each other—like a two-piece jigsaw puzzle. Neither aspect alone provides an adequate picture of light or matter, but taken together they provide us with a model for understanding these elusive but most fundamental constituents of nature. That is what wave-particle duality is all about.

It is not easy to go beyond our commonsense beliefs about the world. Even professional physicists, with years of experience, can find themselves baffled by the strange and mysterious events that take place in the world of quantum physics. The apparent solidity and definitiveness of matter at the macroscopic level makes it exceedingly hard to accept the fact that, deep inside, the constituents of that matter are fuzzy, wave-like objects. These discoveries have forced us to re-examine some of our most basic beliefs about the universe.

One such belief, an integral aspect of Newtonian physics, is that the motion of particles can be predicted. Recall Laplace's quote in Scenic Vista II about a great intelligence who could predict the future with perfect accuracy if, at some instant of time, she knew the positions and velocities of all particles in the universe and all the forces acting on them. While it is not practical to know this information, the fact that it is possible *in principle* led to the "clockwork universe" of the Newtonian worldview.

As we move into the atomic realm, we find that Newtonian physics is not an adequate description of reality. Because of the wave-like properties of matter, we now find that it is *not* possible, not even in principle, to know the position and velocity of a particle with perfect precision. The deterministic, clockwork Newtonian universe must be replaced with a new worldview in which chance and uncertainty play a major role. The quantum world is one of indeterminacy; we cannot, even in principle, predict the future. This is not an issue of just needing to make better measurements; the uncertainty in our knowledge is *inherent* in the waviness of matter, and no amount of improvement in our measuring abilities can get around this fact.

Wave-particle duality is an established and accepted principle of physics as we enter the twenty-first century. But what does it all mean? What does wave-particle duality tell us about the nature of the universe in which we live? Scientists and non-scientists alike, for over 200 years, felt that Newtonian physics provided a fundamental description of reality. The Newtonian worldview provided a familiar and widely-accepted understanding of the place and role of humans in the universe. Wave-particle duality, along with Einstein's relativity, undermines the basic assumptions of the Newtonian worldview, but there have been no fundamental changes in society during the twentieth century. Will the twenty-first century see the development and acceptance of a "quantum worldview" that reflects a different understanding of nature and reality? It is interesting to speculate about what a quantum worldview would be and what implications it would have for society, but the future—as we now know—is uncertain.

We will now leave waves behind for awhile as we head out in entirely new directions. But not forever. The discovery of matter waves has left us with unfinished business, and we will return in Part VI for a more in-depth study of the strange world of quantum physics. There we will find that electrons, and other atomic particles, are described by a completely new kind of wave called a probability wave.

Answers to Selected Exercises and Problems

Chapter 2

1. b) $2\hat{i} - 3\hat{j}$; c) 3.61, 56.3° below x-axis
2. b) $8\hat{i} + 7\hat{j}$; c) 10.6, 41.2° above x-axis
3. b) $-19\hat{i} - 19\hat{j}$; c) 26.9, 45° below $-x$-axis
4. c) 4.71, 35.8° above x-axis   **5.** a) 71.6°;
b) 2.24, 26.6° above $-x$-axis   **6.** a) $-6\hat{i} + 2\hat{j}$
7. a) 3.12, 69.4° below $-x$-axis   **9.** a) 50 m
east of the point that was directly across
from her start   **13.** -15.0 m/s
15. a) 1 m/s^2; b) 1.73 m/s   **17.** a) 0.5 units
uphill; b) 1.67 units away from the floor

Chapter 3

1. a) Bill; b) 20 min   **2.** a) 120 min; b) 50
mph; c) 125 min; d) 48 mph   **3.** a) 2.68
m/s^2; b) 0.273; c) 134 m   **4.** a) 78.4 m;
b) 39.2 s   **5.** a) 20 m/s^2 **6.** 134 m
8. a) 15 m; b) 23 m/s; c) 24 m/s^2 **9.** a) 6 m;
b) 4 m/s; c) 2 m/s^2 **10.** b) 72 m/s; c) -54 m
and 210 m; d) -84 m and 240 m   **14.** 0, 5 m/s,
20 m/s, 30 m/s, 30 m/s   **25.** a) 100 m
28. b) $x = 72.85$ miles   **30.** a) 2.32 m/s

Chapter 5

1. 94.3 N, 58.0° below horizontal   **2.** 57.4 m
3. a) 15.7 N; b) 2.87 m/s; c) 4.34 m/s
4. a) 5.20 m/s^2; b) 1010 kg   **5.** 3.08 m
6. a) (1.0 m/s^2, -0.5 m/s^2); b) (1.0 m/s^2,
0.0 m/s^2)   **7.** a) 6670 N; b) 600 μs
10. a) 3.70 m; b) 6.97 m/s   **11.** a) 3.57×10^{-3} N;

b) 1.04×10^{-2} N   **15.** 4 m/s   **16.** b) 76.5 kg
19. g) 1.73 m/s^2

Chapter 6

1. b) 2 m/s^2 **2.** a) 63.4 ms; b) 783 m/s
3. 5 is slowest; 1, 2, 3, 4 are equal
4. a) $|\vec{v_0}|^2\sin^2\theta/2g$ **5.** a) 5.00 N; b) 30.2
rpm   **6.** 7.27°   **7.** a) 25.5 m; b) 34.4 m
11. 6.87 m   **15.** 24.4 rpm   **18.** c) i) $\sqrt{rg}$;
ii) $\sqrt{rg/2}$ or $\sqrt{3rg/2}$; iii) 0 or $\sqrt{2rg}$;
iv) $\sqrt{3rg}$ **21.** $g \propto 1/r^3$

Chapter 7

1. a) 3.92 N; b) 2.16 m/s^2 **2.** 9800 N
3. a) 36°; b) 66.6 N   **4.** a) 1.83 kg;
b) 1.32 m/s^2 **7.** a) 2.00 m/s^2; b) 99 m
10. $\sqrt{(m_2/m_1)rg}$ **12.** a) 225 N; b) 0.20 m/s
14. e) 13.1 m/s^2

Chapter 8

1. a) 0.2 m/s; b) slides at 0.99 m/s
2. 2.5 mph   **3.** 4.84 m/s   **4.** 0.714 m
7. b) 14.57×10^6 m/s   **8.** 5.48×10^5 m/s
11. b) 13.6 m

Chapter 9

1. a) 4; b) -9; c) 0   **2.** 26.6°   **5.** AB =
4 J, BC = -4 J   **6.** a) 675 kJ; b) 45.9 m

7. a) 392 N/m; b) 17.5 cm; c) 1.10 J
8. a) 49 N; b) 1450 N/m; c) 3.38 cm
9. a) 19.6 kW; b) 58.8 W **12.** a) 2.16 m/s;
b) 0.0058 **13.** a) 225 J; b) 225 N; c) 6.75 kW
15. c) 30.3 m **17.** e) 11 m **19.** c) 16.8 m/s

Chapter 10

1. a) 25.1 m; c) 22.2 m/s **2.** 1.96×10^5 N/m
4. a) 1.40 m/s **5.** 2.00 m/s **6.** a) 15.7 m/s
7. a) 51.0 cm; b) 37.9 cm **10.** b) 15 m
11. 48.2° **15.** f) 0.937 m **16.** c) 22.4
m/s; e) 0.124 N s **19.** g) 1.37×10^8 m/s

Chapter 11

1. 4.68×10^{-24} m/s, 6.8×10^{12} years
2. 0.857 m/s and 2.857 m/s **3.** a) 0; b) 0;
c) 1176 J; d) 0; e) –1176 J; f) 2350 W
9. b) 24.1 kW **11.** 7.94 m/s **13.** b) 0.667

Chapter 12

1. 90.8 min, 7720 m/s **2.** 4.37×10^{11} m;
17,400 m/s **3.** 3.0×10^4 m/s
5. a) 5.34×10^{22} kg **8.** c) 1.36×10^{12} N;
d) 1500 km

Chapter 13

2. c) 5.54 cm; d) 0.445 rad; e) 69.6 cm/s;
f) 875 cm/s²; g) 0.0484 J **3.** a) 0.628 s;
e) 20 cm/s; g) 0.500 J; j) 1.46 cm/s
4. 0.667 s **6.** a) 6.40 cm; b) 160 cm/s²
7. 1.58 Hz **9.** a) 6.03 cm; b) 6.32 s
11. 2.23 cm

Chapter 14

1. a) 42.9 cm; b) 3.21 mm; c) 93.5 GHz
2. 6040 m/s **3.** a) 10 m/s; b) 30° **6.** 882 g

8. a) 628 s⁻¹; b) –π; d) 100 m/s; e) 1.00 m;
f) 6.28 m⁻¹; g) 1 mm; h) –π/2 or 3π/2;
j) –1.00 mm **11.** a) 0.159 W/m²
12. d) 1.2×10^{14}

Chapter 15

1. a) 80 cm; b) 69.1 N **2.** 4.8 cm
3. 0.261 m, 0.556 m, 0.852 m **4.** 198 N
9. 265 N **10.** c) 318 Hz, 424 Hz, and 530 Hz
11. 8.19 m/s²

Chapter 16

1. 500 nm **2.** 0.40 mm **3.** 2.7 mm
4. 7.89 cm **5.** a) Closest is 2.98 m
7. a) 3.00 mm; c) π/2 **9.** 103 nm
10. b) 1333

Chapter 17

1. a) 0.200 mm; b) 9.5 mm **2.** a) 4.0 cm;
b) 4.9 cm **3.** a) 1.258 m; b) 7 **4.** b) 500
nm, 550 nm, 650 nm **5.** c) 1.0 cm
8. b) 1.11×10^7 bits/cm²; c) 115 MB
10. d) $d\sin\theta = m\lambda + d\sin\phi$; f) There are 6
angles, one being 17.0°

Chapter 18

2. a) 3.64×10^6 m/s; b) 1985 m/s
3. 1.8×10^{-5} m **4.** 3.98×10^{-19} J
5. d) 820.6 nm; e) 31.4 cm **9.** ≈165 m/s
10. a) first is 2.41×10^{-21} J; b) first is
7.28×10^4 m/s **11.** e) 4.17×10^{14} Hz to
5.83×10^{14} Hz

Bibliography

The references cited here will give the interested reader an introduction to physics education research on the topic of mechanics. The pedagogical strategies in this text are derived from these research findings.

Arons, Arnold. *A Guide to Introductory Physics Teaching*. John Wiley and Sons, New York, 1990.

Arons, Arnold. "Student Patterns of Thinking and Reasoning, Part 1." *The Physics Teacher*, vol. 21, 1983, p. 576.

Arons, Arnold. "Student Patterns of Thinking and Reasoning, Part 2." *The Physics Teacher*, vol. 22, 1984, p. 21.

Clement, John. "Students' Preconceptions in Introductory Mechanics." *American Journal of Physics,* vol. 50, 1982, p. 66.

Hake, Richard. "Promoting Student Crossover to the Newtonian World." *American Journal of Physics,* vol. 55, 1987, p. 878.

Halloun, Ibrahim and David Hestenes. "The Initial Knowledge State of College Physics Students." *American Journal of Physics,* vol. 53, 1985, p. 1043.

Halloun, Ibrahim and David Hestenes. "Common Sense Concepts About Motion." *American Journal of Physics,* vol. 53, 1985, p. 1056.

McDermott, Lillian, Mark Rosenquist, and Emily Zee. "Student Difficulties in Connecting Graphs and Physics: Examples from Kinematics." *American Journal of Physics,* vol. 55, 1987, p. 503.

McDermott, Lillian, Peter Shaffer, and Mark Somers. "Research as a Guide for Teaching Introductory Mechanics: An Illustration in the Context of the Atwood's Machine." *American Journal of Physics,* vol. 62, 1994, p. 46.

Reif, Fred and Joan Heller. "Knowledge Structure and Problem Solving in Physics." *Educational Psychologist,* vol. 17, 1982, p. 102.

Rosenquist, Mark and Lillian McDermott. "A Conceptual Approach to Teaching Kinematics." *American Journal of Physics,* vol. 55, 1987, p. 407.

Thornton, Ronald and David Sokoloff. "Learning Motion Concepts Using Real-Time Microcomputer-Based Laboratory Tools." *American Journal of Physics*, vol. 58, 1990, p. 858.

Van Heuvelen, Alan. "Learning to Think Like a Physicist: A Review of Research-Based Instructional Strategies." *American Journal of Physics,* vol. 59, 1991, p. 891.

Van Heuvelen, Alan. "Overview, Case Study Physics." *American Journal of Physics,* vol. 59, 1991, p. 898.

Index

STUDENT EVALUATION FORM

Knight *Physics: A Contemporary Perspective*
Chapter 1   Concepts of Motion

Name (Optional): _____   Date: _____

College: _____   Professor: _____

1.   What did you like about this chapter? Why?

2.   Were there any areas of particular difficulty? If yes, please specify those areas, and, if possible, explain why you found them to be difficult.

3.   How do you rate the illustrations and photos in this chapter?
     (High)   5   4   3   2   1   (Low)
     What suggestions do you have for their improvement?

4.   How do you rate the worked examples in this chapter?
     (High)   5   4   3   2   1   (Low)
     Which did you particularly like or dislike? Why?

     Any suggestions for improvement?

Continued on back

5. How do you rate the end-of-chapter problems for this chapter?

(High)   5   4   3   2   1   (Low)

Which did you particularly like or dislike? Why?

Any suggestions for improvement?

6. How do you rate the Student Workbook exercises for this chapter?

(High)   5   4   3   2   1   (Low)

Which did you particularly like or dislike? Why?

Any suggestions for improvement?

7. Please provide any additional comments you think would be helpful to the author for improving or strengthening the material in this chapter.

STUDENT EVALUATION FORM

Knight *Physics: A Contemporary Perspective*
Chapter 2   Vectors and Coordinate Systems

Name (Optional): _____   Date: _____

College: _____   Professor: _____

1.   What did you like about this chapter? Why?

2.   Were there any areas of particular difficulty? If yes, please specify those areas, and, if possible, explain why you found them to be difficult.

3.   How do you rate the illustrations and photos in this chapter?

   (High)   5   4   3   2   1   (Low)

   What suggestions do you have for their improvement?

4.   How do you rate the worked examples in this chapter?

   (High)   5   4   3   2   1   (Low)

   Which did you particularly like or dislike? Why?

   Any suggestions for improvement?

Continued on back

5. How do you rate the end-of-chapter problems for this chapter?

    (High)   5   4   3   2   1   (Low)

    Which did you particularly like or dislike? Why?

    Any suggestions for improvement?

6. How do you rate the Student Workbook exercises for this chapter?

    (High)   5   4   3   2   1   (Low)

    Which did you particularly like or dislike? Why?

    Any suggestions for improvement?

7. Please provide any additional comments you think would be helpful to the author for improving or strengthening the material in this chapter.

THANK YOU FOR YOUR FEEDBACK.
PLEASE GIVE THIS FORM TO YOUR PROFESSOR.
ADDISON-WESLEY PUBLISHING COMPANY

STUDENT EVALUATION FORM

Knight *Physics: A Contemporary Perspective*
Chapter 3   Kinematics: The Mathematics of Motion

Name (Optional): _____   Date: _____

College: _____   Professor: _____

1. What did you like about this chapter? Why?

2. Were there any areas of particular difficulty? If yes, please specify those areas, and, if possible, explain why you found them to be difficult.

3. How do you rate the illustrations and photos in this chapter?
    (High)   5   4   3   2   1   (Low)
    What suggestions do you have for their improvement?

4. How do you rate the worked examples in this chapter?
    (High)   5   4   3   2   1   (Low)
    Which did you particularly like or dislike? Why?

    Any suggestions for improvement?

Continued on back

5. How do you rate the end-of-chapter problems for this chapter?

(High) 5   4   3   2   1   (Low)

Which did you particularly like or dislike? Why?

Any suggestions for improvement?

6. How do you rate the Student Workbook exercises for this chapter?

(High) 5   4   3   2   1   (Low)

Which did you particularly like or dislike? Why?

Any suggestions for improvement?

7. Please provide any additional comments you think would be helpful to the author for improving or strengthening the material in this chapter.

STUDENT EVALUATION FORM

Knight *Physics: A Contemporary Perspective*
Chapter 4   Force and Motion

Name (Optional): _____   Date: _____

College: _____   Professor: _____

1.   What did you like about this chapter? Why?

2.   Were there any areas of particular difficulty? If yes, please specify those areas, and, if possible, explain why you found them to be difficult.

3.   How do you rate the illustrations and photos in this chapter?
     (High)   5   4   3   2   1   (Low)
     What suggestions do you have for their improvement?

4.   How do you rate the worked examples in this chapter?
     (High)   5   4   3   2   1   (Low)
     Which did you particularly like or dislike? Why?

     Any suggestions for improvement?

Continued on back

5. How do you rate the end-of-chapter problems for this chapter?

(High)   5   4   3   2   1   (Low)

Which did you particularly like or dislike? Why?

Any suggestions for improvement?

6. How do you rate the Student Workbook exercises for this chapter?

(High)   5   4   3   2   1   (Low)

Which did you particularly like or dislike? Why?

Any suggestions for improvement?

7. Please provide any additional comments you think would be helpful to the author for improving or strengthening the material in this chapter.

STUDENT EVALUATION FORM

Knight *Physics: A Contemporary Perspective*
Chapter 5   Dynamics I: Newton's Second Law

Name (Optional): _____   Date: _____

College: _____   Professor: _____

1. What did you like about this chapter? Why?

2. Were there any areas of particular difficulty? If yes, please specify those areas, and, if possible, explain why you found them to be difficult.

3. How do you rate the illustrations and photos in this chapter?

    (High)   5   4   3   2   1   (Low)

    What suggestions do you have for their improvement?

4. How do you rate the worked examples in this chapter?

    (High)   5   4   3   2   1   (Low)

    Which did you particularly like or dislike? Why?

    Any suggestions for improvement?

Continued on back

5. How do you rate the end-of-chapter problems for this chapter?

   (High)   5   4   3   2   1   (Low)

   Which did you particularly like or dislike? Why?

   Any suggestions for improvement?

6. How do you rate the Student Workbook exercises for this chapter?

   (High)   5   4   3   2   1   (Low)

   Which did you particularly like or dislike? Why?

   Any suggestions for improvement?

7. Please provide any additional comments you think would be helpful to the author for improving or strengthening the material in this chapter.

STUDENT EVALUATION FORM

Knight *Physics: A Contemporary Perspective*
Chapter 6   Dynamics II: Motion in a Plane

Name (Optional): _____   Date: _____

College: _____   Professor: _____

1. What did you like about this chapter? Why?

2. Were there any areas of particular difficulty? If yes, please specify those areas, and, if possible, explain why you found them to be difficult.

3. How do you rate the illustrations and photos in this chapter?

    (High)   5   4   3   2   1   (Low)

    What suggestions do you have for their improvement?

4. How do you rate the worked examples in this chapter?

    (High)   5   4   3   2   1   (Low)

    Which did you particularly like or dislike? Why?

    Any suggestions for improvement?

Continued on back

5. How do you rate the end-of-chapter problems for this chapter?

(High) 5   4   3   2   1   (Low)

Which did you particularly like or dislike? Why?

Any suggestions for improvement?

6. How do you rate the Student Workbook exercises for this chapter?

(High) 5   4   3   2   1   (Low)

Which did you particularly like or dislike? Why?

Any suggestions for improvement?

7. Please provide any additional comments you think would be helpful to the author for improving or strengthening the material in this chapter.

THANK YOU FOR YOUR FEEDBACK.
PLEASE GIVE THIS FORM TO YOUR PROFESSOR.
ADDISON-WESLEY PUBLISHING COMPANY

STUDENT EVALUATION FORM

Knight *Physics: A Contemporary Perspective*
Chapter 7 **Dynamics III: Newton's Third Law**

Name (Optional): _____   Date: _____

College: _____   Professor: _____

1. What did you like about this chapter? Why?

2. Were there any areas of particular difficulty? If yes, please specify those areas, and, if possible, explain why you found them to be difficult.

3. How do you rate the illustrations and photos in this chapter?
   (High)   5   4   3   2   1   (Low)
   What suggestions do you have for their improvement?

4. How do you rate the worked examples in this chapter?
   (High)   5   4   3   2   1   (Low)
   Which did you particularly like or dislike? Why?

   Any suggestions for improvement?

Continued on back

5. How do you rate the end-of-chapter problems for this chapter?

(High) 5 4 3 2 1 (Low)

Which did you particularly like or dislike? Why?

Any suggestions for improvement?

6. How do you rate the Student Workbook exercises for this chapter?

(High) 5 4 3 2 1 (Low)

Which did you particularly like or dislike? Why?

Any suggestions for improvement?

7. Please provide any additional comments you think would be helpful to the author for improving or strengthening the material in this chapter.

STUDENT EVALUATION FORM

Knight *Physics: A Contemporary Perspective*
Chapter 8   Momentum and Its Conservation

Name (Optional): _____   Date: _____

College: _____   Professor: _____

1.   What did you like about this chapter? Why?

2.   Were there any areas of particular difficulty? If yes, please specify those areas, and, if possible, explain why you found them to be difficult.

3.   How do you rate the illustrations and photos in this chapter?
     (High)   5   4   3   2   1   (Low)
     What suggestions do you have for their improvement?

4.   How do you rate the worked examples in this chapter?
     (High)   5   4   3   2   1   (Low)
     Which did you particularly like or dislike? Why?

     Any suggestions for improvement?

Continued on back

5. How do you rate the end-of-chapter problems for this chapter?

    (High)   5   4   3   2   1   (Low)

    Which did you particularly like or dislike? Why?

    Any suggestions for improvement?

6. How do you rate the Student Workbook exercises for this chapter?

    (High)   5   4   3   2   1   (Low)

    Which did you particularly like or dislike? Why?

    Any suggestions for improvement?

7. Please provide any additional comments you think would be helpful to the author for improving or strengthening the material in this chapter.

STUDENT EVALUATION FORM

Knight *Physics: A Contemporary Perspective*
Chapter 9 Concepts of Energy I: Work and Energy

Name (Optional): _____   Date: _____

College: _____   Professor: _____

1. What did you like about this chapter? Why?

2. Were there any areas of particular difficulty? If yes, please specify those areas, and, if possible, explain why you found them to be difficult.

3. How do you rate the illustrations and photos in this chapter?
   (High)   5   4   3   2   1   (Low)
   What suggestions do you have for their improvement?

4. How do you rate the worked examples in this chapter?
   (High)   5   4   3   2   1   (Low)
   Which did you particularly like or dislike? Why?

   Any suggestions for improvement?

Continued on back

5. How do you rate the end-of-chapter problems for this chapter?

(High)   5   4   3   2   1   (Low)

Which did you particularly like or dislike? Why?

Any suggestions for improvement?

6. How do you rate the Student Workbook exercises for this chapter?

(High)   5   4   3   2   1   (Low)

Which did you particularly like or dislike? Why?

Any suggestions for improvement?

7. Please provide any additional comments you think would be helpful to the author for improving or strengthening the material in this chapter.

STUDENT EVALUATION FORM

Knight *Physics: A Contemporary Perspective*
Chapter 10   Concepts of Energy II: Potential Energy and Conservation

Name (Optional): _____   Date: _____

College: _____   Professor: _____

1.   What did you like about this chapter? Why?

2.   Were there any areas of particular difficulty? If yes, please specify those areas, and, if possible, explain why you found them to be difficult.

3.   How do you rate the illustrations and photos in this chapter?
     (High)   5   4   3   2   1   (Low)
     What suggestions do you have for their improvement?

4.   How do you rate the worked examples in this chapter?
     (High)   5   4   3   2   1   (Low)
     Which did you particularly like or dislike? Why?

     Any suggestions for improvement?

Continued on back

5. How do you rate the end-of-chapter problems for this chapter?

   (High)   5   4   3   2   1   (Low)

   Which did you particularly like or dislike? Why?

   Any suggestions for improvement?

6. How do you rate the Student Workbook exercises for this chapter?

   (High)   5   4   3   2   1   (Low)

   Which did you particularly like or dislike? Why?

   Any suggestions for improvement?

7. Please provide any additional comments you think would be helpful to the author for improving or strengthening the material in this chapter.

THANK YOU FOR YOUR FEEDBACK.
PLEASE GIVE THIS FORM TO YOUR PROFESSOR.
ADDISON-WESLEY PUBLISHING COMPANY

STUDENT EVALUATION FORM

Knight *Physics: A Contemporary Perspective*
Chapter 11   Expanding the Concept of Energy

Name (Optional): _____   Date: _____

College: _____   Professor: _____

1. What did you like about this chapter? Why?

2. Were there any areas of particular difficulty? If yes, please specify those areas, and, if possible, explain why you found them to be difficult.

3. How do you rate the illustrations and photos in this chapter?
    (High)   5   4   3   2   1   (Low)
    What suggestions do you have for their improvement?

4. How do you rate the worked examples in this chapter?
    (High)   5   4   3   2   1   (Low)
    Which did you particularly like or dislike? Why?

    Any suggestions for improvement?

Continued on back

5. How do you rate the end-of-chapter problems for this chapter?

   (High)   5   4   3   2   1   (Low)

   Which did you particularly like or dislike? Why?

   Any suggestions for improvement?

6. How do you rate the Student Workbook exercises for this chapter?

   (High)   5   4   3   2   1   (Low)

   Which did you particularly like or dislike? Why?

   Any suggestions for improvement?

7. Please provide any additional comments you think would be helpful to the author for improving or strengthening the material in this chapter.

STUDENT EVALUATION FORM

Knight *Physics: A Contemporary Perspective*
Chapter 12   Newton's Theory of Gravity

Name (Optional): _____   Date: _____

College: _____   Professor: _____

1. What did you like about this chapter? Why?

2. Were there any areas of particular difficulty? If yes, please specify those areas, and, if possible, explain why you found them to be difficult.

3. How do you rate the illustrations and photos in this chapter?
   (High)   5   4   3   2   1   (Low)
   What suggestions do you have for their improvement?

4. How do you rate the worked examples in this chapter?
   (High)   5   4   3   2   1   (Low)
   Which did you particularly like or dislike? Why?

   Any suggestions for improvement?

Continued on back

5. How do you rate the end-of-chapter problems for this chapter?

    (High)   5   4   3   2   1   (Low)

    Which did you particularly like or dislike? Why?

    Any suggestions for improvement?

6. How do you rate the Student Workbook exercises for this chapter?

    (High)   5   4   3   2   1   (Low)

    Which did you particularly like or dislike? Why?

    Any suggestions for improvement?

7. Please provide any additional comments you think would be helpful to the author for improving or strengthening the material in this chapter.

STUDENT EVALUATION FORM

Knight *Physics: A Contemporary Perspective*
Chapter 13   Oscillations

Name (Optional): _____   Date: _____

College: _____   Professor: _____

1. What did you like about this chapter? Why?

2. Were there any areas of particular difficulty? If yes, please specify those areas, and, if possible, explain why you found them to be difficult.

3. How do you rate the illustrations and photos in this chapter?
    (High)   5   4   3   2   1   (Low)
    What suggestions do you have for their improvement?

4. How do you rate the worked examples in this chapter?
    (High)   5   4   3   2   1   (Low)
    Which did you particularly like or dislike? Why?

    Any suggestions for improvement?

Continued on back

5. How do you rate the end-of-chapter problems for this chapter?

    (High)   5   4   3   2   1   (Low)

    Which did you particularly like or dislike? Why?

    Any suggestions for improvement?

6. How do you rate the Student Workbook exercises for this chapter?

    (High)   5   4   3   2   1   (Low)

    Which did you particularly like or dislike? Why?

    Any suggestions for improvement?

7. Please provide any additional comments you think would be helpful to the author for improving or strengthening the material in this chapter.

THANK YOU FOR YOUR FEEDBACK.
PLEASE GIVE THIS FORM TO YOUR PROFESSOR.
ADDISON-WESLEY PUBLISHING COMPANY

STUDENT EVALUATION FORM

Knight *Physics: A Contemporary Perspective*
Chapter 14   Traveling Waves

Name (Optional): _____   Date: _____

College: _____   Professor: _____

1.   What did you like about this chapter? Why?

2.   Were there any areas of particular difficulty? If yes, please specify those areas, and, if possible, explain why you found them to be difficult.

3.   How do you rate the illustrations and photos in this chapter?
     (High)   5   4   3   2   1   (Low)
     What suggestions do you have for their improvement?

4.   How do you rate the worked examples in this chapter?
     (High)   5   4   3   2   1   (Low)
     Which did you particularly like or dislike? Why?

     Any suggestions for improvement?

Continued on back

5. How do you rate the end-of-chapter problems for this chapter?

(High)   5   4   3   2   1   (Low)

Which did you particularly like or dislike? Why?

Any suggestions for improvement?

6. How do you rate the Student Workbook exercises for this chapter?

(High)   5   4   3   2   1   (Low)

Which did you particularly like or dislike? Why?

Any suggestions for improvement?

7. Please provide any additional comments you think would be helpful to the author for improving or strengthening the material in this chapter.

THANK YOU FOR YOUR FEEDBACK.
PLEASE GIVE THIS FORM TO YOUR PROFESSOR.
ADDISON-WESLEY PUBLISHING COMPANY

STUDENT EVALUATION FORM

Knight *Physics: A Contemporary Perspective*
Chapter 15   Standing Waves

Name (Optional): _____   Date: _____

College: _____   Professor: _____

1. What did you like about this chapter? Why?

2. Were there any areas of particular difficulty? If yes, please specify those areas, and, if possible, explain why you found them to be difficult.

3. How do you rate the illustrations and photos in this chapter?
    (High)   5   4   3   2   1   (Low)
    What suggestions do you have for their improvement?

4. How do you rate the worked examples in this chapter?
    (High)   5   4   3   2   1   (Low)
    Which did you particularly like or dislike? Why?

    Any suggestions for improvement?

Continued on back

5. How do you rate the end-of-chapter problems for this chapter?

(High) 5   4   3   2   1   (Low)

Which did you particularly like or dislike? Why?

Any suggestions for improvement?

6. How do you rate the Student Workbook exercises for this chapter?

(High) 5   4   3   2   1   (Low)

Which did you particularly like or dislike? Why?

Any suggestions for improvement?

7. Please provide any additional comments you think would be helpful to the author for improving or strengthening the material in this chapter.

STUDENT EVALUATION FORM

Knight *Physics: A Contemporary Perspective*
Chapter 16   Interference

Name (Optional): _____   Date: _____

College: _____   Professor: _____

1. What did you like about this chapter? Why?

2. Were there any areas of particular difficulty? If yes, please specify those areas, and, if possible, explain why you found them to be difficult.

3. How do you rate the illustrations and photos in this chapter?
    (High)   5   4   3   2   1   (Low)
    What suggestions do you have for their improvement?

4. How do you rate the worked examples in this chapter?
    (High)   5   4   3   2   1   (Low)
    Which did you particularly like or dislike? Why?

    Any suggestions for improvement?

Continued on back

5. How do you rate the end-of-chapter problems for this chapter?

    (High)   5   4   3   2   1   (Low)

    Which did you particularly like or dislike? Why?

    Any suggestions for improvement?

6. How do you rate the Student Workbook exercises for this chapter?

    (High)   5   4   3   2   1   (Low)

    Which did you particularly like or dislike? Why?

    Any suggestions for improvement?

7. Please provide any additional comments you think would be helpful to the author for improving or strengthening the material in this chapter.

THANK YOU FOR YOUR FEEDBACK.
PLEASE GIVE THIS FORM TO YOUR PROFESSOR.
ADDISON-WESLEY PUBLISHING COMPANY

STUDENT EVALUATION FORM

Knight *Physics: A Contemporary Perspective*
Chapter 17   Diffraction

Name (Optional): _____   Date: _____

College: _____   Professor: _____

1. What did you like about this chapter? Why?

2. Were there any areas of particular difficulty? If yes, please specify those areas, and, if possible, explain why you found them to be difficult.

3. How do you rate the illustrations and photos in this chapter?
    (High)   5   4   3   2   1   (Low)
    What suggestions do you have for their improvement?

4. How do you rate the worked examples in this chapter?
    (High)   5   4   3   2   1   (Low)
    Which did you particularly like or dislike? Why?

    Any suggestions for improvement?

Continued on back

5. How do you rate the end-of-chapter problems for this chapter?

   (High)   5   4   3   2   1   (Low)

   Which did you particularly like or dislike? Why?

   Any suggestions for improvement?

6. How do you rate the Student Workbook exercises for this chapter?

   (High)   5   4   3   2   1   (Low)

   Which did you particularly like or dislike? Why?

   Any suggestions for improvement?

7. Please provide any additional comments you think would be helpful to the author for improving or strengthening the material in this chapter.

THANK YOU FOR YOUR FEEDBACK.
PLEASE GIVE THIS FORM TO YOUR PROFESSOR.
ADDISON-WESLEY PUBLISHING COMPANY

STUDENT EVALUATION FORM

Knight *Physics: A Contemporary Perspective*
Chapter 18   A Closer Look at Light and Matter

Name (Optional): _____   Date: _____

College: _____   Professor: _____

1.   What did you like about this chapter? Why?

2.   Were there any areas of particular difficulty? If yes, please specify those areas, and, if possible, explain why you found them to be difficult.

3.   How do you rate the illustrations and photos in this chapter?

(High)   5   4   3   2   1   (Low)

What suggestions do you have for their improvement?

4.   How do you rate the worked examples in this chapter?

(High)   5   4   3   2   1   (Low)

Which did you particularly like or dislike? Why?

Any suggestions for improvement?

Continued on back

5. How do you rate the end-of-chapter problems for this chapter?

(High) 5 4 3 2 1 (Low)

Which did you particularly like or dislike? Why?

Any suggestions for improvement?

6. How do you rate the Student Workbook exercises for this chapter?

(High) 5 4 3 2 1 (Low)

Which did you particularly like or dislike? Why?

Any suggestions for improvement?

7. Please provide any additional comments you think would be helpful to the author for improving or strengthening the material in this chapter.

THANK YOU FOR YOUR FEEDBACK.
PLEASE GIVE THIS FORM TO YOUR PROFESSOR.
ADDISON-WESLEY PUBLISHING COMPANY